Mathematics
HIGHER LEVEL
for the IB Diploma

Bill Roberts
Sandy MacKenzie

OXFORD
UNIVERSITY PRESS

UNIVERSITY PRESS

Great Clarendon Street, Oxford OX2 6DP

Oxford University Press is a department of the University of Oxford.
It furthers the University's objective of excellence in research,
scholarship, and education by publishing worldwide in

Oxford   New York

Auckland   Cape Town   Dar es Salaam   Hong Kong   Karachi
Kuala Lumpur   Madrid   Melbourne   Mexico City   Nairobi
New Delhi   Shanghai   Taipei   Toronto

With offices in

Argentina   Austria   Brazil   Chile   Czech Republic   France   Greece
Guatemala   Hungary   Italy   Japan   Poland   Portugal   Singapore
South Korea   Switzerland   Thailand   Turkey   Ukraine   Vietnam

Oxford is a registered trade mark of Oxford University Press
in the UK and in certain other countries

British Library Cataloguing in Publication Data

Data available

ISBN-13: 978-0-19-915226-1

10 9 8 7 6 5 4 3

Printed in Great Britain by Bell and Bain Ltd, Glasgow.

Acknowledgments

The Publisher would like to thank the following for permission to reproduce photographs:

p1 Iain Gilfillan; p35 stanford.edu; p39 Erik Von Weber/GettyImages; p58 (left) Adam Hart - Davis /
Science Photo Library; p58 (right) David Brimm / ShutterStock; p108 Mary Evans Picture Library;
p183 (left) American Institute of Physics / Science Photo Library; p183 (right) Science Source /
Science Photo Library; p217 Photo Researchers, Inc.; p337 University of Gdaƒsk; p373 The Print
Collector/Alamy; p403 Science Photo Library / Photolibrary; p446 Ria Novosti/ Science Photo Library
/ Photolibrary; p473 University of Massachusetts Lowell; p595 Science Photo Library / Photolibrary.

Cover image courtesy of Ryan Briscall/Design Pics Inc./Alamy

In a few cases we have been unable to trace the copyright holder prior to publication. If notified
the publisher will be pleased to amend the acknowledgements in any future addition.

Mixed Sources

Product group from well-managed
forests and other controlled sources
www.fsc.org   Cert no. TT-COC-002769
© 1996 Forest Stewardship Council

FSC

About the Authors

Bill Roberts is an experienced educator, having taught mathematics in a number of international schools for the past twenty years. He also works for the International Baccalaureate as an examiner and trainer. Bill is currently based at the University of Newcastle in England.

Sandy MacKenzie is an experienced teacher of mathematics in both the International Baccalaureate and the Scottish systems and works for the International Baccalaureate as an examiner. Sandy is currently Assistant Rector at Morrison's Academy in Scotland.

Dedications

To my father, George Roberts, who encouraged and nurtured my love and appreciation for mathematics.

To Nancy MacKenzie, and the Cairns sisters, Alexia and Rosemary, my three mothers, without whom none of this would have been possible.

Preface

This book is intended primarily for use by students and teachers of Higher Level Mathematics in the International Baccalaureate Diploma Programme, but will also be of use to students on other courses.

Detailed coverage is provided for the core part of this syllabus and provides excellent preparation for the final examination. The book provides guidance on areas of the syllabus that may be examined differently depending on whether a graphical calculator is used or not.

Points of theory are presented and explained concisely and are illustrated by worked examples which identify the key skills and techniques. Where appropriate, information and methods are highlighted and margin notes provide further tips and important reminders. These are supported by varied and graded exercises, which consolidate the theory, thus enabling the reader to practise basic skills and challenging exam-style questions. Each chapter concludes with a review exercise that covers all of the skills within the chapter, with a clear distinction between questions where a calculator is allowed and ones where it is not. The icon ▦ is used to indicate where a calculator may be used, and ▉ indicates where it may not. Many of these questions are from past IB papers and we would like to thank the International Baccalaureate for permission to reproduce these questions.

Throughout the text, we have aimed to produce chapters that have been sequenced in a logical teaching order with major topics grouped together. However, we have also built in flexibility and in some cases the order in which chapters are used can be changed. For example we have split differential and integral calculus, but these can be taught as one section. Although many students will use this book in a teacher-led environment, it has also been designed to be accessible to students for self-study.

The book is accompanied by a CD. As well as containing an electronic version of the entire book, there is a presumed knowledge chapter covering basic skills, revision exercises of the whole syllabus grouped into six sections, and twenty extended exam-style questions.

We would like extend our thanks to family, friends, colleagues and students who supported and encouraged us throughout the process of writing this book and especially to those closest to each of us, for their patience and understanding.

Bill Roberts
Sandy MacKenzie
2007

Contents

1 Trigonometry 1

Although most people connect trigonometry with the study of triangles, it is from the circle that this area of mathematics originates.

The study of trigonometry is not new. Its roots come from the Babylonians around 300 BC. This area of mathematics was further developed by the Ancient Greeks around 100 BC. Hipparchus, Ptolemy and Menelaus are considered to have founded trigonometry as we now know it. It was originally used to aid the study of astronomy.

In the modern world trigonometry can be used to answer questions like "How far apart are each of the 32 pods on the London Eye?" and "What would a graph of someone's height on the London Eye look like?"

1.1 Circle problems

Radians

It is likely that up until now you have measured angles in degrees, but as for most measurements, there is more than one unit that can be used.

Consider a circle with radius 1 unit.

As θ increases, the arc length increases. For a particular value of θ, the arc will be the same length as the radius. When this occurs, the angle is defined to be 1 radian.

The circumference of a circle is given by $C = 2\pi r$, so when $r = 1$, $C = 2\pi$.

As there are 360° at the centre of a circle, and 1 radian is defined to be the angle subtended by an arc of length 1,

$$2\pi \text{ radians} = 360°$$

Hence 1 radian $= \dfrac{360°}{2\pi} \simeq 57.3°$.

Method for converting between degrees and radians

To convert degrees to radians, multiply by $\dfrac{2\pi}{360°} = \dfrac{\pi}{180°}$.

To convert radians to degrees, multiply by $\dfrac{360°}{2\pi} = \dfrac{180°}{\pi}$.

Some angles measured in radians can be written as simple fractions of π.
You must learn these.

Degrees	0°	15°	30°	45°	60°	90°	180°	270°	360°
Radians	0	$\dfrac{\pi}{12}$	$\dfrac{\pi}{6}$	$\dfrac{\pi}{4}$	$\dfrac{\pi}{3}$	$\dfrac{\pi}{2}$	π	$\dfrac{3\pi}{2}$	2π

Where an angle is given without units, assume it is in radians.

Example

Convert $\dfrac{2\pi}{3}$ radians into degrees.

$\dfrac{\pi}{3} = 60°$ (see table) so $\dfrac{2\pi}{3} = 60° \times 2 = 120°$.

Example

Convert 250° into radians.

This is not one of the commonly used angles (nor a multiple), so use the method for converting degrees to radians.

$250° \times \dfrac{\pi}{180°} = \dfrac{25\pi}{18} \simeq 4.36$

Circle sectors and segments

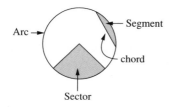

Arc → Segment — chord Sector

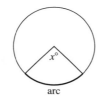

$x°$ arc

Considering the infinite rotational symmetry of the circle,

$$\frac{x°}{360°} = \frac{\text{arc length}}{2\pi r} = \frac{\text{sector area}}{\pi r^2}$$

That is, dividing the angle by 360°, the arc length by the circumference, and the sector area by the circle area gives the same fraction.

This is very useful when solving problems related to circles.

Changing the angles to radians gives formulae for the length of an arc and the area of a sector:

$$\frac{\theta}{2\pi} = \frac{\text{arc length}}{2\pi r}$$

$$\Rightarrow \quad \text{arc length} = r\theta$$

$$\frac{\theta}{2\pi} = \frac{\text{sector area}}{\pi r^2}$$

$$\Rightarrow \quad \text{sector area} = \frac{1}{2}r^2\theta$$

These formulae only work if θ is in radians.

Example

What is the area of the sector shown below?

Sector area $= \dfrac{1}{2}r^2\theta$

$\qquad = \dfrac{1}{2} \times 8^2 \times \dfrac{\pi}{3}$

$\qquad = 33.5 \text{ cm}^2$

Example

The fairground ride shown below moves through an angle of 50° from point A to point B. What is the length of the arc AB?

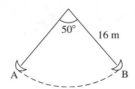

Start by converting 50° into radians. $\quad \theta = \dfrac{50°}{360°} \times 2\pi$

$\qquad\qquad\qquad\qquad\qquad\qquad\qquad = 0.872\ldots$

Hence arc length $= r\theta = 16 \times 0.872\ldots$

$\qquad\qquad\qquad\quad = 13.96 \text{ m}$

Example

What is the volume of water lying in this pipe of radius 2.5 m?

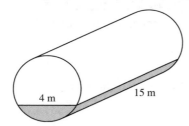

4 m   15 m

In this example, we need to find the area of a segment. The method for doing this is:

Area of segment = Area of sector − Area of triangle ············

> It is important to remember this.

First find the angle at the centre in radians: $\frac{1}{2}\theta = \sin^{-1}\frac{2}{2.5}$

$$= 0.927\ldots$$

2.5 $\frac{1}{2}\theta$

2

Area of triangle $= \frac{1}{2} \times 4 \times 1.5$ ············

> Use Pythagoras to find the height of the triangle.

$$= 3 \text{ m}^2$$

Area of sector $= \frac{1}{2}r^2\theta = \frac{1}{2} \times 2.5^2 \times 0.972\ldots \times 2$

$$= 5.79\ldots \text{ m}^2$$

Area of segment = Area of sector − Area of triangle = $5.79\ldots - 3$

$$= 2.79\ldots \text{ m}^2$$

Volume $= 2.79\ldots \times 15 = 41.9 \text{ m}^3$

Exercise 1

1 Express each angle in degrees.

   a $\frac{3\pi}{4}$ **b** $\frac{\pi}{9}$ **c** $\frac{2\pi}{5}$ **d** $\frac{5\pi}{6}$ **e** $\frac{7\pi}{12}$ **f** $\frac{\pi}{8}$

   g $\frac{11\pi}{18}$ **h** 2   **i** 1.5   **j** 4   **k** 3.6   **l** 0.4

2 Express each angle in radians, giving your answer in terms of π.

   a 30°   **b** 210°   **c** 135°   **d** 315°

   e 240°   **f** 70°   **g** 72°   **h** 54°

3 Express each angle in radians, giving your answer to 3 sf.

   a 35°   **b** 100°   **c** 300°

   d 80°   **e** 132°   **f** 278°

4 Find the area of each shaded sector.

   a

   b

   c
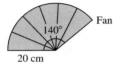

   d

5 Find the length of each arc.

   a

   b

   c

   d

6 Find the perimeter of each shape.

   a

   b

   c
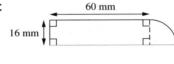

7 The diagram below shows a windscreen wiper cleaning a car windscreen.

   a What is the length of the arc swept out?

   b What area of the windscreen is not cleared?

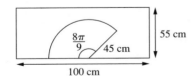

8 Find the area of the shaded segment.

9 What is the area of this shape?

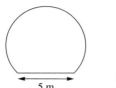

Diameter = 8 m

5 m

10 Radius = 32 cm
Area of sector = 1787 cm²
What is the angle at the centre of the sector?

11 Find the perimeter of this segment.

$\frac{\pi}{4}$

6 cm

12 A sector has an area of 942.5 cm² and an arc length of 62.8 cm. What is the radius of the circle?

r

13 Two circles are used to form the logo for a company as shown below. One circle is of radius 12 cm. The other is of radius 9 cm. Their centres are 15 cm apart. What is the perimeter of the logo?

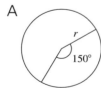

14 What is the ratio of the areas of the major sector in diagram A to the minor sector in diagram B?

A

r

150°

B

r 60°

15 Two circular table mats, each of radius 12 cm, are laid on a table with their centres 16 cm apart. Find
a the length of the common chord
b the area common to the two mats.

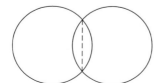

1.2 Trigonometric ratios

This unit circle can be used to define the trigonometric ratios.

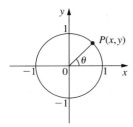

You should already know that for a right-angled triangle

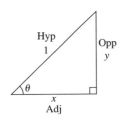

$$\sin \theta = \frac{\text{Opposite}}{\text{Hypotenuse}}$$

$$\cos \theta = \frac{\text{Adjacent}}{\text{Hypotenuse}}$$

$$\tan \theta = \frac{\text{Opposite}}{\text{Adjacent}}$$

The x-coordinate is defined to be $\cos \theta$.

The y-coordinate is defined to be $\sin \theta$.

The results for a right-angled triangle follow from the definitions of the x- and y-coordinates in the unit circle.

$$\tan \theta = \frac{\text{Opp}}{\text{Adj}} \Rightarrow \tan \theta = \frac{y}{x}$$

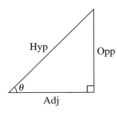

$$\Rightarrow \tan \theta = \frac{\sin \theta}{\cos \theta}$$

This is the definition of $\tan \theta$ and is a useful identity.

Using the definition of $\sin \theta$ and $\cos \theta$ from the unit circle, we can see that these trigonometric ratios are defined not only for acute angles, but for any angle. For example, $\sin 120° = 0.866$ (3 sf).

As the x-coordinate is $\cos \theta$ and the y-coordinate is $\sin \theta$, for obtuse angles $\sin \theta$ is positive and $\cos \theta$ is negative.

> More work will be done on trigonometric identities in Chapter 7.

Exact values

You need to learn sin, cos and tan of the angles given in the table overleaf for non-calculator examinations.

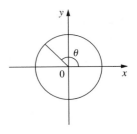

θ (in radians)	0	$\dfrac{\pi}{6}$	$\dfrac{\pi}{4}$	$\dfrac{\pi}{3}$	$\dfrac{\pi}{2}$
θ (in degrees)	0°	30°	45°	60°	90°
$\sin\theta$	0	$\dfrac{1}{2}$	$\dfrac{1}{\sqrt{2}}$	$\dfrac{\sqrt{3}}{2}$	1
$\cos\theta$	1	$\dfrac{\sqrt{3}}{2}$	$\dfrac{1}{\sqrt{2}}$	$\dfrac{1}{2}$	0
$\tan\theta$	0	$\dfrac{1}{\sqrt{3}}$	1	$\sqrt{3}$	undefined

> The last row is given by $\tan\theta = \dfrac{\sin\theta}{\cos\theta}$.

These values can also be remembered using the triangles shown below.

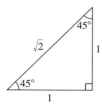 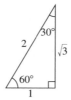

Finding an angle

When solving right-angled triangles, you found an acute angle.

Example

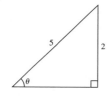

$$\sin\theta = \frac{2}{5}$$

$$\Rightarrow \theta = \sin^{-1}\left(\frac{2}{5}\right)$$

$$\Rightarrow \theta = 23.6°$$

However, $\sin\theta = \dfrac{2}{5}$ has two possible solutions:

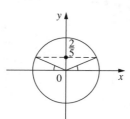

$$\sin\theta = \frac{2}{5}$$

$$\Rightarrow \theta = 23.6° \text{ or } 156.4°$$

> This is recognizing the symmetry of the circle.

Example

Solve $\cos \theta = \dfrac{1}{2}$ for $0 \le \theta < 2\pi$.

$\cos \theta = \dfrac{1}{2}$

$\Rightarrow \theta = \cos^{-1}\left(\dfrac{1}{2}\right)$

$\Rightarrow \theta = \dfrac{\pi}{3}$ or $\theta = 2\pi - \dfrac{\pi}{3}$

$\Rightarrow \theta = \dfrac{\pi}{3}$ or $\theta = \dfrac{5\pi}{3}$

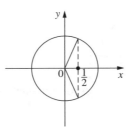

Exercise 2

1 Find the value of each of these.

   a $\sin 150°$ **b** $\sin 170°$ **c** $\cos 135°$ **d** $\cos 175°$

   e $\sin \dfrac{2\pi}{3}$ **f** $\sin \dfrac{3\pi}{4}$ **g** $\cos \dfrac{5\pi}{6}$ **h** $\cos 2.4$(radians)

2 Without using a calculator, find the value of each of these.

   a $\sin \dfrac{\pi}{6}$ **b** $\cos \dfrac{\pi}{3}$ **c** $\tan \dfrac{\pi}{4}$ **d** $\sin \dfrac{2\pi}{3}$

   e $\cos \dfrac{5\pi}{3}$ **f** $\sin 135°$ **g** $\cos 315°$ **h** $\sin 180°$

   i $\cos 180°$ **j** $\cos 270°$

3 Find the possible values of $x°$, given that $0° \le x° < 360°$.

   a $\sin x° = \dfrac{1}{2}$ **b** $\cos x° = \dfrac{1}{3}$ **c** $\sin x° = \dfrac{2}{3}$

   d $\cos x° = \dfrac{1}{6}$ **e** $\sin x° = \dfrac{3}{8}$ **f** $\cos x° = \dfrac{4}{7}$

4 Find the possible values of θ, given that $0 \le \theta < 2\pi$.

   a $\cos \theta = \dfrac{1}{2}$ **b** $\sin \theta = \dfrac{\sqrt{3}}{2}$ **c** $\cos \theta = \dfrac{1}{\sqrt{2}}$

   d $\sin \theta = 2$ **e** $\cos \theta = \dfrac{\sqrt{3}}{2}$ **f** $\sin \theta = \dfrac{2}{7}$

   g $\cos \theta = \dfrac{4}{11}$ **h** $\sin \theta = 0.7$

1.3 Solving triangles

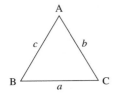

> Vertices are given capital letters. The side opposite a vertex is labelled with the corresponding lower-case letter.

Area of a triangle

We know that the area of a triangle is given by the formula

$A = \dfrac{1}{2} \times$ base $\times$ perpendicular height

To be able to use this formula, it is necessary to know the perpendicular height. This height can be found using trigonometry.

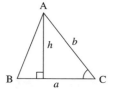

$\sin C = \dfrac{h}{b}$

$\Rightarrow h = b \sin C$

So the area of the triangle is given by

$$\text{Area} = \dfrac{1}{2}\, ab \sin C$$

This formula is equivalent to $\dfrac{1}{2} \times$ one side $\times$ another side $\times$ sine of angle between.

> **Example**
>
> Find the area of this triangle.
>
>
>
> $\text{Area} = \dfrac{1}{2} \times 6 \times 7 \times \sin 40°$
>
> $= 13.5 \text{ cm}^2$

Sine rule

Not all triangle problems can be solved using right-angled trigonometry. A formula called the sine rule is used in these problems.

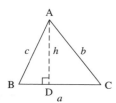

$\sin B = \dfrac{AD}{AB}$

$\Rightarrow AD = AB \times \sin B$

$\Rightarrow AD = c \sin B$

$\sin C = \dfrac{AD}{AC}$

$\Rightarrow AD = AC \times \sin C$

$\Rightarrow AD = b \sin C$

$\Rightarrow c \sin B = b \sin C$

$\Rightarrow \dfrac{b}{\sin B} = \dfrac{c}{\sin C}$

Drawing a line perpendicular to AC from B provides a similar result: $\dfrac{a}{\sin A} = \dfrac{c}{\sin C}$

Putting these results together gives the sine rule:

$$\frac{a}{\sin A} = \frac{b}{\sin B} = \frac{c}{\sin C}$$

Look at this obtuse-angled triangle:

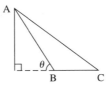

If we consider the unit circle, it is clear that $\sin \theta = \sin(180° - \theta)$ and hence $\sin \theta = \sin B$. So the result is the same.

> This is dealt with in more detail later in the chapter in relation to trigonometric graphs.

Use the sine rule in this form when finding a side:

$$\frac{a}{\sin A} = \frac{b}{\sin B} = \frac{c}{\sin C}$$

Use the sine rule in this form when finding an angle:

$$\frac{\sin A}{a} = \frac{\sin B}{b} = \frac{\sin C}{c}$$

Example

Find x.

$$\frac{c}{\sin C} = \frac{a}{\sin A}$$

$$\Rightarrow \frac{x}{\sin 40°} = \frac{8}{\sin 60°}$$

$$\Rightarrow x = \frac{8 \sin 40°}{\sin 60°}$$

$$\Rightarrow x = 5.94 \text{ m}$$

Example

Find angle P.

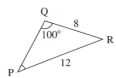

$$\frac{\sin P}{p} = \frac{\sin Q}{q}$$

$$\Rightarrow \frac{\sin P}{8} = \frac{\sin 100°}{12}$$

$$\Rightarrow \sin P = \frac{8 \sin 100°}{12}$$

$$\Rightarrow \sin P = 0.656 \ldots$$

$$\Rightarrow P = 41.0°$$

When the given angle is acute and it is opposite the shorter of two given sides, there are two possible triangles.

Example

In a triangle, angle $A = 40°$, $a = 9$ and $b = 13$. Find angle B.

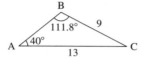

$$\frac{\sin 40°}{9} = \frac{\sin B}{13}$$

$$\Rightarrow \sin B = \frac{13 \sin 40°}{9}$$

$$\Rightarrow \sin B = 0.928 \cdots$$

$$\Rightarrow B = 68.2° \text{ or } B = 180° - 68.2° = 111.8°$$

Hence it is possible to draw two different triangles with this information:

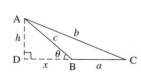

Cosine rule

The sine rule is useful for solving triangle problems but it cannot be used in every situation. If you know two sides and the angle between them, and want to find the third side, the cosine rule is useful.

The cosine rule is

$$a^2 = b^2 + c^2 - 2bc \cos A$$

We can prove this using an acute-angled triangle:

We know that $h^2 = c^2 - (a - x)^2 = c^2 - a^2 + 2ax - x^2$ and $h^2 = b^2 - x^2$.

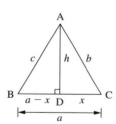

Hence $b^2 - x^2 = c^2 - a^2 + 2ax - x^2$

$$\Rightarrow c^2 = a^2 + b^2 - 2ax$$

Now $\cos C = \dfrac{x}{b}$

$$\Rightarrow x = b \cos C$$

$$\Rightarrow c^2 = a^2 + b^2 - 2ab \cos C$$

Drawing the perpendicular from the other vertices provides different versions of the rule:

$$a^2 = b^2 + c^2 - 2bc \cos A$$
$$b^2 = a^2 + c^2 - 2ac \cos B$$

The proof for an obtuse-angled triangle is similar:

In triangle ABD,   In triangle ACD,

$$h^2 = c^2 - x^2 \qquad h^2 = b^2 - (a + x)^2$$

$$= b^2 - a^2 - 2ax - x^2$$

$$\Rightarrow c^2 - x^2 = b^2 - a^2 - 2ax - x^2$$

$$\Rightarrow c^2 = b^2 - a^2 - 2ax$$

$$\text{Now } \frac{x}{c} = \cos \theta$$
$$= \cos(180° - B) \dotfill$$
$$= -\cos B$$
$$\Rightarrow x = -c \cos B$$

We will return to this later in the chapter.

$$\Rightarrow c^2 = b^2 - a^2 + 2ac \cos B$$
$$\Rightarrow b^2 = a^2 + c^2 - 2ac \cos B$$

This situation is similar to the area of the triangle formula. The different forms do not need to be remembered: it is best thought of as two sides and the angle in between.

Example

Find x.

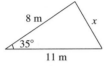

$$x^2 = 8^2 + 11^2 - 2 \times 8 \times 11 \times \cos 35°$$
$$x^2 = 40.829 \ldots$$
$$x = 6.39 \text{ m}$$

Pythagoras' theorem can be considered a special case of the cosine rule. This is the case where $A = 90° \Rightarrow \cos A = 0$.

The cosine rule can be rearranged to find an angle:

$$a^2 = b^2 + c^2 - 2bc \cos A$$
$$\Rightarrow 2bc \cos A = b^2 + c^2 - a^2$$

$$\Rightarrow \cos A = \frac{b^2 + c^2 - a^2}{2bc}$$

This is only one form. It may be useful to re-label the vertices in the triangle.

Example

Find angle A.

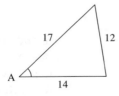

$$\cos A = \frac{b^2 + c^2 - a^2}{2bc}$$
$$\Rightarrow \cos A = \frac{17^2 + 14^2 - 12^2}{2 \times 17 \times 14}$$
$$= 0.716 \ldots$$
$$\Rightarrow A = 44.2°$$

Example

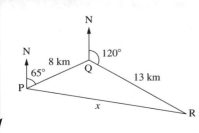

A ship sails on a bearing of 065° for 8 km, then changes direction at Q to a bearing of 120° for 13 km. Find the distance and bearing of R from P.

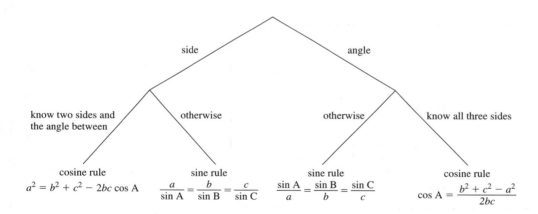

To find the distance x, angle Q is needed.

As the north lines are parallel, we can find angle Q.
So Q = 125°

Using the cosine rule, $x^2 = 8^2 + 13^2 - 2 \times 8 \times 13 \times \cos 125°$
$$x^2 = 352.3 \cdots$$
$$\Rightarrow x = 18.8 \text{ km } (3\text{sf})$$

We can now find the bearing of R from P.

Using the sine rule, $\dfrac{\sin P}{13} = \dfrac{\sin 125°}{18.769\ldots}$

$$\Rightarrow \sin P = \frac{13 \sin 125°}{18.769\ldots} = 0.5673\ldots$$
$$\Rightarrow P = 34.6°$$

Bearing of R from P is $65 + 34.6 = 099.6°$.

Decision making about triangle problems

It is worth remembering that Pythagoras' theorem and right-angled trigonometry can be applied to right-angled triangles, and they should not need the use of the sine rule or the cosine rule.

For non-right angled triangles, use this decision tree.

```
                              side        angle
know two sides and    otherwise     otherwise    know all three sides
the angle between

  cosine rule          sine rule       sine rule        cosine rule
a² = b² + c² − 2bc cos A   a/sinA = b/sinB = c/sinC   sinA/a = sinB/b = sinC/c   cosA = (b²+c²−a²)/2bc
```

cosine rule
$$a^2 = b^2 + c^2 - 2bc \cos A$$

sine rule
$$\frac{a}{\sin A} = \frac{b}{\sin B} = \frac{c}{\sin C}$$

sine rule
$$\frac{\sin A}{a} = \frac{\sin B}{b} = \frac{\sin C}{c}$$

cosine rule
$$\cos A = \frac{b^2 + c^2 - a^2}{2bc}$$

Once two angles in a triangle are known, the third angle can be found by subtracting the other two angles from 180°.

Exercise 3

1 Calculate the area of each triangle.

a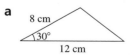

b

c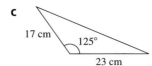

d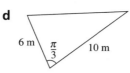

2 Three roads intersect as shown, with a triangular building plot between them. Calculate the area of the building plot.

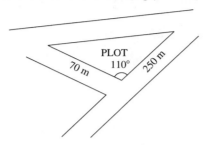

3 A design is created by an equilateral triangle of side 14 cm at the centre of a circle.
   a Find the area of the triangle.
   b Hence find the area of the segments.

4 An extension to a house is built as shown.
What is the volume of the extension?

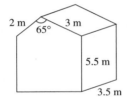

5 Find the area of this campsite.

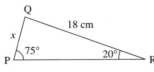

6 Use the sine rule to find the marked side.

a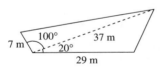

b

c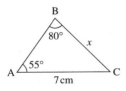

d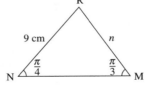

e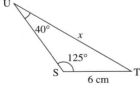

7 Use the sine rule to find the marked angle.

a
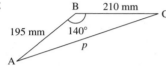

b
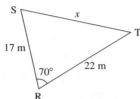

8 Triangle LMN has sides LM = 32 m and MN = 35 m with LNM = 40°
Find the possible values for ∠MLN.

9 Triangle ABC has sides AB = 11 km and BC = 6 km and ∠BAC = 20°.
Calculate ∠BCA.

10 Use the cosine rule to find the marked side.

a
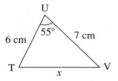

b
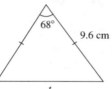

c
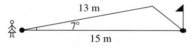

d
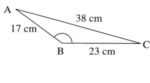

11 A golfer is standing 15 m from the hole. She putts 7° off-line and the ball
travels 13 m. How far is her ball from the hole?

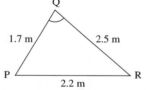

12 Use the cosine rule to find the marked angle.

a
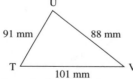

b
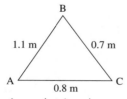

13 Calculate the size of the largest angle in triangle TUV.

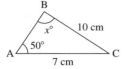

14 Find the size of all the angles in triangle ABC.

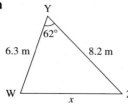

15 Calculate *x* in each triangle.

a
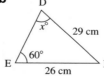

b **c**
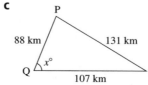

d
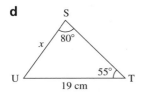

e
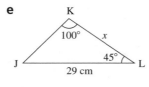

16 A plane flies from New York JFK airport on a bearing of 205° for 200 km. Another plane also leaves from JFK and flies for 170 km on a bearing of 320°. What distance are the two planes now apart?

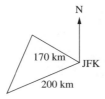

17 Twins Anna and Tanya, who are both 1.75 m tall, both look at the top of Cleopatra's Needle in Central Park, New York. If they are standing 7 m apart, how tall is the Needle?

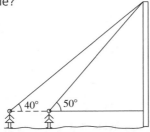

18 Find the size of angle ACE.

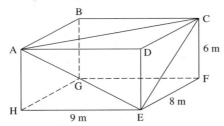

19 Find the area of triangle SWV.

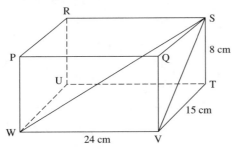

1.4 Trigonometric functions and graphs

$\sin \theta$ is defined as the y-coordinate of points on the unit circle.

θ	0°	30°	45°	60°	90°	180°	270°	360°
$\sin \theta$	0	$\dfrac{1}{2}$	$\dfrac{1}{\sqrt{2}}$	$\dfrac{\sqrt{3}}{2}$	1	0	−1	0

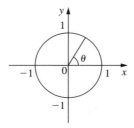

cos θ is defined as the x-coordinate of points on the unit circle.

θ	0°	30°	45°	60°	90°	180°	270°	360°
cos θ	1	$\dfrac{\sqrt{3}}{2}$	$\dfrac{1}{\sqrt{2}}$	$\dfrac{1}{2}$	0	-1	0	1

These functions are plotted below.

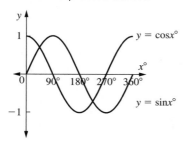

Both graphs only have y-values of $-1 \leq y \leq 1$.

Periodicity

When considering angles in the circle, it is clear that any angle has an equivalent angle in the domain $0 \leq x° < 360°$.

For example, an angle of 440° is equivalent to an angle of 80°.

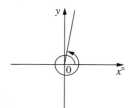

$$440° = 360° + 80°$$

This is also true for negative angles.

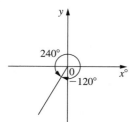

$$-120° = 240°$$

This means that the sine and cosine graphs are infinite but repeat every 360° or 2π.

$y = \sin \theta$

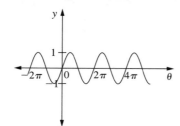

These graphs can be drawn using degrees or radians.

$y = \cos\theta$

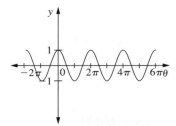

Repeating at regular intervals is known as **periodicity**. The period is the interval between repetitions.

For $y = \sin\theta$ and $y = \cos\theta$, the period is 360° or 2π.

Graph of tan x°

We have defined $\tan\theta$ as $\dfrac{\sin\theta}{\cos\theta}$. This allows us to draw its graph.

θ	0°	30°	45°	60°	90°	180°	270°	360°
$\sin\theta$	0	$\dfrac{1}{2}$	$\dfrac{1}{\sqrt{2}}$	$\dfrac{\sqrt{3}}{2}$	1	0	-1	0
$\cos\theta$	1	$\dfrac{\sqrt{3}}{2}$	$\dfrac{1}{\sqrt{2}}$	$\dfrac{1}{2}$	0	-1	0	1
$\tan\theta$	0	$\dfrac{1}{\sqrt{3}}$	1	$\sqrt{3}$	undefined	0	undefined	0

There is a problem when $x° = 90°, 270° \ldots$ because there is a zero on the denominator. This is undefined (or infinity). Graphically, this creates a **vertical asymptote**. This is created by an x-value where the function is not defined. The definition of an asymptote is that it is a line associated with a curve such that as a point moves along a branch of the curve, the distance between the line and the curve approaches zero. By examining either side of the vertical asymptote, we can obtain the behaviour of the function around the asymptote.

> The vertical asymptote is a line: there are other types of asymptote that we will meet later.

As $x° \rightarrow 90°$ ($x°$ approaches 90°) tan x° increases and approaches ∞ (infinity):

$\tan 85° = 11.4$, $\tan 89° = 57.3$, $\tan 89.9° = 573$ etc.

On the other side of the asymptote, tan x° decreases and approaches $-\infty$:

$\tan 95° = -11.4$, $\tan 91° = -57.3$, $\tan 90.1° = -573$ etc.

The graph of $y = \tan x°$ is shown below.

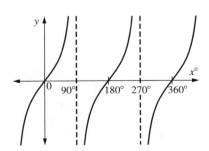

It is clear that this graph is also periodic, and the period is 180°.

Reciprocal trigonometric functions

There are three more trigonometrical functions, defined as the reciprocal trigonometric functions – secant, cosecant and cotangent. Secant is the reciprocal function to cosine, cosecant is the reciprocal function to sine, and cotangent is the reciprocal function to tangent. These are abbreviated as follows:

$$\sec \theta = \frac{1}{\cos \theta}, \quad \csc \theta = \frac{1}{\sin \theta} \text{ (or cosec } \theta) \quad \cot \theta = \frac{1}{\tan \theta}$$

In order to obtain the graph of $f(x) = \csc \theta$, consider the table below.

θ	0°	30°	45°	60°	90°	180°	270°	360°
$\sin \theta$	0	$\frac{1}{2}$	$\frac{1}{\sqrt{2}}$	$\frac{\sqrt{3}}{2}$	1	0	-1	0
$\csc \theta = \dfrac{1}{\sin \theta}$	∞	2	$\sqrt{2}$	$\frac{2}{\sqrt{3}}$	1	∞	-1	∞

The roots (zeros) of the original function become vertical asymptotes in the reciprocal function.

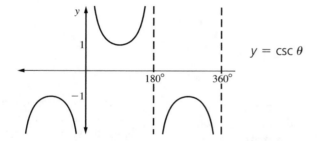

$y = \csc \theta$

This function is also periodic with a period of 360°.

Similarly we can obtain the graphs of $y = \sec \theta$ and $y = \cot \theta$:

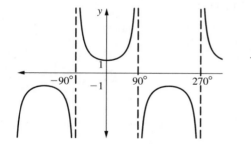

$y = \sec \theta$

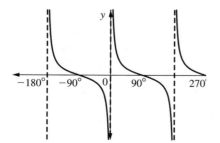

$y = \cot \theta$

The general method for plotting reciprocal graphs will be addressed in Chapter 8.

Composite graphs

Using your graphing calculator, draw the following graphs to observe the effects of the transformations.

1. $y = 2 \sin x°$
   $y = 3 \cos x°$
   $y = 5 \sin x°$
   $y = \dfrac{1}{2} \sin x°$

2. $y = \sin 2x°$
   $y = \cos 3x°$
   $y = \sin 5x°$
   $y = \cos\dfrac{1}{2}x°$

3. $y = -\sin x°$
   $y = -\cos x°$
   $y = -\tan x°$

4. $y = \sin(-x°)$
   $y = \cos(-x°)$
   $y = \tan(-x°)$

5. $y = \sin x° + 2$
   $y = \cos x° + 2$
   $y = \sin x° - 1$
   $y = \cos x° - 1$

6. $y = \sin(x - 30)°$
   $y = \cos(x - 30)°$
   $y = \sin\left(\theta + \dfrac{\pi}{3}\right)$
   $y = \cos\left(\theta + \dfrac{\pi}{3}\right)$

The table summarizes the effects.

y =	Effect	Notes
$A \sin x$, $A \cos x$, $A \tan x$	Vertical stretch	
$\sin Bx$, $\cos Bx$, $\tan Bx$	Horizontal stretch/ compression	This is the only transformation that affects the period of the graph
$-\sin x$, $-\cos x$, $-\tan x$	Reflection in x-axis	
$\sin(-x)$, $\cos(-x)$, $\tan(-x)$	Reflection in y-axis	
$\sin x + C$, $\cos x + C$, $\tan x + C$	Vertical shift	
$\sin(x + D)$, $\cos(x + D)$, $\tan(x + D)$	Horizontal shift	Positive D left, negative D right

Example

Draw the graph of $y = 2\cos\theta + 1$ for $0 \le \theta < 2\pi$.

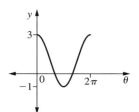

This is a vertical stretch $\times\,2$ and a vertical shift $+1$.

The domain tells you how much of the graph to draw and whether to work in degrees or radians.

Example

Draw the graph of $y = \csc(x - 90)° + 1$ for $0° \le x° < 180°$.

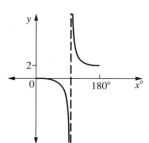

This is a horizontal shift of 90° to the right and a vertical shift $+1$.

Example

Draw the graph of $y = -3\sin 2x°$ for $0° \le x° < 360°$.

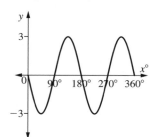

In this case, there are two full waves in 360°. There is a reflection in the x-axis and a vertical stretch $\times\,3$.

Example

What is the equation of this graph?

We assume that, because of the shape, it is either a sine or cosine graph. Since it begins at a minimum point, we will make the assumption that it is a cosine graph. (We could use sine but this would involve a horizontal shift, making the question more complicated.) $y = \cos\theta$

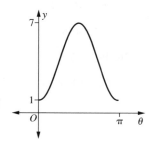

Since it is "upside down" there is a reflection in the x axis $\Rightarrow y = -\cos \theta$
There is a period of π and so there are two full waves in $2\pi \Rightarrow y = -\cos 2\theta$
There is a difference of 6 between the max and min values. There would
normally be a difference of 2 and hence there is a $\times 3$ vertical stretch
$$\Rightarrow y = -3 \cos 2\theta$$
The min and max values are 1, 7 so there is a shift up of 4
$$\Rightarrow y = -3 \cos 2\theta + 4$$

So the equation of this graph is $y = -3 \cos 2\theta + 4$.

Exercise 4

1 What is the period of each function?

a $y = \sin 2x°$ **b** $y = \cos 3x°$ **c** $y = \cos 4\theta + 1$

d $y = \tan 2x°$ **e** $y = \sec x°$ **f** $y = 2 \cos 3x° - 3$

g $y = 5 \csc 2x° + 3$ **h** $y = 7 - 3 \sin 4\theta$ **i** $y = 9 \sin 10x°$

j $y = 8 \tan 60x°$

2 Draw the graphs of these functions for $0° \le x° < 360°$.

a $y = \sin 3x°$ **b** $y = -\cos x°$ **c** $y = \sin(-x°)$

d $y = 4 \csc x$ **e** $y = \tan(x - 30)°$ **f** $y = \sec x° + 2$

3 Draw the graphs of these functions for $0 \le \theta < 2\pi$.

a $y = -\sin 2\theta$ **b** $y = \cot\left(\theta + \dfrac{\pi}{3}\right)$ **c** $y = 4 \cos \theta - 2$

d $y = \csc(-\theta)$ **e** $y = 3 - 5 \sin \theta$

4 Draw the graphs of these functions for $0° \le x° < 180°$.

a $y = \cos 3x°$ **b** $y = 2 \sin 4x°$ **c** $y = 3 \sec 2x°$

d $y = 6 \sin 10x°$ **e** $y = 2 \tan(x + 30)°$ **f** $y = 3 \cos 2x° - 1$

5 Draw the graphs of these functions for $0 \le \theta < \pi$.

a $y = 6 \cos \theta + 2$ **b** $y = 3 \sin 4\theta - 5$ **c** $y = 7 - 4 \cos \theta$

d $y = \cot 3\theta$ **e** $y = \tan\left(2\theta - \dfrac{\pi}{3}\right)$

6 Draw the graph of $y = 4 \sin 2x° - 3$ for $0° \le x° < 720°$.

7 Draw the graph of $y = 6 \cos 30x°$ for $0° \le x° < 12°$.

8 Draw the graph of $y = 8 - 3 \sin 4\theta$ for $0 \le \theta < \dfrac{\pi}{2}$.

9 Find the equation of each graph.

a

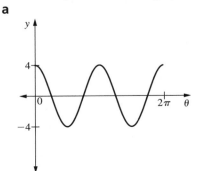

b

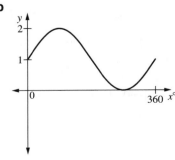

c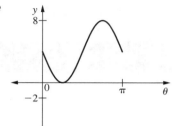

d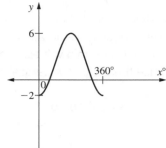

e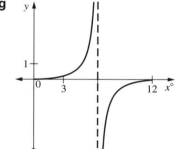

f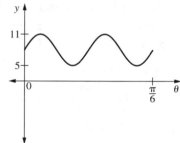

g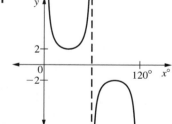

h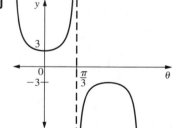

i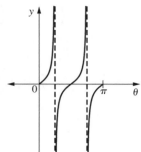

j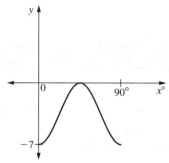

1.5 Related angles

To be able to solve trigonometric equations algebraically we need to consider properties of the trigonometric graphs. Each graph takes a specific y-value for an infinite number of x-values. Within this curriculum, we consider this only within a finite domain. Consider the graphs below, which have a domain $0° \leq x° < 360°$.

$y = \sin x°$ $y = \cos x°$ $y = \tan x°$

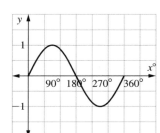 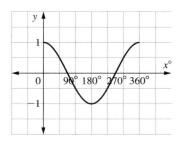 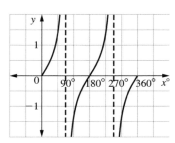

These graphs can be split into four quadrants, each of 90°. We can see that in the first quadrant all three graphs are above the x-axis (positive).

In each of the other three quadrants, only one of the functions is positive.

This is summarised in the following diagram.

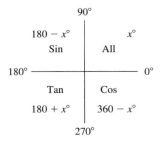

The diagram shows two important features. First, it shows where each function is positive. Second, for every acute angle, there is a related angle in each of the other three quadrants. These related angles give the same numerical value for each trigonometric function, ignoring the sign. This diagram is sometimes known as the **bow-tie diagram**.

By taking an example of 25°, we can see all of the information that the bow-tie diagram provides:

Related angles
25°
$180 - 25 = 155°$
$180 + 25 = 205°$
$360 - 25 = 335°$

Using 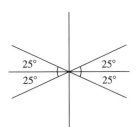 we can say that

$\sin 155° = \sin 25°$
$\sin 205° = -\sin 25°$
$\sin 335° = -\sin 25°$

$\cos 155° = -\cos 25°$
$\cos 205° = -\cos 25°$
$\cos 335° = \cos 25°$

$\tan 155° = -\tan 25°$
$\tan 205° = \tan 25°$
$\tan 335° = -\tan 25°$

Example

Find the exact value of $\cos\dfrac{4\pi}{3}$, $\sin\dfrac{7\pi}{6}$ and $\tan\dfrac{7\pi}{4}$.

This is the bow-tie diagram in radians:

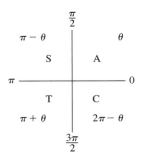

For $\cos\dfrac{4\pi}{3}$, we need to find the related acute angle.

$$\pi + \theta = \frac{4\pi}{3}$$

$$\Rightarrow \theta = \frac{\pi}{3}$$

Since $\dfrac{4\pi}{3}$ is in the third quadrant, $\cos\dfrac{4\pi}{3}$ is negative.

So $\cos\dfrac{4\pi}{3} = -\cos\dfrac{\pi}{3} = -\dfrac{1}{2}$

Considering $\sin\dfrac{7\pi}{6}$:

$$\pi + \theta = \frac{7\pi}{6}$$

$$\Rightarrow \theta = \frac{\pi}{6}$$

Since $\dfrac{7\pi}{6}$ is in the third quadrant, $\sin\dfrac{7\pi}{6}$ is negative.

So $\sin\dfrac{7\pi}{6} = -\sin\dfrac{\pi}{6} = -\dfrac{1}{2}$

Considering $\tan\dfrac{7\pi}{4}$:

$$2\pi - \theta = \frac{7\pi}{4}$$

$$\Rightarrow \theta = \frac{\pi}{4}$$

Since $\dfrac{7\pi}{4}$ is in the fourth quadrant, $\tan\dfrac{7\pi}{4}$ is negative.

So $\tan\dfrac{7\pi}{4} = -\tan\dfrac{\pi}{4} = -1$

Exercise 5

1 Find the exact value of each of these.

   a cos 120°   **b** tan 135°   **c** sin 150°   **d** cos 300°

   e tan 225°   **f** cos 210°   **g** tan 300°   **h** sin 240°

   i cos 330°   **j** cos 150°

2 Find the exact value of each of these.

   a $\tan\dfrac{7\pi}{6}$ **b** $\sin\dfrac{3\pi}{4}$ **c** $\cos\dfrac{11\pi}{6}$ **d** $\tan\dfrac{5\pi}{3}$

   e $\sin\dfrac{5\pi}{4}$ **f** $\tan\dfrac{5\pi}{6}$ **g** $\cos\dfrac{3\pi}{2}$ **h** $\sin\dfrac{5\pi}{3}$

   i $2\sin\dfrac{5\pi}{6}$ **j** $8\cos\dfrac{11\pi}{6}$

3 Express the following angles, using the bow-tie diagram, in terms of the related acute angle.

   a sin 137°   **b** cos 310°   **c** tan 200°

   d sin 230°   **e** cos 157°   **f** tan 146°

   g cos 195°   **h** sin 340°   **i** tan 314°

4 State two possible values for $x°$ given that $0° \leq x° < 360°$.

   a $\sin x° = \dfrac{1}{2}$ **b** $\cos x° = \dfrac{\sqrt{3}}{2}$

   c $\tan x° = \sqrt{3}$ **d** $\tan x° = -1$

5 State two possible values for θ given that $0 \leq \theta < 2\pi$.

   a $\sin\theta = \dfrac{\sqrt{3}}{2}$ **b** $\tan\theta = \dfrac{1}{\sqrt{3}}$

   c $\cos\theta = \dfrac{1}{2}$ **d** $\cos\theta = -\dfrac{\sqrt{3}}{2}$

1.6 Trigonometric equations

We can use related angles to help solve trigonometric equations, especially without a calculator.

Example

Solve $2\sin x° + 3 = 4$ for $0° \leq x° < 360°$.

$2\sin x° + 3 = 4$

$\Rightarrow 2\sin x° = 1$

$\Rightarrow \sin x° = \dfrac{1}{2}$

$\Rightarrow x° = 30°, (180 - 30)°$

$\Rightarrow x° = 30°, 150°$

Thinking of the graph of sin $x°$, it is clear these are the only two answers:

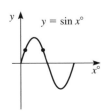

It is very important to take account of the domain.

Example

Solve $2\cos\theta + \sqrt{3} = 0$ for $0 \le \theta < 2\pi$.

$2\cos\theta + \sqrt{3} = 0$

$$\Rightarrow \cos\theta = -\frac{\sqrt{3}}{2}$$

We know that $\cos\dfrac{\pi}{6} = \dfrac{\sqrt{3}}{2}$ and cos is negative in

the second and third quadrants.

$$\Rightarrow \theta = \left(\pi - \frac{\pi}{6}\right), \left(\pi + \frac{\pi}{6}\right)$$

$$\Rightarrow \theta = \frac{5\pi}{6}, \frac{7\pi}{6}$$

Example

Solve $2\cos(3x - 15)° = 1$ for $0° \le x° < 360°$.

$2\cos(3x - 15)° = 1$

$$\Rightarrow \cos(3x - 15)° = \frac{1}{2}$$

$$\Rightarrow (3x - 15)° = 60° \text{ or } 300°$$

$$\Rightarrow 3x° = 75° \text{ or } 315°$$

$$\Rightarrow x° = 25° \text{ or } 105°$$

We know that $3x$ means three full waves in 360° and so the period is 120°.

The other solutions can be found by adding on the period to these initial values:
$x° = 25°, 105°, 145°, 225°, 265°, 345°$

A graphical method can also be used to solve trigonometric equations, using a calculator.

Example

Solve $5\sin x° + 2 = 3$ for $0° \le x° < 360°$.

Using a calculator:

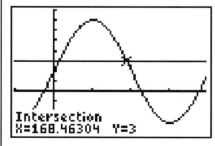

$x° = 11.5°, 168.5°$

Example

Solve $7 - 3 \tan \theta = 11$ for $0 \le \theta < 2\pi$.

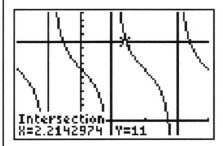

Intersection
X=2.2142974 Y=11

$\theta = 2.21, 5.36$

Example

Solve $5 - 2 \sec x° = 8$ for $0° \le x° < 180°$.

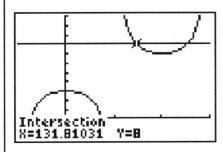

Intersection
X=131.81031 Y=8

Noting the domain, $x° = 131.8°$

The algebraic method can be used in conjunction with a calculator to solve any equation.

Example

Solve $3 \cos 3\theta + 5 = 4$ for $0 \le \theta < 2\pi$.

$3 \cos 3\theta + 5 = 4$
$\Rightarrow 3 \cos 3\theta = -1$
$\Rightarrow \cos 3\theta = -\dfrac{1}{3}$

Use a calculator to find $\cos^{-1}\left(\dfrac{1}{3}\right) = 1.23$

✓S	A
✓T	C

Cos is negative in the second and third quadrants.
$\Rightarrow 3\theta = \pi - 1.23, \pi + 1.23$
$\Rightarrow 3\theta = 1.91, 4.37$
$\Rightarrow \theta = 0.637, 1.46$

Here the period is $\frac{2\pi}{3}$ and hence we can find all six solutions:

$\theta = 0.637, 1.46, 2.73, 3.55, 4.83, 5.65$

Example

Solve $2\sin 2\theta + 3 = 2$ for $-\pi \le \theta < \pi$.

Here we notice that the domain includes negative angles. It is solved in the same way.

$2\sin 2\theta + 3 = 2$

$\Rightarrow 2\sin 2\theta = -1$

$\Rightarrow \sin 2\theta = -\frac{1}{2}$

sin is negative in the third and fourth quadrants:

We know that $\sin\frac{\pi}{6} = \frac{1}{2}$

so the related angles are $\frac{7\pi}{6}$ and $\frac{11\pi}{6}$.

Hence $2\theta = \frac{7\pi}{6}, \frac{11\pi}{6}$

$\Rightarrow \theta = \frac{7\pi}{12}, \frac{11\pi}{12}$

Now we just need to ensure that we have all of the solutions within the domain by using the period. These two solutions are both within the domain. The other two solutions required can be found by subtracting a period:

$\theta = -\frac{5\pi}{12}, -\frac{\pi}{12}, \frac{7\pi}{12}, \frac{11\pi}{12}$

Exercise 6

 1 Solve these for $0° \le x° < 360°$.

    a $\tan x° = \sqrt{3}$ **b** $\cos x° = \frac{1}{2}$ **c** $\sin x° = \frac{\sqrt{3}}{2}$

    d $2\sin x° + 1 = 0$ **e** $2\cos x° = -\sqrt{3}$ **f** $\cos x° + 1 = 0$

    g $4\sin x° - 3 = 1$ **h** $\csc x° = 2$ **i** $6\cot x° - 1 = 5$

 2 Solve these for $0 \le \theta < 2\pi$.

    a $\cos\theta = \frac{\sqrt{3}}{2}$ **b** $\sin\theta = -\frac{1}{2}$ **c** $\tan\theta = -\frac{1}{\sqrt{3}}$

    d $3\tan\theta + 2 = 5$ **e** $4 - 2\sin\theta = 3$ **f** $3\tan\theta = \sqrt{3}$

    g $\sin\left(\theta - \frac{\pi}{6}\right) = \frac{\sqrt{3}}{2}$ **h** $\sqrt{3}\sec\theta = 2$

3 Solve these for $0° \leq x° < 360°$.

   a $\sin 2x° = \dfrac{1}{2}$ **b** $2\cos 3x° = \sqrt{3}$ **c** $6\tan 4x° = 6$

   d $2\cos 2x° = -1$ **e** $4\sin(3x - 15)° = 2\sqrt{3}$ **f** $\sec 3x° = -2$

 4 Solve these for $0 \leq \theta < 2\pi$.

   a $\cos 4\theta = \dfrac{1}{2}$ **b** $\tan\left(2\theta - \dfrac{\pi}{6}\right) = \dfrac{1}{\sqrt{3}}$

   c $4 - 2\sin 5\theta = 3$ **d** $6\cos 2\theta = -\sqrt{27}$

 5 Solve $\sqrt{3}\tan 2x° - 1 = 0$ for $0° \leq x° < 180°$.

 6 Solve $2\sin 4\theta + 1 = 0$ for $0 \leq \theta < \pi$.

 7 Solve $6\sin 30x° - 3 = 0$ for $0° \leq x° < 24°$.

 8 Solve $2\tan x° = \sqrt{12}$ for $-180° \leq x° < 180°$.

9 Solve $6\cos 3\theta + 2 = -1$ for $-\pi \leq \theta < \pi$.

 10 Solve these for $0° \leq x° < 360°$.

   a $3\sin x° = 1$ **b** $4\cos x° = 3$ **c** $5\tan x° - 1 = 7$

   d $6\cos x° - 5 = -1$ **e** $4\sin x° - 3 = 0$ **f** $8\cos(x + 20)° = 5$

   g $3 - 4\cos x° = 2$ **h** $\sqrt{2}\sin x° - 3 = -2$ **i** $9\sin(x - 15)° = -5$

   j $7 - 5\sin x° = 4$ **k** $6\sin 2x° - 5 = -1$ **l** $8\cos 3x° + 5 = 7$

   m $7 + 11\tan 5x° = -9$ **n** $\sec x° = 3$ **o** $\csc x° - 2 = 5$

   p $4\sec x° + 3 = 9$ **q** $6\cot x° - 1 = 8$ **r** $9\sec 4x° + 3 = 21$

   s $4 - 3\sin 30x° = 2$

 11 Solve these for $0 \leq \theta < 2\pi$.

   a $4\sin \theta = 1$ **b** $9\cos \theta = -4$ **c** $8\tan \theta - 2 = 17$

   d $\sqrt{5}\cos \theta - 4 = -3$ **e** $7\cos\left(\theta - \dfrac{\pi}{3}\right) = 4$ **f** $6 - 5\sin \theta = 7$

   g $9 + 5\tan \theta = 23$ **h** $3\cos 3\theta - 1 = 0$ **i** $6\sin 2\theta = -1$

   j $7 - 2\tan 4\theta = 13$ **k** $9 - 4\sin 3\theta = 6$ **l** $8\sec \theta = 19$

   m $1 - 3\csc \theta = 11$ **n** $2 + \cot \theta = 9$ **o** $6\csc 4\theta - 3 = 11$

 12 Solve $8 - 3\cos x° = 7$ for $0° \leq x° < 720°$.

 13 Solve $5 + 2\sin\left(3\theta - \dfrac{\pi}{4}\right) = 6$ for $-\pi \leq \theta < \pi$.

14 The height of a basket on a Ferris wheel is modelled by

$$H(t) = 21 - 18\sin\left(\dfrac{2\pi}{3}t\right)$$

   where H is the height above the ground in metres and t is the time in minutes.
   a How long does it take to make one complete revolution?
   b Sketch the graph of the height of the basket during one revolution.
   c When is the basket at its (i) maximum height (ii) minimum height?

 15 The population of tropical fish in a lake can be estimated using

$$P(t) = 6000 + 1500\cos 15t$$

   where t is the time in years. Estimate the population
   a initially   **b** after 3 years.
   c Find the minimum population estimate and when this occurs.

1.7 Inverse trigonometric functions

In order to solve trigonometric equations, we employed the inverse function.

For example, $\sin x° = \dfrac{1}{2}$

$$\Rightarrow x° = \arcsin\left(\dfrac{1}{2}\right)$$

> arcsin is the inverse sine function (also denoted $\sin^{-1}$).

An inverse function is one which has the opposite effect to the function itself.

For an inverse function, the range becomes the domain and the domain becomes the range.

> More work is done on inverse functions in Chapter 3.

Hence for the inverse of the sine and cosine functions, the domain is $[-1, 1]$.

The graphs of the inverse trigonometric functions are:

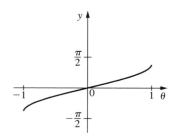

$y = \arcsin\theta$

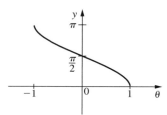

$y = \arccos\theta$

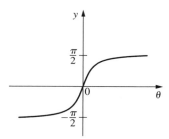

$y = \arctan\theta$

Review exercise

 1 Express in degrees: **a** $\dfrac{\pi}{6}$ **b** $\dfrac{5\pi}{12}$

 2 Express in radians: **a** $120°$ **b** $195°$

 3 Find the area of this segment.

18 cm

70 cm

4 Find the length of this arc

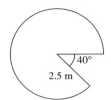

 5 The diagram below shows a circle centre O and radius OA = 5 cm.
The angle AOB = 135°.
Find the shaded area.

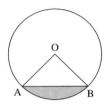

[IB Nov 04 P1 Q9]

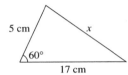 **6** Find x in each triangle.

a

b

c

d

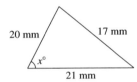
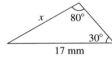

 7 In the triangle ABC, the side AB has length 5 and the angle BAC = 28°. For what range of values of the length of BC will two distinct triangles ABC be possible?

 8 Find the exact value of each of these.

a $\cos\dfrac{2\pi}{3}$ **b** $\sin\dfrac{3\pi}{4}$ **c** $\tan\dfrac{5\pi}{6}$ **d** $\sin\dfrac{7\pi}{6}$

e $\cos\dfrac{7\pi}{4}$ **f** $\sin 300°$ **g** $\tan 240°$ **h** $\cos 135°$

i $\tan 330°$ **j** $\sec 60°$ **k** $\csc 240°$

 9 Sketch each of these graphs.
a $y = 6\cos 2\theta - 1$ **b** $y = 4\sin(x - 30)°$
c $y = 8 - 3\cos 4\theta$ **d** $y = 5\sec\theta$
e $y = \arcsin\theta$

 10 State the equation of the graph.

a

b

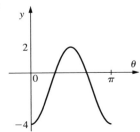
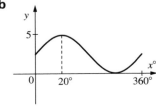

11 Solve these for $0 \leq \theta < 2\pi$.

    a $2 \sin \theta - 1 = 0$ **b** $2 \cos \theta + \sqrt{3} = 0$

    c $6 \tan \theta - 6 = 0$ **d** $2 \sin 4\theta - \sqrt{3} = 0$

    e $\sqrt{3} \tan 2\theta + 1 = 0$

 12 Solve these for $0° \leq x° < 360°$.

    a $8 \tan x° + 8 = 0$ **b** $9 \sin x° = 9$

    c $4 \sin x° + 2 = 0$ **d** $\sqrt{3} \tan x° + 1 = 4$

    e $6 \cos 2x° = 3\sqrt{3}$ **f** $8 \sin 3x° - 4 = 0$

 13 Solve these for $0° \leq x° < 360°$.

    a $7 \cos x° - 3 = 0$ **b** $8 \sin 2x° + 5 = 0$

    c $9 \tan 3x° - 17 = 0$ **d** $3 \sec x° - 7 = 0$

 14 Solve $2 \sin x = \tan x$ for $-\dfrac{\pi}{2} \leq x \leq \dfrac{\pi}{2}$.   [IB May 01 P1 Q2]

 15 The angle θ satisfies the equation $\tan \theta + \cot \theta = 3$ where θ is in degrees. Find all the possible values of θ lying in the interval $[0°, 90°]$.

                                                    [IB May 02 P1 Q10]

 16 The height in cm of a cylindrical piston above its central position is given by

$$h = 16 \sin 4t$$

    where t is the time in seconds, $0 \leq t \leq \dfrac{\pi}{4}$.

    a   What is the height after $\dfrac{1}{2}$ second? `

    b   Find the first time at which the height is 10 cm.

17 Let $f(x) = \sin\left(\arcsin \dfrac{x}{4} - \arccos \dfrac{3}{5}\right)$ for $-4 \leq x \leq 4$.

    a   Sketch the graph of $f(x)$.

    b   On the sketch, clearly indicate the coordinates of the x-intercept, the y-intercept, the minimum point and the endpoints of the curve of $f(x)$.

    c   Solve $f(x) = -\dfrac{1}{2}$.   [IB Nov 03 P1 Q14]

2 Quadratic Equations, Functions and Inequalities

The first reference to quadratic equations appears to be made by the Babylonians in 400 BC, even though they did not actually have the notion of an equation. However, they succeeded in developing an algorithmic approach to solving problems that could be turned into quadratic equations. Most of the problems that the Babylonians worked on involved length, hence they had no concept of a negative answer. The Hindu mathematician, Brahmagupta, undertook more work in the seventh century and he realised that negative quantities were possible and he worked on the idea of letters for unknowns. In the ninth century, in his book *Hisab al-jabr w'al-muqabala*, Al-khwarizmi solved quadratic equations entirely in words. The word "algebra" is derived from the title of his book. It was not until the twelfth century that Abraham bar Hiyya ha-Nasi finally developed a full solution to a quadratic equation.

USSR stamp featuring Al-khwarizmi

2.1 Introduction to quadratic functions

Consider the curve $y = x^2$. What does it look like?

To draw the graph a table of values can be set up on a calculator or drawn on paper.

x	−3	−2	−1	0	1	2	3
x^2	9	4	1	0	1	4	9
y	9	4	1	0	1	4	9

The curve is shown below.

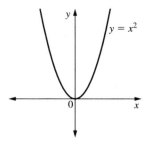

The main features of the curve are:

- It is symmetrical about the y-axis.
- The y-values are always greater than or equal to zero (there are no negative y-values).
- The minimum value is $y = 0$.

Now consider the curve $y = x^2 - x - 2$. This is the table of values:

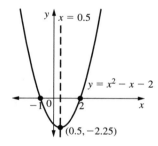

The curve is shown below.

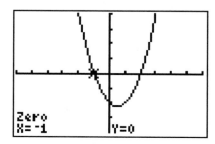

In this case:

- The line of symmetry is $x = 0.5$.
- The curve intersects the x-axis at $x = -1$ and $x = 2$.
- The minimum value of the curve is $y = -2.25$.

This information can also be found on the calculator, and the displays for this are shown below.

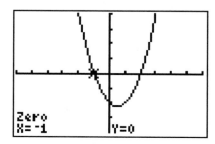

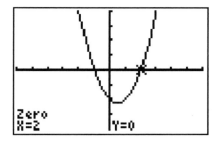

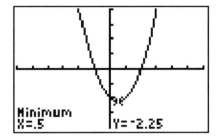

The standard quadratic function is $f(x) = ax^2 + bx + c$ where $a \in \mathbb{R}, a \neq 0$ and $b, c \in \mathbb{R}$. Its graph is known as a **parabola**. However, this is not the only form that produces a parabolic graph. Graphs of the form $y^2 = 2ax$ are also parabolic, and an example is shown below.

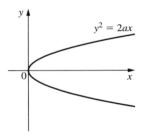

Investigation

Sketch the following 12 curves using your calculator:

$y = x^2 - 5x + 6$ $y = x^2 - 4x + 4$ $y = 2x^2 + 5x + 2$

$y = 2x^2 + 5x - 3$ $y = -x^2 - 6x - 8$ $y = -x^2 - 4x - 6$

$y = x^2 + 7x + 9$ $y = -x^2 + 4x - 4$ $y = 4x^2 - 4x + 1$

$y = -x^2 + 3x - 7$ $y = -4x^2 - 12x - 9$ $y = -x^2 + 4x - 3$

Use the graphs to deduce general rules. Think about about the following points:

- When do these curves have maximum turning points? When do they have minimum turning points?
- What is the connection between the x-value at the turning point and the line of symmetry?
- What is the connection between the x-intercepts of the curve and the line of symmetry?
- Describe the intersection of the curve with the x-axis.

> The maximum or minimum turning point is the point where the curve turns and has its greatest or least value. This will be looked at in the context of other curves in Chapter 8.

From this investigation, it is possible to deduce that:

- If the coefficient of x^2 is positive, the curve has a minimum turning point. If it is negative, the curve has a maximum turning point.
- The x-value where the maximum or minimum turning point occurs is also the line of symmetry.
- The line of symmetry is always halfway between the x-intercepts if the curve crosses the x-axis twice.
- There are three possible scenarios for the intersection of the curve with the x-axis:
  - It intersects twice.
  - It touches the x-axis
  - It does not cross or touch the x-axis at all.

In Chapter 1 transformations of curves were introduced. We will now look at three transformations when applied to the quadratic function.

$y = (x + a)^2$

This translates (shifts) the curve $y = x^2$ left by a units if a is positive or right by a units if a is negative. The curve does not change shape; it merely shifts left or right. Examples are shown below.

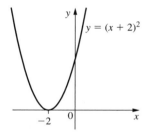
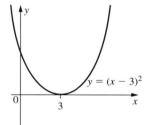

$y = x^2 + a$

This translates (shifts) the curve $y = x^2$ up by a units if a is positive or down by a units if a is negative. Once again, the curve does not change shape; it merely shifts up or down. Examples are shown below.

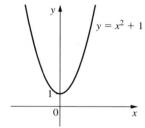
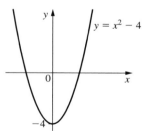

$y = ax^2$

In this case the curve does change shape.

If a is positive this stretches the curve $y = x^2$ parallel to the y-axis by scale factor a. If a is negative the stretch is the same but the curve is also reflected in the x-axis. $y = 2x^2$ and $y = -2x^2$ are shown below.

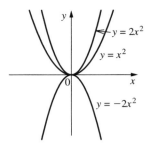

The parabola shape of a quadratic graph occurs in a number of natural settings. One possible example is the displacement-time graph of a projectile, but it must be remembered that this does not take into account the effect of air resistance. A suspension bridge provides a more realistic example. When a flexible chain is hung loosely between two supports, the shape formed by the chain is a curve called a **catenary**. However, if a load is hung from this chain, as happens in a suspension bridge, then the curve produced is in fact a parabola. Hence the shape of the suspending wires on the Golden Gate Bridge in San Francisco is a parabola.

2.2 Solving quadratic equations

Quadratic equations are equations of the form $ax^2 + bx + c = 0$.

There are effectively two methods of solving a quadratic equation.

Solving a quadratic equation by factorisation

This is where the quadratic equation is rearranged to equal zero, the quadratic expression is factorised into two brackets, and then the values of x for which each bracket equals zero are found. The values of x that make the brackets zero are called the **roots** or **zeros** of the equation and are also the x-intercepts of the curve.

> Factorisation was covered in the chapter on presumed knowledge.

Example

Solve the quadratic equation $x^2 - 3x - 10 = 0$.
$$x^2 - 3x - 10 = 0$$
$$\Rightarrow (x - 5)(x + 2) = 0$$
$$\Rightarrow x - 5 = 0 \text{ or } x + 2 = 0$$
$$\Rightarrow x = 5 \text{ or } x = -2$$

Example

Solve the quadratic equation $2x^2 - 5x + 2 = 0$.

$$2x^2 - 5x + 2 = 0$$
$$\Rightarrow (2x - 1)(x - 2) = 0$$
$$\Rightarrow 2x - 1 = 0 \text{ or } x - 2 = 0$$
$$\Rightarrow x = \frac{1}{2} \text{ or } x = 2$$

Using the formula to solve a quadratic equation

Some quadratic equations cannot be factorised, but still have solutions. This is when the quadratic formula is used:

$$x = \frac{-b \pm \sqrt{b^2 - 4ac}}{2a}$$

a, b and c refer to $ax^2 + bx + c = 0$.

$b^2 - 4ac$ is known as the **discriminant**.

We begin by looking at the previous example, which was solved by factorisation, and show how it can be solved using the quadratic formula.

The formula works for all quadratic equations, irrespective of whether they factorise or not, and will be proved later in the chapter.

Example

Solve the quadratic equation $2x^2 - 5x + 2 = 0$ using the quadratic formula.
In this case, $a = 2$, $b = -5$ and $c = 2$.

$$x = \frac{-b \pm \sqrt{b^2 - 4ac}}{2a}$$
$$\Rightarrow x = \frac{5 \pm \sqrt{(-5)^2 - 4(2)(2)}}{2(2)}$$
$$\Rightarrow x = \frac{5 \pm \sqrt{25 - 16}}{4}$$
$$\Rightarrow x = \frac{5 \pm 3}{4}$$
$$\Rightarrow x = 2, \frac{1}{2}$$

Example

Solve the equation $x^2 + 6x - 10 = 0$ using the quadratic formula.

$$x^2 + 6x - 10 = 0$$

In this case, $a = 1$, $b = 6$ and $c = -10$.

$$x = \frac{-b \pm \sqrt{b^2 - 4ac}}{2a}$$
$$\Rightarrow x = \frac{-6 \pm \sqrt{6^2 - 4(1)(-10)}}{2(1)}$$

$$\Rightarrow x = \frac{-6 \pm \sqrt{36 + 40}}{2}$$

$$\Rightarrow x = \frac{-6 \pm \sqrt{76}}{2}$$

$$\Rightarrow x = -7.36, \ 1.36$$

Exercise 1

1 Find the solutions to the following quadratic equations using factorisation.

a $x^2 - 5x + 4 = 0$ **b** $x^2 - x - 6 = 0$

c $2x^2 + 17x + 8 = 0$ **d** $x(x - 1) = x + 3$

2 Find the x-intercepts on the following curves using factorisation.

a $y = x^2 - 7x + 10$ **b** $y = x^2 - 5x - 24$ **c** $y = 2x^2 + 5x - 12$

d $y = 6x^2 + 5x - 6$ **e** $y = 3x^2 - 11x + 6$

3 Find the solutions to the following quadratic equations using the quadratic formula.

a $x^2 - 6x + 6 = 0$ **b** $x^2 - 5x - 5 = 0$ **c** $2x^2 + 7x + 2 = 0$

d $5x^2 + 9x + 2 = 0$ **e** $3x^2 + 5x - 3 = 0$

4 Find the x-intercepts on the following curves using the quadratic formula.

a $y = x^2 + 6x + 3$ **b** $y = x^2 - 4x - 9$ **c** $y = 3x^2 + 7x + 3$

d $y = 2x^2 + 5x - 11$ **e** $y = 1 - 2x^2 - 3x$

2.3 Quadratic functions

We can write quadratic functions in a number of different forms.

Standard form

This is the form $f(x) = ax^2 + bx + c$ where c is the y-intercept because $f(0) = c$. Remember that if a is positive the curve will have a minimum point, and if a is negative the curve will have a maximum turning point. If the curve is given in this form, then to draw it we would usually use a calculator or draw a table of values.

Intercept form

This is the form $f(x) = (ax + b)(cx + d)$ where the quadratic function has been factorised. In this form the x-intercepts are given by $-\dfrac{b}{a}$ and $-\dfrac{d}{c}$ since they are found by letting $f(x) = 0$. Knowing the x-intercepts and the y-intercept, there is normally enough information to draw the curve.

Example

Sketch the curve $f(x) = 2x^2 + 3x + 1$.
The y-intercept is at $f(0) = 1$.
Factorising the function:
$f(x) = (2x + 1)(x + 1)$
So to find the x-intercepts solve $(2x + 1)(x + 1) = 0$.
Hence the x-intercepts are $-\dfrac{1}{2}$ and -1.

The curve has a positive coefficient of x^2, so it has a minimum turning point.
This point occurs halfway between $-\dfrac{1}{2}$ and -1, i.e. at $-\dfrac{3}{4}$.

When $x = -\dfrac{3}{4}$, $f(x) = 2\left(-\dfrac{3}{4}\right)^2 + 3\left(-\dfrac{3}{4}\right) + 1 = -\dfrac{1}{8}$. So the minimum point is $\left(-\dfrac{3}{4}, -\dfrac{1}{8}\right)$.

A sketch of the graph is shown below.

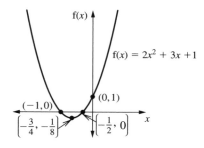

Turning point form

This is when the function is in the form $f(x) = r(x - p)^2 + q$. The graph is the curve $y = x^2$ which has been translated p units to the right, stretched parallel to the y-axis by scale factor r, then translated q upwards.

In this form the maximum or minimum turning point has coordinates (p, q). This is the reason:

If r is positive, then since $(x - p)^2$ is never negative, the least possible value of $f(x)$ is given when $(x - p)^2 = 0$. Hence $f(x)$ has a minimum value of q, which occurs when $x = p$.

If r is negative, then since $(x - p)^2$ is never negative, the greatest possible value of $f(x)$ is given when $(x - p)^2 = 0$. Hence $f(x)$ has a maximum value of q, which occurs when $x = p$.

Remember that $x = p$ is the line of symmetry of the curve.

Completing the square

Writing a quadratic in the form $r(x \pm p)^2 \pm q$ is known as completing the square.

This technique will be needed in later chapters.

This is demostrated in the following examples.

Example

Write the function $f(x) = x^2 + 6x - 27$ in the form $(x + p)^2 - q$.

We know from the expansion of brackets $(x - a)(x - a)$ that this equals $x^2 - 2ax + a^2$. Hence the coefficient of x is always twice the value of p. Therefore $f(x) = (x + 3)^2$ plus or minus a number. To find this we subtract 3^2 because it is not required and then a further 27 is subtracted.

Hence $f(x) = (x + 3)^2 - 9 - 27 = (x + 3)^2 - 36$.

Method for completing the square on $ax^2 + bx + c$

1. Take out a, leaving the constant alone: $a\left(x^2 + \dfrac{b}{a}x\right) + c$

2. Complete the square: $a\left[\left(x + \dfrac{b}{2a}\right)^2 - \left(\dfrac{b}{2a}\right)^2\right] + c$

3. Multiply out the outer bracket.

4. Tidy up the constants.

Example

Complete the square on the function $f(x) = 2x^2 + 3x + 8$.

In this case the coefficient of x^2 is 2 and we need to make it 1: hence the 2 is factorised out, but the constant is left unchanged.

$$f(x) = 2\left(x^2 + \frac{3}{2}x\right) + 8$$

Following the same procedure as the example above:

$$f(x) = 2\left[\left(x + \frac{3}{4}\right)^2 - \frac{9}{16}\right] + 8$$

$$f(x) = 2\left(x + \frac{3}{4}\right)^2 - \frac{9}{8} + 8 = 2\left(x + \frac{3}{4}\right)^2 + \frac{55}{8}$$

Example

Complete the square on the function $f(x) = -x^2 + 6x - 11$.

Step 1 $\Rightarrow f(x) = -(x^2 - 6x) - 11$

Step 2 $\Rightarrow f(x) = -[(x - 3)^2 - 9] - 11$

Step 3 $\Rightarrow f(x) = -(x - 3)^2 + 9 - 11$

Step 4 $\Rightarrow f(x) = -(x - 3)^2 - 2$

Example

Complete the square on the function $f(x) = 2x^2 - 6x + 9$ and hence find the maximum or minimum turning point.

$$f(x) = 2x^2 - 6x + 9$$
$$\Rightarrow f(x) = 2(x^2 - 3x) + 9$$
$$\Rightarrow f(x) = 2\left[\left(x - \frac{3}{2}\right)^2 - \frac{9}{4}\right] + 9$$
$$\Rightarrow f(x) = 2\left(x - \frac{3}{2}\right)^2 - \frac{9}{2} + 9$$
$$\Rightarrow f(x) = 2\left(x - \frac{3}{2}\right)^2 + \frac{9}{2}$$

The curve will have its least value when $x - \frac{3}{2} = 0$, that is $x = \frac{3}{2}$, and this least value will be $\frac{9}{2}$. Hence the curve has a minimum turning point, which is $\left(\frac{3}{2}, \frac{9}{2}\right)$.

Example

Without using a calculator, sketch the curve $y = 2x^2 + 6x - 8$.

From this form of the curve we note that the y-intercept occurs when $x = 0$ and is therefore $y = -8$.

By turning it into intercept form the x-intercepts can be found.

$$y = 2(x^2 + 3x - 4)$$
$$\Rightarrow y = 2(x + 4)(x - 1)$$

Hence the x-intercepts are when $(x + 4)(x - 1) = 0$
$$\Rightarrow x = -4, x = 1$$

Complete the square to transform it into turning point form:

$$y = 2(x^2 + 3) - 8$$
$$\Rightarrow y = 2\left[\left(x + \frac{3}{2}\right)^2 - \frac{9}{4}\right] - 8$$
$$\Rightarrow y = 2\left(x + \frac{3}{2}\right)^2 - \frac{9}{2} - 8$$
$$\Rightarrow y = 2\left(x + \frac{3}{2}\right)^2 - \frac{25}{2}$$

Since the coefficient of x^2 is positive, we know that the curve has a minimum turning point with coordinates $\left(-\frac{3}{2}, -\frac{25}{2}\right)$.

> If the quadratic equation does not factorise, use the formula. If there is no solution the curve is entirely above or below the x-axis.

Hence the curve is:

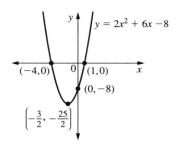

We could sketch the curve from the intercept form alone, but if the curve does not cut the x-axis the turning point form must be used.

Example

By using a method of completing the square without a calculator, sketch the curve $y = 3x^2 - 2x + 4$.

In this form we can see that the curve cuts the y-axis at $(0,4)$.

To find the x-intercepts it is necessary to solve $3x^2 - 2x + 4 = 0$.

Using the quadratic formula: $x = \dfrac{2 \pm \sqrt{4 - 48}}{6} = \dfrac{2 \pm \sqrt{-44}}{6}$

This gives no real roots and hence the curve does not cut the x-axis.

Hence in this situation the only way to find the turning point is to complete the square.

$$y = 3x^2 - 2x + 4$$

$$\Rightarrow y = 3\left(x^2 - \frac{2}{3}\right) + 4$$

$$\Rightarrow y = 3\left[\left(x - \frac{1}{3}\right)^2 - \frac{1}{9}\right] + 4$$

$$\Rightarrow y = 3\left(x - \frac{1}{3}\right)^2 - \frac{1}{3} + 4$$

$$\Rightarrow y = 3\left(x - \frac{1}{3}\right)^2 + \frac{11}{3}$$

Since the coefficient of x^2 is positive, we know that the curve has a minimum turning point with coordinates $\left(\dfrac{1}{3}, \dfrac{11}{3}\right)$.

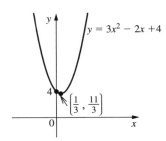

All these curves can be sketched using a calculator and the x-intercepts, y-intercept and the maximum or minimum turning point found.

Using a calculator, sketch the curve $y = x^2 + 5x - 3$ showing the coordinates of the minimum turning point and the x- and y-intercepts.

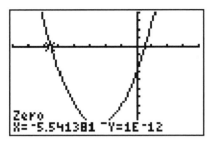

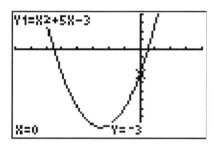

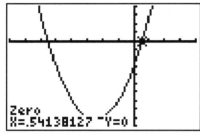

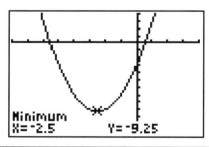

It is also possible to solve quadratic equation using the method of completing the square.

Solve the quadratic equation $x^2 + 6x - 13 = 0$ by completing the square.

$$x^2 + 6x - 13 = 0$$
$$\Rightarrow (x + 3)^2 - 9 - 13 = 0$$
$$\Rightarrow (x + 3)^2 - 22 = 0$$
$$\Rightarrow (x + 3)^2 = 22$$
$$\Rightarrow x + 3 = \pm\sqrt{22}$$
$$\Rightarrow x = -3 \pm\sqrt{22}$$

This method is not often used because the answer looks identical to the answer found using the quadratic formula. In fact the formula is just a generalised form of completing the square, and this is how we prove the quadratic formula.

We begin with $ax^2 + bx + c = 0$

$$\Rightarrow a\left(x^2 + \frac{b}{a}\right) + c = 0$$

$$\Rightarrow a\left[\left(x + \frac{b}{2a}\right)^2 - \frac{b^2}{4a^2}\right] + c = 0 \quad \cdots\cdots\cdots\cdots\cdots\cdots\cdots\cdots\cdots$$

$$\Rightarrow a\left(x + \frac{b}{2a}\right)^2 - \frac{b^2}{4a} + c = 0$$

$$\Rightarrow \left(x + \frac{b}{2a}\right)^2 - \frac{b^2}{4a^2} + \frac{c}{a} = 0$$

$$\Rightarrow \left(x + \frac{b}{2a}\right)^2 + \frac{4ac - b^2}{4a^2} = 0$$

$$\Rightarrow \left(x + \frac{b}{2a}\right)^2 = \frac{b^2 - 4ac}{4a^2}$$

$$\Rightarrow x + \frac{b}{2a} = \frac{\pm\sqrt{b^2 - 4ac}}{2a}$$

$$\Rightarrow x = \frac{-b \pm \sqrt{b^2 - 4ac}}{2a}$$

> From here it should be noted that $x = -\dfrac{b}{2a}$ is the line of symmetry of the curve and the x-value where the maximum or minimum turning point occurs.

Exercise 2

1 Complete the square:
   a $x^2 + 2x + 5$ b $x^2 - 3x + 3$ c $-x^2 + 3x - 5$
   d $3x^2 + 6x - 8$ e $5x^2 + 7x - 3$

2 Complete the square and hence sketch the parabola, showing the coordinates of the maximum or minimum turning point and the x- and y-intercepts.
   a $y = x^2 + 6x + 4$ b $y = x^2 - 4x + 3$ c $y = x^2 + 5x + 2$
   d $y = -x^2 - 4x + 3$ e $y = -x^2 + 8x + 3$ f $y = 2x^2 + 10x - 11$
   g $y = 4x^2 - 3x + 1$ h $y = 3x^2 + 5x + 2$ i $y = -2x^2 + 3x + 4$

3 Draw each of these on a calculator and identify the maximum or minimum turning point, the x-intercepts and the y-intercept.
   a $y = 2x^2 + 5x - 7$ b $y = -x^2 + 5x - 7$
   c $y = 5x^2 + 6x + 16$ d $y = -3x^2 + 5x + 9$

2.4 Linear and quadratic inequalities

Linear inequalities

The equation $y = mx + c$ represents a straight line. We now need to consider what is meant by the inequalities $y > mx + c$ and $y < mx + c$.

Consider the line $y = x - 2$. At any point on the line the value of y is equal to the value of $x - 2$. What happens when we move away from the line in a direction parallel to the y-axis? Clearly the value of $x - 2$ stays the same, but the value of y will increase if we move in the direction of positive y and decrease if we move in the direction of negative y. Hence for all points above the line $y > x - 2$ and for all points below the line $y < x - 2$. This is shown below.

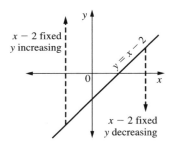

This argument can now be extended to $y = mx + c$. Hence all points above the line fit the inequality $y > mx + c$ and all points below the line fit the inequality $y < mx + c$.

Example

Sketch the region of the x, y plane represented by the inequality $2y + 3x > 4$.

First we consider the line $2y + 3x = 4$.

Rearranging this into the form $y = mx + c$ gives $y = -\dfrac{3}{2}x + 2$.

This line has gradient $-\dfrac{3}{2}$ and a y-intercept of 2. Since "greater than" is required, the area above the line is needed. This is shown below.

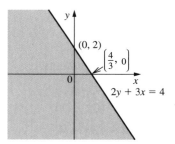

The shading shows the non-required area. The line is not included.

Now we consider how to solve linear inequalities.

Consider the fact that $10 > 9$. If the same number is added to both sides of the inequality, then the inequality remains true. The same is true if the same number is subtracted from both sides of the inequality.

Now consider multiplication. If both sides are multiplied by 3, this gives $30 > 27$, which is still a true statement. If both sides are multiplied by -3 then $-30 > -27$, which is no longer true. Division by a negative number leads to the same problem.

This demonstrates the general result that whenever an inequality is multiplied or divided by a negative number then the inequality sign must reverse.

In all other ways, linear inequalities are solved in the same way as linear equations. We can avoid this problem by ensuring that the coefficient of x always remains positive.

Example

Find the solution set to $2(x - 1) > 3 - 4(1 - x)$.
$$2(x - 1) > 3 - 4(1 - x)$$
$$\Rightarrow 2x - 2 > 3 - 4 + 4x$$
$$\Rightarrow -2x > 1$$
$$\Rightarrow x < -\frac{1}{2}$$

Alternatively:
$$2(x - 1) > 3 - 4(1 - x)$$
$$\Rightarrow 2x - 2 > 3 - 4 + 4x$$
$$\Rightarrow -1 > 2x$$
$$\Rightarrow x < -\frac{1}{2}$$

Quadratic inequalities

Any inequality that involves a quadratic function is called a **quadratic inequality**. Quadratic inequalities are normally solved by referring to the graph. This is best demonstrated by example.

Example

Find the solution set that satisfies $x^2 + 2x - 15 > 0$.

Begin by sketching the curve $y = x^2 + 2x - 15$. This is essential in these questions. However, we are interested only in whether the curve has a maximum or minimum point and where the x-intercepts are.

Since the coefficient of x^2 is positive, the curve has a minimum point and the x-intercepts can be found by factorisation.

$$x^2 + 2x - 15 = 0$$
$$\Rightarrow (x + 5)(x - 3) = 0$$
$$\Rightarrow x = -5, x = 3$$

> If the quadratic formula does not factorise, use the formula or a graphing calculator.

This curve is shown below.

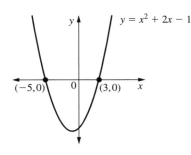

The question now is "When is this curve greater than zero?" The answer to this is when the curve is above the *x*-axis.

This is shown below.

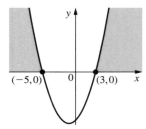

The answer needs to be represented as two inequalities.

Hence the solution set is $x < -5$ and $x > 3$

This can also be done using a calculator to draw the graph, finding the *x*-intercepts and then deducing the inequalities.

Example

Using a calculator, find the solution set to $x^2 - 5x + 3 \geq 0$.
The calculator screen dump is shown below.

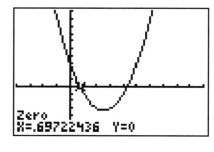

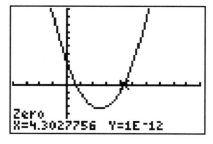

From this we can deduce that the solution set is $x \leq 0.697, x \geq 4.30$.

Example

Find the solution set that satisfies $-x^2 + x + 6 > 0$.

Consider the curve $y = -x^2 + x + 6$. Since the coefficient of x^2 is negative, the curve has a maximum turning point.

Solving $-x^2 + x + 6 = 0$

$\Rightarrow (-x + 3)(x + 2) = 0$

$\Rightarrow x = 3, x = -2$

The curve is shown below.

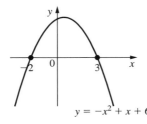

$$y = -x^2 + x + 6$$

Since the question asks when the curve is greater than zero, we need to know when it is above the x-axis. Hence the solution set is $-2 < x < 3$.

In this case the solution can be represented as a single inequality.

Exercise 3

1 Show these inequalities on diagrams.

  a $y > 2x - 5$ **b** $y < -3x + 8$

  c $x + 2y \geq 5$ **d** $3x - 4y + 5 \leq 0$

2 Without using a calculator, find the range(s) of values of x that satisfy these inequalities.

  a $x + 5 > 4 - 3x$ **b** $7x - 5 < 2x + 5$

  c $-2(3x - 1) - 4(x - 2) \leq 12$ **d** $(x - 3)(x + 5) > 0$

  e $(x - 6)(5 - x) < 0$ **f** $(2x + 1)(3 - 4x) \leq 0$

  g $2x^2 - 13x + 21 \leq 0$ **h** $x^2 + 7x + 12 > 0$

  i $x^2 + 3x + 2 < 0$ **j** $-x^2 - x + 6 \geq 0$

  k $x^2 + 12x + 4 \leq -x^2 - 5x - 4$ **l** $4 + 11x + 6x^2 < 0$

3 Using a calculator, find the range(s) of values of x that satisfy these inequalities.

  a $x^2 > 6x - 4$ **b** $x^2 - 5x - 5 > 0$

  c $-4x^2 + 7x - 1 < 0$ **d** $2x^2 + 4x \leq 3 - x^2 - x$

  e $4x^2 \leq 4x + 1$

2.5 Nature of roots of quadratic equations

We now need to take a more in-depth look at quadratic equations.

Consider the following equations and what happens when they are solved using the formula.

a $x^2 + 6x + 3 = 0$

$\Rightarrow x = \dfrac{-6 \pm \sqrt{(6)^2 - 4(1)(3)}}{2(1)}$

$\Rightarrow x = \dfrac{-6 \pm \sqrt{36 - 12}}{2}$

$\Rightarrow x = \dfrac{-6 \pm \sqrt{24}}{2}$

$\Rightarrow x = -0.551, -5.45$

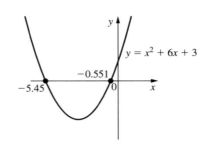

b $x^2 + 6x + 9 = 0$

$\Rightarrow x = \dfrac{-6 \pm \sqrt{(6)^2 - 4(1)(9)}}{2(1)}$

$\Rightarrow x = \dfrac{-6 \pm \sqrt{36 - 36}}{2}$

$\Rightarrow x = \dfrac{-6}{2} = -3$

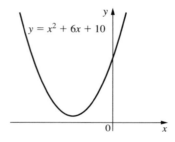

c $x^2 + 6x + 10 = 0$

$\Rightarrow x = \dfrac{-6 \pm \sqrt{(6)^2 - 4(1)(10)}}{2(1)}$

$\Rightarrow x = \dfrac{-6 \pm \sqrt{36 - 40}}{2}$

$\Rightarrow x = \dfrac{-6 \pm \sqrt{-4}}{2}$

This suggests that the roots of quadratic equations can be classified into three categories and that there are conditions for each category.

In case a) we see that the discriminant is greater than zero, and since the square root exists the $\pm$ produces two different roots. Hence case a) is two real distinct roots, and the condition for it to happen is $b^2 - 4ac > 0$.

In case b) we see that the discriminant is equal to zero so there is only one repeated root, which is $x = -\dfrac{b}{2a}$. Hence case b) is two real equal roots, and the condition for it to happen is $b^2 - 4ac = 0$. In this case the maximum or minimum turning point is on the x-axis.

In case c) we see that the discriminant is less than zero and so there are no real answers. There are roots to this equation, which are called **complex roots**, but these will be met formally in Chapter 17. Hence case c) is no real roots and the condition for it to happen is $b^2 - 4ac < 0$.

A summary of this is shown in the table below.

$b^2 - 4ac > 0$	$b^2 - 4ac = 0$	$b^2 - 4ac < 0$

> If a question talks about the condition for real roots then $b^2 - 4ac \geq 0$ is used.

Example

Determine the nature of the roots of $x^2 - 3x + 4 = 0$.
In this case $b^2 - 4ac = 9 - 16 = -7$.
Hence $b^2 - 4ac < 0$ and there are no real roots of the equation.

Example

If $ax^2 - 8x + 2 = 0$ has a repeated root, find the value of a.
This is an alternative way of asking about the conditions for real equal roots.
For this equation to have a repeated root, $b^2 - 4ac = 0$.
$$\Rightarrow 64 - 8a = 0$$
$$\Rightarrow a = 8$$

Example

Show that the roots of $2ax^2 + (a + b)x + \frac{1}{2}b = 0$ are real for all values of a and b $(a, b \in \mathbb{R})$.
The condition for real roots is $b^2 - 4ac \geq 0$.
$$\Rightarrow (a + b)^2 - 4ab \geq 0$$
$$\Rightarrow a^2 + 2ab + b^2 - 4ab \geq 0$$
$$\Rightarrow a^2 - 2ab + b^2 \geq 0$$
$$\Rightarrow (a - b)^2 \geq 0$$
Now the square of any number is always either positive or zero and hence the roots are real irrespective of the values of a and b.

An application of the nature of roots of quadratic equations

Consider the straight line $y = 2x - 1$ and the parabola $y = x^2 - x - 11$. In the figure below we see that the line intersects the curve in two places.

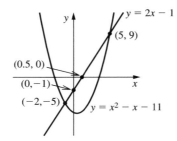

To find the x-coordinates of this point of intersection, we solve the equations simultaneously.

$\Rightarrow 2x - 1 = x^2 - x - 11$

$\Rightarrow x^2 - 3x - 10 = 0$ ....

> This is the resulting quadratic equation.

$\Rightarrow (x - 5)(x + 2) = 0$

$\Rightarrow x = 5, x = -2$

Hence what can be called the resulting quadratic equation has two real distinct roots. This gives a method of finding the different conditions for which a line and a parabola may or may not intersect. There are three possible cases.

If the parabola and the straight line intersect then there are two roots, and hence this is the case of two real different roots i.e. $b^2 - 4ac > 0$ for the resulting quadratic equation. This is shown below.

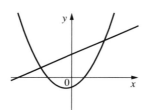

If the straight line is a tangent to the parabola (it touches the curve at only one point), then there is one root which is the point of contact and hence this is the case of real, equal roots i.e. $b^2 - 4ac = 0$ for the resulting equation. This is shown below.

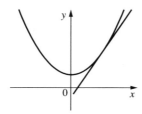

If the parabola and the straight line do not intersect then there are no real roots, and hence this is the case of no real roots i.e. $b^2 - 4ac < 0$ for the resulting quadratic equation. This is shown below.

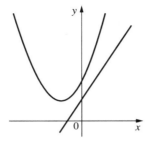

Example

Prove that $y = 4x - 9$ is a tangent to $y = 4x(x - 2)$.
If this is true, then the equation $4x - 9 = 4x(x - 2)$ should have real equal roots, as there is only one point of contact.
$$\Rightarrow 4x^2 - 12x + 9 = 0$$
In this case $b^2 - 4ac = 144 - 144 = 0$.
Hence it does have real equal roots and $y = 4x - 9$ is a tangent to $y = 4x(x - 2)$.

Exercise 4

1 Determine the nature of the roots of the following equations, but do not solve the equations.

   a $x^2 - 3x + 4 = 0$ **b** $2x^2 - 4x - 7 = 0$ **c** $3x^2 - 6x + 4 = 0$

   d $-2x^2 + 7x + 2 = 0$ **e** $4x^2 - 4x + 1 = 0$ **f** $x^2 + 1 = 3x - 4$

2 For what values of p is $4x^2 + px + 49 = 0$ a perfect square?

3 Find the value of q if $2x^2 - 3x + q = 0$ has real equal roots.

4 Prove that $qx^2 + 3x - 6 - 4q = 0$ will always have real roots independent of the value of q.

5 Find a relationship between a and b if the roots of
$$2abx^2 + x\sqrt{a - b} + b^2 - 2a = 0 \text{ are equal.}$$

6 If x is real and $s = \dfrac{4x^2 + 3}{2x - 1}$, prove that $s^2 - 4s - 12 \geq 0$.

7 Find the values of p for which the expression $2p - 3 + 4px - px^2$ is a perfect square.

8 If $x^2 + (3 - 4r)x + 6r^2 - 2 = 0$, show that there is no real value of r.

9 Find the value of m for which the curve $y = 8mx^2 + 3mx + 1$ touches the x-axis.

10 Prove that $y = x - 3$ is a tangent to the curve $y = x^2 - 5x + 6$.

11 For each part of this question, which of the following statements apply?

   i The straight line is a tangent to the curve.

  ii The straight line cuts the curve in two distinct points.

 iii The straight line neither cuts nor touches the curve.

  a Curve: $y = 3x^2 - 4x - 2$ **b** Curve: $y = 7x(x - 1)$

     Line: $y = x - 3$ Line: $y + 2x + 1 = 0$

  c Curve: $y = 9x^2 - 3x + 10$ **d** Curve: $y = (3x - 4)(x + 1)$

     Line: $y = 3(x + 3)$ Line: $y + 10x + 11 = 0$

Review exercise

 1 Express $x(4 - x)$ as the difference of two squares.

 2 Given that $y = x^2 - 2x - 3$ $(x \in \mathbb{R})$, find the set of values of x for which $y < 0$.   [IB May 87 P1 Q9]

 3 Consider the equation $(1 + 2k)x^2 - 10x + k - 2 = 0$, $k \in \mathbb{R}$. Find the set of values of k for which the equation has real roots. [IB Nov 03 P1 Q13]

 4 Solve the following simultaneous equations.

  i $y - x = 2$ **ii** $x + y = 9$

    $2x^2 + 3xy + y^2 = 8$ $x^2 - 3xy + 2y^2 = 0$

 5 The equation $kx^2 - 3x + (k + 2) = 0$ has two distinct real roots. Find the set of possible values of k.   [IB May 01 P1 Q18]

 6 For what values of m is the line $y = mx + 5$ a tangent to the parabola $y = 4 - x^2$?   [IB Nov 00 P1 Q13]

 7 Prove that if $x^2 > k(x - 2)$ for all real x, then $0 < k < 8$.

 8 By letting $y = x^{\frac{1}{4}}$ find the values of x for which $x^{\frac{1}{4}} - 2x^{-\frac{1}{4}} = 1$.

 9 Knowing that the values of x satisfying the equation $2x^2 + kx + k = 0$ are real numbers, determine the range of possible values of $k \in \mathbb{R}$.   [IB Nov 91 P1 Q13]

 10 Express $(2 - x)(x - 5)$ in the form $a - (x - b)^2$, where a and b are constants. State the coordinates of the maximum point on the graph of $y = (2 - x)(x - 5)$ and also state what symmetry the graph has.

 11 William's father is two years older than his mother and his mother's age is the square of his own. The sum of all three ages is 80 years. How old is William?

12

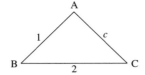

The diagram above shows a triangle ABC, in which AB = 1 unit, BC = 2 units and
AC = c units. Find an expression for cos C in terms of c.

Given that $\cos C > \dfrac{4}{5}$, show that $5c^2 + 4c + 3 < 0$. Find the set of values of c

which satisfy this inequality.

3 Functions

Calculators are an integral part of school mathematics, and this course requires the use of a graphing calculator. Although much of the content of this chapter is "ancient" mathematics in that it has been studied for thousands of years, the use of the calculator in its learning is very recent. Devices for calculating have been around for a long time, the most famous being the abacus. Calculating machines have been built at various times including the difference engine shown below left, built by Charles Babbage in 1822. However, hand-held calculators did not become available until the 1970s, and the first graphing calculator was only produced in 1985. In many ways, the advent of the graphing calculator has transformed the learning of functions and their graphs, and this is a very recent development. What will the next 25 years bring to revolutionise the study of mathematics? Will it be that CAS (computer algebra systems) calculators will become commonplace and an integral part of school mathematics curricula and thus shift the content of these curricula?

3.1 Functions

A function is a mathematical rule. Although the word "function" is often used for any mathematical rule, this is not strictly correct. For a mathematical rule to be a function, each value of *x* can have only one image.

These are arrow diagrams.

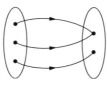

Many to one
(a function)

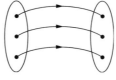

One to one
(a function which
has an inverse)

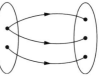

One to many
(not a function)

A simple test can be performed on a graph to find whether it represents a function: if any vertical line cuts more than one point on the graph, it is not a function.

Definitions

Domain – the set of numbers that provide the input for the rule.
Image – the output from the rule of an element in the domain.
Range – the set of numbers consisting of the images of the domain.
Co-domain – a set containing the range.
Function – a rule that links each member of the domain to exactly one member of the range.

> This means that a function may have range $y > 0$ but the co-domain could be stated to be $\mathbb{R}$. The term "co-domain" will not be used on examination papers.

Notation

Functions can be expressed in two forms:

$$f : x \rightarrow 2x + 1$$
$$f(x) = 2x + 1$$

The second form is more common but it is important to be aware of both forms.

Finding an image

To find an image, we substitute the value into the function.

Example

Find $f(2)$ for $f(x) = x^3 - 2x^2 + 7x + 1$.
$$f(2) = 2^3 - 2(2)^2 + 7(2) + 1$$
$$= 8 - 8 + 14 + 1 = 15$$

Domains and ranges are sets of numbers. It is important to remember the notation of the major sets of numbers.

$\mathbb{Z}$ – the set of integers $\{ \ldots, -3, -2, -1, 0, 1, 2, 3, \ldots \}$
$\mathbb{Z}^+$ – the set of positive integers $\{1, 2, 3, \ldots \}$
$\mathbb{N}$ – the set of natural numbers $\{0, 1, 2, 3, \ldots \}$
$\mathbb{Q}$ – the set of rational numbers $\left\{ x : x = \dfrac{p}{q}, p, q \in \mathbb{Z}\ q \neq 0 \right\}$
$\mathbb{R}$ – the set of real numbers.

> Rational numbers are numbers that can be expressed as a fraction of two integers.

If the domain is not stated, it can be assumed to be the set of real numbers. However, if a domain needs to be restricted for the function to be defined, it should always be given. This is particularly true for rational functions, which are covered later in the chapter.

> All numbers on the number line are real numbers (including irrational numbers such as π, $\sqrt{2}$, $\sqrt{3}$).

Example

For $f(t) = t^2 - 3$ with a domain $\{-3, -1, 2, 3\}$, find the range.

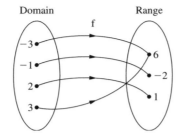

Domain   Range

The range is the set of images $= \{-2, 1, 6\}$.

Use an arrow diagram.

Remember, each value of x has only one image in the range.

Example

For the function $f(x) = 2x + 3, 0 \le x \le 5$
(a) find $f(1)$
(b) sketch the graph of $f(x)$
(c) state the range of this function.

(a) $f(1) = 2(1) + 3 = 5$

(b)

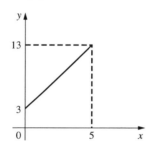

(c) The range is the set of images $= \{3 \le y \le 13\}$.

Example

For $f(x) = x^2 + 5x - 3$, find $f(2)$, $f(2x)$ and $f\left(\dfrac{3}{x}\right)$.

$$f(2) = 2^2 + 5(2) - 3$$
$$= 4 + 10 - 3$$
$$= 11$$

To find $f(2x)$, substitute $2x$ for x in the rule $f(x)$.

So $f(2x) = (2x)^2 + 5(2x) - 3$
$$= 4x^2 + 10x - 3$$

Similarly,

$$f\left(\frac{3}{x}\right) = \left(\frac{3}{x}\right)^2 + 5\left(\frac{3}{x}\right) - 3$$
$$= \frac{9}{x^2} + \frac{15}{x} - 3$$
$$= \frac{9 + 15x - 3x^2}{x^2}$$

Exercise 1

1 For the following functions, find f(4).

    a $f(x) = 3x - 1$ **b** $f(x) = 9 - 2x$

    c $f(x) = x^2 - 3$ **d** $f(x) = \dfrac{24}{x}, x \neq 0$

2 For the following functions, find g(−2).

    a $g(t) = 6t - 5$ **b** $g(t) = 3t^2$

    c $g(t) = t^3 - 4t^2 + 5t + 7$ **d** $g(t) = \dfrac{17 - t}{t^2}, t \neq 0$

3 Draw an arrow diagram for $f(x) = 4x - 3$ with domain {−1, 1, 5} and state the range.

4 Draw an arrow diagram for $g(x) = \dfrac{x - 3}{x}$ with domain {−3, −1, 1, 6} and state the range.

5 Draw the graph of $f(x) = x - 5$ for $0 \leq x \leq 7$ and state the range.

6 Draw the graph of $f(x) = 9 - 2x$ for $-3 \leq x \leq 2$ and state the range.

7 Draw the graph of $g(x) = x^2 - x - 6$ for $0 \leq x \leq 6$ and state the range.

8 Draw the graph of $p(x) = 2x^3 - 7x + 6$ for $-1 \leq x \leq 5$ and state the range.

9 For $f(x) = 2x + 5, x \in \mathbb{R}^+$, what is the range?

10 For each of these graphs, state the domain and range.

    a   **b**

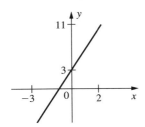 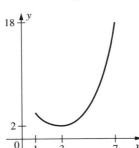

    c

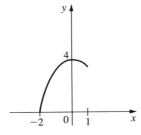

11 For $f(x) = 3x - 2$, find

    a $f(2x)$ **b** $f(-x)$ **c** $f\left(\dfrac{1}{x}\right)$

12 For $g(x) = x^2 - 3x$, find

    a $g(2x)$ **b** $g(x + 4)$ **c** $g(6x)$ **d** $g(2x - 1)$

13 For $h(x) = \dfrac{x}{2 - x}$, find

    a $h(-x)$ **b** $h(4x)$ **c** $h\left(\dfrac{1}{x}\right)$ **d** $h(x + 2)$

14 For $k(x) = x - 9$, find $k(x + 9)$.

3.2 Composite functions

When one function is followed by another, the resultant effect can be expressed as a single function. When functions are combined like this, the resultant function is known as a **composite function**.

Example

Find $h(x) = g(f(x))$ if
$f(x) = 2x$
$g(x) = x + 4$

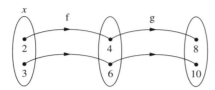

This composite function is f followed by g.
So in this example the effect is $\times 2$ then $+4$.
$h(x) = 2x + 4$

$g(f(x))$ is f followed by g (the order is important).

This was a very simple example, but for more complicated functions it is useful to find the composite function in two steps like this:

$g(f(x)) = g(2x)$ (substituting $2x$ for x in $g(x)$)

     $= 2x + 4$

$f(g(x))$ is sometimes also written $(f \circ g)(x)$

Example

For $f(x) = 2x - 1$ and $g(x) = x^2 + 5x - 2$, find $g(f(x))$.
$$g(f(x)) = g(2x - 1)$$
$$= (2x - 1)^2 + 5(2x - 1) - 2$$
$$= 4x^2 - 4x + 1 + 10x - 5 - 2$$
$$= 4x^2 + 6x - 6$$

It is important to note that the order of the functions is very important.

In general, $g(f(x)) \neq f(g(x))$.

In the example $f(g(x)) = f(x^2 + 5x - 2)$

$$= 2(x^2 + 5x - 2) - 1$$

$$= 2x^2 + 10x - 4 - 1$$
$$= 2x^2 + 10x - 5$$

Example

For $h(\theta) = \sec\theta$ and $p(\theta) = \sin 2\theta - 4\theta$, find $p(h(\theta))$.
$p(h(\theta)) = p(\sec\theta)$
$\qquad = \sin(2\sec\theta) - 4\sec\theta$

Example

For $f(x) = \dfrac{2}{x-1}, x \neq 1$, and $g(x) = x + 4$, find (a) $g(f(4))$ (b) $f(g(x))$

(a) $\quad f(4) = \dfrac{2}{4-1} = \dfrac{2}{3}$

$\qquad g(f(4)) = g\left(\dfrac{2}{3}\right)$

$\qquad\qquad = \dfrac{2}{3} + 4$

$\qquad\qquad = \dfrac{14}{3}$

(b) $\quad f(g(x)) = f(x+4)$

$\qquad\qquad = \dfrac{2}{x+4-1}$

$\qquad\qquad = \dfrac{2}{x+3}$

Note that we did not find $g(f(x))$ here in order to find $g(f(4))$. It is generally easier to just calculate $f(4)$ and then input this value into $g(x)$.

Exercise 2

1 For $f(x) = x + 6$ and $g(x) = 4x$, find
   a $f(g(2))$ **b** $g(f(0))$
   c $f(g(-1))$ **d** $g(f(x))$

2 For $h(x) = 3x + 2$ and $p(x) = x^2 - 3$, find
   a $p(h(2))$ **b** $h(p(-2))$
   c $p(h(-3))$ **d** $h(p(x))$

3 For $f(\theta) = \sin\theta$ and $g(\theta) = \theta + \dfrac{\pi}{3}$, find
   a $g\left(f\left(\dfrac{\pi}{2}\right)\right)$ **b** $f\left(g\left(\dfrac{\pi}{3}\right)\right)$
   c $f(g(-2\pi))$ **d** $f(g(\theta))$

4 For each pair of functions, find (i) $f(g(x))$ and (ii) $g(f(x))$.
   a $f(x) = 2x - 1, g(x) = x^2$ **b** $f(x) = x^2 - 4, g(x) = 3x + 5$
   c $f(x) = x^3, g(x) = x^2 - 6$ **d** $f(x) = \cos x, g(x) = 3x^2$
   e $f(x) = x^3 - x + 7, g(x) = 2x + 3$ **f** $f(x) = x^6 - 2x + 3, g(x) = x^2 + 4$
   g $f(x) = \sin 2x, g(x) = x^2 - 7$

5 For $f(x) = 3x + 4$ and $g(x) = 2x - p$, where p is a constant, find

    a $f(g(x))$ **b** $g(f(x))$ **c** p if $f(g(x)) = g(f(x))$

6 For $f(x) = 6x^2$ and $g(x) = 2x - 3$, find

    a $g(f(x))$ **b** $f(g(x))$ **c** $f(f(x))$ **d** $g(g(x))$

7 For $f(x) = x + \dfrac{\pi}{2}$ and $g(x) = \cos x$, find

    a $g(f(x))$ **b** $f(g(x))$ **c** $f(f(x))$ **d** $g(g(x))$

8 For each pair of functions, find (i) $f(g(x))$ and (ii) $g(f(x))$, in simplest form.

    a $f(x) = \dfrac{2}{x - 3}, x \neq 3, g(x) = 3x + 1$

    b $f(x) = x^2 - 3x, g(x) = \dfrac{3}{x}, x \neq 0$

    c $f(x) = 2 - 5x, g(x) = \dfrac{x}{x + 1}, x \neq -1$

    d $f(x) = \dfrac{2}{3x - 1}, x \neq \dfrac{1}{3}, g(x) = \dfrac{1}{x}, x \neq 0$

    e $f(x) = \dfrac{x}{2 + x}, x \neq -2, g(x) = \dfrac{2}{x}, x \neq 0$

9 For $f(x) = \dfrac{1}{x + 7}, x \neq -7$, and $g(x) = \dfrac{1}{x} - 7, x \neq 0$, find $f(g(x))$ in its simplest form.

3.3 Inverse functions

Consider a function $y = f(x)$. If we put in a single value of x, we find a single value of y. If we were given a single value of y and asked to find the single value of x, how would we do this? This is similar to solving a linear equation for a linear function. The function that allows us to find the value of x is called the **inverse function**. Note that the original function and its inverse function are inverses of each other. We know that addition and subtraction are opposite operations and division is the opposite of multiplication. So, the inverse function of $f(x) = x + 3$ is $f^{-1}(x) = x - 3$.

> $f^{-1}(x)$ is the inverse of $f(x)$.

Note that both the original function $f(x)$ and the inverse function $f^{-1}(x)$ are functions. This means that for both functions there can only be one image for each element in the domain. This can be illustrated in an arrow diagram.

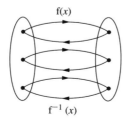

This means that not all functions have an inverse. An inverse exists only if there is a one-to-one correspondence between domain and range in the function.

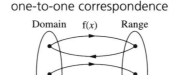

one-to-one correspondence

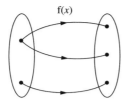

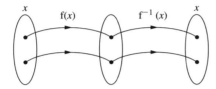

No inverse exists
for the function f(x)

f(x) is not a function

f(x) is a function
and an inverse
function exists

> An inverse does not exist
> for a many-to-one function,
> unless the domain is
> restricted.

When testing whether a mapping was a function, we used a vertical line test. To test if a mapping is a one-to-one correspondence, we can use a vertical and horizontal line test. If the graph crosses any vertical line more than once, the graph is not a function. If the graph crosses any horizontal line more than once, there is no inverse (for that domain).

The arrow diagram for a one-to-one correspondence shows that the range for f(x) becomes the domain for $f^{-1}(x)$. Also, the domain of f(x) becomes the range for $f^{-1}(x)$.

Looking at f(x) and $f^{-1}(x)$ from a composite function view, we have

that is $f^{-1}(f(x)) = x$.

Finding an inverse function

For some functions, the inverse function is obvious, as in the example of $f(x) = x + 3$, which has inverse function $f^{-1}(x) = x - 3$. However, for most functions more thought is required.

Method for finding an inverse function

1. Check that an inverse function exists for the given domain.
2. Rearrange the function so that the subject is x.
3. Interchange x and y.

Example

Find the inverse function for $f(x) = 2x - 1$.
Here there is no domain stated and so we assume that it is for $x \in \mathbb{R}$. $f(x) = 2x - 1$ has a one-to-one correspondence for all real numbers and so an inverse exists.

Let $y = 2x - 1$
So $2x = y + 1$
$$\Rightarrow x = \frac{1}{2}y + \frac{1}{2}$$

Interchanging x and y gives $y = \frac{1}{2}x + \frac{1}{2}$.

So the inverse function is $f^{-1}(x) = \frac{1}{2}x + \frac{1}{2}$.

Example

Find the inverse function for $f(x) = \dfrac{6}{x-2}$ for $x > 2$, $x \in \mathbb{R}$.

An inverse exists as there is a one-to-one correspondence for f(x) when $x > 2$.

$$\text{Let } y = \frac{6}{x-2}$$

$$\text{So } x - 2 = \frac{6}{y}$$

$$\Rightarrow x = \frac{6}{y} + 2$$

$$\Rightarrow x = \frac{2y + 6}{y}$$

Interchanging x and y gives $y = \dfrac{2x + 6}{x}$.

So the inverse function is $f^{-1}(x) = \dfrac{2x + 6}{x}$, $x \in \mathbb{R}^+$.

Example

For $f(x) = x^2$, find $f^{-1}(x)$ for a suitable domain.

Considering the graph of f(x), it is clear that there is not a one-to-one correspondence for all $x \in \mathbb{R}$.

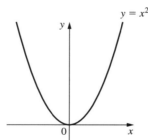

$y = x^2$

However, if we consider only one half of the graph and restrict the domain to $x \in \mathbb{R}$, $x \geq 0$, an inverse does exist.

Let $y = x^2$

Then $x = \sqrt{y}$

Interchanging x and y gives $y = \sqrt{x}$ (positive root only)

$\Rightarrow f^{-1}(x) = \sqrt{x}$, $x \in \mathbb{R}$, $x \geq 0$

> This is important to note: a square root is only a function if only the positive root or only the negative root is considered.

The equation $x^2 + y^2 = 25$ is for a circle with centre the origin, as shown below.

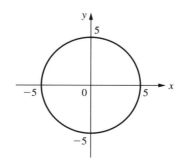

This graph is not a function as there exist vertical lines that cut the graph twice. There are also horizontal lines that cut the circle twice. Even restricting the domain to $0 \le x \le 5$ does not allow there to be an inverse as the original graph is not a function. So it is not possible to find an inverse for $x^2 + y^2 = 25$.

Exercise 3

1 Which of the following have an inverse function for $x \in \mathbb{R}$?
   a $f(x) = 2x - 3$ **b** $f(x) = x^2 - 1$ **c** $f(x) = x^3$ **d** $y = \cos x$

2 For each function f(x), find the inverse function $f^{-1}(x)$.
   a $f(x) = 4x$ **b** $f(x) = x - 5$ **c** $f(x) = x + 6$

   d $f(x) = \dfrac{2}{3}x$ **e** $f(x) = 7 - x$ **f** $f(x) = 9 - 4x$

   g $f(x) = 2x + 9$ **h** $f(x) = x^3 - 6$ **i** $f(x) = 8x^3$

3 What is the largest domain for which f(x) has an inverse function?
   a $f(x) = \dfrac{1}{x - 3}$ **b** $f(x) = \dfrac{2}{x + 4}$ **c** $f(x) = \dfrac{3}{2x - 1}$

   d $f(x) = x^2 - 5$ **e** $f(x) = 9 - x^2$ **f** $f(x) = x^2 - x - 12$

   g $f(x) = \cos x$

4 For each function f(x), (i) choose a suitable domain so that an inverse exists (ii) find the inverse function $f^{-1}(x)$.

   a $f(x) = \dfrac{1}{x - 6}$ **b** $f(x) = \dfrac{3}{x + 7}$ **c** $f(x) = \dfrac{5}{3x - 2}$

   d $f(x) = \dfrac{7}{2 - x}$ **e** $f(x) = \dfrac{8}{4x - 9}$ **f** $f(x) = \dfrac{4}{5x + 6}$

   g $f(x) = 6x^2$ **h** $f(x) = x^2 - 4$ **i** $f(x) = 2x^2 + 3$

   j $f(x) = 16 - 9x^2$ **k** $f(x) = x^4$ **l** $f(x) = 2x^3 - 5$

5 For $f(x) = 3x$ and $g(x) = x - 2$, find
   a $h(x) = f(g(x))$ **b** $h^{-1}(x)$

3.4 Graphs of inverse functions

On a graphing calculator, graph the following functions and their inverse functions and look for a pattern:

1 $f(x) = 2x$, $f^{-1}(x) = \dfrac{1}{2}x$

2 $f(x) = x + 4$, $f^{-1}(x) = x - 4$

3 $f(x) = x - 1$, $f^{-1}(x) = x + 1$

4 $f(x) = 3x - 1$, $f^{-1}(x) = \dfrac{1}{3}x + \dfrac{1}{3}$

5 $f(x) = \dfrac{2}{x - 3}$, $f^{-1}(x) = \dfrac{2 + 3x}{x}$ for $x > 3$

Through this investigation it should be clear that the graph of a function and its inverse are connected. The connection is that one graph is the reflection of the other in the line $y = x$.

This connection should make sense. By reflecting the point (x, y) in the line $y = x$, the image is (y, x). In other words, the domain becomes the range and vice versa. This reflection also makes sense when we remember that $f(f^{-1}(x)) = x$.

Thus if we have a graph (without knowing the equation), we can sketch the graph of the inverse function.

Example

For the graph of $f(x)$ below, sketch the graph of its inverse, $f^{-1}(x)$.

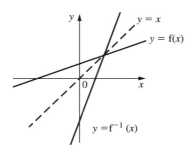

Drawing in the line $y = x$ and then reflecting the graph in this line produces the graph of the inverse function $f^{-1}(x)$.

Exercise 4

1 For each function $f(x)$, find the inverse function $f^{-1}(x)$ and draw the graphs of $y = f(x)$ and $y = f^{-1}(x)$ on the same diagram.
  a $f(x) = 2x$ **b** $f(x) = x + 2$ **c** $f(x) = x - 3$
  d $f(x) = 3x + 1$ **e** $f(x) = 2x - 4$

2 For each function $f(x)$, draw the graph of $y = f(x)$ for $x \geq 0$.
  Find the inverse function $f^{-1}(x)$ and draw it on the same graph.
  a $f(x) = x^2$ **b** $f(x) = 3x^2$
  c $f(x) = x^2 + 4$ **d** $f(x) = 5 - x^2$

3 For each function $f(x)$, draw the graph of $y = f(x)$ for $x \geq 0$.
  Find the inverse function $f^{-1}(x)$ and draw it on the same graph.
  a $f(x) = \dfrac{1}{x + 2}$ **b** $f(x) = \dfrac{1}{x + 5}$ **c** $f(x) = \dfrac{2}{x + 1}$

4 Sketch the graph of the inverse function $f^{-1}(x)$ for each graph.

a

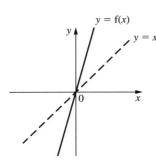

b

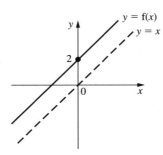

c

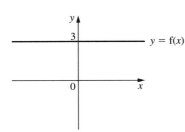

d

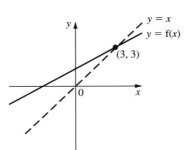

e

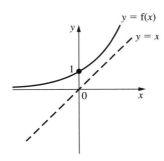

f

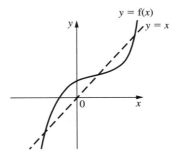

3.5 Special functions

The reciprocal function

The function known as the reciprocal function is $f(x) = \dfrac{1}{x}$. In Chapter 1, we met vertical asymptotes. These occur when a function is not defined (when the denominator is zero). For $f(x) = \dfrac{1}{x}$, there is a vertical asymptote when $x = 0$. To draw the graph, we consider what happens either side of the asymptote. When $x = -0.1, f(-0.1) = -10$. When $x = 0.1, f(0.1) = 10$.

Now consider what happens for large values of x, that is, as $x \rightarrow \pm\infty$.

As $x \rightarrow \infty, \dfrac{1}{x} \rightarrow 0$. As $x \rightarrow -\infty, \dfrac{1}{x} \rightarrow 0$.

So the graph of $f(x) = \dfrac{1}{x}$ is

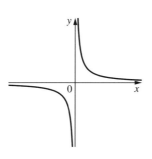

For $x \to \infty, y \to 0$. For large values of x, the graph approaches the x-axis. This is known as a horizontal asymptote. As with vertical asymptotes, this is a line that the graph approaches but does not reach.

It is clear from the graph that this function has an inverse, provided $x \neq 0$.

Let $y = \dfrac{1}{x}$

$\Rightarrow x = \dfrac{1}{y}$

Interchanging y and x, $y = \dfrac{1}{x}$.

Hence $f^{-1}(x) = \dfrac{1}{x}$, $x \neq 0$, $x \in \mathbb{R}$. So this function is the inverse of itself. Hence it has a self-inverse nature and this is an important feature of this function.

The absolute value function

The function denoted $f(x) = |x|$ is known as the **absolute value function**. This function can be described as making every y value positive, that is, ignoring the negative sign. This can be defined strictly as

$$f(x) = \begin{cases} x, & x \geq 0 \\ -x, & x < 0 \end{cases}$$

This is known as a **piecewise function** as it is defined in two pieces.

This is the graph:

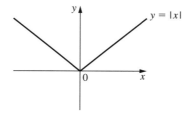

$y = |x|$

The whole graph is contained above the x-axis. Note the unusual "sharp" corner at $x = 0$.

The absolute value can be applied to any function. The effect is to reflect in the x-axis any part of the graph that is below the x-axis while not changing any part above the x-axis.

Example

Sketch the graph of $y = |x - 3|$.

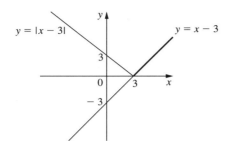

Example

Sketch the graph of $y = |x^2 - 2x - 8|$.

Start by sketching the graph of $y = x^2 - 2x - 8$
$$= (x - 4)(x + 2)$$

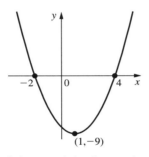

Reflect the negative part of the graph in the x-axis:

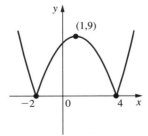

The graph of an absolute value function can be used to solve an equation or inequality.

Example

Solve $|2x - 3| = 5$.

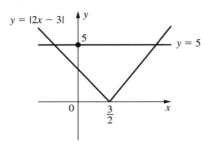

For the negative solution solve

$-(2x - 3) = 5$

$\Rightarrow -2x + 3 = 5$

$\Rightarrow -2x = 2$

$\Rightarrow x = -1$

For the positive solution solve

$2x - 3 = 5$

$\Rightarrow 2x = 8$

$\Rightarrow x = 4$

Example

Solve $|x^2 + x - 6| \leq 4$.

$x^2 + x - 6 = (x + 3)(x - 2)$ so the graph of $y = |x^2 + x - 6|$ is

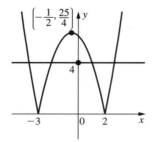

To find the points of intersection of $y = |x^2 + x - 6|$ and $y = 4$ solve

$-x^2 - x + 6 = 4$ and $x^2 + x - 6 = 4$

$\Rightarrow x^2 + x - 2 = 0$ $\Rightarrow x^2 + x - 10 = 0$

$\Rightarrow (x + 2)(x - 1) = 0$ $\Rightarrow x = -3.70$ or $x = 2.70$

$\Rightarrow x = -2$ or $x = 1$ (using the quadratic formula)

Hence $|x^2 + x - 6| \leq 4 \Rightarrow -3.70 \leq x \leq -2$ or $1 \leq x \leq 2.70$

Exercise 5

1 Write $f(x) = |x - 2|$ as a piecewise function.

2 Write $f(x) = |2x + 1|$ as a piecewise function.

3 Write $f(x) = |x^2 - x - 12|$ as a piecewise function.

4 Write $f(x) = |2x^2 - 5x - 3|$ as a piecewise function.

5 Sketch the graph of $y = |x + 4|$.

6 Sketch the graph of $y = |3x|$.

7 Sketch the graph of $y = |3x - 5|$.

8 Sketch the graph of $y = |x^2 + 4x - 12|$.

9 Sketch the graph of $y = |x^2 - 7x + 12|$.

10 Sketch the graph of $y = |x^2 + 5x + 6|$.

11 Sketch the graph of $y = |3x^2 + 5x - 2|$.

12 Solve $|x + 2| = 3$.

13 Solve $|x - 5| = 1$.

14 Solve $|2x + 5| = 3$.

15 Solve $|7 - 2x| = 3$.

16 Solve $|x^2 + x - 6| = 2$.

17 Solve $|2x^2 + x - 10| = 4$.

18 Solve $|x + 2| < 5$.

19 Solve $|2x - 1| \leq 9$.

20 Solve $|9 - 4x| < 1$.

21 Solve $|x^2 + 4x - 12| \leq 7$.

22 Solve $|2x^2 + 5x - 12| < 9$.

3.6 Drawing a graph

In the first two chapters, we covered drawing trigonometric graphs and drawing quadratic graphs. We have now met some of the major features of the graphs, including

- roots – values of x when $y = 0$
- y-intercept – the value of y when $x = 0$
- turning points
- vertical asymptotes – when y is not defined
- horizontal asymptotes – when $x \rightarrow \pm\infty$

> More work will be done on sketching graphs in Chapter 8.

Example

Sketch the graph of $y = x^2 - 4x - 12$, noting the major features.

$x^2 - 4x - 12 = (x + 2)(x - 6)$
so the graph has roots at
$x = -2$ and $x = 6$.
We know the shape of this
function, and that it has a
minimum turning point at
$(2, -16)$ by the symmetry of
the graph.
Setting $x = 0$ gives the
y-intercept as $y = -12$.
This graph has no asymptotes.

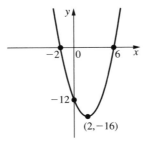

> This process was covered in Chapter 2.

Example

Sketch the graph of $y = \dfrac{2}{x-3}$.

This graph has no roots as the numerator is never zero.

When $x = 0$, $y = -\dfrac{2}{3}$ so $\left(0, -\dfrac{2}{3}\right)$ is the y-intercept.

There is a vertical asymptote when $x - 3 = 0 \Rightarrow x = 3$.

As $x \to \pm\infty$, the denominator becomes very large and so $y \to 0$.

By taking values of x close to the vertical asymptote, we can determine the behaviour of the graph around the asymptote. So, when $x = 2.9, y = -20$.

When $x = 3.1, y = 20$. So the graph is

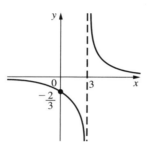

This is an example of a rational function. More work on these is covered on page 79.

All of these features can be identified when sketching a function using a graphing calculator.

Example

Sketch the graph of $y = 10 - 3x - x^2$.

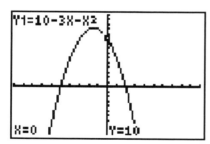

The calculator can be used to calculate points such as intercepts and turning points. The asymptotes (if any) are clear from the graph (as long as an appropriate window is chosen).

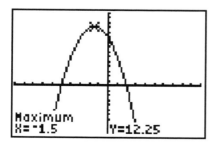

$x = -5$ and $x = 2$ are the roots.

$(0, 10)$ is the y-intercept.

$\left(-\dfrac{3}{2}, \dfrac{49}{4}\right)$ is the maximum turning point.

Many types of function can be sketched using a graphing calculator. Although we study a number of functions in detail, including straight lines, polynomials and trigonometric functions, there are some functions that we only sketch using the calculator (in this course).

Example

Sketch the graph of $y = \dfrac{(x^2 + 5)^{\frac{1}{2}}}{x - 2}$.

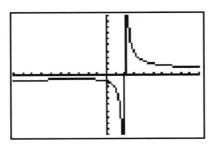

Here we can see that there is a vertical asymptote at $x = 2$, a horizontal asymptote at $y = 0$, there is no turning point, and the y-intercept is $-\dfrac{\sqrt{5}}{2}$.

Exercise 6

1 For each function, sketch the graph of $y = f(x)$, indicating asymptotes, roots, y-intercepts and turning points.

a $f(x) = x + 4$

b $f(x) = 2x - 1$

c $f(x) = 3 - x$

d $f(x) = 7 - 2x$

e $f(x) = x^2 + 7x + 12$

f $f(x) = x^2 - 8x + 12$

g $f(x) = x^2 - 5x - 24$

h $f(x) = 3x^2 + 2x - 8$

i $f(x) = 6x^2 + x - 15$

j $f(x) = 20 + 17x - 10x^2$

k $f(x) = \dfrac{1}{x + 4}$

l $f(x) = \dfrac{3}{x - 2}$

m $f(x) = \dfrac{4}{2x - 1}$

2 Using a graphing calculator, make a sketch of $y = f(x)$, indicating asymptotes, roots, y-intercepts and turning points.

a $f(x) = x^2 - x - 30$

b $f(x) = x^2 + 5x + 3$

c $f(x) = x^2 + 2x + 5$

d $f(x) = \dfrac{6}{2x + 3}$

e $f(x) = \dfrac{5}{(x + 2)(x - 3)}$

f $f(x) = \dfrac{7}{x^2 - 7x + 12}$

g $f(x) = \dfrac{x + 1}{x + 5}$

h $f(x) = \dfrac{x + 3}{x^2 - 3x - 10}$

i $f(x) = \dfrac{x - 2}{x^2 + 6}$

j $f(x) = \dfrac{x^2 + 8}{x - 5}$

k $f(x) = \dfrac{x^2 - x - 6}{x^2 + 10x + 24}$

l $f(x) = \dfrac{(x^2 + 3)^{\frac{3}{2}}(x + 2)}{x + 7}$

m $f(x) = \dfrac{\sin x}{x^2}$

n $f(x) = \dfrac{\cos x^2}{x + 1}$

3.7 Transformations of functions

In Chapter 1, we met trigonometric graphs such as $y = 2 \sin 3x + 1$.

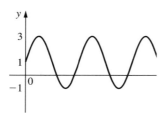

Here the 3 has the effect of producing three waves in 2π (three times as many graphs as $y = \sin x$).

The 2 stretches the graph vertically.

The 1 shifts the graph vertically.

In Chapter 2, we met quadratic graphs such as $y = -(x - 2)^2 + 4$.

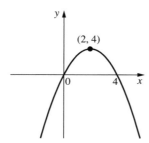

Here the -2 inside the bracket has the effect of shifting $y = x^2$ right.

The -1 in front of the bracket reflects the graph in the x-axis.

The $+4$ shifts the graph vertically.

We can see that there are similar effects for both quadratic and trigonometric graphs. We can now generalize transformations as follows:

For $k\mathrm{f}(x)$, each y-value is multiplied by k and so this creates a vertical stretch.

For $\mathrm{f}(kx)$, each x-value is multiplied by k and so this creates a horizontal stretch.

For $\mathrm{f}(x) + k$, k is added to each y-value and so the graph is shifted vertically.

For $\mathrm{f}(x + k)$, k is added to each x-value and so the graph is shifted horizontally.

For $-\mathrm{f}(x)$, each y-value is multiplied by -1 and so each point is reflected in the x-axis.

For $\mathrm{f}(-x)$, each x-value is multiplied by -1 and so each point in reflected in the y-axis.

General form	Example	Effect
$k\mathrm{f}(x)$	$y = 3 \sin x$	Vertical stretch
$\mathrm{f}(kx)$	$y = \cos 2x$	Horizontal stretch
$\mathrm{f}(x) + k$	$y = x^2 + 5$	Vertical shift $[\,k > 0$ up, $k < 0$ down$]$
$\mathrm{f}(x + k)$	$y = (x + 3)^2$	Horizontal shift $[\,k > 0$ left, $k < 0$ right$]$
$-\mathrm{f}(x)$	$y = -\cos x$	Reflection in x-axis
$\mathrm{f}(-x)$	$y = \sin(-x)$	Reflection in y-axis

Example

Sketch the graph of $y = -2 \cos\left(\theta + \dfrac{\pi}{4}\right) + 1$.

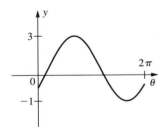

Example

Sketch the graph of $y = (x - 2)^2 - 3$.

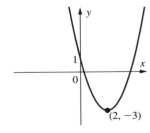

Example

This is a graph of $y = f(x)$.
Draw
(a) $f(x - 2)$
(b) $f(-x)$

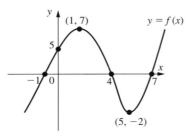

(a) $f(x - 2)$

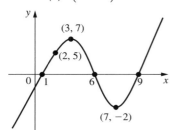

(b) $f(-x)$

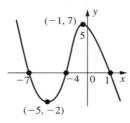

This is a horizontal shift of 2 to the right.

This is a reflection in the y-axis.

The absolute value of a function $|f(x)|$ can be considered to be a transformation of a function (one that reflects any parts below the x-axis). A graphing calculator can also be used to sketch transformations of functions, as shown below.

Example

Given that $f(x) = x^2 - 4$, sketch the graph of (a) $f(x + 3)$ (b) $|f(x)|$
(a)

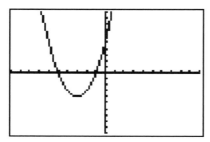

As expected, this is the graph of $f(x)$ shifted 3 places to the left. The calculator cannot find the "new" function, but the answer can be checked once it is found algebraically, if required.

$f(x + 3) = (x + 3)^2 - 4$
$\qquad = x^2 + 6x + 9 - 4$
$\qquad = x^2 + 6x + 5$

(b) The negative part of the curve is reflected in the x-axis, that is, the part defined by $-2 \leq x \leq 2$.

This can be sketched and checked to be the same as $f(x + 3)$.

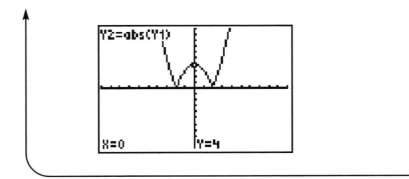

Similarly, the function of the absolute value of x, that is, $f(|x|)$, can be considered to be a transformation of a function. If we consider this as a piecewise function, we know that the absolute value part will have no effect for $x \geq 0$. However, for $x < 0$, the effect will be that it becomes $f(-x)$. This means that the graph of $f(|x|)$ will be the graph of $f(x)$ for $x \geq 0$ and this will then be reflected in the y-axis.

Example

Sketch the graph of $f(x) = x^2 - 4x - 5$ and the graph of $f(|x|)$.
Using a graphing calculator, we can sketch both graphs:

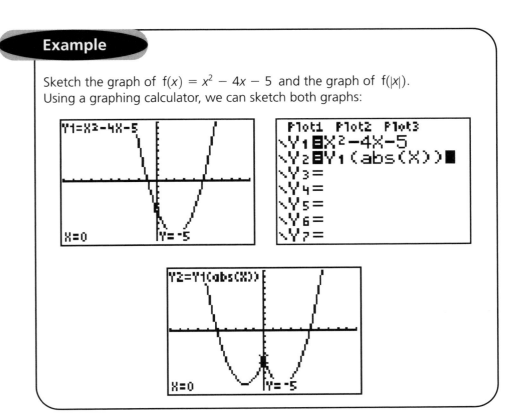

Rational functions

Rational functions are functions of the type $f(x) = \dfrac{g(x)}{h(x)}$ where $g(x)$ and $h(x)$ are polynomials. Here we shall consider functions of the type $\dfrac{a}{px + q}$ and $\dfrac{bx + c}{px + q}$.

Functions of the type $\dfrac{a}{px + q}$

These can be considered to be a transformation of the reciprocal function $f(x) = \dfrac{1}{x}$.

Example

Sketch the graph of $y = \dfrac{2}{x - 3}$.

Comparing this to $f(x) = \dfrac{1}{x}$, $y = 2f(x - 3)$. So its graph is the graph of $y = \dfrac{1}{x}$, stretched vertically $\times 2$ and shifted 3 to the right.

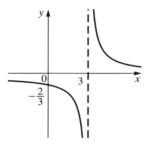

Example

Sketch the graph of $y = \dfrac{5}{2x + 1}$.

To consider this as a transformation, it can be written as $y = \dfrac{\frac{5}{2}}{x + \frac{1}{2}}$.

This is $y = \dfrac{1}{x}$ shifted left $\dfrac{1}{2}$ and vertically stretched by $\dfrac{5}{2}$. However, it is probably easiest just to calculate the vertical asymptote and the y-intercept. Here the vertical asymptote is when $2x + 1 = 0 \Rightarrow x = -\dfrac{1}{2}$. The y-intercept is when $x = 0 \Rightarrow y = 5$.

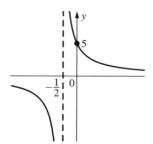

Functions of the type $\dfrac{bx + c}{px + q}$

The shape of this graph is very similar to the previous type but the horizontal asymptote is not the x-axis.

The horizontal asymptote is $y = \dfrac{b}{p}$, as when $x \to \pm\infty$, $y \to \dfrac{b}{p}$.

Example

Sketch the graph of $y = \dfrac{2x + 1}{x - 3}$.

This has a vertical asymptote at $x = 3$.

The horizontal asymptote is $y = 2$. [As $x \to \pm\infty$, $y \to \dfrac{2x}{x} = 2$]

The y intercept is $y = -\dfrac{1}{3}$.

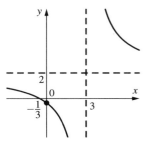

Example

Sketch the graph of $y = \dfrac{7 - 2x}{3x + 5}$.

This has a vertical asymptote at $x = -\dfrac{5}{3}$.

There is a horizontal asymptote at $y = -\dfrac{2}{3}$.

The y-intercept, when $x = 0$, is $y = \dfrac{7}{5}$.

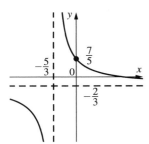

Example

(a) Sketch the curve $f(x) = \dfrac{3}{x-2}$.

(b) Solve $\dfrac{3}{x-2} < 4,\, x > 2$.

(a) For f(x), we know that the graph has a vertical asymptote at $x = 2$ and will have a horizontal asymptote at $y = 0$.

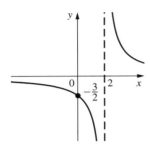

(b) In order to solve this inequality, we first solve $f(x) = 4$

$$\Rightarrow \frac{3}{x-2} = 4$$
$$\Rightarrow 4x - 8 = 3$$
$$\Rightarrow 4x = 11$$
$$\Rightarrow x = \frac{11}{4}$$

Using the graph, the solution to the inequality is $x > \dfrac{11}{4}$.

Exercise 7

1 Sketch (a) $y = 2\cos 3x°$ (b) $y = 4\sin\left(\theta + \dfrac{\pi}{3}\right)$ (c) $y = -3\sin\theta + 2$

2 Sketch (a) $y = 3x^2$ (b) $y = (x-2)^2$ (c) $y = 8 - x^2$

3 For each function f(x), sketch (i) $y = f(x)$ (ii) $y = f(x-2)$ (iii) $y = f(x) - 1$
(iv) $y = -2\,f(x)$

   a $f(x) = x^2$ **b** $f(x) = x^3$ **c** $f(x) = 3x$ **d** $f(x) = 4 - x$

   e $f(x) = \dfrac{1}{x}$ **f** $f(x) = x^2 - 3$ **g** $f(x) = \dfrac{3}{x-4}$ **h** $f(x) = \dfrac{x+2}{x-1}$

4 For each of the functions in question 3, find an expression for $2 - f(x+1)$ algebraically. Sketch each graph of $2 - f(x+1)$ using a graphing calculator, thus checking your answer.

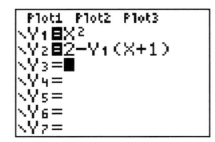

5 For each graph $y = f(x)$, sketch (i) $f(x + 3)$ (ii) $f(-x)$ (iii) $5 - 3f(x)$

a

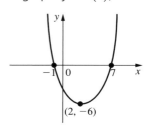

b

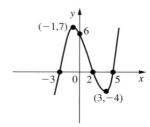

c

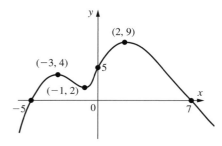

6 Sketch the graph of the rational function $y = f(x)$.

a $f(x) = \dfrac{2}{x}, x \neq 0$

b $f(x) = \dfrac{1}{x - 3}, x \neq 3$

c $f(x) = \dfrac{1}{x + 2}, x \neq -2$

d $f(x) = \dfrac{3}{x - 4}, x \neq 4$

e $f(x) = \dfrac{5}{x + 7}, x \neq -7$

f $f(x) = \dfrac{1}{2x + 1}, x \neq -\dfrac{1}{2}$

g $f(x) = \dfrac{6}{2x - 3}, x \neq \dfrac{3}{2}$

h $f(x) = \dfrac{1}{7 - x}, x \neq 7$

i $f(x) = \dfrac{-4}{x + 5}, x \neq -5$

j $f(x) = \dfrac{3}{8 - 5x}, x \neq \dfrac{8}{5}$

7 Sketch the graph of the rational function $y = g(x)$.

a $g(x) = \dfrac{x + 6}{x - 1}, x \neq 1$

b $g(x) = \dfrac{x - 4}{x + 3}, x \neq -3$

c $g(x) = \dfrac{2x + 1}{x - 6}, x \neq 6$

d $g(x) = \dfrac{8 - x}{x + 2}, x \neq -2$

e $g(x) = \dfrac{9 - x}{3 - x}, x \neq 3$

f $g(x) = \dfrac{5x - 2}{2x + 1}, x \neq -\dfrac{1}{2}$

g $g(x) = \dfrac{10x - 1}{2x + 3}, x \neq -\dfrac{3}{2}$

h $g(x) = \dfrac{7x + 2}{2x - 3}, x \neq \dfrac{3}{2}$

8 Solve the following equations for $x \in \mathbb{R}$.

a $\dfrac{10}{x + 2} = 4$

b $\dfrac{7}{x - 1} = 3$

c $\dfrac{3}{2x + 1} = 5$

d $\dfrac{8}{2 - x} = 6$

e $\dfrac{9}{7 - 2x} = -4$

9 Solve the following inequalities for $x \in \mathbb{R}, x > 0$.

a $\dfrac{1}{x + 5} < 6$

b $\dfrac{1}{x - 3} \leq 4$

c $\dfrac{2}{x + 3} < 7$

d $\dfrac{8}{2x - 1} < 1$

e $\dfrac{8}{3x - 2} < 4$

f $\dfrac{2}{3 - x} \geq -2$

g $\dfrac{x + 2}{x - 1} < 5$

h $\dfrac{6x + 1}{2x - 1} < 5$

10 Use your graphing calculator to draw a sketch of
$f(x) = x^2(x - 2)^3$, $g(x) = f(-x) + 3$ and $h(x) = |g(x)|$.

11 Use your graphing calculator to draw a sketch of $f(x) = \dfrac{x^2 + 1}{x - 3}$ for

$x \neq 3$, $g(x) = f(x - 2)$ and $h(x) = |g(x)|$.

12 Use your graphing calculator to draw a sketch of $p(x) = g(f(x))$ given
$f(x) = x^2 - x - 6$ and $g(x) = 3x - 1$. Hence draw the graph of $p(|x|)$.

Review exercise

1 For $f(x) = \dfrac{3x^2 - 5}{x}$, find f(2).

2 For $g(x) = \dfrac{2x + 1}{x - 2}$ with domain $\{-4, 0, 1, 5\}$, draw an arrow diagram
and state the range.

3 For $f(x) = 7x - 4$, find

  a f(3x)   **b** f(2x − 1)   **c** $f\left(\dfrac{1}{x}\right)$

4 For $f(x) = 8 - 3x$ and $g(x) = \dfrac{x}{x - 1}$, $x \neq 1$ find

  a f(g(x))   **b** g(f(x))   **c** f(f(x))   **d** g(g(x))

5 For each function f(x), choose a suitable domain so that an inverse exists and
find $f^{-1}(x)$.

  a $f(x) = x^2 - 6$ **b** $f(x) = \dfrac{1}{x + 5}$ **c** $f(x) = \dfrac{7}{2x + 3}$

6 Sketch the graph of $f(x) = \dfrac{2}{x + 1}$ and its inverse function $f^{-1}(x)$.

7 For $f(x) = x^2 + 4x - 12$, sketch the graph of $y = |f(x)|$ and $y = f(|x|)$.

8 Solve $|2x + 9| = 7$.

9 Solve $|7 - 5x| < 3$.

10 Sketch the graph of $y = \dfrac{2}{x + 3}$, indicating asymptotes and the y-intercept.

11 Sketch the graph of $y = \dfrac{x + 2}{x^2 + 4x - 9}$, indicating asymptotes, roots, y-intercept
and turning points.

12 Sketch the graph of $y = \dfrac{\cos x}{3x^2}$, indicating asymptotes, roots, y-intercept
and turning points.

13 For $f(x) = 2x + 1$, sketch the graph of
  a f(x − 3)   **b** f(3x)   **c** 4 − 5 f(x)

14 Sketch the graph of each of these rational functions:

  a $f(x) = \dfrac{7}{2x + 1}$ **b** $f(x) = \dfrac{-4}{3x + 2}$ **c** $f(x) = \dfrac{8x - 3}{2x + 1}$

15 Solve $\dfrac{9}{3x - 2} = 5$.

16 Solve $\dfrac{2}{4x - 1} \leq 3$ for $x > 0$.

 17 Use your graphing calculator to draw a sketch of

$$f(x) = \frac{x^3 - 5x + 1}{x + 2}, \ g(x) = f(x + 3) \text{ and } h(x) = |g(x)|.$$

 18 The one-to-one function f is defined on the domain $x > 0$ by $f(x) = \dfrac{2x - 1}{x + 2}$.

    a State the range, A, of f.   **b** Obtain an expression for $f^{-1}(x)$, for $x \in A$.

                                                          [IB May 02 P1 Q15]

 19 Solve the inequality $|x - 2| \geq |2x + 1|$.   [IB May 03 P1 Q13]

 20 A function f is defined for $x \leq 0$ by $f(x) = \dfrac{x^2 - 1}{x^2 + 1}$.

    Find an expression for $f^{-1}(x)$.   [IB May 03 P1 Q17]

21 Let $f : x \to \sqrt{\dfrac{1}{x^2} - 2}$.

    Find

    a the set of real values of x for which   **b** the range of f.

       f is real and finite   [IB May 01 P1 Q5]

22 Let $f(x) = \dfrac{x + 4}{x + 1}, x \neq -1$ and $g(x) = \dfrac{x - 2}{x - 4}, x \neq 4$.

    Find the set of values of x such that $f(x) \leq g(x)$.   [IB Nov 03 P1 Q17]

4 Polynomials

The Italian mathematician Paolo Ruffini, born in 1765, is responsible for synthetic division, also known as Ruffini's rule, a technique used for the division of polynomials that is covered in this chapter.

Ruffini was not merely a mathematician but also held a licence to practise medicine. During the turbulent years of the French Revolution, Ruffini lost his chair of mathematics at the university of Modena by refusing to swear an oath to the republic. Ruffini seemed unbothered by this, indeed the fact that he could no longer teach mathematics meant that he could devote more time to his patients, who meant a lot to him. It also gave him a chance to do further mathematical research. The project he was working on

Paolo Ruffini

was to prove that the quintic equation cannot be solved by radicals. Before Ruffini, no other mathematician published the fact that it was not possible to solve the quintic equation by radicals. For example, Lagrange in his paper *Reflections on the resolution of algebraic equations* said that he would return to this question, indicating that he still hoped to solve it by radicals. Unfortunately, although his work was correct, very few mathematicians appeared to care about this new finding. His article was never accepted by the mathematical community, and the theorem is now credited to being solved by Abel.

4.1 Polynomial functions

Polynomials are expressions of the type $f(x) = ax^n + bx^{n-1} + \ldots + px + c$. These expressions are known as polynomials only when all of the powers of x are positive integers (so no roots, or negative powers). The **degree** of a polynomial is the highest power of x (or whatever the variable is called). We are already familiar with some of these functions, and those with a small degree have special names:

> This chapter treats this topic as if a calculator is not available throughout until the section on using a calculator at the end.

Degree	Form of polynomial	Name of function
1	$ax + b$	Linear
2	$ax^2 + bx + c$	Quadratic
3	$ax^3 + bx^2 + cx + d$	Cubic
4	$ax^4 + bx^3 + cx^2 + dx + e$	Quartic
5	$ax^5 + bx^4 + cx^3 + dx^2 + ex + f$	Quintic

$f(x) = 2x^5 + 3x^2 - 7$ is a polynomial is of degree 5 or quintic function. The coefficient of the leading term is 2, and -7 is the constant term.

Values of a polynomial

We can evaluate a polynomial in two different ways. The first method is to substitute the value into the polynomial, term by term, as in the example below.

> This was covered in Chapter 3.

Example

Find the value of $f(x) = x^3 - 3x^2 + 6x - 4$ when $x = 2$.

Substituting: $f(2) = 2^3 - 3(2)^2 + 6(2) - 4$

$\qquad\qquad = 8 - 12 + 12 - 4$

$\qquad\qquad = 4$

The second method is to use what is known as a **nested scheme**.

This is where the coefficients of the polynomial are entered into a table, and then the polynomial can be evaluated, as shown in the example below.

Example

Using the nested calculation scheme, evaluate the polynomial
$f(x) = 2x^4 - 4x^3 + 5x - 8$ when $x = -2$.

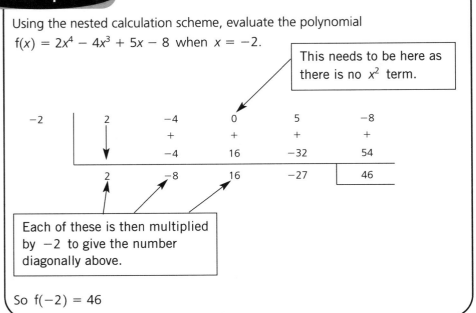

This needs to be here as there is no x^2 term.

Each of these is then multiplied by -2 to give the number diagonally above.

So $f(-2) = 46$

To see why this nested calculation scheme works, consider the polynomial $2x^3 + x^2 - x + 5$.

x	2	1	-1	5
	$\downarrow$	+	+	+
		$2x$	$2x^2 + x$	$2x^3 + x^2 - x$
	2	$2x + 1$	$2x^2 + x - 1$	$2x^3 + x^2 - x + 5$

Example

Find the value of the polynomial $g(x) = x^3 - 7x + 6$ when $x = 2$.

2	1	0	-7	6
	$\downarrow$	+	+	+
		2	4	-6
	1	2	-3	0

Here $g(2) = 0$. This means that $x = 2$ is a root of $g(x) = x^3 - 7x + 6$.

Division of polynomials

This nested calculation scheme can also be used to divide a polynomial by a linear expression. This is known as **synthetic division**.

When we divide numbers, we obtain a quotient and a remainder. For example, in the calculation $603 \div 40 = 15$ R 3, 603 is the dividend, 40 is the divisor, 15 is the quotient and 3 is the remainder.

The same is true for algebraic division. Synthetic division is a shortcut for dividing polynomials by linear expressions – algebraic long division is covered later in the chapter.

Synthetic division works in exactly the same way as the nested calculation scheme. The value of x that is used is the root that the divisor provides. This is best demonstrated by example.

> Synthetic division works only for linear divisors.

Example

Divide $3x^3 - x^2 + 2x - 5$ by $x - 2$ using synthetic division.

We need the value of x such that $x - 2 = 0$, that is, $x = 2$.

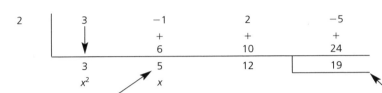

These numbers are the coefficients of the quotient.

This is the remainder.

So $3x^3 - x^2 + 2x - 5 = (x - 2)(3x^2 + 5x + 12) + 19$

This could be checked by expanding the brackets.

Example

Divide $x^3 - 11x + 3$ by $x + 5$.

$$
\begin{array}{r|rrrr}
-5 & 1 & 0 & -11 & 3 \\
   &   & -5 & 25 & -70 \\
\hline
   & 1 & -5 & 14 & -67
\end{array}
$$

So $x^3 - 11x + 3 = (x + 5)(x^2 - 5x + 14) - 67$

Example

Divide $2x^3 + x^2 + 5x - 1$ by $2x - 1$.

Here the coefficient of x in the divisor is not 1.

$$2x - 1 = 0$$
$$\Rightarrow 2\left(x - \frac{1}{2}\right) = 0$$
$$\Rightarrow x = \frac{1}{2}$$

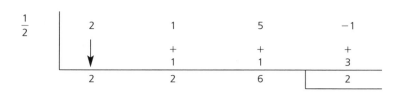

So, from this we can say that

$$2x^3 + x^2 + 5x - 1 = \left(x - \frac{1}{2}\right)(2x^2 + 2x + 6) + 2$$

$$= (2x - 1)(x^2 + x + 3) + 2$$

> In this situation, there will always be a common factor in the quotient. This common factor is the coefficient of x in the divisor.

Exercise 1

1 Evaluate $f(x) = x^4 - 3x^3 + 3x^2 + 7x - 4$ for $x = 2$.

2 Evaluate $g(x) = 7x^3 - 2x^2 - 8x + 1$ for $x = -2$.

3 Evaluate $f(x) = x^6 - 4x^3 - 7x + 9$ for $x = -1$.

4 Find f(4) for each polynomial.

   a $f(x) = x^3 - x^2 + 2x - 5$ **b** $f(x) = 5x^4 - 4x^2 + 8$

   c $f(t) = t^5 - 6t^3 + 7t + 6$ **d** $f(x) = 6 - 7x + 5x^2 - 2x^3$

5 Calculate $f\left(-\frac{1}{2}\right)$ for $f(x) = 6x^3 - 4x^2 - 2x + 3$.

6 Use synthetic division to find the quotient and remainder for each of these calculations.

   a $(x^2 + 6x - 3) \div (x - 2)$ **b** $(x^3 - 4x^2 + 5x - 1) \div (x - 1)$

   c $(2x^3 + x^2 - 8x + 7) \div (x - 6)$ **d** $(x^3 + 5x^2 - x - 9) \div (x + 4)$

   e $(x^4 - 5x^2 + 3x + 7) \div (x + 1)$ **f** $(x^5 - x^2 - 5x - 11) \div (2x - 1)$

   g $(t^3 - 7t + 9) \div (2t + 1)$ **h** $(-3x^4 - 4x^3 - 5x^2 + 13) \div (4x + 3)$

7 Express each function in the form $f(x) = (px - q)Q(x) + R$ where Q(x) is the quotient on dividing f(x) by $(px - q)$ and R is the remainder.

	$f(x)$	$(px - q)$
(a)	$3x^2 - 7x + 2$	$x - 2$
(b)	$x^3 + 6x^2 - 8x + 7$	$x - 5$
(c)	$4x^3 + 7x^2 - 9x - 17$	$x + 3$
(d)	$5x^5 - 4x^3 + 3x - 2$	$x + 4$
(e)	$2x^6 - 5x^4 + 9$	$x + 1$
(f)	$x^3 - 7x^2 + 4x - 2$	$2x - 1$
(g)	$2x^4 - 4x^2 + 11$	$2x + 1$

4.2 Factor and remainder theorems

The remainder theorem

If a polynomial f(x) is divided by $(x - h)$ the remainder is f(h).

Proof

We know that $f(x) = (x - h)Q(x) + R$ where $Q(x)$ is the quotient and R is the remainder.

For $x = h$, $\quad f(h) = (h - h)Q(h) + R$

$$= (0 \times Q(h)) + R$$

$$= R$$

Therefore, $f(x) = (x - h)Q(x) + f(h)$.

The factor theorem

If f(h) = 0 then $(x - h)$ is a factor of f(x).
Conversely, if $(x - h)$ is a factor of f(x) then f(h) = 0.

Proof

For any function $f(x) = (x - h)Q(x) + f(h)$.

If $f(h) = 0$ then $f(x) = (x - h)Q(x)$.

Hence $(x - h)$ is a factor of f(x).

Conversely, if $(x - h)$ is a factor of f(x) then $f(x) = (x - h)Q(x)$.

Hence $f(h) = (h - h)Q(h) = 0$.

Example

Show that $(x + 5)$ is a factor of $f(x) = 2x^3 + 7x^2 - 9x + 30$.

This can be done by substituting $x = -5$ into the polynomial.

$$f(-5) = 2(-5)^3 + 7(-5)^2 - 9(-5) + 30$$
$$= -250 + 175 + 45 + 30$$
$$= 0$$

Since $f(-5) = 0$, $(x + 5)$ is a factor of $f(x) = 2x^3 + 7x^2 - 9x + 30$.

This can also be done using synthetic division. This is how we would proceed if asked to fully factorise a polynomial.

Example

Factorise fully $g(x) = 2x^4 + x^3 - 38x^2 - 79x - 30$.

Without a calculator, we need to guess a possible factor of this polynomial. Since the constant term is -30, we know that possible roots are $\pm 1, \pm 2, \pm 3, \pm 5, \pm 6, \pm 10, \pm 15, \pm 30$.

We may need to try some of these before finding a root. Normally we would begin by trying the smaller numbers.

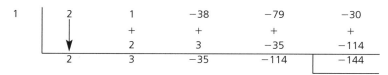

$$
\begin{array}{c|ccccc}
1 & 2 & 1 & -38 & -79 & -30 \\
  &   & + & + & + & + \\
  & \downarrow & 2 & 3 & -35 & -114 \\
\hline
  & 2 & 3 & -35 & -114 & \boxed{-144}
\end{array}
$$

Clearly $x - 1$ is not a factor.

Trying $x = -1$ and $x = 2$ also does not produce a value of 0. So we need to try another possible factor. Try $x + 2$.

$$
\begin{array}{c|ccccc}
-2 & 2 & 1 & -38 & -79 & -30 \\
   &   & + & + & + & + \\
   & \downarrow & -4 & 6 & 64 & 30 \\
\hline
   & 2 & -3 & -32 & -15 & \boxed{0}
\end{array}
$$

So $x + 2$ is a factor.

Now we need to factorise $2x^3 - 3x^2 - 32x - 15$. We know that $x = \pm 1$ do not produce factors so we try $x = -3$.

$$
\begin{array}{c|cccc}
-3 & 2 & -3 & -32 & -15 \\
   &   & + & + & + \\
   & \downarrow & -6 & 27 & 15 \\
\hline
   & 2 & -9 & -5 & \boxed{0}
\end{array}
$$

Hence $g(x) = (x + 3)(x + 2)(2x^2 - 9x - 5)$
$\qquad\qquad = (x + 3)(x + 2)(2x + 1)(x - 5)$

> We do not need to use division methods to factorise a quadratic.

Exercise 2

1 Show that $x - 3$ is a factor of $x^2 + x - 12$.

2 Show that $x - 3$ is a factor of $x^3 + 2x^2 - 14x - 3$.

3 Show that $x - 2$ is a factor of $x^3 - 3x^2 - 10x + 24$.

4 Show that $2x - 1$ is a factor of $2x^3 + 13x^2 + 17x - 12$.

5 Show that $3x + 2$ is a factor of $3x^3 - x^2 - 20x - 12$.

6 Show that $x + 5$ is a factor of $x^4 + 8x^3 + 17x^2 + 16x + 30$.

7 Which of these are factors of $x^3 - 28x - 48$?

   a $x + 1$ **b** $x - 2$

   c $x + 2$ **d** $x - 6$

   e $x - 8$ **f** $x + 4$

8 Factorise fully:

   a $x^3 - x^2 - x + 1$ **b** $x^3 - 7x + 6$

   c $x^3 - 4x^2 - 7x + 10$ **d** $x^4 - 1$

   e $2x^3 - 3x^2 - 23x + 12$ **f** $2x^3 + 21x^2 + 58x + 24$

   g $12x^3 + 8x^2 - 23x - 12$ **h** $x^4 - 7x^2 - 18$

   i $2x^5 + 6x^4 + 7x^3 + 21x^2 + 5x + 15$

   j $36x^5 + 132x^4 + 241x^3 + 508x^2 + 388x - 80$

4.3 Finding a polynomial's coefficients

Sometimes the factor and remainder theorems can be utilized to find a coefficient of a polynomial. This is demonstrated in the following examples.

Example

Find p if $x + 3$ is a factor of $x^3 - x^2 + px + 15$.

Since $x + 3$ is a factor, we know that -3 is a root of the polynomial.

Hence the value of the polynomial is zero when $x = -3$ and so we can use synthetic division to find the coefficient.

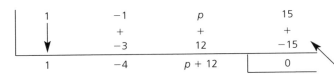

-3	1	-1	p	15
		$+$	$+$	$+$
		-3	12	-15
	1	-4	$p + 12$	0

This is working backwards from the zero.

So $-3(p + 12) = -15$

   $\Rightarrow -3p - 36 = -15$

     $\Rightarrow -3p = 21$

      $\Rightarrow p = -7$

This can also be done by substitution.

If $f(x) = x^3 - x^2 + px + 15$, then $f(-3) = (-3)^3 - (-3)^2 - 3p + 15 = 0$.

So $-27 - 9 - 3p + 15 = 0$

      $\Rightarrow -3p - 21 = 0$

        $\Rightarrow p = -7$

Example

Find p and q if $x + 5$ and $x - 1$ are factors of $f(x) = 2x^4 + 3x^3 + px^2 + qx + 15$, and hence fully factorise the polynomial.

Using synthetic division for each factor, we can produce equations in p and q.

-5	2	3	p	q	15
		$+$	$+$	$+$	$+$
		-10	35	$-5p - 175$	-15
	2	-7	$p + 35$	3	0

So $q - 5p - 175 = 3$

$$\Rightarrow q = 5p + 178$$

1	2	3	p	q	15
		$+$	$+$	$+$	$+$
		2	5	$p + 5$	-15
	2	5	$p + 5$	-15	0

So $q + p + 5 = -15$

$$\Rightarrow q + p = -20$$

Solving $q = 5p + 178$ and $q + p = -20$ simultaneously:

$5p + 178 + p = -20$

$$\Rightarrow 6p = -198$$
$$\Rightarrow p = -33$$

and $q - 33 = -20$

$$\Rightarrow q = 13$$

So $f(x) = 2x^4 + 3x^3 - 33x^2 + 13x + 15$

Now we know that $x + 5$ and $x - 1$ are factors:

-5	2	3	-33	13	15
		$+$	$+$	$+$	$+$
		-10	35	-10	-15
	2	-7	2	3	0

1	2	-7	2	3
		$+$	$+$	$+$
		2	-5	-3
	2	-5	-3	0

Hence $f(x) = (x + 5)(x - 1)(2x^2 - 5x - 3)$

$$\Rightarrow f(x) = (2x + 1)(x + 5)(x - 1)(x - 3)$$

Exercise 3

1 Find the remainder when $x^3 + 2x^2 - 6x + 5$ is divided by $x + 2$.
2 Find the remainder when $5x^4 + 6x^2 - x + 7$ is divided by $2x - 1$.
3 Find the value of p if $(x - 2)$ is a factor of $x^3 - 3x^2 - 10x + p$.
4 Find the value of k if $(x + 5)$ is a factor of $3x^4 + 15x^3 - kx^2 - 9x + 5$.
5 Find the value of k if $(x - 3)$ is a factor of $f(x) = 2x^3 - 9x^2 + kx - 3$ and hence factorise $f(x)$ fully.
6 Find the value of a if $(x + 2)$ is a factor of $g(x) = x^3 + ax^2 - 9x - 18$ and hence factorise $g(x)$ fully.
7 When $x^4 - x^3 + x^2 + px + q$ is divided by $x - 1$, the remainder is zero, and when it is divided by $x + 2$, the remainder is 27. Find p and q.
8 Find the value of k if $(2x + 1)$ is a factor of $f(x) = 2x^3 + 5x^2 + kx - 24$ and hence factorise $f(x)$ fully.
9 Find the values of p and q if $(x + 3)$ and $(x + 7)$ are factors of $x^4 + px^3 + 30x^2 + 11x + q$.
10 The same remainder is found when $2x^3 + kx^2 + 6x + 31$ and $x^4 - 3x^2 - 7x + 5$ are divided by $x + 2$. Find k.

4.4 Solving polynomial equations

In Chapter 2 we solved quadratic equations, which are polynomial equations of degree 2. Just as with quadratic equations, the method of solving other polynomial equations is to make the polynomial equal to 0 and then factorise.

Example

Solve $x^3 + 4x^2 + x - 6 = 0$.

In order to factorise the polynomial, we need a root of the equation. Here the possible roots are $\pm1, \pm2, \pm3, \pm6$. Trying $x = 1$ works:

$$\begin{array}{c|cccc} 1 & 1 & 4 & 1 & -6 \\ & & +1 & +5 & +6 \\ \hline & 1 & 5 & 6 & 0 \end{array}$$

As the remainder is zero, $x - 1$ is a factor.

Hence the equation becomes $(x - 1)(x^2 + 5x + 6) = 0$
$$\Rightarrow (x - 1)(x + 3)(x + 2) = 0$$
$$\Rightarrow x - 1 = 0 \text{ or } x + 3 = 0 \text{ or } x + 2 = 0$$
$$\Rightarrow x = 1 \text{ or } x = -3 \text{ or } x = -2$$

Example

Find the points of intersection of the curve $y = 2x^3 - 3x^2 - 9x + 1$ and the line $y = 2x - 5$.

At intersection, $2x^3 - 3x^2 - 9x + 1 = 2x - 5$

$$\Rightarrow 2x^3 - 3x^2 - 11x + 6 = 0$$

Here the possible roots are $\pm 1, \pm 2, \pm 3, \pm 6$. Trying $x = 3$ works:

3		2	-3	-11	6
			+	+	+
			6	9	-6
		2	3	-2	0

So the equation becomes $(x - 3)(2x^2 + 3x - 2) = 0$

$$\Rightarrow (x - 3)(2x - 1)(x + 2) = 0$$

$$\Rightarrow x = 3 \text{ or } x = \frac{1}{2} \text{ or } x = -2$$

To find the points of intersection, we need to find the y-coordinates.

When $x = 3$

$$y = 2(3) - 5 = 1$$

When $x = \frac{1}{2}$

$$y = 2\left(\frac{1}{2}\right) - 5 = -4$$

When $x = -2$

$$y = 2(-2) - 5 = -9$$

Hence the points of intersection are $(3, 1)$, $\left(\frac{1}{2}, -4\right)$, $(-2, -9)$.

Example

Solve $x^3 - 2x^2 - 5x + 6 \leq 0$ for $x \geq 0$.

This is an inequality which we can solve in the same way as an equation.

First we need to factorise $x^3 - 2x^2 - 5x + 6$.

1		1	-2	-5	6
			+	+	+
			1	-1	-6
		1	-1	-6	0

$$x^3 - 2x^2 - 5x + 6 = (x - 1)(x^2 - x - 6)$$
$$= (x - 1)(x + 2)(x - 3)$$

So the inequality becomes $(x - 1)(x + 2)(x - 3) \leq 0$.

Plotting these points on a graph, and considering points either side of them such as $x = -3, x = 0, x = 2, x = 4$, we can sketch the graph.

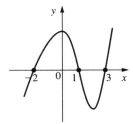

This provides the solution: $1 \le x \le 3$

Exercise 4

1 Show that 5 is a root of $x^3 - x^2 - 17x - 15 = 0$ and hence find the other roots.

2 Show that 2 is a root of $2x^3 - 15x^2 + 16x + 12 = 0$ and hence find the other roots.

3 Show that -3 is a root of $2x^5 - 9x^4 - 34x^3 + 111x^2 + 194x - 120 = 0$ and hence find the other roots.

4 Solve the following equations.

    a $x^3 - 6x^2 + 5x + 12 = 0$ **b** $x^3 + 7x^2 - 4x - 28 = 0$

    c $x^3 + 17x^2 + 75x + 99 = 0$ **d** $x^4 - 4x^3 - 19x^2 + 46x + 120 = 0$

    e $x^4 - 4x^3 - 12x^2 + 32x + 64 = 0$ **f** $2x^3 - 13x^2 - 26x + 16 = 0$

    g $12x^3 - 16x^2 - 7x + 6 = 0$

5 Find where the graph of $f(x) = x^3 - 8x^2 - 11x + 18$ cuts the x-axis.

6 Show that $x^3 - x^2 + 2x - 2 = 0$ has only one root.

7 Find the only root of $x^3 - 3x^2 + 5x - 15 = 0$.

8 $x = 2$ is a root of $g(x) = x^3 - 10x^2 + 31x - p = 0$.

    a Find the value of p.

    b Hence solve the equation $g(x) = 0$.

9 $x = -6$ is a root of $f(x) = 2x^3 - 3x^2 - kx + 42 = 0$.

    a Find the value of k.

    b Hence solve the equation $f(x) = 0$.

10 Solve the following inequalities for $x \ge 0$.

    a $(x + 5)(x - 1)(x - 7) \le 0$

    b $x^3 - 4x^2 - 11x + 30 \le 0$

    c $x^3 + 7x^2 + 4x - 12 \le 0$

    d $x^3 - 9x^2 + 11x + 31 \le 10$

    e $-6x^3 + 207x^2 + 108x - 105 \ge 0$

11 The profit of a football club after a takeover is modelled by
$P = t^3 - 14t^2 + 20t + 120$, where t is the number of years after the takeover.
In which years was the club making a loss?

4.5 Finding a function from its graph

We can find an expression for a function from its graph using the relationship between its roots and factors.

Example

For the graph below, find an expression for the polynomial f(x).

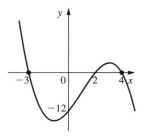

We can see that the graph has roots at $x = -3$, $x = 2$ and $x = 4$.

Hence f(x) has factors $(x + 3)$, $(x - 2)$ and $(x - 4)$.

Since the graph cuts the y-axis at $(0, -12)$, we can find an equation:

$$f(x) = k(x + 3)(x - 2)(x - 4)$$
$$f(0) = k(3)(-2)(-4) = -12$$
$$\Rightarrow 24k = -12$$
$$\Rightarrow k = -\frac{1}{2}$$

Hence $f(x) = -\frac{1}{2}(x + 3)(x - 2)(x - 4)$

Example

For the graph below, find an expression for the polynomial f(x).

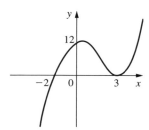

We can see that the graph has roots at $x = -2$ and $x = 3$.

Hence f(x) has factors $(x + 2)$ and $(x - 3)$. Since the graph has a turning point at $(3,0)$, this is a repeated root (in the same way as quadratic functions that have a turning point on the x-axis have a repeated root).

Since the graph cuts the y-axis at $(0,12)$, we can find the equation:

$$f(x) = k(x + 2)(x - 3)^2$$
$$f(0) = k(2)(-3)^2 = 12$$
$$\Rightarrow 18k = 12$$
$$\Rightarrow k = \frac{2}{3}$$

Hence $f(x) = \frac{2}{3}(x + 2)(x - 3)^2$

$$= \frac{2}{3}x^3 - \frac{8}{3}x^2 - 2x + 12$$

Exercise 5

Find an expression for the polynomial $f(x)$ for each of these graphs.

1
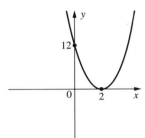

2
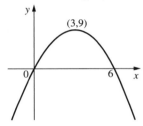

3
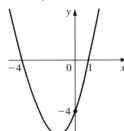

4
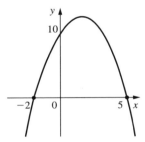

5
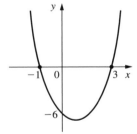

6
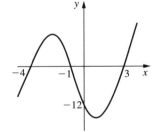

7
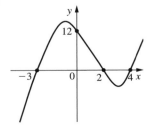

8
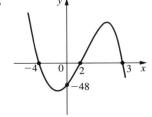

9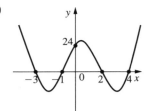

10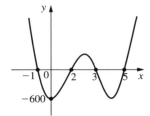

11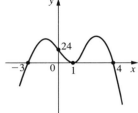

4.6 Algebraic long division

Synthetic division works very well as a "shortcut" for dividing polynomials when the divisor is a linear function. However, it can only be used for this type of division, and in order to divide a polynomial by another polynomial (not of degree 1) it is necessary to use algebraic long division.

This process is very similar to long division for integers.

Example

Find $6087 \div 13$.

$$
\begin{array}{r}
4 \\
13\overline{)6087} \\
52 \\
\hline
8
\end{array}
$$

| There are four 13s in 60 remainder 8 |

This process continues:

$$
\begin{array}{r}
468 \\
13\overline{)6087} \\
52 \\
\hline
88 \\
78 \\
\hline
107 \\
104 \\
\hline
3
\end{array}
$$

| There are four 13s in 60 remainder 8 |

| There are six 13s in 88 remainder 10 |

| There are eight 13s in 107 remainder 3 |

Hence $6087 \div 13 = 468$ R 3.

This can also be expressed as $6087 = 13 \times 468 + 3$.

To perform algebraic division, the same process is employed.

Example

Find $\dfrac{2x^3 - 7x^2 + 6x - 4}{x - 3}$.

We put this into a division format:

$(x - 3)\overline{)2x^3 - 7x^2 + 6x - 4}$

The first step is to work out what the leading term of the divisor x needs to be multiplied by to achieve $2x^3$. The answer to this is $2x^2$ and so this is the first part of the quotient.

$$\begin{array}{r} 2x^2 \\ (x - 3)\overline{)2x^3 - 7x^2 + 6x - 4} \\ \underline{2x^3 - 6x^2 } \\ -x^2 + 6x - 4 \end{array}$$

> Multiplying the divisor by $2x^2$ provides this. This is then subtracted from the dividend.

This process is then repeated: the next part of the quotient is what x needs to be multiplied by to give $-x^2$.

This is continued thus:

$$\begin{array}{r} 2x^2 - x + 3 \\ (x - 3)\overline{)2x^3 - 7x^2 + 6x - 4} \\ \underline{2x^3 + 6x^2 } \\ -x^2 + 6x - 4 \\ \underline{-x^2 + 3x } \\ 3x - 4 \\ \underline{3x - 9} \\ 5 \end{array}$$

So the remainder is 5.

So $\dfrac{2x^3 - 7x^2 + 6x - 4}{x - 3} = 2x^2 - x + 3 + \dfrac{5}{x - 3}.$

Example

Find $\dfrac{3x^4 - 4x^3 + 5x^2 - 7x + 4}{x^2 + 2}$.

> To obtain $3x^4$, x^2 must be multiplied by $3x^2$.

$$\begin{array}{r} 3x^2 \\ (x^2 + 2)\overline{)3x^4 - 4x^3 + 5x^2 - 7x + 4} \\ \underline{3x^4 + 6x^2 } \\ -4x^3 - x^2 - 7x + 4 \end{array}$$

This continues to give:

$$
\begin{array}{r}
3x^2 - 4x - 1 \\
(x^2 + 2)\overline{)3x^4 - 4x^3 + 5x^2 - 7x + 4} \\
\underline{3x^4 \quad\quad + 6x^2} \\
-4x^3 - x^2 - 7x + 4 \\
\underline{-4x^3 \quad\quad\quad - 8x} \\
-x^2 + x + 4 \\
\underline{-x^2 \quad\quad - 2} \\
x + 6
\end{array}
$$

> x^2 is multiplied by $-4x$ to obtain $-4x^3$.
> x^2 is multiplied by -1 to obtain $-x^2$.

So the remainder here is $x + 6$.

Hence $3x^4 - 4x^3 + 5x^2 - 7x + 4 = (3x^2 - 4x - 1)(x^2 + 2) + x + 6$.

Although the process of algebraic long division is not part of this curriculum, it is an important skill that is employed in curve sketching, found in Chapter 8.

Example

Find $\dfrac{5 + 3x^3 + x^6}{x^2 + 4}$.

In this case, the numerator is not presented with the powers in descending order. It is vital that it is rearranged so that its powers are in descending order before dividing. There are also some "missing" powers. These must be put into the dividend with zero coefficients to avoid mistakes being made.

$$(x^2 + 4)\overline{)x^6 + 0x^5 + 0x^4 + 3x^3 + 0x^2 + 0x + 5}$$

Having presented the division as above, the process is the same.

$$
\begin{array}{r}
x^4 \quad\quad - 4x^2 + 3x + 16 \\
(x^2 + 4)\overline{)x^6 + 0x^5 + 0x^4 + 3x^3 + 0x^2 + 0x + 5} \\
\underline{x^6 \quad\quad + 4x^4} \\
-4x^4 + 3x^3 + 0x^2 + 0x + 5 \\
\underline{-4x^4 \quad\quad\quad - 16x^2} \\
3x^3 + 16x^2 + 0x + 5 \\
\underline{3x^3 \quad\quad\quad + 12x} \\
16x^2 - 12x + 5 \\
\underline{16x^2 \quad\quad + 64} \\
-12x - 59
\end{array}
$$

Hence $\dfrac{5 + 3x^3 + x^6}{x^2 + 4} = x^4 - 4x^2 + 3x + 16 - \dfrac{12x + 59}{x^2 + 4}$.

Exercise 6

Use algebraic long division for the following questions.

1 Show that $\dfrac{x^2 + 5x + 6}{x + 2} = x + 3$.

2 Show that $(x^2 - 1) \div (x - 1) = x + 1$.

3 Find $(x^2 - x - 12) \div (x - 4)$.

4 Find $\dfrac{2x^2 - 5x - 3}{x - 3}$.

5 Find the quotient and remainder for $\dfrac{3x^2 - 5x - 7}{x + 2}$.

6 Show that $x = 5$ is a root of $x^3 - 5x^2 + 4x - 20 = 0$.

7 Find $\dfrac{x^3 + 5x^2 - 9x + 4}{x^2 + 5}$, expressing your answer in the form $px + q + \dfrac{ax + b}{x^2 + 5}$.

8 Find $\dfrac{2x^3 - 9x^2 - 8x + 11}{2x^2 + 7}$.

9 Find $\dfrac{3x^4 - 2x^2 + 7}{x + 2}$.

10 Find $\dfrac{2x^5 + 13}{x^2 - 3}$.

11 Find $\dfrac{9 - 5x + 7x^2 - x^4}{2x + 1}$.

12 Find $\dfrac{11 + 5x^3 - x^6}{x^2 + 1}$.

13 Given that $x - 1$ is a factor of f(x), solve $f(x) = x^4 + 4x^3 - 7x^2 - 22x + 24 = 0$.

14 Given that $x^2 + x + 4$ is a factor of g(x), solve
$g(x) = x^5 + 7x^4 + 6x^3 - 4x^2 - 40x - 96 = 0$.

4.7 Using a calculator with polynomials

Everything covered so far in this chapter has been treated as if a calculator were not available. For the polynomials examined so far it has been possible to factorise and hence solve polynomial equations to find the roots. For many polynomials it is not possible to factorise, and the best method of solving these is to employ graphing calculator technology. In these cases, the roots are not exact values and so are given as approximate roots. Also, if a calculator is available, it can help to factorise a polynomial as it removes the need for trial and error to find the initial root.

Example

Sketch the graph of $f(x) = x^3 - 4x^2 + 2x - 1$ and hence find its root(s).

Using a calculator, we can obtain its graph:

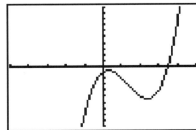

We can see that this cubic function has only one root. We can find this using the calculator. As the calculator uses a numerical process to find the root, it is important to make the left bound and right bound as close as possible to the root.

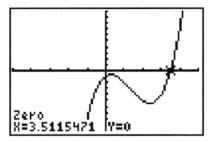

Hence $x = 3.51$.

Example

Factorise fully $f(x) = 6x^3 + 13x^2 - 19x - 12$.

Using the graph of the function,

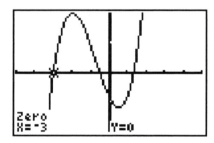

we can see that $x = -3$ is a root of $f(x)$ and so we can use this in synthetic division.

-3	6	13	-19	-12
		+	+	+
		-18	15	12
	6	-5	-4	0

Hence $f(x) = (x + 3)(6x^2 - 5x - 4)$
$= (x + 3)(3x - 4)(2x + 1)$.

Solving a polynomial inequality is also made simple through the use of graphing calculator technology.

Example

(a) Solve $f(x) < 10$ where $f(x) = x^3 + 6x^2 - 7x - 2$.

(b) Find the range of values of a so that there are three solutions to the equation $f(x) = a$.

(a) Here is the graph of $f(x)$:

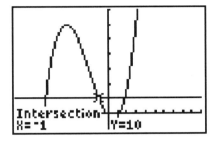

We can find the three points of intersection of $f(x)$ with the line $y = 10$: $x = -6.77, -1, 1.77$

Looking at the graph, it is clear that the solution to the inequality is $x < -6.77$ and $-1 < x < 1.77$.

(b) By calculating the maximum and minimum turning points, we can find the values of a so that there are three solutions. To have three solutions, a must lie between the maximum and minimum values (y-values).

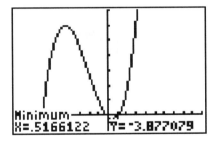

From the calculator, it is clear that $-3.88 < a < 59.9$.

Exercise 7

Use a calculator to solve all of the following equations.

1 $2x^3 - 5x^2 - 4x + 3 = 0$

2 $2x^4 + 9x^3 - 46x^2 - 81x - 28 = 0$

3 $2x^4 + 3x^3 - 15x^2 - 32x - 12 = 0$

4 $3x^3 - 9x^2 + 4x - 12 = 0$

5 $x^3 + 2x^2 - 11x - 5 = 0$

6 $x^3 - 6x^2 + 4x - 7 = 0$

7 $x^4 - x^3 + 5x^2 - 7x + 2 = 0$

8 $x^4 - 22x^2 - 19x + 41 = 0$

9 $x^3 - 9x^2 - 11x + 4 = 2x + 1$

10 For what value of x do the curves $y = 2x^3 - 4x^2 + 3x + 7$ and $y = x^2 - 3x - 5$ meet?

11 State the equation of f(x) from its graph below. Hence find the points of intersection of f(x) and $g(x) = x^2 - 4$.

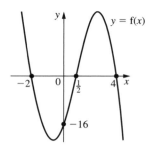

12 Factorise $x^3 - 21x + 20$, by obtaining an initial root using a calculator.

13 Factorise $36x^4 - 73x^2 + 16$, by obtaining an initial root using a calculator.

14 a Solve $f(x) = x^3 + 6x^2 - 7x - 2 < 0$.

　　b Find the values of a so that there are three solutions to the equation $f(x) = a$.

15 a Solve $f(x) = x^3 + 6x^2 - 7x - 2 < 20$.

　　b Find the values of a so that there are three solutions to the equation $f(x) = a$.

Review Exercise

1 Evaluate $2x^3 - 5x^2 + 3x + 7$ when $x = -2$.

2 Find $(3x^4 - 2x^3 + 6x + 1) \div (x - 2)$.

3 Express $f(x) = 2x^5 + 4x^2 - 7$ in the form $Q(x)(2x - 1) + R$ by dividing f(x) by $2x - 1$.

4 Show that $x + 2$ is a factor of $x^3 - 10x^2 + 3x + 54$, and hence find the other factors.

5 Factorise fully $f(x) = 2x^3 - 3x^2 - 17x - 12$.

6 Factorise fully $g(x) = 6x^4 - 19x^3 - 59x^2 + 16x + 20$.

7 Factorise fully $k(x) = x^4 - 3x^3 + x^2 - 15x - 20$.

8 Show that $x = -2$ is a root of $x^3 + 6x^2 - 13x - 42 = 0$, and hence find the other roots.

9 Solve $2x^5 + 5x^4 + x^3 + 34x^2 - 66x + 24 = 0$.

10 Find an expression for f(x) from its graph.

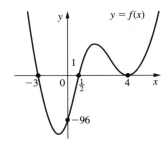

 11 Solve $5x^3 - 6x + 7 = 0$, correct to 3 significant figures.

 12 Solve $x^3 - 7x^2 - 2x + 31 = 0$, correct to 3 significant figures.

 13 Using algebraic long division, find $\dfrac{x^3 - 5x^2 + 6x - 4}{2x + 1}$.

 14 Using algebraic long division, find $(x^4 - 4x^2 + 3x - 5) \div (x^2 + 1)$.

 15 The polynomial $x^3 + ax^2 - 3x + b$ is divisible by $(x - 2)$ and has a remainder 6 when divided by $(x + 1)$. Find the value of a and of b.   [IB May 03 P1 Q4]

 16 The polynomial $f(x) = x^3 + 3x^2 + ax + b$ leaves the same remainder when divided by $x - 2$ as when divided by $x + 1$. Find the value of a.   [IB Nov 01 P1 Q3]

 17 When the polynomial $x^4 + ax + 3$ is divided by $x - 1$, the remainder is 8. Find the value of a.   [IB Nov 02 P1 Q1]

 18 Consider $f(x) = x^3 - 2x^2 - 5x + k$. Find the value of k if $x + 2$ is a factor of f(x).   [IB Nov 04 P1 Q1]

5 Exponential and Logarithmic Functions

In this chapter we will meet logarithms, which have many important applications, particularly in the field of natural science. Logarithms were invented by John Napier as an aid to computation in the 16th century.

John Napier was born in Edinburgh, Scotland, in 1550. Few records exist about John Napier's early life, but it is known that he was educated at St Andrews University, beginning in 1563 at the age of 13. However, it appears that he did not graduate from the university as his name does not appear on any subsequent pass lists. The assumption is that Napier left to study in Europe. There is no record of where he went, but the University of Paris is likely, and there are also indications that he spent time in Italy and the Netherlands.

John Napier

While at St. Andrews University, Napier became very interested in theology and he took part in the religious controversies of the time. He was a devout Protestant, and his most important work, the *Plaine Discovery of the Whole Revelation of St. John* was published in 1593.

It is not clear where Napier learned mathematics, but it remained a hobby of his, with him saying that he often found it hard to find the time to work on it alongside his work on theology. He is best remembered for his invention of logarithms, which were used by Kepler, whose work was the basis for Newton's theory of gravitation. However his mathematics went beyond this and he also worked on exponential expressions for trigonometric functions, the decimal notation for fractions, a mnemonic for formulae used in solving spherical triangles, and "Napier's analogies", two formulae used in solving spherical triangles. He was also the inventor of "Napier's bones", used for mechanically multiplying, dividing and taking square and cube roots. Napier also found exponential epressions for trignometric functions, and introduced the decimal notation for fractions.

We can still sympathize with his sentiments today, when in the preface to the *Mirifici logarithmorum canonis descriptio*, Napier says he hopes that his "logarithms will save calculators much time and free them from the *slippery errors* of calculations".

5.1 Exponential functions

An exponential (or power) function is of the form $y = a^x$.

a is known as the base $(a \neq 1)$.

x is known as the exponent, power or index.

Remember the following rules for indices:

1. $a^p \times a^q = a^{p+q}$
2. $\dfrac{a^p}{a^q} = a^{p-q}$
3. $(a^p)^q = a^{pq}$
4. $\dfrac{1}{a^p} = a^{-p}$
5. $\sqrt[q]{a^p} = a^{\frac{p}{q}}$
6. $a^0 = 1$

Example

Simplify $\dfrac{x^{\frac{1}{5}} \times x^{\frac{2}{5}}}{\sqrt[5]{x^3}}$.

$$\frac{x^{\frac{1}{5}} \times x^{\frac{2}{5}}}{\sqrt[5]{x^3}} = \frac{x^{\frac{3}{5}}}{x^{\frac{3}{5}}} = x^0 = 1$$

Example

Evaluate $8^{-\frac{2}{3}}$ without a calculator.

$$8^{-\frac{2}{3}} = \frac{1}{\sqrt[3]{8^2}} = \frac{1}{2^2} = \frac{1}{4}$$

Graphing exponential functions

Consider the function $y = 2^x$.

x	-2	-1	0	1	2	3	4	5
y	$\dfrac{1}{4}$	$\dfrac{1}{2}$	1	2	4	8	16	32

The y-values double for every integral increase of x.

The first few points are shown in this graph:

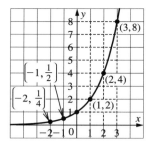

Investigation

Using a graphing calculator, sketch these graphs:

(a) $y = 3^x$ (b) $y = 4^x$ (c) $y = 5^x$ (d) $y = 10^x$

Try to identify a pattern.

The investigation should have revealed that all exponential graphs

1. pass through the point (0,1)
2. have a similar shape
3. are entirely above the x-axis.

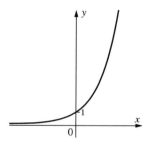

This shape is known as exponential growth, and all graphs of the form $y = a^x$, $a > 1$ have this shape. The domain restriction of $a > 1$ is important. We know that when $a = 1$ the graph is the horizontal line $y = 1$, and below we will see what happens when $0 < a < 1$.

For graphs of the form $y = a^x$, $0 < a < 1$ let us consider $y = \left(\dfrac{1}{2}\right)^x$.

x	-2	-1	0	1	2	3	4	5
y	4	2	1	$\dfrac{1}{2}$	$\dfrac{1}{4}$	$\dfrac{1}{8}$	$\dfrac{1}{16}$	$\dfrac{1}{32}$

The first few points are shown in this graph:

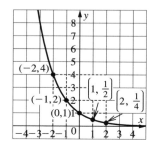

Now $y = \left(\dfrac{1}{2}\right)^x = \dfrac{1}{2^x} = 2^{-x}$. All exponential graphs of the form $y = a^x$, $0 < a < 1$ can

be expressed in this way, and from our knowledge of transformations of functions this is

actually a reflection in the y-axis.

Hence this is the general graph of $y = a^{-x}$, $a > 1$:

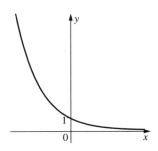

This is known as exponential decay. Exponential decay graphs can be expressed as $y = a^{-x}$, $a > 1$ or as $y = a^x$, $0 < a < 1$ as shown above.

Exercise 1

1 Simplify these.

a $p^4 \times p^5$ **b** $\dfrac{p^7}{p^2}$ **c** $(x^3)^5$ **d** $3y^2 \times 7y^3$

e $(2x^3)^4$ **f** $t^4 \times t^{-2}$ **g** $\dfrac{8p^6}{4p^4}$ **h** $\dfrac{18p^5}{3p^{-2}}$

2 Without using a calculator, evaluate these.

a $16^{\frac{1}{2}}$ **b** $81^{\frac{1}{4}}$ **c** 10^{-1}

d 19^0 **e** $25^{\frac{3}{2}}$ **f** $9^{-\frac{1}{2}}$

g $8^{-\frac{2}{3}}$ **h** $4^{-\frac{3}{2}}$ **i** $\left(\dfrac{1}{27}\right)^{-\frac{2}{3}}$

3 Simplify these.

a $\dfrac{x^5 \times x^3}{x^2}$ **b** $\dfrac{4y^3 \times 2y^6}{6y^5}$

c $\dfrac{5p^3 \times 2p^{-5}}{p^4}$ **d** $\dfrac{t^{\frac{1}{2}} \times t^3}{t^{\frac{3}{2}}}$

e $\dfrac{4m^{\frac{5}{3}} \times 3m^{-\frac{5}{3}}}{2m^{\frac{2}{3}}}$ **f** $3x^2(4x^3 + 5x^{-1})$

g $x^{\frac{1}{2}}(2x^{\frac{1}{2}} + x^{-\frac{1}{2}})$ **h** $(x^{\frac{1}{2}} + x^{-\frac{1}{2}})^2$

4 Draw the graph of each of these.

a $y = 3^x$ **b** $y = 5^x$ **c** $y = 6^x$ **d** $y = 10^x$

e $y = \left(\dfrac{1}{4}\right)^x$ **f** $y = 6^{-x}$ **g** $y = \left(\dfrac{3}{2}\right)^x$ **h** $y = \left(\dfrac{2}{3}\right)^x$

5.2 Logarithmic graphs

In the study of inverse functions, it was found that an inverse function exists only for one-to-one functions. The question is whether there is an inverse function for exponential functions.

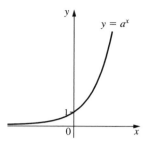

Using the tests of horizontal and vertical lines, it is clear that any of these lines pass through only one point (or no point) on the graph.

So, for any exponential function $y = a^x (a \neq 1)$, an inverse function exists.

In Chapter 3, we also found that the graph of an inverse function is the reflection of the original function in the line $y = x$. Using this, we can find the shape of the inverse function.

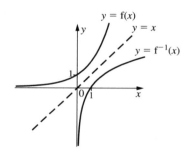

As all exponential graphs have the same shape, all inverse graphs will also have the same shape. These inverse functions are known as **logarithmic functions**.

Logarithmic functions are defined

$$y = a^x \Leftrightarrow x = \log_a y$$

Consider the exponential function $y = 2^x$.

$y = 2^x \Rightarrow x = \log_2 y$

This means that the inverse function is written $y = \log_2 x$.

There are two key features of logarithmic graphs:

1. Logarithmic functions are defined only for $x > 0$.
2. All logarithmic graphs pass through $(1,0)$.

In $\log_2 x$, 2 is known as the base.

This means that we can summarize the domain and range for exponential and logarithmic functions.

	Domain	Range
Exponential	$\mathbb{R}$	$y > 0$
Logarithmic	$x > 0$	$\mathbb{R}$

Interpreting a logarithm

A logarithm can be interpreted by "the answer to a logarithm is a power".

This comes from the definition:

$$\log_a q = p \Leftrightarrow a^p = q$$

So, for example, $\log_2 64 = x \Leftrightarrow 2^x = 64 \Rightarrow x = 6$.

Example

Find $\log_5 125$.

$\log_5 125 = x \Rightarrow 5^x = 125$

$ \Rightarrow x = 3$

> This is asking "What power of 5 gives 125?"

Example

Evaluate (a) $\log_{25} 5$ (b) $\log_5\left(\dfrac{1}{25}\right)$

(a) $\log_{25} 5 = x$

$ \Rightarrow 25^x = 5$

$ \Rightarrow x = \dfrac{1}{2}$

(b) $\log_5\left(\dfrac{1}{25}\right) = x$

$ \Rightarrow 5^x = \dfrac{1}{25}$

$ \Rightarrow x = -2$

There are two important results to remember:

$$\log_a 1 = 0$$
$$\log_a a = 1$$

> These come from $a^0 = 1$ and $a^1 = a$.

Example

Sketch the graph of $y = \log_3 x$.

We know the shape, and that the graph passes through $(1, 0)$.

As the base is 3, we know that $\log_3 3 = 1$, so the graph passes through $(3,1)$.

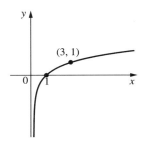

Exercise 2

1 Sketch these graphs.

    a $y = \log_2 x$ **b** $y = \log_4 x$

    c $y = \log_5 x$ **d** $y = \log_{10} x$

2 Sketch these functions on the same graph.

    a $y = 3^x$ and $y = \log_3 x$ **b** $y = 5^x$ and $y = \log_5 x$

3 Without a calculator, evaluate these logarithms.

    a $\log_2 4$ **b** $\log_3 9$ **c** $\log_5 25$ **d** $\log_3 27$ **e** $\log_6 216$

    f $\log_2 32$ **g** $\log_4 64$ **h** $\log_2 64$ **i** $\log_8 64$ **j** $\log_{10} 100$

    k $\log_{10} 1000$ **l** $\log_7 7$ **m** $\log_3 3$ **n** $\log_8 1$

4 Without a calculator, evaluate these logarithms.

    a $\log_8 2$ **b** $\log_9 3$ **c** $\log_4 2$ **d** $\log_{25} 5$

    e $\log_{64} 8$ **f** $\log_{64} 2$ **g** $\log_8 4$ **h** $\log_{16} 8$

5 Without a calculator, evaluate these logarithms.

    a $\log_2\left(\dfrac{1}{16}\right)$ **b** $\log_2\left(\dfrac{1}{8}\right)$ **c** $\log_3\left(\dfrac{1}{9}\right)$ **d** $\log_3\left(\dfrac{1}{27}\right)$

    e $\log_3\left(\dfrac{1}{81}\right)$ **f** $\log_8\left(\dfrac{1}{2}\right)$ **g** $\log_{25}\left(\dfrac{1}{5}\right)$

6 Without a calculator, evaluate these logarithms.

    a $\log_a a$ **b** $\log_a a^2$ **c** $\log_a \sqrt{a}$ **d** $\log_a\left(\dfrac{1}{a}\right)$ **e** $-\log_a a$

5.3 Rules of logarithms

As there are for exponentials, there are rules for logarithms that help to simplify logarithmic expressions.

For exponentials, we have the rules:

1. $a^p \times a^q = a^{p+q}$ 2. $a^p \div a^q = a^{p-q}$ 3. $(a^p)^q = a^{pq}$

The corresponding rules for logarithms are:

> 1. $\log_a xy = \log_a x + \log_a y$ 2. $\log_a\left(\dfrac{x}{y}\right) = \log_a x - \log_a y$
>
> 3. $\log_a x^m = m \log_a x$

Proofs

1. Let $\log_a x = p$ and $\log_a y = q$.

    This means that $x = a^p$ and $y = a^q$.

    Since $xy = a^p \times a^q = a^{p+q}$, $xy = a^{p+q}$.

    By the definition of a logarithm this means that $\log_a xy = p + q$

$$\Rightarrow \log_a xy = \log_a x + \log_a y$$

> It is very important to remember that
> $\log_p a + \log_p b \neq \log_p(a + b)$.

2. Similarly, let $\log_a x = p$.

$$\text{So } \frac{x}{y} = \frac{a^p}{a^q} = a^{p-q}$$

Hence $\log_a\left(\dfrac{x}{y}\right) = p - q$

$$\Rightarrow \log_a\left(\frac{x}{y}\right) = \log_a x - \log_a y$$

3. Again, let $\log_a x = p$.

$$\text{So } x^m = (a^p)^m = a^{mp}$$

Hence $\log_a x^m = mp$

$$= m \log_a x$$

> It is worth noting that
> $$\log_a\left(\frac{1}{x}\right) = -\log_a x.$$

> For these rules to work, the logarithms must have the same base.

Example

Simplify $\log_x 8 + \log_x 3 - \log_x 6$.

$$\log_x 8 + \log_x 3 - \log_x 6 = \log_x\left(\frac{8 \times 3}{6}\right)$$
$$= \log_x 4$$

Example

Simplify $3\log_p 2 - \log_p 12 + 2\log_p 4$

$$3\log_p 2 - \log_p 12 + 2\log_p 4 = \log_p 2^3 - \log_p 12 + \log_p 4^2$$
$$= \log_p 8 - \log_p 12 + \log_p 16$$
$$= \log_p\left(\frac{8 \times 16}{12}\right)$$
$$= \log_p\left(\frac{32}{3}\right)$$

Example

Simplify and evaluate $2\log_{10} 5 + 2\log_{10} 2$.

$$2\log_{10} 5 + 2\log_{10} 2 = \log_{10} 5^2 + \log_{10} 2^2$$
$$= \log_{10} 25 + \log_{10} 4$$
$$= \log_{10} 100$$
$$= 2$$

These rules can also be used to solve equations involving logarithms.

Example

Solve $\log_6(x + 2) + \log_6(x + 1) = 1$ for $x > 0$.

$$\log_6(x + 2) + \log_6(x + 1) = 1$$
$$\Rightarrow \log_6(x + 2)(x + 1) = 1$$
$$\Rightarrow \log_6(x + 2)(x + 1) = \log_6 6 \qquad \boxed{\text{Expressing 1 as a logarithm.}}$$
$$\Rightarrow (x + 2)(x + 1) = 6$$
$$\Rightarrow x^2 + 3x + 2 = 6$$
$$\Rightarrow x^2 + 3x - 4 = 0$$
$$\Rightarrow (x + 4)(x - 1) = 0$$
$$\Rightarrow x = 1 \qquad \boxed{\text{Since } x > 0.}$$

Exercise 3

1 Simplify these.

   a $\log_a 2 + \log_a 9$ **b** $\log_a 5 + \log_a 3$ **c** $\log_a 10 - \log_a 2$

   d $\log_a 8 + \log_a 8$ **e** $\log_a 2 + \log_a 3 + \log_a 4$ **f** $\frac{1}{2}\log_a 16$

   g $5 \log_a 2$ **h** $2 \log_a 6 + \log_a 2 - \log_a 12$ **i** $-3 \log_a 2$

   j $2 \log_a 3 + 3 \log_a 2$ **k** $\log_a 6 - 2 \log_a 2 + \log_a 8$

2 Express each of these as a single logarithm of a number.

   a $1 + \log_3 5$ **b** $\log_2 10 - 2$

   c $5 - 2 \log_2 6$ **d** $\log_a x + 2 \log_a y - 3 \log_a t$

3 Simplify these.

   a $\log_{10} 4 + \log_{10} 125$ **b** $\log_3 63 - \log_3 7$

   c $\log_6 2 + \log_6 3$ **d** $\log_4 36 - \log_4 18$

   e $\log_3 6 + \log_3 12 - \log_3 8$ **f** $\log_6 12 - \log_6\left(\frac{1}{3}\right)$

   g $\frac{1}{2}\log_2 16 - \frac{1}{3}\log_2 8$ **h** $\log_5 64 - 6 \log_5 2$

   i $\log_2 3 + \log_2 2 - \log_2 6 - \log_2 8$ **j** $\log_2\left(\frac{1}{4}\right) - 2 \log_2\left(\frac{1}{8}\right)$

   k $-2 \log_4 8 + \log_4\left(\frac{1}{2}\right)$

4 Simplify these.

   a $\log_a 3 + \log_a x + 2 \log_a x$

   b $\log_a 4 - \log_a 2x$

   c $\log_a(x + 1) - \log_a(x^2 - 1) + 2 \log_a(x - 1)$

   d $3 \log_a(x + 2) - \log_a(3x^2 - 12) + \log_a(x - 2)$

5 Simplify these.

   a $\log_2 4$ **b** $\log_4 2$ **c** $\log_9 27$

   d $\log_{27} 9$ **e** $\log_{10} 100$ **f** $\log_{100} 10$

What is the connection between $\log_x y$ and $\log_y x$?

6 If $\log_a y = \log_a 4 + 3 \log_a x$, express y in terms of x.

7 If $\log_a y = 2 \log_a 3 + 4 \log_a x$, express y in terms of x.

8 If $\log_a y = \log_a p + 5 \log_a x$, express y in terms of p and x.

9 If $2 \log_2 y = \log_2(x + 1) + 3$, show that $y^2 = 8(x + 1)$.

10 Solve for $x > 0$.

    a $\log_a x + \log_a 2 = \log_a 14$ **b** $\log_a x - \log_a 3 = \log_a 19$

    c $\log_a x + 2 \log_a 2 = \log_a 24$ **d** $\log_a x + 2 \log_a 5 = \log_a 225$

    e $\log_a x^2 + \log_a\left(\dfrac{1}{2}\right) = \log_a 32$ **f** $\log_a 6 + \log_a x = \log_a 1$

    g $\dfrac{1}{2} \log_a x + \log_a 6 = \log_a 30$ **h** $2 \log_a x - \log_a x = \log_a 9$

11 Solve for $x > 0$.

    a $\log_a(x + 2) + \log_a(x - 1) = \log_a 4$ **b** $\log_5(x + 1) + \log_5(x - 3) = 1$

    c $\log_4(3x - 1) - \log_4(x - 1) = 1$ **d** $\log_8(x^2 - 1) - \log_8(x - 1) = 2$

    e $\log_{16}(x + 2) - \log_{16}(x - 6) = \dfrac{1}{2}$ **f** $\log_7(2x + 5) - \log_7(x - 5) = \log_7\left(\dfrac{x}{2}\right)$

12 Volume of sound is measured in decibels. The difference in volume between two sounds can be calculated using the formula $d = 50 \log_{10}\left(\dfrac{S_1}{S_2}\right)$ where S_1 and S_2 are sound intensities $(S_1 > S_2)$. The volume of normal conversation is 60 dB and the volume of a car horn is 110 dB. The sound intensity of normal conversation (S_2) is 40 phons. What is the sound intensity of a car horn (S_1)?

5.4 Logarithms on a calculator

The natural base

There is a special base, denoted e, which is known as the **natural base**. The reason why this base is special is covered in Chapter 9. This number e is the irrational number 2.718 …

The exponential function to the base e is $f(x) = e^x$, which is also written exp(x).

$$\exp(x) = e^x$$

There is also a notation for its inverse (logarithmic) function known as the **natural logarithmic function**. It is also sometimes called a Naperian logarithm after John Napier.

$$\log_e x = \ln x$$

The graphs of these functions have the same shape as other logarithmic and exponential graphs.

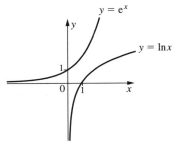

In particular note that $\ln e = 1$.

Calculators perform logarithms in only two bases, 10 and e. For these two logarithms, the base is rarely explicitly stated. For the natural base the notation is ln, and for base 10 it is often just written log x, and the base is assumed to be 10.

Example

Find $\log_{10} 7$ and ln 7.

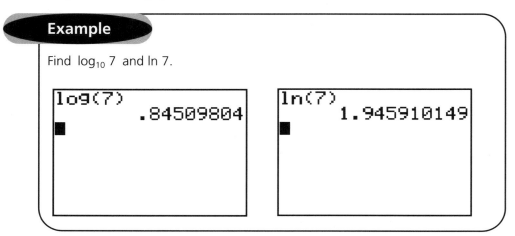

Change of base formula

To find logarithms in other bases, we need to change the base using this formula:

$$\log_a x = \frac{\log_b x}{\log_b a}$$

Proof

$\log_a x = y$

$\Rightarrow x = a^y$

$\Rightarrow \log_b x = \log_b a^y$

$\Rightarrow \log_b x = y \log_b a$

$\Rightarrow y = \dfrac{\log_b x}{\log_b a}$

If we are using this formula to find a logarithm on a calculator, it is often written as

$$\log_a x = \frac{\ln x}{\ln a}$$

Example

Use the change of base formula to evaluate these.
(a) $\log_2 8$ (b) $\log_3 10$

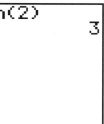

Be careful with brackets.

The change of base formula can be used to sketch any logarithmic function on the calculator.

Example

Sketch $y = \log_6 x$.

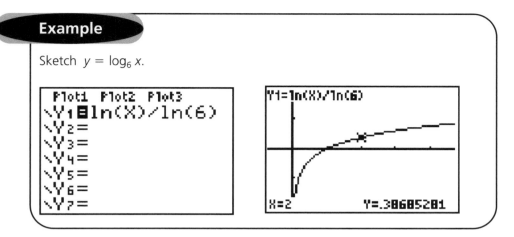

Exercise 3 question 5 asked "What is the connection between $\log_x y$ and $\log_y x$? " The answer is a special case of the change of base formula, namely that

$$\log_x y = \frac{\log_y y}{\log_y x} = \frac{1}{\log_y x}$$

This can be used to help solve equations.

Example

Use the change of base formula to solve $\log_9 x + 2 \log_x 9 = 3$.

This can be changed into $\log_9 x + \dfrac{2}{\log_9 x} = 3$.

Multiplying by $\log_9 x$ gives

$$(\log_9 x)^2 + 2 = 3 \log_9 x$$
$$\Rightarrow (\log_9 x)^2 - 3 \log_9 x + 2 = 0$$
$$\Rightarrow y^2 - 3y + 2 = 0 \quad \cdots\cdots\cdots\cdots\cdots\cdots \boxed{\text{Let } y = \log_9 x}$$
$$\Rightarrow (y - 1)(y - 2) = 0$$
$$\Rightarrow (\log_9 x - 1)(\log_9 x - 2) = 0$$
$$\Rightarrow \log_9 x = 1 \text{ or } \log_9 x = 2$$
$$\Rightarrow x = 9 \text{ or } x = 81$$

Exercise 4

1 Using a calculator, evaluate these.

  a $\log_{10} 1000$ **b** $\log_{10} 8$ **c** $\log_{10} 26$

  d $\log_{10} 3$ **e** $\log_{10}\left(\dfrac{1}{2}\right)$

2 Using a calculator, evaluate these.

  a $\ln 10$ **b** $\ln 9$ **c** $\ln 31$

  d $\ln\left(\dfrac{1}{2}\right)$ **e** $\ln 7.5$ **f** $\ln(0.328)$

3 Use the change of base formula to evaluate these.

   a $\log_4 16$ **b** $\log_2 9$ **c** $\log_3 11$ **d** $\log_4 17$ **e** $\log_5 80$

   f $\log_6 12$ **g** $\log_9 12$ **h** $\log_4 13$ **i** $\log_4\left(\dfrac{1}{2}\right)$ **j** $\log_5\left(\dfrac{1}{6}\right)$

   k $\log_7\left(\dfrac{1}{8}\right)$ **l** $\log_8\left(\dfrac{1}{4}\right)$ **m** $\log_6\left(\dfrac{2}{3}\right)$ **n** $\log_5\left(\dfrac{5}{7}\right)$ **o** $\log_8 9.21$

   p $\log_6 4.38$ **q** $\log_4(0.126)$ **r** $\log_9(0.324)$

4 Use your calculator to sketch these.

   a $y = \log_5 x$ **b** $y = \log_7 x$ **c** $y = \log_9 x$

5 Use the change of base formula to solve these.

   a $\log_4 x + 5 \log_x 4 = 6$

   b $\log_2 x - 6 \log_x 2 = 1$

   c $\log_7 x - 12 \log_x 7 = 4$

5.5 Exponential equations

Logarithms can be used to solve exponential equations and this is one of their greatest applications today. When logarithms were first advanced by Napier, they were used as a computational aid. Exponential equations are ones where we are trying to find the power. The logarithmic rule of $\log_a x^p = p \log_a x$ is particularly useful for these equations.

Method for solving an exponential equation

> **1.** Take natural logs of both sides.
> **2.** "Bring down" the power.
> **3.** Divide the logs.
> **4.** Solve for x.

Example

Solve $2^x = 7$.

$$2^x = 7$$
$$\Rightarrow \ln 2^x = \ln 7$$
$$\Rightarrow x \ln 2 = \ln 7$$
$$\Rightarrow x = \frac{\ln 7}{\ln 2}$$
$$\Rightarrow x = 2.81$$

Example

Solve $e^x = 12$.

$$e^x = 12$$
$$\Rightarrow \ln e^x = \ln 12$$
$$\Rightarrow x \ln e = \ln 12$$
$$\Rightarrow x = \ln 12 \quad \cdots\cdots\cdots \text{Since } \ln e = 1.$$
$$\Rightarrow x = 2.48$$

Example

Solve $3^{2x-1} = 40$.

$$3^{2x-1} = 40$$
$$\Rightarrow \ln 3^{2x-1} = \ln 40$$
$$\Rightarrow (2x - 1)\ln 3 = \ln 40$$
$$\Rightarrow 2x - 1 = \frac{\ln 40}{\ln 3}$$
$$\Rightarrow 2x - 1 = 3.357\ldots$$
$$\Rightarrow x = 2.18 \text{ (3 sf)}$$

Natural logarithm equations can also be solved using a calculator.

Example

Solve $\ln x = 11$.

Remembering this means $\log_e x = 11$, this can be written

$$x = e^{11}$$
$$\Rightarrow x = 59\,900 \text{ (3 sf)}$$

Exponential functions are very important in the study of growth and decay, and are often used as mathematical models.

Example

A population of rats increases according to the formula $R(t) = 8e^{0.22t}$, where t is the time in months.
(a) How many rats were there at the beginning?
(b) How long will it be until there are 80 rats?

(a) When $t = 0$, $R(0) = 8e^0 = 8$

(b) $8e^{0.22t} = 80$

$$\Rightarrow e^{0.22t} = 10$$
$$\Rightarrow \ln e^{0.22t} = \ln 10$$
$$\Rightarrow 0.22t = \ln 10$$
$$\Rightarrow t = \frac{\ln 10}{0.22}$$
$$\Rightarrow t = 10.5 \text{ months}$$

Example

For a radioactive isotope $A = A_0 e^{-kt}$, where A is the mass of isotope in grams, A_0 is the initial mass, and t is time in years.
In 5 years, 40 g of this substance reduced to 34 g.
(a) Find the value of k, correct to 3 sig figs.
(b) Find the half-life of this substance.

(a) $A_0 = 40$, $A = 34$, $t = 5$

$\Rightarrow 40e^{-5k} = 34$

$\Rightarrow e^{-5k} = \dfrac{34}{40}$

$\Rightarrow -5k = \ln\left(\dfrac{34}{40}\right)$

$\Rightarrow k = 0.0325$

(b) The half-life of a radioactive substance is the time taken for only half of the original amount to remain.

i.e. $\dfrac{A}{A_0} = \dfrac{1}{2}$

$\Rightarrow e^{-0.0325t} = \dfrac{1}{2}$

$\Rightarrow -0.0325t = \ln\dfrac{1}{2}$

$\Rightarrow t = 21.3$ years

Example

Solve $(6^x)(3^{2x+1}) = 4^{x+2}$, giving your answer in the form $x = \dfrac{\ln a}{\ln b}$, where $a, b \in \mathbb{Q}$.

$$(6^x)(3^{2x+1}) = 4^{x+2}$$
$$\Rightarrow \ln[(6^x)(3^{2x+1})] = \ln 4^{x+2}$$
$$\Rightarrow x \ln 6 + (2x + 1)\ln 3 = (x + 2)\ln 4$$
$$\Rightarrow x \ln 6 + 2x \ln 3 + \ln 3 = x \ln 4 + 2 \ln 4$$
$$\Rightarrow x \ln 6 + x \ln 9 + \ln 3 = x \ln 4 + \ln 16$$
$$\Rightarrow x(\ln 6 + \ln 9 - \ln 4) = \ln 16 - \ln 3$$
$$\Rightarrow x \ln\left(\dfrac{27}{2}\right) = \ln\left(\dfrac{16}{3}\right)$$
$$\Rightarrow x = \dfrac{\ln\left(\dfrac{16}{3}\right)}{\ln\left(\dfrac{27}{2}\right)}$$

Sometimes exponential equations can be reduced to quadratic form.

Example

Solve $7(3^{x+1}) = 2 + \dfrac{3}{3^x}$, giving the answer in the form $a - \log_3 b$, where $a, b \in \mathbb{Z}$.

$$7(3^{x+1}) = 2 + \dfrac{3}{3^x}$$
$$\Rightarrow 7(3^x)(3^1) = 2 + \dfrac{3}{3^x}$$

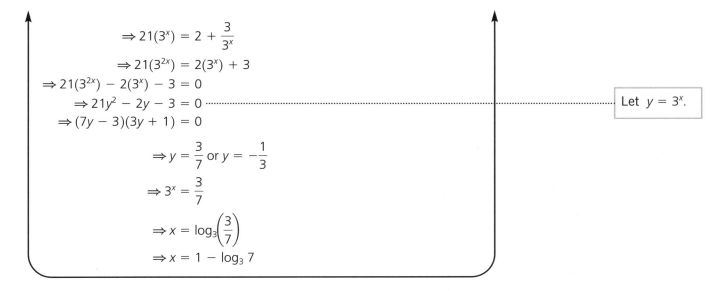

$$\Rightarrow 21(3^x) = 2 + \frac{3}{3^x}$$

$$\Rightarrow 21(3^{2x}) = 2(3^x) + 3$$

$$\Rightarrow 21(3^{2x}) - 2(3^x) - 3 = 0$$

$$\Rightarrow 21y^2 - 2y - 3 = 0 \quad \cdots\cdots\cdots\cdots\cdots\cdots\cdots\cdots\cdots\cdots\cdots\cdots\cdots\cdots \text{Let } y = 3^x.$$

$$\Rightarrow (7y - 3)(3y + 1) = 0$$

$$\Rightarrow y = \frac{3}{7} \text{ or } y = -\frac{1}{3}$$

$$\Rightarrow 3^x = \frac{3}{7}$$

$$\Rightarrow x = \log_3\left(\frac{3}{7}\right)$$

$$\Rightarrow x = 1 - \log_3 7$$

Exercise 5

1 Solve for x.

  a $2^x = 256$ **b** $3^x = 40$ **c** $5^x = 20$

  d $12^x = 6500$ **e** $8^x = 6$ **f** $14^x = 3$

2 Solve for x.

  a $e^x = 12$ **b** $e^x = 30$ **c** $e^x = 270$

  d $4e^x = 18$ **e** $8e^x = 3$

3 Solve for x.

  a $\ln x = 9$ **b** $\ln x = 2$ **c** $\ln x = 10$

  d $\ln x = 16$ **e** $\ln x = 0.2$

4 Find the least positive value of $x \in \mathbb{Z}$ for which the inequality is true.

  a $2^x > 350$ **b** $3^x > 300$ **c** $10^x > 2^{10}$ **d** $5^x > 7200$

5 The number of bacteria in a culture is given by $B(t) = 40e^{0.6t}$, where t is the time in days.

  a How many bacteria are there when $t = 0$?

  b How many bacteria are there after 2 days?

  c How long will it take for the number of bacteria to increase to ten times its original number?

6 According to one mobile phone company, the number of people owning a mobile phone is growing according to the formula $N(t) = 100\,000e^{0.09t}$, where t is time in months. Their target is for 3 million people to own a mobile phone. How long will it be before this target is reached?

7 When a bowl of soup is removed from the microwave, it cools according to the model $T(t) = 80e^{-0.12t}$, t in minutes and T in $^\circ$C.

  a What was its temperature when removed from the microwave?

  b The temperature of the room is 22°C. How long will it be before the soup has cooled to room temperature?

8 A radioactive isotope is giving off radiation and hence losing mass according to the model $M(t) = 2100e^{-0.012t}$, t in years and M in grams.

  a What was its original mass?

  b What will its mass be after 20 years?

  c What is the half-life of the isotope?

9 The height of a satellite orbiting Earth is changing according to the formula
$H(t) = 30000e^{-0.2t}$, t in years and H in km.

a What will be the height of the satellite above the Earth after 2 years?

b When the satellite reaches 320 km from the Earth, it will burn up in the Earth's atmosphere. How long before this happens?

10 Scientists are concerned about the population of cheetahs in a game park in Tanzania. Their study in 2006 produced the model
$P(t) = 220e^{-0.15t}$, where P is the population of cheetahs and t is the time in years from 2006.

a How many cheetahs were there in 2006?

b How many cheetahs do they predict will be in the park in 2015?

c If the population drops to single figures, scientists predict the remaining cheetahs will not survive. When will this take place?

11 The pressure in a boiler is falling according to the formula $P_t = P_0e^{-kt}$, where P_0 is the initial pressure, P_t is the pressure at time t, and t is the time in hours.

a At time zero, the pressure is 2.2 units but 24 hours later it has dropped 1.6 units. Find the value of k to 3 sf.

b If the pressure falls below 0.9 units, the boiler cuts out. How long before it will cut out?

c If the boiler's initial pressure is changed to 2.5 units, how much longer will it be operational?

12 A radioactive substance is losing mass according to the formula $M_t = M_0e^{-kt}$ where M_0 is the initial mass, M_t is the mass after t years.

a If the initial mass is 900 g and after 5 years it has reduced to 850 g, find k.

b What is the half-life of this substance?

13 Solve $3^{x+1} = 2^{2-x}$. Give your answer in the form $x = \dfrac{\ln a}{\ln b}$, where $a, b \in \mathbb{Z}$.

14 Solve $(4^x)(3^{2x+1}) = 6^{x+1}$. Give your answer in the form $x = \dfrac{\ln a}{\ln b}$, where $a, b \in \mathbb{Z}$.

15 Solve $(4^x)(5^{x+1}) = 2^{2x+1}$. Give your answer in the form $x = \dfrac{\ln a}{\ln b}$, where $a, b \in \mathbb{Q}$.

16 Solve $(2^x)(3^{2x+1}) = 4^{x+3}$. Give your answer in the form $x = \dfrac{\ln a}{\ln b}$, where $a, b \in \mathbb{Q}$.

17 Solve $5(2^{x+1}) = 3 + \dfrac{4}{2^x}$, giving your answer in the form $x = a - \log_2 b$, where $a, b \in \mathbb{Z}$.

18 Solve $2(4^{x+1}) = 2 + \dfrac{3}{4^x}$, giving your answer in the form $x = a - \log_4 b$, where $a, b \in \mathbb{Z}$.

19 Solve $2(4^{x+2}) = 12 + \dfrac{5}{4^x}$, giving your answer in the form $x = a - \log_4 b$, where $a, b \in \mathbb{Q}$.

20 Solve $3(6^{x-1}) = 1 + \dfrac{4}{6^x}$, giving your answer in the form $x = a - \log_6 b$, where $a, b \in \mathbb{Z}$.

21 Solve $4^x + 4(2^x) - 5 = 0$.

22 Solve $9^x + 4(3^x) - 12 = 0$.

5.6 Related graphs

Exponential and logarithmic graphs are transformed in the same way as other functions, as studied previously.

We can sketch and interpret related exponential and logarithmic graphs using this information.

Example

Sketch the graph of $y = 4^x - 2$.

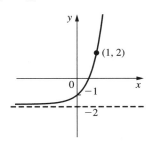

This is a vertical shift downwards of 2 units.

For exponential graphs, we often plot the points for $x = 0$ and $x = 1$.

Example

Sketch the graph of $y = \log_2(x - 1) + 3$.

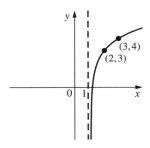

This is a horizontal shift right of 1 unit and a vertical shift of 3 units.

For logarithmic graphs, we often plot the points when $x = 1$ and $x = a$, where a is the base (or their images under transformation). So here $(1, 0) \rightarrow (2, 3)$ and $(2, 1) \rightarrow (3, 4)$

Notice the vertical asymptote has moved to $x = 1$.

Example

Sketch the graph of $y = \log_3 x^2 + 4$.

From log rules, we know that this is the same as $y = 2 \log_3 x + 4$.

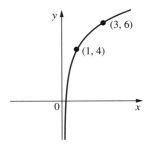

$(1, 0) \rightarrow (1, 4)$
$(3, 1) \rightarrow (3, 6)$

Example

For this graph of $y = ke^{-x}$, what is the value of k?

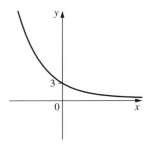

As $(0,3)$ lies on the graph it has been stretched $\times 3$. So $k = 3$.

Example

Part of the graph of $y = p \log_2(x - q)$ is shown. What are the values of p and q?

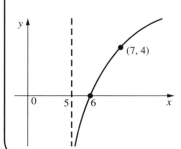

The graph has been shifted 5 places right so $q = 5$.

So for $(7, 4)$, $4 = p \log_2(7 - 5)$
$$= p \log_2 2$$
$$= p$$

So $p = 4$ and $q = 5$

Exercise 6

1 Sketch these graphs.
  a $y = 2^x$ **b** $y = 2^x - 3$ **c** $y = 2^{-x}$ **d** $y = 2^{x+3}$

2 Sketch these graphs.
  a $y = e^x$ **b** $y = 4e^x$ **c** $y = -e^x$ **d** $y = e^{x-2}$

3 Sketch these graphs.
  a $y = 4e^{-x}$ **b** $y = 2e^x - 1$ **c** $y = 3e^{x+2}$ **d** $y = 5e^{x-1} + 3$

4 Sketch these graphs.
  a $y = \log_4 x$ **b** $y = \log_4 x + 2$
  c $y = \log_4(x + 2)$ **d** $y = -\log_4 x$

5 Sketch these graphs.
  a $y = \log_3 x$ **b** $y = \log_3 x^2$

  c $y = \log_3\left(\dfrac{1}{x}\right)$ **d** $y = 3 \log_3(x + 2)$

6 For this graph of $y = ke^x$, what is the value of k?

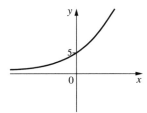

7 For this graph of $y = k \times 2^x + p$, what are the values of k and p?

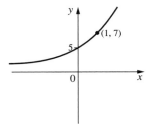

8 The sketch shows the graph of $y = \log_a x$. Find the value of a.

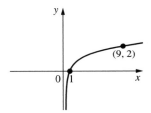

9 The sketch shows the graph of $y = \log_a(x + p)$. Find the values of p and a.

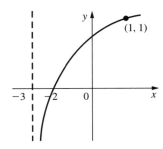

10 The sketch shows part of the graph of $y = a \log_6(x - p)$. Find the values of a and p.

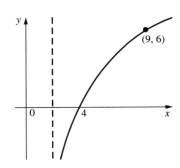

Review exercise

 1 Simplify these.

  a $\dfrac{x^7 \times x^2}{x^3}$ **b** $\dfrac{5p^{\frac{1}{2}} \times 3p^{\frac{3}{4}}}{p^{\frac{3}{2}}}$ **c** $x^{-\frac{1}{2}}(2x^{\frac{1}{2}} + 4x^{-\frac{3}{2}})$

 2 Draw these graphs.

  a $y = 6^x$ **b** $\log_6 x$

 3 Evaluate these.

  a $\log_2 32$ **b** $\log_5 125$ **c** $\log_8 2$ **d** $\log_5\left(\dfrac{1}{25}\right)$

 4 Simplify these.

  a $\log_a 16 + \log_a 3$ **b** $\log_p 8 + \log_p 4 - \log_p 16$ **c** $2\log_a 5 + 1$

 5 Simplify these.

  a $\log_3 x + \log_3 8$ **b** $\log_4 6 + \log_4 2 - \log_4 3$

 6 Solve for $x > 0$.

  a $\log_a x + \log_a 6 = \log_a 54$ **b** $\dfrac{1}{2}\log_a x + \log_a 5 = \log_a 45$

  c $\log_5(x - 1) + \log_5(x + 2) = \log_5 10$ **d** $\log_3(x + 2) - \log_3(x - 1) = 2$

 7 Evaluate these.

  a $\log_4 7$ **b** $\log_9 4$ **c** $\log_{11} 2$ **d** $3\log_8 5$

8 Solve for x.

  a $\ln x = 9$ **b** $\ln x = 17$ **c** $5\ln x = 19$

9 Solve for x.

  a $3^x = 320$ **b** $7^x = 2$ **c** $e^x = 8$ **d** $5e^x = 19$

10 Find the least positive value of $x \in \mathbb{Z}$ for which the inequality is true.

  a $3^x > 190$ **b** $5^x > 2000$ **c** $e^x > 291$

11 Sketch these graphs.

  a $y = 5^x - 2$ **b** $y = 6^{x+3}$ **c** $y = 4^{x-1} + 2$ **d** $y = -3^x - 5$

12 Sketch these graphs.

  a $y = \log_5 x - 2$ **b** $y = \log_7\left(\dfrac{1}{x}\right)$ **c** $y = 5\log_2(x + 1)$

13 For this graph of $y = ke^{-x}$, what is the value of k?

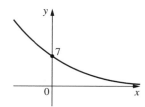

14 The sketch shows part of the graph of $y = \log_p(x + q)$. What are the values of p and q?

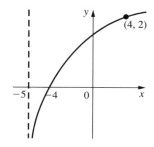

15 A truck has a slow puncture in one of its tyres, causing the pressure to drop. The pressure at time t, P_t, is modelled by $P_t = P_0 e^{-kt}$, where t is in hours and P_0 is the inflation pressure.

a Initially the tyre is inflated to 50 units. After 18 hours, it drops to 16 units. Calculate the value of k.

b The truck will not be allowed to make a journey if the pressure falls below 30 units. If the driver inflates the tyre to 50 units immediately before departure, will he be able to make a round trip that takes 6 hours?

16 Solve $\log_{16} \sqrt[3]{100 - x^2} = \dfrac{1}{2}$.   [IB Nov 03 P1 Q10]

17 Find the exact value of x satisfying the equation $(3^x)(4^{2x+1}) = 6^{x+2}$

Give your answer in the form $\dfrac{\ln a}{\ln b}$, where $a, b \in \mathbb{Z}$.   [IB May 03 P1 Q12]

18 Solve $2(5^{x+1}) = 1 + \dfrac{3}{5^x}$, giving your answer in the form $a + \log_5 b$,

where $a, b \in \mathbb{Z}$.   [IB Nov 03 P1 Q19]

19 Solve the simultaneous equations $\log_x y = 1$ and $xy = 16$ for $x, y > 0$.

20 Solve the simultaneous equations $\log_a(x + y) = 0$ and $2\log_a x = \log(4y + 1)$.

21 Solve the system of simultaneous equations:

$$x + 2y = 5$$
$$4^x = 8^y$$ [IB Nov 98 P1 Q2]

22 If $f(x) = \ln(6x^2 - 5x - 6)$, find

a the exact domain of $f(x)$

b the range of $f(x)$.   [IB Nov 98 P1 Q7]

23 Find all real values of x so that $3^{x^2-1} = (\sqrt{3})^{126}$.   [IB May 98 P1 Q3]

24 a Given that $\log_a b = \dfrac{\log_c b}{\log_c a}$, find the real numbers k and m such that

$\log_9 x^3 = k \log_3 x$ and $\log_{27} 512 = m \log_3 8$.

b Find all values of x for which $\log_9 x^3 + \log_3 x^{\frac{1}{2}} = \log_{27} 512$.

          [IB Nov 97 P1 Q4]

6 Sequences, Series and Binomial Theorem

Pascal's triangle is constructed by adding together the two numbers above as shown (with a 1 on the end of each row). There are many interesting results and applications related to this triangle. For example, notice that the sum of each row is a power of 2.

Pascal's triangle is named after Blaise Pascal, born 1623, a French mathematician who made great contributions to the fields of number theory, geometry and probability. However, it is not universally known as Pascal's triangle as it was not discovered first by him. There is evidence that Chinese and Persian mathematicians independently found the triangle as early as the 11th century. Chia Hsien, Yang Hui and Omar Khayyam are all documented as using the triangle. In fact, there may be reference to the triangle as early as 450 BC by an Indian mathematician who described the "Staircase of Mount Meru". In China the triangle is known as the Chinese triangle, and in Italy it is known as Tartaglia's triangle, named after a 16th century Italian mathematician, Nicolo Tartaglia.

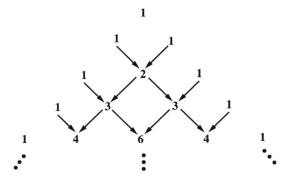

http://www.bath.ac.uk/~ma3mja/history.html

Accessed 14 February 2006

A sequence is defined as an ordered set of objects. In most cases these objects are numbers, but this is not necessarily the case. Sequences and series occur in nature, such as the patterns on snail shells and seed heads in flowers, and in man-made applications such as the world of finance and so are a useful area of study. Whereas a sequence is a list of objects in a definite order, a series is the sum of these objects.

Consider these sequences:

1. △, □, ⬠, ⬡, ...

2. J, A, S, O, N, D, J, ...

3. M, W, F, S, T, T, S, ...

4. Moscow, Los Angeles, Seoul, Barcelona, Atlanta, Sydney, Athens, Beijing, London, ...

5. 2, 4, 6, 8, 10, ...

6. 10, 13, 16, 19, 22, ...

7. 3, 6, 12, 24, 48, ...

8. 1, 3, 7, 15, 31, ...

How can these sequences be described?

Here are some possible descriptions:

1. Plane shapes beginning with triangle, with one vertex (and side) added each time.

2. Initial letter of each month (in English) beginning with July.

3. Initial letter of days, starting with Monday, going forward each time by two days.

4. Olympic cities beginning with Moscow (1980).

5. Even numbers beginning with 2, increasing by 2 each time.

6. Numbers beginning with 10, increasing by 3 each time.

7. Beginning with 3, each term is the previous term multiplied by 2.

8. Beginning with 1, each term is double the previous term plus 1.

It is natural to describe a sequence by the change occurring each time from one term to the next, along with the starting point. Although all of the above are sequences, in this course only the types of which 5, 6 and 7 are examples are studied.

In order to describe sequences mathematically, some notation is required.

---

u_n is known as the nth term of a sequence.
This provides a formula for the general term of a sequence related to its term number, n.
S_n is the notation for the sum of the first n terms.
a or u_1 is commonly used to denote the initial term of a sequence.

In this course, two types of sequence are considered: arithmetic sequences and geometric sequences.

6.1 Arithmetic sequences

An arithmetic sequence, sometimes known as an arithmetic progression, is one where the terms are separated by the same amount each time. This is known as the **common difference** and is denoted by d. Note that for a sequence to be arithmetic, a common difference must exist.

Consider the sequence 5, 7, 9, 11, 13, …

The first term is 5 and the common difference is 2.
So we can say $a = 5$ and $d = 2$.

Sequences can be defined in two ways, explicitly or implicitly. An **implicit expression** gives the result in relation to the previous term, whereas an **explicit expression** gives the result in terms of n. Although it is very easy to express sequences implicitly, it is usually more useful to find an explicit expression in terms of n.

Here, an implicit expression could be $u_n = u_{n-1} + 2$.

nth term   previous term

For an explicit expression consider this table.

n	1	2	3	4	5
u_n	5	7	9	11	13

In this case, $u_n = 2n + 3$.

Compare this with finding the straight line with gradient 2 and y-intercept 3.

It is clear that an arithmetic sequence will be of the form

$$a, a + d, a + 2d, a + 3d, a + 4d, \ldots$$

Hence, the general formula for the nth term of an arithmetic sequence is

$$u_n = a + (n - 1)d$$

Example

Consider the arithmetic sequence 4, 11, 18, 25, 32, …
(a) Find an expression for u_n, the nth term of the sequence.
(b) Find u_{12}, the 12th term of the sequence.
(c) Is (i) 602 (ii) 711 a member of this sequence?

(a) Clearly for this sequence $u_1 = a = 4$ and $d = 7$,
so $u_n = 4 + 7(n - 1)$
$u_n = 4 + 7n - 7$
$u_n = 7n - 3$
(b) Hence $u_{12} = 7 \times 12 - 3$
$= 81$
(c) (i) $7n - 3 = 602$
$\Rightarrow 7n = 605$
$\Rightarrow n = 86.4$
Since n is not an integer, 602 cannot be a term of this sequence.
(ii) $7n - 3 = 711$
$\Rightarrow 7n = 714$
$\Rightarrow n = 102$
Clearly 711 is a member of the sequence, the 102nd term.

Example

What is the nth term of a sequence with $u_7 = 79$ and $u_{12} = 64$?
If $u_7 = 79$, then $a + 6d = 79$.
If $u_{12} = 64$, then $a + 11d = 64$.
Subtracting, $-5d = 15$
$\Rightarrow d = -3$
Now substituting this into $a + 6d = 79$,
$a - 18 = 79$
$\Rightarrow a = 97$
Hence $u_n = 97 - 3(n - 1)$
$= 100 - 3n$
It is easy to verify that this is the correct formula by checking u_{12}.

Example

If $k, 12, k^2 - 6k$ are consecutive terms of an arithmetic sequence, find the possible values of k.

As the sequence is arithmetic, a common difference must exist.

Hence $d = 12 - k$ and $d = k^2 - 6k - 12$.

So $k^2 - 6k - 12 = 12 - k$

$\Rightarrow k^2 - 5k - 24 = 0$

$\Rightarrow (k + 3)(k - 8) = 0$

$\Rightarrow k = -3$ or $k = 8$

Exercise 1

1 Find u_n for these sequences.

  a 5, 7, 9, 11, 13, …

  b 1, 6, 11, 16, 21, …

  c 8, 14, 20, 26, 32, …

  d 60, 51, 42, 33, 24, …

  e 4, 0, -4, -8, -12, …

2 For the sequence 7, 18, 29, 40, 51, … find u_n and u_{20}.

3 For the sequence 200, 310, 420, 530, 640, … find u_n and u_{13}.

4 For the sequence 17, 10, 3, -4, -11, … find u_n and u_{19}.

5 For the sequence $1, \dfrac{3}{2}, 2, \dfrac{5}{2}, 3, \ldots$ find u_n and u_{15}.

6 For 9, 16, 23, 30, 37, … which term is the first to exceed 1000?

7 For 28, 50, 72, 94, 116, … which term is the first to exceed 500?

8 For 160, 154, 148, 142, 136, … which term is the last positive term?

9 Find u_n given $u_5 = 17$ and $u_9 = 33$.

10 Find u_n given $u_4 = 43$ and $u_{10} = 97$.

11 Find u_n given $u_3 = 32$ and $u_7 = 8$.

12 Find u_n given $u_8 = -8$ and $u_{14} = -11$.

13 Given that $k, 8, 7k$ are consecutive terms of an arithmetic sequence, find k.

14 Given that $k - 1, 11, 2k - 1$ are consecutive terms of an arithmetic sequence, find k.

15 Given that $4k - 2, 18, 9k - 1$ are consecutive terms of an arithmetic sequence, find k.

16 Given that $k^2 + 4, 29, 3k$ are consecutive terms of an arithmetic sequence, find k.

6.2 Sum of the first n terms of an arithmetic sequence

An arithmetic series is the sum of an arithmetic sequence.

So for $u_n = 3n + 5$, i.e. 8, 11, 14, 17, 20, …, the arithmetic series is
$S_n = 8 + 11 + 14 + 17 + 20 + \ldots$

So S_5 means $8 + 11 + 14 + 17 + 20 = 70$.

How can a formula for S_n be found?

$$S_n = u_1 + u_2 + \ldots + u_{n-1} + u_n = a + (a + d) + \ldots + [a + (n-2)d] + [a + (n-1)d]$$

Re-ordering,

$$S_n = u_n + u_{n-1} + \ldots + u_2 + u_1 = [a + (n-1)d] + [a + (n-2)d] + \ldots + (a + d) + a$$

Adding,

$$2S_n = 2a + (n-1)d + 2a + (n-1)d + \ldots + 2a + (n-1)d + 2a + (n-1)d$$

$$= n[2a + (n-1)d]$$

$$\Rightarrow S_n = \frac{n}{2}[2a + (n-1)d]$$

This is the formula for S_n. It can be expressed in two ways:

$$S_n = \frac{n}{2}[2a + (n-1)d] \qquad\qquad S_n = \frac{n}{2}[u_1 + u_n]$$

This is because
$$u_n = a + (n-1)d$$

So in the above example, $S_n = \dfrac{n}{2}[16 + 3(n-1)]$

$$= \frac{n}{2}(16 + 3n - 3)$$

$$= \frac{n}{2}(3n + 13)$$

$$= \frac{3}{2}n^2 + \frac{13}{2}n$$

Example

Find a formula for S_n for 7, 15, 23, 31, 39, ... and hence find S_8.
Here $a = 7$ and $d = 8$.
So $S_n = \dfrac{n}{2}[14 + 8(n-1)]$ and $S_8 = 4 \times 64 + 3 \times 8$

$$S_n = \frac{n}{2}(8n + 6) \qquad\qquad S_8 = 256 + 24$$

$$S_n = 4n^2 + 3n \qquad\qquad S_8 = 280$$

Example

Find the number of terms in the series $4 + 10 + 16 + 22 + 28 + \ldots$
required to exceed 500.

$a = 4, d = 6$

$$S_n = \frac{n}{2}[8 + 6(n-1)]$$

$$S_n = 3n^2 + n$$

So $3n^2 + n > 500$

$$\Rightarrow 3n^2 + n - 500 > 0$$

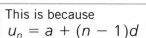

$y = 3n^2 + n - 500$

So $n > 12.7$ (The solution of -13.1 is not valid as n must be positive.)
The number of terms required in the series is 13 (and $S_{13} = 520$).

Example

Given $S_4 = 32$ and $S_7 = 98$, find u_{11}.

$S_4 = \dfrac{4}{2}(2a + 3d) = 32$ $\qquad$ $S_7 = \dfrac{7}{2}(2a + 6d) = 98$

$\Rightarrow 4a + 6d = 32$ $\qquad\qquad$ $\Rightarrow 7a + 21d = 98$

$\Rightarrow 2a + 3d = 16$ ⓘ $\qquad\qquad$ $\Rightarrow a + 3d = 14$ ⓘⓘ

$\qquad\qquad\qquad 2a + 3d = 16$ ⓘ

ⓘ − ⓘⓘ $\quad\underline{a + 3d = 14}$ ⓘⓘ

$\qquad\qquad\qquad\quad a = 2$

Substituting in ⓘⓘ

$a + 3d = 14$

$\Rightarrow 3d = 12$

$\Rightarrow d = 4$

So $u_n = a + (n - 1)d = 2 + 4(n - 1)$

$\qquad u_n = 4n - 2$

$\Rightarrow u_{11} = 44 - 2 = 42$

Exercise 2

1 Find a formula for S_n for these series.

   a $2 + 5 + 8 + 11 + 14 + \ldots$

   b $8 + 10 + 12 + 14 + 16 + \ldots$

   c $80 + 77 + 74 + 71 + 68 + \ldots$

   d $2008 + 1996 + 1984 + 1972 + 1960 + \ldots$

   e $\dfrac{1}{2} + \dfrac{5}{6} + \dfrac{7}{6} + \dfrac{3}{2} + \dfrac{11}{6} + \ldots$

2 Find S_7 for $8 + 15 + 22 + 29 + 36 + \ldots$.

3 For the series obtained from the arithmetic sequence $u_n = 5n - 3$, find S_{12}.

4 Find the sum of the first 20 multiples of 5 (including 5 itself).

5 Find the sum of the multiples of 7 between 100 and 300.

6 Find an expression for the sum of the first n positive integers.

7 Find an expression for the sum of the first n odd numbers.

8 Given that three consecutive terms of an arithmetic sequence add together to make 30 and have a product of 640, find the three terms.

9 Find the number of terms in the arithmetic series $9 + 14 + 19 + 24 + \ldots$ required to exceed 700.

10 Find the greatest possible number of terms in the arithmetic series $18 + 22 + 26 + \ldots$ such that the total is less than 200.

11 What is the greatest total possible (maximum value) of the arithmetic series $187 + 173 + 159 + \ldots$?

12 In an arithmetic progression, the 10th term is twice the 5th term and the 30th term of the sequence is 60.

   a Find the common difference.

   b Find the sum of the 9th to the 20th terms inclusive.

6.3 Geometric sequences and series

An example of a geometric sequence is 4, 8, 16, 32, 64, ...

In a geometric sequence each term is the previous one multiplied by a non-zero constant. This constant is known as the **common ratio**, denoted by r.

The algebraic definition of this is:

$$\frac{u_{n+1}}{u_n} = r \Leftrightarrow \text{the sequence is geometric}$$

Formula for u_n

A geometric sequence has the form

$a, ar, ar^2, ar^3, \ldots$

$$\text{So } u_n = ar^{n-1}$$

Example

Find a formula for u_n for the geometric sequence
6, 12, 24, 48, 96, ...

Here $a = 6$ and $r = 2$.
So $u_n = 6 \times 2^{n-1}$.

Example

Find u_n given that $u_3 = 36$ and $u_5 = 324$.

$$u_3 = ar^2 = 36 \qquad u_5 = ar^4 = 324$$

So $\dfrac{u_5}{u_3} = \dfrac{ar^4}{ar^2} = r^2 = \dfrac{324}{36}$

So $r^2 = 9$

$\Rightarrow r = \pm 3$

$ar^2 = 36$

$\Rightarrow 9a = 36$

$\Rightarrow a = 4$

i.e. $u_n = 4 \times 3^{n-1}$ or $u_n = 4 \times (-3)^{n-1}$

This technique of dividing one term by another is commonly used when solving problems related to geometric sequences.

Example

Given that the following are three consecutive terms of a geometric sequence, find k.

$k - 4, 2k - 2, k^2 + 14$

So $r = \dfrac{2k - 2}{k - 4} = \dfrac{k^2 + 14}{2k - 2}$

$\Rightarrow (2k - 2)^2 = (k - 4)(k^2 + 14)$

$\Rightarrow 4k^2 - 8k + 4 = k^3 + 14k - 4k^2 - 56$

$\Rightarrow k^3 - 8k^2 + 22k - 60 = 0$

$\Rightarrow k = 6$

> Use a calculator to solve the cubic equation.

Sum of a geometric series

As with arithmetic series, a geometric series is the sum of a geometric sequence.

So $S_n = a + ar + ar^2 + ar^3 + \cdots + ar^{n-1}$

$\Rightarrow rS_n = ar + ar^2 + \cdots + ar^{n-1} + ar^n$

$\Rightarrow rS_n = S_n + ar^n - a$

$\Rightarrow rS_n - S_n = ar^n - a$

$\Rightarrow S_n(r - 1) = a(r^n - 1)$

$\Rightarrow S_n = \dfrac{a(r^n - 1)}{r - 1}$

The formula for S_n can be expressed in two ways:

$$S_n = \frac{a(r^n - 1)}{r - 1} \quad \text{or} \quad S_n = \frac{a(1 - r^n)}{1 - r}$$

Example

Find S_n for $4 - 8 + 16 - 32 + 64\ldots$

Here $a = 4$ and $r = -2$.

So $S_n = \dfrac{4((-2)^n - 1)}{-2 - 1}$

$= -\dfrac{4}{3}((-2)^n - 1)$

Exercise 3

1 Find the 6th term and the nth term for these geometric sequences.

   a 8, 4, 2, …
   b 80, 20, 5, …
   c 2, 6, 18, …
   d 5, −10, 20, …
   e 100, −50, 25, …
   f $u_1 = 12, r = 2$
   g $a = 6, r = 5$

2 Find the sum of the first eight terms for each of the sequences in question 1. Also find the sum to n terms of these numerical sequences.

3 Find the sum to n terms of these geometric sequences.

a $x + x^2 + x^3 + \cdots$

b $1 - x + x^2 - \cdots$

c $1 - 3x + 9x^2 - 27x^3 + \cdots$

4 Find the general term, u_n, of the geometric sequence that has:

a $u_3 = 20$ and $u_6 = 160$

b $u_2 = 90$ and $u_5 = \dfrac{10}{3}$

c $u_2 = -12$ and $u_5 = 324$

d $u_2 = -\dfrac{1}{2}$ and $u_7 = 512$

5 Given these three consecutive terms of a geometric sequence, find k.

a $k - 4, k + 8, 5k + 4$

b $k - 1, 2 - 2k, k^2 - 1$

c $\dfrac{k}{2}, k + 8, k^2$

6 Find the first term in this geometric sequence that exceeds 500.

$2, 4, 8, 16, \ldots$

7 If $a = 8$ and $r = 4$, find the last term that is less than 8000.

8 For the geometric series $3 + 6 + 12 + 24 + \cdots$, how many terms are required for a total exceeding 600?

9 The first two terms of a geometric series have a sum of -4. The fourth and fifth terms have a sum of 256. Find the first term and the common ratio of the series.

6.4 Sum of an infinite series

In order to consider infinite series, it is first important to understand the ideas of convergence and divergence. If two (or more) things converge, then they move towards each other. In a sequence or series, this means that successive terms become closer and closer together; to test convergence, the gap between the terms is examined.

Consider these three series:

1. $24 + 12 + 6 + 3 + \cdots$

2. $5 + 10 + 20 + 40 + \cdots$

3. $4 + 7 + 10 + 13 + \cdots$

In the first (geometric) series, the gap between the terms narrows so the series is said to be **convergent**.

In the second (geometric) series, the gap between the terms widens and will continue to increase, so the series is said to be **divergent**.

The third series is arithmetic and so the gap between the terms remains the same throughout, known as the common difference. Although the gap remains constant, the series continues to increase in absolute size towards infinity and hence all arithmetic series are divergent.

Plotting a graph of the above series can help to visualize what is happening with these series.

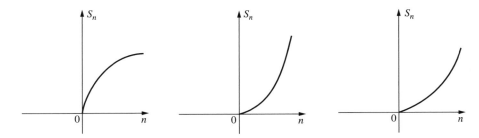

All of the above series are infinite but only the first series converges. Finding the infinite sum of a divergent series does not make any sense, and hence in order to find the sum to infinity of a series, the series must converge.

In order to find a result for the sum of an infinite series, it is important to understand the concept of a **limit**.

The concept of a limit is not particularly easy to define. The formal definition can be stated as

"A number or point L that is approached by a function $f(x)$ as x approaches a if, for every positive number ε, there exists a number δ such that $|f(x) - L| < \varepsilon$ if $0 < |x - a| < \delta$."

This is not necessarily helpful in visualizing the meaning of the term. A more informal viewpoint may help.

Consider Freddie Frog, who gets tired very quickly. Freddie hops 2 metres on his first hop. On his second hop, he is tired and can hop only half the distance, 1 metre. This continues, and each time he can hop only half the distance of his previous hop.

Consider Freddie trying to hop across a 4 metre road:

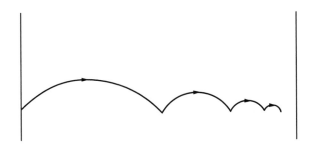

With each hop, he gets closer to the other side, but will he ever make it across the road? The distance that he has hopped can be considered to be

$$2 + 1 + \frac{1}{2} + \frac{1}{4} + \frac{1}{8} + \frac{1}{16} + \ldots$$

It is clear that he is getting very close to a distance of 4 metres but, as each hop is only half of his previous hop (and therefore half of the remaining distance), he will never actually reach 4 metres. In this situation, 4 metres is considered to be the limit of the distance hopped.

A limit is a value that a function or series approaches and becomes infinitesimally close to but will never reach. This idea has been covered in Chapter 3 – a horizontal asymptote is a value that a function approaches as x becomes large but never reaches. It is said that a series converges to a limit.

The notation for this is $\lim_{n \to \infty} S_n = L$ where L is the limit.

The formal definition for a limit in relation to functions given above is also true for series.

$\lim_{n \to \infty} S_n = L$ is true provided that S_n can be made as close to L as required by choosing n sufficiently large. In mathematical notation this can be stated "Given any number $\varepsilon > 0$, there exists an integer N such that $|S_n - L| < \varepsilon$ for all $n \geq N$".

Returning to the consideration of geometric series, will all infinite series converge to a limit? It is clear that the above series describing the frog does converge to a limit. However, consider the series

$1 + 10 + 100 + 1000 + 10000 + \ldots$

It is immediately clear that this series will continue to grow, and the gap between terms will continue to grow.

This is the key to understanding whether a series will converge – the gap between successive terms. If this gap is decreasing with each term, then the series will ultimately converge. Hence for the sum of a geometric series to converge, the common ratio must be reducing the terms. Putting this into mathematical notation,

a series will only converge if $|r| < 1$

If $|r| < 1$, then r^n will have a limit of zero as n becomes very large.

Considering the formula for the sum of n terms of a geometric series,

$$S_n = \frac{a(1 - r^n)}{1 - r} = \frac{a - ar^n}{1 - r}$$

For large values of n (as n approaches ∞), $ar^n \to 0$ if $|r| < 1$.

Think of 0.1^{10}, 0.1^{50} etc.

So when $n \to \infty$, the sum becomes

$$S_\infty = \frac{a}{1 - r} \quad (|r| < 1)$$

This formula can be used to find the limit of a convergent series, also known as the **sum to infinity** or infinite sum.

Example

Show that the sum to infinity of $8 + 2 + \frac{1}{2} + \frac{1}{8} + \frac{1}{32} + \ldots$ exists and find this sum.

Clearly this series converges as $|r| = \frac{1}{4} < 1$.

So $S_\infty = \frac{a}{1 - r} = \frac{8}{\frac{3}{4}} = \frac{32}{3}$.

A recurring decimal such as $5.\dot{8} = 5.8888888\ldots$ can be considered to be an infinite geometric series as it is $5 + \dfrac{8}{10} + \dfrac{8}{100} + \dfrac{8}{1000} + \ldots$. This means that the formula for the sum to infinity can be used to find an exact (fractional) value for the decimal. This is demonstrated by example.

Example

Find the exact value of the recurring decimal $1.\dot{2}$.

This can be considered as $1 + \dfrac{2}{10} + \dfrac{2}{100} + \dfrac{2}{1000} + \ldots$

So the decimal part is a geometric series with $a = \dfrac{2}{10}$ and $r = \dfrac{1}{10}$.

Hence a limit exists since $r = \dfrac{1}{10} < 1$.

$$S_\infty = \frac{a}{1 - r} = \frac{\dfrac{2}{10}}{\dfrac{9}{10}} = \frac{2}{9}.$$

So we can write $1.\dot{2} = 1 + \dfrac{2}{9} = \dfrac{11}{9}$.

Exercise 4

Determine whether the series below converge. If they do, find the sum to infinity.

1 $20 + 10 + 5 + \cdots$

2 $81 + 27 + 9 + \cdots$

3 $4 + 12 + 36 + \cdots$

4 $-64 + 40 - 25 + \cdots$

5 $8 - 12 + 18 - \cdots$

Find the sum to infinity for the geometric series with:

6 $a = 6, r = \dfrac{1}{2}$

7 $a = 100, r = \dfrac{2}{3}$

8 $a = 60, r = -\dfrac{1}{5}$

9 $a = 9, r = -\dfrac{3}{4}$

Find the range of values of x for which the following series converge.

10 $1 + x + x^2 + x^3 + \ldots$

11 $4x - 4 + \dfrac{4}{x} - \dfrac{4}{x^2} + \ldots$

Find the exact value of these recurring decimals.

12 $6.\dot{4}$

13 $2.\dot{1}\dot{6}$

14 $7.3\dot{4}$

15 Find the sum
  a of the even numbers from 50 to 100 inclusive
  b of the first ten terms of the geometric series that has a first term of 16 and a common ratio of 1.5
  c to infinity of the geometric series whose second term is $\frac{2}{3}$ and third term $\frac{1}{2}$.

6.5 Applications of sequences and series

Although sequences and series occur naturally and in many applications, these mostly involve more complicated series than met in this course. Most common examples of geometric series at this level model financial applications and population.

Example

Katherine receives €200 for her twelfth birthday and opens a bank account that provides 5% compound interest per annum (per year). Assuming she makes no withdrawals nor any further deposits, how much money will she have on her eighteenth birthday?

This can be considered as a geometric series with $a = 200$ and $r = 1.05$. The common ratio is 1.05 because 5% is being added to 100%, which gives $105\% = 1.05$.

So in six years the balance will be $u_7 = 200 \times 1.05^6$
$$\Rightarrow u_7 = 268.02$$

So she will have €268.02 on her eighteenth birthday.

> u_1 is the first term. After six years the balance will be u_7.

If Katherine receives €200 on every birthday following her twelfth, how much will she have by her eighteenth?

After one year the balance will be $u_2 = 1.05 \times 200 + 200$.

After two years, the balance will be $u_3 = 1.05u_2 + 200$.

This can be expressed as $u_3 = 1.05(1.05 \times 200 + 200) + 200$

$$= 1.05^2 \times 200 + 1.05 \times 200 + 200$$
$$= 200(1.05^2 + 1.05 + 1)$$

So $u_7 = 200(1.05^6 + 1.05^5 + ... + 1.05 + 1)$

The part in brackets is a geometric series with $a = 1$ and $r = 1.05$.

So the sum in brackets is $\dfrac{1(1 - 1.05^7)}{1 - 1.05}$
$$= 8.1420 ...$$

Hence the balance on her eighteenth birthday will be
$200 \times 8.1420 ... = $ €1628.40.

Exercise 5

1 In his training, Marcin does 10 sit-ups one day, then 12 sit-ups the following day. If he continues to do 2 more each day, how long before he completes 1000 sit-ups?

2 Karen invests $2000 in an account paying 8% per year. How much will be in the account after 4 years?

3 Anders invests 50 000 DKr (Danish kroner) at 12% per year. How much is it worth after 6 years?

4 A kind benefactor sets up a prize in an international school. The benefactor donates £10 000. The school invests the money in an account paying 5% interest. If £750 is paid out annually, for how long can the full prize be given out?

5 If Yu wants to invest 50 000 yen with a return of 20 000 yen over 8 years, what % rate must she find?

6 What initial investment is required to produce a final balance of £12 000 if invested at 8% per year over 4 years?

7 In a Parisian sewer, the population of rats increases by 12% each month.
   a If the initial population is 10 000, how many rats will there be after 5 months?
   b How long before there are 50 000 rats?

8 The number of leopards in a Kenyan national park has been decreasing in recent years. There were 300 leopards in 2000 and the population has decreased at a rate of 9% annually.
   a What was the population in 2005?
   b When will the population drop below 100?

9 Each time a ball bounces, it reaches 85% of the height reached on the previous bounce. It is dropped from a height of 5 metres.
   a What height does the ball reach after its third bounce?
   b How many times does it bounce before the ball can no longer reach a height of 1 metre?

6.6 Sigma notation

Sigma is the Greek letter that corresponds to S in the Roman alphabet, and is written σ or Σ. The σ form is often used in statistics but the capital form Σ is used to denote a sum of discrete elements. This notation is a useful shorthand rather than writing out a long string of numbers. It is normally used on the set of integers.

Consider $1 + 2 + 3 + 4 + 5 + 6 + 7$

This can be written as
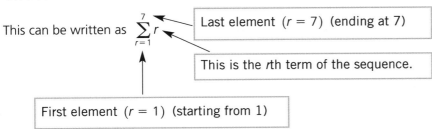

First element ($r = 1$) (starting from 1)

Last element ($r = 7$) (ending at 7)

This is the rth term of the sequence.

Similarly,

$$2 + 5 + 8 + 11 + 14 = \sum_{r=1}^{5} 3r - 1$$

We know that $3r - 1$ is the rth term because when $r = 1, 3r - 1 = 2$ and when $r = 2, 3r - 1 = 5$, etc.

Both arithmetic and geometric series can be expressed using this notation.

Example

Consider the arithmetic series
$100 + 96 + 92 + \ldots + 60$
Express the series using sigma notation.
This has $a = 100$ and $d = -4$. So the corresponding sequence has the general term
$$u_n = 100 - 4(n - 1)$$
$$= 104 - 4n$$

This series can be expressed as $\displaystyle\sum_{r=1}^{11} 104 - 4r$.

The first n terms (i.e. S_n) could be expressed as $\displaystyle\sum_{r=1}^{n} 104 - 4r$.

Example

Consider the geometric series
$4 + 8 + 16 + 32 + \ldots$
Express the series using sigma notation.

This can be expressed as $\displaystyle\sum_{r=1}^{n} 4 \times 2^{r-1}$ or $\displaystyle\sum_{r=1}^{n} 2^{r+1}$.

Example

Express the sum $16 + 4 + 1 + \dfrac{1}{4} + \ldots$ using sigma notation.

The infinite sum $16 + 4 + 1 + \dfrac{1}{4} + \ldots$ can be expressed as $\displaystyle\sum_{r=1}^{\infty} 16 \times \left(\dfrac{1}{4}\right)^{r-1}$

There are results that we can use with sigma notation that help to simplify expressions. These are presented here without proof but are proved in Chapter 18.

Result 1

$$\sum_{r=1}^{n} 1 = \underbrace{1 + 1 + \cdots + 1}_{n \text{ times}} = n$$

$$\sum_{r=1}^{n} a = an \text{ where } a \text{ is a constant.}$$

This should be obvious as $\sum_{r=1}^{n} a = a + a + a + \cdots + a = na$

We can also see that a constant can be removed outside a sum:

$$\sum_{r=1}^{n} a = a \sum_{r=1}^{n} 1$$

Result 2

$$\sum_{r=1}^{n} r = \frac{n(n + 1)}{2}$$

This is the sum of the first n natural numbers, and each of these sums is also a triangular number (because that number of objects can be arranged as a triangle).

For example, $\sum_{r=1}^{2} r = \frac{2(2 + 1)}{2} = 3$

$$\sum_{r=1}^{3} r = \frac{3(3 + 1)}{2} = 6$$

Result 3

$$\sum_{r=1}^{n} r^2 = \frac{1}{6}n(n + 1)(2n + 1)$$

These three results can be used to simplify other sigma notation sums. Note that they apply only to sums beginning with $r = 1$. If the sums begin with another value the question becomes more complicated, and these are not dealt with in this curriculum.

Example

Simplify $\sum_{r=1}^{n} 4r^2 - 3r - 5$ and hence find $\sum_{r=1}^{6} 4r^2 - 3r - 5$.

$$\sum_{r=1}^{n} 4r^2 - 3r - 5 = 4\sum_{r=1}^{n} r^2 - 3\sum_{r=1}^{n} r - 5\sum_{r=1}^{n} 1$$

$$= 4 \times \frac{1}{6}n(n + 1)(2n + 1) - 3\frac{n(n + 1)}{2} - 5n$$

$$= \frac{1}{6}n[(4n + 4)(2n + 1) - 9(n + 1) - 30]$$

$$= \frac{1}{6}n(8n^2 + 12n + 4 - 9n - 9 - 30)$$

$$= \frac{1}{6}n(8n^2 + 3n - 35)$$

Hence $\displaystyle\sum_{r=1}^{6} 4r^2 - 3r - 5 = \frac{1}{6} \times 6(8 \times 6^2 + 3 \times 6 - 35)$

$$= 288 + 18 - 35$$

$$= 271$$

Exercise 6

1 Evaluate

a $\displaystyle\sum_{r=1}^{5} 3r - 2$ **b** $\displaystyle\sum_{i=4}^{7} 2i^2$ **c** $\displaystyle\sum_{k=3}^{8} 5k^2 - 3k$

2 Express each of these sums in sigma notation.

a $4 + 8 + 12 + 16 + 20$

b $-2 + 3 + 8 + 13 + \cdots + (5n + 3)$

c $9 + 13 + 17 + 21 + \cdots$

3 Use the results for $\displaystyle\sum_{r=1}^{n} 1$, $\displaystyle\sum_{r=1}^{n} r$ and $\displaystyle\sum_{r=1}^{n} r^2$ to simplify these.

a $\displaystyle\sum_{r=1}^{n} 6r - 2$ **b** $\displaystyle\sum_{k=1}^{n} 2k^2 - k + 3$ **c** $\displaystyle\sum_{k=1}^{2n} 9 - k^2$ **d** $\displaystyle\sum_{r=1}^{k+1} 7r - 3$

6.7 Factorial notation

Sigma notation is a method used to simplify and shorten sums of numbers. There are also ways to shorten multiplication, one of which is factorial notation.

A factorial is denoted with an exclamation mark ! and means the product of all the positive integers up to that number.

$$n! = 1 \times 2 \times \cdots \times (n - 1) \times n$$

So $5! = 1 \times 2 \times 3 \times 4 \times 5 = 120$.

It is important to be able to perform arithmetic with factorials, as demonstrated in the examples below.

It is worth noting that 0! is defined to be 1.

Example

Simplify $\dfrac{8!}{5!}$.

It should be obvious that $8! = 8 \times 7 \times 6 \times 5!$ so $\dfrac{8!}{5!}$ can be simplified to

$$\frac{8 \times 7 \times 6 \times 5!}{5!} = 8 \times 7 \times 6 = 336$$

Example

Simplify $\dfrac{n!}{(n-2)!}$

Using the same rationale as above,

$$\dfrac{n!}{(n-2)!} = \dfrac{n \times (n-1) \times (n-2)!}{(n-2)!} = n(n-1)$$

Example

Simplify $n! - (n-2)!$
This can be factorised with a common factor of $(n-2)!$
So $n! - (n-2)! = (n-2)![n(n-1) - 1]$
$$\qquad\qquad\qquad = (n-2)!\,(n^2 - n - 1)$$

Permutations and combinations

Factorial notation is used most commonly with counting methods known as permutations and combinations.

Factorials can be used to determine the number of ways of arranging n objects.

Consider four people standing in a line: Anna, Julio, Mehmet and Shobana. How many different orders can they stand in?

The different ways can be listed systematically:

A	J	M	S
A	J	S	M
A	M	J	S
A	M	S	J
A	S	M	J
A	S	J	M
J	A	M	S
J	A	S	M
J	M	A	S
J	M	S	A
J	S	A	M
J	S	M	A
M	A	J	S
M	A	S	J
M	J	A	S
M	J	S	A
M	S	A	J
M	S	J	A
S	A	J	M
S	A	M	J
S	J	A	M
S	J	M	A
S	M	A	J
S	M	J	A

There are clearly 24 possibilities. This comes as no surprise as this can be considered as having 4 ways of choosing position 1, then for each choice having 3 ways of choosing position 2, and for each choice 2 ways of choosing position 3, leaving only 1 choice for position 4 each time.

This is equivalent to having $4 \times 3 \times 2 \times 1$ possibilities.

> So $n!$ is the number of ways of arranging n objects in order.

Consider a bag with five balls in it, labelled A, B, C, D and E.

If two balls are chosen from the bag at random, there are 10 possible arrangements:

A&B   A&C   A&D   A&E   B&C   B&D   B&E   C&D   C&E   D&E

If the order that the balls come out in matters, there would be 20 possible outcomes.

| AB | AC | AD | AE | BC | BD | BE | CD | CE | DE |
| BA | CA | DA | EA | CB | DB | EB | DC | EC | ED |

The first type are known as **combinations** (where order does not matter) and the second type as **permutations** (where order is important).

The formula for the number of combinations when choosing r objects at random from n objects is $\dfrac{n!}{r!\,(n-r)!}$.

> There are two notations for combinations, ^nC_r or $\dbinom{n}{r}$.

$$\text{So } \binom{n}{r} = \frac{n!}{r!\,(n-r)!}$$

The formula for the number of permutations when choosing r objects at random from n objects is very similar:

$$^nP_r = \frac{n!}{(n-r)!}$$

This makes it clear that the $r!$ in the combinations formula removes the duplication of combinations merely in a different order. This topic is further developed in Chapter 20 in its application to probability.

> It is important that we recognize whether we are working with a permutation or a combination, i.e. does order matter?

Example

How many 5 letter words (arrangements of letters) can be made from the letters of EIGHTYFOUR?

Here, the order of the letters matters so the number of words will be given by $^{10}P_5$.

$$^{10}P_5 = \frac{10!}{5!} = 30240$$

Example

How many different hockey teams (11 players) can be chosen from a squad of 15? Here, the order in which the players are chosen is unimportant.

So the number of different teams is $^{15}C_{11} = \dfrac{15!}{11!\,(15-11)!} = 1365$.

Many calculators have in-built formulae for permutations and combinations.

Pascal's triangle revisited

Here is Pascal's triangle.

$$
\begin{array}{c}
1 \\
1 \quad 1 \\
1 \quad 2 \quad 1 \\
1 \quad 3 \quad 3 \quad 1 \\
1 \quad 4 \quad 6 \quad 4 \quad 1 \\
1 \quad 5 \quad 10 \quad 10 \quad 5 \quad 1 \\
1 \quad 6 \quad 15 \quad 20 \quad 15 \quad 6 \quad 1
\end{array}
$$

Notice that this could also be written as

Row 1 $\qquad \dbinom{1}{0}\dbinom{1}{1}$

Row 2 $\qquad \dbinom{2}{0}\dbinom{2}{1}\dbinom{2}{2}$

Row 3 $\qquad \dbinom{3}{0}\dbinom{3}{1}\dbinom{3}{2}\dbinom{3}{3}$

Row 4 $\qquad \dbinom{4}{0}\dbinom{4}{1}\dbinom{4}{2}\dbinom{4}{3}\dbinom{4}{4}$ etc.

So Pascal's triangle is also given by the possible combinations in each row n. This leads to recognizing some important results about combinations.

Result 1

$$
\binom{n}{0} = \binom{n}{n} = 1
$$

This is fairly obvious from the definition of $^{n}C_{r}$.

$$
\binom{n}{0} = \frac{n!}{n!\,0!} = \frac{n!}{n!} = 1 \qquad\qquad \binom{n}{n} = \frac{n!}{n!\,0!} = \frac{n!}{n!} = 1
$$

Result 2

$$\binom{n}{1} = \binom{n}{n-1} = n$$

$$\binom{n}{1} = \frac{n!}{(n-1)!\,1!} = \frac{n!}{(n-1)!} = n \qquad\qquad \binom{n}{n-1} = \frac{n!}{(n-1)!\,1!} = \frac{n!}{(n-1)!} = n$$

The above two results and the symmetry of Pascal's triangle lead to result 3.

Result 3

$$\binom{n}{r} = \binom{n}{n-r}$$

This is again easy to show:

$$\binom{n}{r} = \frac{n!}{r!\,(n-r)!} = \frac{n!}{[n-(n-r)]!\,(n-r)!} = \frac{n!}{(n-r)!\,r!} = \binom{n}{n-r}$$

Result 4

$$\binom{n}{r-1} + \binom{n}{r} = \binom{n+1}{r}$$

This is the equivalent statement of saying that to obtain the next row of Pascal's triangle, add the two numbers above.

The proof of result 4 is as follows.

$$\binom{n}{r-1} + \binom{n}{r}$$

$$= \frac{n!}{(r-1)!\,(n-r+1)!} + \frac{n!}{r!\,(n-r)!}$$

$$= \frac{r \cdot n!}{r!\,(n-r+1)!} + \frac{(n-r+1)\cdot n!}{r!\,(n-r+1)!}$$

$$= \frac{r \cdot n! + (n+1)\cdot n! - r\cdot n!}{r!\,(n-r+1)!}$$

$$= \frac{(n+1)!}{r!\,(n-r+1)!}$$

$$= \binom{n+1}{r}$$

Example

Solve

$$\binom{n + 1}{1} + \binom{n + 1}{2} = 66$$

From result 4

$$\Rightarrow \binom{n + 2}{2} = 66$$

$$\Rightarrow \frac{(n + 2)!}{(n + 2 - 2)! \, 2!} = 66$$

$$\Rightarrow \frac{(n + 2)!}{n! \times 2} = 66$$

$$\Rightarrow (n + 2)(n + 1) = 132$$

$$\Rightarrow n^2 + 3n - 130 = 0$$

$$\Rightarrow (n + 13)(n - 10) = 0$$

$$\Rightarrow n = 10$$

> Remember that n must be positive.

Exercise 7

1 Evaluate the following:

   a $^6P_2$ b $^8P_3$ c $^8C_3$ d $\binom{9}{5}$ e $\binom{8}{4}$

2 How many different 4 letter words (arrangements where order matters) can be made from the letters A, E, I, O, U, Y?

3 How many different committees of 9 can be made from 14 people?

4 A grade 5 class has 11 students.

   a If the teacher lines them up, how many different orders can there be?

   b If 3 students are selected as president, secretary and treasurer of the Eco-Club, how many different ways can this be done?

   c If 7 students are chosen for a mini-rugby match, how many different teams are possible?

5 In the UK national lottery, 6 balls are chosen at random from 49 balls. In the Viking lottery operated in Scandinavia, 6 balls are chosen from 48 balls. How many more possible combinations result from the extra ball?

6 The EuroMillions game chooses 5 numbers at random from 50 balls and then 2 more balls known as lucky stars from balls numbered 1–9. How many possible combinations are there for the jackpot prize (5 numbers plus 2 lucky stars)?

7 José is choosing his 11 players for a soccer match. Of his squad of 20, one player is suspended. He has three players whom he always picks (certainties). How many possible teams can he create?

8 How many 3-digit numbers can be created from the digits 2, 3, 4, 5, 6 and 7 if each digit may be used

   a any number of times   b only once.

9 Solve these equations.

   a $\binom{n}{2} = 15$ b $\binom{n}{3} = 10$ c $\binom{2n}{2} = 28$ d $\binom{2n}{2} = 66$

10 Solve these equations.

a $\dbinom{n}{n-2} = 6$ **b** $\dbinom{n}{n-2} = 45$ **c** $\dbinom{n}{n-3} = 84$

11 Find a value of n that satisfies each equation.

a $\dbinom{n}{1} + \dbinom{n}{2} = 28$ **b** $\dbinom{n+2}{2} + \dbinom{n+2}{3} = 20$ **c** $\dbinom{2n}{3} + \dbinom{2n}{4} = 35$

6.8 Binomial theorem

The binomial theorem is a result that provides the expansion of $(x + y)^n$.

Consider the expansions of $(x + y)^1$, $(x + y)^2$ and $(x + y)^3$.

$(x + y)^1 = 1x + 1y$ $(x + y)^2 = 1x^2 + 2xy + 1y^2$ $(x + y)^3 = (x + y)(x^2 + 2xy + y^2)$

$$= 1x^3 + 3x^2y + 3xy^2 + 1y^3$$

Notice that the coefficients are the same as the numbers in Pascal's triangle.

Similarly, $(x + y)^4 = 1x^4 + 4x^3y + 6x^2y^2 + 4xy^3 + 1y^4$

From Pascal's triangle, this could be rewritten

$$(x + y)^4 = \dbinom{4}{0}x^4 + \dbinom{4}{1}x^3y + \dbinom{4}{2}x^2y^2 + \dbinom{4}{3}xy^3 + \dbinom{4}{4}y^4$$

This leads to the general expansion

$$(x + y)^n = \dbinom{n}{0}x^n + \dbinom{n}{1}x^{n-1}y + \dbinom{n}{2}x^{n-2}y^2 + \cdots + \dbinom{n}{n-1}xy^{n-1} + \dbinom{n}{n}y^n$$

This can be shortened to

$$(x + y)^n = \sum_{r=0}^{n} \dbinom{n}{r}x^{n-r}y^r$$

This result is stated here without proof; the proof is presented in Chapter 18.

A useful special case is the expansion of $(1 + x)^n$

$$(1 + x)^n = 1 + nx + \frac{n(n - 1)}{2!}x^2 + \frac{n(n - 1)(n - 2)}{3!}x^3 + \ldots$$

Example

Using the binomial theorem, expand $(2x + 3)^5$.

This can be written $(2x + 3)^5 = \sum_{r=0}^{5} \dbinom{5}{r}(2x)^{5-r}3^r$

So

$$(2x + 3)^5 = \dbinom{5}{0}2^5x^53^0 + \dbinom{5}{1}2^4x^43^1 + \dbinom{5}{2}2^3x^33^2 + \dbinom{5}{3}2^2x^23^3 + \dbinom{5}{4}2^1x^13^4 + \dbinom{5}{5}2^0x^03^5$$

$$= 32x^5 + 5 \times 16 \times 3x^4 + 10 \times 8 \times 9x^3 + 10 \times 4 \times 27x^2$$
$$+ 5 \times 2 \times 81x + 243$$

$$= 32x^5 + 240x^4 + 720x^3 + 1080x^2 + 810x + 243$$

Example

Using the result for $(1 + x)^n$, find the expansion of $(4 + 3x)^4$.

$$(4 + 3x)^4 = 4\left(1 + \frac{3}{4}x\right)^4$$

Using the result for $(1 + x)^n$, this becomes

$$4\left[1 + 4\left(\frac{3}{4}x\right) + \frac{4 \times 3}{2!}\left(\frac{3}{4}x\right)^2 + \frac{4 \times 3 \times 2}{3!}\left(\frac{3}{4}x\right)^3 + \frac{4 \times 3 \times 2 \times 1}{4!}\left(\frac{3}{4}x\right)^4\right]$$

$$= 4\left(1 + 3x + \frac{27}{8}x^2 + \frac{27}{16}x^3 + \frac{81}{256}x^4\right)$$

$$= 4 + 12x + \frac{27}{2}x^2 + \frac{27}{4}x^3 + \frac{81}{64}x^4$$

Example

Expand $\left(x - \frac{4}{x}\right)^3$.

This can be rewritten as $\displaystyle\sum_{r=0}^{3}\binom{3}{r}x^{3-r}(-1)^r\left(\frac{4}{x}\right)^r$.

Before expanding, it is often useful to simplify this further.

So $\displaystyle\sum_{r=0}^{3}\binom{3}{r}x^{3-r}(-1)^r\left(\frac{4}{x}\right)^r = \sum_{r=0}^{3}\binom{3}{r}x^{3-r}4^r(-1)^r x^{-r}$

$$= \sum_{r=0}^{3}\binom{3}{r}4^r(-1)^r x^{3-2r}$$

Expanding gives

$$\left(x - \frac{4}{x}\right)^3 = x^3 + \binom{3}{1}4^1(-1)^1 x^1 + \binom{3}{2}4^2(-1)^2 x^{-1} + 4^3(-1)^3 x^{-3}$$

$$= x^3 - 12x + 48x^{-1} - 64x^{-3}$$

Example

What is the coefficient of x^2 in the expansion of $\left(3x - \frac{2}{5x}\right)^8$?

Rewriting using sigma notation,

$$\left(3x - \frac{2}{5x}\right)^8 = \sum_{r=0}^{8}\binom{8}{r}3^{8-r}x^{8-r}(-1)^r\left(\frac{2}{5}\right)^r(x^{-1})^r$$

$$= \sum_{r=0}^{8}\binom{8}{r}3^{8-r}(-1)^r\left(\frac{2}{5}\right)^r x^{8-2r}$$

For the x^2 term, it is clear that $8 - 2r = 2$

$$\Rightarrow r = 3$$

Hence the term required is $\binom{8}{3}3^5(-1)^3\left(\frac{2}{5}\right)^3 x^2$.

So the coefficient is $56 \times 243 \times (-1) \times \dfrac{8}{125}$

$$= -\frac{108\,864}{125}$$

This method of finding the required term is very useful, and avoids expanding large expressions.

Example

Find the term independent of x in the expansion of $(2 + x)\left(2x + \dfrac{1}{x}\right)^5$.

For this to produce a term independent of x, the expansion of $\left(2x + \dfrac{1}{x}\right)^5$

must have a constant term or a term in x^{-1}.

$$\left(2x + \frac{1}{x}\right)^5 = \sum_{r=0}^{5} \binom{5}{r} 2^{5-r} x^{5-r} (x^{-1})^r$$

So the power of x is given by $5 - 2r$. This cannot be zero for positive integer values of r. Hence the required coefficient is given by

$$5 - 2r = -1$$
$$\Rightarrow r = 3$$

The required term is therefore given by $(2 + x)(\cdots + 10 \times 2^2 \times x^{-1} + \ldots)$.
So the term independent of x is 40.

Example

Find the term independent of x in the expansion of $(2x + 1)^7\left(x - \dfrac{2}{x}\right)^5$.

This is the product of two expansions, which need to be considered separately at first.

$$(2x + 1)^7 = \sum_{r=0}^{7} \binom{7}{r} 2^{7-r} x^{7-r} 1^r \text{ and } \left(x - \frac{2}{x}\right)^5 = \sum_{k=0}^{5} \binom{5}{k} x^{5-k}(-1)^k 2^k x^{-k}$$

So the general terms are $\binom{7}{r} 2^{7-r} x^{7-r}$ and $\binom{5}{k}(-1)^k 2^k x^{5-2k}$.

For the term independent in x, that is x^0, the general terms need to multiply together to make x^0.

So $x^{7-r} \cdot x^{5-2k} = x^0$
$$\Rightarrow 7 - r + 5 - 2k = 0$$
$$\Rightarrow r = 12 - 2k$$

This type of equation is often best solved using a tabular method (there is often more than one solution).

r	$12 - 2k$	k
0	12	0
1	10	1
2	8	2
3	6	3
4	4	4
5	2	5
6		
7		

So the three scenarios that give terms independent of x when the brackets are multiplied are:

$k = 3, r = 6$ $k = 4, r = 4$ $k = 5, r = 2$

$\binom{7}{6}2^1 \times \binom{5}{3}(-1)^3 2^3$ $\binom{7}{4}2^3 \times \binom{5}{4}(-1)^4 2^4$ $\binom{7}{2}2^5 \times \binom{5}{5}(-1)^5 2^5$

$= 7 \times 2 \times 10$ $= 35 \times 8 \times 5$ $= 21 \times 32 \times 1$

$\quad\quad \times -1 \times 8$ $\times 1 \times 16$ $\times -1 \times 32$

$= -1120$ $= 22400$ $= -21504$

So the term independent of x in the expansion is $\quad -1120 + 22400 - 21504$

$$= -224$$

Example

Expand $(2 + x)^5$ and hence find 1.9^5.

$(2 + x)^5 = 32 + 80x + 80x^2 + 40x^3 + 10x^4 + x^5$

So 1.9^5 can be considered to be when $x = -0.1$ in the above expansion.

So $1.9^5 = 32 + 80(-0.1) + 80(-0.1)^2 + 40(-0.1)^3 + 10(-0.1)^4 + (-0.1)^5$

$\quad\quad\quad = 32 - 8 + 0.8 - 0.04 + 0.001 - 0.00001$

$\quad\quad\quad = 24.76099$

Exercise 8

1 Use the binomial theorem to expand the following expressions.

  a $(a + b)^4$ **b** $(3x + 2)^6$ **c** $(1 - x)^4$ **d** $(2p - 3q)^5$

2 Expand the following using the binomial theorem.

  a $\left(x + \dfrac{1}{x}\right)^3$ **b** $\left(x + \dfrac{2}{x}\right)^5$ **c** $\left(x - \dfrac{1}{x}\right)^6$ **d** $\left(2t - \dfrac{1}{4t}\right)^4$

3 Expand $(1 + 3x + x^2)^3$ by considering it as $([1 + 3x] + x^2)^3$.

4 What is the coefficient of:

  a x^3 in the expansion of $(x + 2)^5$

  b x^5 in the expansion of $(x + 5)^8$

  c x^2 in the expansion of $(x - 4)^6$

  d x^3 in the expansion of $(2x + 9)^5$

  e x in the expansion of $(8 - x)^9$

  f x^3 in the expansion of $\left(x + \dfrac{1}{x}\right)^7$

  g x^2 in the expansion of $\left(x - \dfrac{2}{x}\right)^4$

  h the term independent of x in the expansion of $\left(2x - \dfrac{3}{x}\right)^8$.

5 What is the coefficient of:

  a x^3 in the expansion of $(x + 1)^5(2x + 1)^4$

  b x^6 in the expansion of $(x - 2)^4(x + 4)^6$

c x in the expansion of $(3 + x)^3(1 - 2x)^5$

d x^2 in the expansion of $(x^2 + x - 3)^4$.

6 Expand these expressions.

a $(x + 5)^3(x - 4)^4$ **b** $\left(x + \dfrac{1}{x}\right)^3(x - 2)^3$ **c** $\left(x + \dfrac{1}{x}\right)^4\left(x - \dfrac{2}{x}\right)^3$

7 What is the coefficient of:

a x in the expansion of $(x + 1)^4\left(x + \dfrac{1}{x}\right)^3$

b x^3 in the expansion of $(2x + 3)^5\left(x - \dfrac{1}{x}\right)^4$

c x^{-10} in the expansion of $\left(x + \dfrac{1}{x}\right)^7\left(x - \dfrac{4}{x}\right)^5$.

8 Find the term independent of p in the expansion of $\left(2p^2 - \dfrac{1}{p}\right)^5\left(p + \dfrac{2}{p}\right)^4$.

9 Calculate the following correct to three significant figures.

a 1.01^4 **b** 0.8^7 **c** 7.94^8

10 For small values of x, any terms with powers higher than 3 are negligible for the expression $(x^2 - x + 5)^2(x - 2)^7$.

Find the approximate expression, $ax^2 + bx + c$, for this expansion.

11 In the expansions $\left(px + \dfrac{q}{x}\right)^6$ and $\left(px^2 - \dfrac{q}{x}\right)^4$, the constant terms are equal.

Show that this is never true for $p, q \in \mathbb{R}, p, q \neq 0$.

Review exercise

 1 Find the coefficient of x^3 in the binomial expansion of $\left(1 - \dfrac{1}{2}x\right)^8$.

[IB Nov 02 P1 Q3]

 2 The nth term u_n of a geometric sequence is given by $u_n = 3(4)^{n+1}, n \in \mathbb{Z}^+$.

a Find the common ratio r.

b Hence, or otherwise, find S_n, the sum of the first n terms of this sequence.

[IB May 01 P1 Q7]

 3 Consider the arithmetic series $2 + 5 + 8 + \cdots$.

a Find an expression for S_n, the sum of the first n terms.

b Find the value of n for which $S_n = 1365$.   [IB May 02 P1 Q1]

 4 A geometric sequence has all positive terms. The sum of the first two terms is 15 and the sum to infinity is 27. Find the value of

a the common ratio

b the first term.   [IB May 03 P1 Q1]

 5 The sum of the first n terms of a series is given by $S_n = 2n^2 - n$, where $n \in \mathbb{Z}^+$.

a Find the first three terms of the series.

b Find an expression for the nth term of the series, giving your answer in terms of n.   [IB Nov 04 P1 Q3]

6 Consider the infinite geometric series

$$1 + \left(\frac{2x}{3}\right) + \left(\frac{2x}{3}\right)^2 + \left(\frac{2x}{3}\right)^3 + \dots$$

  a For what values of x does the series converge?

  b Find the sum of the series if $x = 1.2$. [IB Nov 01 P1 Q4]

 7 An arithmetic sequence has 5 and 13 as its first two terms respectively.

  a Write down, in terms of n, an expression for the nth term, a_n.

  b Find the number of terms of the sequence which are less than 400.

  [IB Nov 99 P1 Q1]

 8 The coefficient of x in the expansion of $\left(x + \dfrac{1}{ax^2}\right)^7$ is $\dfrac{7}{3}$. Find the possible values of a.

  [IB Nov 00 P1 Q12]

 9 The sum of an infinite geometric sequence is $13\dfrac{1}{2}$, and the sum of the first three terms is 13. Find the first term. [IB Nov 00 P1 Q15]

 10 a $x + 1, 3x + 1, 6x - 2$ are the first three terms of an arithmetic sequence. For what value of n does S_n, the sum of the first n terms, first exceed 100?

  b The sum of the first three terms of a positive geometric sequence is 315 and the sum of the 5th, 6th and 7th terms is 80 640. Identify the first term and the common ratio.

 11 The first four terms of an arithmetic sequence are $2, a - b, 2a + b + 7$ and $a - 3b$, where a and b are constants. Find a and b. [IB Nov 03 P1 Q9]

 12 a Find the expansion of $(2 + x)^5$, giving your answer in ascending powers of x.

  b By letting $x = 0.01$ or otherwise, find the **exact** value of 2.01^5.

  [IB Nov 04 P1 Q8]

 13 The first three terms of a geometric sequence are also the first, eleventh and sixteenth terms of an arithmetic sequence.
The terms of the geometric sequence are all different.
The sum to infinity of the geometric sequence is 18.

  a Find the common ratio of the geometric sequence, clearly showing all working.

  b Find the common difference of the arithmetic sequence.

  [IB May 05 P2 Q4]

 14 a An arithmetic progression is such that the sum of the first 8 terms is 424, and the sum of the first 10 terms is 650. Find the fifth term.

  b A 28.5 m length of rope is cut into pieces whose lengths are in arithmetic progression with a common difference of d m. Given that the lengths of the shortest and longest pieces are 1 m and 3.75 m respectively, find the number of pieces and the value of d.

  c The second and fourth terms of a geometric progression are 24 and 3.84 respectively. Given that all terms are positive, find

   i the sum, to the nearest whole number, of the first 5 terms

   ii the sum to infinity.

 15 Determine the coefficients of $\dfrac{1}{x}$ and $\dfrac{1}{x^3}$ in the expansion $\left(2x + \dfrac{1}{x}\right)^7$.

16 The constant in the expansions of $\left(kx^2 + \dfrac{6}{x^2}\right)^4$ and $\left(kx^3 + \dfrac{p}{x^3}\right)^6$ are equal, and k and p are both greater than zero. Express k in terms of p.

17 Find the constant term in the expansion of $\left(3x^2 + \dfrac{2}{x^6}\right)^{12}$ giving your answer as an integer.

18 What is the coefficient of x^9 in the expansion of $(x^2 - 2x + 1)^3\left(3x + \dfrac{2}{x}\right)^5$?

19 Simplify $\displaystyle\sum_{k=1}^{n} 6 - 5k^2$ and hence find $\displaystyle\sum_{k=1}^{8} 6 - 5k^2$.

20 Solve $\dbinom{n+1}{n-2} = 165$.

7 Trigonometry 2

Consider the equation below. If all angles are in radians

$$\arctan(1) + \arctan(2) + \arctan(3) = \pi$$

Can you prove this?

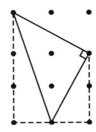

The mathematics behind the fact

The hint is in the diagram, since we can check that the angles of the three triangles at their common vertex add up to π. Is it possible to find a similar proof for the following equation?

$$\arctan\left(\frac{1}{2}\right) + \arctan\left(\frac{1}{3}\right) = \frac{\pi}{4}$$

http://www.math.hmc.edu/funfacts/ffiles/20005.2.shtml

Accessed 1 Dec 06

In Chapter 1, we met the trigonometric functions and their graphs. In this chapter we will meet some trigonometric identities. These can be used to solve problems and are also used to prove other trigonometric results.

7.1 Identities

An identity is a result or equality that holds true regardless of the value of any of the variables within it.

> $\equiv$ means "is identical to" although the equals sign is often still used in identity work.

We met an example of a trigonometric identity in Chapter 1: $\tan\theta \equiv \dfrac{\sin\theta}{\cos\theta}$

Pythagorean identities

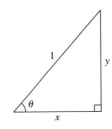

Applying Pythagoras' theorem to the right-angled triangle obtained from the unit circle,

$$y^2 + x^2 = 1 \Rightarrow \sin^2 \theta + \cos^2 \theta = 1$$

This resulted from the definition of $\sin \theta$ and $\cos \theta$ from the unit circle. We can also see that it is true from a right-angled triangle.

We know that $\sin \theta = \dfrac{y}{r}$ and $\cos \theta = \dfrac{x}{r}$.

By Pythagoras' theorem, $x^2 + y^2 = r^2$.

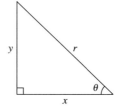

So $\sin^2 \theta + \cos^2 \theta = \left(\dfrac{y}{r}\right)^2 + \left(\dfrac{x}{r}\right)^2$

$$= \dfrac{y^2 + x^2}{r^2}$$

$$= \dfrac{r^2}{r^2}$$

$$= 1$$

> This is an identity and so strictly should be written $\sin^2 \theta + \cos^2 \theta \equiv 1$.

Finding $\cos \theta$ given $\sin \theta$

We now know that $\sin^2 \theta + \cos^2 \theta = 1$.

This is often expressed as $\sin^2 \theta = 1 - \cos^2 \theta$ or $\cos^2 \theta = 1 - \sin^2 \theta$.

Example

If $\sin \theta = \dfrac{1}{4}$, find possible values of $\cos \theta$.

Since $\sin \theta = \dfrac{1}{4}$, $\sin^2 \theta = \dfrac{1}{16}$

$\Rightarrow \cos^2 \theta = 1 - \dfrac{1}{16}$

$\qquad = \dfrac{15}{16}$

$\Rightarrow \cos \theta = \pm \dfrac{\sqrt{15}}{4}$

The information could also be displayed in a triangle, and $\cos \theta$ calculated that way.

> As we do not know which quadrant θ lies in, $\cos \theta$ could be positive or negative.

Example

If $\cos \theta = -\dfrac{1}{2}$, find possible values of $\sin \theta$.

$\cos^2 \theta = \dfrac{1}{4}$

$\Rightarrow \sin^2 \theta = 1 - \dfrac{1}{4}$

$\qquad = \dfrac{3}{4}$

$\Rightarrow \sin \theta = \pm \dfrac{\sqrt{3}}{2}$

Taking this identity of $\sin^2 \theta + \cos^2 \theta = 1$, we can create other identities that are useful in trigonometric work.

By dividing both sides by $\cos^2 \theta$,

$$\frac{\sin^2 \theta}{\cos^2 \theta} + \frac{\cos^2 \theta}{\cos^2 \theta} = \frac{1}{\cos^2 \theta}$$

So

$$\tan^2 \theta + 1 \equiv \sec^2 \theta$$

Similarly, by dividing both sides by $\sin^2 \theta$ we obtain

$$1 + \cot^2 \theta \equiv \csc^2 \theta$$

Example

Simplify $\dfrac{\sin \theta}{1 + \cos \theta} + \dfrac{1 + \cos \theta}{\sin \theta}$.

$\dfrac{\sin \theta}{1 + \cos \theta} + \dfrac{1 + \cos \theta}{\sin \theta} = \dfrac{\sin^2 \theta}{\sin \theta (1 + \cos \theta)} + \dfrac{(1 + \cos \theta)^2}{\sin \theta (1 + \cos \theta)}$

$\qquad = \dfrac{\sin^2 \theta + 1 + 2 \cos \theta + \cos^2 \theta}{\sin \theta (1 + \cos \theta)}$

$\qquad = \dfrac{2 + 2 \cos \theta}{\sin \theta (1 + \cos \theta)}$

$\qquad = \dfrac{2}{\sin \theta}$

Identities are often used to simplify expressions.

Questions involving simplification can also be presented as proving another identity.

Example

Prove that $\dfrac{\sin^4 \theta}{\tan \theta} \equiv \sin \theta \cos \theta - \sin \theta \cos^3 \theta$.

$$\text{LHS} \equiv \frac{\sin^4 \theta}{\tan \theta}$$

$$\equiv \frac{\sin^4 \theta \cos \theta}{\sin \theta}$$

$$\equiv \sin^3 \theta \cos \theta$$

$$\equiv \sin \theta \sin^2 \theta \cos \theta$$

$$\equiv \sin \theta \cos \theta (1 - \cos^2 \theta)$$

$$\equiv \sin \theta \cos \theta - \sin \theta \cos^3 \theta$$

$$\equiv \text{RHS}$$

When faced with this type of question, the strategy is to take one side (normally the left-hand side) and work it through to produce the right-hand side.

Example

Solve $3 \sec^2 \theta + 2 \tan \theta - 4 = 0$ for $0 \le \theta < 2\pi$.

Using $\sec^2 \theta \equiv \tan^2 \theta + 1$ this becomes

$$3(\tan^2 \theta + 1) + 2 \tan \theta - 4 = 0$$

$$\Rightarrow 3 \tan^2 \theta + 2 \tan \theta - 1 = 0$$

By using the identity, the equation has been transformed into one that involves only one type of trigonometric function, $\tan \theta$.

This is a quadratic equation where the variable is $\tan \theta$. It can be solved using factorisation or the quadratic formula.

Here, $3 \tan^2 \theta + 2 \tan \theta - 1 = 0$

$$\Rightarrow (3 \tan \theta - 1)(\tan \theta + 1) = 0$$

$$\Rightarrow \tan \theta = \frac{1}{3} \text{ or } \tan \theta = -1$$

$$\Rightarrow \theta = 0.322, 3.46 \text{ or } \theta = \frac{3\pi}{4}, \frac{7\pi}{4}$$

Example

Solve $\tan \theta + 3 \cot \theta = 5 \sec \theta$ for $0 \le \theta < 2\pi$.

Here we can use the definitions of each of these three trigonometric functions to simplify the equation.

$$\frac{\sin \theta}{\cos \theta} + \frac{3 \cos \theta}{\sin \theta} = \frac{5}{\cos \theta}$$

$$\Rightarrow \frac{\sin \theta - 5}{\cos \theta} = \frac{-3 \cos \theta}{\sin \theta}$$

$$\Rightarrow \sin^2 \theta - 5 \sin \theta = -3 \cos^2 \theta$$
$$\Rightarrow \sin^2 \theta - 5 \sin \theta = -3(1 - \sin^2 \theta)$$
$$\Rightarrow 2 \sin^2 \theta + 5 \sin \theta - 3 = 0$$
$$\Rightarrow (2 \sin \theta - 1)(\sin \theta + 3) = 0$$
$$\Rightarrow \sin \theta = \frac{1}{2} \text{ or } \sin \theta = -3$$

Since $\sin \theta = -3$ has no solution, the solution to the equation is $\theta = \dfrac{\pi}{6}, \dfrac{5\pi}{6}$.

Exercise 1

These Pythagorean identities can also be used to help solve trigonometric equations.

1 For the given values of $\sin \theta$, give possible values of $\cos \theta$.

   a $\sin \theta = \dfrac{1}{2}$ **b** $\sin \theta = \dfrac{1}{\sqrt{2}}$ **c** $\sin \theta = \dfrac{5}{7}$ **d** $\sin \theta = 3$

2 For the given values of $\cos \theta$, give possible values of $\sin \theta$.

   a $\cos \theta = \dfrac{\sqrt{3}}{2}$ **b** $\cos \theta = \dfrac{4}{5}$ **c** $\cos \theta = 4$ **d** $\cos \theta = 0.2$

3 Prove the following to be true, using trigonometric identities.

   a $\cos^3 \theta \tan \theta \equiv \sin \theta - \sin^3 \theta$

   b $\cos^5 \theta \equiv \cos \theta - 2 \sin^2 \theta \cos \theta$

   c $(4 \sin \theta + 3 \cos \theta)^2 + (3 \sin \theta - 4 \cos \theta)^2 \equiv 25$

   d $(\sec \theta - 1)^2 - 2 \sec \theta \equiv \tan^2 \theta$

4 If $\cot^2 \theta = 8$, what are possible values for $\sin \theta$?

5 Simplify these.

   a $\dfrac{6 - 6 \cos^2 \theta}{2 \sin \theta}$ **b** $\dfrac{\sin^3 \theta + \sin \theta \cos^2 \theta}{\cos \theta}$

   c $\dfrac{\tan^2 \theta - \sec^2 \theta}{\csc \theta}$ **d** $\dfrac{7 \cos^4 \theta + 7 \sin^2 \theta \cos^2 \theta}{\sin^2 \theta}$

6 Prove that $\tan^2 \phi + \cot^2 \phi \equiv \sec^2 \phi + \csc^2 \phi - 2$.

7 Solve these following equations for $0 \leq \theta < 2\pi$.

   a $\csc^2 \theta + \cot^2 \theta = 5$ **b** $\sin^2 \theta + 2 \cos \theta - 1 = 0$

   c $3 \tan \theta = 4 \sec^2 \theta - 5$ **d** $\cot \theta + \tan \theta = 2$

   e $6 \cot^2 \theta + 13 \csc \theta - 2 = 0$ **f** $\cos \theta - 2 \sin^2 \theta = -1$

7.2 Compound angle (addition) formulae

These formulae allow the expansion of expressions such as $\sin (A + B)$.

It is very important to recognize that $\sin (A + B) \neq \sin A + \sin B$.

This becomes clear by taking a simple example:

$$\sin 90° = \sin(30 + 60)° = 1$$

But $\sin 30° + \sin 60° = \dfrac{1}{2} + \dfrac{\sqrt{3}}{2} \approx 1.37 \ (\neq 1)$

The six addition formulae are:

$$\sin(A + B) \equiv \sin A \cos B + \cos A \sin B$$
$$\sin(A - B) \equiv \sin A \cos B - \cos A \sin B$$
$$\cos(A + B) \equiv \cos A \cos B - \sin A \sin B$$
$$\cos(A - B) \equiv \cos A \cos B + \sin A \sin B$$
$$\tan(A + B) \equiv \frac{\tan A + \tan B}{1 - \tan A \tan B}$$
$$\tan(A - B) \equiv \frac{\tan A - \tan B}{1 + \tan A \tan B}$$

The formula for $\cos(A - B)$ can be proved as follows:

Take two points on the unit circle, as shown.

These two points have coordinates

M($\cos B$, $\sin B$) and N($\cos A$, $\sin A$).

The square of the distance from
M to N is given by Pythagoras' theorem:

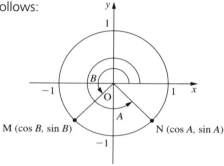

$$MN^2 = (\cos A - \cos B)^2 + (\sin A - \sin B)^2$$
$$= \cos^2 A - 2\cos A \cos B + \cos^2 B + \sin^2 A - 2 \sin A \sin B + \sin^2 B$$
$$= 1 + 1 - 2\cos A \cos B - 2 \sin A \sin B$$
$$= 2 - 2(\cos A \cos B + \sin A \sin B)$$

Using the cosine rule in triangle MON:
$$MN^2 = 1^2 + 1^2 - 2 \times 1 \times 1 \times \cos(A - B)$$
$$= 2 - 2 \cos(A - B)$$

Angle MON = angle A − angle B

Equating these two results for MN^2:

$$2 - 2 \cos(A - B) = 2 - 2(\cos A \cos B + \sin A \sin B)$$
$$\cos(A - B) = \cos A \cos B + \sin A \sin B$$

The other formulae are easily proved from this one. The starting point for each is given below:

$$\cos(A + B) = \cos[A - (-B)]$$

$$\sin(A - B) = \cos\left[\frac{\pi}{2} - (A - B)\right] \qquad \sin(A + B) = \sin[A - (-B)]$$

$$\tan(A - B) = \frac{\sin(A - B)}{\cos(A - B)} \qquad \tan(A + B) = \tan[A - (-B)]$$

Example

Using $15° = 60° - 45°$, find the exact value of $\cos 15°$.

$$\cos 15° = \cos(60 - 45)°$$
$$= \cos 60° \cos 45° + \sin 60° \sin 45°$$
$$= \left(\frac{1}{2} \times \frac{1}{\sqrt{2}}\right) + \left(\frac{\sqrt{3}}{2} \times \frac{1}{\sqrt{2}}\right)$$
$$= \frac{1}{2\sqrt{2}} + \frac{\sqrt{3}}{2\sqrt{2}}$$
$$= \frac{1 + \sqrt{3}}{2\sqrt{2}}$$

This is a typical non-calculator question.

Example

Prove that $2 \cos\left(\theta + \dfrac{\pi}{3}\right) = \cos \theta - \sqrt{3} \sin \theta$.

$$\text{LHS} = 2 \cos\left(\theta + \frac{\pi}{3}\right)$$
$$= 2\left(\cos \theta \cos \frac{\pi}{3} - \sin \theta \sin \frac{\pi}{3}\right)$$
$$= 2\left(\frac{1}{2}\cos \theta - \frac{\sqrt{3}}{2}\sin \theta\right)$$
$$= \cos \theta - \sqrt{3} \sin \theta = \text{RHS}$$

Example

Simplify $\dfrac{\tan 96° - \tan 51°}{1 + \tan 96° \tan 51°}$.

Recognizing this is of the form $\tan(A - B)$, it can be written

$$\frac{\tan 96° - \tan 51°}{1 + \tan 96° \tan 51°} = \tan(96 - 51)°$$
$$= \tan 45°$$
$$= 1$$

Example

In the diagram below, find the exact value of $\cos A\hat{B}C$.

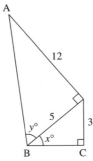

In the diagram, $\hat{ABC} = x° + y°$.

Using Pythagoras' theorem, we can calculate BC to be 4 and AB to be 13.

Hence $\cos x° = \dfrac{4}{5}$, $\cos y° = \dfrac{5}{13}$

$\qquad \sin x° = \dfrac{3}{5}$, $\sin y° = \dfrac{12}{13}$

So $\cos \hat{ABC} = \cos(x + y)°$

$\qquad\qquad = \cos x° \cos y° - \sin x° \sin y°$

$\qquad\qquad = \left(\dfrac{4}{5} \times \dfrac{5}{13}\right) - \left(\dfrac{3}{5} \times \dfrac{12}{13}\right)$

$\qquad\qquad = \dfrac{20}{65} - \dfrac{36}{65}$

$\qquad\qquad = -\dfrac{16}{65}$

For questions of this type it is worth remembering the Pythagorean triples (and multiples thereof) such as:

$\qquad$ 3, 4, 5
$\qquad$ 5, 12, 13
$\qquad$ 8, 15, 17
$\qquad$ 7, 24, 25

These identities can also be employed to solve equations, as in the following examples.

Example

Solve $\cos \theta = \sin\left(\theta + \dfrac{\pi}{3}\right)$ for $0 \le \theta < 2\pi$.

$\qquad\qquad \cos \theta = \sin \theta \cos \dfrac{\pi}{3} + \cos \theta \sin \dfrac{\pi}{3}$

$\qquad \Rightarrow \cos \theta = \dfrac{1}{2}\sin \theta + \dfrac{\sqrt{3}}{2}\cos \theta$

$\Rightarrow \dfrac{2 - \sqrt{3}}{2}\cos \theta = \dfrac{1}{2}\sin \theta$

$\qquad \Rightarrow \tan \theta = 2 - \sqrt{3}$

$\qquad\qquad \Rightarrow \theta = 0.262,\ 3.40$

The difference between this type of question and an identity should be noted. An identity holds true for all values of θ, whereas this equation is true only for certain values of θ.

Example

Solve $\cos(45 - x)° = \sin(30 + x)°$ for $0° \le x° < 360°$.

$\cos 45° \cos x° + \sin 45° \sin x° = \sin 30° \cos x° + \cos 30° \sin x°$

$\qquad \Rightarrow \dfrac{1}{\sqrt{2}}\cos x° + \dfrac{1}{\sqrt{2}}\sin x° = \dfrac{1}{2}\cos x° + \dfrac{\sqrt{3}}{2}\sin x°$

$\qquad\qquad \Rightarrow \dfrac{\sqrt{2} - 1}{2}\cos x° = \dfrac{\sqrt{3} - \sqrt{2}}{2}\sin x°$

$\qquad\qquad\qquad \Rightarrow \tan x° = \dfrac{\sqrt{2} - 1}{\sqrt{3} - \sqrt{2}}$

$\qquad\qquad\qquad\qquad \Rightarrow x° = 52.5°,\ 232.5°$

Exercise 2

1 By considering $75° = 30° + 45°$, find these.
   a $\sin 75°$ **b** $\cos 75°$ **c** $\tan 75°$

2 Find $\sin 15°$ by calculating
   a $\sin(60 - 45)°$ **b** $\sin(45 - 30)°$

3 Find the exact value of $\cos 105°$.

4 Find $\cos\dfrac{11\pi}{12}$ by using the fact that $\dfrac{11\pi}{12} = \dfrac{\pi}{4} + \dfrac{2\pi}{3}$.

5 Prove that $4\sin\left(\theta - \dfrac{\pi}{6}\right) \equiv 2\sqrt{3}\sin\theta - 2\cos\theta$.

6 Prove that $\sin(x + y)\sin(x - y) \equiv \sin^2 x - \sin^2 y$.

7 Prove that $\csc\left(\dfrac{\pi}{2} + \theta\right) \equiv \sec\theta$.

8 Simplify these.

   a $\cos 310° \cos 40° + \sin 310° \sin 40°$

   b $\sin\dfrac{\pi}{2}\cos\dfrac{11\pi}{6} + \sin\dfrac{11\pi}{6}\cos\dfrac{\pi}{2}$

9 A and B are acute angles as shown.

 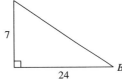

   Find these.
   a $\sin(A + B)$ **b** $\cos(A - B)$ **c** $\tan(B - A)$

10 In the diagram below, find $\cos P\hat{Q}R$.

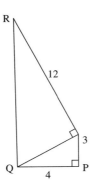

11 In the diagram below, find $\sin A\hat{B}C$.

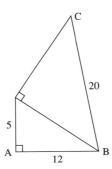

12 In the diagram below, find $\tan P\hat{Q}R$.

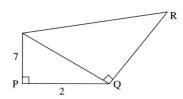

13 A is acute, B is obtuse, $\sin A = \dfrac{3}{7}$ and $\sin B = \dfrac{2}{3}$. Without using a calculator, find the possible values of $\sin(A + B)$ and $\cos(A + B)$.

14 Solve $\cos(45 - x)° = \sin(30 + x)°$ for $0° \leq x° < 360°$.

15 Show that $\cos\left(\dfrac{\pi}{4} + \theta\right) - \sin\left(\dfrac{\pi}{4} - \theta\right) = 0$.

16 The gradients of OP and OQ are $\dfrac{1}{3}$ and 3 respectively.

Find $\cos P\hat{O}Q$.

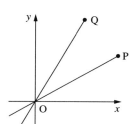

17

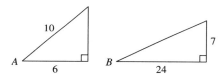

For the triangles above, and by considering $2A = A + A$, find these.
a $\sin 2A$ **b** $\cos 2A$ **c** $\sin 2B$ **d** $\cos 2B$

18 Solve the following equations for $0° \leq x° < 360°$.
a $\sin(x + 30)° = 2\cos x°$
b $\cos(x + 45)° = \sin(x + 45)°$
c $6\sin x° = \cos(x + 30)°$

19 Solve the following equations for $0 \leq \theta < 2\pi$.

a $\sin\theta + \cos\theta = \cos\left(\theta - \dfrac{\pi}{6}\right)$

b $\tan\left(\theta + \dfrac{\pi}{4}\right) = \sin\dfrac{\pi}{3}$

c $\cos\left(x + \dfrac{2\pi}{3}\right) = \sin\left(x + \dfrac{3\pi}{4}\right)$

7.3 Double angle formulae

It is useful to consider the special cases of addition formulae that are the double angle formulae.

$$\begin{aligned}
\sin 2A &= \sin(A + A) \\
&= \sin A \cos A + \cos A \sin A \\
&= 2 \sin A \cos A
\end{aligned}$$

$$\begin{aligned}
\cos 2A &= \cos(A + A) \\
&= \cos A \cos A - \sin A \sin A \\
&= \cos^2 A - \sin^2 A
\end{aligned}$$

$$\begin{aligned}
\tan 2A &= \tan(A + A) \\
&= \frac{\tan A + \tan A}{1 - \tan A \tan A} \\
&= \frac{2 \tan A}{1 - \tan^2 A}
\end{aligned}$$

It is often useful to rearrange $\cos 2A$ by using the identity $\sin^2 A + \cos^2 A = 1$.

$$\begin{aligned}
\text{So } \cos 2A &= \cos^2 A - \sin^2 A \\
&= 1 - \sin^2 A - \sin^2 A \\
&= 1 - 2 \sin^2 A
\end{aligned}$$

$$\begin{aligned}
\text{Or } \cos 2A &= \cos^2 A - \sin^2 A \\
&= \cos^2 A - (1 - \cos^2 A) \\
&= 2 \cos^2 A - 1
\end{aligned}$$

In this chapter we have met the Pythagorean identities, compound angle formulae and double angle formulae. These are all summarized below.

Pythagorean identities

$$\sin^2 \theta + \cos^2 \theta \equiv 1$$
$$\tan^2 \theta + 1 \equiv \sec^2 \theta$$
$$1 + \cot^2 \theta \equiv \csc^2 \theta$$

Compound angle formulae

$$\sin(A + B) \equiv \sin A \cos B + \cos A \sin B$$
$$\sin(A - B) \equiv \sin A \cos B - \cos A \sin B$$
$$\cos(A + B) \equiv \cos A \cos B - \sin A \sin B$$
$$\cos(A - B) \equiv \cos A \cos B + \sin A \sin B$$
$$\tan(A + B) \equiv \frac{\tan A + \tan B}{1 - \tan A \tan B}$$
$$\tan(A - B) \equiv \frac{\tan A - \tan B}{1 + \tan A \tan B}$$

Double angle formulae

$$\sin 2A = 2 \sin A \cos A$$
$$\begin{aligned}
\cos 2A &= \cos^2 A - \sin^2 A \\
&= 2 \cos^2 A - 1 \\
&= 1 - 2 \sin^2 A
\end{aligned}$$
$$\tan 2A = \frac{2 \tan A}{1 - \tan^2 A}$$

Example

By considering $\dfrac{\pi}{2}$ as $2\left(\dfrac{\pi}{4}\right)$, use the double angle formulae to find these.

(a) $\sin\dfrac{\pi}{2}$ (b) $\cos\dfrac{\pi}{2}$ (c) $\tan\dfrac{\pi}{2}$

(a) $\sin\dfrac{\pi}{2} = 2\sin\dfrac{\pi}{4}\cos\dfrac{\pi}{4}$

$\qquad\quad = 2 \times \dfrac{1}{\sqrt{2}} \times \dfrac{1}{\sqrt{2}}$

$\qquad\quad = 2 \times \dfrac{1}{2}$

$\qquad\quad = 1$

(b) $\cos\dfrac{\pi}{2} = \cos^2\dfrac{\pi}{4} - \sin^2\dfrac{\pi}{4}$

$\qquad\quad = \left(\dfrac{1}{\sqrt{2}}\right)^2 - \left(\dfrac{1}{\sqrt{2}}\right)^2$

$\qquad\quad = 0$

(c) $\tan\dfrac{\pi}{2} = \dfrac{2\tan\dfrac{\pi}{4}}{1 - \tan^2\dfrac{\pi}{4}}$

$\qquad\quad = \dfrac{2}{1 - 1}$

$\qquad\quad = \dfrac{2}{0}$

$\qquad\quad = \infty$

Example

Find $\cos 2\theta$.

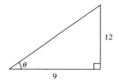

Clearly $\cos\theta = \dfrac{9}{15}$

So $\cos 2\theta = 2\cos^2\theta - 1$

$\qquad\quad = 2\left(\dfrac{9}{15}\right)^2 - 1$

$\qquad\quad = \dfrac{162}{225} - 1$

$\qquad\quad = -\dfrac{63}{225}$

Example

Find an expression for $\sin 3\theta$ in terms of $\sin \theta$.

$$\begin{aligned}
\sin 3\theta &= \sin(2\theta + \theta) \\
&= \sin 2\theta \cos \theta + \cos 2\theta \sin \theta \\
&= 2 \sin \theta \cos \theta \cos \theta + (1 - 2 \sin^2 \theta) \sin \theta \\
&= 2 \sin \theta \cos^2 \theta + \sin \theta - 2 \sin^3 \theta \\
&= 2 \sin \theta(1 - \sin^2 \theta) + \sin \theta - 2 \sin^3 \theta \\
&= 2 \sin \theta - 2 \sin^3 \theta + \sin \theta - 2 \sin^3 \theta \\
&= 3 \sin \theta - 4 \sin^3 \theta
\end{aligned}$$

Example

Find an expression for $\cos 4\theta$ in terms of $\cos \theta$.

Here we can use the double angle formula – remember that the double angle formulae do not apply only to 2θ; they work for any angle that is twice the size of another angle.

$$\begin{aligned}
\cos 4\theta &= 2 \cos^2 2\theta - 1 \\
&= 2(2 \cos^2 \theta - 1)^2 - 1 \\
&= 2(4 \cos^4 \theta - 4 \cos^2 \theta + 1) - 1 \\
&= 8 \cos^4 \theta - 8 \cos^2 \theta + 1
\end{aligned}$$

Example

Prove the identity $\dfrac{1 - \tan^2 \theta}{1 + \tan^2 \theta} \equiv \cos 2\theta$.

$$\begin{aligned}
\text{LHS} &= \frac{1 - \tan^2 \theta}{1 + \tan^2 \theta} \\
&\equiv \frac{1 - \tan^2 \theta}{\sec^2 \theta} \\
&\equiv \cos^2 \theta\left(1 - \frac{\sin^2 \theta}{\cos^2 \theta}\right) \\
&\equiv \cos^2 \theta - \sin^2 \theta \\
&\equiv \cos 2\theta = \text{RHS}
\end{aligned}$$

Example

Prove the identity $\tan 3\alpha + \tan \alpha \equiv \dfrac{\sin 4\alpha}{\cos 3\alpha \cos \alpha}$.

In this example, it is probably easiest to begin with the right-hand side and show that it is identical to the left-hand side. Although it involves what appears to be a double angle (4α), in fact it is not useful to use the double angle formulae.

To recognize this, it is important to look at the other side of the identity and realize what the goal is.

$$RHS = \frac{\sin 4\alpha}{\cos 3\alpha \cos \alpha}$$

$$\equiv \frac{\sin(3\alpha + \alpha)}{\cos 3\alpha \cos \alpha}$$

$$= \frac{\sin 3\alpha \cos \alpha + \cos 3\alpha \sin \alpha}{\cos 3\alpha \cos \alpha}$$

$$= \frac{\sin 3\alpha \cos \alpha}{\cos 3\alpha \cos \alpha} + \frac{\cos 3\alpha \sin \alpha}{\cos 3\alpha \cos \alpha}$$

$$= \tan 3\alpha + \tan \alpha = LHS$$

Exercise 3

1 Use the double angle formulae to find these.

a $\sin\dfrac{2\pi}{3}$ **b** $\cos\dfrac{5\pi}{3}$ **c** $\tan\dfrac{2\pi}{3}$

2 Find $\sin 2\theta$.

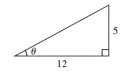

3 Find $\cos 2\theta$.

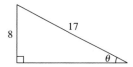

4 Given that θ is an acute angle with $\tan \theta = \dfrac{1}{2}$, calculate the exact value of these.

a $\sin \theta$ **b** $\cos \theta$

5 Find an expression for $\cos 5\theta$ in terms of $\sin \theta$ and $\cos \theta$.

6 Find an expression for $\sin 6\theta$ in terms of $\sin \theta$.

7 Find the exact value of $\cos 2\theta$ given that $\sin \theta = \dfrac{12}{13}$ (θ is not acute).

8 Prove that $\tan \theta \equiv \dfrac{1 - \cos 2\theta}{\sin 2\theta}$.

9 Prove that $\dfrac{1 + \cos 2y}{\sin 2y} \equiv \dfrac{\sin 2y}{1 - \cos 2y}$.

10 Prove that $\cos \phi + \sin \phi \equiv \dfrac{\cos 2\phi}{\cos \phi - \sin \phi}$.

11 Show that $\cos 3\theta - \sin 4\theta = 4\cos^3 \theta(1 - 2\sin \theta) + \cos \theta(4\sin \theta - 3)$.

12 Prove that $\dfrac{\cos A + \sin A}{\cos A - \sin A} \equiv \sec 2A + \tan 2A$.

13 Prove that $\sin 2\alpha \equiv \dfrac{2\tan \alpha}{1 + \tan^2 \alpha}$.

7.4 Using double angle formulae

Half angle formulae

It is useful to rearrange the double angle formulae to obtain formulae for a half angle. We have seen that double angle formulae can be applied to different angles.

So we can find expressions for $\cos\frac{1}{2}\theta$ and $\sin\frac{1}{2}\theta$.

These formulae are particularly applied when integrating trigonometric functions (see Chapter 15).

$\cos 2\theta = 2\cos^2\theta - 1$ can be rearranged to

$2\cos^2\theta = \cos 2\theta + 1$

$\Rightarrow \cos^2\theta = \frac{1}{2}(\cos 2\theta + 1)$

$\Rightarrow \cos\theta = \pm\sqrt{\frac{1}{2}(\cos 2\theta + 1)}$

$$\text{So } \cos\frac{\theta}{2} = \pm\sqrt{\frac{1}{2}(\cos\theta + 1)}$$

Similarly

$1 - 2\sin^2\theta = \cos 2\theta$

$\Rightarrow 2\sin^2\theta = 1 - \cos 2\theta$

$\Rightarrow \sin^2\theta = \frac{1}{2}(1 - \cos 2\theta)$

$\Rightarrow \sin\theta = \pm\sqrt{\frac{1}{2}(1 - \cos 2\theta)}$

$$\text{So } \sin\frac{\theta}{2} = \pm\sqrt{\frac{1}{2}(1 - \cos\theta)}$$

Example

Find the exact value of $\cos 15°$.

$\cos 15° = \sqrt{\frac{1}{2}(\cos 30° + 1)}$

$= \sqrt{\frac{1}{2}\left(\frac{\sqrt{3}}{2} + 1\right)}$

$= \sqrt{\frac{1}{2}\left(\frac{2 + \sqrt{3}}{2}\right)}$

$= \sqrt{\frac{2 + \sqrt{3}}{4}}$

$= \frac{\sqrt{2 + \sqrt{3}}}{2}$

Trigonometric equations involving double angles

We covered basic equations involving double angles that can be solved without using the double angle formulae in Chapter 1.

Example

Solve $\sin 2x° = \dfrac{\sqrt{3}}{2}$ for $0° \leq x° < 360°$.

$\sin 2x° = \dfrac{\sqrt{3}}{2}$

$\Rightarrow 2x° = 60°, 120°$

$\quad \Rightarrow x° = 30°, 60°, 210°, 240°$

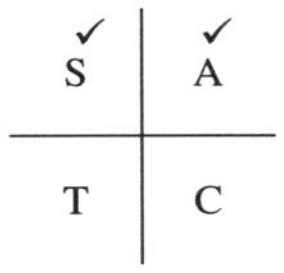

$\begin{array}{c|c} \overset{\checkmark}{S} & \overset{\checkmark}{A} \\ \hline T & C \end{array}$

However, if there is another trigonometric term involved (and we cannot use a calculator to solve the equation), then factorisation methods need to be employed.

Example

Solve $\sin 2\theta - \sin \theta = 0$ for $0 \leq \theta < 2\pi$.

$$\sin 2\theta - \sin \theta = 0$$
$$\Rightarrow 2 \sin \theta \cos \theta - \sin \theta = 0$$
$$\Rightarrow \sin \theta (2 \cos \theta - 1) = 0$$
$$\Rightarrow \sin \theta = 0 \text{ or } 2 \cos \theta - 1 = 0$$
$$\Rightarrow \theta = 0, \pi \quad \Rightarrow \cos \theta = \frac{1}{2}$$
$$\Rightarrow \theta = \frac{\pi}{3}, \frac{5\pi}{3}$$

So $\theta = 0, \dfrac{\pi}{3}, \pi$ or $\dfrac{5\pi}{3}$.

Example

Solve $\cos 2x° - 5 \cos x° = 2$ for $0° \leq x° < 360°$.

$$\cos 2x° - 5 \cos x° = 2$$
$$\Rightarrow 2 \cos^2 x° - 1 - 5 \cos x° = 2$$
$$\Rightarrow 2 \cos^2 x° - 5 \cos x° - 3 = 0$$
$$\Rightarrow (2 \cos x° + 1)(\cos x° - 3) = 0$$
$$\Rightarrow 2 \cos x° + 1 = 0 \text{ or } \cos x° - 3 = 0$$
$$\Rightarrow \cos x° = -\frac{1}{2} \text{ or } \cos x° = 3$$
$$\Rightarrow x° = 120°, 240°$$

$\cos x° = 3$ has no solution.

Example

Solve $3 \cos 2\theta + 11 \sin \theta = -4$ for $0 \le \theta < 2\pi$.

$$\Rightarrow 3(1 - 2 \sin^2 \theta) + 11 \sin \theta + 4 = 0$$
$$\Rightarrow 3 - 6 \sin^2 \theta + 11 \sin \theta + 4 = 0$$
$$\Rightarrow 6 \sin^2 \theta - 11 \sin \theta - 7 = 0$$
$$\Rightarrow (3 \sin \theta - 7)(2 \sin \theta + 1) = 0$$

$$\Rightarrow \sin \theta = \frac{7}{3} \text{ or } \sin \theta = -\frac{1}{2}$$
$$\Rightarrow \theta = \frac{7\pi}{6}, \frac{11\pi}{6}$$

> The form of $\cos 2\theta$ required is determined by the other term in the equation (sin or cos).

> $\sin \theta = \frac{7}{3}$ has no solution.

An equation of this type is often solved using a calculator (when it is available). In fact, in some cases this is the only appropriate method.

Example

Solve $\sin 3x° + \cos 2x° = 1$ for $0° \le x° < 180°$.

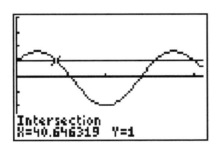

Intersection
X=40.646319 Y=1

$x° = 0°, 40.6°, 139°$

Exercise 4

1 Prove $\tan\left(\dfrac{\theta}{2}\right) \equiv \csc \theta - \cot \theta$.

2 Prove $2 \sin^2 \dfrac{\theta}{2} \equiv 1 - \cos \theta$.

3 Find the value of $\sin 75°$, using a half angle formula.

4 Find the exact value of $\cos \dfrac{\pi}{8}$, using a half angle formula.

5 Solve these equations for $0 \le \theta < 2\pi$.

  a $\sin 2\theta = \dfrac{\sqrt{3}}{2}$ **b** $6 \cos 2\theta + 1 = 4$

6 Solve these equations for $0° \le x° < 360°$.

  a $\sin 2x° - \cos x° = 0$ **b** $\sin 2x° - 4 \sin x° = 0$
  c $\cos 2x° - \cos x° + 1 = 0$ **d** $\cos 2x° - 4 \sin x° + 5 = 0$
  e $\cos 2x° + 5 \cos x° - 2 = 0$ **f** $\cos 2x° + 3 \cos x° - 1 = 0$
  g $2 \cos 2x° + \cos x° - 1 = 0$ **h** $\cos 2 = 7 \sin x° + 4$

7 Solve these equations for $0 \le \theta < 2\pi$.

  a $\cos 2\theta + \cos \theta = 0$ **b** $\sin 2\theta + \sin \theta = 0$

  c $2\cos 2\theta + 1 = 0$ **d** $\sin 2\theta - \cos \theta = 0$

  e $\cos 2\theta = \cos \theta$ **f** $\cos 2\theta = 4\cos \theta + 5$

  g $\cos 2\theta = \sin \theta + 1$

8 Solve these equations for $-\pi \le \theta < \pi$.

  a $\sin 2\theta = 2\sin \theta$ **b** $\cos 2\theta = 1 - 3\cos \theta$

  c $2\cos \theta + 2 = \cos 2\theta + 4$ **d** $\cos 2\theta + 3\cos \theta + 2 = 0$

9 Solve these equations for $0 \le \theta < 2\pi$. Give your answers to 3 sf.

  a $6\cos 2\theta - 5\cos \theta + 4 = 0$ **b** $2\cos 2\theta - 3\sin \theta + 1 = 0$

7.5 Wave function

This refers to functions of the type $f(x) = a\cos x° + b\sin x°$, where there are both sine and cosine terms in one function.

What does the graph of this type of function look like?

This is the graph of $y = 2\cos x° + 3\sin x°$:

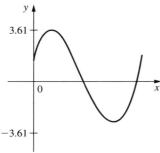

It appears to have all of the properties of a single trigonometric function, in that it is a periodic wave with symmetrical features. As its maximum and minimum values are numerically the same, we can conclude that there is a stretch factor involved. As the graph begins neither on the x-axis nor at a maximum/minimum, there must be a horizontal shift.

This suggests a function of the form $k\cos(x - \alpha)°$.

We can check this by finding k and $\alpha°$.

> Although this part of trigonometry is not explicitly stated as part of the IB HL syllabus, it is really an application of compound angle formulae and is worth studying.

Example

Express $f(x) = 2\cos x° + 3\sin x°$ in the form $k\cos(x - \alpha)°$.

We know that $k\cos(x - \alpha)° = k\cos x° \cos \alpha° + k\sin x° \sin \alpha°$, so for f(x) to be expressed in this form

$2\cos x° + 3\sin x° = k\cos x° \cos \alpha° + k\sin x° \sin \alpha°$

Comparing the two sides, we can conclude that

$k\sin \alpha° = 3$

$k\cos \alpha° = 2$

By squaring and adding, we can find k:

$k^2 \sin^2 \alpha° + k^2 \cos^2 \alpha° = 3^2 + 2^2$

$\Rightarrow k^2(\sin^2 \alpha° + \cos^2 \alpha°) = 13$

$\Rightarrow k^2 = 13$

$\Rightarrow k = \sqrt{13}$

By dividing, we can find $\alpha°$:

$$\frac{k \sin \alpha°}{k \cos \alpha°} = \tan \alpha° = \frac{3}{2}$$

$$\Rightarrow \alpha° = \tan^{-1}\frac{3}{2}$$

$$\Rightarrow \alpha° = 56.3°$$

It is vital to consider which quadrant $\alpha°$ lies in. This is best achieved using the diagram.

S	A
T	C

In this case, we know that

$k \sin \alpha° = +$

$k \cos \alpha° = +$

S ✓	A ✓✓
T	C ✓

Since $k > 0$ (always), we can see that the quadrant with two ticks in it is the first quadrant.

So here $\alpha°$ is acute.

Hence $f(x) = 2 \cos x° + 3 \sin x° = \sqrt{13} \cos(x - 56.3)°$.

Method for wave function

To express a function of the form $f(\theta) = a \sin \theta + b \cos \theta$ as a single trigonometric function:

1. Expand the desired form using the compound formula (if no form is given, use $k \cos(x - \alpha)°$).
2. Compare the two sides to find $k \sin \alpha°$ and $k \cos \alpha°$ (write them in this order).
3. Square and add to find k^2.
4. Divide to find $\tan \alpha°$.
5. Use the positivity diagram to find $\alpha°$.

Example

Express $3 \cos x° - 4 \sin x°$ as $k \cos(x - \alpha)°$, $k > 0$, $0° \leq \alpha° < 360°$.

Let $3 \cos x° - 4 \sin x° = k \cos(x - \alpha)°$
$$= k \cos x° \cos \alpha° + k \sin x° \sin \alpha°$$

So $k \sin \alpha° = -4$

$\quad k \cos \alpha° = 3$

Squaring and adding:

$$k^2 \sin^2 \alpha° + k^2 \cos^2 \alpha° = (-4)^2 + 3^2$$

$$k^2 = 25$$

$$k = 5$$

Dividing:

$$\frac{k \sin \alpha°}{k \cos \alpha°} = \tan \alpha° = -\frac{4}{3}$$

$$\tan^{-1}\left(\frac{4}{3}\right) = 53.1°$$

S	A ✓
T ✓	C ✓✓

So $\alpha° = 360° - 53.1°$

$\quad\quad = 306.9°$

Hence $3 \cos x° - 4 \sin x° = 5 \cos(x - 306.9)°$

It is not always the form $k\cos(x - \alpha)°$ that is required, but the method is precisely the same.

Example

Express $\cos\theta + \sqrt{3}\sin\theta$ as $k\sin(\theta - \alpha)$, $k > 0$, $0 \le \alpha < 2\pi$.

Let $\cos\theta + \sqrt{3}\sin\theta = k\sin(\theta - \alpha)$
$$= k\sin\theta\cos\alpha - k\cos\theta\sin\alpha$$

So $-k\sin\alpha = 1$
$\Rightarrow k\sin\alpha = -1$
and $k\cos\alpha = \sqrt{3}$

Squaring and adding:
$$k^2\sin^2\alpha + k^2\cos^2\alpha = (-1)^2 + \sqrt{3}^2$$
$$k^2 = 4$$
$$k = 2$$

Dividing:
$$\frac{k\sin\alpha}{k\cos\alpha} = \tan\alpha = -\frac{1}{\sqrt{3}}$$

$$\tan^{-1}\left(\frac{1}{\sqrt{3}}\right) = \frac{\pi}{6}$$

S	A ✓
T ✓	C ✓✓

So $\alpha = 2\pi - \dfrac{\pi}{6}$

$\quad = \dfrac{11\pi}{6}$

Hence $\cos\theta + \sqrt{3}\sin\theta = 2\sin\left(\theta - \dfrac{11\pi}{6}\right)$

This method works in the same way for functions involving a multiple angle (as long as both the sine and cosine parts have the same multiple angle).

Example

Find the maximum value of $f(\theta) = \cos 2\theta + \sin 2\theta$ and the smallest possible positive value of θ where this occurs.

First, express this as a single trigonometric function. Any of the four forms can be chosen, but the simplest is generally $k\cos(2\theta - \alpha)$.

Let $f(\theta) = 1\cos 2\theta + 1\sin 2\theta = k\cos(2\theta - \alpha)$
$$= k\cos 2\theta\cos\alpha + k\sin 2\theta\sin\alpha$$

So $k\sin\alpha = 1$
$\quad k\cos\alpha = 1$

Squaring and adding:

$$k^2 \sin^2 \alpha + k^2 \cos^2 \alpha = 1^2 + 1^2$$
$$k^2 = 2$$
$$k = \sqrt{2}$$

Dividing:

$$\frac{k \sin \alpha}{k \cos \alpha} = \tan \alpha = \frac{1}{1} = 1$$

$$\tan^{-1}(1) = \frac{\pi}{4}$$

So $\alpha = \dfrac{\pi}{4}$

Hence $\cos 2\theta + \sin 2\theta = \sqrt{2} \cos\left(2\theta - \dfrac{\pi}{4}\right)$

The maximum value of $f(\theta)$ is $\sqrt{2}$. This normally occurs when $\theta = 0$ for $\cos \theta$ and so here

$$2\theta - \frac{\pi}{4} = 0$$
$$\Rightarrow 2\theta = \frac{\pi}{4}$$
$$\Rightarrow \theta = \frac{\pi}{8}$$

There will be an infinite number of maximum points but this is the first one with a positive value of θ.

This method can also be used to solve equations (if no calculator is available).

Example

Solve $\sqrt{3} \cos 2\theta - \sin 2\theta - 1 = 0$ by first expressing $\sqrt{3} \cos 2\theta - \sin 2\theta$ in the form $k \cos(2\theta - \alpha)$, $0 \le \theta \le 2\pi$

Let $\sqrt{3} \cos 2\theta - \sin 2\theta = k \cos(2\theta - \alpha)$
$$= k \cos 2\theta \cos \alpha + k \sin 2\theta \sin \alpha$$

So $k \sin \alpha = -1$
   $k \cos \alpha = \sqrt{3}$

Squaring and adding:

$$k^2 \sin^2 \alpha + k^2 \cos^2 \alpha = (-1)^2 + \sqrt{3}^2$$
$$k^2 = 4$$
$$k = 2$$

Dividing:

$$\frac{k \sin \alpha}{k \cos \alpha} = \tan \alpha = \frac{-1}{\sqrt{3}}$$

$$\tan^{-1}\left(\frac{1}{\sqrt{3}}\right) = \frac{\pi}{6}$$

So $\alpha = 2\pi - \dfrac{\pi}{6}$

$\quad = \dfrac{11\pi}{6}$

Hence $\sqrt{3}\cos 2\theta - \sin 2\theta - 1 = 2\cos\left(2\theta - \dfrac{11\pi}{6}\right) - 1$

The equation becomes

$2\cos\left(2\theta - \dfrac{11\pi}{6}\right) - 1 = 0$

$\Rightarrow \cos\left(2\theta - \dfrac{11\pi}{6}\right) = \dfrac{1}{2}$

$\Rightarrow 2\theta - \dfrac{11\pi}{6} = \dfrac{\pi}{3}, \dfrac{5\pi}{3}$

$\Rightarrow 2\theta = \dfrac{13\pi}{6}, \dfrac{21\pi}{6}$

$\Rightarrow \theta = \dfrac{13\pi}{12}, \dfrac{21\pi}{12}$

There will be two more solutions one period away (π radians).

So $\theta = \dfrac{\pi}{12}, \dfrac{9\pi}{12}, \dfrac{13\pi}{12}, \dfrac{21\pi}{12}$

Exercise 5

1 Express each of these in the form $k\cos(x - \alpha)°$, where $k > 0$, $0° \le \alpha° < 360°$.

  a $6\cos x° + 8\sin x°$ **b** $5\cos x° + 12\sin x°$

  c $\cos x° - 3\sin x°$ **d** $\sin x° - 2\cos x°$

2 Express each of these in the form $k\cos(\theta - \alpha)$, where $k > 0$, $0 \le \alpha < 2\pi$.

  a $\sqrt{3}\cos\theta - \sin\theta$ **b** $\cos\theta - \sin\theta$

  c $-\cos\theta - 2\sin\theta$ **d** $\sqrt{3}\sin\theta - \cos\theta$

3 Express each of these in the form $k\cos(x + \alpha)°$, where $k > 0$, $0° \le \alpha° < 360°$.

  a $15\cos x° - 8\sin x°$ **b** $2.5\cos x° - 3.5\sin x°$

4 Express each of these in the form $k\sin(\theta + \alpha)$, where $k > 0$, $0 \le \alpha < 2\pi$.

  a $\sqrt{3}\cos\theta - \sin\theta$ **b** $\cos\theta - \sin\theta$

5 Express each of these in the form $k\sin(x - \alpha)°$, where $k > 0$, $0° \le \alpha° < 360°$.

  a $-\sin x° - 3\cos x°$ **b** $\sqrt{3}\cos x° - \sin x°$

6 Express each of these as a single trigonometric function.

  a $\cos 2x° + \sin 2x°$ **b** $\cos 3x° - \sqrt{3}\sin 3x°$

  c $\sqrt{27}\cos\theta - 3\sin\theta$ **d** $\sin 30\theta - \cos 30\theta$

7 Without using a calculator, state the maximum and minimum values of each function, and the corresponding values of $x°$, for $0° \le x° < 360°$.

  a $f(x) = 5\cos x° - 5\sin x°$

  b $f(x) = \sqrt{3}\cos x° - \sin x° + 5$

  c $f(x) = 7\sin 2x° - 7\cos 2x° + 1$

8 State the maximum and minimum values of each function, and the corresponding values of θ, for $0 \leq \theta < 2\pi$.

  a $f(\theta) = \sqrt{8} \sin \theta - 2 \cos \theta$

  b $f(\theta) = -\sqrt{3} \cos 3\theta - \sin 3\theta$

  c $f(\theta) = \sqrt{6} \sin 8\theta - \sqrt{2} \cos 8\theta - 3$

9 By expressing the left-hand side of each equation as a single trigonometric function, solve the equation for $0 \leq \theta < 2\pi$.

  a $\cos \theta - \sin \theta = -1$ **b** $\sqrt{3} \cos \theta - \sin \theta = \sqrt{3}$

  c $\cos \theta - \sin \theta = 1$ **d** $\cos 4\theta - \sqrt{3} \sin 4\theta = 1$

10 By expressing the left-hand side of each equation as a single trigonometric function, solve the equation for $0° \leq x° < 360°$.

  a $3 \sin x° - 5 \cos x° = 4$ **b** $8 \cos 3x° + 15 \sin 3x° = 13$

Review exercise

 1 If $\sin x° = \dfrac{1}{2}$, what are possible values for $\cos x°$?

 2 For the billiards shot shown in the diagram,

  a prove that

    i $\sin q° = \sin 2p°$

    ii $\cos q° = -\cos 2p°$

  b Find the value of

    i $\sin q°$ **ii** $\cos q°$

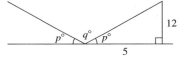

 3 Show that, for triangle KLM, $m = \dfrac{k \sin(\alpha + \beta)}{\sin \alpha}$.

 4 Prove that $\dfrac{1 + \sin 2\phi}{\cos 2\phi} \equiv \dfrac{\cos \phi + \sin \phi}{\cos \phi - \sin \phi}$.

 5 Prove that $2 \csc x \equiv \tan\left(\dfrac{x}{2}\right) + \cot\left(\dfrac{x}{2}\right)$.

 6 Solve the equation $\cos \theta + \cos 3\theta + \cos 5\theta = 0$ for $0 \leq \theta < \pi$.

 7 The function f is defined on the domain $[0, \pi]$ by $f(\theta) = 4 \cos \theta + 3 \sin \theta$.

  a Express $f(\theta)$ in the form $R \cos(\theta - \alpha)$ where $R > 0, 0 \leq \alpha < 2\pi$.

  b Hence, or otherwise, write down the value of θ for which $f(\theta)$ takes its maximum value.   [IB May 02 P1 Q12]

 8 Find all the values of θ in the interval $[0, \pi]$ that satisfy the equation $\cos \theta = \sin^2 \theta$.   [IB May 03 P1 Q2]

 9 Use the fact that $-15° = (45 - 60)°$ to find $\sin(-15)°$.

 10 In the following diagram, find $\cos(\alpha + \beta)$.

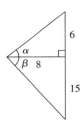

11 Find csc 22.5° by using a half angle formula.

 12 Find tan 2θ.

7

5

 13 Express $\sqrt{48}\cos\theta - 8\sin\theta$ in the form $k\sin(\theta - \alpha)$ where

$k > 0, 0 \le \alpha < 2\pi$.

14 Express $6\cos 8x° - 6\sin 8x°$ in the form $k\cos(8x - \alpha)°$. Hence solve

$6\cos 8x° - 6\sin 8x° + 7 = 1$ for $0° \le x° < 90°$.

15 Prove that $\tan 3\theta \equiv \dfrac{3\tan\theta - \tan^3\theta}{1 + 3\tan^2\theta}$.

16 K is the point with coordinates $\left(\sin\left(\theta + \dfrac{\pi}{6}\right), \cos\left(\theta - \dfrac{\pi}{6}\right)\right)$ and L has coordinates

$\left(\sin\left(\theta - \dfrac{\pi}{6}\right), \cos\left(\theta + \dfrac{\pi}{6}\right)\right)$. Find, in its simplest form, an expression for the

gradient of the line KL.

17 The rotor blade of a helicopter is modelled using the following diagram, where
TUVW is a square.

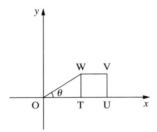

a Show that the area of OTUVW is

$$A = \sin^2\theta + \frac{1}{4}\sin 2\theta$$

b Express the area in the form $k\cos(2\theta - \alpha) + p$.

c Hence find the maximum area of the rotor blade, and the smallest positive value
of θ when this maximum occurs.

 8 Differential Calculus 1 – Introduction

The ideas that are the basis for calculus have been with us for a very long time. Between 450 BC and 225 BC, Greek mathematicians were working on problems that would find their absolute solution with the invention of calculus. However, the main developments were much more recent; it was not until the 16th century that major progress was made by mathematicians such as Fermat, Roberval and Cavalieri. In the 17th century, calculus as it is now known was developed by Sir Isaac Newton and Gottfried Wilhelm von Leibniz.

Sir Isaac Newton famously "discovered" gravity when an apple fell on his head.

Sir Isaac Newton   **Gottfried Wilhelm von Leibniz**

Consider the graph of a quadratic, cubic or trigonometric function.

Differential calculus is a branch of mathematics that is concerned with rate of change. In a graph, the rate of change is the gradient. Although linear functions have a constant gradient, most functions have changing gradients. Being able to find a pattern for the gradient of curves is the aim of differentiation. Differentiation is the process used to find rate of change.

The gradient of a straight line is constant.

For example, in the diagram below, the gradient $= 2$.

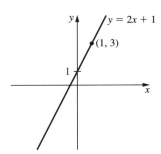

However, when a curve is considered, it is obvious that the gradient is constantly changing.

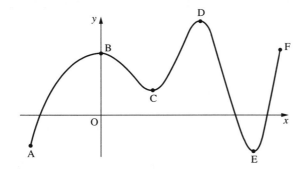

The sections AB, CD, EF have positive gradient (the function is increasing) and sections BC, DE have negative gradient (the function is decreasing). The question we need to answer is: how do we measure the gradient of a curve?

8.1 Differentiation by first principles

We know $\text{gradient} = \dfrac{\Delta y}{\Delta x}$, and one method of finding the gradient of a straight line is to use $\dfrac{y_2 - y_1}{x_2 - x_1}$.

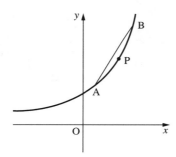

Consider the coordinates $(x, f(x))$ and $(x + h, f(x + h))$ – the gap between the x-coordinates is h. This can be used to find an approximation for the gradient at P as seen in the diagram.

$$\text{Gradient} = \frac{y_2 - y_1}{x_2 - x_1} = \frac{f(x + h) - f(x)}{x + h - x} = \frac{f(x + h) - f(x)}{h}$$

This is calculating the gradient of the chord AB shown in the diagram. As the chord becomes smaller, the end-points of the curve are getting closer together, and h becomes smaller. Obviously, this approximation becomes more accurate as h becomes smaller. Finally h becomes close to zero and the chord's length becomes so small that it can be considered to be the same as the point P.

The gradient of a function, known as the **derivative**, with notation $f'(x)$ is defined:

$$f'(x) = \lim_{h \to 0} \frac{f(x + h) - f(x)}{h}$$

The notation $\lim_{h \to 0}$ means the limit as h tends to zero. This is the value to which the expression converges as h becomes infinitesimally small.

The idea of a limit is similar to sum of a infinite series met in Chapter 6 and also to horizontal asymptotes in Chapter 3.

Example

Find the derivative of $f(x) = 3x^2$.

$f(x + h) = 3(x + h)^2 = 3x^2 + 6hx + 3h^2$

$f(x + h) - f(x) = 3x^2 + 6hx + 3h^2 - 3x^2 = 6hx + 3h^2$

$\dfrac{f(x + h) - f(x)}{h} = \dfrac{6hx + 3h^2}{h} = 6x + 3h$

So $f'(x) = \lim_{h \to 0} \dfrac{f(x + h) - f(x)}{h} = 6x$ (as $3h \to 0$)

Hence at any point on the curve, the gradient is given by $6x$.

This process is known as differentiation by first principles.

Example

Find the derivative of $f(x) = x^3$.

$f(x + h) = (x + h)^3 = x^3 + 3x^2h + 3xh^2 + h^3$

$f(x + h) - f(x) = x^3 + 3x^2h + 3xh^2 + h^3 - x^3 = 3x^2h + 3xh^2 + h^3$

$\dfrac{f(x + h) - f(x)}{h} = \dfrac{3x^2h + 3xh^2 + h^3}{h} = 3x^2 + 3xh + h^2$

So $f'(x) = \lim_{h \to 0} \dfrac{f(x + h) - f(x)}{h} = 3x^2$ (as $3xh + h^2 \to 0$)

Example

Find the derivative of $f(x) = -7x$.

$f(x + h) = -7(x + h) = -7x - 7h$

$f(x + h) - f(x) = -7x - 7h - (-7x) = -7h$

$\dfrac{f(x + h) - f(x)}{h} = \dfrac{-7h}{h} = -7$

So $f'(x) = \lim_{h \to 0} \dfrac{f(x + h) - f(x)}{h} = -7$

Example

Find the derivative of $f(x) = 5$.

$f(x + h) = 5$

$f(x + h) - f(x) = 5 - 5 = 0$

$\dfrac{f(x + h) - f(x)}{h} = \dfrac{0}{h} = 0$

So $f'(x) = \lim\limits_{h \to 0} \dfrac{f(x + h) - f(x)}{h} = 0$

What happens in a sum or difference of a set of functions? Consider the sum of the previous three examples, i.e. $f(x) = x^3 - 7x + 5$.

Example

Find the derivative of $f(x) = x^3 - 7x + 5$.

$f(x + h) = (x + h)^3 - 7(x + h) + 5$

$\qquad = x^3 + 3x^2h + 3xh^2 + h^3 - 7x - 7h + 5$

$f(x + h) - f(x) = x^3 + 3x^2h + 3xh^2 + h^3 - 7x - 7h + 5 - x^3 + 7x - 5$

$\qquad = 3x^2h + 3xh^2 + h^3 - 7h$

$\dfrac{f(x + h) - f(x)}{h} = \dfrac{3x^2h + 3xh^2 + h^3 - 7h}{h}$

$\qquad = 3x^2 + 3xh + h^2 - 7$

So $f'(x) = \lim\limits_{h \to 0} \dfrac{f(x + h) - f(x)}{h} = 3x^2 - 7$ (as $3xh + h^2 \to 0$)

This demonstrates that differentiation of a function containing a number of terms can be differentiated term by term.

Exercise 1

Find $f'(x)$ using the method of differentiation from first principles:

1 $f(x) = 5x$ **2** $f(x) = 8x$ **3** $f(x) = -2x$

4 $f(x) = x^2$ **5** $f(x) = x^3$ **6** $f(x) = x^4$

7 $f(x) = 2x^2$ **8** $f(x) = 5x^2$ **9** $f(x) = 4x^3$

10 $f(x) = 9$ **11** $f(x) = \dfrac{3}{x}$ **12** $f(x) = x^2 + 4$

13 $f(x) = 8 - 3x$ **14** $f(x) = x^2 - 4x + 9$ **15** $f(x) = 2x - \dfrac{1}{x}$

8.2 Differentiation using a rule

Looking at the patterns in Exercise 1, it should be obvious that for:

$$f(x) = ax^n$$
$$f'(x) = anx^{n-1}$$

Multiply by the power and subtract 1 from the power

This rule can be used to perform differentiation.

In particular, notice that $f(x) = ax$ gives $f'(x) = a$.

Unless specifically required, differentiation by first principles is not used – the above rule makes the process much shorter and easier.

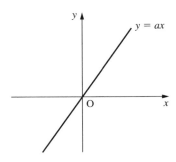

This is no surprise – the gradient of a linear function is constant.

Also, $f(x) = k$ gives $f'(x) = 0$.

The gradient of a horizontal line is zero.

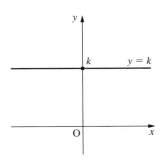

Differential calculus was developed by two mathematicians, Isaac Newton and Gottfried Leibniz. There are two commonly used notations:

Functional / Newtonian notation	Geometrical / Leibniz notation
$f(x) =$	$y =$
Derivative $f'(x) =$	$\dfrac{dy}{dx} =$

Either notation can be used, and both will appear in questions.

Example

Differentiate $y = 5x^3 - \dfrac{4}{x}$.

Simplifying, $y = 5x^3 - 4x^{-1}$

$\dfrac{dy}{dx} = 5 \cdot 3x^{(3-1)} - (-4)x^{(-1-1)}$

$\qquad = 15x^2 + 4x^{-2}$

$\qquad = 15x^2 + \dfrac{4}{x^2}$

As with first principles, we can differentiate a sum by differentiating term by term.

Example

Find the derivative of $f(x) = \dfrac{2}{\sqrt[3]{x}}$.

$f(x) = 2x^{-\frac{1}{3}}$

$f'(x) = -\dfrac{2}{3}x^{-\frac{4}{3}}$

Example

Differentiate $y = \dfrac{2x(x-5)}{\sqrt{x}}$.

$y = \dfrac{2x^2 - 10x}{x^{\frac{1}{2}}}$

$\quad = 2x^{\frac{3}{2}} - 10x^{\frac{1}{2}}$

$\dfrac{dy}{dx} = 2 \cdot \dfrac{3}{2}x^{\frac{1}{2}} - 10 \cdot \dfrac{1}{2}x^{-\frac{1}{2}}$

$\quad = 3x^{\frac{1}{2}} - 5x^{-\frac{1}{2}}$

Sometimes it is necessary to simplify the function before differentiating.

Example

Find $g'(4)$ for $g(x) = x^2 - x - 6$.

Here we are evaluating the derivative when $x = 4$.

First differentiate: $\qquad\qquad g'(x) = 2x - 1$

Then substitute $x = 4$: $\qquad\quad g'(4) = 2 \cdot 4 - 1 = 7$

So the gradient of $g(x)$ at $x = 4$ is 7.

Example

Find the coordinates of the points where the gradient is -2 for

$$f(x) = \frac{1}{3}x^3 - \frac{1}{2}x^2 - 8x + 7.$$

Here we are finding the points on the curve where the derivative is -2.

First differentiate: $f'(x) = x^2 - x - 8$

Then solve the equation $f'(x) = x^2 - x - 8 = -2$

$$\Rightarrow x^2 - x - 6 = 0$$
$$\Rightarrow (x + 2)(x - 3) = 0$$
$$\Rightarrow x = -2 \text{ or } \Rightarrow x = 3$$

At $x = -2$, $y = \frac{55}{3}$ and at $x = 3$, $y = -\frac{25}{2}$

So the coordinates required are $\left(-2, \frac{55}{3}\right)$ and $\left(3, -\frac{25}{2}\right)$.

Exercise 2

1 Differentiate these functions.

   a $f(x) = 9x^2$ **b** $f(x) = 10x^3$ **c** $f(x) = 6x^4$

   d $f(x) = -3x^5$ **e** $f(x) = 12$ **f** $f(x) = 7x$

   g $f(x) = 11x$ **h** $f(x) = 8x - 9$ **i** $f(x) = \dfrac{4}{x^2}$

   j $f(x) = 5\sqrt{x}$ **k** $y = x^2 + 5x + 6$ **l** $f(x) = \dfrac{5}{\sqrt{x^5}}$

   m $y = x^3 + 5x^2 - 7x - 4$ **n** $y = 6x^2 - \dfrac{2}{x}$ **o** $y = \sqrt[4]{x}$

   p $y = \sqrt[3]{x^5}$ **q** $y = \dfrac{4x(x^2 - 3)}{3x^2}$ **r** $y = \dfrac{3x^2(x^3 - 3)}{5\sqrt{x}}$

2 Find $f'(3)$ for $f(x) = x^2 - 4x + 9$.

3 Find $g'(6)$ for $g(x) = \dfrac{4 - x^2}{x}$.

4 Find the gradient of $y = x^3 - 6x + 9$ when $x = 2$.

5 Find the gradient of $y = \dfrac{4x^2 - 9}{\sqrt{x}}$ when $x = 16$.

6 Find the coordinates of the point where the gradient is 4 for
   $f(x) = x^2 - 6x + 12$.

7 Find the coordinates of the points where the gradient is 2 for
   $f(x) = \dfrac{2}{3}x^3 - \dfrac{9}{2}x^2 - 3x + 8$.

8.3 Gradient of a tangent

A tangent is a straight line that touches a curve (or circle) at one point.

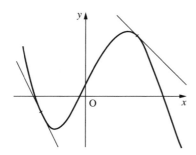

Differentiation can be used to find the value of the gradient at any particular point on the curve. At this instant the value of the gradient of the curve is the same as the gradient of the tangent to the curve at that point.

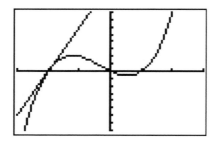

Finding the gradient using a graphing calculator

Using a graphing calculator, the value of the gradient at any point can be calculated.

For example, for $y = x^2 - x - 6$, at $x = -1$

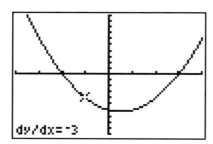

dy/dx=-3

This is helpful, especially for checking answers. However, we often need the derivative function and so need to differentiate by hand. The calculator can only find the gradient using a numerical process and is unable to differentiate algebraically.

Tangents and normals

The gradient at a point is the same as the gradient of the tangent to the curve at that point. Often it is necessary to find the equation of the tangent to the curve.

Method for finding the equation of a tangent

1. Differentiate the function.
2. Substitute the required value to find the gradient.
3. Find the y-coordinate (if not given).
4. Find the equation of the tangent using this gradient and the point of contact using $y - y_1 = m(x - x_1)$.

Example

Find the equation of the tangent to $y = x^2 - x - 6$ at $x = -1$.

Differentiating,

$\dfrac{dy}{dx} = 2x - 1$ and so at $x = -1$, $\dfrac{dy}{dx} = 2 \times (-1) - 1 = -3$

The point of contact is when $x = -1$, and so $y = (-1)^2 - (-1) - 6 = -4$,
i.e. $(-1, -4)$

Using $y - y_1 = m(x - x_1)$, the equation of the tangent is

$y - (-4) = -3 \, (x - -1) \Rightarrow y = -3x - 7$

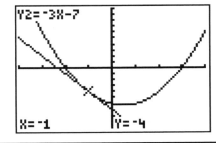

The normal to a curve is also a straight line. The normal to the curve is perpendicular to the curve at the point of contact (therefore it is perpendicular to the tangent).

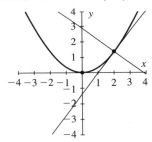

Finding the equation of a normal to a curve is a very similar process to finding the tangent.

Method for finding the equation of a normal

1. Differentiate the function.
2. Substitute the required value to find the gradient.
3. Find the gradient of the perpendicular using $m_1 m_2 = -1$.
4. Find the y-coordinate (if not given).
5. Find the equation of the normal using this gradient and the point of contact using $y - y_1 = m(x - x_1)$.

Example

Find the equation of the tangent, and the equation of the normal, to
$y = x^2 - 9x - 12$ at $x = 3$.

Using the method, $\dfrac{dy}{dx} = 2x - 9$.

At $x = 3$, $\dfrac{dy}{dx} = 2(3) - 9 = -3$.

At $x = 3$, $y = 3^2 - 9(3) - 12 = -30$.

So the equation of the tangent is $y + 30 = -3(x - 3)$

$$\Rightarrow y = -3x - 21$$

The equation of the normal uses the same point but the gradient is different.

Using $m_1m_2 = -1$, the gradient of the normal is $\frac{1}{3}$.

Using $y - y_1 = m(x - x_1)$ the equation of the normal is $y + 30 = \frac{1}{3}(x - 3)$

$$\Rightarrow y = \frac{1}{3}x - 31$$

> To find the perpendicular gradient turn the fraction upside down and change the sign.

Example

Find the equation of the tangent, and the equation of the normal, to $y = x^3 - 1$ where the curve crosses the x-axis.

The curve crosses the x-axis when $y = 0$. So $x^3 = 1 \Rightarrow x = 1$, i.e. $(1, 0)$

Differentiating, $\frac{dy}{dx} = 3x^2$.

At $x = 1$, $\frac{dy}{dx} = 3(1)^2 = 3$.

Using $y - y_1 = m(x - x_1)$, the equation of the tangent is
$y = 3(x - 1) = 3x - 3$.

Then the gradient of the normal will be $-\frac{1}{3}$.

Using $y - y_1 = m(x - x_1)$ the equation of the normal is $y = -\frac{1}{3}(x - 1)$

$$\Rightarrow y = -\frac{1}{3}x + \frac{1}{3}$$

Exercise 3

1 Find the equation of the tangent and the equation of the normal to:

   a $y = 3x^2$ at $x = 1$ **b** $y = x^2 - 3x$ at $x = 2$

   c $y = x^4$ at $x = -1$ **d** $y = \sqrt{x}$ at $x = 9$

   e $y = \frac{4}{x^2}$ at $x = 1$ **f** $y = 20 - 3x^2$ at $x = -3$

2 The curve $y = (x^2 + 3)(x - 1)$ meets the x-axis at A and the y-axis at B. Find the equation of the tangents at A and B.

3 Find the equation of the normal to $y = \frac{16}{x^3}$ at $x = 2$.

4 The tangent at P(1, 0) to the curve $y = x^3 + x^2 - 2$ meets the curve again at Q. Find the coordinates of Q.

5 Find the equation of the tangent to $y = 9 - 2x - 2x^2$ at $x = -1$.

6 Find the equations of the tangents to the curve $y = (2x + 1)(x - 1)$ at the points where the curve cuts the x-axis. Find the point of intersection of these tangents.

7 Find the equations of the tangents to the curve $y = 3x^2 - 5x - 9$ at the points of intersection of the curve and the line $y = 6x - 5$.

8 Find the equation of the normal to $y = x^2 - 3x + 2$, which has a gradient of 3.

9 Find the equations for the tangents at the points where the curves $y = x^2 - x - 6$ and $y = -x^2 + x + 6$ meet.

10 For $y = x^2 - 7$, find the equation of the tangent at $x = 1$.

For $y = x^2 - x - 12$, find the equation of the normal to the curve at $x = 4$.

Now find the area of the triangle formed between these two lines and the y-axis.

Tangents on a graphing calculator

It is possible to draw the tangent to a curve using a graphing calculator.

To find the equation of the tangent to $y = x^2 - x - 12$ at $x = 3$, the calculator can draw the tangent and provide the equation of the tangent.

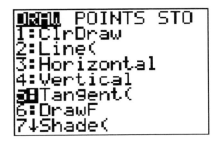
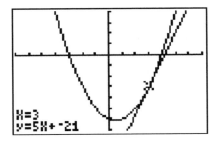

8.4 Stationary points

The gradient of a curve is constantly changing. In some regions, the function is increasing, in others it is decreasing, and at other points it is stationary.

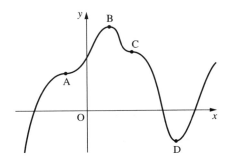

At points A, B, C and D, the tangent to the curve is horizontal and so the gradient is zero. These points are known as **stationary points**. Often these points are very important to find, particularly when functions are used to model real-life situations.

For example, a stone is thrown and its height, in metres, is given by $h(t) = 4t - t^2$, $0 \leq t \leq 4$.

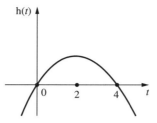

$h'(t) = 4 - 2t$ and so $h'(t) = 0$ when $4 - 2t = 0$

$$\text{i.e. } t = 2$$

So the maximum height of the stone is given by $h(2) = 4$ metres, which is the point where the gradient is zero. We have met this concept before as maximum and minimum turning points in Chapters 2, 3 and 4, and these are in fact examples of stationary points.

- Stationary points are when $\dfrac{dy}{dx} = 0$.
- Stationary points are coordinate points.
- The x-coordinate is when the stationary point occurs.
- The y-coordinate is the stationary value.

Note that the maximum turning point is not necessarily the maximum value of that function. Although it is the maximum value in that region (a local maximum) there may be greater values. For example, for the cubic function $y = x^3 + 2x^2 - 3x + 4$ the greatest value is not a turning point as it tends to infinity in the positive x-direction.

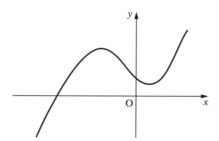

Method for finding stationary points

1. Differentiate the function.
2. Solve the equation $\dfrac{dy}{dx} = 0$.
3. Find the y-coordinate of each stationary point.

Example

Find the stationary points of $y = x^3 - 7x^2 - 5x + 1$.

Differentiating, $\dfrac{dy}{dx} = 3x^2 - 14x - 5$

When $\dfrac{dy}{dx} = 0$, $3x^2 - 14x - 5 = 0$

So $(3x + 1)(x - 5) = 0$

$\Rightarrow x = -\dfrac{1}{3}$ or $x = 5$

When $x = -\dfrac{1}{3}, y = \dfrac{50}{27}$ and when $x = 5, y = -74$

So the stationary points are $\left(-\dfrac{1}{3}, \dfrac{50}{27}\right)$ and $(5, -74)$.

Determining the nature of stationary points

There are four possible types of stationary point.

Maximum turning point	Minimum turning point	Rising point of inflexion	Falling point of inflexion

A stationary point of inflexion is when the sign of the gradient does not change either side of the stationary point.

There are two methods for testing the nature of stationary points.

Method 1 — Using the signs of f′(x)

Here the gradient immediately before and after the stationary point is examined. This is best demonstrated by example.

Example

Find the stationary points of $y = 2x^3 + 3x^2 - 36x + 5$ and determine their nature.

Using the steps of the method suggested above,

1. $\dfrac{dy}{dx} = 6x^2 + 6x - 36$

2. $\dfrac{dy}{dx} = 0$

$\Rightarrow 6x^2 + 6x - 36 = 0$

$\Rightarrow 6(x + 3)(x - 2) = 0$

$\Rightarrow x = -3$ or $x = 2$

3. When $x = -3$, $y = 2(-3)^3 + 3(-3)^2 - 36(-3) + 5$
$$= 86$$

When $x = 2$, $y = 2(2)^3 + 3(2)^2 - 36(2) + 5$
$$= -39$$

Therefore the coordinates of the stationary points are $(-3, 86)$ and $(2, -39)$.

To find the nature of the stationary points, we can examine the gradient before and after $x = -3$ and $x = 2$ using a table of signs.

This means the negative side of -3	This means the positive side of -3

$x =$	-3^-	-3	-3^+	2^-	2	2^+
$\dfrac{dy}{dx}$	$+$	0	$-$	$-$	0	$+$
Shape	/	—	\	\	—	/

We can choose values either side of the stationary point to test the gradient either side of the stationary point. This is the meaning of the notation -3^+ and -3^-. -3^+ means taking a value just on the positive side of -3, that is slightly higher than -3. -3^- means taking a value just on the negative side of -3, that is slightly lower than -3.

It is important to be careful of any vertical asymptotes that create a discontinuity.

So for -3^-, $x = -4$ could be used and so $\dfrac{dy}{dx} = 6(-4 + 3)(-4 - 2)$. What is important is whether this is positive or negative. The brackets are both negative and so the gradient is positive. A similar process with, say $x = -2$, $x = 1$, $x = 3$, fills in the above table.

This provides the shape of the curve around each stationary value and hence the nature of each stationary point.

So $(-3, 86)$ is a maximum turning point and $(2, -39)$ is a minimum turning point. Strictly these should be known as a **local maximum** and a **local minimum** as they are not necessarily the maximum or minimum values of the function – these would be called the **global** maximum or minimum.

Example

Find the stationary point for $y = x^3$ and determine its nature.

1. $\dfrac{dy}{dx} = 3x^2$

2. $\dfrac{dy}{dx} = 0$ when $3x^2 = 0$
$$\Rightarrow x = 0$$

3. When $x = 0$, $y = 0$, i.e. (0, 0)

4.

$x =$	0^-	0	0^+
$\dfrac{dy}{dx}$	$+$	0	$+$
Shape	/	———	/

So the stationary point (0, 0) is a rising point of inflexion.

Method 2 — Using the sign of $f''(x)$

When a function is differentiated a second time, the rate of change of the gradient of the function is found. This is known as the **concavity** of the function.

> The two notations used here for the second derivative are:
>
> $f''(x)$ in functional notation and $\dfrac{d^2y}{dx^2}$ in Leibniz notation.
>
> This Leibniz notation arises from differentiating $\dfrac{dy}{dx}$ again.
>
> This is $\dfrac{d}{dx}\left(\dfrac{dy}{dx}\right) = \dfrac{d^2y}{dx^2}$.

For a section of curve, if the **gradient** is increasing then it is said to be concave up.

The curve is getting less steep in this section, i.e. it is becoming less negative and so is increasing.

Similarly, if the gradient is decreasing it is said to be concave down.

Looking at the sign of $\dfrac{d^2y}{dx^2}$ can help us determine the nature of stationary points.

Consider a minimum turning point:

At the turning point, $\dfrac{dy}{dx} = 0$, although the gradient is zero, the gradient is increasing

(moving from negative to positive) and so $\dfrac{d^2y}{dx^2}$ is positive.

Consider a maximum turning point:

At the turning point, $\dfrac{dy}{dx} = 0$, although the gradient is zero, the gradient is decreasing (moving from positive to negative) and so $\dfrac{d^2y}{dx^2}$ is negative.

At a point of inflexion, $\dfrac{d^2y}{dx^2}$ is zero.

This table summarizes the nature of stationary points in relation $\dfrac{dy}{dx}$ and $\dfrac{d^2y}{dx^2}$.

	Maximum turning point	Minimum turning point	Rising point of inflexion	Falling point of inflexion
$\dfrac{dy}{dx}$	0	0	0	0
$\dfrac{d^2y}{dx^2}$	−	+	0	0

This method is often considered more powerful than method 1 (when the functions become more complicated). For examination purposes, it is always best to use the second derivative to test nature. However, note that for stationary points of inflexion, it is still necessary to use a table of signs.

Although the table above is true, it is unfortunately not the whole picture. A positive or negative answer for $\dfrac{d^2y}{dx^2}$ provides a conclusive answer to the nature of a stationary point.

$\dfrac{d^2y}{dx^2} = 0$ is not quite as helpful. In most cases, this will mean that there is a stationary

point of inflexion. However, this needs to be tested using a table of signs as it is possible that it will in fact be a minimum or maximum turning point. A table of signs is also required to determine whether a stationary point of inflexion is rising or falling. See the second example below for further clarification.

Example

Find the stationary points of $y = x + \dfrac{4}{x}$.,

1. $\dfrac{dy}{dx} = 1 - 4x^{-2}$

2. This is stationary when $\dfrac{dy}{dx} = 0$.

   So $1 - 4x^{-2} = 0$

   $\Rightarrow \dfrac{4}{x^2} = 1$

   $\Rightarrow x^2 = 4$

   $\Rightarrow x = -2$ or $x = 2$

3. When $x = -2$, $y = -4$, i.e. $(-2, -4)$ and when $x = 2$, $y = 4$, i.e. $(2, 4)$

4. To test the nature using the second derivative,

   $\dfrac{d^2y}{dx^2} = 8x^{-3} = \dfrac{8}{x^3}$

   At $x = -2$, $\dfrac{d^2y}{dx^2} = \dfrac{8}{-8} = -1$ and since this is negative, this is a maximum turning point.

   At $x = 2$, $\dfrac{d^2y}{dx^2} = \dfrac{8}{8} = 1$ and since this is positive, this is a minimum turning point.

   So stationary points are $(-2, -4)$, a local maximum, and $(2, 4)$, a local minimum.

Example

Find the stationary point(s) of $y = x^4$.

1. $\dfrac{dy}{dx} = 4x^3$

2. Stationary when $\dfrac{dy}{dx} = 0$,

   So $4x^3 = 0$

   $\Rightarrow x^3 = 0$

   $\Rightarrow x = 0$

3. When $x = 0$, $y = 0$, i.e. $(0, 0)$.

4. To test the nature using the second derivative,

$$\frac{d^2y}{dx^2} = 12x^2$$

At $x = 0$, $\frac{d^2y}{dx^2} = 12 \times 0^2 = 0$

As $\frac{d^2y}{dx^2} = 0$ for this stationary point, no assumptions can be made about its

nature and so a table of signs is needed.

$x =$	0^-	0	0^+
$\dfrac{dy}{dx}$	$-$	0	$+$
Shape	\	—	/

Hence $(0, 0)$ is a minimum turning point. This can be verified with a calculator.

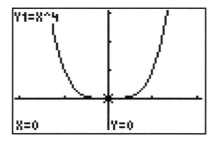

This is an exceptional case, which does not often occur. However, be aware of this "anomaly".

Exercise 4

1 Find the stationary points and determine their nature using a table of signs.

   a $f(x) = x^2 - 8x + 3$

   b $y = x^3 - 12x + 7$

   c $f(x) = 5x^4$

   d $y = (3x - 4)(x + 2)$

   e $f(x) = 4x + \dfrac{1}{x}$

2 Find the stationary points and determine their nature using the second derivative.

   a $y = 2x^2 + 8x - 5$

   b $y = (4 - x)(x + 6)$

   c $f(x) = x(x - 4)^2$

   d $y = 2x^3 - 9x^2 + 12x + 5$

   e $f(x) = 3x^5$

3 Find the stationary points and determine their nature using either method.

a $f(x) = \dfrac{1}{3}x^3 - 2x^2 + 3x - 4$

b $y = (2x - 5)^2$

c $f(x) = 16x - \dfrac{1}{x^2}$

d $y = x^6$

e $y = x^5 - 2x^3 + 5x^2 + 2$

4 Find the distance between the turning points of the graph of
$y = -(x^2 - 4)(x^2 + 2)$.

8.5 Points of inflexion

The concavity of a function is determined by the second derivative.

$f''(x) > 0$	Concave up
$f''(x) < 0$	Concave down

So what happens when $f''(x) = 0$?

We know that when $f'(x) = 0$ and $f''(x) = 0$, there is a stationary point – normally a stationary point of inflexion (with the exceptions as previously discussed).

In fact, apart from the previously noted exceptions, whenever $f''(x) = 0$ it is known as a **point of inflexion**. The type met so far are stationary points of inflexion when the gradient is also zero (also known as horizontal points of inflexion).

However, consider the curve:

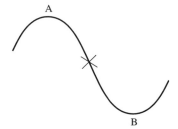

Looking at the gradient between the turning points, it is constantly changing, the curve becoming steeper and then less steep as it approaches B. So the rate of change of gradient is negative (concave down) around A and then positive (concave up) around B.

Clearly the rate of change of gradient $\left(\dfrac{d^2y}{dx^2}\right)$ must be zero at some point between A

and B. This is the steepest part of the curve between A and B, and it is this point that is known as a point of inflexion. This is clearly not stationary. So a point of inflexion can now be defined to be a point where the concavity of the graph changes sign.

If $\dfrac{d^2y}{dx^2} = 0$, there is a point of inflexion:

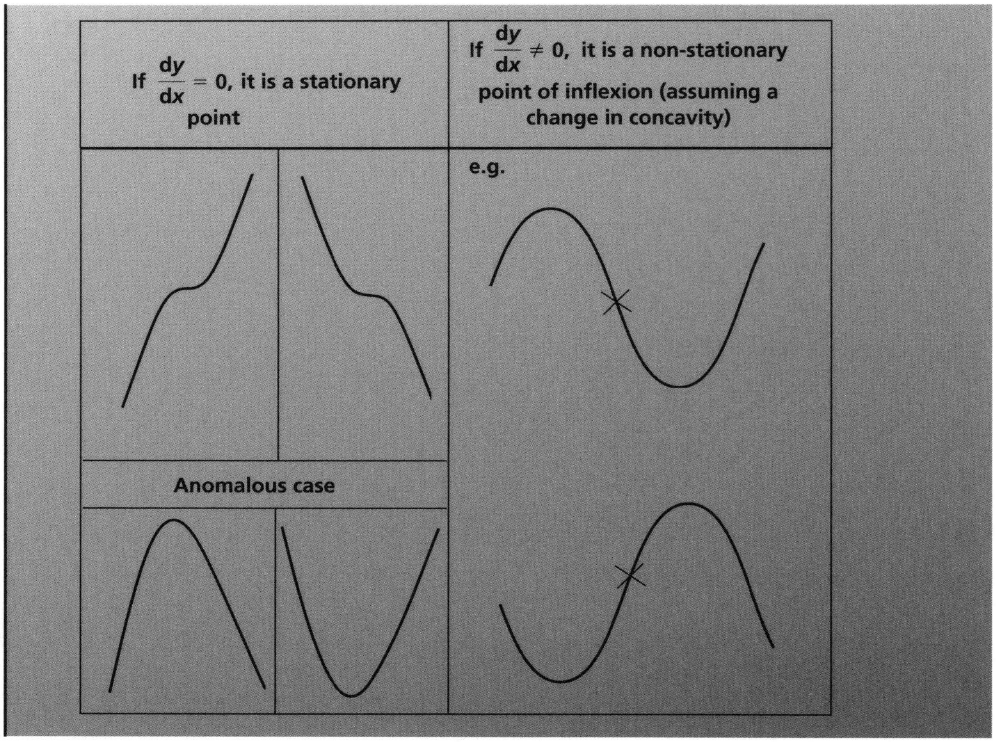

If $\dfrac{dy}{dx} = 0$, it is a stationary point	If $\dfrac{dy}{dx} \neq 0$, it is a non-stationary point of inflexion (assuming a change in concavity)
	e.g.
Anomalous case	

Method for finding points of inflexion

1. Differentiate the function twice to find $\dfrac{d^2y}{dx^2}$.

2. Solve the equation $\dfrac{d^2y}{dx^2} = 0$.

3. Find the y-coordinate of each point.

4. Test the concavity around this point, i.e. $\dfrac{d^2y}{dx^2}$ must change sign.

Example

Find the points of inflexion of the curve $f(x) = x^5 - 15x^3$ and determine whether they are stationary.

1. $f'(x) = 5x^4 - 45x^2$

   $f''(x) = 20x^3 - 90x$

2. For points of inflexion, $f''(x) = 0$

   So $20x^3 - 90x = 0$

   $\Rightarrow 10x(2x^2 - 9) = 0$

   $\Rightarrow x = 0$ or $x^2 = \dfrac{9}{2}$

   $\Rightarrow x = 0$ or $x = \dfrac{\pm 3}{\sqrt{2}}$

3. $f\left(-\dfrac{3}{\sqrt{2}}\right) = 100.2324$

   $f(0) = 0$

   $f\left(\dfrac{3}{\sqrt{2}}\right) = -100.2324$

4.

$x =$	$-\dfrac{3}{\sqrt{2}}^{-}$	$-\dfrac{3}{\sqrt{2}}$	$-\dfrac{3}{\sqrt{2}}^{+}$	0^{-}	0	0^{+}
$f''(x)$	$-$	0	$+$	$+$	0	$-$

$x =$	$\dfrac{3}{\sqrt{2}}^{-}$	$\dfrac{3}{\sqrt{2}}$	$\dfrac{3}{\sqrt{2}}^{+}$
$f''(x)$	$-$	0	$+$

There is a change in concavity (the sign of the second derivative changes) around each point. So each of these three points is a point of inflexion.

To test whether each point is stationary, consider $f'(x)$.

$$f'\left(\frac{3}{\sqrt{2}}\right) = 5\left(\frac{3}{\sqrt{2}}\right)^4 - 45\left(\frac{3}{\sqrt{2}}\right)^2 = -\frac{405}{4}$$

Hence $x = \dfrac{3}{\sqrt{2}}$ provides a non-stationary point of inflexion.

$$f'\left(-\frac{3}{\sqrt{2}}\right) = 5\left(-\frac{3}{\sqrt{2}}\right)^4 - 45\left(-\frac{3}{\sqrt{2}}\right)^2 = -\frac{405}{4}$$

Hence $x = -\dfrac{3}{\sqrt{2}}$ also provides a non-stationary point of inflexion.

$$f'(0) = 5(0)^4 - 45(0)^2 = 0$$

Hence $x = 0$ provides a stationary point of inflexion.

The three points of inflexion are $\left(-\dfrac{3}{\sqrt{2}}, 100.2324\right)$, $(0, 0)$ and

$\left(\dfrac{3}{\sqrt{2}}, -100.2324\right)$. This can be verified on a calculator.

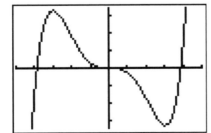

Exercise 5

1 Find the points of inflexion for the following functions and determine whether they are stationary.

  a $f(x) = x^5 - \dfrac{40}{3}x^3$

  b $f(x) = x^3 + 3x^2 - 6x + 7$

  c i $y = x^2 - x + 18$

    ii $y = 5x^2 - 9$

    iii $f(x) = ax^2 + bx + c$

  Make a general statement about quadratic functions.

2 Find the points of inflexion for the following functions and determine whether they are stationary.

  a $y = 4x^3$

  b $f(x) = x^3 - 3x^2 + 7$

  c $y = -x^3 - 6x^2 + 8x - 3$

  d $f(x) = ax^3 + bx^2 + cx + d$

  Make a general statement about cubic functions.

3 Find the points of inflexion for the following functions and determine whether they are stationary.

  a $f(x) = x^4 - 6x^2 + 8$

  b $y = 3x^4 + 5x^3 - 3x^2 + 7x + 3$

  c $f(x) = x^5 - 3x^4 + 5x^3$

  d $y = x^4 - 3$

4 Find the equation of the tangent to $y = x^3 - 9x^2 + 6x + 9$ at the point of inflexion.

5 For the graph of $y = 2x^3 - 12x^2 + 5x - 3$, find the distance between the point of inflexion and the root.

8.6 Curve sketching

Bringing together knowledge of functions, polynomials and differentiation, it is now possible to identify all the important features of a function and hence sketch its curve.

The important features of a graph are:

- **Vertical asymptotes** (where the function is not defined)
  This is usually when the denominator is zero.

- **Intercepts**
  These are when $x = 0$ and $y = 0$.

- **Stationary points and points of inflexion**
  Determine when $\dfrac{dy}{dx} = 0$ and when $\dfrac{d^2y}{dx^2} = 0$.

- **Behaviour as** $x \to \pm\infty$
  This provides horizontal and oblique asymptotes.

With the exception of oblique asymptotes, all of the necessary concepts have been met in this chapter, and in Chapters 1, 2, 3 and 4.

Oblique asymptotes

In Chapter 3 we met horizontal asymptotes. These occur where $x \to \pm\infty$. This is also true for oblique asymptotes.

Consider the function $y = 2 + \dfrac{3}{x - 1}$.

It is clear that as $x \to \pm\infty$, the $\dfrac{3}{x - 1}$ becomes negligible and so $y \to 2$.

Hence $y = 2$ is a horizontal asymptote for this function.

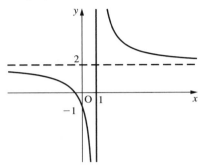

Now consider the function $y = 2x - 1 + \dfrac{5}{3x + 2}$.

In a similar way, the fractional part $\left(\dfrac{5}{3x + 2}\right)$ tends to zero as $x \to \pm\infty$ and so $y \to 2x - 1$. This means that $y = 2x - 1$ is an **oblique asymptote** (also known as a slant asymptote).

Method for sketching a function

1. Find the vertical asymptotes (where the function is not defined).
2. If it is an improper rational function (degree of numerator $\geq$ degree of denominator), divide algebraically to produce a proper rational function.
3. Consider what happens for very large positive and negative values of x. This will provide horizontal and oblique asymptotes.
4. Find the intercepts with the axes.
   These are when $x = 0$ and $y = 0$.
5. Find the stationary points and points of inflexion (and their nature).
   Determine when $\dfrac{dy}{dx} = 0$ and when $\dfrac{d^2y}{dx^2} = 0$.
6. Sketch the curve, ensuring that all of the above important points are annotated on the graph.

Example

Find the asymptotes of $y = \dfrac{-2x^2 - 7x - 1}{x - 4}$.

Clearly the function is not defined at $x = 4$ and so this is a vertical asymptote.

Dividing,
$$
\begin{array}{r}
-2x - 15 \\
x - 4 \overline{) -2x^2 - 7x - 1} \\
\underline{-2x^2 + 8x} \\
-15x - 1 \\
\underline{-15x + 60} \\
-61
\end{array}
$$

Hence $y = -2x - 15 - \dfrac{61}{x - 4}$ and so as $x \to \pm\infty$, $y \to -2x - 15$.

Therefore $y = -2x - 15$ is an oblique asymptote.

This is clear from the graph:

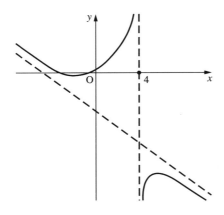

Example

Sketch the graph of $y = \dfrac{x^2}{x + 1}$, identifying all asymptotes, intercepts, stationary points, and non-horizontal points of inflexion.

There is a vertical asymptote at $x = -1$

Dividing,

$$
\begin{array}{r}
x - 1 \\
(x + 1) \overline{)\, x^2 } \\
\underline{x^2 + x} \\
- x \\
\underline{-x - 1} \\
1
\end{array}
$$

This gives $y = x - 1 + \dfrac{1}{x + 1}$.

So $y = x - 1$ is an oblique asymptote.

Putting $x = 0$ and $y = 0$ gives an intercept at $(0, 0)$. There are no other roots.

Preparing for differentiation, $y = x - 1 + \dfrac{1}{x + 1}$

$$= x - 1 + (x + 1)^{-1}$$

Differentiating, $\dfrac{dy}{dx} = 1 - (x + 1)^{-2}$

Stationary when $\dfrac{dy}{dx} = 0$, i.e. when $\dfrac{1}{(x + 1)^2} = 1$

$$\Rightarrow x + 1 = -1 \text{ or } x + 1 = 1$$

$$\Rightarrow x = -2 \text{ or } x = 0$$

Substituting into the original function provides the coordinates $(0, 0)$ and $(-2, -4)$.

For the nature of these stationary points,

$$\frac{d^2y}{dx^2} = 2(x + 1)^{-3} = \frac{2}{(x + 1)^3}$$

At $x = 0$, $\frac{d^2y}{dx^2} = 2 > 0$ so $(0, 0)$ is a local minimum turning point.

At $x = -2$, $\frac{d^2y}{dx^2} = -2 < 0$ so $(-2, -4)$ is a local maximum point.

As $\frac{d^2y}{dx^2} = \frac{2}{(x + 1)^3} \neq 0$ $\forall x \in \mathbb{R}$, there are no non-horizontal points of inflexion.

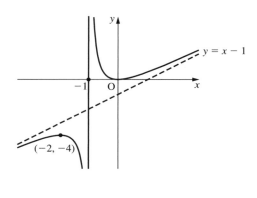

> The notation $\forall$ means "for all". So $\forall\ x \in \mathbb{R}$, means for all real values of x.

All of these features can be checked using a graphing calculator, if it is available. In some cases, an examination question may expect the use of a graphing calculator to find some of these important points, particularly stationary points.

For example, the calculator provides this graph for the above function:

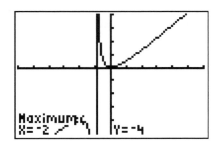

In fact, it is possible to be asked to sketch a graph that would be difficult without use of the calculator. Consider the next example.

Example

$$\frac{x^2}{9} - \frac{y^2}{4} = 36$$

In order to graph this function using a calculator, we need to rearrange into a $y =$ form.

$$\frac{x^2}{9} - \frac{y^2}{4} = 36$$

$$\Rightarrow \frac{y^2}{4} = \frac{x^2}{9} - 36$$

$$\Rightarrow y^2 = \frac{4x^2}{9} - 144$$

$$\Rightarrow y = \pm\sqrt{\frac{4x^2}{9} - 144}$$

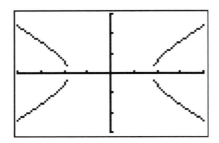

Clearly this function is not defined for a large section (in fact $-18 < x < 18$).

The oblique asymptotes are not immediately obvious. Rearranging the equation to give $\frac{y}{x} = \pm\sqrt{\frac{4}{9} - \frac{144}{x^2}}$ makes it clearer.

As $x \to \pm\infty$, $\frac{y}{x} \to \pm\frac{2}{3}$

$$\Rightarrow y \to \pm\frac{2}{3}x$$

Exercise 6

Find all asymptotes (vertical and non-vertical) for these functions.

1 $y = \dfrac{2}{x}$ 　　　　**2** $y = \dfrac{x}{x-3}$ 　　　　**3** $y = \dfrac{x^2+5}{x}$

4 $f(x) = \dfrac{x+3}{x-2}$ 　　　**5** $f(x) = \dfrac{x^2+1}{x+2}$ 　　　**6** $y = \dfrac{2x^2+3x-5}{x-3}$

7 $f(x) = \dfrac{x^3 - 2x^2 + 3x + 5}{x^2 + 4}$ 　　　**8** $y = \dfrac{x^3 - 4x}{x^2 + 1}$

9 $y = \dfrac{5x}{(x-1)(x-4)}$ 　　　**10** $f(x) = \dfrac{x^2+6}{x^2-1}$

11 $y = \dfrac{3x^2+8}{x^2-9}$ 　　　**12** $y = \dfrac{4x^3+9}{x^2-x-6}$

Sketch the graphs of these functions, including asymptotes, stationary points and intercepts.

13 $y = \dfrac{x-1}{x+1}$ 　　　**14** $y = \dfrac{x-1}{x(x+1)}$ 　　　**15** $y = \dfrac{x}{x+4}$

16 $y = \dfrac{2x^2}{x + 1}$ **17** $y = \dfrac{x}{x^2 - 1}$ **18** $y = \dfrac{x^2}{1 - x}$

19 $y = \dfrac{(2x + 5)(x - 4)}{(x + 2)(x - 3)}$ **20** $y = \dfrac{1}{x^2 - x - 12}$

21 $y = \dfrac{4}{x^2 - 2}$ **22** $\dfrac{x^2}{16} - \dfrac{y^2}{9} = 25$

8.7 Sketching the graph of the derived function

Given the graph of the original function it is sometimes useful to consider the graph of its derivative. For example, non-horizontal points of inflexion now become obvious from the graph of the derived function, since they become stationary points. Horizontal points of inflexion are already stationary.

If the original function is known, then it is straightforward to sketch the graph of the derived function. This can be done by:

a) finding the derivative and sketching it

b) using a graphing calculator to sketch the derived function.

Example

Sketch the derivative of $y = x^3 - 6x^2 + 3x - 5$.

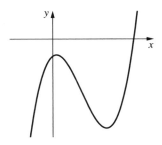

Using method a)

Differentiating gives $\dfrac{dy}{dx} = 3x^2 - 12x + 3$

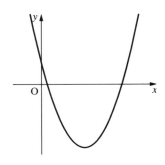

Using method b)

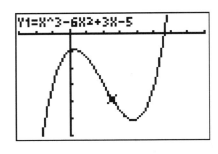

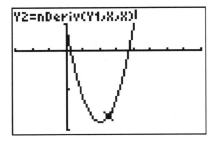

These methods are possible only if the function is known.

If the function is not known, the gradient of the graph needs to be examined.

Example

Sketch the graph of the derivative of this graph.

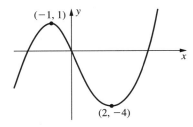

Note where the graph is increasing, stationary and decreasing. Stationary points on the original graph become roots of the derived function $\left(\dfrac{dy}{dx} = 0\right)$, increasing regions are above the x-axis $\left(\dfrac{dy}{dx} > 0\right)$, and decreasing regions are below the x-axis $\left(\dfrac{dy}{dx} < 0\right)$. So the graph becomes

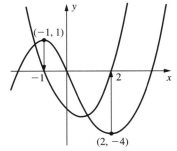

With polynomial functions, the degree of a derived function is always one less than the original function.

We can consider this process in reverse, and draw a possible graph of the original function, given the graph of the derived function. In order to do this note that:

1. roots of the derived function are stationary points on the original graph
2. stationary points on the derived function graph are points of inflexion on the original graph.

Example

Given this graph of the derived function $y = f'(x)$, sketch a possible graph of the original function $y = f(x)$.

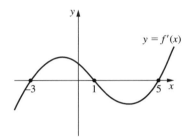

We can see that there are roots on this graph and hence stationary points on the original at $x = -3$, $x = 1$, and $x = 5$.
It is helpful to consider the gradient of the original to be able to draw a curve.

$x =$	$\rightarrow$	-3	$\rightarrow$	1	$\rightarrow$	5	$\rightarrow$
$f'(x)$	$-$	0	$+$	0	$-$	0	$+$

So a possible graph of the original function $y = f(x)$ is

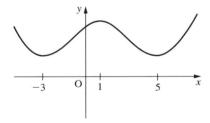

We cannot determine the y-values of the points on the curve but we can be certain of the shape of it. This will be covered in further detail in Chapter 14.

Sketching the reciprocal function

Sometimes we are asked to sketch the graph of a reciprocal function, i.e. $\dfrac{1}{f(x)}$.

If $f(x)$ is known and a calculator is used, this is easy. However, if it is not known, then we need to consider the following points.

1. At $f(x) = 0$, $\dfrac{1}{f(x)} \to \infty$. Hence roots on the original graph become vertical asymptotes on the reciprocal graph.

2. At a vertical asymptote, $f(x) \to \infty \Rightarrow \dfrac{1}{f(x)} = 0$. Hence vertical asymptotes on $f(x)$ become roots on the reciprocal graph.

3. Maximum turning points become minimum turning points and minimum turning points become maximum turning points. The x-value of the turning point stays the same but the y-value is reciprocated.

4. If $f(x)$ is above the x-axis, $\dfrac{1}{f(x)}$ is also above the x-axis, and if $f(x)$ is below the x-axis, $\dfrac{1}{f(x)}$ is also below the x-axis.

We will now demonstrate this by example.

Example

If $f(x) = \sin x$, $0 \le x \le 2\pi$ draw $\dfrac{1}{f(x)}$.

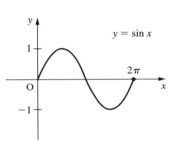
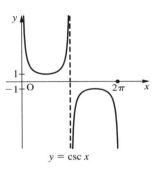

This is the standard way of drawing the curves of $y = \sec x$, $y = \csc x$ and $y = \cot x$.

Example

For the following graph of $y = f(x)$, draw the graph of $y = \dfrac{1}{f(x)}$.

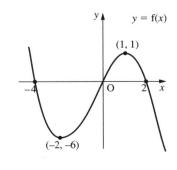
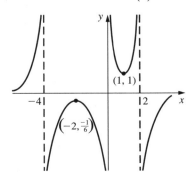

Exercise 7

1 Sketch the graph of the derived function of the following:

a $y = 4x$ **b** $y = -x$ **c** $y = 4$

d $y = x^2$ **e** $y = -4x^2$ **f** $y = x^2 + x - 7$

g $y = x^3$ **h** $y = \frac{1}{3}x^3 + 2x^2 + 3x - 8$ **i** $y = \frac{1}{4}x^4$

2 Sketch the graph of the derived function of the following:

a
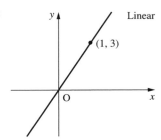

b
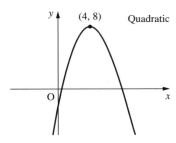

c
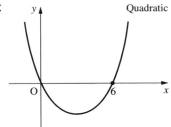

d
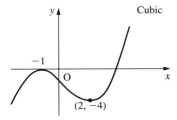

e
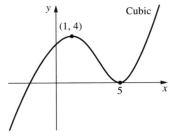

f
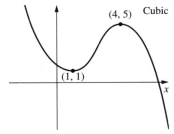

g
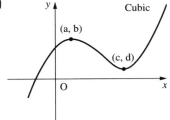

h
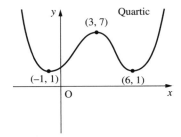

i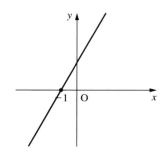

Quartic

(2, 3)

O

(−2, −4)

j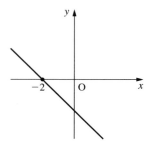

Fifth degree

(3, 7)

(−1, 3)

1

(6, −4)

3 Sketch a possible graph of the original function $y = f(x)$, given the derived function graph $y = f'(x)$ in each case.

a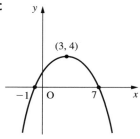

−1   O

b

−2   O

c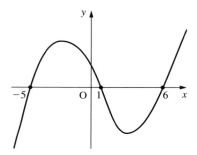

(3, 4)

−1   O   7

d

−5   O   1   6

4 For the following functions, draw the graph of $\dfrac{1}{f(x)}$

a $f(x) = 2x - 1$ **b** $f(x) = (x - 2)^2$

c $f(x) = x^3 - 2x^2 - 5x + 6$ **d** $f(x) = 2x^3 - 15x^2 + 24x + 16$

e $f(x) = e^x$ **f** $f(x) = \ln x$

g $f(x) = \cos x,\ 0 \le x \le 2\pi$ **h** $f(x) = \dfrac{6}{(x - 3)}$

5 For the following graphs of $y = f(x)$, sketch the graph of the reciprocal function $y = \dfrac{1}{f(x)}$.

a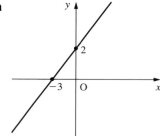

2

−3   O

b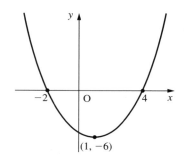

−2   O   4

(1, −6)

c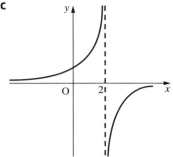

d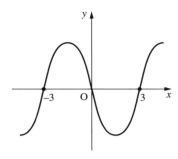

e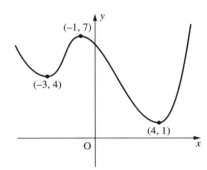

f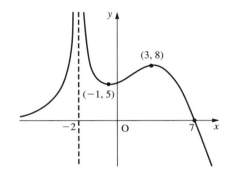

Review exercise

 1 Differentiate $f(x) = x^3 - 4x + 5$ using first principles.

 2 For $f(x) = \sqrt{x} + \dfrac{2}{x^2}$, find $f'(4)$.

3 Given that $y = \dfrac{3x - x^9}{2\sqrt{x}}$, find $\dfrac{dy}{dx}$.

 4 A function is defined as $f(x) = 2x + 3 + \dfrac{64}{x^2}$. Find values of x for which the function is increasing.

 5 Given that $f(x) = 5x^2 - 1$ and $g(x) = 3x + 2$, find $h(x) = f(g(x))$. Hence find $h'(x)$.

 6 Find the positive value of x for which the gradient of the tangent is -6 for $y = 6x - x^3$. Hence find the equation of the tangent at this point.

 7 Sketch the graph of the derivative for the graph below.

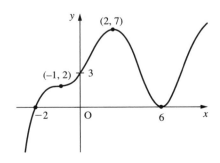

8 Find the stationary points of $y = (4x - 1)(2x^2 - 2)$ and investigate their nature.

9 Find the equation of the tangent to $y = x^5 - 3$ at $x = -1$ and find the equation of the normal to $y = 9 - x^2$ at $x = -1$. Find the point where these lines cross.

10 Sketch the graph of $y = \dfrac{x^3}{x^2 - x - 6}$, including all asymptotes, stationary points and intercepts.

11 Find the equations of all the asymptotes of the graph of $y = \dfrac{x^2 - 5x - 4}{x^2 - 5x + 4}$

[IB Nov 02 p1 Q4]

12 The line $y = 16x - 9$ is a tangent to the curve $y = 2x^3 + ax^2 + bx - 9$ at the point (1, 7). Find the values of a and b. [IB Nov 01 p1 Q7]

13 For the following graphs of $y = f(x)$, draw the graphs of $y = \dfrac{1}{f(x)}$.

a

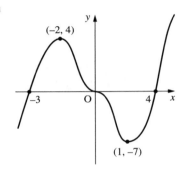

b

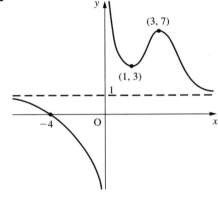

9 Differentiation 2 — Further Techniques

Leonhard Euler is considered to be one of the most important mathematicians of all time. He was born on 15 April 1707 in Basel, Switzerland, and died on 18 September 1783 in St Petersburg, Russia, although he spent much of his life in Berlin. Euler's mathematical discoveries are in many branches of mathematics including number theory, geometry, trigonometry, mechanics, calculus and analysis. Some of the best-known notation was created by Euler including the notation f(x) for a function, e for the base of natural logs, i for the square root of -1, π for pi, Σ for summation and many others. Euler enjoyed his work immensely, writing in 1741, "The King calls me his professor, and I think I am the happiest man in the world." Even on his dying day he continued to enjoy mathematics, giving a mathematics lesson to his grandchildren and doing some work on the motion of ballons.

In Chapter 8, the basic concepts of differentiation were covered. However, the only functions that we differentiated all reduced to functions of the form $y = ax^n + \cdots + k$. In this chapter, we will meet and use further techniques to differentiate other functions. These include trigonometric, exponential and logarithmic functions, functions that are given implicitly, and functions that are the product or quotient of two (or more) functions.

9.1 Differentiating trigonometric functions

What is the derivative of sin x?

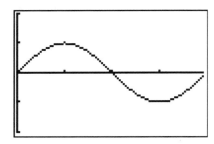

Using our knowledge of sketching the derived function, we know that the graph must be of this form:

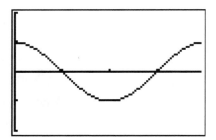

We can use a calculator to draw the derivative graph as above.

This graph looks very much like the cosine function. We now need to see if it is.

In order to prove this, there are two results that need to be investigated. First, we need to consider what happens to $\dfrac{\sin h}{h}$ for small values of h. The calculator can be used to investigate this:

X	Y₁	
0	ERROR	
.01	.99998	
.02	.99993	
.03	.99985	
.04	.99973	
.05	.99958	
.06	.9994	
X=0		

It is clear that, as $h \to 0$, $\dfrac{\sin h}{h} \to 1$.

Second, we also need to investigate $\dfrac{\cos h - 1}{h}$ for small values of h.

X	Y₂	
0	ERROR	
.01	-.005	
.02	-.01	
.03	-.015	
.04	-.02	
.05	-.025	
.06	-.03	
X=0		

It is clear that, as $h \to 0$, $\dfrac{\cos h - 1}{h} \to 0$.

Now we can use differentiation by first principles to find the derivative of $f(x) = \sin x$.

$f(x + h) = \sin(x + h)$

$f(x + h) - f(x) = \sin(x + h) - \sin x$

$\qquad = \sin x \cos h + \cos x \sin h - \sin x$

$\qquad = \sin x(\cos h - 1) + \cos x \sin h$

$\dfrac{f(x + h) - f(x)}{h} = \dfrac{\sin x(\cos h - 1)}{h} + \dfrac{\cos x \sin h}{h}$

So $\displaystyle\lim_{h \to 0} \dfrac{f(x + h) - f(x)}{h} = 0 + \cos x$ using the above results.

Hence $\displaystyle\lim_{h \to 0} \dfrac{f(x + h) - f(x)}{h} = \cos x$

Therefore if $f(x) = \sin x$, $f'(x) = \cos x$.

What about the derivative of $\cos x$?

Examining the graph of the derived function using the calculator, this would appear to be $-\sin x$.

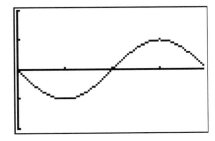

> When dealing with trigonometric functions it is vital that radians are used. This is because of the results that we investigated above.
>
> In degrees, $\displaystyle\lim_{h \to 0} \dfrac{\sin h}{h} \neq 1$ as seen below.
>
> | X | Y1 | |
> |---|---|---|
> | 0 | ERROR | |
> | .01 | .01745 | |
> | .02 | .01745 | |
> | .03 | .01745 | |
> | .04 | .01745 | |
> | .05 | .01745 | |
> | .06 | .01745 | |
>
> X=0
>
> So in degrees, the derivative of $\sin x$ is not $\cos x$.
>
> Therefore, for calculus, we must always use radians.

Again, we can use using differentiation by first principles to find the derivative of $f(x) = \cos x$.

$\qquad f(x + h) = \cos(x + h)$

$f(x + h) - f(x) = \cos(x + h) - \cos x$

$\qquad\qquad = \cos x \cos h - \sin x \sin h - \cos x$

$\qquad\qquad = \cos x(\cos h - 1) - \sin x \sin h$

$\dfrac{f(x + h) - f(x)}{h} = \dfrac{\cos x(\cos h - 1)}{h} - \dfrac{\sin x \sin h}{h}$

So $\displaystyle\lim_{h \to 0} \dfrac{f(x + h) - f(x)}{h} = 0 - \sin x$ using the previous results.

Hence $\displaystyle\lim_{h \to 0} \dfrac{f(x + h) - f(x)}{h} = -\sin x$

Therefore if $f(x) = \cos x$, $f'(x) = -\sin x$.

Summarizing:

$$y = \sin x \qquad\qquad y = \cos x$$
$$\frac{dy}{dx} = \cos x \qquad\qquad \frac{dy}{dx} = -\sin x$$

It is clear that these are connected (as the two functions themselves are). Starting with $\sin x$, repeated differentiation gives $\cos x$, $-\sin x$, $-\cos x$ and then back to $\sin x$. This cycle can be remembered by

$$\text{Differentiate} \quad \left\downarrow \begin{array}{c} S \\ C \\ -S \\ -C \end{array} \right.$$

We now have the derivatives of two of the six trigonometric functions. The other four functions are all defined in terms of $\sin x$ and $\cos x$ (remembering $\tan x = \dfrac{\sin x}{\cos x}$) and so this information provides the derivatives of the other four functions.

Proofs of these require the use of rules that have not yet been covered, and hence these are to be found later in the chapter. However, the results are shown below.

$$y = \tan x \qquad y = \csc x \qquad\qquad y = \sec x \qquad\qquad y = \cot x$$
$$\frac{dy}{dx} = \sec^2 x \qquad \frac{dy}{dx} = -\csc x \cot x \qquad \frac{dy}{dx} = \sec x \tan x \qquad \frac{dy}{dx} = -\csc^2 x$$

Example

Find the derivative of $y = \cos x - \sec x$.

$$\frac{dy}{dx} = -\sin x - \sec x \tan x$$

Example

Find the derivative of $y = 8 \sin x$.

$$\frac{dy}{dx} = 8 \cos x$$

Exercise 1

Find the derivative of each of these.

1 $y = \tan x + 3$ **2** $y = \sin x - \csc x$

3 $y = \sin x + 6x^2$ **4** $y = 5 \cos x$

5 $y = 7 \cot x$ **6** $y = -3 \sec x$

7 $y = 9x^2 - 4 \cos x$ **8** $y = 7x - 5 \sin x - \sec x$

9.2 Differentiating functions of functions (chain rule)

The chain rule is a very useful and important rule for differentiation. This allows us to differentiate composite functions. First consider $y = (ax + b)^n$.

Investigation

Consider these functions:

1 $y = (2x + 1)^2$ **2** $y = (2x + 1)^3$ **3** $y = (3x - 2)^4$
4 $y = (3x - 2)^5$ **5** $y = (4 - x)^2$ **6** $y = (4 - x)^3$

Using knowledge of the binomial theorem and differentiation, find the derivatives of the above functions. Factorise the answers.

You should have noticed a pattern that will allow us to take a "shortcut", which we always use, when differentiating this type of function.

This is that for functions of the form $y = (ax + b)^n$,

$$\frac{dy}{dx} = an(ax + b)^{n-1}$$

This is a specific case of a more general rule, known as the chain rule, which can be stated as:

$$\frac{dy}{dx} = \frac{dy}{du} \cdot \frac{du}{dx}$$

> We can consider this as differentiating the bracket to the power n and then multiplying by the derivative of the bracket.

> This is not due to cancelling!

This is where y is a function of a function. This means that we can consider y as a function of u and u as a function of x.

Proof

Consider $y = g(u)$ where $u = f(x)$.

If δx is a small increase in x, then we can consider δu and δy as the corresponding increases in u and y.

Then, as $\delta x \to 0$, δu and δy also tend to zero.

We know from Chapter 8 that $\dfrac{dy}{dx} = \lim\limits_{\delta x \to 0}\left(\dfrac{\delta y}{\delta x}\right)$

$$= \lim_{\delta x \to 0}\left(\frac{\delta y}{\delta u} \cdot \frac{\delta u}{\delta x}\right)$$

$$= \lim_{\delta x \to 0}\left(\frac{\delta y}{\delta u}\right) \cdot \lim_{\delta x \to 0}\left(\frac{\delta u}{\delta x}\right)$$

$$= \lim_{\delta u \to 0}\left(\frac{\delta y}{\delta u}\right) \cdot \lim_{\delta x \to 0}\left(\frac{\delta u}{\delta x}\right)$$

So $\dfrac{dy}{dx} = \dfrac{dy}{du} \cdot \dfrac{du}{dx}$

The use of this rule is made clear in the following examples.

Example

Differentiate $y = (3x - 4)^4$.

Let $u = 3x - 4$ and $y = u^4$.

Then $\dfrac{dy}{du} = 4u^3$ and $\dfrac{du}{dx} = 3$.

Hence $\dfrac{dy}{dx} = 4u^3 \cdot 3 = 12u^3$.

Substituting back for x gives $\dfrac{dy}{dx} = 12(3x - 4)^3$.

Example

Differentiate $y = (5 - 2x)^7$.

Let $u = 5 - 2x$ and $y = u^7$.

Then $\dfrac{dy}{du} = 7u^6$ and $\dfrac{du}{dx} = -2$.

Hence $\dfrac{dy}{dx} = 7u^6 \cdot -2 = -14u^6$.

Substituting back for x gives $\dfrac{dy}{dx} = -14(5 - 2x)^6$.

We will now apply the chain rule to other cases of a function of a function.

Example

Differentiate $y = \sin 4x$.

Let $u = 4x$ and $y = \sin u$.

Then $\dfrac{dy}{du} = \cos u$ and $\dfrac{du}{dx} = 4$.

Hence $\dfrac{dy}{dx} = 4 \cos u$.

Substituting back for x gives $\dfrac{dy}{dx} = 4 \cos 4x$.

Example

Differentiate $y = \cos^2(3x)$.

Remember that this means $y = (\cos 3x)^2$.
Here there is more than one composition and so the chain rule must be extended to:

$$\frac{dy}{dx} = \frac{dy}{du} \cdot \frac{du}{dv} \cdot \frac{dv}{dx}$$

Let $v = 3x$ and $u = \cos v$ and $y = u^2$.

Then $\dfrac{dy}{du} = 2u$, $\dfrac{du}{dv} = -\sin v$ and $\dfrac{dv}{dx} = 3$.

Hence $\dfrac{dy}{dx} = -6u \sin v$.

Substituting back for x gives $\dfrac{dy}{dx} = -6 \cos 3x \sin 3x$.

This is the formal version of the working for chain rule problems. In practice, the substitution is often implied, as shown in the following examples. However, it is important to be able to use the formal substitution, both for more difficult chain rule examples and as a skill for further techniques in calculus.

Example

$f(x) = 2(7 - 3x)^6 - \tan 2x$

$f'(x) = 12(7 - 3x)^5 \cdot (-3) - 2 \sec^2 2x$ ....

$\quad = -36(7 - 3x)^5 - 2 \sec^2 2x$

> This working is sufficient and is what is usually done.

Example

$f(x) = \tan(3x^2 - 4) + \dfrac{2}{\sqrt{2x - 1}} = \tan(3x^2 - 4) + 2(2x - 1)^{-\frac{1}{2}}$

$f'(x) = [\sec^2(3x^2 - 4) \cdot 6x] + \left[2 \cdot -\dfrac{1}{2}(2x - 1)^{-\frac{3}{2}} \cdot 2 \right]$

$\quad = 6x \sec^2(3x^2 - 4) - \dfrac{2}{(2x - 1)^{\frac{3}{2}}}$

Exercise 2

Differentiate the following:

1 $f(x) = (x + 4)^2$ **2** $f(x) = (2x + 3)^2$ **3** $f(x) = (3x - 4)^2$

4 $f(x) = (5x - 2)^4$ **5** $f(x) = (5 - x)^3$ **6** $f(x) = (7 - 2x)^4$

7 $y = (9 - 4x)^5$ **8** $y = 4(2x + 3)^6$ **9** $y = (3x + 8)^{\frac{1}{2}}$

10 $y = (2x - 9)^{\frac{5}{3}}$ **11** $y = \sqrt[3]{6x - 5}$ **12** $y = \dfrac{1}{\sqrt{3x - 2}}$

13 $f(x) = \dfrac{4}{5x - 4}$ **14** $f(x) = \dfrac{7}{3 - 8x}$ **15** $P = \dfrac{3}{(4 - 3k)^2}$

16 $N = \dfrac{5}{\sqrt{(8 - 5p)^3}}$ **17** $y = \sin 4x$ **18** $y = \cos 3x$

19 $y = -\sin\dfrac{1}{2}x$ **20** $y = \tan 6x$ **21** $y = \sec 9x$

22 $y = 6x + \cot 3x$

23 $y = \csc 2x + (3x + 2)^4$

24 $y = \sin 5x - \dfrac{4}{\sqrt{(3x + 4)^5}}$

25 $y = \sin^3 x$

26 $y = \tan^2(4x)$

27 $y = 3x^4 - \cos^3 x$

28 $y = \dfrac{2}{(3x - 4)^5} - \sec^2(2x)$

29 $y = \cos\left(3x - \dfrac{\pi}{4}\right)$

30 $y = \tan(\sqrt{x + 1})$

9.3 Differentiating exponential and logarithmic functions

Investigation

Draw graphs of: (a) $y = 10^x$ (b) $y = 5^x$ (c) $y = 3^x$ (d) $y = 2^x$

For each graph, draw the derivative graph, using a graphing calculator.

You should notice that the derivative graph is of a similar form to the original, that is, it is an exponential graph.

For $y = 3^x$, the derivative graph is just above the original.

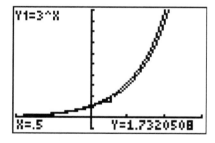

For $y = 2^x$, the derivative graph is below the original.

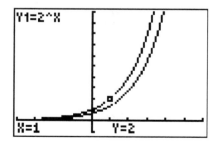

This suggests that there is a function for which the derivative graph is identical to the original graph and that the base of this function lies between 2 and 3. What is this base?

This question was studied for many years by many mathematicians including Leonhard Euler, who first used the symbol e. The answer is that this base is e. Check that $y = e^x$ produces its own graph for the derived function on your calculator. Remember that e = 2.71828... is an irrational number.

During the study of exponential functions in Chapter 5, we met the natural exponential function, $y = e^x$. The significance of this function becomes clearer now: the derivative of e^x is itself.

$$\frac{d}{dx}(e^x) = e^x$$

Below is a formal proof of this.

Proof of derivative of e^x

Consider the curve $y = a^x$.

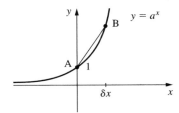

Gradient of the chord $AB = \dfrac{a^{\delta x} - 1}{\delta x}$

Gradient at $A = \displaystyle\lim_{\delta x \to 0}\left(\dfrac{a^{\delta x} - 1}{\delta x}\right)$

Now consider two general points on the exponential curve.

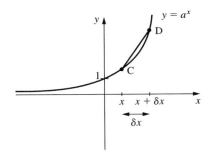

Gradient $CD = \dfrac{a^{x + \delta x} - a^x}{x + \delta x - x}$

Gradient at $C = \displaystyle\lim_{\delta x \to 0}\dfrac{a^x(a^{\delta x} - 1)}{\delta x}$

Hence the gradient at C is a^x multiplied by the gradient at A.

But when $a \equiv e$, the gradient at $A = 1$ (this can be checked on a graphing calculator).
Gradient at $C = e^x \cdot 1$

Hence $\dfrac{d}{dx}(e^x) = e^x$

> This is another property of the curve $y = e^x$. At $(0,1)$ its gradient is 1.

Example

Differentiate $y = e^{5x}$.

This can be considered as $y = e^u$ where $u = 5x$.

Since $\dfrac{du}{dx} = 5$ and $\dfrac{dy}{du} = e^u$

$\dfrac{dy}{dx} = e^u \cdot 5$

$\phantom{\dfrac{dy}{dx}} = 5e^{5x}$

Example

$y = e^{x^2 - 1}$

$\Rightarrow \dfrac{dy}{dx} = 2xe^{x^2 - 1}$

The result for e^x can be combined with the chain rule to create a general rule for differentiating exponential functions.

$$\frac{d}{dx}(e^{f(x)}) = f'(x)e^{f(x)}$$

This now allows us to differentiate the inverse function of $y = e^x$, known as the **natural logarithmic function** $y = \ln x$.

As $y = e^x$ and $y = \ln x$ are inverse functions, from Chapter 5 we know that $e^{\ln x} = x$.

We can differentiate both sides of this equation (with respect to x).

$$\frac{d}{dx}(e^{\ln x}) = \frac{d}{dx}(x)$$

$$\Rightarrow e^{\ln x} \cdot \frac{d}{dx}(\ln x) = 1$$

$$\Rightarrow \frac{d}{dx}(\ln x) = \frac{1}{e^{\ln x}}$$

$$\Rightarrow \frac{d}{dx}(\ln x) = \frac{1}{x}$$

This is a very important result.

$$\frac{d}{dx}(\ln x) = \frac{1}{x}$$

The result for $\ln x$ can be combined with the chain rule to create this general result:

If $\quad y = \ln(f(x))$

Then $\dfrac{dy}{dx} = \dfrac{1}{f(x)} \cdot f'(x)$

$\phantom{Then \dfrac{dy}{dx}} = \dfrac{f'(x)}{f(x)}$

This result is particularly important for integration in Chapter 15.

Example

Differentiate $y = \ln(4x)$.

$$\frac{dy}{dx} = 4 \cdot \frac{1}{4x} = \frac{1}{x}$$

Example

Differentiate $y = \ln(\sin x)$.

$$\frac{dy}{dx} = \frac{1}{\sin x} \cdot \cos x$$
$$= \cot x$$

Example

Differentiate $y = \ln(3x - 2)$.

$$\frac{dy}{dx} = \frac{1}{3x - 2} \cdot 3$$
$$= \frac{3}{3x - 2}$$

Using the results for e^x and $\ln x$ helps us generalize so that we can find the derivatives of any exponential or logarithmic function.

To find out how to differentiate $y = a^x$ we first consider $y = 4^x$.

Since $e^{\ln 4} = 4$, we can rewrite this function as $y = (e^{\ln 4})^x$.

$$y = e^{x \ln 4}$$

$$\Rightarrow \frac{dy}{dx} = \ln 4 e^{x \ln 4}$$

$$\Rightarrow \frac{dy}{dx} = \ln 4 \cdot 4^x$$

In general, $a^x = (e^{\ln a})^x = e^{x \ln a}$

and so $\dfrac{dy}{dx} = \ln a \cdot e^{x \ln a} = \ln a \cdot a^x$

$$\frac{d}{dx}(a^x) = \ln a \cdot a^x$$

We will now look at $y = \log_a x$. In this case the change of base formula will help.

$$\log_a x = \frac{\ln x}{\ln a} = \frac{1}{\ln a} \cdot \ln x$$

Differentiating gives us

$$\frac{d}{dx}(\log_a x) = \frac{1}{\ln a} \cdot \frac{1}{x}$$

$$= \frac{1}{x \ln a}$$

$$\frac{d}{dx}(\log_a x) = \frac{1}{x \ln a}$$

Considering these two general results, it is clear that the results for e^x and $\ln x$ are actually just special cases.

Example

Differentiate $y = 3 \log_7 x$.

$$\frac{dy}{dx} = \frac{3}{x \ln 7}$$

Example

Differentiate $y = k \cdot 6^{2x}$.

$$\frac{dy}{dx} = 2k \ln 6 \cdot 6^{2x}$$

Sometimes it is useful to use laws of logarithms to assist the differentiation.

Example

Differentiate $y = \ln\left(\frac{2x + 1}{x - 4}\right)$.

We can consider this as $y = \ln(2x + 1) - \ln(x - 4)$.

So $\dfrac{dy}{dx} = \dfrac{2}{2x + 1} - \dfrac{1}{x - 4}$

$$= \frac{2(x - 4) - (2x + 1)}{(2x + 1)(x - 4)}$$

$$= \frac{-9}{(2x + 1)(x - 4)}$$

Exercise 3

Differentiate the following:

1 $f(x) = e^{3x}$ **2** $f(x) = e^{7x}$ **3** $f(x) = -e^{4x}$

4 $f(x) = \dfrac{2}{e^{5x}}$ **5** $f(x) = -\dfrac{6}{e^{9x}}$ **6** $f(x) = e^{x^2}$

7 $f(x) = e^{2x+3}$ **8** $f(x) = \ln 3x$ **9** $f(x) = \ln 7x$

10 $f(x) = -2 \ln 4x$ **11.** $f(x) = \ln(2x^2 + 4)$ **12** $y = 4^x$

13 $y = 10^x$ **14** $y = 6 \cdot 5^x$ **15** $y = e^{3x} - 3^x$

16 $y = \ln 2x - 2^x$ **17** $y = \log_2 x$ **18** $y = \log_8 x$

19 $y = 4^x - \log_5 x$ **20** $y = e^{4x} - \sin 2x + \ln x$ **21** $y = \ln(\cos x)$

22 $y = \ln(\tan x)$ **23** $y = \tan(\ln x)$

9.4 Product rule

Using the chain rule, $y = (2x + 3)^4$ can be differentiated without multiplying out the brackets first. However, this does not really help to differentiate $y = (3x + 2)(2x - 3)^3$ without some unpleasant simplification. Equally, we cannot currently differentiate $y = e^x \sin x$. These functions are products of two functions, and to be able to differentiate these we need to use the product rule.

> For $y = uv$ where u and v are functions of x,
> $$\frac{dy}{dx} = v\frac{du}{dx} + u\frac{dv}{dx}$$

This is sometimes remembered in the shortened form $\dfrac{dy}{dx} = v\,du + u\,dv$

Proof

Consider $y = uv$ where u and v are functions of x.

If δx is a small increase in x, and δu, δv and δy are the corresponding increases in u, v and y, then

$$y + \delta y = (u + \delta u)(v + \delta v) = uv + u\delta v + v\delta u + \delta u\delta v$$

As $y = uv$, $\delta y = u\delta v + v\delta u + \delta u\delta v$

So $\dfrac{\delta y}{\delta x} = u\dfrac{\delta v}{\delta x} + v\dfrac{\delta u}{\delta x} + \delta u\dfrac{\delta v}{\delta x}$

Now when $\delta x \to 0$, $\dfrac{\delta y}{\delta x} \to \dfrac{dy}{dx}$, $\dfrac{\delta u}{\delta x} \to \dfrac{du}{dx}$, $\dfrac{\delta v}{\delta x} \to \dfrac{dv}{dx}$, $\delta u \to 0$

Therefore $\dfrac{dy}{dx} = \lim\limits_{\delta x \to 0}\left(\dfrac{\delta y}{\delta x}\right)$

$$\Rightarrow \frac{dy}{dx} = u\frac{dv}{dx} + v\frac{du}{dx} + 0$$

$$\Rightarrow \frac{dy}{dx} = u\frac{dv}{dx} + v\frac{du}{dx}$$

Example

Differentiate $y = (3x + 2)(2x - 3)^3$.

Let $y = uv$ where $u = 3x + 2$ and $v = (2x - 3)^3$.

$$\frac{du}{dx} = 3 \qquad \frac{dv}{dx} = 3(2x - 3)^2 \cdot 2$$
$$= 6(2x - 3)^2$$

$$\frac{dy}{dx} = (3x + 2) \cdot 6(2x - 3)^2 + 3(2x - 3)^3$$
$$= 3(2x - 3)^2[2(3x + 2) + 2x - 3]$$
$$= 3(2x - 3)^2(8x + 1)$$

> There are often common factors which can be used to simplify the answer.

Example

Differentiate $y = e^x \sin x$.

Let $y = uv$ where $u = e^x$ and $v = \sin x$.

$$\frac{du}{dx} = e^x \qquad \frac{dv}{dx} = \cos x$$

$$\frac{dy}{dx} = e^x \sin x + e^x \cos x$$
$$= e^x(\sin x + \cos x)$$

> This is the mechanics of the solution. It is not absolutely necessary for it to be shown as part of the solution.

Example

Differentiate $y = 4x^2 \ln x$.

Let $y = uv$ where $u = 4x^2$ and $v = \ln x$.

$$\frac{du}{dx} = 8x \qquad \frac{dv}{dx} = \frac{1}{x}$$

$$\frac{dy}{dx} = 8x \ln x + 4x^2 \cdot \frac{1}{x}$$
$$= 8x \ln x + 4x$$
$$= 4x(2 \ln x + 1)$$

Example

Differentiate $y = (2x + 1)^3 e^{2x} \cos x$.

This example is a product of three functions. We need to split it into two parts and then further split the second part.

Let $y = uv$ where $u = (2x + 1)^3$ and $v = e^{2x} \cos x$.

$$\frac{du}{dx} = 3(2x + 1)^2 \cdot 2 \qquad \text{For } \frac{dv}{dx} \text{ we need to use the product}$$
$$\text{rule again.}$$

$$= 6(2x + 1)^2$$

Let $v = fg$ where $f = e^{2x}$ and $g = \cos x$

$$\frac{df}{dx} = 2e^{2x} \qquad \frac{dg}{dx} = -\sin x$$

$$\frac{dv}{dx} = 2e^{2x} \cos x - e^{2x} \sin x$$

$$= e^{2x}(2 \cos x - \sin x)$$

$$\frac{dy}{dx} = 6(2x + 1)^2 e^{2x} \cos x + (2x + 1)^3 e^{2x}(2 \cos x - \sin x)$$

$$= (2x + 1)^2 e^{2x}[6 \cos x + (2x + 1)(2 \cos x - \sin x)]$$

Exercise 4

Find the derivative of each of these.

1 $y = x^2 \sin x$ **2** $y = x^3 \cos x$ **3** $y = 3x^2 e^x$

4 $y = e^{3x} \sin x$ **5** $y = \ln x \sin x$ **6** $y = \sin x \cos x$

7 $y = \sin 3x \cos 2x$ **8** $y = x^2(x - 1)^2$ **9** $y = x^3(x - 2)^4$

10 $y = 2x^3(3x + 2)^2$ **11** $y = (x - 2)(2x + 1)^3$

12 $y = (x + 5)^2(3x - 2)^4$ **13** $y = (5 - 2x)^3(3x + 4)^2$

14 $y = (3x + 4)^3 \sin x$ **15** $y = 5^x \cos x$

16 $y = x^3 \log_6 x$ **17** $y = e^{4x} \sec 3x$

18 $y = (2x - 1)^3 \csc 3x$ **19** $y = 4^x \log_8 x$

20 $y = x \ln(2x + 3)$ **21** $y = 4x^2 \ln(x^2 + 2x + 5)$

22 $y = e^{3x} \sec\left(2x - \dfrac{\pi}{4}\right)$ **23** $y = \dfrac{3}{x^4} \tan\left(3x + \dfrac{\pi}{2}\right)$

24 $y = x^2 \ln x \sin x$ **25** $y = e^{3x}(x + 2)^2 \tan x$

9.5 Quotient rule

This rule is used for differentiating a quotient (one function divided by another) such as $y = \dfrac{u}{v}$.

Consider the function $y = \dfrac{(3x - 4)^2}{x^2} = x^{-2}(3x - 4)^2$

We can differentiate this using the product rule.

Let $y = uv$ where $u = (3x - 4)^2$ and $v = x^{-2}$.

$$\frac{du}{dx} = 6(3x - 4) \qquad \frac{dv}{dx} = -2x^{-3}$$

$$\frac{dy}{dx} = 6x^{-2}(3x - 4) - 2x^{-3}(3x - 4)^2$$

$$= 2x^{-3}(3x - 4)[3x - (3x - 4)]$$

$$= 2x^{-3}(3x - 4)(4)$$

$$= \frac{8(3x - 4)}{x^3}$$

In this case it was reasonably easy to rearrange into a product, but that is not always so. Generally, it is not wise to use the product rule for differentiating quotients as it often leads to an answer that is difficult to simplify, and hence a rule for differentiating quotients would be useful.

Consider $y = \dfrac{u}{v}$ where u and v are functions of x.

This can be written as $y = u \cdot \dfrac{1}{v}$.

$$\frac{d}{dx}\left(\frac{1}{v}\right) = -v^{-2} \cdot \frac{d}{dx}(v)$$

> This is a slightly different application of the chain rule.

$$= -\frac{1}{v^2} \cdot \frac{dv}{dx}$$

Using the product rule:

$$\frac{dy}{dx} = \frac{1}{v} \cdot \frac{du}{dx} + u \cdot -\frac{1}{v^2} \cdot \frac{dv}{dx}$$

$$= \frac{v}{v^2} \cdot \frac{du}{dx} + -\frac{u}{v^2} \cdot \frac{dv}{dx}$$

$$= \frac{v\dfrac{du}{dx} - u\dfrac{dv}{dx}}{v^2}$$

This is the quotient rule:

$$\frac{dy}{dx} = \frac{v\dfrac{du}{dx} - u\dfrac{dv}{dx}}{v^2}$$

> This is often remembered as
> $$\frac{dy}{dx} = \frac{v\,du - u\,dv}{v^2}$$
> The numerator of the quotient rule is very similar to the product rule but the sign is different.

Example

Differentiate $y = \dfrac{(3x - 4)^2}{x^2}$ using the quotient rule.

Let $u = (3x - 4)^2 \qquad v = x^2 \qquad v^2 = x^4$

$$\frac{du}{dx} = 6(3x - 4) \qquad \frac{dv}{dx} = 2x$$

> Once again, this is the mechanics of the solution and so does not necessarily need to be shown.

So $\dfrac{dy}{dx} = \dfrac{6x^2(3x - 4) - 2x(3x - 4)^2}{x^4}$

$$= \frac{2x(3x - 4)[3x - (3x - 4)]}{x^4}$$

$$= \frac{2(3x - 4)(4)}{x^3}$$

$$= \frac{8(3x - 4)}{x^3}$$

Example

Differentiate $y = \dfrac{e^{2x}}{\sin x}$.

Let $u = e^{2x}$ $\quad v = \sin x$ $\quad v^2 = \sin^2 x$

$$\frac{du}{dx} = 2e^{2x} \quad \frac{dv}{dx} = \cos x$$

So $\dfrac{dy}{dx} = \dfrac{2e^{2x} \sin x - e^{2x} \cos x}{\sin^2 x}$

$\qquad = \dfrac{e^{2x}(2 \sin x - \cos x)}{\sin^2 x}$

Example

Differentiate $y = \dfrac{\ln x}{(2x - 5)^3}$.

Let $u = \ln x$ $\quad v = (2x - 5)^3$ $\quad v^2 = (2x - 5)^6$

$$\frac{du}{dx} = \frac{1}{x} \quad\quad \frac{dv}{dx} = 6(2x - 5)^2$$

So $\dfrac{dy}{dx} = \dfrac{\dfrac{1}{x}(2x - 5)^3 - 6 \ln x (2x - 5)^2}{(2x - 5)^6}$

$\qquad = \dfrac{\dfrac{1}{x}(2x - 5) - 6 \ln x}{(2x - 5)^4}$

$\qquad = \dfrac{2x - 5 - 6x \ln x}{x(2x - 5)^4}$

Exercise 5

Use the quotient rule to differentiate these.

1 $f(x) = \dfrac{e^x}{\cos x}$ 　　**2** $f(x) = \dfrac{6x^2}{x + 3}$ 　　**3** $f(x) = \dfrac{7x}{\tan x}$

4 $f(x) = \dfrac{\ln x}{4x}$ 　　**5** $f(x) = \dfrac{e^x}{x - 4}$ 　　**6** $f(x) = \dfrac{x + 3}{x - 3}$

7 $f(x) = \dfrac{\sqrt{x+9}}{x^2}$

8 $f(x) = \dfrac{4^x}{\sqrt{x}}$

9 $f(x) = \dfrac{2x}{\sqrt{x-1}}$

10 $y = \dfrac{e^{3x}}{9x^2}$

11 $y = \dfrac{\log_6 x}{x+6}$

12 $y = \dfrac{\ln x}{\ln(x-4)}$

13 $y = \dfrac{e^x}{e^x - e^{-x}}$

14 $y = \dfrac{\sin 2x}{e^{6x}}$

15 $y = \dfrac{4(3x-2)^5}{(2x+3)^3}$

16 $y = \dfrac{x \sin x}{e^x}$

17 $y = \dfrac{x^2 e^{3x}}{(x+5)^2}$

18 $y = \dfrac{\sec\left(x + \dfrac{\pi}{4}\right)}{e^{2x}}$

19 $y = \dfrac{\cot\left(2x - \dfrac{\pi}{3}\right)}{\ln(3x+1)}$

20 Use the quotient rule to prove the results for $\tan x$, $\csc x$, $\sec x$ and $\cot x$. You need to remember that $\tan x = \dfrac{\sin x}{\cos x}$.

> Also consider how you could have proved these two results using only the chain rule.

9.6 Implicit differentiation

This is the differentiation of functions that are stated implicitly. Until now we have mostly considered functions that are stated explicitly, that is, $y = \ldots$

Functions defined implicitly have equations that are not in the form $y = \ldots$ Some of these equations are easily made explicit (such as $2x + 3y = 5$) but others are more difficult to rearrange. Some of these implicit equations may be familiar, such as the circle equation $(x-4)^2 + (y+3)^2 = 36$. Differentiating implicit functions does not require any further mathematical techniques than those covered so far. The key concept utilized in implicit differentiation is the chain rule.

Method for implicit differentiation

> **1.** Differentiate each term, applying the chain rule to functions of the variable.
>
> **2.** Rearrange the answer to the form $\dfrac{dy}{dx}$.

Example

Find $\dfrac{dy}{dx}$ and $\dfrac{d^2y}{dx^2}$ for $3x^2 + y^2 = 7$

Differentiating with respect to x:

$6x + 2y\dfrac{dy}{dx} = 0$

$\Rightarrow 2y\dfrac{dy}{dx} = -6x$

> It is possible to rearrange this function to an explicit form. However, unless you are told otherwise, it is often better to leave it in this form and differentiate implicitly.

> Applying the chain rule gives
>
> $\dfrac{d}{dx}(y^2) = 2y \cdot \dfrac{dy}{dx}$.

$$\Rightarrow \frac{dy}{dx} = -\frac{3x}{y}$$

Note that the answer contains both x and y.

We can now find the second derivative by differentiating this again.

Using the quotient rule:

$$\frac{d^2y}{dx^2} = \frac{-3y + 3x\dfrac{dy}{dx}}{y^2}$$

$$= \frac{-3y + 3x\left(-\dfrac{3x}{y}\right)}{y^2}$$

$$= \frac{-3y - \dfrac{9x^2}{y}}{y^2}$$

$$= \frac{-3y^2 - 9x^2}{y^3}$$

In a case like this it is important to be able to explicitly state $\dfrac{dy}{dx} = \ldots$ so that the second derivative can be found, but this is not always the situation.

We would usually leave the answer in this form. However, if we wanted $\dfrac{d^2y}{dx^2}$ as a function of x, we could proceed as follows:

$$\frac{d^2y}{dx^2} = \frac{-3(7 - 3x^2) - 9x^2}{(7 - 3x^2)^{\frac{3}{2}}}$$

$$= \frac{-21}{(7 - 3x^2)^{\frac{3}{2}}}$$

We know that
$$3x^2 + y^2 = 7$$
$$\Rightarrow y^2 = 7 - 3x^2$$

Example

Find $\dfrac{dy}{dx}$ for $6 \sin x - e^{3x}y^3 = 9$.

Differentiating with respect to x:

$$6 \cos x - \left(3e^{3x}y^3 + e^{3x}3y^2\frac{dy}{dx}\right) = 0$$

$$\Rightarrow 3e^{3x}y^2\left(y + \frac{dy}{dx}\right) = 6 \cos x$$

$$\Rightarrow y + \frac{dy}{dx} = \frac{2 \cos x}{e^{3x}y^2}$$

$$\Rightarrow \frac{dy}{dx} = \frac{2 \cos x}{e^{3x}y^2} - y$$

Use the product rule to differentiate $e^{3x}y^3$.

Example

Find $\dfrac{dQ}{dp}$ for $\sin \pi p + \dfrac{2Q}{(p+3)^2} = \ln p$.

Differentiating with respect to p:

$$\pi \cos \pi p + \dfrac{2\dfrac{dQ}{dp}(p+3)^2 - 2(p+3) \cdot 2Q}{(p+3)^4} = \dfrac{1}{p}$$

$$\Rightarrow \dfrac{2\dfrac{dQ}{dp}(p+3)^2 - 4Q(p+3)}{(p+3)^4} = \dfrac{1}{p} - \pi \cos \pi p$$

$$\Rightarrow \dfrac{2\dfrac{dQ}{dp}}{(p+3)^2} - \dfrac{4Q}{(p+3)^3} = \dfrac{1}{p} - \pi \cos \pi p$$

$$\Rightarrow \dfrac{2\dfrac{dQ}{dp}}{(p+3)^2} = \dfrac{1}{p} - \pi \cos \pi p + \dfrac{4Q}{(p+3)^3}$$

$$\Rightarrow 2\dfrac{dQ}{dp} = \dfrac{(p+3)^2}{p} - \pi(p+3)^2 \cos \pi p + \dfrac{4Q}{p+3}$$

$$\Rightarrow \dfrac{dQ}{dp} = \dfrac{(p+3)^2}{2p} - \dfrac{\pi}{2}(p+3)^2 \cos \pi p + \dfrac{2Q}{p+3}$$

Example

Find the equations of the tangents to $3x^2y - y^2 = 27$ when $x = 2$.

Differentiating with respect to x:

$$6xy + 3x^2\dfrac{dy}{dx} - 2y\dfrac{dy}{dx} = 0$$

$$\Rightarrow \dfrac{dy}{dx}(3x^2 - 2y) = -6xy$$

$$\Rightarrow \dfrac{dy}{dx} = \dfrac{-6xy}{3x^2 - 2y}$$

To find $\dfrac{dy}{dx}$ we now require the y-coordinates. So, from the formula

$3x^2y - y^2 = 27$, when $x = 2$, we find

$$12y - y^2 = 27$$
$$\Rightarrow y^2 - 12y + 27 = 0$$
$$\Rightarrow (y - 3)(y - 9) = 0$$
$$\Rightarrow y = 3, y = 9$$

At $(2, 3)$

$$\dfrac{dy}{dx} = \dfrac{-6 \cdot 2 \cdot 3}{12 - 6}$$

$$= \dfrac{-36}{6}$$

$$= -6$$

At $(2, 9)$

$$\dfrac{dy}{dx} = \dfrac{-6 \cdot 2 \cdot 9}{12 - 18}$$

$$= \dfrac{-108}{-6}$$

$$= 18$$

So the equation of the tangent is

$y - 3 = -6(x - 2)$

$\Rightarrow y = -6x + 15$

So the equation of the tangent is

$y - 9 = 18(x - 2)$

$\Rightarrow y = 18x - 27$

Some questions will require the second derivative to be found, and a result to be shown to be true that involves $\dfrac{d^2y}{dx^2}, \dfrac{dy}{dx}$ and y. In the examples so far we have found $\dfrac{dy}{dx} = \ldots$ and then differentiated this again with respect to x to find $\dfrac{d^2y}{dx^2}$. With other questions, it is best to leave the result as an implicit function and differentiate for a second time, implicitly. The following two examples demonstrate this.

Example

Show that $x^2\dfrac{d^2y}{dx^2} + 4x\dfrac{dy}{dx} + 2y = -\dfrac{1}{x}$ for $x^2y + x \ln x = 6x$.

Differentiating with respect to x:

$2xy + x^2\dfrac{dy}{dx} + \ln x + \dfrac{1}{x} \cdot x = 6$

$\Rightarrow 2xy + x^2\dfrac{dy}{dx} + \ln x + 1 = 6$

Differentiating again with respect to x:

$2y + 2x\dfrac{dy}{dx} + 2x\dfrac{dy}{dx} + x^2\dfrac{d^2y}{dx^2} + \dfrac{1}{x} = 0$

$\Rightarrow x^2\dfrac{d^2y}{dx^2} + 4x\dfrac{dy}{dx} + 2y = -\dfrac{1}{x}$

Example

Given that $e^x y = \cos x$, show that $2y + 2\dfrac{dy}{dx} + \dfrac{d^2y}{dx^2} = 0$.

Differentiating with respect to x:

$e^x y + e^x\dfrac{dy}{dx} = -\sin x$

Differentiating again with respect to x:

$e^x y + e^x\dfrac{dy}{dx} + e^x\dfrac{dy}{dx} + e^x\dfrac{d^2y}{dx^2} = -\cos x$

From the original function, $-\cos x = -e^x y$

So we have $e^x y + e^x\dfrac{dy}{dx} + e^x\dfrac{dy}{dx} + e^x\dfrac{d^2y}{dx^2} = -e^x y$

237

$$\Rightarrow e^x y + 2e^x \frac{dy}{dx} + e^x \frac{d^2y}{dx^2} = -e^x y$$

$$\Rightarrow 2e^x y + 2e^x \frac{dy}{dx} + e^x \frac{d^2y}{dx^2} = 0$$

Dividing by e^x (since $e^x \neq 0 \ \forall \, x \in \mathbb{R}$):

$$\Rightarrow 2y + 2\frac{dy}{dx} + \frac{d^2y}{dx^2} = 0$$

Exercise 6

1 Find $\dfrac{dy}{dx}$ for:

   a $x^3 + xy = 4$ **b** $4x^2 + y^2 = 9$ **c** $y^3 - \sqrt{x} = 0$

   d $(x + 3)(y + 2) = \ln x$ **e** $xy = y^2 - 7$ **f** $e^y = (x - y)^2$

   g $e^{2x}y^3 = 9 - \sin 3x$ **h** $y = \cos(x + y)$ **i** $x^4 = y \ln y$

   j $(x + y)^3 = e^y$ **k** $\dfrac{(x + y)^4}{y} = 8x - e^x$

2 Find $\dfrac{dy}{dx}$ and $\dfrac{d^2y}{dx^2}$ for:

   a $4y + 3y^2 = x^2$ **b** $4xy = \sin x - y$ **c** $xe^y = 8$

3 For the function defined implicitly by $x^4 - 2xy + y^2 = 4$, find the equations of the tangents at $x = 1$.

4 Show that $2y + \dfrac{dy}{dx} + \dfrac{d^2y}{dx^2} = 0$ for $e^x y = \sin x$.

5 Given that $xy = \sin x$, show that $x^2 \dfrac{d^2y}{dx^2} + 2x \dfrac{dy}{dx} + x^2 y = 0$.

6 Show that $x^3 \dfrac{d^2y}{dx^2} + x^2 \dfrac{dy}{dx} + xy = -2$ for $xy = \ln x$.

7 Given that $e^{2x}y = \ln x$, show that $x^2 e^{2x}\left(4y + 4\dfrac{dy}{dx} + \dfrac{d^2y}{dx^2}\right) = -1$.

9.7 Differentiating inverse trigonometric functions

In order to find the derivative of $\sin^{-1} x$ (or $\arcsin x$), we apply implicit differentiation.

Consider $y = \sin^{-1} x$

   $\Rightarrow x = \sin y$

Differentiating with respect to x:

$$1 = \cos y \cdot \frac{dy}{dx}$$

$$\Rightarrow \frac{dy}{dx} = \frac{1}{\cos y}$$

$$\Rightarrow \frac{dy}{dx} = \frac{1}{\sqrt{1-x^2}} \quad \cdots\cdots\cdots\cdots\cdots$$

This is because $\sin y = x \Rightarrow \sin^2 y = x^2$

$$\Rightarrow 1 - \cos^2 y = x^2$$

$$\Rightarrow \cos^2 y = 1 - x^2$$

$$\Rightarrow \cos y = \sqrt{1-x^2}$$

For $y = \sin^{-1} x \quad \dfrac{dy}{dx} = \dfrac{1}{\sqrt{1-x^2}}$

We can now consider $y = \sin^{-1}\left(\dfrac{x}{a}\right)$

$$\Rightarrow \frac{x}{a} = \sin y$$

$$\Rightarrow x = a \sin y$$

Differentiating with respect to x:

$$1 = a \cos y \frac{dy}{dx}$$

$$\Rightarrow \frac{dy}{dx} = \frac{1}{a \cos y}$$

$$\Rightarrow \frac{dy}{dx} = \frac{1}{a\sqrt{1 - \dfrac{x^2}{a^2}}} \quad \cdots\cdots\cdots$$

This is because $\sin y = \dfrac{x}{a} \Rightarrow 1 - \cos^2 y = \dfrac{x^2}{a^2}$

$$\Rightarrow \cos^2 y = 1 - \frac{x^2}{a^2}$$

$$= \frac{\sqrt{a^2}}{a\sqrt{a^2 - x^2}}$$

$$= \frac{1}{\sqrt{a^2 - x^2}}$$

For $y = \sin^{-1}\left(\dfrac{x}{a}\right) \quad \dfrac{dy}{dx} = \dfrac{1}{\sqrt{a^2 - x^2}}$

Similarly $\dfrac{dy}{dx}$ can be obtained for $y = \cos^{-1}(x)$ and $y = \cos^{-1}\left(\dfrac{x}{a}\right)$.

For $y = \cos^{-1} x \quad \dfrac{dy}{dx} = \dfrac{-1}{\sqrt{1-x^2}}$

For $y = \cos^{-1}\left(\dfrac{x}{a}\right) \quad \dfrac{dy}{dx} = \dfrac{-1}{\sqrt{a^2 - x^2}}$

Now consider $y = \tan^{-1}(x)$

$$\Rightarrow x = \tan y$$

Differentiating with respect to x:

$$1 = \sec^2 y \cdot \frac{dy}{dx}$$

$$\Rightarrow \frac{dy}{dx} = \frac{1}{\sec^2 y}$$

$$\Rightarrow \frac{dy}{dx} = \frac{1}{x^2 + 1}$$

> Remember that
> $\sec^2 x = \tan^2 x + 1$
>
> So $\sec^2 y = \tan^2 y + 1$
> $\qquad\quad = x^2 + 1$

For $y = \tan^{-1}x$ $\qquad \dfrac{dy}{dx} = \dfrac{1}{1 + x^2}$

A similar result can be found for $\tan^{-1}\left(\dfrac{x}{a}\right)$.

For $y = \tan^{-1}\left(\dfrac{x}{a}\right)$ $\qquad \dfrac{dy}{dx} = \dfrac{a}{a^2 + x^2}$

Example

Differentiate $y = \cos^{-1}\left(\dfrac{x}{3}\right)$.

$$\frac{dy}{dx} = \frac{-1}{\sqrt{9 - x^2}}$$

Example

Differentiate $y = \sin^{-1}(4x)$.

$$\frac{dy}{dx} = \frac{1}{\sqrt{\dfrac{1}{16} - x^2}}$$

$$= \frac{1}{\sqrt{\dfrac{1}{16}}\sqrt{1 - 16x^2}}$$

$$= \frac{4}{\sqrt{1 - 16x^2}}$$

> In this case $a = \dfrac{1}{4}$.

We could also consider these examples to be applications of the chain rule. This may be easier and shorter (but both methods are perfectly valid). This is demonstrated below.

$$y = \sin^{-1}(4x)$$

Then $\dfrac{dy}{dx} = \dfrac{1}{\sqrt{1 - (4x)^2}} \cdot 4$

$$= \dfrac{4}{\sqrt{1 - 16x^2}}$$

In some cases it is not possible to use the stated results, and the chain rule must be applied.

Example

Differentiate $y = \tan^{-1}\sqrt{x}$.

Here we must use the chain rule.

$$\dfrac{dy}{dx} = \dfrac{1}{1 + (\sqrt{x})^2} \cdot \dfrac{1}{2}x^{-\frac{1}{2}}$$

$$= \dfrac{1}{2\sqrt{x}(1 + x)}$$

Exercise 7

Differentiate the following functions.

1 $y = \sin^{-1}\left(\dfrac{x}{5}\right)$

2 $y = \cos^{-1}\left(\dfrac{x}{8}\right)$

3 $y = \tan^{-1}\left(\dfrac{x}{10}\right)$

4 $y = \sin^{-1}\left(\dfrac{2x}{3}\right)$

5 $y = \cos^{-1}(3x)$

6 $y = \tan^{-1}\left(\dfrac{e^x}{2}\right)$

7 $y = \cos^{-1}\sqrt{x + 4}$

8 $y = \tan^{-1}(2x - 1)$

9 $y = \sin^{-1}(\ln 5x)$

9.8 Summary of standard results

This chapter has covered a variety of techniques including the chain rule, product rule, quotient rule and implicit differentiation. These have produced a number of standard results, which are summarized below.

$y =$	$\dfrac{dy}{dx}$
$\sin x$	$\cos x$
$\cos x$	$-\sin x$
$\tan x$	$\sec^2 x$
$\csc x$	$-\csc x \cot x$
$\sec x$	$\sec x \tan x$
$\cot x$	$-\csc^2 x$
e^x	e^x
$\ln x$	$\dfrac{1}{x}$
a^x	$a^x \ln a$
$\log_a x$	$\dfrac{1}{x \ln a}$
$\sin^{-1}(x)$	$\dfrac{1}{\sqrt{1 - x^2}}$
$\cos^{-1}(x)$	$\dfrac{-1}{\sqrt{1 - x^2}}$
$\tan^{-1}(x)$	$\dfrac{1}{1 + x^2}$
$\sin^{-1}\left(\dfrac{x}{a}\right)$	$\dfrac{1}{\sqrt{a^2 - x^2}}$
$\cos^{-1}\left(\dfrac{x}{a}\right)$	$\dfrac{-1}{\sqrt{a^2 - x^2}}$
$\tan^{-1}\left(\dfrac{x}{a}\right)$	$\dfrac{a}{a^2 + x^2}$

Within this chapter and Chapter 8, we have covered all of the differentiation techniques and skills for IB Higher Level. One of the key skills in an examination is to be able to identify which technique is required to solve a particular problem. Exercise 8 contains a mixture of examples that require the knowledge and use of standard results and the above techniques.

Exercise 8

Differentiate the following functions using the appropriate techniques and results.

1 $f(x) = x^2 - 5x + 9$ **2** $y = (2x - 7)^3$ **3** $f(x) = \cos 8x - \sqrt{9x}$

4 $y = \sec x - e^{5x}$ **5** $f(x) = x^3 e^{-4x}$ **6** $y = x^2 \ln x$

7 $f(x) = \dfrac{\sin 3x}{e^x}$ **8** $y = \dfrac{x \sin x}{e^{4x}}$ **9** $f(x) = 3^x \sin x$

10 $y = \dfrac{\log_2 x}{(x - 4)^3}$ **11** $f(x) = \dfrac{x^2 \ln x}{x + 9}$ **12** $y = 3 \cos 2x \sin 4x$

13 $y = 6 \sin^{-1} 2x$ **14** $f(x) = \dfrac{\cos^{-1} x}{3x^2}$ **15** $y = x \sin x \ln x$

16 $y = \dfrac{\ln(\cot x)}{e^x}$ **17** Find $f'(2)$ for $f(x) = \dfrac{x^3(x - 7)^2}{(2x - 1)^3}$.

18 Find $f'(4)$ for $f(x) = \dfrac{1}{x}\tan^{-1}\!\left(\dfrac{x}{4}\right)$.   **19** Find $\dfrac{dy}{dx}$ for $x^2 y + e^x y^2 = 9$.

20 Find $\dfrac{dy}{dx}$ for $x^4 y^3 - y \sin x = 2$.

9.9 Further differentiation problems

The techniques covered in this chapter can also be combined to solve differentiation problems of various types, including equations of tangents and normals, and stationary points. Problems of this type are given in Exercise 9.

Example

Find the stationary point for $\dfrac{y}{e^x} = x$ and determine its nature.

In this case, it is easiest to consider this as $y = xe^x$.

Using the product rule, $\dfrac{dy}{dx} = e^x x + e^x$

$$= e^x(x + 1)$$

For stationary points, $\dfrac{dy}{dx} = 0$

Hence $e^x(x + 1) = 0$

$$\Rightarrow x = -1$$

At $x = -1$,

$$y = e^{-1} \cdot -1$$

$$= -\frac{1}{e}$$

Hence the stationary point is $\left(-1, -\dfrac{1}{e}\right)$

$$\frac{d^2y}{dx^2} = e^x(x + 1) + e^x$$

$$= e^x(x + 2)$$

So, at $x = -1, \dfrac{d^2y}{dx^2} = e^{-1}(1)$

$= \dfrac{1}{e}$

So $\dfrac{d^2y}{dx^2} > 0$, therefore $\left(-1, -\dfrac{1}{e}\right)$ is a local minimum turning point.

Exercise 9

1 Find the gradient of the tangent to $y = \tan^{-1} 3x$ where $x = \dfrac{1}{\sqrt{3}}$.

2 Find the gradient of the tangent to $y = \ln\sqrt{1 - \cos 2x}$ where $x = \dfrac{\pi}{4}$.

3 Given $y = \dfrac{4^x}{e^x(x + 2)}$, find the rate of change where $x = 2$.

4 Find the equation of the tangents to $x^2y + y^2 = 6$ at $x = 1$.

5 Find the gradient of the tangent to $2x \ln x - y \ln y = 2e(1 - e)$ at the point (e, e^2).

6 Find the value of $\dfrac{d^2y}{dx^2}$ when $x = \pi$ for $\dfrac{y}{x} = \dfrac{\sin 2x}{e^x}$.

7 Find the stationary points of $y = \dfrac{x^2}{e^x}$.

8 Find the stationary points of $y = 4x^2 \ln x, x > 0$.

9 Find the stationary points, and their nature, of the curve given by $y = \dfrac{x^3}{e^x}$.

10 Show that the gradient of the tangent to the curve given by
$\dfrac{xy}{\pi} + \sin x \ln x = \cos x + 1$ at $x = \pi$ is $\ln \pi$.

Review exercise

 1 Differentiate these functions.

    a $y = 5(3x - 2)^4$ **b** $f(x) = \dfrac{7}{\sqrt{(3 - 2x^2)}}$ **c** $y = 6t - \sec 3t$

    d $f(x) = 6e^{8x}$ **e** $y = \ln 6x - 3^x$

 2 Differentiate these functions.

    a $y = e^{4x} \sin 3x$ **b** $y = \ln(x \sin x)$ **c** $f(x) = \dfrac{e^{5x}}{\sqrt{x + 4}}$

    d $y = \ln\left(\dfrac{3x + 4}{2x - 1}\right)$ **e** $y = \log_{10}\left(\dfrac{e^{2x} \cos 3x}{(x + 4)^2}\right)$

 3 Find $\dfrac{dy}{dx}$ for: **a** $4y^2 - 3x^2y = 5$ **b** $x^3 = y \ln x$

4 Find $\dfrac{d^2y}{dx^2}$ for $x^2 \sin x - e^xy = 7$.

5 Differentiate $y = 2\tan^{-1}\left(\dfrac{1 + \cos x}{\sin x}\right)$.

6 Find the exact value of the gradient of the tangent to $y = \dfrac{1}{x \sin x}$ where $x = \dfrac{\pi}{4}$.

 7 Find the gradient of the tangent to $3x^2 + 4y^2 = 7$ at the point where $x = 1$ and $y > 0$. [IB May 01 P1 Q4]

 8 A curve has equation $xy^3 + 2x^2y = 3$.
Find the equation of the tangent to this curve at the point $(1, 1)$. [IB May 02 P1 Q17]

 9 A curve has equation $x^3y^2 = 8$.
Find the equation of the normal to the curve at the point $(2,1)$. [IB May 03 P1 Q10]

 10 Find the stationary points of $y = x^2 \tan^{-1} x$.

 11 Find the stationary points of $y = \dfrac{e^{2x} \sin x}{x + 1}$ for $0 < x < \pi$.

 12 Show that the point $P(2, -2)$ lies on the circle with equation $(x + 2)^2 + (y - 2)^2 = 32$ and the parabola with equation $y^2 = 12 - 4x$.
Also show that these curves share a common tangent at P, and state the equation of this tangent.

 13 If $y = \ln(2x - 1)$, find $\dfrac{d^2y}{dx^2}$. [IB Nov 04 P1 Q5]

 14 Consider the function $f(t) = 3 \sec 2t + 5t$.
  a Find $f'(t)$.
  b Find the **exact** values of
    i $f(\pi)$
    ii $f'(\pi)$. [IB Nov 03 P1 Q8]

 15 Consider the equation $2xy^2 = x^2y + 3$.
  a Find y when $x = 1$ and $y < 0$.
  b Find $\dfrac{dy}{dx}$ when $x = 1$ and $y < 0$. [IB Nov 03 P1 Q15]

10 Differentiation 3 – Applications

Differential calculus is widely used in both the natural sciences and the human sciences. In physics, if we want to investigate the speed of a body falling under gravity, the force that will give a body a certain acceleration and hence a certain velocity, or the rate of decay of a radioactive material, then differential calculus will help us. In chemistry, we determine rates of reaction using calculus, and in biology a problem such as the rate of absorption of aspirin into the bloodstream as a function of time would have its solution based on differential calculus.

Differentiation may also be applied to a large number of problems that deal with the issue of extremes; this could include the biggest, the smallest, the greatest or the least. These maximum or minimum amounts may be described as values for which a certain rate of change (increase or decrease) is zero, i.e. stationary points. For example, it is possible to determine how high a projectile will go by finding the point at which its change of altitude with respect to time, that is, its velocity, is equal to zero.

10.1 Optimization problems

We have already met the idea of stationary points on a curve, which can give rise to local maxima and minima. This idea of something having a maximum and a minimum value can be used in a variety of situations. If we need to find out when a quantity is as small as possible or as large as possible, given that we can model the situation mathematically, then we can use differential calculus.

Imagine a car manufacturing company that is aiming its new model at the cheaper end of the market. One of the jobs of the marketing department in this company is to decide how much to sell each car for. If they decide to sell at just above cost price, then the company will only make a small amount of profit per car, but provided all other features of the marketing are correct the company will sell a large number. If they decide to charge a higher price, then the company will make more profit per car, but will probably sell fewer cars. Hence the marketing department need to find the right price to charge that will maximize the company's profit.

Obviously to maximize a function f(x) you are looking for the greatest value within a given region. Similarly to minimize a function, you are looking for the smallest. This may or may not be a stationary point. Many economists and engineers are faced with problems such as these, and this area of study is known as **optimization**. For the purposes of this course the greatest and least values will always occur at a stationary point.

Because it is based on "real life" situations, the problem is not always as simple as it might first seem. At its most basic level, optimization is just applying differentiation to a formula given in a "real life" scenario. However, in more complex questions there are often a number of steps that may need to precede this.

Method for solving optimization problems

1 Draw a diagram and write down the formula suggested by the question.
2 If the formula involves three variables, find a link between two of them.
3 Now substitute into the original formula. We now have a formula that links two variables.
4 Differentiate.
5 Make the equation equal to zero and solve.
6 Find the other value(s) by substitution.
7 Check whether it is a maximum or minimum point.

We will now demonstrate this with a number of examples.

> Always check that it is the two variables that the question is talking about!

Example

In design technology class, Ayesha is asked to make a box in the shape of a cuboid from a square sheet of card with each edge being 1.5 metres long. To do this she removes a square from each corner of the card. What is the biggest box that she can make?

The diagram below shows the cardboard with a square of side x metres removed from each corner, and the box into which it is made.

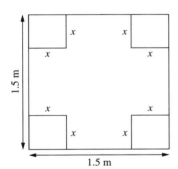 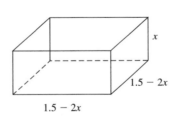

Step 1. Volume of the box, $V = x(1.5 - 2x)^2$. In this case the formula has only two variables: hence steps 2 and 3 can be ignored.
Step 4.

$$V = x(2.25 - 6x + 4x^2)$$
$$= 2.25x - 6x^2 + 4x^3$$
$$\Rightarrow \frac{dV}{dx} = 2.25 - 12x + 12x^2$$

Step 5. This is stationary when $2.25 - 12x + 12x^2 = 0$.

Using a calculator to solve the quadratic equation gives $x = \dfrac{1}{4}$ or $x = \dfrac{3}{4}$.

Step 6. $V = \dfrac{1}{4}$ or 0

For step 7, to test that this is indeed a maximum, we differentiate $\dfrac{dV}{dx}$ and apply the second derivative test.

$$\dfrac{d^2V}{dx^2} = -12 + 24x$$

When $x = \dfrac{1}{4}$, $\dfrac{d^2V}{dx^2} = -6$ and when $x = \dfrac{3}{4}$, $\dfrac{d^2V}{dx^2} = 6$. Hence the maximum value occurs when $x = \dfrac{1}{4}$ and the maximum volume is $\dfrac{1}{4}$m³.

This question can also be done on a calculator by inputting the curve $y = x(1.5 - 2x)^2$ and finding the maximum value.
The calculator display is shown below.

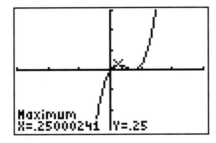

Hence the maximum volume is $\dfrac{1}{4}$m³.

Example

A farmer wishes to fence in part of his field as a safe area for his sheep. The shape of the area is a rectangle, but he has only 100 m of fencing. What are the dimensions of the safe area that will make it as large as possible?
The area is a rectangle with dimension x m by y m.
This is shown below.

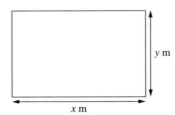

Step 1. Area = xy, i.e. $A = xy$. Hence we have a formula with three variables.
Step 2. We know that the farmer has 100 m of fencing, hence $2x + 2y = 100$.

Therefore $y = \dfrac{100 - 2x}{2} = 50 - x$

Step 3. Find a formula for the area in terms of x. In this case it does not matter if we substitute for x or for y, as we need to find both in the end.

So $A = x(50 - x)$

$\Rightarrow A = 50x - x^2$

Step 4. $\dfrac{dA}{dx} = 50 - 2x$

Step 5. This is stationary when $50 - 2x = 0$

$$\Rightarrow x = 25$$

Step 6. $y = 25$ and therefore $A = 25^2 = 625$

For step 7, to test that this is indeed a maximum, we differentiate $\dfrac{dA}{dx}$ and apply the second derivative test.

$$\dfrac{d^2A}{dx^2} = -2$$

Since it is negative, the area of 625 m^2 is a maximum value. Hence the maximum value is given when $x = y = 25$ m. This should come as no surprise, as the maximum area of any rectangle is when it is a square.

> Even though it is obvious that the value given is a maximum, it is still important that we demonstrate it.

Example

The diagram shows a solid body made from a cylinder fixed to a cuboid. The cuboid has a square base with each edge measuring $4x$ cm and a height of x cm. The cylinder has a height of h cm, and the base of the cylinder fits exactly on the cuboid with no overlap. Given that the total volume of the solid is 80 cm^3, find the minimum surface area.

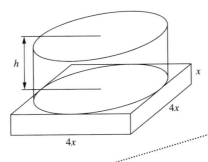

Step 1.
(Surface area) $A = \overbrace{4x^2 + 4x^2 + 4x^2 + 4x^2} + \overbrace{16x^2 + 16x^2}$
$\qquad + \underbrace{\pi(2x)^2 - \pi(2x)^2} + \underbrace{2\pi(2x)h}$

> These are the areas of the four sides of the cuboid.

> This is the area of the top and bottom.

> These are the areas of the circles. You need to add on the top one, but subtract the bottom, as the area showing is a square minus a circle. The radius of the circle is $2x$.

Hence we have a formula with three variables:

$A = 48x^2 + 4\pi xh$

Step 2. We know that the volume of the body is 80 cm^3, hence:

$16x^3 + \pi(2x)^2 h = 80$

$$\Rightarrow h = \dfrac{80 - 16x^3}{4\pi x^2}$$

> This is the curved surface area of the cylinder.

Step 3.

$$A = 48x^2 + 4\pi x\left(\dfrac{80 - 16x^3}{4\pi x^2}\right)$$

$$A = 48x^2 + \frac{80}{x} - 16x^2$$

$$\Rightarrow A = 32x^2 + \frac{80}{x}$$

Step 4.

$$\frac{dA}{dx} = 64x - \frac{80}{x^2}$$

Step 5. This is stationary when

$$64x - \frac{80}{x^2} = 0$$

$$\Rightarrow 64x^3 = 80$$

$$\Rightarrow x = \sqrt[3]{\frac{80}{64}} = 1.07\ldots$$

Step 6 allows us to find that $A = 111\ldots$ and $h = 4.11\ldots$

For step 7, to test that this is indeed a minimum, differentiate $\dfrac{dA}{dx}$ and apply the second derivative test.

$$\frac{dA}{dx} = 64x - 80x^{-2}$$

$$\frac{d^2A}{dx^2} = 64 + \frac{160}{x^3}$$

So when $x = 1.07\ldots$, $\dfrac{d^2A}{dx^2} = 192\ldots$. Since it is positive the area of 111 cm^2 is a minimum value.

Example

The marketing director for MacKenzie Motors has calculated that the profit made on each car is given by the formula $P = xy - 5000y$ where x is the selling price of each car (in thousands of pounds sterling) and y is the number of cars sold, which is affected by the time of year. Given that x and y are related by the formula $y = \sin x$, find the maximum profit that can be made.

Step 1. $P = xy - 5000y$.

Step 2. The link between two of the variables is $y = \sin x$.

Step 3. $P = x \sin x - 5000 \sin x$.

Steps 4, 5 and 6. This can be differentiated, but the resulting equation will need to be solved on a calculator. Hence it will be more effective to find the maximum value of the curve at this stage. Because this is a sinusoidal curve, we need to find where the first maximum occurs.

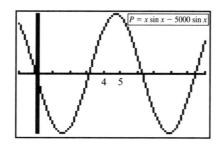

Hence we know the maximum value of $P = 4995.28\ldots$ and occurs when $x = 4.71$.

For step 7 it is not sensible to do a second derivative test. Instead we put in a sketch of the curve above and state that the y-values either side of the maximum are less than the maximum. It is important that actual values are given.

When $x = 4.70$, $P = 4994.91\ldots$

and when $x = 4.72$, $P = 4995.13\ldots$ which are both less than $4995.28\ldots$ and hence the maximum value of $P = £5000$.

Exercise 1

1 The amount of power a car engine produces is related to the speed at which the car is travelling. The actual relationship is given by the formula $P = 15v + \dfrac{7500}{v}$. Find the speed when the car is working most efficiently, i.e. when the power is the least.

2 A cuboid has a square base and a total surface area of $300\ \text{cm}^2$. Find the dimensions of the cuboid for the volume to be a maximum.

3 The base for a table lamp is in the shape of a cylinder with one end open and one end closed. If the volume of the base needs to be $1000\ \text{cm}^3$, find the radius of the base such that the amount of material used is a minimum.

4 For stacking purposes, a manufacturer of jewellery boxes needs to make them in the shape of a cuboid where the length of the box must be three times the width. The box must have a capacity of $400\ \text{cm}^3$. Find the dimensions of the box that would have the smallest surface area.

5 The diagram below shows a rectangle with an equilateral triangle on top. If the perimeter of the shape is 28 cm, find the length of the sides of the rectangle such that the area of the shape is a maximum.

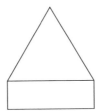

6 One of the clients of a packaging company is a soup manufacturer who needs tin cans manufactured. To maximize the profit, the surface area of the can should be as small as possible. Given that the can must hold 0.25 litres and is cylindrical, find the minimum surface area.

7 Consider the semicircle below. It has diameter XY and the point A is any point on the arc XY. The point A can move but it is required that $XA + AY = 25$. Find the maximum area of the triangle XAY.

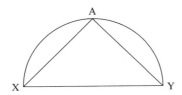

8 A cone is cut from a sphere as shown below. The radius of the sphere is r and x is the distance of the base of the cone from the centre of the sphere.

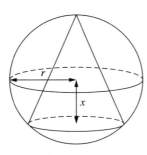

a Prove that the volume V of the cone is $V = \dfrac{\pi}{3}(r - x)(r + x)^2$.

b Find the height of the cone when the volume of the cone is a maximum.

9 A courier company requires that parcels be secured by three pieces of string. David wants to send a parcel in the shape of a cuboid. The cuboid has square ends. The square is of side x cm and the length of the parcel is y cm. This is shown in the diagram below.

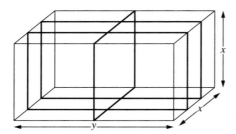

Given that the total length of string used is 900 cm, find the volume of the parcel in terms of x. Hence find the values of x and y for which the volume has a stationary value and determine whether this is a maximum or a minimum.

10 An open container is made from four pieces of sheet metal. The two end pieces are both isosceles triangles with sides of length $13x$, $13x$ and $24x$ as shown below. The other two pieces that make up the container are rectangles of length y and width $13x$. The total amount of sheet metal used is $900\ \text{cm}^2$.

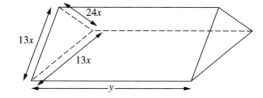

a Show that $y = \dfrac{450 - 60x^2}{13x}$.

b Find the volume of the container in terms of x.

c Find the value of x for which the volume of the container, V, has a stationary value and determine whether this is a maximum or a minimum.

11 In an intensive care unit in hospital, the drug adrenalin is used to stabilize blood pressure. The amount of the drug in the bloodstream y at any time t is the combination of two functions, $P(t)$ and $Q(t)$, which takes into consideration the fact that the drug is administered into the body repeatedly. Researchers at the hospital have found that $P(t) = e^{-t} \sin t$, while the manufacturers of the drug have found that $Q(t) = \cos t + 2$. The researchers have also found that $y = P(t) + Q(t)$. Given that the drug is initially administered at time $t = 0$, find the first two times when the quantity of drug in the bloodstream is the greatest, and verify using differentiation that these are in fact maximum values.

12 The population P, in thousands, of mosquitoes in the Kilimanjaro region of Tanzania over a 30-day period in May is affected by two variables: the average daily rainfall, r, and the average daily temperature, θ. The rainfall is given by the function $r = t \cos t + 6$ and the temperature is given by $\theta = e^{\frac{t}{10}} + 7$, where t is the time in days. It has been found that $P = r + \theta$. Find the minimum number of mosquitoes after the first five days of May and verify that it is a minimum.

13 In Japan the running cost of a car in yen per hour, Y, is dependent on its average speed v. This is given by the formula $Y = 6 + \dfrac{v^2 + 1}{v - 1}$, where v is the speed in tens of kilometres per hour. Write down the cost of a journey of 200 km covered at an average speed of 50 km h^{-1} and find the speed that would make the cost of this journey a minimum.

14 A new car hire company, Bob's Rentals, has just opened and wants to make the maximum amount of profit. The amount of profit, P, is dependent on two factors: x, the number of tens of cars rented; and y, the distance travelled by each car. The profit is given by the formula $P = xy + 40$, where $y = \sin x$. Find the smallest number of cars that the company needs to rent to give the maximum amount of profit.

15 In a triangle PQR, angle PQR is $90°$, PQ $= 7$ cm and PR $= 25$ cm. The rectangle QFGH is such that its vertices F, G and H lie on QR, PR and PQ respectively. This is shown below.

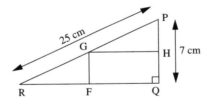

a Given that QH $= x$ cm and GH $= y$ cm, find a relationship between x and y.
b Hence express the area of the rectangle in terms of x only.
c Calculate the maximum value of this area as x varies.

10.2 Rates of change of connected variables

Consider a variable x. If the rate of change of the variable x is 2 ms^{-1}, then what we mean is that the rate of change of x with respect to time is 2, i.e. $\dfrac{dx}{dt} = 2$. The units of the variable give this information. Now, sometimes we want to find the rate of change with respect to time of a variable that is connected to x, say y, where $y = x^2$, at the point when $x = 10$. To do this, we use the chain rule $\dfrac{dy}{dx} = \dfrac{dy}{du} \times \dfrac{du}{dx}$. In this case we would use the formula $\dfrac{dy}{dt} = \dfrac{dx}{dt} \times \dfrac{dy}{dx}$.

Since $\dfrac{dy}{dx} = 2x$

$\Rightarrow \dfrac{dy}{dx} = 20.$

Therefore $\dfrac{dy}{dt} = 2 \times 20 = 40 \text{ ms}^{-1}.$

This is known as a **connected rate of change**.

Method for finding connected rates of change

This occurs when a question asks for a rate of change of one quantity but does not give a direct equation, and hence it is necessary to make a connection to another equation.

1 Write down the rate of change required by the question.
2 Write down the rate of change given by the question.
3 Write down an expression that connects the rate of change required and the one given.
4 This connection produces a third rate of change, which needs to be calculated. Find an equation that will give this new rate of change.
5 Differentiate the new equation.
6 Multiply the two formulae together and substitute to find the required rate of change.

Example

The radius, r, of an ink spot is increasing at the rate of 2 mms^{-1}. Find the rate at which the area, A, is increasing when the radius is 8 mm.

Step 1. The rate of change required is $\dfrac{dA}{dt}$.

Step 2. The known rate of change is $\dfrac{dr}{dt} = 2$.

Step 3. The connection is $\dfrac{dA}{dt} = \dfrac{dr}{dt} \times \dfrac{dA}{dr}$.

Step 4. We now need an equation linking A and r. For a circle $A = \pi r^2$.

Step 5. $\dfrac{dA}{dr} = 2\pi r$

Step 6. $\dfrac{dA}{dt} = \dfrac{dr}{dt} \times \dfrac{dA}{dr}$

$\Rightarrow \dfrac{dA}{dt} = 2 \times 2\pi r$

$\Rightarrow \dfrac{dA}{dt} = 4\pi r$

$\Rightarrow \dfrac{dA}{dt} = 4 \times \pi \times 8$

$\Rightarrow \dfrac{dA}{dt} = 32\pi$

Before we proceed with further examples we need to establish the result that $\dfrac{dx}{dy} = \dfrac{1}{\dfrac{dy}{dx}}$

We know that $\dfrac{dy}{dx} = \lim\limits_{\delta x \to 0} \dfrac{\delta y}{\delta x}.$

Since $\dfrac{\delta y}{\delta x}$ is a fraction $\dfrac{dx}{dy} = \lim\limits_{\delta x \to 0} \dfrac{1}{\dfrac{\delta y}{\delta x}}$.

However, as $\delta y \to 0$, $\delta x \to 0$.

Therefore $\dfrac{dx}{dy} = \dfrac{1}{\lim\limits_{\delta x \to 0}\left(\dfrac{\delta y}{\delta x}\right)}$, giving the result that

$$\dfrac{dx}{dy} = \dfrac{1}{\dfrac{dy}{dx}}$$

> Remember δy and δx are numbers whereas $\dfrac{dy}{dx}$ is a notation.

Example

A spherical balloon is blown up so that its volume, V, increases at a constant rate of $3 \ \text{cm}^3 \text{s}^{-1}$. Find the equation for the rate of increase of the radius r.

Step 1. The rate of change required is $\dfrac{dr}{dt}$.

Step 2. The known rate of change is $\dfrac{dV}{dt} = 3$.

Step 3. The connection is $\dfrac{dr}{dt} = \dfrac{dV}{dt} \times \dfrac{dr}{dV}$.

Step 4. We now need an equation linking V and r. For a sphere $V = \dfrac{4}{3}\pi r^3$. It is much easier to find $\dfrac{dV}{dr}$ than $\dfrac{dr}{dV}$ and hence we use the rule above.

Step 5. $\dfrac{dV}{dr} = 4\pi r^2$.

Step 6. $\dfrac{dr}{dt} = \dfrac{dV}{dt} \times \dfrac{dr}{dV}$

$\dfrac{dr}{dt} = 3 \times \dfrac{1}{4\pi r^2} = \dfrac{3}{4\pi r^2}$

Example

The surface area, A, of a cube is increasing at a rate of $20 \ \text{cm}^2 \text{s}^{-1}$. Find the rate of increase of the volume, V, of the cube when the edge of the cube is 10 cm.

Step 1. The rate of change required is $\dfrac{dV}{dt}$.

Step 2. The known rate of change is $\dfrac{dA}{dt} = 20$.

Step 3. The connection is $\dfrac{dV}{dt} = \dfrac{dA}{dt} \times \dfrac{dV}{dA}$.

A formula linking volume and area is not very straightforward, and nor is differentiating it. Hence we now connect the volume and area using the length of an edge, x.

This gives $\dfrac{dV}{dA} = \dfrac{dV}{dx} \times \dfrac{dx}{dA}$

> This is rather different from what has been asked previously, where questions have required that we differentiate the dependent variable directly with respect to what is known as the independent variable. In the case of $\dfrac{dy}{dx}$, y is the dependent variable and x is the independent variable. However in a case like this the independent variable is t since all the other variables are dependent on this. Which variable is the independent one is not always immediately obvious.

which leads to the formula $\dfrac{dV}{dt} = \dfrac{dA}{dt} \times \dfrac{dV}{dx} \times \dfrac{dx}{dA}$.

Step 4. We now need equations linking V and x and A and x. For a cube $V = x^3$ and $A = 6x^2$.

Step 5. $\dfrac{dV}{dx} = 3x^2$ and $\dfrac{dA}{dx} = 12x$

Step 6. $\dfrac{dV}{dt} = \dfrac{dA}{dt} \times \dfrac{dV}{dx} \times \dfrac{dx}{dA} = 20 \times 3x^2 \times \dfrac{1}{12x} = 5x$

So when $x = 10$, $\dfrac{dV}{dt} = 50 \text{ cm}^3\,\text{s}^{-1}$.

Now steps 5 and 6 can be done in an alternative way.

Step 5. $\dfrac{dV}{dx} = 3x^2$ and $\dfrac{dA}{dx} = 12x$

When $x = 10$, $\dfrac{dV}{dx} = 300$ and $\dfrac{dA}{dx} = 120$.

Step 6. $\dfrac{dV}{dt} = \dfrac{dA}{dt} \times \dfrac{dV}{dx} \times \dfrac{dx}{dA} = 20 \times 300 \times \dfrac{1}{120} = 50 \text{ cm}^3\,\text{s}^{-1}$

The method to use is personal choice, although if a formula is required, then the first method must be used.

Example

Water is being poured into a cone, with its vertex pointing downwards. This is shown below. The cone is initially empty and water is poured in at a rate of $25 \text{ cm}^3\,\text{s}^{-1}$. Find the rate at which the depth of the liquid is increasing after 30 seconds.

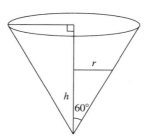

Step 1. The rate of change required is $\dfrac{dh}{dt}$, where h is the depth of the liquid.

Step 2. The known rate of change is $\dfrac{dV}{dt} = 25$.

Step 3. The connection is $\dfrac{dh}{dt} = \dfrac{dV}{dt} \times \dfrac{dh}{dV}$.

Step 4. We need to find a formula linking V and h. For a cone $V = \dfrac{1}{3}\pi r^2 h$. We now need to find a formula that connects h and r. From the diagram

$\tan 60° = \dfrac{r}{h}$

$\Rightarrow r = h\sqrt{3}$

$\Rightarrow V = \dfrac{1}{3}\pi \times 3h^2 \times h$

$$\Rightarrow V = \pi h^3$$

Step 5. $\dfrac{dV}{dh} = 3\pi h^2$.

Step 6. $\dfrac{dh}{dt} = \dfrac{dV}{dt} \times \dfrac{dh}{dV}$

$$\Rightarrow \dfrac{dh}{dt} = 25 \times \dfrac{1}{3\pi h^2} = \dfrac{25}{3\pi h^2}$$

After 30 seconds, the volume will be $30 \times 25 = 750$ cm³. Using $V = \pi h^3$
gives $750 = \pi h^3$

$$\Rightarrow h = 6.20\ldots$$

$$\Rightarrow \dfrac{dh}{dt} = \dfrac{25}{3\pi \times 6.20^2} = 0.0689\ \text{cm s}^{-1}$$

Example

A point P moves in such a way that its coordinates at any time t are given by
$x = te^{2\sin t}$ and $y = e^{-2\cos t}$. Find the gradient of the curve after 5 seconds.

Step 1. The rate of change required is $\dfrac{dy}{dx}$.

Step 2 is not required.

Step 3. The connection is $\dfrac{dy}{dx} = \dfrac{dy}{dt} \times \dfrac{dt}{dx}$.

Step 4 is not required.

Step 5. $x = te^{2\sin t}$

$$\Rightarrow \dfrac{dx}{dt} = e^{2\sin t} + 2t\cos t e^{2\sin t}$$

When $t = 5$,

$$\dfrac{dx}{dt} = e^{2\sin 5} + 2 \times 5\cos 5 e^{2\sin 5}$$

$$= 0.563\ldots$$

and

$$y = e^{-2\cos t}$$

$$\Rightarrow \dfrac{dy}{dt} = 2\sin t e^{-2\cos t}$$

When $t = 5$,

$$\dfrac{dy}{dt} = 2\sin 5 e^{-2\cos 5}$$

$$= -1.08\ldots$$

Step 6.

$$\dfrac{dy}{dx} = \dfrac{dy}{dt} \times \dfrac{dt}{dx}$$

$$\Rightarrow \dfrac{dy}{dx} = -1.09 \times \dfrac{1}{0.564}$$

$$\Rightarrow \dfrac{dy}{dx} = -1.93$$

Exercise 2

1 The surface area of a sphere is given by the formula $A = 4\pi r^2$, where r is the radius. Find the value of $\dfrac{dA}{dr}$ when $r = 3$ cm. The rate of increase of the radius is 4 cm s^{-1}. Find the rate of increase of the area when $r = 4$ cm.

2 Orange juice is being poured into an open beaker, which can be considered to be a cylinder, at a rate of $30 \text{ cm}^3 \text{ s}^{-1}$. The radius of the cylinder is 8 cm. Find the rate at which the depth of the orange juice is increasing.

3 The cross-sectional area of a trough is an isosceles triangle of height 36 cm and base 30 cm. The trough is 3 m long. This is shown below.

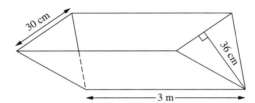

If water flows into the trough at a rate of $500 \text{ cm}^3 \text{ s}^{-1}$, find the rate at which the water level is increasing when the height is h.

4 The population, P, of termites varies with time t hours according to the formula $P = N_0 e^{3m}$ where N_0 is the initial population of termites and m is a variable given by $m = 3e^{\sin t}$. Find the rate of change of the termite population after 6 hours, giving your answer in terms of N_0.

5 A wine glass has been made such that when the depth of wine is x, the volume of wine, V, is given by the formula $V = 3x^3 - \dfrac{1}{3x}$. Alexander pours wine into the glass at a steady rate, and at the point when its depth is 4 cm, the level is rising at a rate of 1.5 cm s^{-1}. Find the rate at which the wine is being poured into the glass.

6 Bill, who is 1.85 m tall, walks directly away from a street lamp of height 6 m on a level street at a velocity of 2.5 m s^{-1}. Find the rate at which the length of his shadow is increasing when he is 4 m away from the foot of the lamp.

7 Consider the segment of a circle of fixed radius 8 cm. If the angle θ increases at a rate of 0.05 radians per second, find the rate of increase of the area of the segment when $\theta = 1.5$ radians.

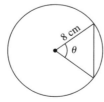

8 An empty hollow cone of radius a and height $4a$ is held vertex downwards and water is poured in at a rate of $8\pi \text{ cm}^3 \text{ s}^{-1}$. Find the rate at which the depth of water is increasing after 25 seconds.

9 A point P moves in such a way that its coordinates at any time t are given by
$x = \dfrac{1}{1 + t^2}$ and $y = \tan^{-1} t$. Find the gradient of the line OP after 3 seconds.

10 A circular disc of radius r rolls, without slipping, along the x-axis. The plane of the disc remains in the plane Oxy. A point P is fixed on the circumference of the disc and is initially at O. When the disc is rolled through θ radians, the point of contact is now Q and the length of the arc PQ is now the same as OQ. This is shown below.

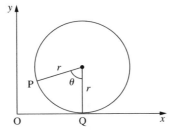

a Find the coordinates of P in terms of r and θ.

b As the disc rolls, the point P traces out a curve. Using connected rates of change, show that the gradient of the curve is $\cot\dfrac{\theta}{2}$.

11 A balloon is blown up so that its surface area is increasing at a rate of $25\ \text{cm}^2\,\text{s}^{-1}$. What is the rate of increase of the volume when its radius is 8 cm? Assume the balloon is spherical at all times.

12 A point moves on a curve such that $x = e^{3t}\cos 3t$ and $y = e^{3t}\sin 3t$, where t is the time taken. Show that the gradient at any time t is given by the formula

$$\frac{dy}{dx} = \tan\left(3t + \frac{\pi}{4}\right).$$

13 Water is being poured into a cone, with its vertex pointing downwards. This is shown below.

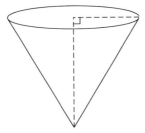

The cone is initially empty and water is poured in at a rate of $2\pi\sqrt{3}\ \text{cm}^3\,\text{s}^{-1}$. Find the rate of increase of:

a the radius of the circular surface of the water after 4.5 seconds

b the area of the circular surface of the water after 4.5 seconds.

10.3 Displacement, velocity and acceleration

This is another important application of differential calculus, which goes back to the basic definitions met in Chapter 8. Usually we define s as the displacement, v as the velocity, and a as the acceleration. If we consider a body moving 100 m in 25 seconds, some very basic knowledge will tell us that its average speed is $\dfrac{100}{25}\ \text{ms}^{-1}$, i.e. total distance travelled divided by total time taken. However, unless the body keeps a constant speed we have no idea what the velocity was after 4 seconds. In order to deal with this we now deal with velocity in a different way. By definition, velocity is the rate of change of displacement with respect to time. In questions like this displacement and distance are often used interchangeably, but because direction matters, we technically need to use a vector quantity and hence we usually talk about displacement. The distinction between vector and scalar quantities is discussed in Chapter 12. Since the differential operator $\frac{d}{dt}$ means "rate of change with respect to time", we now find that velocity, $v = \frac{ds}{dt}$. Using a similar argument, acceleration $a = \frac{dv}{dt}$ or alternatively $\frac{d^2s}{dt^2}$. Hence if we have a displacement formula as a function of time, we can now work out its velocity and acceleration. However, what happens when acceleration is related to displacement?

We know that $a = \frac{dv}{dt}$. From the work on connected rates of change, $\frac{dv}{dt}$ could be written as $\frac{ds}{dt} \times \frac{dv}{ds}$. However, $\frac{ds}{dt}$ is actually v. Hence $a = v\frac{dv}{ds}$. This now gives a formula that links velocity and displacement.

To summarize:

Quantity	Notation
Velocity	$\frac{ds}{dt}$
Acceleration	$\frac{dv}{dt}$ or $\frac{d^2s}{dt^2}$ or $v\frac{dv}{ds}$

Example

If the displacement of a particle is given by the formula $s = 3t^3 - 20t^2 + 40t$, find:
a) the displacement after 3 seconds
b) the formula for the velocity at any time t
c) the values of t when the particle is not moving
d) the initial velocity of the particle
e) the formula for the acceleration at any time t
f) the initial acceleration of the particle.

a) When $t = 3$, $s = 3 \times 27 - 20 \times 9 + 40 \times 3 = 21$ m.
b) To find the velocity, differentiate the formula for s.

$$v = \frac{ds}{dt} = 9t^2 - 40t + 40$$

Hence $v = 9t^2 - 40t + 40$.
c) When the particle is not moving $v = 0$

$$\Rightarrow 9t^2 - 40t + 40 = 0$$

$$\Rightarrow \frac{40 \pm \sqrt{40^2 - 4 \times 9 \times 40}}{2 \times 9} = 0$$

$$\Rightarrow t = 2.92 \text{ secs or } t = 1.52 \text{ secs}$$

d) The initial velocity occurs when $t = 0$

$$\Rightarrow v = 40 \text{ ms}^{-1}$$

e) To find the acceleration, differentiate the formula for v

$$a = \frac{dv}{dt} = 18t - 40$$

f) The initial acceleration occurs when $t = 0$.

$$\Rightarrow a = -40 \text{ m s}^{-2}$$

The negative sign means it is a deceleration rather than an acceleration.

Example

A boat travels with variable speed. Its displacement at any time t is given by $s = 2t^3 - 8t^2 + 8t$. After how long in the journey:
a) is its displacement a maximum and what is its displacement at that point?
b) is its velocity a minimum and what is its velocity at that point?

a) To find the maximum displacement we differentiate:

$$s = 2t^3 - 8t^2 + 8t$$

$$v = \frac{ds}{dt} = 6t^2 - 16t + 8$$

For a maximum displacement

$$v = 6t^2 - 16t + 8 = 0$$

The maximum displacement occurs when the velocity is zero

$$\Rightarrow 3t^2 - 8t + 4 = 0$$

$$\Rightarrow (3t - 2)(t - 2) = 0$$

$$\Rightarrow t = \frac{2}{3} \text{ or } t = 2$$

Now $\dfrac{d^2s}{dt^2} = 12t - 16$.

Hence when $t = 2$, $\dfrac{d^2s}{dt^2} = 8$, which is positive and therefore a minimum.

When $t = \dfrac{2}{3}$, $\dfrac{d^2s}{dt^2} = -8$, which is negative and therefore a maximum.

When $t = \dfrac{2}{3}$, $s = \dfrac{64}{27}$ m.

b) To find the minimum velocity we use $\dfrac{dv}{dt}$, i.e. $\dfrac{d^2s}{dt^2}$.

$$v = 6t^2 - 16t + 8$$

$$\frac{dv}{dt} = 12t - 16$$

For a minimum velocity $12t - 16 = 0$

$$\Rightarrow t = \frac{4}{3} \text{ sec}$$

When $t = \dfrac{4}{3}$, $v = 6\left(\dfrac{4}{3}\right)^2 - 16\left(\dfrac{4}{3}\right) + 8 = -\dfrac{8}{3} \text{ m s}^{-1}$.

To test whether it is a minimum we find $\dfrac{d^2v}{dt^2}$.

Since $\dfrac{d^2v}{dt^2} = 12$, it is a minimum.

Example

The displacement of a particle is given by the equation $s = 5\cos\dfrac{\pi}{3}t + 10\sin\dfrac{\pi}{3}t$.

a) Give a formula for its velocity at any time t.
b) What is its initial velocity?
c) What is the minimum displacement of the particle?
d) Give a formula for its acceleration at any time t.

a) $s = 5\cos\dfrac{\pi}{3}t + 10\sin\dfrac{\pi}{3}t$

$$\frac{ds}{dt} = v = \frac{-5\pi}{3}\sin\frac{\pi}{3}t + \frac{10\pi}{3}\cos\frac{\pi}{3}t$$

b) The initial velocity is when $t = 0$.

$$v = \frac{-5\pi}{3}\sin 0 + \frac{10\pi}{3}\cos 0$$

$$\Rightarrow v = \frac{10\pi}{3}\ \text{ms}^{-1}$$

c) The minimum displacement occurs when $\dfrac{ds}{dt} = 0$, i.e. $v = 0$.

$$\frac{-5\pi}{3}\sin\frac{\pi}{3}t + \frac{10\pi}{3}\cos\frac{\pi}{3}t = 0$$

$$\Rightarrow \frac{\sin\dfrac{\pi}{3}t}{\cos\dfrac{\pi}{3}t} = \frac{\dfrac{10\pi}{3}}{\dfrac{5\pi}{3}}$$

$$\Rightarrow \tan\frac{\pi}{3}t = 2$$

$$\Rightarrow \frac{\pi}{3}t = 1.10\ldots,\ 4.24\ldots$$

$$\Rightarrow t = 1.05\ldots,\ 4.05\ldots$$

> This equation has multiple solutions, but we only need to consider the first two positive solutions as after this it will just give repeated maxima and minima.

$$\frac{d^2s}{dt^2} = \frac{-5\pi^2}{9}\cos\frac{\pi}{3}t - \frac{10\pi^2}{9}\sin\frac{\pi}{3}t$$

Now when, $t = 1.05\ldots$, $\dfrac{d^2s}{dt^2}$ is negative and when $t = 4.05\ldots$, $\dfrac{d^2s}{dt^2}$ is positive.

Hence $t = 4.05\ldots$ gives the minimum displacement. In this case $s = -11.2$ m.

d) The acceleration is given by $\dfrac{dv}{dt}$, i.e. $\dfrac{d^2s}{dt^2}$.

Hence $a = \dfrac{-5\pi^2}{9}\cos\dfrac{\pi}{3}t - \dfrac{10\pi^2}{9}\sin\dfrac{\pi}{3}t$.

Example

The displacement of a particle is given by the formula $s = \dfrac{\ln t}{t^2}$.

Find:
a) the formula for the velocity of the particle at any time t
b) the velocity of the particle after 3 seconds
c) the formula for the acceleration of the particle
d) the acceleration after 2 seconds.

a) $v = \dfrac{ds}{dt} = \dfrac{t^2 \dfrac{1}{t} - 2t \ln t}{t^4} = \dfrac{1 - 2 \ln t}{t^3}$

b) When $t = 3$, $v = \dfrac{1 - 2 \ln 3}{27} = -0.0443 \text{ ms}^{-1}$.

> The negative sign means the velocity is in the opposite direction.

c) $a = \dfrac{dv}{dt} = \dfrac{d^2s}{dt^2} = \dfrac{t^3 \left(\dfrac{-2}{t} \right) - 3t^2(1 - 2 \ln t)}{t^6}$

$\Rightarrow a = \dfrac{t^2(-2 - 3 + 6 \ln t)}{t^6}$

$\Rightarrow a = \dfrac{6 \ln t - 5}{t^4}$

d) When $t = 2$, $a = \dfrac{6 \ln 2 - 5}{16} = -0.0526 \text{ ms}^{-2}$.

Example

If the velocity of a particle is proportional to the square of the displacement travelled, prove that the acceleration is directly proportional to the cube of the displacement.

$v \propto s^2$

$\Rightarrow v = ks^2$

We know that the acceleration, a, is given by $v\dfrac{dv}{ds}$.

Now $\dfrac{dv}{ds} = 2ks$.

Therefore

$a = ks^2 \times 2ks$

$\quad = 2k^2 s^3$

Therefore the acceleration is directly proportional to the cube of the displacement as $2k^2$ is a constant.

Exercise 3

1 The displacement, s, travelled in metres by a bicycle moving in a straight line is dependent on the time, t, and is connected by the formula $s = 4t - t^3$.

a Find the velocity and the acceleration of the cyclist when $t = \dfrac{1}{2}$ sec.

b At what time does the cyclist stop?

2 If $v = 16t - 6t^2$ and the body is initially at O, find:

a the velocity when $t = 2$ secs

b an expression for the acceleration at any time t

c the acceleration when $t = 3$ secs.

3 The velocity of a car is dependent on time and is given by the formula $v = (1 - 2t)^2$.

a Find the acceleration of the car after t seconds.

b When does the car first stop?

c What is the acceleration at the instant when the car stops?

4 The displacement of a particle is given by the formula $s = \dfrac{t \sin t}{t - 1}$

$(t \in \mathbb{R}, t \neq 1)$. Find:

a a general formula for the velocity v

b the velocity when $t = 2$ secs

c a general formula for the acceleration a.

5 If the velocity of a particle is inversely proportional to the square root of the displacement travelled, prove that the acceleration is inversely proportional to the square of the displacement.

6 If the velocity of a particle is given by $v = e^{2s} \cos 2s$, show that the acceleration is $e^{4s}(\cos 4s - \sin 4s + 1)$.

7 A particle is moving along a straight line such that its displacement at any time t is given by the formula $s = 2 \cos 2t + 6 \sin 2t$.

a Show that the acceleration is directly proportional to the displacement.

b Using the compound angle formula $R \cos(2t + \alpha)$, where $R > 0$ and $0 \leq \alpha \leq \dfrac{\pi}{2}$, show that the velocity is periodic and find the period.

8 The displacement of a particle is given by the formula $s = \dfrac{e^{2t}}{t^2 - 1}$, $t \in \mathbb{R}, t \neq 1$.

Find:

a a formula for the velocity of the particle at any time t

b the velocity of the particle after 2 seconds

c a formula for the acceleration of the particle

d the acceleration after 2 seconds in terms of e.

9 The velocity of a particle is given by $v^2 = \dfrac{6s^2}{\sqrt{s^2 - 1}}$. Show that the acceleration of the particle is given by the formula $a = \dfrac{12s^3 - 12s + 1}{2(s^2 - 1)^{\frac{3}{2}}}$.

10 David is visiting the fairground and his favourite ride is the big wheel. At any time t his horizontal displacement is given by the formula $s = 3 \sin(kt + c)$, where k and c are constants.

a Find his horizontal velocity at any time t.

b Find a general formula for the time when he first reaches his maximum horizontal velocity.

c Given that he has a horizontal displacement of $\dfrac{3}{\sqrt{2}}$ m after 10 seconds and a horizontal velocity of $\dfrac{3k\sqrt{3}}{2}$ m s^{-1} after 15 seconds, find the value of the acceleration after 20 seconds.

11 For a rocket to leave the earth's atmosphere, its displacement from the earth's surface increases exponentially with respect to time and is given by the formula $s = te^{kt^2}$ (for $t > 0$). Find:
   a the value of k, given that when $t = 10$ seconds, $s = 3000$ m
   b a general formula for the velocity at any time t
   c a general formula for the acceleration at any time t
   d the time when numerically the acceleration is twice the velocity $(t > 0)$.

12 The displacement of the East African mosquito has been modelled as a formula related to time, which is $s = \ln\left(\dfrac{t^2}{t-1}\right)$. However, this formula is not totally successful, and works only for certain values of t. The maximum value of t is 20 seconds. The minimum value of t is the minimum point of the curve.
   a Sketch the curve on a calculator and find the minimum value of t.
   b Find the velocity of the mosquito at any time t and state any restrictions on the time t.
   c Find the acceleration of the mosquito at any time t, stating any restrictions on t.
   d Find the velocity and acceleration of the mosquito after 10 seconds.

13 The displacement of a train at any time t is given by the formula $s^2 + 2st - 2t^2 = 4$. Find:
   a the velocity in terms of s and t
   b the acceleration in terms of s and t
   c the relationship between the displacement and the time when the velocity has a stationary value.

Review exercise

1 An airplane is flying at a constant speed at a constant altitude of 3 km in a straight line that will take it directly over an observer at ground level. At a given instant the observer notes that the angle θ is $\dfrac{1}{3}\pi$ radians and is increasing at $\dfrac{1}{60}$ radians per second. Find the speed, in kilometres per hour, at which the airplane is moving towards the observer.   [IB Nov 03 P1 Q20]

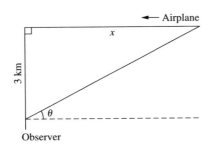

2 Particle A moves in a straight line starting at O with a velocity in metres per second given by the formula $v_A = t^2 + 3t - 4$. Particle B also moves in a straight line starting at O with a velocity in metres per second given by the formula $v_B = 2te^{0.5t} - 3t^2$. Find:

a the acceleration of particle A when $t = 5$

b the times when the particles have the same velocity

c the maximum and minimum velocity of particle B in the range $0 \leq t \leq 3$.

3 Air is pumped into a spherical ball, which expands at a rate of 8 cm³ per second ($8 \text{ cm}^3 \text{ s}^{-1}$). Find the exact rate of increase of the radius of the ball when the radius is 2 cm. [IB Nov 02 P1 Q16]

4 For a regular hexagon of side a cm and a circle of radius b cm, the sum of the perimeter of the hexagon and the circumference of the circle is 300 cm. What are the values of a and b if the sum of the areas is a minimum?

5 An astronaut on the moon throws a ball vertically upwards. The height, s metres, of the ball after t seconds is given by the equation

$s = 40t + 0.5at^2$, where a is a constant. If the ball reaches its maximum height when $t = 25$, find the value of a. [IB May 01 P1 Q17]

6 A manufacturer of cans for Lite Lemonade needs to make cans that hold 500 ml of drink. A can is manufactured from a sheet of aluminium, and the area of aluminium used to make the cans needs to be a minimum.

a If the radius of the can is r and the height of the can is h, find an expression for the area A of aluminium needed to make one can.

b Hence find the radius of the can such that the surface area is a minimum.

c Find the surface area of this can.

7 A particle moves such that its displacement at any time t hours is given by the function $f(t) = 3t^2 \sin 5t$, $t > 0$. Find:

a the velocity at any time t

b the time when the particle first comes to rest

c the time when the particle first has its maximum velocity.

8 A rectangle is drawn so that its lower vertices are on the x-axis and its upper vertices are on the curve $y = \sin x$, where $0 \leq x \leq \pi$.

a Write down an expression for the area of the rectangle.

b Find the maximum area of the rectangle. [IB May 00 P1 Q17]

9 A triangle has two sides of length 5 cm and 8 cm. The angle θ between these two sides is changing at a rate of $\dfrac{\pi}{30}$ radians per minute. What is the rate of change of the area of the triangle when $\theta = \dfrac{\pi}{3}$?

10 A particle moves along a straight line. When it is a distance s from a fixed point O, where $s > 1$, the velocity v is given by $v = \dfrac{3s + 2}{2s - 1}$. Find the acceleration when $s = 2$. [IB May 99 P1 Q20]

11 The depth h of the water at a certain point in the ocean at time t hours is given by the function $h = 2 \cos 3t - 3 \cos 2t + 6 \cos t + 15$, $t > 0$.

a Find the first time when the depth is a maximum.

b How long will it be before the water reaches its maximum depth again?

12 A square-based pyramid has a base of length x cm. The height of the pyramid is h cm. If the rate of change of x is 3 cm s^{-1}, and the rate of change of h is 2 cm s^{-1}, find the rate of change of the volume V when $x = 8$ cm and $h = 12$ cm.

 13

A company makes channelling from a rectangular sheet of metal of width $2x$. A cross-section of a channel is shown in the diagram, where AB + BC + CD = $2x$. The depth of the channel is h. AB and CD are inclined to the line BC at an angle θ. Find:

a the length of BC in terms of x, h and θ

b the area of the cross-section

c the maximum value of the cross-section as θ varies.

 14 A drop of ink is placed on a piece of absorbent paper. The ink makes a cirular mark, which starts to increase in size. The radius of the circular mark is given by the formula $r = \dfrac{4(1 + t^4)}{8 + t^4}$, where r is the radius in centimetres of the circular mark and t is the time in minutes after the ink is placed on the paper.

a Find t when $r = \dfrac{17}{6}$.

b Find a simplified expression, in terms of t, for the rate of change of the radius.

c Find the rate of change of the area of the circular mark when $r = \dfrac{17}{6}$.

d Find the value of t when the rate of change of the radius starts to decrease, that is, find the value of t, $t > 0$, at the point of inflexion on the curve $r = \dfrac{4(1 + t^4)}{8 + t^4}$. [IB May 98 P2 Q2]

11 Matrices

The concept of matrices and determinants was probably first understood by the Babylonians, who were certainly studying systems of linear equations. However, it was *Nine Chapters on the Mathematical Art*, written during the Han Dynasty in China between 200 BC and 100 BC, which gave the first known example of matrix methods as set up in the problem below.

There are three types of corn, of which three bundles of the first, two of the second, and one of the third make 39 measures. Two of the first, three of the second and one of the third make 34 measures. One of the first, two of the second and three of the third make 26 measures. How many measures of corn are contained in one bundle of each type?

The author of the text sets up the coefficients of the system of three linear equations in three unknowns as a table on a "counting board" (see Matrix 1). The author now instructs the reader to multiply the middle column by 3 and subtract the right column *as many times as possible*. The right column is then subtracted *as many times as possible* from 3 times the first column (see Matrix 2). The left-most column is then multiplied by 5 and the middle column is subtracted *as many times as possible* (see Matrix 3).

> This chapter will reveal that we now write linear equations as the rows of a matrix rather than columns, but the method is identical.

1	2	3
2	3	2
3	1	1
26	34	39

Matrix 1

0	0	3
4	5	2
8	1	1
39	24	39

Matrix 2

0	0	3
0	5	2
36	1	1
99	24	39

Matrix 3

Looking at the left-hand column, the solution can now be found for the third type of corn. We can now use the middle column and substitution to find the value for the second type of corn and finally the right-hand column to find the value for the first type of corn. This is basically the method of Gaussian elimination, which did not become well known until the early 19th century and is introduced in this chapter.

11.1 Introduction to matrices

Definitions

Elements: The numbers or symbols in a matrix.

Matrix: A rectangular array of numbers called entries or elements.

Row: A horizontal line of elements in the matrix.

Column: A vertical line of elements in the matrix.

Order: The size of the matrix. A matrix of order $m \times n$ has m rows and n columns.

Hence the matrix $A = \begin{pmatrix} 2 & 4 & -1 \\ 3 & 7 & 1 \end{pmatrix}$ has six elements, two rows, three columns, and its order is 2×3.

> A matrix is usually denoted by a capital letter.
>
> A square matrix is one that has the same number of rows as columns.

The most elementary form of matrix is simply a collection of data in tabular form like this:

Sales of	Week		
	1	**2**	**3**
Butter	75	70	82
Cheese	102	114	100
Milk	70	69	72

This data can be represented using the matrix $\begin{pmatrix} 75 & 70 & 82 \\ 102 & 114 & 100 \\ 70 & 69 & 72 \end{pmatrix}$.

Operations

Equality

Two matrices are equal if they are of the same order and their corresponding elements are equal.

> **Example**
>
> Find the value of a if $\begin{pmatrix} 2 & 5 \\ -3 & 4 \end{pmatrix} = \begin{pmatrix} a & 5 \\ -3 & 4 \end{pmatrix}$.
>
> Clearly in this case, $a = 2$.

Addition and subtraction

To add or subtract two or more matrices, they must be of the same order. We add or subtract corresponding elements.

Example

Evaluate $\begin{pmatrix} 2 & 4 & 3 \\ -1 & 3 & 7 \end{pmatrix} + \begin{pmatrix} 6 & -2 & 7 \\ 4 & -4 & -2 \end{pmatrix}$.

In this case the answer is $\begin{pmatrix} 8 & 2 & 10 \\ 3 & -1 & 5 \end{pmatrix}$.

> If the question appears on a calculator paper and does not involve variables, then a calculator can be used to do this.

We can now return to the example of a matrix given at the beginning of the chapter where the table

	Week		
Sales of	**1**	**2**	**3**
Butter	75	70	82
Cheese	102	114	100
Milk	70	69	72

can be represented as the matrix $\begin{pmatrix} 75 & 70 & 82 \\ 102 & 114 & 100 \\ 70 & 69 & 72 \end{pmatrix}$.

If this were to represent the sales in one shop, and the matrix $\begin{pmatrix} 79 & 78 & 79 \\ 97 & 101 & 109 \\ 81 & 75 & 74 \end{pmatrix}$ represents the sales in another branch of the shop, then adding the matrices together would give the total combined sales i.e. $\begin{pmatrix} 154 & 148 & 161 \\ 199 & 215 & 209 \\ 151 & 144 & 146 \end{pmatrix}$.

Multiplication by a scalar

The scalar outside the matrix multiplies every element of the matrix.

> This can be done by calculator, but is probably easier to do mentally.

Example

Evaluate $-3\begin{pmatrix} 1 & 2 & -4 \\ 3 & 1 & 2 \\ -1 & 3 & 5 \end{pmatrix}$.

In this case the answer is $\begin{pmatrix} -3 & -6 & 12 \\ -9 & -3 & -6 \\ 3 & -9 & -15 \end{pmatrix}$.

Multiplication of matrices

To multiply two matrices there are a number of issues we need to consider. In matrix multiplication we multiply each row by each column, and hence the number of columns in the first matrix must equal the number of rows in the second matrix. Multiplying an

$m \times n$ matrix by an $n \times p$ matrix is possible because the first matrix has n columns and the second matrix has n rows. If this is not the case, then the multiplication cannot be carried out. The answer matrix has the same number of rows as the first matrix and the same number of columns as the second.

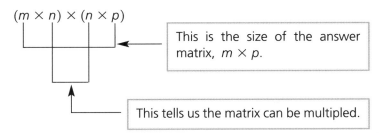

$$(m \times n) \times (n \times p)$$

This is the size of the answer matrix, $m \times p$.

This tells us the matrix can be multipled.

To find the element in the first row and first column of the answer matrix we multiply the first row by the first column. The operation is the same for all other elements in the answer. For example, the answer in the second row, third column of the answer comes from multiplying the second row of the first matrix by the third column of the second matrix.

The matrices $A = \begin{pmatrix} a & b & c \\ d & e & f \end{pmatrix}$ and $B = \begin{pmatrix} p & q \\ r & s \\ t & u \end{pmatrix}$ are 2×3 and 3×2, so they can be

multiplied to find AB, which will be a 2×2 matrix.

In this case $AB = \begin{pmatrix} ap + br + ct & aq + bs + cu \\ dp + er + ft & dq + es + fu \end{pmatrix}$. BA could also be found, and would

be a 3×3 matrix.

If we consider the case of $C = \begin{pmatrix} p & q \\ r & s \end{pmatrix}$ and $D = \begin{pmatrix} t \\ u \\ v \end{pmatrix}$, then it is not possible to find

either CD or DC since we have a 2×2 matrix multiplied by a 3×1 matrix or a 3×1 matrix multiplied by a 2×2 matrix.

Example

Evaluate $\begin{pmatrix} 3 & 6 \\ -2 & 4 \\ 1 & -5 \\ 3 & -7 \end{pmatrix} \begin{pmatrix} 5 & -1 \\ 2 & 3 \end{pmatrix}$.

We have a 4×2 matrix multiplied by a 2×2 matrix, and hence they can be multiplied and the answer will be a 4×2 matrix.

$$\begin{pmatrix} 3 & 6 \\ -2 & 4 \\ 1 & -5 \\ 3 & -7 \end{pmatrix} \begin{pmatrix} 5 & -1 \\ 2 & 3 \end{pmatrix} = \begin{pmatrix} (3 \times 5) + (6 \times 2) & (3 \times -1) + (6 \times 3) \\ (-2 \times 5) + (4 \times 2) & (-2 \times -1) + (4 \times 3) \\ (1 \times 5) + (-5 \times 2) & (1 \times -1) + (-5 \times 3) \\ (3 \times 5) + (-7 \times 2) & (3 \times -1) + (-7 \times 3) \end{pmatrix}$$

$$= \begin{pmatrix} 27 & 15 \\ -2 & 14 \\ -5 & -16 \\ 1 & -24 \end{pmatrix}$$

Example

If $A = \begin{pmatrix} 3 & -1 \\ -4 & 5 \end{pmatrix}$ find A^2.

As with algebra, A^2 means $A \times A$.

Hence $A^2 = \begin{pmatrix} 3 & -1 \\ -4 & 5 \end{pmatrix}\begin{pmatrix} 3 & -1 \\ -4 & 5 \end{pmatrix} = \begin{pmatrix} 13 & -8 \\ -32 & 29 \end{pmatrix}$

Example

Find the value of x and of y if $\begin{pmatrix} 3 & -1 \\ x & 2 \end{pmatrix}\begin{pmatrix} 2 & 6 \\ y & -1 \end{pmatrix} = \begin{pmatrix} 3 & 19 \\ -4 & -32 \end{pmatrix}$.

Multiplying the left-hand side of the equation

$$\Rightarrow \begin{pmatrix} 6 - y & 19 \\ 2x + 2y & 6x - 2 \end{pmatrix} = \begin{pmatrix} 3 & 19 \\ -4 & -32 \end{pmatrix}$$

Equating elements:

$$6 - y = 3$$
$$\Rightarrow y = 3$$
$$2x + 2y = -4$$
$$\Rightarrow 2x + 6 = -4$$
$$\Rightarrow x = -5$$

We now check $6x - 2 = -32$, which is true.

Example

If $A = \begin{pmatrix} 5 & 9 \\ -2 & 1 \end{pmatrix}$, $B = \begin{pmatrix} 2 & -5 \\ 1 & -2 \end{pmatrix}$ and $C = \begin{pmatrix} 1 & k \\ 0 & 2 \end{pmatrix}$, find $-2A(B - 2C)$.

We begin by finding $B - 2C = \begin{pmatrix} 2 & -5 \\ 1 & -2 \end{pmatrix} - \begin{pmatrix} 2 & 2k \\ 0 & 4 \end{pmatrix} = \begin{pmatrix} 0 & -5 - 2k \\ 1 & -6 \end{pmatrix}$

and $\qquad -2A = \begin{pmatrix} -10 & -18 \\ 4 & -2 \end{pmatrix}$

Hence $\qquad -2A(B - 2C) = \begin{pmatrix} -10 & -18 \\ 4 & -2 \end{pmatrix}\begin{pmatrix} 0 & -5 - 2k \\ 1 & -6 \end{pmatrix}$

$$= \begin{pmatrix} -18 & 158 + 20k \\ -2 & -8k - 8 \end{pmatrix}$$

Commutativity

Commutativity is when the result of an operation is independent of the order in which the elements are taken. Matrix multiplication is not commutative because in general $AB \neq BA$. In many cases multiplying two matrices is only possible one way or, if it possible both ways, the matrices are of different orders. Only in the case of a square matrix is it possible to multiply both ways and gain an answer of the same order, and even then the answers are often not the same.

Example

If $A = \begin{pmatrix} 2 & -1 \\ 3 & 4 \end{pmatrix}$ and $B = \begin{pmatrix} 3 & -2 \\ 3 & 4 \end{pmatrix}$, find:

a) AB
b) BA

a) $AB = \begin{pmatrix} 2 & -1 \\ 3 & 4 \end{pmatrix}\begin{pmatrix} 3 & -2 \\ 3 & 4 \end{pmatrix} = \begin{pmatrix} 3 & -8 \\ 21 & 10 \end{pmatrix}$

b) $BA = \begin{pmatrix} 3 & -2 \\ 3 & 4 \end{pmatrix}\begin{pmatrix} 2 & -1 \\ 3 & 4 \end{pmatrix} = \begin{pmatrix} 0 & -11 \\ 18 & 13 \end{pmatrix}$

Hence if we have a matrix A and multiply it by a matrix X, then we need to state whether we want XA or AX as they are often not the same thing. To do this we introduce the terms pre- and post-multiplication. If we **pre-multiply** a matrix A by X we are finding XA, but if we **post-multiply** a matrix A by X we are finding AX.

Identity matrix

Under the operation of multiplication, the identity matrix is one that fulfils the following properties. If A is any matrix and I is the identity matrix, then $A \times I = I \times A = A$. In other words, if any square matrix is pre- or post-multiplied by the identity matrix, then the answer is the original matrix. This is similar to the role that 1 has in multiplication of real numbers, where $1 \times x = x \times 1 = x$ for $x \in \mathbb{R}$.

$\begin{pmatrix} 1 & 0 \\ 0 & 1 \end{pmatrix}$ is the identity matrix for 2×2 matrices and $\begin{pmatrix} 1 & 0 & 0 \\ 0 & 1 & 0 \\ 0 & 0 & 1 \end{pmatrix}$ is the identity matrix for 3×3 matrices.

Only square matrices have an identity of this form.

Zero matrix

Under the operation of addition, the matrix that has the identity property is the zero matrix. This is true for a matrix of any size.

For a 2×2 matrix the zero matrix is $\begin{pmatrix} 0 & 0 \\ 0 & 0 \end{pmatrix}$ and for a 2×3 matrix it is $\begin{pmatrix} 0 & 0 & 0 \\ 0 & 0 & 0 \end{pmatrix}$.

The role of the zero matrix is similar to the role that 0 has in addition of real numbers, where $0 + x = x + 0 = x$ for $x \in \mathbb{R}$. If we multiply by a zero matrix, the answer will be the zero matrix.

Associativity

Matrix multiplication is associative. This means that $(AB)C = A(BC)$.

We will prove this for 2×2 matrices. The method of proof is the same for any three matrices that will multiply.

Let $A = \begin{pmatrix} a & b \\ c & d \end{pmatrix}$, $B = \begin{pmatrix} e & f \\ g & h \end{pmatrix}$ and $C = \begin{pmatrix} i & j \\ k & l \end{pmatrix}$.

$$A(BC) = \begin{pmatrix} a & b \\ c & d \end{pmatrix} \left[\begin{pmatrix} e & f \\ g & h \end{pmatrix} \begin{pmatrix} i & j \\ k & l \end{pmatrix} \right]$$

$$= \begin{pmatrix} a & b \\ c & d \end{pmatrix} \left[\begin{pmatrix} ei + fk & ej + fl \\ gi + hk & gj + hl \end{pmatrix} \right]$$

$$= \begin{pmatrix} aei + afk + bgi + bhk & aej + afl + bgj + bhl \\ cei + cfk + dgi + dhk & cej + cfl + dgj + dhl \end{pmatrix}$$

$$(AB)C = \left[\begin{pmatrix} a & b \\ c & d \end{pmatrix} \begin{pmatrix} e & f \\ g & h \end{pmatrix} \right] \begin{pmatrix} i & j \\ k & l \end{pmatrix}$$

$$= \left[\begin{pmatrix} ae + bg & af + bh \\ ce + dg & cf + dh \end{pmatrix} \right] \begin{pmatrix} i & j \\ k & l \end{pmatrix}$$

$$= \begin{pmatrix} aei + bgi + afk + bhk & aej + bgj + afl + + bhl \\ cei + dgi + cfk + + dhk & cej + dgj + cfl + + dhl \end{pmatrix}$$

$$= \begin{pmatrix} aei + afk + bgi + bhk & aej + afl + bgj + bhl \\ cei + cfk + dgi + dhk & cej + cfl + dgj + dhl \end{pmatrix}$$

Hence matrix multiplication on 2×2 matrices is associative.

Distributivity

Matrix multiplication is distributive across addition. This means that $A(B + C) = AB + AC$.

We will prove this for 2×2 matrices. The method of proof is the same for any three matrices that will multiply.

Let $A = \begin{pmatrix} a & b \\ c & d \end{pmatrix}$, $B = \begin{pmatrix} e & f \\ g & h \end{pmatrix}$ and $C = \begin{pmatrix} i & j \\ k & l \end{pmatrix}$.

$$A(B + C) = \begin{pmatrix} a & b \\ c & d \end{pmatrix} \left[\begin{pmatrix} e & f \\ g & h \end{pmatrix} + \begin{pmatrix} i & j \\ k & l \end{pmatrix} \right]$$

$$= \begin{pmatrix} a & b \\ c & d \end{pmatrix} \left[\begin{pmatrix} e + i & f + j \\ g + k & h + l \end{pmatrix} \right]$$

$$= \begin{pmatrix} ae + ai + bg + bk & af + aj + bh + bl \\ ce + ci + dg + dk & cf + cj + dh + dl \end{pmatrix}$$

$$AB + AC = \begin{pmatrix} a & b \\ c & d \end{pmatrix}\begin{pmatrix} e & f \\ g & h \end{pmatrix} + \begin{pmatrix} a & b \\ c & d \end{pmatrix}\begin{pmatrix} i & j \\ k & l \end{pmatrix}$$

$$= \begin{pmatrix} ae + bg & af + bh \\ ce + dg & cf + dh \end{pmatrix} + \begin{pmatrix} ai + bk & aj + bl \\ ci + dk & cj + dl \end{pmatrix}$$

$$= \begin{pmatrix} ae + ai + bg + bk & af + aj + bh + bl \\ ce + ci + dg + dk & cf + cj + dh + dl \end{pmatrix}$$

Hence matrix multiplication is distributive over addition in 2×2 matrices.

Exercise 1

1 What is the order of each of these matrices?

a $(1 \quad 2 \quad -3)$

b $\begin{pmatrix} 2 & 6 & -1 \\ 4 & 3 & 7 \end{pmatrix}$

c $\begin{pmatrix} 2 & 6 & -1 \\ -3 & 3 & 7 \\ 1 & -3 & 2 \end{pmatrix}$

d $\begin{pmatrix} 1 \\ k \\ 6k \\ 2 \end{pmatrix}$

2 Alan, Bill and Colin buy magazines and newspapers each week. The tables below show their purchases in three consecutive weeks.

Week 1

	Magazines	Newspapers
Alan	3	1
Bill	2	2
Colin	4	4

Week 2

	Magazines	Newspapers
Alan	1	2
Bill	4	1
Colin	0	1

Week 3

	Magazines	Newspapers
Alan	4	2
Bill	1	0
Colin	1	1

Write each of these in matrix form. What operation do you need to perform on the matrices to find the total number of magazines and the total number of newspapers bought by each of the men? What are these numbers?

3 Simplify these.

a $4\begin{pmatrix} 2 & 1 & 3 \\ 5 & -2 & 3 \\ 7 & -4 & 1 \end{pmatrix}$

b $-6\begin{pmatrix} 3 & 4 \\ -1 & 2 \\ -3 & -4 \end{pmatrix}$

c $k\begin{pmatrix} 3 & 6 \\ -4 & -1 \\ 12 & 4 \end{pmatrix}$

d $(k - 1)\begin{pmatrix} 3 & 2 \\ -1 & 0 \end{pmatrix}$

4 Find the unknowns in these equations.

a $\begin{pmatrix} 2 & 4 & -1 \\ 3 & 1 & k \end{pmatrix} = \begin{pmatrix} 2 & 4 & -1 \\ 3 & 1 & 6 \end{pmatrix}$

b $\begin{pmatrix} 2 & 5 \\ 7 & k \end{pmatrix} = \begin{pmatrix} 2 & 5 \\ 7 & k^2 \end{pmatrix}$

c $\begin{pmatrix} 2 & 1 \\ 3 & 7 \\ 1 & k^2 \end{pmatrix} = \begin{pmatrix} 2 & k \\ 3 & 7 \\ 1 & k \end{pmatrix}$

d $3\begin{pmatrix} 2 & k \\ -1 & 3 \end{pmatrix} = \begin{pmatrix} 6 & k^2 \\ -3 & 9 \end{pmatrix}$

e $\frac{1}{2}\begin{pmatrix} 2 & 3 \\ -1 & \frac{3}{2} \end{pmatrix} = \frac{k}{2}\begin{pmatrix} 4 & 6 \\ -2 & 3k \end{pmatrix}$

f $\begin{pmatrix} 2 & 3 \\ 2 & -k \end{pmatrix} + \begin{pmatrix} 3 & 2 \\ k & 1 \end{pmatrix} = \begin{pmatrix} 5 & 5 \\ 7 & 6 - 2k \end{pmatrix}$

g $\begin{pmatrix} 3 & -4 \\ 4 & x \end{pmatrix} + \begin{pmatrix} 1 & y \\ 5 & 1 \end{pmatrix} = \begin{pmatrix} 4 & 7 \\ -6 & 1 \end{pmatrix}$

5 For $P = \begin{pmatrix} 2 & -3 \\ 1 & 0 \end{pmatrix}$, $Q = \begin{pmatrix} -4 \\ 1 \end{pmatrix}$, $R = \begin{pmatrix} -3 & 4 \\ 2 & -3 \\ 4 & 8 \end{pmatrix}$, $S = \begin{pmatrix} 8 & 1 \\ -1 & 4 \end{pmatrix}$,

$T = \begin{pmatrix} 8 & -1 \\ -4 & -3 \\ -4 & 7 \end{pmatrix}$ and $U = \begin{pmatrix} 1 & 0 \\ 0 & 1 \end{pmatrix}$, find, if possible:

a $P + Q$ **b** $P + S$ **c** $Q + R$

d $R - T$ **e** $S - U$ **f** $P + S - U$

g $2R + 3T$ **h** $-P - 2S + 3U$

6 Multiply these matrices.

a $\begin{pmatrix} 2 & -3 \\ 4 & 1 \end{pmatrix}\begin{pmatrix} -3 & 7 \\ 6 & 1 \end{pmatrix}$

b $(2 \quad -4)\begin{pmatrix} 3 \\ 8 \end{pmatrix}$

c $\begin{pmatrix} 3 & -4 \\ 2 & -4 \\ 1 & 7 \end{pmatrix}\begin{pmatrix} 5 & 7 \\ 1 & -2 \end{pmatrix}$

d $\begin{pmatrix} 3 & 5 & -2 \\ 2 & 5 & 2 \\ 1 & -4 & -3 \end{pmatrix}\begin{pmatrix} 2 & 2 & -2 \\ 3 & 7 & -1 \\ -4 & 1 & 0 \end{pmatrix}$

e $\begin{pmatrix} 2 & k \\ -1 & k \end{pmatrix}\begin{pmatrix} 3 & 1 \\ k & 2 \end{pmatrix}$

f $(1 \quad 2 \quad -1 \quad k)\begin{pmatrix} 7 & 0 & k & 2 \\ 3 & -k & 2 & k + 1 \\ 6 & 5 & 2k & 3 \\ 3k - 4 & 2 & 0 & -2k \end{pmatrix}$

7 If $A = \begin{pmatrix} 3 & 4 \\ -2 & 5 \end{pmatrix}$, $B = \begin{pmatrix} 3 & 4 \\ 9 & -2 \end{pmatrix}$ and $C = \begin{pmatrix} 1 & k \\ 0 & 1 \end{pmatrix}$, find:

a AB **b** $A(BC)$ **c** $(AB)C$ **d** $C(AB)$

e $3BC$ **f** $(A - B)C$ **g** $(2A + B)(A - C)$ **h** $3(A + B)(A - B)$

8 Find the values of x and y.

a $\begin{pmatrix} 2 & x \\ -1 & 2 \end{pmatrix}\begin{pmatrix} 1 & y \\ 3 & 2 \end{pmatrix} = \begin{pmatrix} 8 & 6 \\ 5 & 3 \end{pmatrix}$

b $\begin{pmatrix} 3 & 7 \\ -2 & 5 \end{pmatrix}\begin{pmatrix} x \\ y \end{pmatrix} = \begin{pmatrix} 7 \\ 5 \end{pmatrix}$

c $\begin{pmatrix} 2 & 4 \\ 0 & -1 \end{pmatrix}\begin{pmatrix} 3 & -7 \\ y & 4 \end{pmatrix} = \begin{pmatrix} x & 2 \\ -1 & -4 \end{pmatrix}$

d $\begin{pmatrix} 3 & x \\ 1 & -3 \end{pmatrix}\begin{pmatrix} 4 & 1 \\ y & 4 \end{pmatrix} = \begin{pmatrix} 0 & 19 \\ 13 & -11 \end{pmatrix}$

9 If $A = \begin{pmatrix} 4 & -1 \\ 3 & 0 \end{pmatrix}$ find A^2 and A^3.

10 a The table below shows the number of men, women and children dieting in a school on two consecutive days.

	Men	Women	Children
Day 1	3	2	4
Day 2	5	7	2

Write this in matrix form, calling the matrix A.

b The minimum number of calories to stay healthy is shown in the table below.

	Calories
Men	1900
Women	1300
Children	1100

Write this in matrix form, calling the matrix B.

c Evaluate the matrix AB and explain the result.

d If $C = \begin{pmatrix} 1 \\ 1 \\ 1 \end{pmatrix}$, $D = (1 \quad 1 \quad 1)$, and $E = (1 \quad 1)$ find:

i EAB

ii AC

iii DB

In each case, explain the meaning of the result.

11 The table below shows the numbers of games won, drawn and lost for five soccer teams.

	Won	Drawn	Lost
Absolutes	3	4	7
Brilliants	6	2	6
Charismatics	10	1	3
Defenders	3	9	2
Extras	8	3	3

a Write this as matrix P.

b If a team gains 3 points for a win, 1 point for a draw and no points for losing, write down a matrix Q that when multiplied by P will give the total points for each team.

c Find this matrix product.

12 Given that $N = \begin{pmatrix} 1 & 3 \\ -2 & 2 \end{pmatrix}$ find:

a $2N - 3I$ **b** $N^2 - 2I$ **c** $N^2 - 3N + 2I$

13 If $A = \begin{pmatrix} 2 & -1 \\ 3 & 4 \end{pmatrix}$ and $B = \begin{pmatrix} m & 2 \\ n & -3 \end{pmatrix}$ find the values of m and n such that the multiplication of A and B is commutative.

14 If $M = \begin{pmatrix} 4 & 2 \\ 1 & -3 \end{pmatrix}$ and $M^2 - M - 4I = \begin{pmatrix} k & 0 \\ 0 & k \end{pmatrix}$, find the value of k.

15 If $M^2 = 2M + I$ where M is any 2×2 matrix, show that
$M^4 = 2M^2 + 8M + 5I$.

16 Given that $A = \begin{pmatrix} 2 & -1 \\ -3 & 4 \end{pmatrix}$ and $A^2 - 6A + cI = 0$, find the value of c.

17 If $A = \begin{pmatrix} 1 & 3 & c \\ 2 & 0 & 1 \end{pmatrix}$, $B = \begin{pmatrix} 1 & c \\ 3 & 0 \\ -c & 5 \end{pmatrix}$ and $C = \begin{pmatrix} 3 & -6 \\ 2 & \frac{1}{2} \end{pmatrix}$, find the value of c

such that $AB = 2C$.

18 If $P = \begin{pmatrix} 1 & 4 & -3 \\ c & 2 & 0 \end{pmatrix}$ and $Q = \begin{pmatrix} 5 & 1 \\ 2 & 0 \\ 3 & c \end{pmatrix}$, find the products PQ and QP.

19 Find the values of x and y for which the following pairs of matrices are commutative.

a $X = \begin{pmatrix} x & 2 \\ 1 & -2 \end{pmatrix}$ and $Y = \begin{pmatrix} 3 & y \\ 5 & 1 \end{pmatrix}$

b $X = \begin{pmatrix} 3 & y \\ x & -2 \end{pmatrix}$ and $Y = \begin{pmatrix} -1 & y \\ 2 & 1 \end{pmatrix}$

c $X = \begin{pmatrix} 8 & y \\ 3 & -2 \end{pmatrix}$ and $Y = \begin{pmatrix} x & y \\ 1 & 1 \end{pmatrix}$

20 Find in general form the 2×2 matrix A that commutes with $\begin{pmatrix} 1 & 1 \\ 0 & 1 \end{pmatrix}$.

11.2 Determinants and inverses of matrices

Finding inverse matrices

If we think of a matrix A multiplied by a matrix B to give the identity matrix, then the matrix B is called the **inverse** of A and is denoted by A^{-1}. If A is any matrix and I is the identity matrix, then the inverse fulfils the following property:

$$A \times A^{-1} = A^{-1} \times A = I$$

Consider a 2×2 matrix $A = \begin{pmatrix} a & b \\ c & d \end{pmatrix}$. Let its inverse be $A^{-1} = \begin{pmatrix} p & q \\ r & s \end{pmatrix}$ so that
$\begin{pmatrix} a & b \\ c & d \end{pmatrix}\begin{pmatrix} p & q \\ r & s \end{pmatrix} = \begin{pmatrix} 1 & 0 \\ 0 & 1 \end{pmatrix}$.

Equating elements: $ap + br = 1$ (i)
$\qquad\qquad\qquad aq + bs = 0$ (ii)
$\qquad\qquad\qquad cp + dr = 0$ (iii)
$\qquad\qquad\qquad cq + ds = 1$ (iv)
$\quad\;$ (i) $\times d \Rightarrow adp + bdr = d$ (v)
$\quad$ (iii) $\times b \Rightarrow bcp + bdr = 0$ (vi)

$(v) - (vi) \Rightarrow adp - bcp = d$

$$\Rightarrow p = \frac{d}{ad - bc}$$

Using a similar method we find

$$q = \frac{-b}{ad - bc}$$

$$r = \frac{-c}{ad - bc}$$

$$\text{and } s = \frac{a}{ad - bc}$$

$$\text{so } A^{-1} = \begin{pmatrix} \dfrac{d}{ad - bc} & q = \dfrac{-b}{ad - bc} \\ q = \dfrac{-c}{ad - bc} & q = \dfrac{a}{ad - bc} \end{pmatrix}$$

$$= \frac{1}{ad - bc}\begin{pmatrix} d & -b \\ -c & a \end{pmatrix}$$

> $ad - bc$ is known as the **determinant** of a 2×2 matrix.

Provided the determinant does not equal zero, the matrix has an inverse. A matrix where $ad - bc = 0$ is called a **singular matrix** and if $ad - bc \neq 0$ then it is called a **non-singular matrix**.

> The notation for the determinant of matrix A is Det(A) or $|A|$ and for the matrix above is written as $\begin{vmatrix} a & b \\ c & d \end{vmatrix}$.

Example

Find the determinant of the matrix $B = \begin{pmatrix} 3 & 1 \\ 2 & 6 \end{pmatrix}$.

Det(B) $= (3)(6) - (2)(1) = 16$

Method to find the inverse of a 2×2 matrix

If a calculator cannot be used then:

1. Evaluate the determinant to check the matrix is non-singular and divide each element by the determinant.
2. Interchange the elements a and d in the leading diagonal.
3. Change the signs of the remaining elements b and c.

On a calculator paper where no variables are involved, a calculator should be used.

Example

Find the inverse of $M = \begin{pmatrix} 3 & 7 \\ -2 & -5 \end{pmatrix}$.

Det(M) $= -15 - (-14) = -1$ so M^{-1} exists.

$$M^{-1} = \frac{1}{-1}\begin{pmatrix} -5 & -7 \\ 2 & 3 \end{pmatrix} = \begin{pmatrix} 5 & 7 \\ -2 & -3 \end{pmatrix}$$

It is not part of this syllabus to find the inverse of a 3×3 matrix by hand, but you need to be able to do this on a calculator, and you need to be able to verify that a particular matrix is the inverse of a given matrix.

Example

Find the inverse of $\begin{pmatrix} 1 & 3 & 1 \\ 3 & -1 & 1 \\ 1 & -2 & 1 \end{pmatrix}$.

On a calculator this appears as:

```
MATRIX[A] 3 ×3
[ 1      3      1     ]
[ 3     -1      1     ]
[ 1     -2      1     ]




3,3=1
```

and the answer is:

```
[A]⁻¹
  [[-.1  .5   -.4]
   [.2   0    -.2]
   [.5   -.5  1  ]]
■
```

Example

Verify that $\begin{pmatrix} \dfrac{4}{7} & \dfrac{-6}{7} & \dfrac{-11}{7} \\ \dfrac{3}{7} & \dfrac{-8}{7} & \dfrac{-10}{7} \\ \dfrac{-1}{7} & \dfrac{5}{7} & \dfrac{8}{7} \end{pmatrix}$ is the inverse of $\begin{pmatrix} 2 & 1 & 4 \\ 2 & -3 & -1 \\ -1 & 2 & 2 \end{pmatrix}$.

If we multiply these together then the answer is:

```
[A]*[B]
        [[1  0  0]
         [0  1  0]
         [0  0  1]]
```

Since this is the identity matrix, the matrices are inverses of each other.

If $A = \begin{pmatrix} 0 & 0 & 2m \\ 0 & 1 & 0 \\ m & 0 & 0 \end{pmatrix}$ and $A^{-1} = \begin{pmatrix} 0 & 0 & 1 \\ 0 & 1 & 0 \\ n & 0 & 0 \end{pmatrix}$, find the values of m and n.

$$\begin{pmatrix} 0 & 0 & 2m \\ 0 & 1 & 0 \\ m & 0 & 0 \end{pmatrix}\begin{pmatrix} 0 & 0 & 1 \\ 0 & 1 & 0 \\ n & 0 & 0 \end{pmatrix} = \begin{pmatrix} 1 & 0 & 0 \\ 0 & 1 & 0 \\ 0 & 0 & 1 \end{pmatrix}$$

$$\Rightarrow \begin{pmatrix} 2mn & 0 & 0 \\ 0 & 1 & 0 \\ 0 & 0 & m \end{pmatrix} = \begin{pmatrix} 1 & 0 & 0 \\ 0 & 1 & 0 \\ 0 & 0 & 1 \end{pmatrix}$$

so $m = 1$

and $2mn = 1 \Rightarrow n = \dfrac{1}{2}$

General results for inverse matrices

If $AB = I$ then $B = A^{-1}$ and $A = B^{-1}$.

That is, the matrices are inverses of each other.

Proof

Let $AB = I$.

If we pre-multiply both sides of the equation by A^{-1} then

$(A^{-1}A)B = A^{-1}I$

$\quad \Rightarrow IB = A^{-1}I$

$\quad \Rightarrow B = A^{-1}$

Similarly if we post-multiply both sides of the equation by B^{-1} then

$A(BB^{-1}) = IB^{-1}$

$\quad \Rightarrow AI = IB^{-1}$

$\quad \Rightarrow A = B^{-1}$

$$(AB)^{-1} = B^{-1}A^{-1}$$

Proof

We begin with $B^{-1}A^{-1}$.

Pre-multiplying by AB:

$(AB)(B^{-1}A^{-1}) = ABB^{-1}A^{-1}$

$\qquad\qquad\quad = AIA^{-1}$

$\qquad\qquad\quad = AA^{-1}$

$\qquad\qquad\quad = I$

Therefore the inverse of AB is $B^{-1}A^{-1}$, so

$$(AB)^{-1} = B^{-1}A^{-1}$$

Finding the determinant of a 3×3 matrix

This is done by extracting the 2×2 determinants from the 3×3 determinant, and although it can be done on a calculator very easily, it is important to know how to do this by hand.

> 3×3 determinants are particularly important when we come to work with vector equations of planes in Chapter 13.

Method

If the row and column through a particular entry in the 3×3 determinant are crossed out, four entries are left that form a 2×2 determinant, and this is known as the 2×2 determinant through that number. However, there is a slight complication. Every entry in a 3×3 determinant has a sign associated with it, which is not the sign of the entry itself. These are the signs:

$$\begin{vmatrix} + & - & + \\ - & + & - \\ + & - & + \end{vmatrix}$$

Therefore in the determinant $\begin{vmatrix} 6 & 2 & 3 \\ 4 & -1 & -2 \\ -3 & 0 & 5 \end{vmatrix}$ the 2×2 determinant through 6 is

$$\begin{vmatrix} -1 & -2 \\ 0 & 5 \end{vmatrix}.$$

In the same determinant $\begin{vmatrix} 6 & 2 & 3 \\ 4 & -1 & -2 \\ -3 & 0 & 5 \end{vmatrix}$, the 2×2 determinant through 4 is

$$-\begin{vmatrix} 2 & 3 \\ 0 & 5 \end{vmatrix}.$$

To find the determinant of a 3×3 matrix, we extract three 2×2 determinants and then evaluate these as before. It is usual to extract the 2×2 determinants from the top row of the 3×3 determinant.

Example

Without using a calculator, evaluate $\begin{vmatrix} 1 & 6 & 9 \\ 2 & -2 & 5 \\ -8 & 1 & 4 \end{vmatrix}$.

$$\begin{vmatrix} 1 & 6 & 9 \\ 2 & -2 & 5 \\ -8 & 1 & 4 \end{vmatrix} = 1\begin{vmatrix} -2 & 5 \\ 1 & 4 \end{vmatrix} - 6\begin{vmatrix} 2 & 5 \\ -8 & 4 \end{vmatrix} + 9\begin{vmatrix} 2 & -2 \\ -8 & 1 \end{vmatrix}$$

$$= (-8 - 5) - 6(8 + 40) + 9(2 - 16)$$

$$= -13 - 288 - 126 = -427$$

Example

Find the values of y for which the matrix $M = \begin{pmatrix} 1 & 4 & 2 \\ -3 & 2y & 9 \\ y^2 & 1 & -1 \end{pmatrix}$ is singular.

For the matrix to be singular $\text{Det}(M) = 0$.

$$\Rightarrow \begin{vmatrix} 2y & 9 \\ 1 & -1 \end{vmatrix} - 4\begin{vmatrix} -3 & 9 \\ y^2 & -1 \end{vmatrix} + 2\begin{vmatrix} -3 & 2y \\ y^2 & 1 \end{vmatrix} = 0$$

$$\Rightarrow (-2y - 9) - 4(3 - 9y^2) + 2(-3 - 2y^3) = 0$$

$$\Rightarrow -4y^3 + 36y^2 - 2y - 27 = 0$$

We solve this using a calculator.

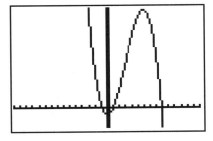

$$\Rightarrow y = -0.805, \ 0.947 \text{ or } 8.86$$

General results for determinants

$$\text{Det}(AB) = \text{Det}(A) \times \text{Det}(B)$$

This is a useful result, which can save time, and is proved below for 2×2 matrices. The proof for 3×3 matrices would be undertaken in exactly the same way.

Proof

Let $A = \begin{pmatrix} a & b \\ c & d \end{pmatrix}$ and $B = \begin{pmatrix} p & q \\ r & s \end{pmatrix}$.

$$AB = \begin{pmatrix} ap + br & aq + bs \\ cp + dr & cq + ds \end{pmatrix}$$

$$\begin{aligned} \text{Det}(AB) &= (ap + br)(cq + ds) - (aq + bs)(cp + dr) \\ &= acpq + adps + bcqr + bdrs - acpq - adqr - bcps - bdrs \\ &= adps + bcqr - adqr - bcps \end{aligned}$$

Now $\text{Det}(A) = ad - bc$ and $\text{Det}(B) = ps - qr$

So $\text{Det}(A) \times \text{Det}(B) = (ad - bc)(ps - qr)$
$$= adps - adqr - bcps + bcqr$$

Hence $\text{Det}(AB) = \text{Det}(A) \times \text{Det}(B)$

Example

If $A = \begin{pmatrix} 3 & 1 & 2 \\ 2 & -4 & 3 \\ 6 & 2 & 4 \end{pmatrix}$ and $B = \begin{pmatrix} 1 & 2 & -4 \\ 3 & 7 & 2 \\ -1 & 4 & 3 \end{pmatrix}$, find Det(AB). If possible find $(AB)^{-1}$.

Using a calculator, $\text{Det}(A) = 0$ and $\text{Det}(B) = -85$. Hence $\text{Det}(AB) = 0$. Since $\text{Det}(AB) = 0$, the matrix is singular so AB has no inverse.

The area of a triangle with vertices (x_1, y_1), (x_2, y_2) and (x_3, y_3) is $\dfrac{1}{2}\begin{vmatrix} 1 & 1 & 1 \\ x_1 & x_2 & x_3 \\ y_1 & y_2 & y_3 \end{vmatrix}$.

Proof

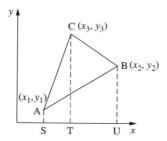

Considering the diagram shown above:

$$\text{Area ABC} = \text{area SATC} + \text{area TCBU} - \text{area SABU}$$
$$= \frac{1}{2}(y_1 + y_3)(x_3 - x_1) + \frac{1}{2}(y_2 + y_3)(x_2 - x_3) - \frac{1}{2}(y_1 + y_2)(x_2 - x_1)$$
$$= \frac{1}{2}(x_3y_1 + x_3y_3 - x_1y_1 - x_1y_3 + x_2y_2 + x_2y_3 - x_3y_2 - x_3y_3 - x_2y_1$$
$$- x_2y_2 + x_1y_1 + x_1y_2)$$
$$= \frac{1}{2}(x_2y_3 - x_3y_2 - x_1y_3 + x_3y_1 + x_1y_2 - x_2y_1)$$
$$= \frac{1}{2}\begin{vmatrix} 1 & 1 & 1 \\ x_1 & x_2 & x_3 \\ y_1 & y_2 & y_3 \end{vmatrix}$$

If A, B and C lie on a straight line (are collinear), then the area of the triangle is zero, and this is one possible way to show that three points are collinear.

Example

Without using a calculator, find the area of the triangle PQR whose vertices have coordinates (1, 2), (2, −3) and (5, −1).

$$\text{Area} = \frac{1}{2}\begin{vmatrix} 1 & 1 & 1 \\ 1 & 2 & 5 \\ 2 & -3 & -1 \end{vmatrix}$$

$$= \frac{1}{2}[(-2 + 15) - (-1 - 10) + (-3 - 4)]$$

$$= \frac{1}{2}(13 + 11 - 7) = \frac{17}{2} \text{ units}^2$$

Exercise 2

1 Find the inverse of each of these matrices.

a $\begin{pmatrix} 1 & -2 \\ 3 & 5 \end{pmatrix}$
b $\begin{pmatrix} 2 & 7 \\ -3 & 10 \end{pmatrix}$
c $\begin{pmatrix} 4 & 8 \\ -9 & 1 \end{pmatrix}$

d $\begin{pmatrix} 3 & 8 & -1 \\ 2 & 5 & 3 \\ 8 & -4 & 2 \end{pmatrix}$
e $\begin{pmatrix} 6 & 5 & 7 \\ -9 & 2 & 6 \\ 0 & 5 & -3 \end{pmatrix}$
f $\begin{pmatrix} 4 & 6 & -6 \\ 2 & 5 & 1 \\ 8 & 3 & 8 \end{pmatrix}$

g $\begin{pmatrix} k & 5 \\ 3k & 1 \end{pmatrix}$
h $\begin{pmatrix} 3k & k \\ 2k - 1 & k + 2 \end{pmatrix}$

2 If $P = \begin{pmatrix} 3 & -1 \\ 2 & 4 \end{pmatrix}$, $Q = \begin{pmatrix} 4 & -1 \\ -2 & 1 \end{pmatrix}$, $R = \begin{pmatrix} 3 & -3 \\ 2 & 4 \end{pmatrix}$, $S = \begin{pmatrix} 4 & 7 \\ 9 & 1 \end{pmatrix}$,

$PX = Q$, $QY = R$ and $RZ = S$, find the matrices X, Y and Z.

3 Evaluate these determinants.

a $\begin{vmatrix} 4 & 7 \\ 6 & 3 \end{vmatrix}$
b $\begin{vmatrix} 6 & -1 \\ -4 & 8 \end{vmatrix}$

c $\begin{vmatrix} 3 & 2 & -1 \\ 1 & 6 & 3 \\ 2 & 1 & 4 \end{vmatrix}$
d $\begin{vmatrix} 2 & 5 & -3 \\ 7 & 3 & 6 \\ 2 & 1 & -5 \end{vmatrix}$

4 Expand and simplify these.

a $\begin{vmatrix} \cos\theta & \sin\theta \\ -\sin\theta & \cos\theta \end{vmatrix}$
b $\begin{vmatrix} \sin\theta & \cos\theta \\ \sin\theta & -\cos\theta \end{vmatrix}$
c $\begin{vmatrix} x + 1 & x - 1 \\ x & x - 2 \end{vmatrix}$

d $\begin{vmatrix} 0 & a & c \\ a & 0 & b \\ c & b & 0 \end{vmatrix}$
e $\begin{vmatrix} \cos\theta & \sin\theta & \tan\theta \\ \tan\theta & \cos\theta & \sin\theta \\ \sin\theta & \tan\theta & \cos\theta \end{vmatrix}$
f $\begin{vmatrix} y - 1 & 0 & y + 1 \\ 1 & 1 & -1 \\ y + 1 & 1 & 1 \end{vmatrix}$

g $\begin{vmatrix} 1 & 1 & 1 \\ b^2 & a^2 & a^2 \\ a & b & a \end{vmatrix}$

5 Using determinants, find the area of the triangle PQR, where P, Q and R are the points:

a $(1, 4), (3, -2), (4, -1)$

b $(3, 7), (-4, -4), (1, 5)$

c $(-3, 5), (9, 1), (5, -1)$

6 Using determinants, determine whether each set of points is collinear.

  a $(1, 2), (6, 7), (3, 4)$

  b $(1, 3), (5, -2), (7, -3)$

  c $(2, 3), (5, 18), (-3, -22)$

7 Verify that these matrices are the inverses of each other.

  a $\begin{pmatrix} 1 & 2 & 0 \\ -1 & 1 & 0 \\ 2 & 5 & 1 \end{pmatrix}$ and $\begin{pmatrix} \dfrac{1}{3} & \dfrac{-2}{3} & 0 \\ \dfrac{1}{3} & \dfrac{1}{3} & 0 \\ \dfrac{-7}{3} & \dfrac{-1}{3} & 1 \end{pmatrix}$

  b $\begin{pmatrix} 1 & -1 & 3 \\ 2 & 0 & 4 \\ 6 & -2 & 22 \end{pmatrix}$ and $\begin{pmatrix} \dfrac{1}{2} & 1 & \dfrac{-1}{4} \\ \dfrac{-5}{4} & \dfrac{1}{4} & \dfrac{1}{8} \\ \dfrac{-1}{4} & \dfrac{-1}{4} & \dfrac{1}{8} \end{pmatrix}$

  c $\begin{pmatrix} 1 & -2 & 1 \\ 3 & 1 & 2 \\ -1 & 4 & 1 \end{pmatrix}$ and $\begin{pmatrix} \dfrac{-7}{16} & \dfrac{3}{8} & \dfrac{-5}{16} \\ \dfrac{-5}{16} & \dfrac{1}{8} & \dfrac{1}{16} \\ \dfrac{13}{16} & \dfrac{-1}{8} & \dfrac{7}{16} \end{pmatrix}$

8 Find the value of k for which the matrix $\begin{pmatrix} k^2 + k + 2 & k^2 & 0 \\ k + 4 & 2 & k^2 \\ 1 & 1 & 1 \end{pmatrix}$ is singular.

9 If $A = \begin{pmatrix} 1 & 3 \\ k & -1 \end{pmatrix}$ and $B = \begin{pmatrix} 4 & -2 \\ 1 & -k \end{pmatrix}$ verify that $(AB)^{-1} = B^{-1}A^{-1}$.

10 Find the values of c for which the matrix $\begin{pmatrix} 4 & -2 & 6 \\ 1 & c & 9 \\ 0 & 3 & c \end{pmatrix}$ is singular.

11 Find the values of y such that $\begin{vmatrix} y & 2y \\ y & 1 \end{vmatrix} = \begin{vmatrix} 2 & 3 & -1 \\ 4 & 2 & 0 \\ 1 & 5 & 1 \end{vmatrix}$.

12 M is the matrix $\begin{pmatrix} 0 & 0 & 1 \\ 1 & x & 0 \\ x^2 & 0 & x \end{pmatrix}$ and N is the matrix $\begin{pmatrix} -y & 0 & 1 \\ 1 & y & -y \\ y^2 & 0 & 0 \end{pmatrix}$. By evaluating the product MN, find the values of x and y for which M is the inverse of N.

13 If $2A - 3BX = B$, where A, B and X are 2×2 matrices, find

  a X in terms of A and B

  b X given that $B^{-1}A = 2I$, where I is the identity matrix.

14 The matrix $M = \begin{pmatrix} x - 3 & x - 1 \\ x + 1 & x + 3 \end{pmatrix}$.

  a Show that $\mathrm{Det}(M)$ is independent of x.

  b Find M^{-1}.

11.3 Solving simultaneous equations in two unknowns

The techniques of solving two simultaneous equations in two unknowns have been met before, but it is worth looking at the different cases and then examining how we can use matrices to solve these.

When we have two linear equations there are three possible scenarios, which are shown in the diagrams below.

The lines intersect, giving a unique solution

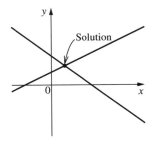

In this case solving the pair of simultaneous equations using a method of elimination or substitution will give the unique solution.

The lines are parallel, giving no solution

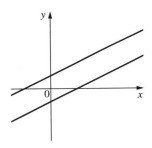

This occurs when we attempt to eliminate a variable and find we have a constant equal to zero.

Example

Determine whether the following equations have a solution.

$x + 5y = 7$ equation (i)

$-3x - 15y = 16$ equation (ii)

$3(i) + (ii) \Rightarrow 0 = 37$

This is not possible, so there is no solution.

The lines are the same, giving infinite solutions

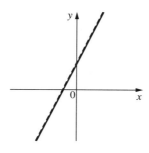

The lines are coincident.

In this case, when we try to eliminate a variable we find $0 = 0$, but we can actually give a solution.

Example

Find the solution to:

$4x - y = 5$ equation (i)
$16x - 4y = 20$ equation (ii)

If we do 4(i) − (ii) we find $0 = 0$.

Hence the solution can be written as $y = 4x - 5$.

If we are asked to show that simultaneous equations are consistent, this means that they either have a unique solution or infinite solutions. If they are inconsistent, they have no solution.

Using matrices to solve simultaneous equations in two unknowns

This is best demonstrated by example.

Example

Find the solution to these simultaneous equations.

$3x + 2y = 4$
$x - y = 5$

This can be represented in matrix form as

$$\begin{pmatrix} 3 & 2 \\ 1 & -1 \end{pmatrix}\begin{pmatrix} x \\ y \end{pmatrix} = \begin{pmatrix} 4 \\ 5 \end{pmatrix}$$

To solve this we pre-multiply both sides by the inverse matrix. This can be calculated either by hand or by using a calculator.

Hence $\begin{pmatrix} \frac{1}{5} & \frac{2}{5} \\ \frac{1}{5} & \frac{-3}{5} \end{pmatrix}\begin{pmatrix} 3 & 2 \\ 1 & -1 \end{pmatrix}\begin{pmatrix} x \\ y \end{pmatrix} = \begin{pmatrix} \frac{1}{5} & \frac{2}{5} \\ \frac{1}{5} & \frac{-3}{5} \end{pmatrix}\begin{pmatrix} 4 \\ 5 \end{pmatrix}$

$$\Rightarrow \begin{pmatrix} 1 & 0 \\ 0 & 1 \end{pmatrix}\begin{pmatrix} x \\ y \end{pmatrix} = \begin{pmatrix} \frac{1}{5} & \frac{2}{5} \\ \frac{1}{5} & \frac{-3}{5} \end{pmatrix}\begin{pmatrix} 4 \\ 5 \end{pmatrix}$$

$$\Rightarrow \begin{pmatrix} x \\ y \end{pmatrix} = \begin{pmatrix} \dfrac{14}{5} \\ \dfrac{-11}{5} \end{pmatrix}$$

So $x = \dfrac{14}{5}$, $y = -\dfrac{11}{5}$

If we are asked to find when equations have no unique solution, then we would need to show the matrix is singular. However, if we need to distinguish between the two cases here, i.e. no solution or infinite solutions, then we need to use Gaussian elimination.

Example

Show that the following system of equations does not have a unique solution.

$2x - 3y = 7$
$6x - 9y = 20$

This can be represented in matrix form as

$$\begin{pmatrix} 2 & -3 \\ 6 & -9 \end{pmatrix} \begin{pmatrix} x \\ y \end{pmatrix} = \begin{pmatrix} 7 \\ 20 \end{pmatrix}$$

If $A = \begin{pmatrix} 2 & -3 \\ 6 & -9 \end{pmatrix}$ then $\text{Det}(A) = -18 + 18 = 0$.

Hence the matrix is singular and the system of equations does not have a unique solution.

We can also write the simultaneous equations as what is called an **augmented matrix** and solve from here. This is effectively a neat way of representing elimination, but becomes very helpful when we deal with three equations in three unknowns. We have demonstrated this in the example on the previous page.

Example

The augmented matrix looks like this.

$$\begin{pmatrix} 3 & 2 & \vdots & 4 \\ 1 & -1 & \vdots & 5 \end{pmatrix}$$

We now conduct row operations on the augmented matrix to find a solution.

Changing Row 1 to Row 1 − 3 (Row 2) ⋯⋯⋯⋯⋯⋯⋯⋯⋯⋯⋯⋯

> This is called Gaussian elimination or row reduction.

$$\Rightarrow \begin{pmatrix} 0 & 5 & \vdots & -11 \\ 1 & -1 & \vdots & 5 \end{pmatrix}$$

> We are trying to make the first element zero.

This is the same as

$$\begin{pmatrix} 0 & 5 \\ 1 & -1 \end{pmatrix} \begin{pmatrix} x \\ y \end{pmatrix} = \begin{pmatrix} -11 \\ 5 \end{pmatrix}$$

So the first row gives:

$5y = -11$

$\Rightarrow y = -\dfrac{11}{5}$

We now substitute in the second row:

$$x - y = 5$$

$$\Rightarrow x + \frac{11}{5} = 5$$

$$\Rightarrow x = \frac{14}{5}$$

Example

Determine whether the following set of equations has no solution or infinite solutions.

$$x - 3y = 7$$
$$-3x + 9y = 15$$

The augmented matrix for these equations is

$$\begin{pmatrix} 1 & -3 & \vdots & 7 \\ -3 & 9 & \vdots & 15 \end{pmatrix}$$

Changing Row 1 to 3 (Row 1) + Row 2

$$\Rightarrow \begin{pmatrix} 0 & 0 & \vdots & 36 \\ -3 & 9 & \vdots & 15 \end{pmatrix}$$

Hence we have a case of $0 = 36$, which is inconsistent, so the equation has no solution.

> Had the equation had infinite solutions we would have had a whole line of zeros.

Exercise 3

1 Use the inverse matrix to solve these equations.

a $3x + y = 4$
    $x - 2y = 7$

b $3p - 5q = 7$
    $p - 2q = 7$

c $3y + 1 - 2x = 0$
    $4x + 3y - 4 = 0$

d $y = 3x - 4$
    $3y = 7x + 2$

2 Using a method of row reduction, solve these pairs of simultaneous equations.

a $x - 5y = 7$
    $3x + 5y = 10$

b $a - 3b = 8$
    $2a + 5b = 7$

c $2x + 3y - 8 = 0$
    $x - 2y + 7 = 0$

d $y = \frac{3}{2}x - 9$
    $y = 4x - 1$

3 Use the method of inverse matrices or row reduction to uniquely solve the following pairs of simultaneous equations. In each case state any restrictions there may be on the value of k.

a $(2k + 1)x - y = 1$
    $(k + 1)x - 2y = 3$

b $x - ky = 3$
    $kx - 3y = 3$

c $y + (2k - 1)x - 1 = 0$
    $5y - kx + 7 = 0$

d $y = kx - 4$
    $(k + 1)y = -3x + 10$

4 By evaluating the determinant, state whether the simultaneous equations have a unique solution.

a $3x + 2y = -7$
$6x - 4y = 14$

b $8x + 7y = 15$
$3x - 8y = 13$

c $y = 2x - 5$
$2y - 4x = -10$

d $(3k - 1)y - x = 5$
$4y - (k + 1)x = 11$

5 Determine the value of c for which the simultaneous equations have no solution. What can you say about the lines in each case?

a $cy + (3c - 1)x = 7$
$cy = -2x + 3c$

b $cx - (2c - 4)y = 15$
$(c + 1)x - 2cy = 9$

6 State with a reason which of these pairs of equations are consistent.

a $y - 3x = 7$
$\dfrac{y}{2} - 5x = 9$

b $2y - 3x = -7$
$y = \dfrac{3}{2}x - \dfrac{14}{4}$

c $y + 2x - 3 = 0$
$x = -\dfrac{y}{2} + \dfrac{6}{4}$

7 Find the value of p for which the lines are coincident.
$2x - 4y - 2p = 0$
$px - 6y - 9 = 0$
$x - 2y - p = 0$

8 Find the value of λ for which the equations are consistent and in this case find the corresponding values of y and x.
$4x + \lambda y = 10$
$3x - y = 4$
$4x + 6y = -2$

11.4 Solving simultaneous equations in three unknowns

We will see in Chapter 13 that an equation of the form $ax + by + cz = d$ is the equation of a plane. Because there are three unknowns, to solve these simultaneously we need three equations. It is important at this stage to consider various scenarios, which, like lines, lead to a unique solution, infinite solutions or no solution.

Unique solution

The three planes intersect in a point.

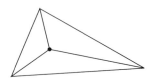

In this case solving the three equations simultaneously using any method will give the unique solution.

No solution

The three planes are parallel.

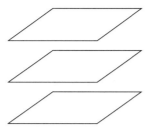

Two planes are coincident and the third plane is parallel.

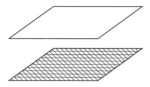

Two planes meet in a line and the third plane is parallel to the line of intersection

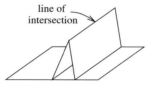

Two planes are parallel and the third plane cuts the other two.

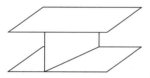

Infinite solutions

Three planes are coincident. In this case the solution is a plane of solutions.

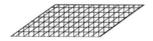

Two planes are coincident and the third plane cuts the other two in a line. In this case the solution is a line of solutions.

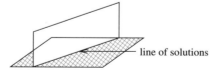

All three meet in a common line. In this case the solution is a line of solutions.

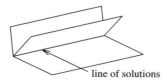

How do we know which case we have? This works in exactly the same way as a pair of simultaneous equations in two unknowns when we consider the augmented matrix.

If there are four zeros in a row of the augmented matrix there are infinite solutions. It does not matter which row it is. Therefore the augmented matrix $\begin{pmatrix} 3 & 2 & -1 & | & 4 \\ 4 & 2 & 6 & | & 1 \\ 0 & 0 & 0 & | & 0 \end{pmatrix}$ would produce infinite solutions.

If there are three zeros in a row of the augmented matrix there are no solutions. Again it does not matter which row it is, but the three zeros have to be the first three entries in the row. Therefore the augmented matrix $\begin{pmatrix} 3 & 2 & -1 & | & 4 \\ 4 & 2 & 6 & | & 1 \\ 0 & 0 & 0 & | & 4 \end{pmatrix}$ would produce no solution.

All other augmented matrices will produce a unique solution. For example $\begin{pmatrix} 3 & 2 & -1 & | & 4 \\ 4 & 2 & 6 & | & 1 \\ 0 & 4 & 0 & | & 0 \end{pmatrix}$ has a unique solution even though there is one row with three zeros in it. This highlights the fact that the position of the three zeros is important.

Elimination

To do this we eliminate one variable using two pairs of equations, leaving us with a pair of simultaneous equations in two unknowns.

Solving simultaneous equations in three unknowns

We are often told which method to use, but if not:
1 If a unique solution is indicated, any method can be used to find it.
2 If we want to distinguish between unique and non-unique solutions, then checking whether the matrix is singular is the easiest method.
3 If we want to establish that there is no solution or find the infinite solutions, then row operations are usually the easiest.

Example

Solve these equations.

$4x + 2y + z = 0$ equation (i)
$3x - 7y - 2z = 20$ equation (ii)
$x + y + 4z = 6$ equation (iii)

$2\text{(i)} + \text{(ii)} \Rightarrow 11x - 3y = 20$ equation (iv)
$2\text{(ii)} + \text{(iii)} \Rightarrow 7x - 13y = 46$ equation (v)

$13\text{(iv)} - 3\text{(v)} \Rightarrow 122x = 122$
$\Rightarrow x = 1$

Substitute in equation (iv):
$11 - 3y = 20$
$\Rightarrow y = -3$

Substituting x and y in equation (iii):
$1 - 3 + 4z = 6$
$\Rightarrow z = 2$

Substitution

To do this we make one variable the subject of one equation, substitute this in the other two equations and then solve the resulting pair in the usual way.

Example

Solve these equations.

$3x + 4y - z = -2$ equation (i)
$2x + 5y + 2z = 7$ equation (ii)
$x - 3y - z = 1$ equation (iii)

Rearranging equation (i) gives $z = 3x + 4y + 2$.

Substituting in equation (ii):
$2x + 5y + 6x + 8y + 4 = 7$
$\Rightarrow 8x + 13y = 3$ equation (iv)

Substituting in equation (iii):
$x - 3y - 3x - 4y - 2 = 1$
$\Rightarrow -2x - 7y = 3$ equation (v)

Rearranging equation (iv) gives $x = \dfrac{3 - 13y}{8}$ equation (vi)

Substituting in equation (v):
$-2\left(\dfrac{3 - 13y}{8}\right) - 7y = 3$
$\Rightarrow -6 + 26y - 56y = 24$
$\Rightarrow y = -1$

Substituting in equation (vi):
$\Rightarrow x = \dfrac{3 + 13}{8} = 2$

Substituting in equation (iii):
$2 + 3 - z = 1$
$\Rightarrow z = 4$

Using inverse matrices

In this case we write the equations in matrix form and then multiply each side of the equation by the inverse matrix. In other words, if $AX = B$ then $X = A^{-1}B$.

> Remember that we must pre-multiply by A^{-1}.

Example

Solve these equations using the inverse matrix.
$2x + y - 3z = -6$
$x + y + 4z = 19$
$2x + y - 5z = -14$

Writing these equations in matrix form:

$$\begin{pmatrix} 2 & 1 & -3 \\ 1 & 1 & 4 \\ 2 & 1 & -5 \end{pmatrix}\begin{pmatrix} x \\ y \\ z \end{pmatrix} = \begin{pmatrix} -6 \\ 19 \\ -14 \end{pmatrix}$$

$$\Rightarrow \begin{pmatrix} x \\ y \\ z \end{pmatrix} = \begin{pmatrix} 4.5 & -1 & -3.5 \\ -6.5 & 2 & 5.5 \\ 0.5 & 0 & -0.5 \end{pmatrix}\begin{pmatrix} -6 \\ 19 \\ -14 \end{pmatrix} = \begin{pmatrix} 3 \\ 0 \\ 4 \end{pmatrix}$$

As with lines, if the matrix is singular, then the system of equations has no solution or infinite solutions. Further work using one of the other methods is necessary to distinguish between the two cases.

Example

Does the following system of equations have a unique solution?

$$x + 3y - 4z = 2$$
$$2x - y + 5z = 1$$
$$3x - 5y + 14z = 7$$

Writing these equations in matrix form:

$$\begin{pmatrix} 1 & 3 & -4 \\ 2 & -1 & 5 \\ 3 & -5 & 14 \end{pmatrix} \begin{pmatrix} x \\ y \\ z \end{pmatrix} = \begin{pmatrix} 2 \\ 1 \\ 7 \end{pmatrix}$$

If we let $A = \begin{pmatrix} 1 & 3 & -4 \\ 2 & -1 & 5 \\ 3 & -5 & 14 \end{pmatrix}$ then $\text{Det}(A) = 0$ (using a calculator) and hence the system of equations does not have no unique solution.

Using row operations

The aim is to produce as many zeros in a row as possible.

Example

Find the unique solution to this system of equations using row operations.

$$x - 3y + 2z = -3$$
$$2x + 4y - 3z = 11$$
$$x + y + 2z = 1$$

The augmented matrix is:

$$\begin{pmatrix} 1 & -3 & 2 & | & -3 \\ 2 & 4 & -3 & | & 11 \\ 1 & 1 & 2 & | & 1 \end{pmatrix}$$

$$\Rightarrow \begin{pmatrix} 1 & -3 & 2 & | & -3 \\ 2 & 4 & -3 & | & 11 \\ 0 & -2 & 7 & | & -9 \end{pmatrix}$$

Change Row 3 to 2 (Row 3) − Row 2.

$$\Rightarrow \begin{pmatrix} 1 & -3 & 2 & | & -3 \\ 0 & 10 & -7 & | & 17 \\ 0 & -2 & 7 & | & -9 \end{pmatrix}$$

Change Row 2 to Row 2 − 2 (Row 1).

$$\Rightarrow \begin{pmatrix} 1 & -3 & 2 & | & -3 \\ 0 & 10 & -7 & | & 17 \\ 0 & 0 & 28 & | & -28 \end{pmatrix}$$

Change Row 3 to 5 (Row 3) + Row 2.

We always begin by producing a pair of zeros above each other in order that when we carry out more row operations on those two lines, we do not go around in circles creating zeros by eliminating ones we already have. Hence the final row operation must use Rows 2 and 3.

We cannot produce any more zeros, so the equation will have a unique solution and we now solve using the rows.

From Row 3
$$28z = -28$$
$$\Rightarrow z = -1$$

From Row 2
$$10y - 7z = 17$$
$$\Rightarrow y = 1$$

From Row 1
$$x - 3y + 2z = -3$$
$$x - 3 - 2 = -3$$
$$\Rightarrow x = 2$$

However, the strength of row operations is in working with infinite solutions and no solution.

In the case of no solution, row operations will produce a line of three zeros.

Example

Verify that this system of equations has no solution.
$$x - 3y + 4z = 5$$
$$2x - y + 3z = 7$$
$$3x - 9y + 12z = 14$$

The augmented matrix is:
$$\begin{pmatrix} 1 & -3 & 4 & | & 5 \\ 2 & -1 & 3 & | & 7 \\ 3 & -9 & 12 & | & 14 \end{pmatrix}$$

$$\Rightarrow \begin{pmatrix} 1 & -3 & 4 & | & 5 \\ 2 & -1 & 3 & | & 7 \\ 0 & 0 & 0 & | & -1 \end{pmatrix}$$

Change Row 3 to Row 3 − 3 (Row 1).

Since there is a line of three zeros, the system of equations is inconsistent and has no solution.

In the case of infinite solutions we will get a line of four zeros, but as we saw previously there are two possibilities for the solution. In Chapter 13 the format of these solutions will become clearer, but for the moment, if a line of zeros is given, the answer will be dependent on one parameter. A plane of solutions can occur only if the three planes are coincident – that is, the three equations are actually the same – and in this case the equation of the plane is actually the solution.

Example

Show that this system of equations has infinite solutions and find the general form of these solutions.
$$2x + y - 3z = 1$$
$$2x + 2y - 4z = 5$$
$$6x + 3y - 9z = 3$$

The augmented matrix is: $\begin{pmatrix} 2 & 1 & -3 & | & 1 \\ 2 & 2 & -4 & | & 5 \\ 6 & 3 & -9 & | & 3 \end{pmatrix}$

$\Rightarrow \begin{pmatrix} 0 & 0 & 0 & | & 0 \\ 2 & 2 & -4 & | & 5 \\ 6 & 3 & -9 & | & 3 \end{pmatrix}$ ....

> Change Row 1 to
> $3\,(\text{Row 1}) - \text{Row 3}$.

Hence the system has infinite solutions. We now eliminate one of the other variables.

$\Rightarrow \begin{pmatrix} 0 & 0 & 0 & | & 0 \\ 0 & 3 & -3 & | & 12 \\ 6 & 3 & -9 & | & 3 \end{pmatrix}$ ....

> Change Row 2 to
> $3\,(\text{Row 2}) - \text{Row 3}$.

This can be read as $\begin{pmatrix} 0 & 0 & 0 \\ 0 & 3 & -3 \\ 6 & 3 & -9 \end{pmatrix}\begin{pmatrix} x \\ y \\ z \end{pmatrix} = \begin{pmatrix} 0 \\ 12 \\ 3 \end{pmatrix}$.

> By letting $y = \lambda$ it is clear that we can get the solutions for x and z in terms of λ.

Hence if we let $y = \lambda$, then $3\lambda - 3z = 12 \Rightarrow z = \lambda - 4$.

If we now substitute these into the equation $6x + 3y - 9z = 3$ we will find x.

$6x + 3\lambda - 9(\lambda - 4) = 3$
$\Rightarrow 6x + 3\lambda - 9\lambda + 36 = 3$
$\Rightarrow x = \dfrac{6\lambda - 33}{6} = \dfrac{2\lambda - 11}{2}$

> This is the parametric equation of a line, and we will learn in Chapter 13 how to write this in other forms.

Hence the general solution to the equation is $x = \dfrac{2\lambda - 11}{2}$, $y = \lambda$, $z = \lambda - 4$.

> This is a line of solutions. By looking at the original equations we can see that the third equation is three times the first equation, and hence this is a case of two planes being coincident and the third plane cutting these two in a line.

Example

Show that the following system of equations has infinite solutions and find the general form of these solutions.

$x + y - 3z = 2$
$2x + 2y - 6z = 4$
$4x + 4y - 12z = 8$

The augmented matrix is: $\begin{pmatrix} 1 & 1 & -3 & | & 2 \\ 2 & 2 & -6 & | & 4 \\ 4 & 4 & -12 & | & 8 \end{pmatrix}$

$\Rightarrow \begin{pmatrix} 0 & 0 & 0 & | & 0 \\ 2 & 2 & -6 & | & 4 \\ 4 & 4 & -12 & | & 8 \end{pmatrix}$ ....

> Change Row 1 to
> $2\,(\text{Row 1}) - \text{Row 2}$.

Hence it is clear that the system has infinite solutions. In this case we cannot eliminate another variable, because any more row operations will eliminate all the variables.

This can be read as $\begin{pmatrix} 0 & 0 & 0 \\ 2 & 2 & -6 \\ 4 & 4 & -12 \end{pmatrix} \begin{pmatrix} x \\ y \\ z \end{pmatrix} = \begin{pmatrix} 0 \\ 4 \\ 8 \end{pmatrix}$.

If we look back at the original equations, we can see that they are in fact the same equation, and hence we have the case of three coincident planes, which leads to a plane of solutions. The plane itself is the solution to the equations, i.e. $x + y - 3z = 2$.

Example

Determine what type of solutions the following system of equations has, and explain the arrangement of the three planes represented by these equations.

$2x + 3y - 2z = 1$
$4x + 6y - 4z = 2$
$6x + 9y - 6z = 4$

The augmented matrix for this system is $\begin{pmatrix} 2 & 3 & -2 & | & 1 \\ 4 & 6 & -4 & | & 2 \\ 6 & 9 & -6 & | & 4 \end{pmatrix}$.

In this case if we perform row operations we get conflicting results.

Changing Row 1 to 2 (Row 1) − Row 2 gives $\begin{pmatrix} 0 & 0 & 0 & | & 0 \\ 4 & 6 & -4 & | & 2 \\ 6 & 9 & -6 & | & 4 \end{pmatrix}$, which implies the system has infinite solutions.

However, changing Row 1 to 3 (Row 1) − Row 3 gives $\begin{pmatrix} 0 & 0 & 0 & | & -1 \\ 4 & 6 & -4 & | & 2 \\ 6 & 9 & -6 & | & 4 \end{pmatrix}$, which implies the system has no solution.

If we now look back at the equations we can see that the first and second equations are multiples of each other. In the third equation the coefficients of x, y and z are multiples of those coefficients in the first and second equations. This means we have two coincident planes and a parallel plane.

> The reasoning behind this will be explained in Chapter 13.

Hence the system actually has no solution. This is also obvious from the initial equations since we clearly do not have three coincident planes.

Row operations on a calculator

A calculator is capable of doing this, but there are a number of points that need to be made. Obviously, if the question appears on a non-calculator paper then this is not an option. However, if a calculator is allowed then it is useful. To find a unique solution we put the 3×4 augmented matrix into the calculator as usual.

Example

Use a calculator to find the solution to this system of equations.

$$x + 2y + z = 3$$
$$3x + y - z = 2$$
$$x + 4y + 2z = 4$$

The augmented matrix for this system is $\begin{pmatrix} 1 & 2 & 1 & | & 3 \\ 3 & 1 & -1 & | & 2 \\ 1 & 4 & 2 & | & 4 \end{pmatrix}$

The calculator display is shown below:

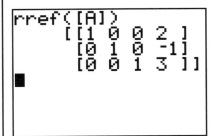

This can be read as $\begin{pmatrix} 1 & 0 & 0 \\ 0 & 1 & 0 \\ 0 & 0 & 1 \end{pmatrix}\begin{pmatrix} x \\ y \\ z \end{pmatrix} = \begin{pmatrix} 2 \\ -1 \\ 3 \end{pmatrix}$

Hence $x = 2, y = -1$, and $z = 3$.

If we put in a system of equations that has no solution then the line of three zeros will occur. The line of four zeros will occur if the system has infinite solutions. The calculator will not find the line or plane of solutions, but it will certainly make it easier.

Example

Find the general solution to this system of equations.

$$x + 3y + z = 4$$
$$2x - y + 2z = 3$$
$$x - 4y + z = -1$$

The augmented matrix for this system is $\begin{pmatrix} 1 & 3 & 1 & | & 4 \\ 2 & -1 & 2 & | & 3 \\ 1 & -4 & 1 & | & -1 \end{pmatrix}$

The calculator display is shown below:

```
rref([A])
[[1 0 1 1.85714...
 [0 1 0 .714285...
 [0 0 0 0        ...
```

The line of four zeros at the bottom indicates the infinite solutions. If we rewrite this in the form

$$\begin{pmatrix} 1 & 0 & 1 \\ 0 & 1 & 0 \\ 0 & 0 & 0 \end{pmatrix} \begin{pmatrix} x \\ y \\ z \end{pmatrix} = \begin{pmatrix} 1.85\ldots \\ 0.714\ldots \\ 0 \end{pmatrix}$$

we can see that $y = 0.714$, but that x and z cannot be solved uniquely. Hence if we let $x = t$, then $t + z = 1.86$

$$\Rightarrow z = -t + 1.86$$

Hence the general solution to the equations is $x = t$, $y = 0.714$ and $z = -t + 1.86$. This is a case of three planes meeting in a line.

Exercise 4

1 Using elimination, solve these systems of equations.

a $6x + 8y + 5z = 1$
$3x + 5y + 3z = 3$
$2x + 3y + 2z = -1$

b $4x + 7y + 3z = 2$
$2x + 5y + 2z = -2$
$5x + 13y + 5z = 0$

c $x - 2y = 10$
$3x - y + z = 7$
$2x - y + z = 5$

d $4x + 8y + 3z = 6$
$3x + 5y + z = 3$
$4x + y + 4z = 15$

2 Using substitution, solve these systems of equations.

a $2x + 2y - z = -6$
$3x + 7y + 2z = 13$
$2x + 5y + 2z = 12$

b $2x + y - 2z = -11$
$x - 3y + 8z = 27$
$3x - 2y + z = -4$

c $3x + y + 4z = 6$
$2x - y - 2z = -7$
$x + 2y + 6z = 13$

d $2x + 3y + 4z = 1$
$4x - y - 6z = 9$
$x - 2y - 8z = -2$

3 Using a method of inverse matrices, solve these systems of equations.

a $x + 3y + 6z = 1$
$2x + 6y + 9z = 4$
$3x + 3y + 12z = 5$

b $x + 3y + 2z = 1$
$4x - y - 6z = 12$
$2x + y - 5z = 10$

c $2x - y - 2z = 4$
$4x + y - 3z = 9$
$6x - 2y - 3z = 7$

d $3x - y + z = 1$
$2x + 3y + 5z = 3$
$x - 2y - 5z = 6$

4 By evaluating the determinant, state whether each system of equations has a unique solution or not.

a $x + 2y - 5z = 15$
$2x - 3y + 7z = -1$
$3x + y - 2z = 12$

b $2x - y + 3z = 12$
$x - y + 7z = 15$
$3x - y - z = 7$

c $x + y - z = 4$
$2x + 2y - 3z = 4$
$3x + 3y - 2z = 8$

d $2x + y - 2z = 4$
$4x + 2y - 4z = 8$
$x + y - z = 2$

5 Using row operations, solve these systems of equations.

a $3x + 4y + 7z = 0$
$2x - y + 4z = 3$
$x + 2y + 5z = 2$

b $2x + y + 3z = 1$
$3x - 4y - 2z = 9$
$x - y - 2z = 0$

c $x - 3y + 2z = 13$
$2x - y + 2z = 8$
$3x + 3y + 2z = -1$

d $2x - y + 3z = 6$
$6x - y + z = 3$
$10x + 3y - 2z = 0$

6 Using a calculator, solve these systems of equations.

a $x + 2y - 3z = 5$
$2x - y + 2z = 7$
$3x + 2y + 5z = 9$

b $3x - 3y + z = 8$
$2x + y - 2z = 5$
$3x + 4y + z = 1$

c $3x + 3y - 6z = 2$
$6x - 8y + z = 8$
$x - y + 3z = 0$

d $3x + 2y - z = 1$
$6x - y + 3z = 7$
$11x + y + 2z = 8$

7 Without using a calculator, solve the following equations where possible.

a $x + 3y + 2z = 1$
$x + y - 2z = 4$
$x + 7y + 10z = -5$

b $2x + y + 3z = 4$
$-x + 2y + z = 2$
$x + 3y + 4z = 6$

c $3x - y + 4z = 1$
$x + 2y - 3z = 4$
$x - 5y + 10z = -7$

d $2x + y + 3z = 4$
$4x + 2y + 6z = 8$
$6x + 3y + 9z = 12$

e $x + 2y - z = 1$
$2x - y + 3z = 4$
$5x - 5y + 5z = 9$

f $-x - y + z = 4$
$2x - y + z = 8$
$y - z = 3$

g $3x + y - 2z = 4$
$x + y + z = 4$
$x + 3y + 6z = 10$

h $3x + y - 2z = 4$
$2x - y + z = 10$
$x - 4y + 4z = 12$

8 Using a calculator, state whether the following equations have a unique solution, no solution or infinite solutions. If the solution is unique, state it, and if the solution is infinite, give it in terms of one parameter.

a $2x + y - 2z = 4$
$x + 3y - 2z = 7$
$x + 8y - 4z = 17$

b $3x + 5y - 2z = 2$
$x + 7y + 4z = 1$
$3x - 3y + 2z = -2$

c $x + 4y - z = 4$
$-x + 7y = 10$
$3x + y - 2z = 7$

d $3x + y - 2z = 4$
$x + y + z = 4$
$x + 3y + 6z = 10$

e $x - y + 5z = 4$
$2x + y - z = 8$
$3y - 11z = 0$

f $x + y - z = 6$
$2x + y - z = 7$
$x + y - 5z = 18$

9 a Find the inverse of $\begin{pmatrix} 1 & -3 & -3 \\ 2 & -1 & 3 \\ 3 & -9 & 9 \end{pmatrix}$.

b Hence solve this system of equations.

$x - 3y - 3z = 2$
$2x - y + 3z = 1$
$3x - 9y + 9z = 4$

10 a Find the determinant of the matrix $\begin{pmatrix} 1 & 9 & 5 \\ 1 & 3 & 2 \\ 1 & 1 & 1 \end{pmatrix}$.

b Find the value of c for which this system of equations can be solved.

$$\begin{pmatrix} 1 & 9 & 5 \\ 1 & 3 & 2 \\ 1 & 1 & 1 \end{pmatrix}\begin{pmatrix} x \\ y \\ z \end{pmatrix} = \begin{pmatrix} 3 \\ c \\ 2 \end{pmatrix}$$

c Using this value of c, give the general solution to the system of equations.

11 Without finding the solution, show that this system of equations has a unique solution.

$$2x - 5y + z = 4$$
$$x - y + 3z = -4$$
$$4x + 4y + 3z = 1$$

12 If the following system of equations does not have a unique solution, state the relationship between a and b.

$$ax - 3y + 2z = 4$$
$$2x - 5y + z = 9$$
$$2x - by + 4z = -1$$

13 a Let $M = \begin{pmatrix} 1 & k & -3 \\ 2 & 4 & -5 \\ 3 & -1 & k \end{pmatrix}$. Find Det M.

b Find the value of k for which this system of equations does not have a unique solution.

$$x + ky - 3z = 1$$
$$2x + 4y - 5z = 2$$
$$3x - y + kz = 3$$

14 Find the value of a for which this system of equations is consistent.

$$x - 3y + z = 3$$
$$x + 5y - 2z = 1$$
$$16x - 2z = a$$

Review exercise

 1 If P is an $m \times n$ matrix and Q is an $n \times p$ matrix find the orders of the matrices R and S such that $3P(-4Q + 2R) = 5S$.

 2 Given that $A = \begin{pmatrix} 3 & -2 \\ -3 & 4 \end{pmatrix}$ and $I = \begin{pmatrix} 1 & 0 \\ 0 & 1 \end{pmatrix}$, find the values of λ for which $(A - \lambda I)$ is a singular matrix. [IB May 03 P1 Q5]

3 a Find the inverse of the matrix $A = \begin{pmatrix} k & -1 \\ 1 & k \end{pmatrix}$, where $k \in \mathbb{R}$.

b Hence or otherwise, solve the simultaneous equations

$$kx - y = 2k$$
$$x + ky = 1 - k^2$$

[IB Nov 97 P1 Q12]

4 If $A = \begin{pmatrix} x & 4 \\ 4 & 2 \end{pmatrix}$ and $B = \begin{pmatrix} 2 & y \\ 8 & 4 \end{pmatrix}$, find the values of x and y, given that

$AB = BA$. [IB Nov 01 P1 Q6]

 5 Consider this system of equations.

$x + (k + 3)y + 5z = 0$
$x + 3y + (k + 1)z = k + 2$
$x + y + kz = 2k - 1$

a Write the system in matrix form $AX = B$ where $X = \begin{pmatrix} x \\ y \\ z \end{pmatrix}$.

b Find the value of k for which the determinant of A is zero.
c Find the value of z in terms of k.
d Describe the solutions to the system of equations.

 6 Given the following two matrices, $M = \begin{pmatrix} 1 & -1 & -2 \\ 1 & 1 & -2 \\ 1 & 2 & a \end{pmatrix}$ and

$M^{-1} = \dfrac{1}{2}\begin{pmatrix} b & -5 & 4 \\ -1 & 1 & 0 \\ 1 & -3 & 2 \end{pmatrix}$, find the values of a and b.

[IB Nov 98 P1 Q14]

 7 a Find the relationship between p, q and r such that the following system of equations has a solution.

$2x - y - 3z = p$
$3x + y + 4z = q$
$-3x - 6y - 21z = r$

b If $p = 3$ and $q = -1$, find the solution to the system of equations. Is this solution unique?

8 Let $A = \begin{pmatrix} 2 & 6 \\ k & -1 \end{pmatrix}$ and $B = \begin{pmatrix} h & 3 \\ -3 & 7 \end{pmatrix}$, where h and k are integers. Given

that $Det\ A = Det\ B$ and that $Det\ AB = 256h$,

a show that h satisfies the equation $49h^2 - 130h + 81 = 0$
b hence find the value of k. [IB May 06 P1 Q17]

 9 If $M = \begin{pmatrix} 2 & -1 \\ 3 & 1 \end{pmatrix}$, $I = \begin{pmatrix} 1 & 0 \\ 0 & 1 \end{pmatrix}$ and $M^2 = pM + qI$, find the values of p

and q.

10 a Find the values of c for which $M = \begin{pmatrix} c^3 & 3 & 8 \\ c & 2 & 2 \\ 1 & 3 & 1 \end{pmatrix}$ is singular.

b Find A where $A = \begin{pmatrix} 8 & 3 & 8 \\ 2 & 2 & 2 \\ 1 & 3 & 1 \end{pmatrix}\begin{pmatrix} 1 & 2 & 3 \\ 4 & 5 & 6 \\ 2 & 1 & 1 \end{pmatrix}$.

c Explain why A is singular.

11 The system of equations represented by the following matrix equation has an infinite number of solutions.

$$\begin{pmatrix} 2 & -1 & -9 \\ 1 & 2 & 3 \\ 2 & 1 & -3 \end{pmatrix} \begin{pmatrix} x \\ y \\ z \end{pmatrix} = \begin{pmatrix} 7 \\ 1 \\ k \end{pmatrix}$$

Find the value of k. [IB May 00 P1 Q6]

 12 The variables x, y and z satisfy the simultaneous equations

$$x - 2y + z = 3$$
$$2x + 3y + 4z = 5$$
$$-x + 9y + z = c$$

where c is a constant.

a Show that these equations do not have a unique solution.

b Find the value of c that makes these equations consistent.

c For this value of c, find the general solution to these equations.

 13 a Find the values of a and b given that the matrix $A = \begin{pmatrix} a & -4 & -6 \\ -8 & 5 & 7 \\ -5 & 3 & 4 \end{pmatrix}$ is

the inverse of the matrix $B = \begin{pmatrix} 1 & 2 & -2 \\ 3 & b & 1 \\ -1 & 1 & -3 \end{pmatrix}$.

b For the values of a and b found in part **a**, solve the system of linear equations

$$x + 2y - 2z = 5$$
$$3x + by + z = 0$$
$$-x + y - 3z = a - 1$$

[IB Nov 99 P1 Q12]

 14 Show that the following system of equations has a solution only when $p - 2q + r = 0$.

$$3a - 5b + c = p$$
$$2a + b - 4c = q$$
$$-a + 7b - 9c = r$$

15 Find the value of a for which the following system of equations does not have a unique solution.

$$4x - y + 2z = 1$$
$$2x + 3y = -6$$
$$x - 2y + az = \frac{7}{2}$$

[IB May 99 P1 Q6]

 16 Given that $P = \begin{pmatrix} 2 & -1 & 3 \\ a & 2 & b \\ 4 & 0 & 0 \end{pmatrix}$, $Q = \begin{pmatrix} 0 & -2 & 1 \\ 3 & 0 & 0 \\ 0 & 1 & -1 \end{pmatrix}$ and

$R = \begin{pmatrix} -3 & -1 & -1 \\ 6 & -9 & 5 \\ 0 & -8 & 4 \end{pmatrix}$, find the values of a and b such that $PQ = R$.

 17 Given the two sets of equations,

$$\begin{aligned} x_1 &= 3z_1 - 2z_2 + 5z_3 & y_1 &= x_1 + 4x_2 - 3x_3 \\ x_2 &= 4z_1 + 5z_2 - 9z_3 & y_2 &= 3x_1 - 5x_2 - 7x_3 \\ x_3 &= z_1 - 6z_2 + 9z_3 & y_3 &= 2x_1 + 2x_2 - x_3 \end{aligned}$$

use matrix methods to obtain three equations that express y_1, y_2 and y_3 directly in terms of z_1, z_2 and z_3.

12 Vector Techniques

Bernard Bolzano was born in 1781 in Prague in what is now the Czech Republic. During Bolzano's early life there were two major influences. The first was his father, who was active in caring for others and the second were the monks who taught him, who were required to take a vow which committed them to take special care of young people. In the year 1799–1800 Bolzano undertook mathematical research with Frantisek Josef Gerstner and contemplated his future. The result of this was that in the autumn of 1800, he went to Charles University to study theology. During this time he also continued to work on mathematics and prepared a doctoral thesis on geometry which led to him publishing a work on the foundations of elementary geometry, *Betrachtungen über einige Gegenstände der Elementargoemetrie* in 1804. In this book Bolzano considers points, lines and planes as undefined elements, and defines operations on them. These are key ideas in the concept of linear space, which then led to the concept of vectors.

Bernard Bolzano

Following this, Bolzano entered two competitions for chairs at the Charles University in Prague. One was for the chair of mathematics and the other for the new chair in the philosophy of religion. Bolzano was placed first in both competitions, but the university gave him the chair in the philosophy of religion. In many ways this was the wrong decision, given the way he was brought up with a belief in social justice and pacifism and the fact he was a free thinker. His appointment was viewed with suspicion by the Austrian rulers in Vienna. He criticised the discrimination of the Czech-speaking Bohemians by the German-speaking Bohemians, against their Czech fellow citizens and the anti-Semitism displayed by both the German and Czech Bohemians. It came as no surprise that Bolzano was suspended from his position in December 1819 after pressure from the Austrian government. He was also suspended from his professorship, put under house arrest, had his mail censored, and was not allowed to publish. He was then tried by the Church, and was required to recant his supposed heresies. He refused to do so and resigned his chair at the university. From 1823 he continued to study, until in the winter of 1848 he contracted a cold which, given the poor condition of his lungs, led to his death.

12.1 Introduction to vectors

Physical quantities can be classified into two different kinds:

(i) scalar quantities, often called **scalars**, which have magnitude, but no associated direction

(ii) vector quantities, often called **vectors**, which have a magnitude and an associated direction.

So travelling 20 m is a scalar quantity and is called distance whereas travelling 20 m due north is a vector quantity and is called displacement.

Vector notation

Vectors can be represented in either two or three dimensions, and are described through components. Hence if we want to move from the point (1, 2) on the Cartesian plane to the point (3, 5) we do this by stating we move 2 in the positive x-direction and 3 in the positive y-direction. There are two possible notations for this, column vector notation and unit vector notation.

Column vector notation

In two dimensions a vector can be represented as $\begin{pmatrix} x \\ y \end{pmatrix}$ and in three dimensions as $\begin{pmatrix} x \\ y \\ z \end{pmatrix}$.

The conventions for x and y in terms of positive and negative are the same as in the standard two-dimensional Cartesian plane. In three dimensions this is also true, but we need to define what a standard three-dimensional plane looks like. There are three different versions, which are all rotations of each other. In all cases they obey what can be called the "right-hand screw rule". This means that if a screw were placed at the origin and turned with a screwdriver in the right hand from the positive x-axis to the positive y-axis, then it would move in the direction of the positive z-axis. This is shown below.

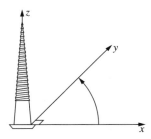

The axes are always drawn like this in this book. Different orientations may be used on IB examination papers.

Unit vector notation

The column vector $\begin{pmatrix} 2 \\ 3 \\ -2 \end{pmatrix}$ can be represented as $2\mathbf{i} + 3\mathbf{j} - 2\mathbf{k}$ using unit vectors.

Here the unit vectors **i**, **j** and **k** are vectors of magnitude 1 in the directions x, y and z respectively. These are shown in the diagram below.

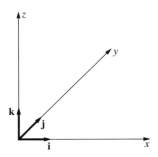

So the vector $2\mathbf{i} + 3\mathbf{j} - 2\mathbf{k}$ means 2 along the x-axis, 3 along the y-axis and -2 along the z-axis.

Hence a vector represents a change in position.

Position vectors, free vectors and tied vectors

A vector can be written as a position vector, a free vector or a tied vector.

A **position vector** is one that specifies a particular position in space relative to the origin. For example, in the diagram, the position vector of A is $\overrightarrow{OA} = \begin{pmatrix} 2 \\ 1 \end{pmatrix}$.

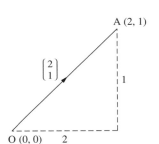

> Point A with coordinates (2, 1) has position vector $\overrightarrow{OA} = \begin{pmatrix} 2 \\ 1 \end{pmatrix}$.

Now if we talk about a vector $\mathbf{a} = \begin{pmatrix} 2 \\ 1 \end{pmatrix}$ then this can be anywhere in space and is therefore a **free vector**.

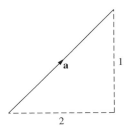

A vector $\overrightarrow{AB} = \begin{pmatrix} 2 \\ 1 \end{pmatrix}$ is a **tied vector** since it is specified as the vector that goes from A to B.

There is no advantage to one notation over the other. Both are used in IB examinations and it is probably best to work in the notation given in the question.

$\overrightarrow{OA} = \begin{pmatrix} 2 \\ 1 \end{pmatrix}$ means A is 2 units to the right of O and 1 unit above it.

This is true for the position vector of any point.

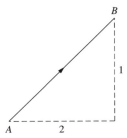

It is obviously possible that $\overrightarrow{OA} = \mathbf{a} = \begin{pmatrix} 2 \\ 1 \end{pmatrix}$ and that $\overrightarrow{AB} = \mathbf{a} = \begin{pmatrix} 2 \\ 1 \end{pmatrix}$, but it must be understood that $\overrightarrow{OA}$ and $\mathbf{a}$ and $\overrightarrow{AB}$ are slightly different concepts.

Whenever vectors are printed in books or in examination papers, free vectors are always written in bold, for example $\mathbf{a}$, but in any written work they are written with a bar underneath, $\underline{a}$. Position vectors and tied vectors are always written as the start and end points of the line representing the vector with an arrow above them, for example $\overrightarrow{OA}$, $\overrightarrow{AB}$.

Forming a tied vector

We now know the vector $\overrightarrow{AB}$ means the vector that takes us from A to B. If we consider A to be the point $(1, 2, -7)$ and B to be the point $(3, 1, 6)$, then to get from A to B we need to move 2 along the x-axis, -1 along the y-axis and 13 along the z-axis. This is shown in the diagram below.

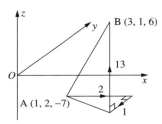

So $\overrightarrow{AB} = \begin{pmatrix} 2 \\ -1 \\ 13 \end{pmatrix}$.

> To find $\overrightarrow{BA}$ we subtract the coordinates of B from those of A.

More commonly, we think of $\overrightarrow{AB}$ as being the coordinates of A subtracted from the coordinates of B.

Example

If A has coordinates $(2, -3, 1)$ and B has coordinates $(3, -4, 1)$ find:

a) $\overrightarrow{AB}$

b) $\overrightarrow{BA}$

a) To get from $(2, -3, 1)$ to $(3, -4, 1)$ we go 1 in the x-direction, -1 in the y-direction and 0 in the z-direction. Alternatively, $\overrightarrow{AB} = \begin{pmatrix} 3 - 2 \\ -4 - (-3) \\ 1 - 1 \end{pmatrix} = \begin{pmatrix} 1 \\ -1 \\ 0 \end{pmatrix}$.

b) Similarly with $\overrightarrow{BA}$, we go -1 in the x-direction, 1 in the y-direction and 0 in the z-direction. Alternatively, $\overrightarrow{BA} = \begin{pmatrix} 2 - 3 \\ -3 - (-4) \\ 1 - 1 \end{pmatrix} = \begin{pmatrix} -1 \\ 1 \\ 0 \end{pmatrix}$.

Notice that in the example

$$\overrightarrow{AB} = -\overrightarrow{BA}$$

This is always true.

The magnitude of a vector

The magnitude (sometimes called the **modulus**) of a vector is the length of the line representing the vector. To calculate this we use Pythagoras' theorem.

> ### Example
>
> Find the magnitude of the vector $\mathbf{a} = \begin{pmatrix} 5 \\ 12 \end{pmatrix}$.
>
> Consider the vector $\mathbf{a} = \begin{pmatrix} 5 \\ 12 \end{pmatrix}$ in the diagram below.
>
>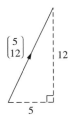
>
> The magnitude is given by the length of the hypotenuse. $|\mathbf{a}| = \sqrt{5^2 + 12^2} = 13$

$|\mathbf{a}|$ means the magnitude of $\mathbf{a}$.

In three dimensions this becomes a little more complicated.

Example

Find the magnitude of the vector $\begin{pmatrix} 3 \\ 4 \\ 7 \end{pmatrix}$.

The vector $\begin{pmatrix} 3 \\ 4 \\ 7 \end{pmatrix}$ is shown in the diagram below. OA is the magnitude of the vector.

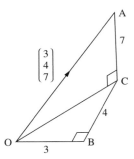

We know by Pythagoras' theorem that $OA = \sqrt{OC^2 + AC^2}$.
Applying Pythagoras' theorem again, $OC^2 = OB^2 + BC^2$.
Hence

$$OA = \sqrt{OB^2 + BC^2 + AC^2}$$
$$\Rightarrow OA = \sqrt{3^2 + 4^2 + 7^2}$$
$$\Rightarrow OA = \sqrt{74}$$

Multiplying a vector by a scalar

When we multiply a vector by a scalar we just multiply each component by the scalar. Hence the vector changes in magnitude, but not in direction. For example $2\begin{pmatrix} 3 \\ 4 \end{pmatrix} = \begin{pmatrix} 6 \\ 8 \end{pmatrix}$ has the same direction as $\begin{pmatrix} 3 \\ 4 \end{pmatrix}$ but has twice the magnitude. This is shown in the diagram below.

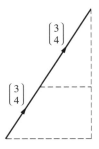

> There is no symbol for multiplication in this case. This is important as the symbols $\cdot$ and $\times$ have specific meanings in vectors.

In general, $c\begin{pmatrix} 3 \\ 4 \end{pmatrix} = \begin{pmatrix} 3c \\ 4c \end{pmatrix}$.

Example

If A has coordinates $(2, 3, -1)$ and B has coordinates $(7, 6, -1)$, find these vectors.

a) $\overrightarrow{AB}$

b) $2\overrightarrow{BA}$

c) $p\overrightarrow{AB}$

(a) $\overrightarrow{AB} = \begin{pmatrix} 7 - 2 \\ 6 - 3 \\ -1 - (-1) \end{pmatrix} = \begin{pmatrix} 5 \\ 3 \\ 0 \end{pmatrix}$

(b) $\overrightarrow{BA} = \begin{pmatrix} 2 - 7 \\ 3 - 6 \\ -1 - (-1) \end{pmatrix} = \begin{pmatrix} -5 \\ -3 \\ 0 \end{pmatrix}$

$2\overrightarrow{BA} = 2\begin{pmatrix} -5 \\ -3 \\ 0 \end{pmatrix} = \begin{pmatrix} -10 \\ -6 \\ 0 \end{pmatrix}$

(c) $p\overrightarrow{AB} = p\begin{pmatrix} 5 \\ 3 \\ 0 \end{pmatrix} = \begin{pmatrix} 5p \\ 3p \\ 0 \end{pmatrix}$

Equal vectors

Vectors are equal if they have the same direction and magnitude.

Example

Find the values of a, b and c for which the vectors $3\begin{pmatrix} a - 1 \\ 2b + 3 \\ c \end{pmatrix}$ and $\begin{pmatrix} 2a - 2 \\ b + 1 \\ 5c - 2 \end{pmatrix}$ are equal.

If they are equal then

$3(a - 1) = 2a - 2$

$\Rightarrow 3a - 3 = 2a - 2 \Rightarrow a = 1$

$3(2b + 3) = b + 1$

$\Rightarrow 6b + 9 = b + 1 \Rightarrow b = -\dfrac{8}{5}$

$3c = 5c - 2 \Rightarrow c = 1$

Negative vectors

A negative vector has the same magnitude as the positive vector but the opposite direction. Hence if $\mathbf{a} = \begin{pmatrix} 2 \\ -3 \end{pmatrix}$ then $-\mathbf{a} = \begin{pmatrix} -2 \\ 3 \end{pmatrix}$.

Zero vectors

A zero or null vector is a vector with zero magnitude and no directional property. It is denoted by $\begin{pmatrix} 0 \\ 0 \end{pmatrix}$ or $0\mathbf{i} + 0\mathbf{j}$ in two dimensions. Adding a vector and its negative vector gives the zero vector, i.e. $\mathbf{a} + (-\mathbf{a}) =$ zero vector.

> **Example**
>
> If $\mathbf{a} = 2\mathbf{i} + 4\mathbf{j} - 7\mathbf{k}$ and $2\mathbf{a} + \mathbf{b} = 0\mathbf{i} + 0\mathbf{j} + 0\mathbf{k}$, find $\mathbf{b}$.
> $2\mathbf{a} = 4\mathbf{i} + 8\mathbf{j} - 14\mathbf{k}$
> $\mathbf{b} = -2\mathbf{a} = -(4\mathbf{i} + 8\mathbf{j} - 14\mathbf{k}) = -4\mathbf{i} - 8\mathbf{j} + 14\mathbf{k}$

In the case of an equation like this the zero vector $0\mathbf{i} + 0\mathbf{j} + 0\mathbf{k}$ could just be written as 0.

If $2\mathbf{a} + \mathbf{b} = 0$ then $\mathbf{b} = -2\mathbf{a}$

Unit vectors

A unit vector is a vector of magnitude one. To find this we divide by the magnitude of the vector. If $\mathbf{n}$ is the vector then the notation for the unit vector is $\hat{\mathbf{n}}$.

> **Example**
>
> Find a unit vector parallel to $\mathbf{m} = 3\mathbf{i} + 5\mathbf{j} - 2\mathbf{k}$.
> The magnitude of $3\mathbf{i} + 5\mathbf{j} - 2\mathbf{k}$ is $\sqrt{3^2 + 5^2 + (-2)^2} = \sqrt{38}$.
> Hence the required vector is $\hat{\mathbf{m}} = \dfrac{1}{\sqrt{38}}(3\mathbf{i} + 5\mathbf{j} - 2\mathbf{k})$.

Parallel vectors

Since parallel vectors must have the same direction, the vectors must be scalar multiples of each other.

So in two dimensions $\begin{pmatrix} 3 \\ -2 \end{pmatrix}$ is parallel to $\begin{pmatrix} 15 \\ -10 \end{pmatrix}$ since $\begin{pmatrix} 15 \\ -10 \end{pmatrix} = 5\begin{pmatrix} 3 \\ -2 \end{pmatrix}$.

In three dimensions $4\mathbf{i} + 2\mathbf{j} - 5\mathbf{k}$ is parallel to $-12\mathbf{i} - 6\mathbf{j} + 15\mathbf{k}$ since $-12\mathbf{i} - 6\mathbf{j} + 15\mathbf{k} = -3(4\mathbf{i} + 2\mathbf{j} - 5\mathbf{k})$.

> **Example**
>
> Find the value of k for which the vectors $\begin{pmatrix} 4 \\ -2 \\ 8 \end{pmatrix}$ and $\begin{pmatrix} 12 \\ -6 \\ k \end{pmatrix}$ are parallel.
>
> $\begin{pmatrix} 4 \\ -2 \\ 8 \end{pmatrix} = 2\begin{pmatrix} 2 \\ -1 \\ 4 \end{pmatrix}$ and $\begin{pmatrix} 12 \\ -6 \\ k \end{pmatrix} = 6\begin{pmatrix} 2 \\ -1 \\ \frac{k}{6} \end{pmatrix}$
>
> Hence these vectors are parallel when $\dfrac{k}{6} = 4 \Rightarrow k = 24$.

Perpendicular vectors

In the two-dimensional case we use the property that with perpendicular lines the product of the gradients is -1.

Example

Find a vector perpendicular to $\begin{pmatrix} 3 \\ -1 \end{pmatrix}$.

From the diagram below we can see that the line representing this vector has a gradient of $-\dfrac{1}{3}$.

Hence the line representing the perpendicular vector will have a gradient of 3.

Therefore a perpendicular vector is $\begin{pmatrix} 1 \\ 3 \end{pmatrix}$. ·············

> There are an infinite number of perpendicular vectors.

In three dimensions this is more complicated and will be dealt with later in the chapter.

Exercise 1

1 Find the values of a, b and c.

a $\begin{pmatrix} 2 \\ 3 \\ 4 \end{pmatrix} = 2\begin{pmatrix} a \\ b+1 \\ c-2 \end{pmatrix}$ **b** $\begin{pmatrix} 1 \\ b \\ -2 \end{pmatrix} = \begin{pmatrix} 3a \\ 2b^2 \\ c+6 \end{pmatrix}$ **c** $3\begin{pmatrix} a \\ b-1 \\ 4 \end{pmatrix} = 4\begin{pmatrix} 2-a \\ 2b+3 \\ 3 \end{pmatrix}$

2 If the position vector of P is $\mathbf{i} + \mathbf{j}$ and the position vector of Q is $2\mathbf{i} - 3\mathbf{j}$, find:

   a $\overrightarrow{PQ}$ **b** $\left| \overrightarrow{PQ} \right|$

3 If the position vector of A is $\begin{pmatrix} 2 \\ 3 \end{pmatrix}$ and the position vector of B is $\begin{pmatrix} -1 \\ 5 \end{pmatrix}$, find:

   a $\overrightarrow{AB}$ **b** $\left| \overrightarrow{AB} \right|$

4 Write down a vector that is parallel to the line $y = 3x + 5$.

5 Find the magnitude of these vectors.

   a $\mathbf{m} = 3\mathbf{i} + 5\mathbf{j}$ **b** $\overrightarrow{OP} = \begin{pmatrix} 2 \\ -7 \end{pmatrix}$ **c** $\begin{pmatrix} x \\ y \end{pmatrix} = \begin{pmatrix} -3 \\ -9 \end{pmatrix}$

   d $\mathbf{a} = 2\mathbf{i} - 4\mathbf{j} + 3\mathbf{k}$ **e** $\begin{pmatrix} x \\ y \\ z \end{pmatrix} = \begin{pmatrix} 4 \\ -1 \\ 2 \end{pmatrix}$ **f** $\overrightarrow{OA} = 2\mathbf{i} - 7\mathbf{j} - 2\mathbf{k}$

6 State which of the following vectors are parallel to $\begin{pmatrix} 1 \\ -12 \\ -16 \end{pmatrix}$.

   a $\begin{pmatrix} 1 \\ -12 \\ -16 \end{pmatrix}$ **b** $\dfrac{1}{3}\mathbf{i} - \mathbf{j} - \dfrac{4}{3}\mathbf{k}$ **c** $\begin{pmatrix} -5 \\ -15 \\ -20 \end{pmatrix}$ **d** $0.5p(2\mathbf{i} - 6\mathbf{j} + 8\mathbf{k})$

7 Find the values of c for which the vectors are parallel.

    a $\mathbf{i} + 2\mathbf{j} - 3\mathbf{k}$ and $3\mathbf{i} + c\mathbf{j} - 9\mathbf{k}$ **b** $\begin{pmatrix} 14 \\ -35 \\ c \end{pmatrix}$ and $\begin{pmatrix} 18 \\ -45 \\ -9 \end{pmatrix}$

    c $\begin{pmatrix} 4t \\ -8t \\ 10t \end{pmatrix}$ and $\begin{pmatrix} ct \\ -12t \\ 15t \end{pmatrix}$

8 A two-dimensional vector has a modulus of 13. It makes an angle of $60°$ with the x-axis and an angle of $30°$ with the y-axis. Find an exact value for this vector.

9 Find a unit vector in the direction of $\begin{pmatrix} -5 \\ -6 \\ 1 \end{pmatrix}$.

10 A, B, C and D have position vectors given by
$\mathbf{i} + 3\mathbf{j} - 2\mathbf{k}$, $2\mathbf{i} + \mathbf{j} + 4\mathbf{k}$, $3\mathbf{i} + 2\mathbf{j} + 7\mathbf{k}$ and $3\mathbf{i} + 5\mathbf{j} + 4\mathbf{k}$. Determine which of the following pairs of lines are parallel.
    a AB and CD   **b** BC and CD   **c** BC and AD

11 A triangle has its vertices at the points P (1, 2), Q (3, 5) and R(−1, −1). Find the vectors $\overrightarrow{PQ}$, $\overrightarrow{QR}$ and $\overrightarrow{PR}$, and the modulus of each of these vectors.

12 A parallelogram has coordinates P(0, 1, 4), Q(4, −1, 3), R(x, y, z) and S(−1, 5, 6).
    a Find the coordinates of R.
    b Find the vectors $\overrightarrow{PQ}$, $\overrightarrow{QR}$, $\overrightarrow{SR}$ and $\overrightarrow{RP}$.
    c Find the magnitude of each of the vectors in part **b**.
    d Hence write down the unit vectors in the directions of $\overrightarrow{PQ}$, $\overrightarrow{QR}$, $\overrightarrow{SR}$ and $\overrightarrow{RP}$.

13 If PQRST is a pentagon, show that $\overrightarrow{PQ} + \overrightarrow{QR} + \overrightarrow{RS} = \overrightarrow{PT} + \overrightarrow{TS}$.

14 Consider the hexagon shown.

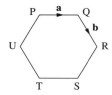

    Find the vector represented by $\overrightarrow{SU}$.

15 The vector $\begin{pmatrix} a - b \\ b \end{pmatrix}$ is parallel to the x-axis and the vector $\begin{pmatrix} 2a - b \\ a + b \end{pmatrix}$ is parallel to $\begin{pmatrix} 2 \\ 1 \end{pmatrix}$. Find the values of a and b.

16 The vector $(p + 2q - r)\mathbf{i} - (2p - q + 6r)\mathbf{j} + (3p - 2q - r)\mathbf{k}$ is parallel to $\mathbf{i} + \mathbf{j}$ and the vector $(p + r)\mathbf{i} - (-q - 2r)\mathbf{j} + (p + 2q - r)\mathbf{k}$ is parallel to the z-axis. Find the values of p, q and r.

17 If $\overrightarrow{OP} = (3x + 2y)\mathbf{p} + (x - y + 3)\mathbf{q}$, $\overrightarrow{OQ} = (x - y + 2)\mathbf{p} - (2x + y + 1)\mathbf{q}$ and $2\overrightarrow{OP} = 3\overrightarrow{OQ}$, where vectors $\mathbf{p}$ and $\mathbf{q}$ are non-parallel vectors, find the values of x and y.

12.2 A geometric approach to vectors

To add two vectors we just add the *x*-components, the *y*-components and the *z*-components. To subtract two vectors we subtract the *x*-components, the *y*-components and the *z*-components.

> **Example**
>
> If $\mathbf{a} = 2\mathbf{i} - 6\mathbf{j} + 12\mathbf{k}$ and $\mathbf{b} = -\mathbf{i} - 6\mathbf{j} + 7\mathbf{k}$ find $\mathbf{a} + \mathbf{b}$.
>
> $$\mathbf{a} + \mathbf{b} = (2 - 1)\mathbf{i} + (-6 - 6)\mathbf{j} + (12 + 7)\mathbf{k}$$
> $$= \mathbf{i} - 12\mathbf{j} + 19\mathbf{k}$$

> **Example**
>
> If $\overrightarrow{OA} = \begin{pmatrix} -3 \\ 4 \\ 2 \end{pmatrix}$ and $\overrightarrow{OB} = \begin{pmatrix} 2 \\ -4 \\ -7 \end{pmatrix}$ find $\overrightarrow{OA} - \overrightarrow{OB}$.
>
> $$\overrightarrow{OA} - \overrightarrow{OB} = \begin{pmatrix} -3 \\ 4 \\ 2 \end{pmatrix} - \begin{pmatrix} 2 \\ -4 \\ -7 \end{pmatrix} = \begin{pmatrix} -3 - 2 \\ 4 - (-4) \\ 2 - (-7) \end{pmatrix} = \begin{pmatrix} -5 \\ 0 \\ 9 \end{pmatrix}$$

We can also look at adding and subtracting vectors geometrically.

Vector addition

Let the vectors $\mathbf{p}$ and $\mathbf{q}$ be represented by the lines $\overrightarrow{AB}$ and $\overrightarrow{BC}$ respectively as shown in the diagram.

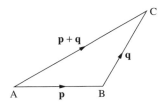

Then the vector represented by the line $\overrightarrow{AC}$ is defined as the sum of $\mathbf{p}$ and $\mathbf{q}$ and is written as $\mathbf{p} + \mathbf{q}$. This is sometimes called the **triangle law** of vector addition.

Alternatively it can also be represented by a parallelogram. In this case let $\mathbf{p}$ and $\mathbf{q}$ be represented by $\overrightarrow{AB}$ and $\overrightarrow{AD}$.

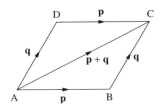

It should be noted that since $\overrightarrow{DC}$ is the same in magnitude and direction as $\overrightarrow{AB}$, the line $\overrightarrow{DC}$ can also represent the vector **p**. Similarly $\overrightarrow{BC}$ can represent the vector **q**.

Comparing this with the triangle, it is clear that the diagonal $\overrightarrow{AC}$ can represent the sum $\overrightarrow{AB} + \overrightarrow{BC}$. This is known as the **parallelogram law** of vector addition.

This also shows that vector addition is commutative.

$$\mathbf{q} + \mathbf{p} = \overrightarrow{AD} + \overrightarrow{DC} = \overrightarrow{AC} = \overrightarrow{AB} + \overrightarrow{BC} = \mathbf{p} + \mathbf{q}$$

Put very simply, vector addition can be thought of as getting from the start point to the end point by any route. Hence in the case of the triangle the route along two sides is the same as the route along the third side because they start and finish at the same point.

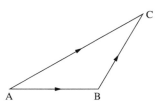

Hence $\boxed{\overrightarrow{AB} + \overrightarrow{BC} = \overrightarrow{AC}}$

This concept can be extended to more than two vectors. To get from A to F we can either go directly from A to F or we can go via B, C, D, and E. Hence

$$\overrightarrow{AF} = \overrightarrow{AB} + \overrightarrow{BC} + \overrightarrow{CD} + \overrightarrow{DE} + \overrightarrow{EF}.$$

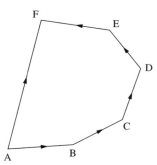

Example

A quadrilateral has coordinates $A(1, -1, 4)$, $B(3, 2, 5)$, $C(1, 2, 0)$, $D(-1, 2, -1)$. Show that:

a) $\overrightarrow{AB} + \overrightarrow{BC} = \overrightarrow{AC}$

b) $\overrightarrow{AB} + \overrightarrow{BC} + \overrightarrow{CD} = \overrightarrow{AD}$

a) $\overrightarrow{AB} = \begin{pmatrix} 2 \\ 3 \\ 1 \end{pmatrix}$, $\overrightarrow{BC} = \begin{pmatrix} -2 \\ 0 \\ -5 \end{pmatrix}$, $\overrightarrow{AC} = \begin{pmatrix} 0 \\ 3 \\ -4 \end{pmatrix}$

$$\overrightarrow{AB} + \overrightarrow{BC} = \begin{pmatrix} 2 \\ 3 \\ 1 \end{pmatrix} + \begin{pmatrix} -2 \\ 0 \\ -5 \end{pmatrix} = \begin{pmatrix} 0 \\ 3 \\ -4 \end{pmatrix} = \overrightarrow{AC}$$

b) $\overrightarrow{CD} = \begin{pmatrix} -2 \\ 0 \\ -1 \end{pmatrix}$, $\overrightarrow{AD} = \begin{pmatrix} -2 \\ 3 \\ -5 \end{pmatrix}$

$$\overrightarrow{AB} + \overrightarrow{BC} + \overrightarrow{CD} = \begin{pmatrix} 2 \\ 3 \\ 1 \end{pmatrix} + \begin{pmatrix} -2 \\ 0 \\ -5 \end{pmatrix} + \begin{pmatrix} -2 \\ 0 \\ -1 \end{pmatrix} = \begin{pmatrix} -2 \\ 3 \\ -5 \end{pmatrix} = \overrightarrow{AD}$$

Example

If A has coordinates $(0, -1, 2)$ and B has coordinates $(2, -3, 5)$, find:
a) the position vector of the point C, the midpoint of AB
b) the position vector of the point D which divides the line AB in the ratio of $1 : 2$.

a)

$$\overrightarrow{AB} = \begin{pmatrix} 2 - 0 \\ -3 - (-1) \\ 5 - 2 \end{pmatrix} = \begin{pmatrix} 2 \\ -2 \\ 3 \end{pmatrix}$$

$$\overrightarrow{AC} = \frac{1}{2}\overrightarrow{AB} = \frac{1}{2}\begin{pmatrix} 2 \\ -2 \\ 3 \end{pmatrix} = \begin{pmatrix} 1 \\ -1 \\ \frac{3}{2} \end{pmatrix}$$

The diagram shows that $\overrightarrow{OC} = \overrightarrow{OA} + \overrightarrow{AC} = \begin{pmatrix} 0 \\ -1 \\ 2 \end{pmatrix} + \begin{pmatrix} 1 \\ -1 \\ \frac{3}{2} \end{pmatrix} = \begin{pmatrix} 1 \\ -2 \\ \frac{7}{2} \end{pmatrix}.$

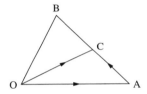

b) Since the line AB is divided in the ratio of $1 : 2$ the point D is $\frac{1}{3}$ the way along the line.

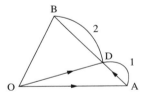

$$\overrightarrow{AD} = \frac{1}{3}\overrightarrow{AB} = \frac{1}{3}\begin{pmatrix} 2 \\ -2 \\ 3 \end{pmatrix} = \begin{pmatrix} \frac{2}{3} \\ \frac{-2}{3} \\ \frac{3}{3} \end{pmatrix} = \begin{pmatrix} \frac{2}{3} \\ \frac{-2}{3} \\ 1 \end{pmatrix}$$

Therefore the position vector of D is

$$\overrightarrow{OD} = \overrightarrow{OA} + \overrightarrow{AD} = \begin{pmatrix} 0 \\ -1 \\ 2 \end{pmatrix} + \begin{pmatrix} \frac{2}{3} \\ \frac{-2}{3} \\ 1 \end{pmatrix} = \begin{pmatrix} \frac{2}{3} \\ \frac{-5}{3} \\ 3 \end{pmatrix}.$$

Vector subtraction

We can now use this principle to look at the subtraction of two vectors. We can first consider **a** − **b** to be the same as **a** + (−**b**).

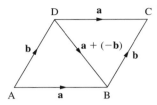

a + (−**b**) is the same as $\overrightarrow{DC}$ + $\overrightarrow{CB}$ and hence the diagonal $\overrightarrow{DB}$ can represent **a** + (−**b**) Thus in terms of the parallelogram one diagonal represents the addition of two vectors and the other the subtraction of two vectors. This explains geometrically why to find $\overrightarrow{AB}$ we subtract the coordinates of A from the coordinates of B.

Alternatively if we consider the triangle below we can see that $\overrightarrow{AB}$ = $\overrightarrow{AO}$ + $\overrightarrow{OB}$. Hence $\overrightarrow{AB}$ = −$\overrightarrow{OA}$ + $\overrightarrow{OB}$ = −**a** + **b** = **b** − **a**.

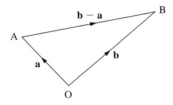

Example

The diagram shows quadrilateral ABCD, where $\overrightarrow{AB}$ = 2**p**, $\overrightarrow{DC}$ = **p** and $\overrightarrow{AD}$ = **q**.
a) What type of quadrilateral is ABCD?
b) Find these in terms of **p** and **q**.

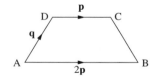

　i) $\overrightarrow{BC}$

　ii) $\overrightarrow{DB}$

　iii) $\overrightarrow{AC}$

a) Since AB and DC are parallel and AD is not parallel to BC, the shape is a trapezium.
b) (i) $\overrightarrow{BC}$ = $\overrightarrow{BA}$ + $\overrightarrow{AD}$ + $\overrightarrow{DC}$
　　　　= −2**p** + **q** + **p** = **q** − **p**
　(ii) $\overrightarrow{DB}$ = $\overrightarrow{DC}$ + $\overrightarrow{CB}$
　　　　= **p** + [−(**q** − **p**)] = 2**p** − **q**
　(iii) $\overrightarrow{AC}$ = $\overrightarrow{AD}$ + $\overrightarrow{DC}$
　　　　= **q** + **p**

Example

PQRS is a quadrilateral where X and Y are the midpoints of PQ and RS respectively.

Show that $\overrightarrow{PS} + \overrightarrow{QR} = 2\overrightarrow{XY}$.

Since $\overrightarrow{PX} = \overrightarrow{XQ}$, $\overrightarrow{XQ} - \overrightarrow{PX} = 0$

$\Rightarrow \overrightarrow{XQ} + \overrightarrow{XP} = 0$

Similarly $\overrightarrow{YR} + \overrightarrow{YS} = 0 \Rightarrow \overrightarrow{RY} + \overrightarrow{SY} = 0$

Now $\overrightarrow{XY} = \overrightarrow{XP} + \overrightarrow{PS} + \overrightarrow{SY}$ and

$\overrightarrow{XY} = \overrightarrow{XQ} + \overrightarrow{QR} + \overrightarrow{RY}$

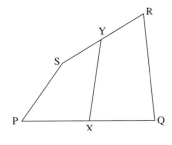

Hence $2\overrightarrow{XY} = \overrightarrow{XP} + \overrightarrow{PS} + \overrightarrow{SY} + \overrightarrow{XQ} + \overrightarrow{QR} + \overrightarrow{RY}$

$\Rightarrow 2\overrightarrow{XY} = \overrightarrow{PS} + \overrightarrow{QR}$

Exercise 2

1 If $\mathbf{a} = 2\mathbf{i} - \mathbf{j} + 5\mathbf{k}$, $\mathbf{b} = \mathbf{i} + 5\mathbf{j} - 6\mathbf{k}$ and $\mathbf{c} = 3\mathbf{i} - 6\mathbf{j} - 8\mathbf{k}$, find:

  a $\mathbf{a} + \mathbf{b}$ **b** $\mathbf{a} + \mathbf{b} + \mathbf{c}$ **c** $\mathbf{b} - \mathbf{c}$ **d** $2\mathbf{a} + \mathbf{b} + 4\mathbf{c}$

  e $3\mathbf{a} - 3\mathbf{b} - 2\mathbf{c}$ **f** $-2\mathbf{a} + 3\mathbf{b} + 7\mathbf{c}$ **g** $m\mathbf{a} + 20m\mathbf{b} - 3m\mathbf{c}$

2 If $\mathbf{a} = \mathbf{i} + \mathbf{j}$, $\mathbf{b} = 2\mathbf{i} - 3\mathbf{j}$ and $\mathbf{c} = 4\mathbf{i} + 7\mathbf{j}$, find:

  a $\mathbf{a} + 2\mathbf{b} + 3\mathbf{c}$ **b** $|\mathbf{a} - 2\mathbf{b} - 3\mathbf{c}|$

  c the angle that $\mathbf{a} - \mathbf{b} + \mathbf{c}$ makes with the x-axis.

3 Vectors $\mathbf{a}$, $\mathbf{b}$, $\mathbf{c}$ and $\mathbf{d}$ are given by $\mathbf{a} = \begin{pmatrix} 1 \\ 1 \\ 1 \end{pmatrix}$, $\mathbf{b} = \begin{pmatrix} 2 \\ 4 \\ 0 \end{pmatrix}$, $\mathbf{c} = \begin{pmatrix} 5 \\ -1 \\ 3 \end{pmatrix}$ and

$\mathbf{d} = \begin{pmatrix} 7 \\ 5 \\ q \end{pmatrix}$. If $\mathbf{b} - \mathbf{a}$ is parallel to $\mathbf{c} - \mathbf{d}$, find the value of q. Also find the

ratio of their moduli.

4 In the triangle shown, $\overrightarrow{OA} = \mathbf{a}$ and $\overrightarrow{OB} = \mathbf{b}$ and C is the midpoint of AB.

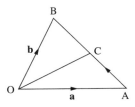

Find:

  a $\overrightarrow{AB}$ **b** $\overrightarrow{AC}$ **c** $\overrightarrow{CB}$

  d Hence, using two different methods, find $\overrightarrow{OC}$.

5 If the position vector of A is $\begin{pmatrix} 4 \\ -16 \end{pmatrix}$ and the position vector of B is $\begin{pmatrix} -1 \\ 4 \end{pmatrix}$,

find:

a $\overrightarrow{AB}$ **b** $|\overrightarrow{AB}|$

c the position vector of the midpoint of AB

d the position vector of the point dividing AB in the ratio 2 : 3.

6 If the position vector of P is $\begin{pmatrix} -7 \\ -13 \end{pmatrix}$ and the position vector of Q is $\begin{pmatrix} -1 \\ -3 \end{pmatrix}$,

find:

a $\overrightarrow{PQ}$ **b** $|\overrightarrow{PQ}|$

c the position vector of the midpoint of $\overrightarrow{PQ}$

d the position vector of the point dividing $\overrightarrow{PQ}$ in the ratio 1 : 7.

7 In the triangle shown, $\overrightarrow{OA} = \mathbf{a}$ and $\overrightarrow{OB} = \mathbf{b}$. The point C lies on AB such that $AC : CB = 1 : k$ where k is a constant.
Find:

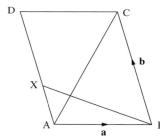

a $\overrightarrow{AC}$ **b** $\overrightarrow{BC}$ **c** $\overrightarrow{BA}$ **d** $\overrightarrow{OC}$

8 ABCD is the parallelogram shown, where $\overrightarrow{AB} = \mathbf{a}$ and $\overrightarrow{BC} = \mathbf{b}$.
$\overrightarrow{AX} = \dfrac{1}{3}\overrightarrow{AD}$.

a Find:

i) $\overrightarrow{CD}$ ii) $\overrightarrow{CA}$ iii) $\overrightarrow{BD}$ iv) $\overrightarrow{AX}$ v) $\overrightarrow{XD}$

b If $\mathbf{a} = \begin{pmatrix} 2k \\ 3c \end{pmatrix}$ and $\mathbf{b} = \begin{pmatrix} 4k \\ c \end{pmatrix}$, find $\overrightarrow{AC}$ in

terms of k and c.

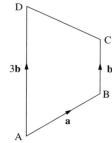

9 The trapezium shown has $\overrightarrow{AB} = \mathbf{a}$, $\overrightarrow{BC} = \mathbf{b}$

and $\overrightarrow{AD} = 3\mathbf{b}$. E and F are points on BC such that $BE : EF : FC = m : n : 3$.
Find:

a $\overrightarrow{BE}$ **b** $\overrightarrow{EF}$

c $\overrightarrow{CF}$ **d** $\overrightarrow{AF}$

e $\overrightarrow{ED}$

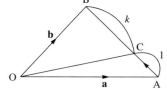

10 The cuboid ABCDEFGH shown has $\overrightarrow{AB} = \mathbf{a}$, $\overrightarrow{AD} = \mathbf{b}$ and $\overrightarrow{AE} = \mathbf{c}$. Find in terms of **a**, **b** and **c**:

a $\overrightarrow{BC}$ **b** $\overrightarrow{FH}$

c $\overrightarrow{AH}$ **d** $\overrightarrow{AG}$

e $\overrightarrow{BH}$

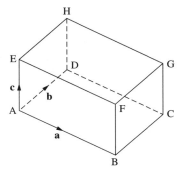

11 If PQR is a triangle and S is the midpoint of PQ, show that
$\overrightarrow{RQ} + \overrightarrow{PR} = 2\overrightarrow{PS}$.

12 ABCDEFGH is a regular octagon in which $\overrightarrow{AB} = \mathbf{a}$, $\overrightarrow{BC} = \mathbf{b}$ and $\overrightarrow{CD} = \mathbf{c}$
and $\overrightarrow{DE} = \mathbf{d}$. Find in terms of **a**, **b**, **c** and **d**:

a $\overrightarrow{DG}$ **b** $\overrightarrow{AH}$ **c** $\overrightarrow{FA}$

13 T, U and V are the midpoints of the sides PQ, QR and PR of a triangle. Show
that $\overrightarrow{OP} + \overrightarrow{OQ} + \overrightarrow{OR} = \overrightarrow{OT} + \overrightarrow{OU} + \overrightarrow{OV}$, where O is the origin.

14 OABC is a rhombus, where O is the origin, $\overrightarrow{OA} = \mathbf{a}$ and $\overrightarrow{OC} = \mathbf{c}$.

a Find $\overrightarrow{AB}$, $\overrightarrow{BC}$, $\overrightarrow{AC}$ and $\overrightarrow{OB}$ in terms of **a** and **c**.

b What is the relationship between **c** + **a** and **c** − **a**?

12.3 Multiplication of vectors

When we multiply two vectors there are two possible answers. One answer is a scalar
and the other is a vector. Hence one is called the **scalar product** and one is called the
vector product. We use a "dot" to signify the scalar product and a "cross" to signify
the vector product. It is quite common therefore to refer to the "dot product" and
"cross product".

The reason why there need to be two cases is best seen through physics. Consider the
concept of force multiplied by displacement. In one context this gives the work done,
which is a scalar quantity. In another context it gives the moment of a force (the turning
effect), which is a vector quantity. Hence in the physical world there are two possibilities,
and both need to be accounted for in the mathematical world.

Scalar product

The scalar product or dot product of two vectors **a** and **b** inclined at an angle of θ is
written as $\mathbf{a} \cdot \mathbf{b}$ and equals $|\mathbf{a}||\mathbf{b}| \cos \theta$.

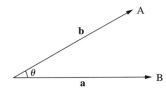

"Inclined at an angle of θ"
means the angle between
the two vectors is θ.

The following results are important.

Parallel vectors

If two vectors are parallel, then $\mathbf{a} \cdot \mathbf{b} = |\mathbf{a}||\mathbf{b}| \cos 0$ or $|\mathbf{a}||\mathbf{b}| \cos \pi$

So $\mathbf{a} \cdot \mathbf{b} = \pm|\mathbf{a}||\mathbf{b}|$

Perpendicular vectors

If two vectors are perpendicular, then $\mathbf{a} \cdot \mathbf{b} = |\mathbf{a}||\mathbf{b}| \cos \dfrac{\pi}{2}$

So $\mathbf{a} \cdot \mathbf{b} = 0$

Commutativity

We know that $\mathbf{a} \cdot \mathbf{b} = |\mathbf{a}||\mathbf{b}| \cos \theta$ and $\mathbf{b} \cdot \mathbf{a} = |\mathbf{b}||\mathbf{a}| \cos \theta$.

Since $|\mathbf{a}||\mathbf{b}| \cos \theta = |\mathbf{b}||\mathbf{a}| \cos \theta$ then $\mathbf{a} \cdot \mathbf{b} = \mathbf{b} \cdot \mathbf{a}$.

The scalar product is commutative.

Distributivity

The scalar product is distributive across addition.

This means $\mathbf{a} \cdot (\mathbf{b} + \boldsymbol{\rho}) = \mathbf{a} \cdot \mathbf{b} + \mathbf{a} \cdot \boldsymbol{\rho}$.

Proof

Consider the diagram below, where $\overrightarrow{OA} = \mathbf{a}$, $\overrightarrow{OB} = \mathbf{b}$ and $\overrightarrow{BC} = \boldsymbol{\rho}$. $\hat{AOB} = \theta$, $\hat{AOC} = \alpha$, and the angle $\overrightarrow{BC}$ makes with $\overrightarrow{BD}$, which is parallel to $\overrightarrow{OA}$, is β.

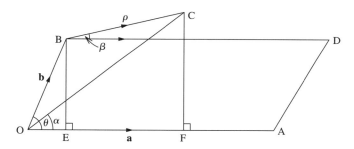

Clearly $\overrightarrow{OC} = \mathbf{b} + \boldsymbol{\rho}$

Now $\mathbf{a} \cdot \mathbf{b} + \mathbf{a} \cdot \boldsymbol{\rho} = |\overrightarrow{OA}||\overrightarrow{OB}| \cos \theta + |\overrightarrow{OA}||\overrightarrow{BC}| \cos \beta$

$$= |\overrightarrow{OA}|\left(|\overrightarrow{OE}| + |\overrightarrow{EF}|\right)$$

$$= |\overrightarrow{OA}||\overrightarrow{OF}|$$

and $\mathbf{a} \cdot (\mathbf{b} + \boldsymbol{\rho}) = |\overrightarrow{OA}||\overrightarrow{OC}| \cos \alpha = |\overrightarrow{OA}||\overrightarrow{OF}|$

So $\mathbf{a} \cdot (\mathbf{b} + \boldsymbol{\rho}) = \mathbf{a} \cdot \mathbf{b} + \mathbf{b} \cdot \boldsymbol{\rho}$, proving the scalar product is distributive over addition.

This result can be extended to any number of vectors.

Scalar product of vectors in component form

Let $\mathbf{p} = a_1\mathbf{i} + b_1\mathbf{j} + c_1\mathbf{k}$ and $\mathbf{q} = a_2\mathbf{i} + b_2\mathbf{j} + c_2\mathbf{k}$

Then $\mathbf{p} \cdot \mathbf{q} = (a_1\mathbf{i} + b_1\mathbf{j} + c_1\mathbf{k}) \cdot (a_2\mathbf{i} + b_2\mathbf{j} + c_2\mathbf{k})$

$$= (a_1a_2\mathbf{i} \cdot \mathbf{i} + b_1b_2\mathbf{j} \cdot \mathbf{j} + c_1c_2\mathbf{k} \cdot \mathbf{k})$$

$$+ (a_1b_2\mathbf{i} \cdot \mathbf{j} + b_1c_2\mathbf{j} \cdot \mathbf{k} + c_1a_2\mathbf{k} \cdot \mathbf{i} + b_1a_2\mathbf{j} \cdot \mathbf{i} + c_1b_2\mathbf{k} \cdot \mathbf{j} + a_1c_2\mathbf{i} \cdot \mathbf{k})$$

We need to look at what happens with various combinations of $\mathbf{i}$, $\mathbf{j}$ and $\mathbf{k}$:

$\mathbf{i} \cdot \mathbf{i} = \mathbf{j} \cdot \mathbf{j} = \mathbf{k} \cdot \mathbf{k} = (1)(1) \cos 0 = 1$

$\mathbf{i} \cdot \mathbf{j} = \mathbf{j} \cdot \mathbf{i} = \mathbf{i} \cdot \mathbf{k} = \mathbf{k} \cdot \mathbf{i} = \mathbf{j} \cdot \mathbf{k} = \mathbf{k} \cdot \mathbf{j} = (1)(1) \cos \dfrac{\pi}{2} = 0$

Hence

$$\mathbf{p} \cdot \mathbf{q} = a_1a_2 + b_1b_2 + c_1c_2$$

So there are two ways of calculating the scalar product. We normally use this form as we rarely know the angle between two vectors.

Example

If $\mathbf{a} = 3\mathbf{i} + \mathbf{j} - \mathbf{k}$ and $\mathbf{b} = 2\mathbf{i} - \mathbf{j} + 6\mathbf{k}$, find $\mathbf{a} \cdot \mathbf{b}$.

$\mathbf{a} \cdot \mathbf{b} = (3)(2) + (1)(-1) + (-1)(6)$
$\phantom{\mathbf{a} \cdot \mathbf{b}} = 6 - 1 - 6$
$\phantom{\mathbf{a} \cdot \mathbf{b}} = -1$

Example

Given that $\mathbf{x} \cdot \mathbf{p} = \mathbf{q} \cdot \mathbf{x}$ show that $\mathbf{x}$ is perpendicular to $\mathbf{p} - \mathbf{q}$.

$ \mathbf{x} \cdot \mathbf{p} = \mathbf{q} \cdot \mathbf{x}$
$\Rightarrow \mathbf{x} \cdot \mathbf{p} - \mathbf{q} \cdot \mathbf{x} = 0$
$ \mathbf{x} \cdot \mathbf{p} - \mathbf{x} \cdot \mathbf{q} = 0$ .... | Since the scalar product is commutative
$ \mathbf{x} \cdot (\mathbf{p} - \mathbf{q}) = 0$ .... | Using the distributive law
Since the scalar product is zero, $\mathbf{x}$ and $\mathbf{p} - \mathbf{q}$ are perpendicular.

Example

Show that the triangle ABC with vertices A(1, 2, 3), B(2, −1, 4) and C(3, −3, 2) is not right-angled.

We first write down the vectors representing each side.

$$\overrightarrow{AB} = \begin{pmatrix} 1 \\ -3 \\ 1 \end{pmatrix}$$

$$\overrightarrow{BC} = \begin{pmatrix} 1 \\ -2 \\ -2 \end{pmatrix}$$

$$\overrightarrow{AC} = \begin{pmatrix} 2 \\ -5 \\ -1 \end{pmatrix}$$

Now $\overrightarrow{AB} \cdot \overrightarrow{BC} = \begin{pmatrix} 1 \\ -3 \\ 1 \end{pmatrix} \cdot \begin{pmatrix} 1 \\ -2 \\ -2 \end{pmatrix} = (1)(1) + (-3)(-2) + (1)(-2) = 1 + 6 - 2 = 5$

$\overrightarrow{BC} \cdot \overrightarrow{AC} = \begin{pmatrix} 1 \\ -2 \\ -2 \end{pmatrix} \cdot \begin{pmatrix} 2 \\ -5 \\ -1 \end{pmatrix} = (1)(2) + (-2)(-5) + (-2)(-1) = 2 + 10 + 2 = 14$

$\overrightarrow{AB} \cdot \overrightarrow{AC} = \begin{pmatrix} 1 \\ -3 \\ 1 \end{pmatrix} \cdot \begin{pmatrix} 2 \\ -5 \\ -1 \end{pmatrix} = (1)(2) + (-3)(-5) + (1)(-1) = 2 + 15 - 1 = 16$

Since none of the scalar products equals zero, none of the sides are at right angles to each other, and hence the triangle is not right-angled.

Example

In the triangle ABC shown, prove that $\mathbf{c} \cdot \mathbf{c} = (\mathbf{a} + \mathbf{b}) \cdot (\mathbf{a} + \mathbf{b})$.

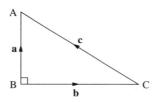

Using the scalar product

$$|\mathbf{c}|^2 = \mathbf{a} \cdot \mathbf{a} + \mathbf{a} \cdot \mathbf{b} + \mathbf{b} \cdot \mathbf{a} + \mathbf{b} \cdot \mathbf{b}$$

$$\Rightarrow |\mathbf{c}|^2 = |\mathbf{a}|^2 + 0 + 0 + |\mathbf{b}|^2$$

$$\Rightarrow AC^2 = AB^2 + BC^2$$

This is Pythagoras' theorem and hence the relationship is proven.

Since AB and BC are at right angles to each other

Angle between two vectors

If we draw two intersecting vectors, there are two possible angles where one is the supplement of the other.

Is the angle between the vectors α or β?

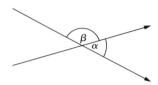

> Remember that supplement means "subtract from 180°" or "subtract from π" depending on whether we are working in radians or degrees.

There is a convention for this. The angle between two vectors is the angle between their directions when those directions both converge or both diverge from a point. Hence in this case we require α.

Now we know which angle to find, we can find it using the two formulae for scalar product.

Example

Find the angle θ between $\mathbf{a} = 3\mathbf{i} - 2\mathbf{j} + 4\mathbf{k}$ and $\mathbf{b} = 2\mathbf{i} + 4\mathbf{j} + 7\mathbf{k}$.

We know that $\mathbf{a} \cdot \mathbf{b} = (3)(2) + (-2)(4) + (4)(7) = 6 - 8 + 28 = 26$

So $|\mathbf{a}||\mathbf{b}| \cos \theta = 26$

Now $|\mathbf{a}| = \sqrt{3^2 + (-2)^2 + 4^2} = \sqrt{9 + 4 + 16} = \sqrt{29}$

And $|\mathbf{b}| = \sqrt{2^2 + 4^2 + 7^2} = \sqrt{4 + 16 + 49} = \sqrt{69}$

Hence $\cos \theta = \dfrac{26}{\sqrt{29}\sqrt{69}}$

$\Rightarrow \theta = 54.5°$

Example

If the angle between the vectors $\mathbf{a} = \begin{pmatrix} 3 \\ -1 \\ -1 \end{pmatrix}$ and $\mathbf{b} = \begin{pmatrix} 2 \\ -1 \\ x \end{pmatrix}$ is 60°, find the values of x.

We know that $\mathbf{a} \cdot \mathbf{b} = (3)(2) + (-1)(-1) + (-1)(x) = 6 + 1 - x = 7 - x$

So $7 - x = |\mathbf{a}||\mathbf{b}| \cos 60°$

Now $|\mathbf{a}| = \sqrt{3^2 + (-1)^2 + (-1)^2} = \sqrt{9 + 1 + 1} = \sqrt{11}$

And $|\mathbf{b}| = \sqrt{2^2 + (-1)^2 + x^2} = \sqrt{4 + 1 + x^2} = \sqrt{5 + x^2}$

Hence $7 - x = \sqrt{11}\sqrt{5 + x^2} \cos 60° = \dfrac{\sqrt{11}\sqrt{5 + x^2}}{2}$

We can to use a calculator to solve this equation.

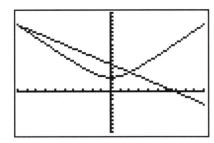

The answer is $x = 2.01$ or -10.0.

Exercise 3

1 Given that $\mathbf{a} = \begin{pmatrix} 3 \\ 1 \end{pmatrix}$, $\mathbf{b} = \begin{pmatrix} 2 \\ 5 \end{pmatrix}$ and $\mathbf{c} = \begin{pmatrix} 7 \\ -3 \end{pmatrix}$, find:

  a $\mathbf{a} \cdot \mathbf{b}$ **b** $\mathbf{b} \cdot \mathbf{c}$ **c** $\mathbf{a} \cdot (\mathbf{b} + \mathbf{c})$ **d** $\mathbf{b} \cdot \mathbf{i}$

  e $\mathbf{a} \cdot (\mathbf{c} - \mathbf{b})$ **f** $3\mathbf{a} \cdot \mathbf{c}$ **g** $\mathbf{a} \cdot (2\mathbf{b} + \mathbf{c})$ **h** $\mathbf{b} \cdot (2\mathbf{a} - 3\mathbf{c})$

2 Given that $\mathbf{a} = 2\mathbf{i} + 3\mathbf{j} - 4\mathbf{k}$, $\mathbf{b} = \mathbf{i} - 5\mathbf{j} - 2\mathbf{k}$ and $\mathbf{c} = -2\mathbf{i} - 2\mathbf{j} + \mathbf{k}$, find:

  a $\mathbf{a} \cdot \mathbf{b}$ **b** $\mathbf{c} \cdot \mathbf{b}$ **c** $\mathbf{b} \cdot \mathbf{b}$ **d** $\mathbf{a} \cdot (\mathbf{c} - \mathbf{b})$

  e $(\mathbf{a} + \mathbf{b}) \cdot \mathbf{i}$ **f** $\mathbf{b} \cdot (\mathbf{a} + 2\mathbf{b})$ **g** $\mathbf{c} \cdot (\mathbf{a} - 2\mathbf{b})$ **h** $\mathbf{b} \cdot \mathbf{a} + \mathbf{b} \cdot \mathbf{c}$

3 Calculate the angle between each pair of vectors.

  a $\mathbf{a} = 2\mathbf{i} - 4\mathbf{j} + 5\mathbf{k}$, $\mathbf{b} = \mathbf{i} + 3\mathbf{j} + 8\mathbf{k}$

  b $\mathbf{a} = \begin{pmatrix} 2 \\ 5 \end{pmatrix}$, $\mathbf{b} = \begin{pmatrix} 3 \\ -1 \end{pmatrix}$

  c $\mathbf{a} = \begin{pmatrix} 3 \\ -3 \\ 4 \end{pmatrix}$, $\mathbf{b} = \begin{pmatrix} 1 \\ -4 \\ 3 \end{pmatrix}$

  d $\mathbf{a} = \mathbf{i} + 2\mathbf{k}$, $\mathbf{b} = \mathbf{j} - \mathbf{k}$

  e $\mathbf{a} = \begin{pmatrix} 3 \\ -1 \\ 4 \end{pmatrix}$, $\mathbf{b} = \begin{pmatrix} 4 \\ 0 \\ 0 \end{pmatrix}$

  f $\mathbf{a} = \begin{pmatrix} 2t \\ t \\ -3t \end{pmatrix}$, $\mathbf{b} = \begin{pmatrix} -\dfrac{1}{t} \\ \dfrac{2}{t} \\ -\dfrac{3}{t} \end{pmatrix}$

4 Find $\mathbf{p} \cdot \mathbf{q}$ and the cosine of the angle θ between $\mathbf{p}$ and $\mathbf{q}$ if
$\mathbf{p} = -\mathbf{i} + 3\mathbf{j} - 2\mathbf{k}$ and $\mathbf{q} = \mathbf{i} + \mathbf{j} - 6\mathbf{k}$.

5 Find which of the following vectors are perpendicular to each other.

  $\mathbf{a} = 3\mathbf{i} + 2\mathbf{j} - \mathbf{k}$, $\mathbf{b} = 2\mathbf{i} - \mathbf{j} - 4\mathbf{k}$, $\mathbf{c} = 3\mathbf{i} + 2\mathbf{j} + \mathbf{k}$,

  $\mathbf{d} = -36\mathbf{i} + 27\mathbf{j} - 54\mathbf{k}$, $\mathbf{e} = \mathbf{i} + 2\mathbf{j}$, $\mathbf{f} = 4\mathbf{i} - 3\mathbf{j} + 6\mathbf{k}$

6 Find the value of λ if the following vectors are perpendicular.

  a $a = \begin{pmatrix} 2 \\ 1 \\ \lambda \end{pmatrix}$, $b = \begin{pmatrix} -1 \\ -1 \\ -2 \end{pmatrix}$

b $\mathbf{a} = 2\mathbf{i} + 5\mathbf{j} - 2\mathbf{k}, \mathbf{b} = \mathbf{i} + 4\mathbf{j} - \lambda\mathbf{k}$

c $\mathbf{a} = \lambda\mathbf{i} - 3\mathbf{k}, \mathbf{b} = 2\mathbf{i} + \mathbf{j} + 5\mathbf{k}$

d $\mathbf{a} = \begin{pmatrix} \lambda \\ 1 \\ 3 \end{pmatrix}, \mathbf{b} = \begin{pmatrix} \lambda \\ \lambda \\ -2 \end{pmatrix}$

7 Show that the triangle ABC is not right-angled, given that A has coordinates (2, −1, 2), B (3, 3, −1) and C (−2, 1, −4).

8 Find a unit vector that is perpendicular to $\overrightarrow{PQ}$ and to $\overrightarrow{PR}$, where

$\overrightarrow{PQ} = \mathbf{i} + \mathbf{j} + 2\mathbf{k}$ and $\overrightarrow{QR} = -\mathbf{i} - 2\mathbf{j} + \mathbf{k}$.

9 If the angle between the vectors $\mathbf{p} = \begin{pmatrix} 2 \\ 1 \\ -1 \end{pmatrix}$ and $\mathbf{q} = \begin{pmatrix} -3 \\ x \\ 4 \end{pmatrix}$ is 80°, find the possible values of x.

10 Taking O as the origin on a cube OABCDEFG of side 2 cm as shown, find the angle between the diagonals OF and AG.

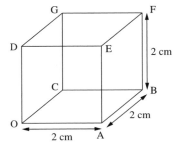

11 A quadrilateral ABCD has coordinates A (0, 0, 1), B (1, 1, 3), C (3, 0, 6) and D (2, −1, 4). Show that the quadrilateral is a parallelogram.

12 A quadrilateral ABCD has coordinates A (1, 2, −1), B (2, 3, 0), C (3, 5, 3) and D (−2, 3, 2). Show that the diagonals of the quadrilateral are perpendicular. Hence state, giving a reason, whether or not the quadrilateral is a rhombus.

13 Using the scalar product, prove that the diagonals of this rhombus are perpendicular to one another.

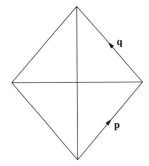

14 In the quadrilateral PQRS shown, prove that

$\overrightarrow{PR} \cdot \overrightarrow{QS} = \overrightarrow{PQ} \cdot \overrightarrow{RS} + \overrightarrow{QR} \cdot \overrightarrow{PS}$.

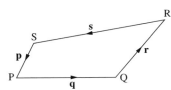

15 If $\mathbf{a} \cdot \mathbf{b} = \mathbf{a} \cdot \mathbf{c}$, show that $\mathbf{a}$ is perpendicular to $\mathbf{b} - \mathbf{c}$.

16 If A, B, C, D are four points such that $\overrightarrow{BC} + \overrightarrow{DA} = 0$, prove that ABCD is a parallelogram. If $\overrightarrow{AB} \cdot \overrightarrow{BC} = 0$, state with a reason whether the parallelogram is a rhombus, a rectangle or a square.

17 Given that $\mathbf{a}$ and $\mathbf{b}$ are non-zero vectors show that if $|\mathbf{a}| = |\mathbf{b}|$ then $\mathbf{a} + \mathbf{b}$ and $\mathbf{a} - \mathbf{b}$ are perpendicular.

18 If **a** and **b** are perpendicular vectors, show that:
$$(\mathbf{a} + \mathbf{b}) \cdot (\mathbf{a} + \mathbf{b}) = (\mathbf{a} - \mathbf{b}) \cdot (\mathbf{a} - \mathbf{b})$$

19 Triangle ABC is right-angled at B. Show that $\overrightarrow{AB} \cdot \overrightarrow{AC} = |\overrightarrow{AB}|^2$.

20 Given that **a**, **b** and **c** are non-zero vectors, $\mathbf{a} \neq \mathbf{b} \neq \mathbf{c}$ and
$\mathbf{a} \cdot (\mathbf{b} + \mathbf{c}) = \mathbf{b} \cdot (\mathbf{a} - \mathbf{c})$, show that $\mathbf{c} \cdot (\mathbf{a} + \mathbf{b}) = 0$.

Vector product

> The vector product or cross product of two vectors **a** and **b** inclined at an angle θ is written as $\mathbf{a} \times \mathbf{b}$ and equals $|\mathbf{a}||\mathbf{b}| \sin \theta \hat{\mathbf{n}}$, which is a vector quantity.

Hence it is a vector of magnitude $|\mathbf{a}||\mathbf{b}| \sin \theta$ in the direction of **n** where **n** is perpendicular to the plane containing **a** and **b**.

> Remember that $\hat{\mathbf{n}}$ is a unit vector.

Now obviously $\hat{\mathbf{n}}$ can have one of two directions. This is decided again by using a "right-hand screw rule" in the sense that the direction of **n** is the direction of a screw turned from **a** to **b** with the right hand. This is shown in the diagram below.

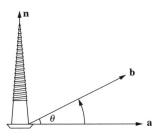

In other words $\mathbf{a} \times \mathbf{b} = |\mathbf{a}||\mathbf{b}| \sin \theta \hat{\mathbf{n}}$ where $\hat{\mathbf{n}}$ is a unit vector perpendicular to both **a** and **b**.

> If you are unsure, try this with a screwdriver and a screw!

The following results are important.

> The fact that the vector product of **a** and **b** is perpendicular to both **a** and **b** is very important when it comes to the work that we will do with planes in Chapter 13.

Parallel vectors

If two vectors are parallel, then $\mathbf{a} \times \mathbf{b} = |\mathbf{a}||\mathbf{b}| \sin 0 \hat{\mathbf{n}}$ or $|\mathbf{a}||\mathbf{b}| \sin \pi \hat{\mathbf{n}}$
So $\mathbf{a} \times \mathbf{b} = 0$

> Remember that for parallel vectors, $\mathbf{a} \cdot \mathbf{b} = |\mathbf{a}||\mathbf{b}|$

Perpendicular vectors

If two vectors are perpendicular, then $\mathbf{a} \times \mathbf{b} = |\mathbf{a}||\mathbf{b}| \sin \frac{\pi}{2} \hat{\mathbf{n}}$
So $\mathbf{a} \times \mathbf{b} = |\mathbf{a}||\mathbf{b}|\hat{\mathbf{n}}$

> Remember that for perpendicular vectors, $\mathbf{a} \cdot \mathbf{b} = 0$

Commutativity

We know $\mathbf{a} \times \mathbf{b} = |\mathbf{a}||\mathbf{b}| \sin \theta \hat{\mathbf{n}}_1$ and that
$\mathbf{b} \times \mathbf{a} = |\mathbf{b}||\mathbf{a}| \sin \theta \hat{\mathbf{n}}_2$.

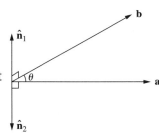

By thinking of the right-hand screw rule we can see that $\hat{\mathbf{n}}_1$ and $\hat{\mathbf{n}}_2$ must be in opposite directions.

Hence $\mathbf{a} \times \mathbf{b} = -\mathbf{b} \times \mathbf{a}$.

The vector product is not commutative.

Distributivity

The vector product is distributive across addition.

$\mathbf{r} \times (\mathbf{p} + \mathbf{q}) = \mathbf{r} \times \mathbf{p} + \mathbf{r} \times \mathbf{q}$.

Proof

Consider two vectors $\mathbf{p}$ and $\mathbf{q}$ with the third vector $\mathbf{r}$ which is perpendicular to both $\mathbf{p}$ and $\mathbf{q}$. Hence the plane containing $\mathbf{p}$ and $\mathbf{q}$ also contains $\mathbf{p} + \mathbf{q}$, and $\mathbf{r}$ is perpendicular to that plane.

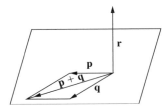

$\mathbf{r} \times \mathbf{p}, \mathbf{r} \times \mathbf{q}$ and $\mathbf{r} \times (\mathbf{p} + \mathbf{q})$ must also lie in this plane as all three are vectors that are perpendicular to $\mathbf{r}$.

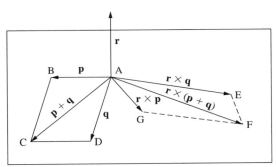

Now $\mathbf{r} \times \mathbf{q} = \overrightarrow{AE}$ which is a vector of magnitude $|\mathbf{r}||\mathbf{q}| \sin 90° = |\mathbf{r}||\mathbf{q}|$

Similarly $\mathbf{r} \times \mathbf{p} = \overrightarrow{AG}$ which is a vector of magnitude $|\mathbf{r}||\mathbf{p}| \sin 90° = |\mathbf{r}||\mathbf{p}|$

And $\mathbf{r} \times (\mathbf{p} + \mathbf{q}) = \overrightarrow{AF}$ which is a vector of magnitude $|\mathbf{r}||(\mathbf{p} + \mathbf{q})| \sin 90° = |\mathbf{r}||(\mathbf{p} + \mathbf{q})|$

Hence the sides of the quadrilateral AEFG are $|\mathbf{r}|$ times the lengths of the sides in quadrilateral ABCD. Since the angles in both figures are the same ($\mathbf{r} \times \mathbf{p}$ is a 90° rotation of $\mathbf{p}$, $\mathbf{r} \times \mathbf{q}$ is a 90° rotation of $\mathbf{q}$, and $\mathbf{r} \times (\mathbf{p} + \mathbf{q})$ is a 90° rotation of $\mathbf{p} + \mathbf{q}$), ABCD and AEFG are both parallelograms.

Now we know that $\overrightarrow{AF} = \overrightarrow{AE} + \overrightarrow{EF}$

Hence $\mathbf{r} \times (\mathbf{p} + \mathbf{q}) = \mathbf{r} \times \mathbf{p} + \mathbf{r} \times \mathbf{q}$, and we have proved the distributive law when $\mathbf{r}$ is perpendicular to $\mathbf{p}$ and $\mathbf{q}$.

Now let us consider the case where $\mathbf{r}$ is not perpendicular to $\mathbf{p}$ and $\mathbf{q}$. In this case we need to form the plane perpendicular to $\mathbf{r}$ with vectors $\mathbf{p}$ and $\mathbf{q}$ inclined at different angles to $\mathbf{r}$. $\mathbf{p}_1, \mathbf{q}_1$ and $(\mathbf{p}_1 + \mathbf{q}_1)$ are the projections of $\mathbf{p}, \mathbf{q}$ and $(\mathbf{p} + \mathbf{q})$ on this plane. θ is the angle between $\mathbf{r}$ and $\mathbf{p}$.

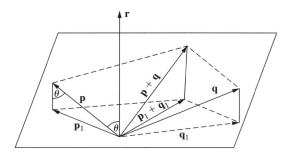

Now $\mathbf{r} \times \mathbf{p} = |\mathbf{r}||\mathbf{p}| \sin \theta \hat{\mathbf{n}}$ and $\mathbf{r} \times \mathbf{p}_1 = |\mathbf{r}||\mathbf{p}_1| \sin 90° \hat{\mathbf{n}}$, where $\hat{\mathbf{n}}$ is a vector perpendicular to $\mathbf{r}$, $\mathbf{p}$ and $\mathbf{p}_1$.

Also $|\mathbf{p}_1| = |\mathbf{p}| \sin \theta$ and hence $\mathbf{r} \times \mathbf{p}_1 = |\mathbf{r}||\mathbf{p}| \sin \theta \hat{\mathbf{n}} = \mathbf{r} \times \mathbf{p}$.

Using an identical method, $\mathbf{r} \times \mathbf{q}_1 = \mathbf{r} \times \mathbf{q}$ and $\mathbf{r} \times (\mathbf{p}_1 + \mathbf{q}_1) = \mathbf{r} \times (\mathbf{p} + \mathbf{q})$.

Since $\mathbf{r}$ is perpendicular to $\mathbf{p}_1$, $\mathbf{q}_1$ and $\mathbf{p}_1 + \mathbf{q}_1$ we can use the distributive law that we proved earlier, that is, $\mathbf{r} \times (\mathbf{p}_1 + \mathbf{q}_1) = \mathbf{r} \times \mathbf{p}_1 + \mathbf{r} \times \mathbf{q}_1$.

> The angle between $\mathbf{r}$ and $\mathbf{p}_1$ is 90° because this is how we set up the plane in the beginning.

Hence $\mathbf{r} \times (\mathbf{p} + \mathbf{q}) = \mathbf{r} \times \mathbf{p} + \mathbf{r} \times \mathbf{q}$ and we have proved that the distributive law also holds when $\mathbf{r}$ is not perpendicular to $\mathbf{p}$ and $\mathbf{q}$.

Vector product of vectors in component form

Let $\mathbf{p} = a_1\mathbf{i} + b_1\mathbf{j} + c_1\mathbf{k}$ and $\mathbf{q} = a_2\mathbf{i} + b_2\mathbf{j} + c_2\mathbf{k}$

Then $\mathbf{p} \times \mathbf{q} = (a_1\mathbf{i} + b_1\mathbf{j} + c_1\mathbf{k}) \times (a_2\mathbf{i} + b_2\mathbf{j} + c_2\mathbf{k})$

$$= (a_1a_2\mathbf{i} \times \mathbf{i} + b_1b_2\mathbf{j} \times \mathbf{j} + c_1c_2\mathbf{k} \times \mathbf{k}) + (a_1b_2\mathbf{i} \times \mathbf{j} + b_1c_2\mathbf{j} \times \mathbf{k} + c_1a_2\mathbf{k} \times \mathbf{i} + b_1a_2\mathbf{j} \times \mathbf{i} + c_1b_2\mathbf{k} \times \mathbf{j} + a_1c_2\mathbf{i} \times \mathbf{k})$$

We need to look at what happens with various combinations of $\mathbf{i}$, $\mathbf{j}$ and $\mathbf{k}$:

$\mathbf{i} \times \mathbf{i} = \mathbf{j} \times \mathbf{j} = \mathbf{k} \times \mathbf{k} = 1 \times 1 \times \sin 0° \times$ perpendicular vector $= 0$

For the others the answer will always be

$1 \times 1 \times \sin \dfrac{\pi}{2} \times$ perpendicular vector $=$ perpendicular vector

From the definition of the unit vectors the perpendicular vector to $\mathbf{i}$ and $\mathbf{j}$ is $\mathbf{k}$, to $\mathbf{j}$ and $\mathbf{k}$ is $\mathbf{i}$, and to $\mathbf{i}$ and $\mathbf{k}$ is $\mathbf{j}$. The only issue is whether it is positive or negative, and this can be determined by the "right-hand screw rule". A list of results is shown below.

$\mathbf{i} \times \mathbf{j} = \mathbf{k}$ $\mathbf{j} \times \mathbf{i} = -\mathbf{k}$

$\mathbf{j} \times \mathbf{k} = \mathbf{i}$ $\mathbf{k} \times \mathbf{j} = -\mathbf{i}$

$\mathbf{k} \times \mathbf{i} = \mathbf{j}$ $\mathbf{i} \times \mathbf{k} = -\mathbf{j}$

Hence $\mathbf{p} \times \mathbf{q} = a_1b_2\mathbf{k} + b_1c_2\mathbf{i} + c_1a_2\mathbf{j} - b_1a_2\mathbf{k} - c_1b_2\mathbf{i} - a_1c_2\mathbf{j}$

$$= (b_1c_2 - c_1b_2)\mathbf{i} + (c_1a_2 - a_1c_2)\mathbf{j} + (a_1b_2 - b_1a_2)\mathbf{k}$$

$$= (b_1c_2 - c_1b_2)\mathbf{i} - (a_1c_2 - c_1a_2)\mathbf{j} + (a_1b_2 - b_1a_2)\mathbf{k}$$

This can be written as a determinant:

$$\mathbf{p} \times \mathbf{q} = \begin{vmatrix} \mathbf{i} & \mathbf{j} & \mathbf{k} \\ a_1 & b_1 & c_1 \\ a_2 & b_2 & c_2 \end{vmatrix}$$

So there are two ways of calculating the vector product. We normally use this determinant form as we rarely know the angle between two vectors.

We can use the vector product to calculate the angle between two vectors, but unless there is a good reason, we would normally use the scalar product. One possible reason would be if we were asked to find the sine of the angle between the vectors.

Example

If $\mathbf{a} = \mathbf{i} + 3\mathbf{j} + 2\mathbf{k}$ and $\mathbf{b} = 2\mathbf{i} - 4\mathbf{j} + \mathbf{k}$, find
a) the unit vector perpendicular to both $\mathbf{a}$ and $\mathbf{b}$
b) the sine of the angle between $\mathbf{a}$ and $\mathbf{b}$.

a) $\mathbf{a} \times \mathbf{b} = \begin{vmatrix} \mathbf{i} & \mathbf{j} & \mathbf{k} \\ 1 & 3 & 2 \\ 2 & -4 & 1 \end{vmatrix}$

$$= \mathbf{i}[3 - (-8)] - \mathbf{j}[1 - 4] + \mathbf{k}[(-4) - 6]$$
$$= 11\mathbf{i} + 3\mathbf{j} - 10\mathbf{k}$$

Hence the unit vector is

$$\frac{1}{\sqrt{11^2 + 3^2 + (-10)^2}}(11\mathbf{i} + 3\mathbf{j} - 10\mathbf{k}) = \frac{1}{\sqrt{230}}(11\mathbf{i} + 3\mathbf{j} - 10\mathbf{k})$$

b) We know that $\mathbf{a} \times \mathbf{b} = |\mathbf{a}||\mathbf{b}| \sin\theta\hat{\mathbf{n}}$

$$\Rightarrow 11\mathbf{i} + 3\mathbf{j} - 10\mathbf{k} = \sqrt{1^2 + 3^2 + 2^2}\sqrt{2^2 + (-4)^2 + 1^2}$$
$$\times \sin\theta \times \frac{1}{\sqrt{230}}(11\mathbf{i} + 3\mathbf{j} - 10\mathbf{k})$$

$$\Rightarrow 1 = \sqrt{14}\sqrt{21} \times \sin\theta \times \frac{1}{\sqrt{230}}$$

$$\Rightarrow \sin\theta = \frac{\sqrt{230}}{\sqrt{14}\sqrt{21}} = \sqrt{\frac{230}{294}} = \sqrt{\frac{115}{147}}$$

Example

A, B and C are the points (2, 5, 6), (3, 8, 9), and (1, 1, 0) respectively. Find the unit vector that is perpendicular to the plane ABC.

The plane ABC must contain the vectors $\overrightarrow{AB}$ and $\overrightarrow{BC}$. Hence we need a vector perpendicular to two other vectors. This is the definition of the cross product.

Now $\overrightarrow{AB} = \begin{pmatrix} 3 - 2 \\ 8 - 5 \\ 9 - 6 \end{pmatrix} = \begin{pmatrix} 1 \\ 3 \\ 3 \end{pmatrix}$ and $\overrightarrow{BC} = \begin{pmatrix} 1 - 3 \\ 1 - 8 \\ 0 - 9 \end{pmatrix} = \begin{pmatrix} -2 \\ -7 \\ -9 \end{pmatrix}$

Therefore the required vector is

$$\overrightarrow{AB} \times \overrightarrow{BC} = \begin{vmatrix} \mathbf{i} & \mathbf{j} & \mathbf{k} \\ 1 & 3 & 3 \\ -2 & -7 & -9 \end{vmatrix}$$

$$= \mathbf{i}[(-27) - (-21)] - \mathbf{j}[(-9) - (-6)] + \mathbf{k}[(-7) - (-6)]$$
$$= -6\mathbf{i} + 3\mathbf{j} - \mathbf{k}$$

Hence the unit vector is

$$\frac{1}{\sqrt{(-6)^2 + 3^2 + (-1)^2}}(-6\mathbf{i} + 3\mathbf{j} - \mathbf{k}) = \frac{1}{\sqrt{46}}(-6\mathbf{i} + 3\mathbf{j} - \mathbf{k})$$

Example

If $\mathbf{a} = \begin{pmatrix} 2 \\ 3 \\ -1 \end{pmatrix}$, $\mathbf{b} = \begin{pmatrix} 6 \\ -3 \\ 2 \end{pmatrix}$ and $\mathbf{c} = \begin{pmatrix} 4 \\ 3 \\ -1 \end{pmatrix}$, find $\mathbf{c} \cdot \mathbf{a} \times \mathbf{b}$.

In an example like this it is important to remember that we have to do the vector product first, because if we calculated the scalar product first we would end up trying to find the vector product of a scalar and a vector, which is not possible.

$$\mathbf{a} \times \mathbf{b} = \begin{vmatrix} \mathbf{i} & \mathbf{j} & \mathbf{k} \\ 2 & 3 & -1 \\ 6 & -3 & 2 \end{vmatrix}$$

$$= \mathbf{i}[6 - 3] - \mathbf{j}[4 - (-6)] + \mathbf{k}[(-6) - 18]$$

$$= \begin{pmatrix} 3 \\ -10 \\ -24 \end{pmatrix}$$

Therefore $\mathbf{c} \cdot \mathbf{a} \times \mathbf{b} = \begin{pmatrix} 4 \\ 3 \\ -1 \end{pmatrix} \cdot \begin{pmatrix} 3 \\ -10 \\ -24 \end{pmatrix} = 12 - 30 + 24 = 6$

Example

Show that $(\mathbf{a} + \mathbf{b}) \times (\mathbf{a} - \mathbf{b}) = 2(\mathbf{b} \times \mathbf{a})$.

Consider the left-hand side.
$(\mathbf{a} + \mathbf{b}) \times (\mathbf{a} - \mathbf{b}) = (\mathbf{a} \times \mathbf{a}) + (\mathbf{b} \times \mathbf{a}) - (\mathbf{a} \times \mathbf{b}) - (\mathbf{b} \times \mathbf{b})$
Now $\mathbf{a} \times \mathbf{a} = \mathbf{b} \times \mathbf{b} = 0$
Hence $(\mathbf{a} + \mathbf{b}) \times (\mathbf{a} - \mathbf{b}) = (\mathbf{b} \times \mathbf{a}) - (\mathbf{a} \times \mathbf{b})$
Now we know $\mathbf{a} \times \mathbf{b}$ and $\mathbf{b} \times \mathbf{a}$ are the same in magnitude but in opposite directions.
Therefore $(\mathbf{a} + \mathbf{b}) \times (\mathbf{a} - \mathbf{b}) = 2(\mathbf{b} \times \mathbf{a})$

Application of vector product

We will see that a very important use of vector products is in the representation of planes, which will be dealt with in Chapter 13. However, there are two other applications that are useful to know.

Area of a parallelogram

The area of a parallelogram = base × height

$$= AD \times h$$

$$= (AD)(AB \sin \theta)$$

$$= |\overrightarrow{AD} \times \overrightarrow{AB}|$$

The area of a parallelogram is the magnitude of the vector product of two adjacent sides.

Example

Find the area of the parallelogram ABCD where A has coordinates $(2, 3, -1)$, B $(3, -2, -1)$, C $(4, -5, 1)$ and D $(3, 0, 1)$.

We first need to find the vectors representing a pair of adjacent sides.

$$\overrightarrow{AB} = \begin{pmatrix} 3 \\ -2 \\ -1 \end{pmatrix} - \begin{pmatrix} 2 \\ 3 \\ -1 \end{pmatrix} = \begin{pmatrix} 1 \\ -5 \\ 0 \end{pmatrix} \text{ and } \overrightarrow{AD} = \begin{pmatrix} 3 \\ 0 \\ 1 \end{pmatrix} - \begin{pmatrix} 2 \\ 3 \\ -1 \end{pmatrix} = \begin{pmatrix} 1 \\ -3 \\ 2 \end{pmatrix}$$

Now $\overrightarrow{AB} \times \overrightarrow{AD} = \begin{vmatrix} \mathbf{i} & \mathbf{j} & \mathbf{k} \\ 1 & -5 & 0 \\ 1 & -3 & 2 \end{vmatrix}$

$$= \mathbf{i}[(-10) - 0] - \mathbf{j}[2 - 0] + \mathbf{k}[(-3) - (-5)]$$

$$= -10\mathbf{i} - 2\mathbf{j} + 2\mathbf{k}$$

Now the area of the parallelogram ABCD is

$$|\overrightarrow{AB} \times \overrightarrow{AD}| = \sqrt{(-10)^2 + (-2)^2 + 2^2} = \sqrt{108} \text{ units}^2$$

Area of a triangle

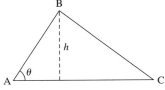

The area of triangle ABC = ½ base × height

$$= \tfrac{1}{2}(AC)(h)$$

$$= \tfrac{1}{2}(AC)(AB \sin \theta)$$

$$= \tfrac{1}{2}|\overrightarrow{AC} \times \overrightarrow{AB}|$$

The area of a triangle is half the magnitude of the vector product of two sides.

This is consistent with the idea that the area of a triangle is half the area of a parallelogram.

Example

Find the area of the triangle ABC with coordinates $A(1, 3, -1)$, $B(-2, 1, -4)$ and $C(4, 3, -3)$.

We begin by finding $\overrightarrow{AB}$ and $\overrightarrow{BC}$.

Now $\overrightarrow{AB} = \begin{pmatrix} -2 - 1 \\ 1 - 3 \\ -4 - (-1) \end{pmatrix} = \begin{pmatrix} -3 \\ -2 \\ -3 \end{pmatrix}$ and $\overrightarrow{BC} = \begin{pmatrix} 4 - (-2) \\ 3 - 1 \\ -3 - (-4) \end{pmatrix} = \begin{pmatrix} 6 \\ 2 \\ 1 \end{pmatrix}$

Hence $\overrightarrow{AB} \times \overrightarrow{BC} = \begin{vmatrix} \mathbf{i} & \mathbf{j} & \mathbf{k} \\ -3 & -2 & -3 \\ 6 & 2 & 1 \end{vmatrix}$

$= \mathbf{i}[(-2) - (-6)] - \mathbf{j}[(-3) - (-18)] + \mathbf{k}[(-6) - (-12)]$
$= 4\mathbf{i} - 15\mathbf{j} + 6\mathbf{k}$

Now the area of triangle ABC is

$\frac{1}{2}|\overrightarrow{AB} \times \overrightarrow{BC}| = \frac{1}{2}\sqrt{4^2 + (-15)^2 + 6^2} = \frac{\sqrt{277}}{2}$ units²

Exercise 4

1 If $\mathbf{a} = \mathbf{i} + 2\mathbf{j} - 3\mathbf{k}$ and $\mathbf{b} = 2\mathbf{i} - 4\mathbf{j} - \mathbf{k}$, find:
   a $\mathbf{a} \times \mathbf{b}$
   b $\mathbf{a} \times (\mathbf{a} + \mathbf{b})$
   c $\mathbf{b} \times (\mathbf{a} - \mathbf{b})$
   d $\mathbf{a} \times (3\mathbf{a} - 2\mathbf{b})$
   e $(2\mathbf{a} + \mathbf{b}) \times \mathbf{a}$
   f $(\mathbf{a} + 2\mathbf{b}) \times (2\mathbf{a} + \mathbf{b})$
   g $\mathbf{a} \cdot (\mathbf{a} \times \mathbf{b})$

2 Find the value of $|\mathbf{a} \times \mathbf{b}|$ for the given modulus of $\mathbf{a}$, modulus of $\mathbf{b}$ and angle between the vectors $\mathbf{a}$ and $\mathbf{b}$.
   a $|\mathbf{a}| = 3$, $|\mathbf{b}| = 7$, $\theta = 60°$
   b $|\mathbf{a}| = 9$, $|\mathbf{b}| = \sqrt{13}$, $\theta = 120°$
   c $|\mathbf{a}| = \sqrt{18}$, $|\mathbf{b}| = 2$, $\theta = 135°$

3 If OPQ is a triangle, show that $|\overrightarrow{OP} \times \overrightarrow{OQ}| = |\overrightarrow{OP} \times \overrightarrow{PQ}|$.

4 If $\mathbf{a} \times \mathbf{b} = 0$, show that $\mathbf{a} = k\mathbf{b}$ where k is a scalar.

5 Given that $\mathbf{b} \times \mathbf{c} = \mathbf{c} \times \mathbf{a}$ show that $\mathbf{a} + \mathbf{b}$ is parallel to $\mathbf{c}$.

6 Consider two vectors $\mathbf{a}$ and $\mathbf{b}$. If $\mathbf{a} \cdot \mathbf{a} \times \mathbf{b} = 0$ write down the angle between $\mathbf{a}$ and $\mathbf{b}$.

7 If $\mathbf{a} = \begin{pmatrix} 3 \\ -4 \\ 3 \end{pmatrix}$ and $\mathbf{b} = \begin{pmatrix} 4 \\ 2 \\ 3 \end{pmatrix}$, find
   a the unit vector $\hat{\mathbf{n}}$ perpendicular to both $\mathbf{a}$ and $\mathbf{b}$
   b the sine of the angle θ between $\mathbf{a}$ and $\mathbf{b}$.

8 If $\mathbf{a} = \mathbf{i} - 3\mathbf{j} - \mathbf{k}$ and $\mathbf{b} = \mathbf{i} + 2\mathbf{k}$, find
   a the unit vector $\hat{\mathbf{n}}$ perpendicular to both $\mathbf{a}$ and $\mathbf{b}$
   b the sine of the angle θ between $\mathbf{a}$ and $\mathbf{b}$.

9 P, Q and R are the points (0, 0, 3), (3, 4, 6) and (0, −1, 0) respectively. Find the unit vector that is perpendicular to the plane PQR.

10 Two sides of a triangle are represented by the vectors $(\mathbf{i} + \mathbf{j} - \mathbf{k})$ and $(5\mathbf{i} + 2\mathbf{j} + 2\mathbf{k})$. Find the area of the triangle.

11 Relative to the origin the points A, B and C have position vectors

$$\begin{pmatrix} 0 \\ 1 \\ -2 \end{pmatrix}, \begin{pmatrix} 2 \\ -2 \\ 1 \end{pmatrix} \text{ and } \begin{pmatrix} 4 \\ -1 \\ -3 \end{pmatrix} \text{ respectively. Find the area of the triangle ABC.}$$

12 The triangle ABC has its vertices at the points A (0, 1, 2), B (0, 0, 1) and C (2, 6, 3). Find the area of the triangle ABC.

13 Given that $\overrightarrow{AB} = \mathbf{i} - 4\mathbf{j}$, $\overrightarrow{AC} = 3\mathbf{i} - \mathbf{j} - 2\mathbf{k}$, $\overrightarrow{AP} = 2\overrightarrow{AB}$ and $\overrightarrow{AQ} = 4\overrightarrow{AC}$, find the area of triangle APQ.

14 A parallelogram OABC has one vertex O at the origin and the vertices A and B at the points (3, 4, 0) and (0, 5, 5) respectively. Find the area of the parallelogram OABC.

15 A parallelogram PQRS has vertices at P(0, 2, −1), Q(2, −3, −7) and R(−1, 0, −4). Find the area of the parallelogram PQRS.

16 A parallelogram PQRS is such that $\overrightarrow{PX} = 5\mathbf{i} - \mathbf{j} + 2\mathbf{k}$ and $\overrightarrow{PY} = -3\mathbf{i} - 7\mathbf{j} + \mathbf{k}$, where $\overrightarrow{PQ} = 5\overrightarrow{PX}$ and Y is the midpoint of $\overrightarrow{PS}$.

Find the vectors representing the sides $\overrightarrow{PQ}$ and $\overrightarrow{PS}$ and hence calculate the area of the parallelogram.

17 If $\mathbf{a} = \begin{pmatrix} 1 \\ 6 \\ 0 \end{pmatrix}$, $\mathbf{b} = \begin{pmatrix} 4 \\ 7 \\ -1 \end{pmatrix}$ and $\mathbf{c} = \begin{pmatrix} 3 \\ 4 \\ -1 \end{pmatrix}$, determine whether or not

$\mathbf{a} \times (\mathbf{b} \times \mathbf{c}) = (\mathbf{a} \times \mathbf{b}) \times \mathbf{c}$.

18 If $\mathbf{a} \times \mathbf{b} = \mathbf{a} \times \mathbf{c}$, show that the vector $\mathbf{c} - \mathbf{b}$ is parallel to $\mathbf{a}$.

Review exercise

1 The points P, Q, R, S have position vectors **p**, **q**, **r**, **s** given by

$$\mathbf{p} = \mathbf{i} + 2\mathbf{k}$$
$$\mathbf{q} = -1.2\mathbf{j} + 1.4\mathbf{k}$$
$$\mathbf{r} = -5\mathbf{i} - 6\mathbf{j} + 8\mathbf{k}$$
$$\mathbf{s} = \mathbf{j} - 7\mathbf{k}$$

respectively. The point X lies on PQ produced and is such that PX = 5PQ, and the point Y is the midpoint of PR.
a Show that XY is not perpendicular to PY.
b Find the area of the triangle PXY.
c Find a vector perpendicular to the plane PQR.
d Find the cosine of the acute angle between PS and RS.

 2 Let $\mathbf{a} = \begin{pmatrix} 2 \\ 1 \\ 0 \end{pmatrix}$, $\mathbf{b} = \begin{pmatrix} -1 \\ p \\ 6 \end{pmatrix}$ and $\mathbf{c} = \begin{pmatrix} 2 \\ -4 \\ 3 \end{pmatrix}$

a Find $\mathbf{a} \times \mathbf{b}$
b Find the value of p, given that $\mathbf{a} \times \mathbf{b}$ is parallel to **c**.   [IB May 06 P1 Q11]

3 The point A is given by the vector $\begin{pmatrix} 1 - m \\ 2 + m \\ 3 + m \end{pmatrix}$ and the point B by $\begin{pmatrix} 1 - 2m \\ 2 + 2m \\ 3 + 2m \end{pmatrix}$,

relative to O. Show that there is no value of m for which $\overrightarrow{OA}$ and $\overrightarrow{OB}$ are perpendicular.

 4 If **a** and **b** are unit vectors and θ is the angle between them, express $|\mathbf{a} - \mathbf{b}|$ in terms of θ. [IB May 93 P1 Q12]

 5 A circle has a radius of 5 units with a centre at (3, 2). A point P on the circle has coordinates (x, y). The angle that this radius makes with the horizontal is θ. Give a vector expression for $\overrightarrow{OP}$.

 6 Given two non-zero vectors **a** and **b** such that $|\mathbf{a} + \mathbf{b}| = |\mathbf{a} - \mathbf{b}|$, find the value of $\mathbf{a} \cdot \mathbf{b}$. [IB Nov 02 P1 Q18]

 7 Show that the points A(0, 1, 3), B(5, 3, 2) and C(15, 7, 0) are collinear (that is, they lie on the same line).

 8 Find the angle between the vectors $\mathbf{v} = \mathbf{i} + \mathbf{j} + 2\mathbf{k}$ and $\mathbf{w} = 2\mathbf{i} + 3\mathbf{j} + \mathbf{k}$. Give your answer in radians. [IB May 02 P1 Q5]

 9 The circle shown has centre D, and the points A, B and C lie on the circumference of the circle. The radius of the circle is 1 unit.

Given that $\overrightarrow{DB} = \mathbf{a}$ and $\overrightarrow{DC} = \mathbf{b}$, show that $\hat{ACB} = 90°$.

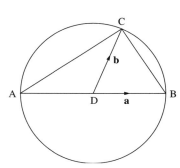

 10 Let α be the angle between **a** and **b**, where

$\mathbf{a} = (\cos\theta)\mathbf{i} + (\sin\theta)\mathbf{j}$, $\mathbf{b} = (\sin\theta)\mathbf{i} + (\cos\theta)\mathbf{j}$ and $0 < \theta < \dfrac{\pi}{4}$. Express α in terms of θ. [IB Nov 00 P1 Q11]

 11 The points X and Y have coordinates (1, 2, 3) and (2, −1, 0) respectively. $\overrightarrow{OA} = 2\overrightarrow{OX}$ and $\overrightarrow{OB} = 3\overrightarrow{OY}$. OABC is a parallelogram.
  a Find the coordinates of A, B and C.
  b Find the area of the parallelogram OABC.
  c Find the position vector of the point of intersection of $\overrightarrow{OB}$ and $\overrightarrow{AC}$.
  d The point E has position vector **k**. Find the angle between $\overrightarrow{AE}$ and $\overrightarrow{BE}$.
  e Find the area of the triangle ABE.

 12 Given $\mathbf{u} = 3\mathbf{i} - 2\mathbf{j} + 5\mathbf{k}$ and $\mathbf{v} = \mathbf{i} + 4\mathbf{j} + m\mathbf{k}$, and that the vector $(2\mathbf{u} - 3\mathbf{v})$ has magnitude $\sqrt{265}$, find the value of m. [IB Nov 93 P1 Q8]

13 Vectors, Lines and Planes

Stefan Banach was born in Kraków, Poland (at the time part of the Austro-Hungarian Empire) on 30 March 1892. During his early life Banach was brought up by Franciszka Plowa, who lived in Kraków with her daughter Maria. Maria's guardian was a French intellectual Juliusz Mien, who quickly recognised Banach's talents. Mien gave Banach a general education, including teaching him to speak French, and in general gave him an appreciation for education. On leaving school Banach chose to study engineering and went to Lvov, which is now in the Ukraine, where he enrolled in the Faculty of Engineering at Lvov Technical University.

Stefan Banach

By chance, Banach met another Polish mathematician, Steinhaus, in 1916. Steinhaus told Banach of a mathematical problem that he was working on without making much headway and after a few days Banach had the main idea for the required counter-example, which led to Steinhaus and Banach writing a joint paper. This was Banach's first paper and it was finally published in 1918. From then on he continued to publish and in 1920 he went to work at Lvov Technical University. He initially worked as an assistant lecturing in mathematics and gaining his doctorate, but was promoted to a full professorship in 1924. Banach worked in an unconventional manner and was often found doing mathematics with his colleagues in the cafés of Lvov. Banach was the founder of modern functional analysis, made major contributions to the theory of topological vector spaces and defined axiomatically what today is called a Banach space, which is a real or complex normed vector space.

At the beginning of the second world war Soviet troops occupied Lvov, but Banach was allowed to continue at the university, and he became the Dean of the Faculty of Science. However, the Nazi occupation of Lvov in June 1941 ended his career in university and by the end of 1941 Banach was working in the German institute dealing with infectious diseases, feeding lice. This was to be his life during the remainder of the Nazi occupation of Lvov in July 1944. Once the Soviet troops retook Lvov, Banach renewed his contacts, but by this time was seriously ill. Banach had planned to take up the chair of mathematics at the Jagiellonian University in Kraków, but he died of lung cancer in Lvov in 1945.

13.1 Equation of a straight line

Vector equation of a straight line

In both two and three dimensions a line is described as passing through a fixed point and having a specific direction, or as passing through two fixed points.

Vector equation of a line passing through a fixed point parallel to a given vector

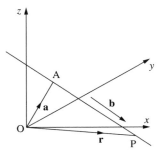

Consider the diagram.

The fixed point A on the line has position vector $\overrightarrow{OA} = \mathbf{a}$.

The line is parallel to vector $\mathbf{b}$.

Any point $P(x, y, z)$ on the line has position vector $\overrightarrow{OP} = \mathbf{r}$.

Hence $\overrightarrow{AP}$ has direction $\mathbf{b}$ and a variable magnitude: that is, $\overrightarrow{AP} = \lambda\mathbf{b}$, where λ is a variable constant.

Using vector addition, $\overrightarrow{OP} = \overrightarrow{OA} + \overrightarrow{AP}$ so $\mathbf{r} = \mathbf{a} + \lambda\mathbf{b}$.

This is the vector equation of a line, and $\mathbf{b}$ is known as the **direction vector** of the line.

Alternatively, if $\mathbf{a} = \begin{pmatrix} a_1 \\ a_2 \\ a_3 \end{pmatrix}$ and $\mathbf{b} = \begin{pmatrix} b_1 \\ b_2 \\ b_3 \end{pmatrix}$ then we could write the vector equation of a

line as: $\begin{pmatrix} x \\ y \\ z \end{pmatrix} = \begin{pmatrix} a_1 \\ a_2 \\ a_3 \end{pmatrix} + \lambda \begin{pmatrix} b_1 \\ b_2 \\ b_3 \end{pmatrix}$.

If we put in values for λ we then get the position vectors of points that lie on the line.

Vector equation of a line passing through two fixed points

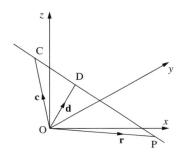

Consider the diagram.

One fixed point on the line is C, which has position vector $\overrightarrow{OC} = \mathbf{c}$, and the second fixed point on the line is D, which has position vector $\overrightarrow{OD} = \mathbf{d}$. Again any point P($x$, y, z) on the line has position vector $\overrightarrow{OP} = \mathbf{r}$.

Hence $\overrightarrow{CP}$ has direction $\mathbf{d} - \mathbf{c}$ and variable magnitude: that is, $\overrightarrow{CP} = \lambda(\mathbf{d} - \mathbf{c})$.

Therefore $\mathbf{r} = \mathbf{c} + \lambda(\mathbf{d} - \mathbf{c})$.

This is the vector equation of a line passing though two fixed points.

> We rarely use the vector equation in this form. It is usually easier to find the direction vector then use the form $\mathbf{r} = \mathbf{a} + \lambda\mathbf{b}$.

Example

Find the vector equation of the line passing through the point A (1, 3, 4) parallel to the vector $\mathbf{i} - \mathbf{j} + \mathbf{k}$.

The position vector of A is $\overrightarrow{OA} = \mathbf{i} + 3\mathbf{j} + 4\mathbf{k}$.

Hence the vector equation of the line is $\mathbf{r} = \mathbf{i} + 3\mathbf{j} + 4\mathbf{k} + \lambda(\mathbf{i} - \mathbf{j} + \mathbf{k})$.

> It is essential not to forget the "$\mathbf{r}$" in the equation.

Example

Find the vector equation of the line passing through A (3, 2) parallel to the vector $\begin{pmatrix} 4 \\ -1 \end{pmatrix}$.

Even though this is in two dimensions it works in exactly the same way, since we are assuming that the z-component of the vector is zero.

The position vector of A is $\overrightarrow{OA} = \begin{pmatrix} 3 \\ 2 \end{pmatrix}$.

Hence the vector equation of the line is $\mathbf{r} = \begin{pmatrix} 3 \\ 2 \end{pmatrix} + \lambda\begin{pmatrix} 4 \\ -1 \end{pmatrix}$.

Example

Find the vector equation of the line passing though A(2, 3, −1) and B(4, −2, 2).

There are a number of ways of tackling this question.

Method 1

If we let A have position vector $\mathbf{a}$ and let B have position vector $\mathbf{b}$, then we can say that $\mathbf{r} = \mathbf{a} + \lambda(\mathbf{b} - \mathbf{a})$.

Now $\mathbf{a} = \begin{pmatrix} 2 \\ 3 \\ -1 \end{pmatrix}$ and $\mathbf{b} = \begin{pmatrix} 4 \\ -2 \\ 2 \end{pmatrix}$

Hence $\mathbf{r} = \begin{pmatrix} 2 \\ 3 \\ -1 \end{pmatrix} + \lambda \begin{pmatrix} 4 - 2 \\ -2 - 3 \\ 2 - (-1) \end{pmatrix}$

$\Rightarrow \mathbf{r} = \begin{pmatrix} 2 \\ 3 \\ -1 \end{pmatrix} + \lambda \begin{pmatrix} 2 \\ -5 \\ 3 \end{pmatrix}$

Method 2

We could use **b** as a position vector instead of **a**, giving $\mathbf{r} = \begin{pmatrix} 4 \\ -2 \\ 2 \end{pmatrix} + \lambda \begin{pmatrix} 2 \\ -5 \\ 3 \end{pmatrix}$.

Method 3

We could begin by forming the direction vector.

The direction vector is $\mathbf{b} - \mathbf{a} = \begin{pmatrix} 4 \\ -2 \\ 2 \end{pmatrix} - \begin{pmatrix} 2 \\ 3 \\ -1 \end{pmatrix} = \begin{pmatrix} 2 \\ -5 \\ 3 \end{pmatrix}$.

> We could also use $\mathbf{a} - \mathbf{b}$ as the direction vector.

Hence the vector equation of the line is $\mathbf{r} = \begin{pmatrix} 2 \\ 3 \\ -1 \end{pmatrix} + \lambda \begin{pmatrix} 2 \\ -5 \\ 3 \end{pmatrix}$

or $\mathbf{r} = \begin{pmatrix} 4 \\ -2 \\ 2 \end{pmatrix} + \lambda \begin{pmatrix} 2 \\ -5 \\ 3 \end{pmatrix}$.

> These two equations are equivalent.

Parametric equations of a straight line

The parametric equations of a straight line are when the vector equation is expressed in terms of the parameter λ. This form is often used when we are doing calculations using the vector equation of a line.

> λ is not always used as the parameter. Other common letters are s, t, m, n and μ.

The way to find the parametric equations is shown in the following example.

Example

Give the vector equation $\mathbf{r} = \mathbf{i} + 2\mathbf{j} - 4\mathbf{k} + \lambda(6\mathbf{i} - 7\mathbf{j} + 3\mathbf{k})$ in parametric form.

$\mathbf{r} = \mathbf{i} + 2\mathbf{j} - 4\mathbf{k} + \lambda(6\mathbf{i} - 7\mathbf{j} + 3\mathbf{k})$ can be rewritten as

$x\mathbf{i} + y\mathbf{j} + z\mathbf{k} = \mathbf{i} + 2\mathbf{j} - 4\mathbf{k} + \lambda(6\mathbf{i} - 7\mathbf{j} + 3\mathbf{k})$

$= (1 + 6\lambda)\mathbf{i} + (2 - 7\lambda)\mathbf{j} + (-4 + 3\lambda)\mathbf{k}$

Equating components gives the parametric equations:

$x = 1 + 6\lambda$
$y = 2 - 7\lambda$
$z = -4 + 3\lambda$

Example

Find the parametric equations of the line passing through the points $A(-2, 1, 3)$ and $B(1, -1, 4)$.

We begin by forming the vector equation.

The direction vector of this equation is $\overrightarrow{AB} = \begin{pmatrix} 1 - (-2) \\ -1 - 1 \\ 4 - 3 \end{pmatrix} = \begin{pmatrix} 3 \\ -2 \\ 1 \end{pmatrix}$.

Hence the vector equation is $\mathbf{r} = \begin{pmatrix} 1 \\ -1 \\ 4 \end{pmatrix} + \mu \begin{pmatrix} 3 \\ -2 \\ 1 \end{pmatrix}$

$$\Rightarrow \begin{pmatrix} x \\ y \\ z \end{pmatrix} = \begin{pmatrix} 1 \\ -1 \\ 4 \end{pmatrix} + \mu \begin{pmatrix} 3 \\ -2 \\ 1 \end{pmatrix} = \begin{pmatrix} 1 + 3\mu \\ -1 - 2\mu \\ 4 + \mu \end{pmatrix}$$

Equating components gives the parametric equations:

$x = 1 + 3\mu$
$y = -1 - 2\mu$
$z = 4 + \mu$

Example

Show that the point $(3, 6, -1)$ fits on the line $\mathbf{r} = \begin{pmatrix} 1 \\ 0 \\ -1 \end{pmatrix} + \lambda \begin{pmatrix} 1 \\ 3 \\ 0 \end{pmatrix}$ but $(2, 3, 0)$ does not.

To do this it is easiest to use the parametric forms of the equation.

Therefore $\begin{pmatrix} x \\ y \\ z \end{pmatrix} = \mathbf{r} = \begin{pmatrix} 1 \\ 0 \\ -1 \end{pmatrix} + \lambda \begin{pmatrix} 1 \\ 3 \\ 0 \end{pmatrix} = \begin{pmatrix} 1 + \lambda \\ 3\lambda \\ -1 \end{pmatrix}$

The parametric equations of this line are:

$x = 1 + \lambda$
$y = 3\lambda$
$z = -1$

Using $x = 1 + \lambda$ we can see that if $x = 3$ then $\lambda = 2$.
We now check this is consistent with the values of y and z.
When $\lambda = 2$, $y = 6$ and $z = -1$. ·· z is not dependent on λ; it is always -1.
Hence $(3, 6, -1)$ lies on the line.

Now if $x = 2$, $\lambda = 1$.
Since $y = 3\lambda$, $y = 3$.
However, $z = -1$ and hence $(2, 3, 0)$ does not lie on the line.

Example

Find the coordinates of the point where the line passing through $A(2, 3, -4)$ parallel to $-\mathbf{i} - 3\mathbf{j} + \mathbf{k}$ crosses the xy-plane.

The easiest way to solve a problem like this is to put the equation in parametric form.

The vector equation of the line is $\mathbf{r} = 2\mathbf{i} + 3\mathbf{j} - 4\mathbf{k} + \lambda(-\mathbf{i} - 3\mathbf{j} + \mathbf{k})$.

Hence $x\mathbf{i} + y\mathbf{j} + z\mathbf{k} = 2\mathbf{i} + 3\mathbf{j} - 4\mathbf{k} + \lambda(-\mathbf{i} - 3\mathbf{j} + \mathbf{k})$

$$= (2 - \lambda)\mathbf{i} + (3 - 3\lambda)\mathbf{j} + (-4 + \lambda)\mathbf{k}$$

Therefore the parametric equations of the line are:

$x = 2 - \lambda$
$y = 3 - 3\lambda$
$z = -4 + \lambda$

This line will cross the xy-plane when $z = 0$.

Hence $\lambda = 4$, giving $x = -2$ and $y = -9$.

Thus the coordinates of the point of intersection are $(-2, -9, 0)$.

Cartesian equations of a straight line

Consider these parametric equations of a straight line:

$x = 3\mu + 1$
$y = 2\mu - 1$
$z = 4 + \mu$

If we now isolate μ we find:

$$\frac{x - 1}{3} = \frac{y + 1}{2} = \frac{z - 4}{1} (= \mu)$$

These are known as the **Cartesian equations of a line** and are in fact a three-dimensional version of $y = mx + c$. In a general form this is written as $\dfrac{x - x_0}{l} = \dfrac{y - y_0}{m} = \dfrac{z - z_0}{n}$

where (x_0, y_0, z_0) are the coordinates of a point on the line and $\begin{pmatrix} l \\ m \\ n \end{pmatrix}$ is the direction vector of the line.

Example

A line is parallel to the vector $2\mathbf{i} - \mathbf{j} + 2\mathbf{k}$ and passes through the point $(2, -3, 5)$. Find the vector equation of the line, the parametric equations of the line, and the Cartesian equations of the line.

The vector equation of the line is given by $\mathbf{r} = 2\mathbf{i} - 3\mathbf{j} + 5\mathbf{k} + s(2\mathbf{i} - \mathbf{j} + 2\mathbf{k})$.

To form the parametric equations we write

$$x\mathbf{i} + y\mathbf{j} + z\mathbf{k} = 2\mathbf{i} - 3\mathbf{j} + 5\mathbf{k} + s(2\mathbf{i} - \mathbf{j} + 2\mathbf{k})$$
$$= (2 + 2s)\mathbf{i} + (-3 - s)\mathbf{j} + (5 + 2s)\mathbf{k}$$

Equating the coefficients of x, y and z gives the parametric equations:

$x = 2 + 2s$
$y = -3 - s$
$z = 5 + 2s$

Eliminating s gives the Cartesian equations:

$$\frac{x - 2}{2} = \frac{y + 3}{-1} = \frac{z - 5}{2} (= s)$$

Sometimes we need to undo the process, as shown in the next example.

Example

Convert the Cartesian equations $\dfrac{2x - 5}{2} = \dfrac{3 - y}{3} = \dfrac{-2z + 5}{5}$ to parametric and vector form.

We begin by writing $\dfrac{2x - 5}{2} = \dfrac{3 - y}{3} = \dfrac{-2z + 5}{5} = t$, where t is a parameter.

Hence $x = \dfrac{2t + 5}{2}$

$\qquad y = 3 - 3t$

$\qquad z = \dfrac{5t - 5}{-2}$

These are the parametric equations of the line.

Now $\begin{pmatrix} x \\ y \\ z \end{pmatrix} = \begin{pmatrix} \dfrac{2t + 5}{2} \\ 3 - 3t \\ \dfrac{5t - 5}{-2} \end{pmatrix}$

$\Rightarrow \begin{pmatrix} x \\ y \\ z \end{pmatrix} = \begin{pmatrix} \dfrac{5}{2} \\ 3 \\ \dfrac{5}{2} \end{pmatrix} + t \begin{pmatrix} 1 \\ -3 \\ -\dfrac{5}{2} \end{pmatrix}$ .... Separating the parameter t

Hence $\mathbf{r} = \begin{pmatrix} 2.5 \\ 3 \\ 2.5 \end{pmatrix} + t \begin{pmatrix} 1 \\ -3 \\ -2.5 \end{pmatrix}$ is the vector equation of the line.

Exercise 1

1 Find the vector equation of the line that is parallel to the given vector and passes through the given point.

a Vector $\begin{pmatrix} 1 \\ -2 \\ -1 \end{pmatrix}$, point $(0, 2, -3)$

b Vector $\mathbf{i} - 4\mathbf{j} - 2\mathbf{k}$, point $(1, -2, 0)$

c Vector $\begin{pmatrix} 0 \\ -5 \\ 12 \end{pmatrix}$, point $(4, 4, 3)$

d Vector $3\mathbf{i} + 6\mathbf{j} - \mathbf{k}$, point $(5, 2, 1)$

e Vector $2\mathbf{i} - \mathbf{j}$, point $(-3, -1)$

f Vector $\begin{pmatrix} 4 \\ -7 \end{pmatrix}$, point $(-5, 1)$

2 Find the vector equation of the line passing through each pair of points.

a $(2, 1, 2)$ and $(-2, 4, 3)$ **b** $(-3, 1, 0)$ and $(4, -1, 2)$

c $(2, -2, 3)$ and $(0, 7, -3)$ **d** $(3, 4, -2)$ and $(2, -5, -1)$

e $(4, -3)$ and $(1, -3)$

3 Write down equations, in vector form, in parametric form and in Cartesian form, for the line passing through point A with position vector **a** and direction vector **b**.

a $\mathbf{a} = \mathbf{i} - 2\mathbf{j} - 4\mathbf{k}$ $\mathbf{b} = 3\mathbf{i} + \mathbf{j} - 5\mathbf{k}$

b $\mathbf{a} = \begin{pmatrix} -3 \\ -2 \\ 3 \end{pmatrix}$ $\mathbf{b} = \begin{pmatrix} 4 \\ -7 \\ 3 \end{pmatrix}$

c $\mathbf{a} = \mathbf{j} + \mathbf{k}$ $\mathbf{b} = \mathbf{i} - 3\mathbf{k}$

d $\mathbf{a} = \begin{pmatrix} 4 \\ 1 \\ 0 \end{pmatrix}$ $\mathbf{b} = \begin{pmatrix} -1 \\ 2 \\ 2 \end{pmatrix}$

4 Convert these vector equations to parametric and Cartesian form.

a $\mathbf{r} = \begin{pmatrix} 1 \\ -1 \\ 2 \end{pmatrix} + \lambda \begin{pmatrix} -2 \\ 3 \\ -2 \end{pmatrix}$

b $\mathbf{r} = 2\mathbf{i} - 5\mathbf{j} - \mathbf{k} + \mu(3\mathbf{i} - \mathbf{j} + 4\mathbf{k})$

c $\mathbf{r} = \begin{pmatrix} 2 \\ 8 \\ -1 \end{pmatrix} + m \begin{pmatrix} 4 \\ -7 \\ 6 \end{pmatrix}$

d $\mathbf{r} = \mathbf{i} - \mathbf{j} + 7\mathbf{k} + n(2\mathbf{i} - 3\mathbf{j} - \mathbf{k})$

e $\mathbf{r} = \begin{pmatrix} 4 \\ 6 \end{pmatrix} + s \begin{pmatrix} 3 \\ -5 \end{pmatrix}$

f $\mathbf{r} = \mathbf{i} - 6\mathbf{j} + t(2\mathbf{i} - 5\mathbf{j})$

5 Convert these parametric equations to vector form.

a $x = 3\lambda - 7$ **b** $x = -\mu - 4$
     $y = \lambda + 6$ $y = 3\mu - 5$
     $z = 2\lambda + 4$ $z = 5\mu + 1$

c $x = 5m + 4$
   $y = 3m$
   $z = 4m - 3$

d $x = -4$
   $y = 2n - 1$
   $z = 5 - 2n$

6 Convert these Cartesian equations to vector form.

a $\dfrac{x - 3}{4} = \dfrac{y + 5}{3} = \dfrac{z + 1}{-3}$

b $\dfrac{2x - 5}{4} = \dfrac{3y + 3}{-4} = \dfrac{z + 1}{2}$

c $\dfrac{2 - 5x}{-4} = \dfrac{3y + 5}{-6} = \dfrac{2z + 7}{3}$

d $\dfrac{6x + 1}{4} = \dfrac{4 - 3y}{-2} = \dfrac{4 - z}{6}$

e $5 - 3x = \dfrac{2 + 3y}{-2} = \dfrac{3z - 1}{-3}$

f $\dfrac{x - 5}{7} = \dfrac{3 - 7y}{4}; z = 2$

7 Determine whether the given point lies on the line.

a $(-11, 4, -11)$ $\mathbf{r} = \mathbf{i} + \mathbf{j} - 2\mathbf{k} + \lambda(4\mathbf{i} - 2\mathbf{j} + 3\mathbf{k})$

b $(7, -16, 15)$ $x = 2\mu - 3, y = 4 - 4\mu, z = 3\mu$

c $(4, 3, 9)$ $x = 5 - m, y = 2m + 1, z = 3 + 5m$

d $(11, 33, 12)$ $\dfrac{x + 4}{3} = \dfrac{y + 2}{7} = \dfrac{2z + 1}{5}$

e $\left(8, 10, \dfrac{7}{3}\right)$ $\dfrac{2x - 4}{3} = \dfrac{3y - 5}{7} = \dfrac{3z + 1}{2}$

f $\left(-\dfrac{1}{2}, -\dfrac{5}{2}, -3\right)$ $\dfrac{2x + 5}{2} = \dfrac{3 - 2y}{4}; z = -3$

8 The Cartesian equation of a line is given by $\dfrac{3x - 5}{6} = \dfrac{2 - y}{3} = \dfrac{3z + 1}{2}$.

Find the vector equation of the parallel line passing through the point with coordinates $(3, 7, -1)$ and find the position vector of the point on this line where $z = 0$.

9 Find the coordinates of the points where the line $\mathbf{r} = \begin{pmatrix} -3 \\ 5 \\ -3 \end{pmatrix} + s\begin{pmatrix} 2 \\ 4 \\ -1 \end{pmatrix}$

intersects the xy-, the yz- and the xz-plane.

10 Write down a vector equation of the line passing through A and B if

a $\overrightarrow{OA}$ is $2\mathbf{i} - \mathbf{j} + 5\mathbf{k}$ and $\overrightarrow{OB}$ is $8\mathbf{i} - 3\mathbf{j} + 7\mathbf{k}$

b A and B have coordinates $(2, 6, 7)$ and $(4, 4, 5)$.
   In each case, find the coordinates of the points where the line crosses the xy-plane, the yz-plane and the xz-plane.

11 Find where the line $\dfrac{3x + 9}{4} = \dfrac{2 - y}{-1} = \dfrac{4 - 3z}{5}$ intersects the xy-, the

yz- and the xz-plane.

13.2 Parallel, intersecting and skew lines

When we have two lines there are three possible scenarios. These are shown below.

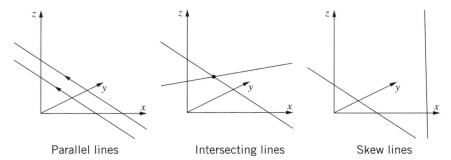

| Parallel lines | Intersecting lines | Skew lines |

Skew lines can only exist in three (or more) dimensions.

It is important to consider these three different cases.

Parallel lines

This is the simplest case, and here the direction vector of one line will be the same as or a multiple of the other. It is only the direction vectors of the line that we need to consider here.

Example

State whether the lines $\mathbf{r} = 2\mathbf{i} - 5\mathbf{j} - 6\mathbf{k} + s(-\mathbf{i} + 3\mathbf{j} + 2\mathbf{k})$ and
$\mathbf{r} = 3\mathbf{i} - 6\mathbf{j} + 6\mathbf{k} + t(4\mathbf{i} - 12\mathbf{j} - 8\mathbf{k})$ are parallel or not, giving a reason.

Since $4\mathbf{i} - 12\mathbf{j} - 8\mathbf{k} = -4(-\mathbf{i} + 3\mathbf{j} + 2\mathbf{k})$ the lines are clearly parallel.

Example

Show that the following lines are parallel.

$$\text{Line 1} \begin{cases} x = 3s - 1 \\ y = 8s + 3 \\ z = 6s + 1 \end{cases} \quad \text{Line 2} \begin{cases} x = 1.5t + 4 \\ y = 4t - 7 \\ z = 3t - 4 \end{cases}$$

In parametric form, all we need to consider are the coefficients of the parameter as these are the direction vectors of the lines.

Hence the direction vector of Line 1 is $\begin{pmatrix} 3 \\ 8 \\ 6 \end{pmatrix}$ and the direction vector of Line 2 is $\begin{pmatrix} 1.5 \\ 4 \\ 3 \end{pmatrix}$.

Since $\begin{pmatrix} 3 \\ 8 \\ 6 \end{pmatrix} = 2\begin{pmatrix} 1.5 \\ 4 \\ 3 \end{pmatrix}$ the lines are parallel.

Example

Show that the lines $\dfrac{x - 3}{2} = \dfrac{1 - y}{4} = \dfrac{2z + 1}{3}$ and $\dfrac{2x + 1}{4} = \dfrac{y + 3}{-4} = \dfrac{4z - 5}{6}$ are parallel.

It is tempting to just look at the denominators and see if they are multiples of one another, but in this case it will give the wrong result.

For the denominators to be the direction vectors of the line they must be in a form with positive unitary coefficients of x, y and z.

Hence Line 1 is $\dfrac{x-3}{2} = \dfrac{y-1}{-4} = \dfrac{z+\frac{1}{2}}{\frac{3}{2}}$ and Line 2 is $\dfrac{x+\frac{1}{2}}{2} = \dfrac{y+3}{-4} = \dfrac{z-\frac{5}{4}}{\frac{6}{4}}$.

By comparing the denominators, which are now the direction vectors of the line, we see they are the same, as $\dfrac{6}{4} = \dfrac{3}{2}$, and hence the lines are parallel.

This method can also be used to find whether two lines are coincident, that is, they are actually one line.

To show two lines are coincident

To do this we need to show that the lines are parallel and that they have a point in common.

Example

Show that the lines $\mathbf{r} = \begin{pmatrix} 1 \\ 2 \\ -3 \end{pmatrix} + s \begin{pmatrix} -1 \\ -1 \\ 4 \end{pmatrix}$ and $\mathbf{r} = \begin{pmatrix} 3 \\ 4 \\ -11 \end{pmatrix} + t \begin{pmatrix} 5 \\ 5 \\ -20 \end{pmatrix}$ are coincident.

Since $-5 \begin{pmatrix} -1 \\ -1 \\ 4 \end{pmatrix} = \begin{pmatrix} 5 \\ 5 \\ -20 \end{pmatrix}$ the lines are parallel.

We know the point $(1, 2, -3)$ lies on the first line. We now test whether it lies on the second line.

The parametric equations for the second line are:

$x = 3 + 5t$

$y = 4 + 5t$

$z = -11 - 20t$

Letting $3 + 5t = 1$

$\Rightarrow t = -\dfrac{2}{5}$

Substituting $t = -\dfrac{2}{5}$ into the equations for y and z gives $y = 4 + 5\left(-\dfrac{2}{5}\right) = 2$

and $z = -11 - 20\left(-\dfrac{2}{5}\right) = -3$.

Hence the point $(1, 2, -3)$ lies on both lines and the lines are coincident.

Intersecting and skew lines

These two cases are treated as a pair. In neither case can the lines be parallel, so we first need to check that the direction vectors are not the same or multiples of each other. Provided the lines are not parallel, then they either have a common point, in which case they intersect, or they do not, in which case they are skew.

Method

> 1. Check the vectors are not parallel and put each line in parametric form. Make sure the parameters for each line are different.
> 2. Assume that they intersect, and equate the x-values, the y-values and the z-values.
> 3. Solve a pair of equations to calculate the values of the parameter.
> 4. Now substitute into the third equation. If the parameters fit, the lines intersect, and if they do not, the lines are skew. To find the coordinates of intersection, substitute either of the calculated parameters into the parametric equations.

Example

Do the lines $\mathbf{r} = \mathbf{k} + \lambda(\mathbf{i} - \mathbf{j} - 3\mathbf{k})$ and $\mathbf{r} = 2\mathbf{i} + \mathbf{j} + \mu(3\mathbf{j} + 5\mathbf{k})$ intersect or are they skew? If they intersect, find the point of intersection.

Step 1

The direction vectors are not equal or multiples of each other, and hence the lines are not parallel. The parametric equations are:

$$x = \lambda \qquad\qquad x = 2$$
$$y = -\lambda \qquad\qquad y = 1 + 3\mu$$
$$z = 1 - 3\lambda \qquad\qquad z = 5\mu$$

Step 2

Equating values of x: $\lambda = 2$ $\qquad$ equation (i)
Equating values of y: $-\lambda = 1 + 3\mu$ $\qquad$ equation (ii)
Equating values of z: $1 - 3\lambda = 5\mu$ $\qquad$ equation (iii)

Step 3

Solve equations (i) and (ii).

Substituting $\lambda = 2$ from equation (i) into equation (ii) gives $-2 = 1 + 3\mu$
$\Rightarrow \mu = -1$.

Step 4

Substitute $\lambda = 2$ and $\mu = -1$ into equation (iii).
$$1 - 3(2) = 5(-1)$$
$$\Rightarrow -5 = -5$$

Hence the lines intersect. Substituting either the value of λ or the value of μ into the equations will give the coordinates of the point of intersection.

Using $\mu = -1 \Rightarrow x = 2$, $y = 1 + 3(-1) = -2$ and $z = 5(-1) = -5$.

Hence the coordinates of the point of intersection are $(2, -2, -5)$.

Example

Do the lines $\mathbf{r} = -3\mathbf{j} - 2\mathbf{k} + \lambda(2\mathbf{i} + \mathbf{j} + \mathbf{k})$ and
$\mathbf{r} = \mathbf{i} + 2\mathbf{j} - 5\mathbf{k} + \mu(\mathbf{i} - \mathbf{j} + 2\mathbf{k})$ intersect or are they skew? If they intersect find the point of intersection.

Step 1

The lines are not parallel. The parametric equations are:

$$x = 2\lambda \qquad\qquad x = 1 + \mu$$
$$y = -3 + \lambda \qquad\qquad y = 2 - \mu$$
$$z = -2 + \lambda \qquad\qquad z = -5 + 2\mu$$

Step 2

Equating values of x: $2\lambda = 1 + \mu$ $\qquad$ equation (i)
Equating values of y: $-3 + \lambda = 2 - \mu$ $\qquad$ equation (ii)
Equating values of z: $-2 + \lambda = -5 + 2\mu$ $\qquad$ equation (iii)

Step 3

Solve equations (i) and (ii).

$2\lambda - \mu = 1$ equation (i)

$\lambda + \mu = 5$ equation (ii)

Adding equations (i) and (ii) $\Rightarrow 3\lambda = 6 \Rightarrow \lambda = 2$

Substitute into equation (i) $\Rightarrow 2(2) = 1 + \mu \Rightarrow \mu = 3$

Step 4

Substitute $\lambda = 2$ and $\mu = 3$ into equation (iii).

$-2 + 2 = -5 + 2(3) \Rightarrow 0 = 1$, which is not possible.

Hence the values of λ and μ do not fit equation (iii), and the lines are skew.

Example

a) Given that the lines $\mathbf{r}_1 = \begin{pmatrix} 1 \\ -1 \\ 3 \end{pmatrix} + m\begin{pmatrix} 1 \\ -1 \\ 1 \end{pmatrix}$ and $\mathbf{r}_2 = \begin{pmatrix} 2 \\ 4 \\ 6 \end{pmatrix} + n\begin{pmatrix} 2 \\ 1 \\ 3 \end{pmatrix}$ intersect

at point P, find the coordinates of P.

(b) Show that the point A(3, −3, 5) and the point B(0, 3, 3) lie on $\mathbf{r}_1$ and $\mathbf{r}_2$ respectively.

(c) Hence find the area of triangle APB.

a) **Step 1**

The direction vectors are not equal or multiples of each other and hence the lines are not parallel. The parametric equations are:

$$x = 1 + m \qquad\qquad x = 2 + 2n$$
$$y = -1 - m \qquad\qquad y = 4 + n$$
$$z = 3 + m \qquad\qquad z = 6 + 3n$$

Step 2

Equating values of x: $1 + m = 2 + 2n$ $\qquad$ equation (i)

Equating values of y: $-1 - m = 4 + n$ $\qquad$ equation (ii)

Equating values of z: $3 + m = 6 + 3n$ $\qquad$ equation (iii)

Step 3

Solve equations (i) and (ii).

$m - 2n = 1$ equation (i)

$m + n = -5$ equation (ii)

Equation (ii) − Equation (i) $\Rightarrow 3n = -6 \Rightarrow n = -2$

Substitute into equation (i) $\Rightarrow m - 2(-2) = 1 \Rightarrow m = -3$

Step 4

In this case we do not need to prove that the lines intersect because we are told in the question. Substituting either the value of n or the value of m into the equations will give the coordinates of the point of intersection.

Using $n = -2 \Rightarrow x = 2 + 2(-2) = -2$, $y = 4 + (-2) = 2$ and

$z = 6 + 3(-2) = 0$.

Hence the coordinates of P are $(-2, 2, 0)$.

b) If the point (3, −3, 5) lies on $\mathbf{r}_1$ then there should be a consistent value for m in the parametric equations of $\mathbf{r}_1$.

$3 = 1 + m \Rightarrow m = 2$

If $m = 2$, $y = -1 - 2 = -3$ and $z = 3 + 2 = 5$. Hence A lies on $\mathbf{r}_1$.

If the point (0, 3, 3) lies on $\mathbf{r}_2$ then there should be a consistent value for n in the parametric equations of $\mathbf{r}_2$.

$0 = 2 + 2n \Rightarrow n = -1$

If $n = -1$, $y = 4 - 1 = 3$ and $z = 6 - 3 = 3$. Hence B lies on $\mathbf{r}_1$.

c) $\overrightarrow{AP} = \begin{pmatrix} -2 \\ 2 \\ 0 \end{pmatrix} - \begin{pmatrix} 3 \\ -3 \\ 5 \end{pmatrix} = \begin{pmatrix} -5 \\ 5 \\ -5 \end{pmatrix}$ and $\overrightarrow{BP} = \begin{pmatrix} -2 \\ 2 \\ 0 \end{pmatrix} - \begin{pmatrix} 0 \\ 3 \\ 3 \end{pmatrix} = \begin{pmatrix} -2 \\ -1 \\ -3 \end{pmatrix}$

$\Rightarrow \overrightarrow{AP} \times \overrightarrow{BP} = \begin{vmatrix} \mathbf{i} & \mathbf{j} & \mathbf{k} \\ -5 & 5 & -5 \\ -2 & -1 & -3 \end{vmatrix}$

$= \mathbf{i}[(-15) - 5] - \mathbf{j}[15 - 10] + \mathbf{k}[5 - (-10)]$

$= -20\mathbf{i} - 5\mathbf{j} + 15\mathbf{k}$

Area of triangle ABP $= \dfrac{1}{2}|\overrightarrow{AP} \times \overrightarrow{BP}|$

$= \dfrac{1}{2}\sqrt{(-20)^2 + (-5)^2 + 15^2} = \dfrac{\sqrt{650}}{2} = \dfrac{5\sqrt{26}}{2}$ units^2

Example

Find the position vector of the point of intersection of the line $\mathbf{r} = -\mathbf{i} - 2\mathbf{j} + 5\mathbf{k} + \lambda(2\mathbf{i} - 3\mathbf{j} + \mathbf{k})$ and the perpendicular that passes through the point $P(1, 2, -5)$. Hence find the shortest distance from $(1, 2, -5)$ to the line.

Any point on the line $\mathbf{r}$ is given by the parametric equations

$x = -1 + 2\lambda$
$y = -2 - 3\lambda$
$z = 5 + \lambda$

This could be point X as shown in the diagram.

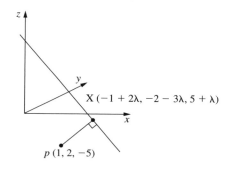

Hence the direction vector $\overrightarrow{PX}$ is

$(-1 + 2\lambda - 1)\mathbf{i} + (-2 - 3\lambda - 2)\mathbf{j} + (5 + \lambda - (-5))\mathbf{k}$

$= (-2 + 2\lambda)\mathbf{i} + (-4 - 3\lambda)\mathbf{j} + (10 + \lambda)\mathbf{k}$.

If $\overrightarrow{PX}$ is perpendicular to $\mathbf{r}$ then the scalar product of $\overrightarrow{PX}$ and the direction vector of $\mathbf{r}$ must be zero.

$\Rightarrow [(-2 + 2\lambda)\mathbf{i} + (-4 - 3\lambda)\mathbf{j} + (10 + \lambda)\mathbf{k}] \cdot [2\mathbf{i} - 3\mathbf{j} + \mathbf{k}] = 0$

$\Rightarrow -4 + 4\lambda + 12 + 9\lambda + 10 + \lambda = 0$

$\Rightarrow 14\lambda = -18$

$\Rightarrow \lambda = -\dfrac{9}{7}$

Substituting $\lambda = -\dfrac{9}{7}$ into the parametric equations will give the point of intersection.

Hence: $x = -1 - \dfrac{18}{7} = \dfrac{-25}{7}$

$y = -2 + \dfrac{27}{7} = \dfrac{13}{17}$

$z = 5 - \dfrac{9}{7} = \dfrac{26}{7}$

Therefore the position vector of the point is $\dfrac{-25}{7}\mathbf{i} + \dfrac{13}{7}\mathbf{j} + \dfrac{26}{7}\mathbf{k}$.

The shortest distance is the distance between $\left(\dfrac{-25}{7}, \dfrac{13}{7}, \dfrac{26}{7}\right)$ and $(1, 2, -5)$

Using Pythagoras' theorem, distance

$= \sqrt{\left(\dfrac{-25}{7} - 1\right)^2 + \left(\dfrac{13}{7} - 2\right)^2 + \left(\dfrac{26}{7} - (-5)\right)^2}$

$= \sqrt{\left(\dfrac{-32}{7}\right)^2 + \left(\dfrac{-1}{7}\right)^2 + \left(\dfrac{61}{7}\right)^2} = \dfrac{\sqrt{678}}{7}$

Finding the angle between two lines

Here we are interested in the direction vectors of the two lines, and it is the angle between these two vectors that we require.

Method

1 Find the direction vector of each line.
2 Apply the scalar product rule and hence find the angle.

Example

Find the angle θ between the lines

$\mathbf{r} = \begin{pmatrix} 1 \\ -1 \\ 4 \end{pmatrix} + \lambda \begin{pmatrix} 2 \\ 1 \\ -1 \end{pmatrix}$ and $\mathbf{r} = \begin{pmatrix} 2 \\ 3 \\ -2 \end{pmatrix} + \lambda \begin{pmatrix} 3 \\ -2 \\ 1 \end{pmatrix}$.

Step 1

The required direction vectors are $\begin{pmatrix} 2 \\ 1 \\ -1 \end{pmatrix}$ and $\begin{pmatrix} 3 \\ -2 \\ 1 \end{pmatrix}$.

Step 2

$\begin{pmatrix} 2 \\ 1 \\ -1 \end{pmatrix} \cdot \begin{pmatrix} 3 \\ -2 \\ 1 \end{pmatrix} = \sqrt{2^2 + 1^2 + (-1)^2}\sqrt{3^2 + (-2)^2 + 1^2}\cos\theta$

$\Rightarrow (2)(3) + (1)(-2) + (-1)(1) = \sqrt{6}\sqrt{14}\cos\theta$

$\Rightarrow \cos\theta = \dfrac{3}{\sqrt{6}\sqrt{14}}$

$\Rightarrow \theta = 70.9°$

Exercise 2

1 Determine whether the following pairs of lines are parallel, coincident, skew or intersecting. If they intersect, give the position vector of the point of intersection.

a $\mathbf{r} = \begin{pmatrix} 1 \\ 1 \\ 0 \end{pmatrix} + \lambda \begin{pmatrix} 2 \\ 3 \\ -3 \end{pmatrix}$ and $\mathbf{r} = \begin{pmatrix} -3 \\ 7 \\ -6 \end{pmatrix} + \mu \begin{pmatrix} 2 \\ 4 \\ -3 \end{pmatrix}$

b $\mathbf{r} = \begin{pmatrix} 1 \\ 3 \\ 3 \end{pmatrix} + \lambda \begin{pmatrix} 1 \\ -2 \\ 3 \end{pmatrix}$ and $\mathbf{r} = \begin{pmatrix} 0 \\ 0 \\ 5 \end{pmatrix} + \mu \begin{pmatrix} 2 \\ 1 \\ 1 \end{pmatrix}$

c $\mathbf{r} = \mathbf{i} - 2\mathbf{j} + 3\mathbf{k} + m(2\mathbf{i} - \mathbf{j} + 3\mathbf{k})$ and
$\mathbf{r} = 3\mathbf{i} + 2\mathbf{j} - 2\mathbf{k} + n(-8\mathbf{i} + 4\mathbf{j} - 12\mathbf{k})$

d $\mathbf{r} = \begin{pmatrix} 1 \\ 5 \\ -1 \end{pmatrix} + s \begin{pmatrix} 4 \\ -1 \\ 2 \end{pmatrix}$ and $x = 1 + 2t, y = 3 - 3t, z = 2t$

e $\begin{aligned} x &= 2 + 3\lambda \\ y &= 1 + \lambda \\ z &= -2 + 8\lambda \end{aligned}$ and $\begin{aligned} x &= 4 - 2\mu \\ y &= 2 + 2\mu \\ z &= 6 + 16\mu \end{aligned}$

f $x - 2 = \dfrac{y - 3}{2} = \dfrac{z + 1}{-3}$ and $-\dfrac{x}{2} = \dfrac{y + 1}{-4} = \dfrac{z - 5}{6}$

g $\mathbf{r} = \mathbf{i} - 2\mathbf{j} + 5\mathbf{k} + m(\mathbf{i} - \mathbf{j} + 3\mathbf{k})$ and $\dfrac{3 - x}{2} = \dfrac{2y + 3}{2} = \dfrac{3z - 5}{-2}$

2 Determine whether the given points lie on the given lines. If not, find the shortest distance from the point to the line.

a $(0, 0, 1)$ $\qquad$ $\mathbf{r} = \mathbf{i} + \mathbf{j} - \mathbf{k} + \lambda(2\mathbf{i} - \mathbf{j} + \mathbf{k})$

b $(1, 2, -1)$ $\qquad$ $\mathbf{r} = \begin{pmatrix} 1 \\ 0 \\ 3 \end{pmatrix} + \mu \begin{pmatrix} 2 \\ 1 \\ 2 \end{pmatrix}$

c $(1, -1, 3)$ $\qquad$ $\mathbf{r} = \mathbf{i} - 2\mathbf{j} + 3\mathbf{k} + t(\mathbf{i} + \mathbf{k})$

d $(-1, -2, 0)$ $\qquad$ $\mathbf{r} = \begin{pmatrix} 3 \\ 1 \\ 1 \end{pmatrix} + s \begin{pmatrix} 1 \\ 2 \\ -1 \end{pmatrix}$

e $\left(4, 0, -\dfrac{7}{2}\right)$ $\qquad$ $\dfrac{x + 2}{3} = \dfrac{2 - y}{1} = \dfrac{2z + 3}{-2}$

f $(3, 2, 1)$ $\qquad$ $x = 2 - m, y = 3m + 1, z = 1 - 2m$

3 The points A(0, 1, −2), B(3, 5, 5) and D(1, 3, −3) are three vertices of a parallelogram ABCD. Find vector and Cartesian equations for the sides AB and AD and find the coordinates of C.

4 Find the acute angle between each of these pairs of lines.

a $\mathbf{r} = \begin{pmatrix} -2 \\ 1 \\ -1 \end{pmatrix} + s \begin{pmatrix} 2 \\ 6 \\ -1 \end{pmatrix}$ and $\mathbf{r} = \begin{pmatrix} 3 \\ -1 \\ -1 \end{pmatrix} + t \begin{pmatrix} 2 \\ -4 \\ -3 \end{pmatrix}$

b $\mathbf{r} = 2\mathbf{i} + 2\mathbf{j} - \mathbf{k} + m(\mathbf{i} - 3\mathbf{j} - \mathbf{k})$ and
$\mathbf{r} = 5\mathbf{i} + 3\mathbf{j} + 3\mathbf{k} + n(-\mathbf{i} - 3\mathbf{j} - 2\mathbf{k})$

c $\mathbf{r} = \begin{pmatrix} 2 \\ -1 \\ -1 \end{pmatrix} + s \begin{pmatrix} 2 \\ -3 \\ -4 \end{pmatrix}$ and $x = 1 + t, y = 2 + 5t, z = 1 - t$

d $\begin{aligned} x &= -1 + \lambda \\ y &= 2 + 5\lambda \\ z &= -2 + 3\lambda \end{aligned}$ and $\begin{aligned} x &= 1 + \mu \\ y &= 2 - 3\mu \\ z &= 3 + 5\mu \end{aligned}$

e $\dfrac{x - 3}{4} = \dfrac{y + 2}{3} = \dfrac{z - 2}{-3}$ and $-\dfrac{2x + 1}{3} = \dfrac{3y + 1}{5} = \dfrac{z - 1}{3}$

5 Two lines have equations $\mathbf{r} = \begin{pmatrix} 3 \\ 2 \\ -1 \end{pmatrix} + \lambda \begin{pmatrix} 2 \\ -3 \\ 2 \end{pmatrix}$ and $\mathbf{r} = \begin{pmatrix} -1 \\ 1 \\ 1 \end{pmatrix} + \mu \begin{pmatrix} 1 \\ 2 \\ a \end{pmatrix}$.

Find the value of a for which the lines intersect and the position vector of the point of intersection.

6 a Show that the points whose position vectors are $\begin{pmatrix} 9 \\ 10 \end{pmatrix}$ and $\begin{pmatrix} -3 \\ -8 \end{pmatrix}$ lie on

the line with equation $\mathbf{r} = \begin{pmatrix} 1 \\ -2 \end{pmatrix} + s \begin{pmatrix} 2 \\ 3 \end{pmatrix}$.

b Obtain, in parametric form, an equation of the line that passes through the

point with position vector $\begin{pmatrix} -3 \\ 8 \end{pmatrix}$ and is perpendicular to the given line.

7 The position vectors of the points A and B are given by $\overrightarrow{OA} = 2\mathbf{i} - 3\mathbf{j} + \mathbf{k}$

and $\overrightarrow{OB} = 3\mathbf{i} + 5\mathbf{j} - 2\mathbf{k}$, where O is the origin.
a Find a vector equation of the straight line passing through A and B.
b Given that this line is perpendicular to the vector $p\mathbf{i} - 2\mathbf{j} + 6\mathbf{k}$, find the value of p.

13.3 Equation of a plane

Definition of a plane

In simple terms a plane can be described as an infinite flat surface such that an infinite straight line joining any two points on it will lie entirely in it. This surface can be vertical, horizontal or sloping. The position of a plane can be described by giving

1. three non-collinear points
2. a point and two non-parallel lines
3. a vector perpendicular to the plane at a given distance from the origin
4. a vector perpendicular to the plane and a point that lies in the plane.

The vector equation of a plane is described as being a specific distance away from the origin and perpendicular to a given vector. The reason behind this is that it makes the vector equation of a plane very straightforward. We will first look at this form and then see how we can derive the other forms from this.

Scalar product form of the vector equation of a plane

Consider the plane below, distance d from the origin, which is perpendicular to the unit vector $\hat{\mathbf{n}}$, where $\hat{\mathbf{n}}$ is directed away from O. X is the point of intersection of the perpendicular with the plane, P is any point (x, y, z), and $\overrightarrow{OP} = \mathbf{r}$.

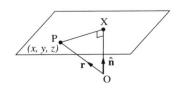

$$\overrightarrow{OX} = d\hat{\mathbf{n}}$$

We also know that $\overrightarrow{OX}$ and $\overrightarrow{PX}$ are perpendicular and thus $\overrightarrow{OX} \cdot \overrightarrow{PX} = 0$.

Now $\overrightarrow{PX} = d\hat{\mathbf{n}} - \mathbf{r}$

Hence $d\hat{\mathbf{n}} \cdot (d\hat{\mathbf{n}} - \mathbf{r}) = 0$

$\Rightarrow d^2\hat{\mathbf{n}} \cdot \hat{\mathbf{n}} - d\hat{\mathbf{n}} \cdot \mathbf{r} = 0$

$\Rightarrow d(\mathbf{r} \cdot \hat{\mathbf{n}}) = d^2$ ⋯⋯⋯⋯⋯⋯⋯⋯⋯⋯⋯⋯⋯⋯⋯ Since $\hat{\mathbf{n}} \cdot \hat{\mathbf{n}} = 1$ and scalar product is commutative

$\Rightarrow \mathbf{r} \cdot \hat{\mathbf{n}} = d$

> $\mathbf{r} \cdot \hat{\mathbf{n}} = d$ is the standard equation of a plane, where $\mathbf{r}$ is the position vector of any point on the plane, $\hat{\mathbf{n}}$ is the unit vector perpendicular to the plane, and d is the distance of the plane from the origin.

Example

Write down the vector equation of the plane in scalar product form.

A plane is perpendicular to the vector $\dfrac{1}{5\sqrt{2}}(3\mathbf{i} - 4\mathbf{j} + 5\mathbf{k})$ and 6 units away from the origin.

Since the perpendicular vector is a unit vector the vector equation of the plane is

$$\mathbf{r} \cdot \frac{1}{5\sqrt{2}}(3\mathbf{i} - 4\mathbf{j} + 5\mathbf{k}) = 6$$

or $\mathbf{r} \cdot (3\mathbf{i} - 4\mathbf{j} + 5\mathbf{k}) = 30\sqrt{2}$

We can see that any equation of the form $\mathbf{r} \cdot \mathbf{n} = D$ represents a plane perpendicular to $\mathbf{n}$, known as the direction normal. If we want the plane in the form $\mathbf{r} \cdot \hat{\mathbf{n}} = d$, we divide by the magnitude of $\mathbf{n}$.

Example

Find the distance of the plane $\mathbf{r} \cdot \begin{pmatrix} 3 \\ 2 \\ -4 \end{pmatrix} = 8$ from the origin, and the unit vector perpendicular to the plane.

The magnitude of the direction vector is $\sqrt{3^2 + 2^2 + (-4)^2} = \sqrt{29}$

$$\Rightarrow \mathbf{r} \cdot \begin{pmatrix} \dfrac{3}{\sqrt{29}} \\ \dfrac{2}{\sqrt{29}} \\ \dfrac{-4}{\sqrt{29}} \end{pmatrix} = \frac{8}{\sqrt{29}}$$

Hence the distance from the origin is $\dfrac{8}{\sqrt{29}}$, and the unit vector perpendicular to the plane is $\begin{pmatrix} \dfrac{3}{\sqrt{29}} \\ \dfrac{2}{\sqrt{29}} \\ \dfrac{-4}{\sqrt{29}} \end{pmatrix}$.

It is not common to be told the distance of the plane from the origin; it is much more likely that a point on the plane will be given. In this case we can still use the scalar product form of the vector equation, as shown in the next example.

Example

A plane is perpendicular to the line $\mathbf{i} - \mathbf{j} - 2\mathbf{k}$ and passes through the point $(-1, 2, 5)$. Find the vector equation of the plane in scalar product form.

The equation of the plane must be of the form $\mathbf{r} \cdot (\mathbf{i} - \mathbf{j} - 2\mathbf{k}) = D$.

Also, the position vector of the point must fit the equation.

$$\Rightarrow (-\mathbf{i} + 2\mathbf{j} + 5\mathbf{k}) \cdot (\mathbf{i} - \mathbf{j} - 2\mathbf{k}) = D$$

$$\Rightarrow D = -1 - 2 - 10 = -13$$

Therefore the vector equation of the plane is $\mathbf{r} \cdot (\mathbf{i} - \mathbf{j} - 2\mathbf{k}) = -13$.

If we are required to find the distance of the plane from the origin or the unit vector perpendicular to the plane, we just divide by the magnitude of the direction normal.

In this case $\hat{\mathbf{n}} = \dfrac{1}{\sqrt{6}}(\mathbf{i} - \mathbf{j} - 2\mathbf{k})$ and $d = \dfrac{-13}{\sqrt{6}}$.

Cartesian equation of a plane

The Cartesian equation is of the form $ax + by + cz = d$ and is found by putting $\mathbf{r} = \begin{pmatrix} x \\ y \\ z \end{pmatrix}$.

If we have the equation $\mathbf{r} \cdot (2\mathbf{i} - 4\mathbf{j} - \mathbf{k}) = 6$, then $(x\mathbf{i} + y\mathbf{j} + z\mathbf{k}) \cdot (2\mathbf{i} - 4\mathbf{j} - \mathbf{k}) = 6$.

Therefore the Cartesian equation of the plane is $2x - 4y - z = 6$.

Example

Convert $x + 3y - 2z = 9$ into scalar product form.

$$\begin{pmatrix} x \\ y \\ z \end{pmatrix} \cdot \begin{pmatrix} 1 \\ 3 \\ -2 \end{pmatrix} = 9$$

Thus $\mathbf{r} \cdot \begin{pmatrix} 1 \\ 3 \\ -2 \end{pmatrix} = 9$

Parametric form of the equation of a plane

The parametric form is used to describe a plane passing through a point and containing two lines. Consider the diagram.

The vectors $\mathbf{m}$ and $\mathbf{n}$ are not parallel and lie on the plane. The point A, whose position vector is $\mathbf{a}$, also lies in the plane.

Let P be any point on the plane with position vector $\mathbf{r}$.

Hence $\overrightarrow{AP} = \lambda\mathbf{m} + \mu\mathbf{n}$, where λ and μ are parameters.

Thus $\mathbf{r} = \mathbf{a} + \overrightarrow{AP} = \mathbf{a} + \lambda\mathbf{m} + \mu\mathbf{n}$.

$\mathbf{r} = \mathbf{a} + \lambda\mathbf{m} + \mu\mathbf{n}$, where $\mathbf{a}$ is the position vector of a point and $\mathbf{m}$ and $\mathbf{n}$ are direction vectors of lines, is the parametric form of the equation of a plane.

It is not easy to work in this form, so we need to be able to convert this to scalar product form.

In any plane, the direction normal is perpendicular to any lines in the plane. Hence to find the direction normal, we need a vector that is perpendicular to two lines in the plane. From Chapter 12, we remember that this is the definition of the vector product.

Method for converting the parametric form to scalar product form

1. Using the direction vectors of the lines, find the perpendicular vector using the vector product. This gives the plane in the form $\mathbf{r} \cdot \mathbf{n} = D$.
2. To find D, substitute in the coordinates of the point.

Example

Convert the parametric form of the vector equation $\mathbf{r} = \begin{pmatrix} -1 \\ -1 \\ 2 \end{pmatrix} + \lambda\begin{pmatrix} -2 \\ 0 \\ 1 \end{pmatrix} + \mu\begin{pmatrix} -2 \\ 3 \\ 0 \end{pmatrix}$

to scalar product form.

In this equation the direction vectors of the lines are $\begin{pmatrix} -2 \\ 0 \\ 1 \end{pmatrix}$ and $\begin{pmatrix} -2 \\ 3 \\ 0 \end{pmatrix}$.

Hence the direction normal of the plane is

$$\begin{vmatrix} \mathbf{i} & \mathbf{j} & \mathbf{k} \\ -2 & 0 & 1 \\ -2 & 3 & 0 \end{vmatrix} = \mathbf{i}(0 - 3) - \mathbf{j}(0 + 2) + \mathbf{k}(-6 - 0)$$

$$= \begin{pmatrix} -3 \\ -2 \\ -6 \end{pmatrix}$$

Therefore the scalar product form of the vector equation of the plane is $\mathbf{r} \cdot \begin{pmatrix} -3 \\ -2 \\ -6 \end{pmatrix} = D$.

$\begin{pmatrix} -1 \\ -1 \\ 2 \end{pmatrix}$ fits this $\Rightarrow \begin{pmatrix} -1 \\ -1 \\ 2 \end{pmatrix} \cdot \begin{pmatrix} -3 \\ -2 \\ -6 \end{pmatrix} = D$

$$\Rightarrow D = 3 + 2 - 12 = -7$$

Therefore the vector equation of the plane in scalar product form is $\mathbf{r} \cdot \begin{pmatrix} -3 \\ -2 \\ -6 \end{pmatrix} = -7$.

This could also be written in the form $\mathbf{r} \cdot \hat{\mathbf{n}} = d$:

$$\mathbf{r} \cdot \begin{pmatrix} \dfrac{-3}{7} \\ \dfrac{-2}{7} \\ \dfrac{-6}{7} \end{pmatrix} = -1$$

Example

Find the equation of the plane that contains the lines

$$\mathbf{r}_1 = 2\mathbf{i} + \mathbf{j} + 3\mathbf{k} + \lambda(3\mathbf{i} - 2\mathbf{j} + 4\mathbf{k})$$

$$\mathbf{r}_2 = -\mathbf{j} + \mathbf{k} + \mu(\mathbf{i} + \mathbf{j} + \mathbf{k})$$

and passes through the point (2, 1, 3). Give the answer in scalar product form and in Cartesian form.

The direction vectors of the lines are $3\mathbf{i} - 2\mathbf{j} + 4\mathbf{k}$ and $\mathbf{i} + \mathbf{j} + \mathbf{k}$.

Hence the direction normal of the plane is

$$\begin{vmatrix} \mathbf{i} & \mathbf{j} & \mathbf{k} \\ 3 & -2 & 4 \\ 1 & 1 & 1 \end{vmatrix} = \mathbf{i}(-2 - 4) - \mathbf{j}(3 - 4) + \mathbf{k}(3 + 2)$$

$$= -6\mathbf{i} + \mathbf{j} + 5\mathbf{k}$$

Therefore the vector equation of the plane is of the form $\mathbf{r} \cdot (-6\mathbf{i} + \mathbf{j} + 5\mathbf{k}) = D$.

The vector $2\mathbf{i} + \mathbf{j} + 3\mathbf{k}$ fits this. Hence $(2\mathbf{i} + \mathbf{j} + 3\mathbf{k}) \cdot (-6\mathbf{i} + \mathbf{j} + 5\mathbf{k}) = D$

$$\Rightarrow D = -12 + 1 + 15 = 4$$

Therefore the vector equation of the plane is $\mathbf{r} \cdot (-6\mathbf{i} + \mathbf{j} + 5\mathbf{k}) = 4$.

This could also be written in the form $\mathbf{r} \cdot \hat{\mathbf{n}} = d$ as

$$\mathbf{r} \cdot \left(-\frac{6}{\sqrt{62}}\mathbf{i} + \frac{1}{\sqrt{62}}\mathbf{j} + \frac{5}{\sqrt{62}}\mathbf{k}\right) = \frac{4}{\sqrt{62}}$$

To find the Cartesian equation we use $\mathbf{r} \cdot (-6\mathbf{i} + \mathbf{j} + 5\mathbf{k}) = 4$ and replace $\mathbf{r}$ with $x\mathbf{i} + y\mathbf{j} + z\mathbf{k}$.

Hence $(x\mathbf{i} + y\mathbf{j} + z\mathbf{k}) \cdot (-6\mathbf{i} + \mathbf{j} + 5\mathbf{k}) = 4$

$$\Rightarrow -6x + y + 5z = 4$$

Example

Find the scalar product form of the vector equation of the plane passing through the points $A(1, -1, -3)$, $B(2, -1, 1)$ and $C(-3, -1, 2)$.

We can use the same method to do this, since we can form the direction vectors of two lines in the plane.

$$\overrightarrow{AB} = (2 - 1)\mathbf{i} + (-1 + 1)\mathbf{j} + (1 + 3)\mathbf{k} = \mathbf{i} + 4\mathbf{k}$$

$$\overrightarrow{BC} = (-3 - 2)\mathbf{i} + (-1 + 1)\mathbf{j} + (2 - 1)\mathbf{k} = -5\mathbf{i} + \mathbf{k}$$

Hence the direction normal of the plane is

$$\begin{vmatrix} \mathbf{i} & \mathbf{j} & \mathbf{k} \\ 1 & 0 & 4 \\ -5 & 0 & 1 \end{vmatrix} = \mathbf{i}(0 - 0) - \mathbf{j}(1 + 20) + \mathbf{k}(0 - 0) = -21\mathbf{j}$$

Therefore the vector equation of the plane is of the form $\mathbf{r} \cdot (-21\mathbf{j}) = D$.
$\overrightarrow{OA} = \mathbf{i} - \mathbf{j} - 3\mathbf{k}$ fits this $\Rightarrow (\mathbf{i} - \mathbf{j} - 3\mathbf{k}) \cdot (-21\mathbf{j}) = D$

$$\Rightarrow D = 21$$

Therefore the vector equation of the plane is $\mathbf{r} \cdot (-21\mathbf{j}) = 21$ or, in the form $\mathbf{r} \cdot \hat{\mathbf{n}} = d$, $\mathbf{r} \cdot \mathbf{j} = -1$.

Example

Find the equation of the plane that is parallel to the plane $\mathbf{r} = 2\mathbf{i} + \mathbf{j} - \mathbf{k} + m(\mathbf{i} + 2\mathbf{j}) + n(\mathbf{i} - \mathbf{j} - 3\mathbf{k})$ and passes through the point $A(3, 2, -2)$. Hence find the distance of this plane from the origin.

The direction vectors of two lines parallel to the plane are $\mathbf{i} + 2\mathbf{j}$ and $\mathbf{i} - \mathbf{j} - 3\mathbf{k}$.

Hence the direction normal of the plane is

$$\begin{vmatrix} \mathbf{i} & \mathbf{j} & \mathbf{k} \\ 1 & 2 & 0 \\ 1 & -1 & -3 \end{vmatrix} = \mathbf{i}(-6 - 0) - \mathbf{j}(-3 - 0) + \mathbf{k}(-1 - 2)$$

$$= -6\mathbf{i} + 3\mathbf{j} - 3\mathbf{k} = 3(-2\mathbf{i} + \mathbf{j} - \mathbf{k})$$

Therefore the scalar product form of the vector equation of the plane is $\mathbf{r} \cdot (-2\mathbf{i} + \mathbf{j} - \mathbf{k}) = D$.

$\overrightarrow{OA} = 3\mathbf{i} + 2\mathbf{j} - 2\mathbf{k}$ fits this $\Rightarrow (3\mathbf{i} + 2\mathbf{j} - 2\mathbf{k}) \cdot (-2\mathbf{i} + \mathbf{j} - \mathbf{k}) = D$

$\Rightarrow D = -6 + 2 + 2 = -2$

Therefore the scalar product form of the vector equation of the plane is $\mathbf{r} \cdot (-2\mathbf{i} + \mathbf{j} - \mathbf{k}) = -2$.

In the form $\mathbf{r} \cdot \hat{\mathbf{n}} = d$ this is $\mathbf{r} \cdot \left(-\dfrac{2}{\sqrt{6}}\mathbf{i} + \dfrac{1}{\sqrt{6}}\mathbf{j} - \dfrac{1}{\sqrt{6}}\mathbf{k}\right) = \dfrac{-2}{\sqrt{6}}$. ....

Hence the distance of this plane from the origin is $\left|\dfrac{-2}{\sqrt{6}}\right| = \dfrac{2}{\sqrt{6}}$.

> The negative sign means that this plane is on the opposite side of the origin to the one where d is positive.

Example

The plane π_1 has equation $\mathbf{r} \cdot \begin{pmatrix} 2 \\ 0 \\ -1 \end{pmatrix} = 7$.

a) Find the distance of π_1 from the origin.
b) Find the equation of the plane π_2, which is parallel to π_1 and passes through the point A$(-1, 2, 5)$.
c) Find the distance of π_2 from the origin.
d) Hence find the distance between π_1 and π_2.

a) Putting π_1 in the form $\mathbf{r} \cdot \hat{\mathbf{n}} = d \Rightarrow \mathbf{r} \cdot \begin{pmatrix} \frac{2}{\sqrt{5}} \\ 0 \\ \frac{-1}{\sqrt{5}} \end{pmatrix} = \frac{7}{\sqrt{5}}$

Hence the distance of π_1 from the origin is $\dfrac{7}{\sqrt{5}}$.

b) Since π_1 and π_2 are parallel, they have the same direction normals.

Hence the equation of π_2 is $\mathbf{r} \cdot \begin{pmatrix} 2 \\ 0 \\ -1 \end{pmatrix} = D$.

$\overrightarrow{OA} = \begin{pmatrix} -1 \\ 2 \\ 5 \end{pmatrix}$ fits this $\Rightarrow \begin{pmatrix} -1 \\ 2 \\ 5 \end{pmatrix} \cdot \begin{pmatrix} 2 \\ 0 \\ -1 \end{pmatrix} = D$

$\Rightarrow D = -2 - 5 = -7$

Therefore π_2 has equation $\mathbf{r} \cdot \begin{pmatrix} 2 \\ 0 \\ -1 \end{pmatrix} = -7$.

c) We now put π_2 in the form $\mathbf{r} \cdot \hat{\mathbf{n}} = d \Rightarrow \mathbf{r} \cdot \begin{pmatrix} \frac{2}{\sqrt{5}} \\ 0 \\ \frac{-1}{\sqrt{5}} \end{pmatrix} = \frac{-7}{\sqrt{5}}$

Hence the distance of π_2 from the origin is $\left| \dfrac{-7}{\sqrt{5}} \right| = \dfrac{7}{\sqrt{5}}$.

d) Hence the distance between π_1 and π_2 is $\dfrac{7}{\sqrt{5}} + \dfrac{7}{\sqrt{5}} = \dfrac{14}{\sqrt{5}}$.

> We add the distances here because the planes are on opposite sides of the origin. Had they been on the same side of the origin, we would have subtracted them.

Lines and planes

Sometimes problems are given that refer to lines and planes. These are often solved by remembering that if a line and a plane are parallel then the direction vector of the line is perpendicular to the direction normal of the plane.

Example

Find the equation of the line that is perpendicular to the plane $\mathbf{r} \cdot (\mathbf{i} + 2\mathbf{j} - \mathbf{k}) = 4$ and passes through the point $(1, 3, -3)$.

Since the line is perpendicular to the plane, the direction vector of the line is the direction normal to the plane. This is shown in the diagram.

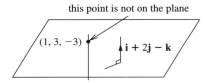

this point is not on the plane

$(1, 3, -3)$ $\mathbf{i} + 2\mathbf{j} - \mathbf{k}$

Hence the vector equation of the line is $\mathbf{r} = \mathbf{i} + 3\mathbf{j} - 3\mathbf{k} + \lambda(\mathbf{i} + 2\mathbf{j} - \mathbf{k})$.

Example

Determine whether the line $\mathbf{r} = \begin{pmatrix} 3 \\ -1 \\ 0 \end{pmatrix} + m \begin{pmatrix} 2 \\ 1 \\ 4 \end{pmatrix}$ intersects with, is parallel to, or

is contained in the plane $\mathbf{r} \cdot \begin{pmatrix} 5 \\ -2 \\ -2 \end{pmatrix} = 17$.

We first test whether the direction vector of the line is perpendicular to the direction normal of the plane. If it is, then the line is either parallel to or contained in the plane. If not, it will intersect the plane.

> We will deal with how to find this point of intersection later in the chapter.

Hence we find the scalar product of $\begin{pmatrix} 2 \\ 1 \\ 4 \end{pmatrix} \cdot \begin{pmatrix} 5 \\ -2 \\ -2 \end{pmatrix} = 10 - 2 - 8 = 0$.

Since the scalar product is zero, the line is either parallel to the plane or contained in the plane. To find out which case this is, we find if they have a point in common. Knowing that the point $(3, -1, 0)$ lies on the line, we now test whether it lies in the plane.

$$\begin{pmatrix} 3 \\ -1 \\ 0 \end{pmatrix} \cdot \begin{pmatrix} 5 \\ -2 \\ -2 \end{pmatrix} = 15 + 2 + 0 = 17$$

Hence the line is contained in the plane.

Exercise 3

1 Find the vector equation of each of these planes in the form $\mathbf{r} \cdot \hat{\mathbf{n}} = d$.

a Perpendicular to the vector $\begin{pmatrix} 1 \\ 4 \\ -1 \end{pmatrix}$ and 3 units away from the origin

b Perpendicular to the vector $\mathbf{i} - 3\mathbf{j} + 5\mathbf{k}$ and containing the point $(3, 2, -1)$

c Perpendicular to the line $\mathbf{r} = \begin{pmatrix} 1 \\ 3 \\ -2 \end{pmatrix} + m\begin{pmatrix} 2 \\ 0 \\ 1 \end{pmatrix}$ and 6 units away from the origin

d Perpendicular to the line $\mathbf{r} = 3\mathbf{i} - 2\mathbf{j} + 4\mathbf{k} + t(-\mathbf{i} - 4\mathbf{j} + \mathbf{k})$ and containing the point $(2, 3, -1)$

e Perpendicular to the line $\mathbf{r} = -\mathbf{i} - \mathbf{j} + 2\mathbf{k} + \mu(\mathbf{i} - 8\mathbf{j} + 3\mathbf{k})$ and passing through the point $A(3, 2, -2)$

f Containing the lines $\mathbf{r} = (\mathbf{i} + 2\mathbf{j}) + s(\mathbf{i} + 2\mathbf{j} - 3\mathbf{k})$ and $\mathbf{r} = (\mathbf{i} + 5\mathbf{j} - 9\mathbf{k}) + t(-\mathbf{i} + \mathbf{j} - 6\mathbf{k})$

g Passing through the points $A(1, 3, -2)$, $B(-6, 1, 0)$ and $C(-4, -3, -1)$

h Containing the lines $\mathbf{r} = \begin{pmatrix} -1 \\ 4 \\ -2 \end{pmatrix} + \lambda\begin{pmatrix} 7 \\ 3 \\ -2 \end{pmatrix}$ and $\dfrac{x + 6}{-2} = \dfrac{y - 1}{4} = z = 0$

i Passing through the points A, B and C with position vectors

$$\overrightarrow{OA} = \begin{pmatrix} 4 \\ 1 \\ 0 \end{pmatrix}, \overrightarrow{OB} = \begin{pmatrix} 1 \\ 0 \\ 6 \end{pmatrix} \text{ and } \overrightarrow{OC} = \begin{pmatrix} 3 \\ -3 \\ 4 \end{pmatrix}$$

j Passing through the origin and perpendicular to $\mathbf{r} = (2 - 3\mu)\mathbf{i} + (-3 - 4\mu)\mathbf{j} + (\mu - 6)\mathbf{k}$

k Passing through the point $(5, 0, 5)$ and parallel to the plane $\mathbf{r} \cdot (3\mathbf{i} - 2\mathbf{j}) = 1$

l Passing through the origin and containing the line $\mathbf{r} = 3\mathbf{i} + \lambda(4\mathbf{j} + 7\mathbf{k})$

2 Convert these equations of planes into scalar product form.

a $r = 2i - k + \lambda(2i) + \mu(3i - 2j + 5k)$

b $r = (1 + 2s - 3t)i + (2 - 5s)j + (3 - 2s + t)k$

c $r = \begin{pmatrix} 1 \\ 4 \\ 0 \end{pmatrix} + \lambda\begin{pmatrix} 2 \\ -1 \\ 3 \end{pmatrix} + \mu\begin{pmatrix} 3 \\ 4 \\ 3 \end{pmatrix}$

d $r = (1 + 2m - n)i + (3 - m - 4n)j + (2 - m - 5n)k$

3 Convert these equations to Cartesian form.

a $r \cdot (i - 2j + 7k) = 9$

b $r \cdot \begin{pmatrix} 4 \\ -1 \\ 0 \end{pmatrix} = -6$

c $r = (i - j + 5k) + \lambda(3i - j + 4k) + \mu(i - 3j - 3k)$

d $r = (1 - 3s - 2t)i + (3 + 4s)j + (2 - 3s - t)k$

4 Find the scalar product form of the vector equation of the plane that is perpendicular to the line $r_1 = (i - 2j + 4k) + s(i - 3j + 7k)$ and contains the line $r_2 = (-3i + k) + s(2i - 3j - k)$.

5 Find the unit vectors perpendicular to these planes.

a $r \cdot (i + 2j - 5k) = 13$

b $4x - y + 5z = 8$

c $r = (i + 3j) + m(2i - j) + n(4i - j + 3k)$

d $5x = 2y - 4z + 15$

6 Show that the planes $r \cdot (i - 2j + 15k) = 19$ and $r \cdot (3i + 9j + k) = 9$ are perpendicular.

7 A plane passes through the points $(1, 1, -2)$ and $(3, 2, 0)$, and is parallel to the vectors $i + 2j + k$ and $2i - j - 2k$. Find the vector equation of the plane in scalar product form.

8 Two planes π_1 and π_2 have vector equations $r \cdot (3i + j - 3k) = 7$ and $r \cdot (3i + j - 3k) = 15$. Explain why π_1 and π_2 are parallel and hence find the distance between them.

9 **a** Show that the line L whose vector equation is
$r = 3i - 2j + 8k + \lambda(i - j + 4k)$ is parallel to the plane π_1 whose vector equation is $r \cdot (i + 5j + k) = 8$.
b Find the equation of plane π_2 that contains the line L and is parallel to π_1.
c Find the distance of π_1 and π_2 from the origin and hence determine the distance between the planes.

10 Find the equation of plane P_1 that contains the point with position vector $2i - 3j + 5k$ and is parallel to the plane P_2 with equation $r \cdot (2i + 5j - 6k) = 14$. Find the distances of P_1 and P_2 from the origin and hence determine the perpendicular distance between P_1 and P_2.

11 A plane passes through the point with position vector $\mathbf{i} + \mathbf{j}$ and is parallel to the lines $\mathbf{r}_1 = 2\mathbf{i} - 3\mathbf{j} + \lambda(\mathbf{i} + \mathbf{k})$ and $\mathbf{r}_2 = 5\mathbf{j} - \mathbf{k} + \mu(\mathbf{i} - 3\mathbf{j} + \mathbf{k})$. Find the vector equation of the plane in scalar product form. Is either of the given lines contained in the plane?

12 A plane passes through the three points whose position vectors are:

$\mathbf{a} = \mathbf{i} + 3\mathbf{j} + 5\mathbf{k}$

$\mathbf{b} = \mathbf{i} - \mathbf{j} + 5\mathbf{k}$

$\mathbf{c} = -4\mathbf{i} + 2\mathbf{j} - 3\mathbf{k}$

Find a vector equation of this plane in the form $\mathbf{r} \cdot \hat{\mathbf{n}} = d$ and hence write down the distance of the plane from the origin.

13 A plane passes through A, B and C, where $\overrightarrow{OA} = \begin{pmatrix} 4 \\ 7 \\ 0 \end{pmatrix}$, $\overrightarrow{OB} = \begin{pmatrix} -2 \\ 1 \\ 3 \end{pmatrix}$ and

$\overrightarrow{OC} = \begin{pmatrix} 4 \\ 1 \\ 9 \end{pmatrix}$. Find the vector equation of the plane in scalar product form.

Hence find the distance of the plane from the origin.

14 Find a vector equation of the line through the point $(4, 3, 7)$ that is perpendicular to the plane $\mathbf{r} \cdot (2\mathbf{i} + 2\mathbf{j} - 5\mathbf{k}) = 9$.

15 a Show that the line L with equation $x + 4 = y = \dfrac{z - 5}{2}$ is parallel to the plane P_1 with equation $\mathbf{r} \cdot (\mathbf{i} + \mathbf{j} - \mathbf{k}) = 8$.

  b Find a vector equation of the plane P_2 in scalar product form that contains the line L and is parallel to P_1.

  c Find the distances of P_1 and P_2 from the origin and hence find the distance between them.

16 Determine whether the given lines are parallel to, contained in, or intersect the plane $\mathbf{r} \cdot (2\mathbf{i} + \mathbf{j} - 3\mathbf{k}) = 5$.

  a $\mathbf{r} = 3\mathbf{i} - 2\mathbf{j} + 7\mathbf{k} + \lambda(-2\mathbf{i} + 3\mathbf{j} - 3\mathbf{k})$

  b $\mathbf{r} = (2\mathbf{i} + 3\mathbf{j}) + s(3\mathbf{i} + 2\mathbf{k})$

  c $\dfrac{x - 1}{2} = \dfrac{2y + 3}{4} = \dfrac{6z - 1}{-12}$

  d $x = 2 - 3t, y = -t, z = 3t - 4$

13.4 Intersecting lines and planes

The intersection of a line and a plane

In this situation there are three possible cases.

1. If the line is parallel to the plane there is no intersection.
2. If the line is contained in the plane, there are an infinite number of solutions, i.e. a line of solutions, which are given by the parametric equations of the line.
3. If the line intersects the plane, there is one solution. These are shown in the diagrams.

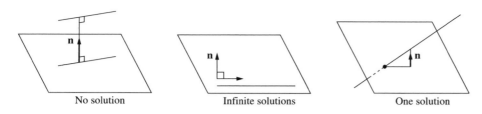

No solution   Infinite solutions   One solution

It is this third case we are interested in.

Method

1. Find an expression for any point on the line, which must fit the equation of the plane.
2. Find the value of the parameter, say λ.
3. Hence find the coordinates of the point of intersection.

Example

Find the point of intersection between the line $\mathbf{r} = (\mathbf{i} + 6\mathbf{j} - 5\mathbf{k}) + \lambda(\mathbf{i} - \mathbf{j} + \mathbf{k})$ and the plane $\mathbf{r} \cdot (3\mathbf{i} + 2\mathbf{j} + \mathbf{k}) = 5$.

Any point on the line is given by the parametric equations of the line:

$x = 1 + \lambda$
$y = 6 - \lambda$
$z = -5 + \lambda$

Substituting these into the equation of the plane:

$$[(1 + \lambda)\mathbf{i} + (6 - \lambda)\mathbf{j} + (-5 + \lambda)\mathbf{k}] \cdot [3\mathbf{i} + 2\mathbf{j} + \mathbf{k}] = 5$$
$$\Rightarrow 3(1 + \lambda) + 2(6 - \lambda) + 1(-5 + \lambda) = 5$$
$$\Rightarrow 3 + 3\lambda + 12 - 2\lambda - 5 + \lambda = 5$$
$$\Rightarrow 2\lambda = -5$$
$$\Rightarrow \lambda = -\frac{5}{2}$$

Hence at the point of intersection:

$$x = 1 - \frac{5}{2} = -\frac{3}{2}$$

$$y = 6 + \frac{5}{2} = \frac{17}{2}$$

$$z = -5 - \frac{5}{2} = -\frac{15}{2}$$

The point of intersection is $\left(-\frac{3}{2}, \frac{17}{2}, \frac{-15}{2}\right)$.

The intersection of two and three planes

This is directly related to the work done in Chapter 11, where the different cases were considered. Many of these questions are best solved using a method of inverse matrices or row reduction, but we will consider the situation of two planes intersecting in a line from a vector point of view.

Since the line of intersection of two planes is contained in both planes, it is perpendicular to both direction normals. This is shown in the diagram.

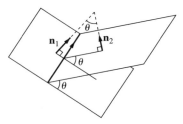

Method

> **1.** Find the vector product of the direction normals. This gives the direction vector of the line.
>
> **2.** Write the equations of the planes in Cartesian form.
>
> **3.** We now assume that $z = 0$ since the line has to intersect this plane.
>
> **4.** Solving simultaneously gives a point on the line.
>
> **5.** Write down a vector equation of the line.

Example

Find a vector equation of the line of intersection of the planes $\mathbf{r} \cdot (2\mathbf{i} + \mathbf{j} - 3\mathbf{k}) = 8$ and $\mathbf{r} \cdot (3\mathbf{i} - \mathbf{j} + 2\mathbf{k}) = 7$.

The direction vector of the line is given by

$$\begin{vmatrix} \mathbf{i} & \mathbf{j} & \mathbf{k} \\ 2 & 1 & -3 \\ 3 & -1 & 2 \end{vmatrix} = \mathbf{i}(2 - 3) - \mathbf{j}(4 + 9) + \mathbf{k}(-2 - 3) = -\mathbf{i} - 13\mathbf{j} - 5\mathbf{k}$$

The Cartesian equations of the plane are $2x + y - 3z = 8$ and $3x - y + 2z = 7$.

Assuming $z = 0$ gives $2x + y = 8$ and $3x - y = 7$.

$2x + y = 8$ equation (i)

$3x - y = 7$ equation (ii)

Equation (i) + equation (ii) $\Rightarrow x = 3$

Substituting in equation (i) $\Rightarrow y = 2$

Thus a point on the line is $(3, 2, 0)$.

Therefore an equation of the line of intersection is
$\mathbf{r} = 3\mathbf{i} + 2\mathbf{j} + \lambda(-\mathbf{i} - 13\mathbf{j} - 5\mathbf{k})$.

Finding the angle between two planes

The angle between two planes is the same as the angle between the direction normals of the planes.

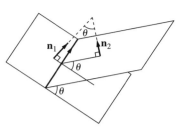

Method

> Apply the scalar product to the direction normals of the planes and find the angle.

Example

Find the angle θ between the planes $\mathbf{r} \cdot (\mathbf{i} + \mathbf{j} + \mathbf{k}) = 3$ and $\mathbf{r} \cdot (2\mathbf{i} + 2\mathbf{j} - \mathbf{k}) = 1$.

$$(\mathbf{i} + \mathbf{j} + \mathbf{k}) \cdot (2\mathbf{i} + 2\mathbf{j} - \mathbf{k}) = \sqrt{3}\sqrt{9} \cos \theta$$

$$\Rightarrow 2 + 2 - 1 = 3\sqrt{3} \cos \theta$$

$$\Rightarrow \cos \theta = \frac{1}{\sqrt{3}}$$

$$\Rightarrow \theta = 54.7°$$

Finding the angle between a line and a plane

The angle between a line and a plane is the complement of the angle between the line and the direction normal.

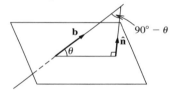

Method

> 1. Use the scalar product to find the angle between the direction normal of the plane and the direction vector of the line.
> 2. Subtract the angle in Step 1 from $90°$ to find the required angle.

Example

Find the angle between the line $\mathbf{r} = 3\mathbf{k} + \lambda(7\mathbf{i} - \mathbf{j} + 4\mathbf{k})$ and the plane $\mathbf{r} \cdot (2\mathbf{i} - 5\mathbf{j} - 2\mathbf{k}) = 8$.

$$(7\mathbf{i} - \mathbf{j} + 4\mathbf{k}) \cdot (2\mathbf{i} - 5\mathbf{j} - 2\mathbf{k}) = \sqrt{66}\sqrt{33} \cos \theta$$

$$\Rightarrow 14 + 5 - 8 = \sqrt{66}\sqrt{33} \cos \theta$$

$$\Rightarrow \cos \theta = \frac{11}{\sqrt{66}\sqrt{33}}$$

$$\Rightarrow \theta = 76.4°$$

Therefore the required angle is $90° - 76.4° = 13.6°$.

Exercise 4

1 Find the point of intersection between the line and plane.

a $x - 3 = 2y + 5 = 4 - 3z$ and $3x + 2y - 6z = 32$

b $\mathbf{r} = \begin{pmatrix} 3 \\ 1 \\ -2 \end{pmatrix} + s\begin{pmatrix} 2 \\ -1 \\ 0 \end{pmatrix}$ and $\mathbf{r} \cdot \begin{pmatrix} 2 \\ 2 \\ 1 \end{pmatrix} = 7$

c $\mathbf{r} = \mathbf{i} - 5\mathbf{k} + \mu(\mathbf{i} - \mathbf{j} - 2\mathbf{k})$ and $3x - 4y + z = 8$

d $x = 1 + 2m, y = m - 3, z = 1$ and $\mathbf{r} \cdot \begin{pmatrix} 2 \\ 3 \\ -1 \end{pmatrix} = 14$

e $\mathbf{r} = 4\mathbf{i} - \mathbf{j} + \lambda(6\mathbf{i} - \mathbf{j} + 3\mathbf{k})$ and

$\mathbf{r} = 2\mathbf{k} + m(\mathbf{i} - 4\mathbf{k}) + n(3\mathbf{i} - \mathbf{j} + 6\mathbf{k})$

f $x = 4 - 3n, y = n - 6, z = 1 + 2n$ and $x + y - 2z = 20$

2 Find the acute angle between the two planes.

a $\mathbf{r} \cdot (3\mathbf{i} - 4\mathbf{j} + 9\mathbf{k}) = 10$ and $\mathbf{r} \cdot (3\mathbf{i} - 6\mathbf{j} + 2\mathbf{k}) = 15$

b $2x + 3y - 4z = 15$ and $x + y - 4z = 17$

c $\mathbf{r} \cdot \begin{pmatrix} 1 \\ 0 \\ -3 \end{pmatrix} = 1$ and $2x - y - 5z = 6$

d $x - y - 9z = 6$ and $y = 3x - z + 5$

e $\mathbf{r} \cdot (\mathbf{i} - \mathbf{j} - 5\mathbf{k}) = 2$ and

$\mathbf{r} = (1 + m + n)\mathbf{i} + (2 - m + 3n)\mathbf{j} + (4 - 5m)\mathbf{k}$

f The plane perpendicular to the line $\mathbf{r} = 4\mathbf{i} - \mathbf{j} - 5\mathbf{k} + s(\mathbf{i} - 2\mathbf{j} + \mathbf{k})$ and passing though the point $(4, -1, -5)$, and the plane $\mathbf{r} \cdot (\mathbf{j} - 5\mathbf{k}) = 14$

3 Find the acute angle between the line and plane.

a $\mathbf{r} = \mathbf{i} - 2\mathbf{j} + \lambda(3\mathbf{i} + 5\mathbf{j} + 9\mathbf{k})$ and $\mathbf{r} \cdot (2\mathbf{i} - 3\mathbf{j} + 2\mathbf{k}) = 7$

b $\mathbf{r} = \begin{pmatrix} 1 \\ 2 \\ -3 \end{pmatrix} + t\begin{pmatrix} 3 \\ 1 \\ -4 \end{pmatrix}$ and $x + 3y - 4z = 5$

c $\dfrac{x - 2}{5} = \dfrac{2y + 7}{5} = \dfrac{3 - 2z}{4}$ and $2x - 4y - 9z = 0$

d $x = \lambda - 1, y = \dfrac{1}{2}\lambda + 3, z = 2 + 3\lambda$ and $\mathbf{r} \cdot (4\mathbf{i} - 5\mathbf{k}) = -6$

e $\dfrac{x + 4}{7} = 2y - 1, z = -2$ and $\mathbf{r} \cdot \begin{pmatrix} 4 \\ -5 \\ 1 \end{pmatrix} = 13$

f $\mathbf{r} = \mathbf{i} - 2\mathbf{j} - \mathbf{k} + m(3\mathbf{i} - 2\mathbf{j} + 3\mathbf{k})$ and

$\mathbf{r} = 2\mathbf{i} - \mathbf{j} + 4\mathbf{k} + \lambda(\mathbf{i} - 2\mathbf{j} + 6\mathbf{k}) + \mu(2\mathbf{i} - \mathbf{j} - \mathbf{k})$

4 Find the vector equation of the line of intersection of the following pairs of planes.

a $\mathbf{r} \cdot (3\mathbf{i} - 4\mathbf{j} + 2\mathbf{k}) = 8$ and $\mathbf{r} \cdot (\mathbf{i} + 3\mathbf{j} - 5\mathbf{k}) = 7$

b $x - y - 2z = 4$ and $3x + y - 5z = 4$

c $\mathbf{r} \cdot (\mathbf{i} - 3\mathbf{j} + 2\mathbf{k}) = -2$ and $2x - y + 4z = 11$

d $\mathbf{r} \cdot (\mathbf{i} - \mathbf{j} + \mathbf{k}) = 7$ and $2x - y + 4z = 8$

e $\mathbf{r} \cdot (3\mathbf{i} - \mathbf{j} + 5\mathbf{k}) = 10$ and
$\mathbf{r} = (1 - 2m + n)\mathbf{i} + (3 - 4m + n)\mathbf{j} + (3m + 2n)\mathbf{k}$

f $\mathbf{r} = (1 - 2\lambda + 3\mu)\mathbf{i} + (2\lambda - \mu)\mathbf{j} + (6 - 5\mu)\mathbf{k}$ and
$\mathbf{r} = (2 - 3s)\mathbf{i} + (1 - 4t)\mathbf{j} + (s - 2t)\mathbf{k}$

5 Prove that the line $\mathbf{r} = 3\mathbf{i} - 5\mathbf{j} + \lambda(4\mathbf{i} - 5\mathbf{j} - 3\mathbf{k})$ is parallel to the intersection of the planes $\mathbf{r} \cdot (3\mathbf{i} + 3\mathbf{j} - \mathbf{k}) = 2$ and $\mathbf{r} \cdot (\mathbf{i} + 2\mathbf{j} - 2\mathbf{k}) = 17$.

6 Find the acute angle between the plane defined by the points $(0, 1, -1)$, $(2, 1, 0)$ and $(3, -2, 0)$, and the plane defined by the points $(0, -1, 2)$, $(1, -1, 1)$ and $(-2, -3, 0)$.

7 a Find the equation of the straight line which passes through the point $P(1, 2, -3)$ and is perpendicular to the plane $3x + y - 3z = -5$.

b Calculate the coordinates of the point Q, which is the point of intersection of the line and the plane.

8 The vector equation of a plane is given by $\mathbf{r} \cdot (\mathbf{i} - 5\mathbf{j} + 6\mathbf{k}) = 6$.

a Given that the point $A(a, 3a, 2a)$ lies on the plane, find the value of a.

b If B has coordinates $(1, 4, -1)$, find the acute angle between the direction normal of the plane and $\overrightarrow{AB}$.

c Hence find the perpendicular distance of B from the direction normal passing through A.

Review exercise

1 The points A, B, C, D have position vectors given by

$$\mathbf{a} = 3\mathbf{i} + 2\mathbf{k}$$
$$\mathbf{b} = 2\mathbf{i} + \mathbf{j} + 5\mathbf{k}$$
$$\mathbf{c} = -\mathbf{i} - \mathbf{j} + 2\mathbf{k}$$
$$\mathbf{d} = 3\mathbf{i} + \mathbf{j} + 2\mathbf{k}$$

respectively. Find

a a unit vector perpendicular to the plane ABC

b a vector equation, in the form $\mathbf{r} \cdot \hat{\mathbf{n}} = d$, of the plane parallel to ABC and passing through D

c the acute angle between the line BD and the perpendicular to the plane ABC.

2 The points A, B, C, D have position vectors given by

$$\mathbf{a} = \mathbf{i} + \mathbf{j} + \mathbf{k}$$
$$\mathbf{b} = \frac{3}{2}\mathbf{i} + \frac{5}{2}\mathbf{j} + \frac{1}{2}\mathbf{k}$$
$$\mathbf{c} = 5\mathbf{i} + 3\mathbf{j} + \mathbf{k}$$
$$\mathbf{d} = 2\mathbf{i} - 2\mathbf{j} - 3\mathbf{k}$$

respectively. The point P lies on AB produced and is such that $\overrightarrow{AP} = 2\overrightarrow{AB}$, and the point Q is the midpoint of AC.

  a Show that $\overrightarrow{PQ}$ is perpendicular to $\overrightarrow{AQ}$.

  b Find the area of the triangle APQ.

  c Find a unit vector perpendicular to the plane ABC.

  d Find the equations of the lines AD and BD in Cartesian form.

  e Find the acute angle between the lines AD and BD.

 3 a The plane π_1 has equation $\mathbf{r} = \begin{pmatrix} 2 \\ 1 \\ 1 \end{pmatrix} + \lambda \begin{pmatrix} -2 \\ 1 \\ 8 \end{pmatrix} + \mu \begin{pmatrix} 1 \\ -3 \\ -9 \end{pmatrix}$.

  The plane π_2 has equation $\mathbf{r} = \begin{pmatrix} 2 \\ 0 \\ 1 \end{pmatrix} + s \begin{pmatrix} 1 \\ 2 \\ 1 \end{pmatrix} + t \begin{pmatrix} 1 \\ 1 \\ 1 \end{pmatrix}$.

  i For points that lie in π_1 and π_2, show that $\lambda = \mu$.

  ii Hence, or otherwise, find a vector equation of the line of intersection of π_1 and π_2.

  b The plane π_3 contains the line $\dfrac{2 - x}{3} = \dfrac{y}{-4} = z + 1$ and is perpendicular to $3\mathbf{i} - 2\mathbf{j} + \mathbf{k}$. Find the Cartesian equation of π_3.

  c Find the intersection of π_1, π_2 and π_3.   [IB May 05 P2 Q3]

4 The line L has equation

$$\mathbf{r} = \begin{pmatrix} 20 \\ 1 \\ 4 \end{pmatrix} + \lambda \begin{pmatrix} 2 \\ 1 \\ 0 \end{pmatrix} \quad \text{where } \lambda \in \mathbb{R}$$

  a Show that L lies in the plane whose equation is $\mathbf{r} \cdot \begin{pmatrix} -1 \\ 2 \\ 4 \end{pmatrix} = -2$.

  b Find the position vector of P, the foot of the perpendicular from the origin O to L.

  c Find an equation of the plane containing the origin and L.

  d Find the position vector of the point where L meets the plane π whose equation is $\mathbf{r} \cdot \begin{pmatrix} 3 \\ -4 \\ 1 \end{pmatrix} = 7$.

 5 The line L has equation $\mathbf{r} = \begin{pmatrix} 1 \\ -1 \\ 2 \end{pmatrix} + t \begin{pmatrix} 0 \\ 4 \\ -1 \end{pmatrix}$ and the plane π has equation

$$\mathbf{r} \cdot \begin{pmatrix} 0 \\ -1 \\ 1 \end{pmatrix} = 5.$$

  a Find the coordinates of the point of intersection, P, of L and π.

  b The point Q has coordinates (6, 1, 2), and R is the foot of the perpendicular from Q to π. Find the coordinates of R.

c Find a vector equation for the line PR.

d Find the angle PQR.

 6 The point $A(2, 5, -1)$ is on the line L, which is perpendicular to the plane with equation $x + y + z - 1 = 0$.

   a Find the Cartesian equation of the line L.

   b Find the point of intersection of the line L and the plane.

   c The point A is reflected in the plane. Find the coordinates of the image of the point A.

   d Calculate the distance from the point $B(2, 0, 6)$ to the line L.   [IB Nov 03 P2 Q1]

 7 Show that the planes $\mathbf{r}_1 \cdot (\mathbf{i} + 2\mathbf{j} - \mathbf{k}) = 4$, $\mathbf{r}_2 \cdot (\mathbf{i} + 3\mathbf{j} - 2\mathbf{k}) = 7$ and

$\mathbf{r}_3 \cdot (\mathbf{i} + 4\mathbf{j} - 3\mathbf{k}) = 10$ intersect in a line and find the vector equation of the line.

 8 Consider the points $A(1, 2, 1)$, $B(0, -1, 2)$, $C(1, 0, 2)$ and $D(2, -1, -6)$.

   a Find the vectors $\overrightarrow{AB}$ and $\overrightarrow{BC}$.

   b Calculate $\overrightarrow{AB} \times \overrightarrow{BC}$.

   c Hence, or otherwise, find the area of triangle ABC.

   d Find the equation of the plane P containing the points A, B and C.

   e Find a set of parametric equations for the line through the point D and perpendicular to the plane P.

   f Find the distance from the point D to the plane P.

   g Find a unit vector which is perpendicular to the plane P.

   h The point E is a reflection of D in the plane P. Find the coordinates of E.

           [IB Nov 99 P2 Q2]

 9 In a particular situation $\overrightarrow{OA} = 2\mathbf{i} - 3\mathbf{j} + 8\mathbf{k}$ and $\overrightarrow{OB} = \mathbf{j} + 9\mathbf{k}$. The plane π is given by the equation $x + 3y + z = 1$.

   a Determine whether or not A and B lie in the plane π.

   b Find the Cartesian equation of the line AB.

   c Find the angle between AB and the normal to the plane π at A.

   d Hence find the perpendicular distance from B to this normal.

   e Find the equation of the plane that contains AB and is perpendicular to π.

 10 The equations of two lines L_1 and L_2 are

L_1: $\mathbf{r} = \mathbf{i} - 20\mathbf{j} - 13\mathbf{k} + t(7\mathbf{i} + \mathbf{k})$, where t is a scalar;

L_2: $\dfrac{x + 30}{2} = \dfrac{y + 39}{1} = \dfrac{2 - 3z}{2}$.

The equations of two planes P_1 and P_2 are

P_1: $6x + 3y - 2z = 12$;

P_2: $\mathbf{r} = 22\mathbf{i} + \lambda(\mathbf{i} + \mathbf{j}) + \mu(\mathbf{i} - 2\mathbf{k})$.

   a Find the vector cross product $(\mathbf{i} + \mathbf{j}) \times (\mathbf{i} - 2\mathbf{k})$.

   b i Write down vectors $\mathbf{n}_1$, $\mathbf{n}_2$ that are normal to the planes P_1, P_2 respectively.

    ii Hence, or otherwise, find the acute angle between the planes correct to the nearest tenth of a degree.

  c Show that L_2 is normal to P_1.

  d **i** Find the coordinates of the point of intersection of L_2 and P_1.

    ii Hence, or otherwise, show that the two lines and the two planes all have a point in common. [IB Nov 98 P2 Q1]

 11 A plane π_1 has equation $\mathbf{r} \cdot (3\mathbf{i} + \mathbf{j}) = -13$.

  a Find, in vector form, an equation for the line passing through the point A with position vector $2\mathbf{i} + \mathbf{j} + 4\mathbf{k}$ and normal to the plane π_1.

  b Find the position vector of the foot B of the perpendicular from A to the plane π_1.

  c Find the sine of the angle between OB and the plane π_1.

  The plane π_2 has equation $\mathbf{r} \cdot (\mathbf{i} + \mathbf{j} + \mathbf{k}) = 5$.

  d Find the position vector of the point P where both the planes π_1, π_2 intersect with the plane perpendicular to the x-axis which passes through the origin.

  e Find the position vector of the point Q where both the planes π_1, π_2 intersect with the plane perpendicular to the y-axis which passes through the origin.

  f Find the vector equation of the line PQ.

 12 The equations of the planes P_1 and P_2 are given by

$$P_1: \mathbf{r} \cdot (3\mathbf{i} - \mathbf{j} + 2\mathbf{k}) = -1$$

$$P_2: \mathbf{r} \cdot (-2\mathbf{i} + \mathbf{j} - 5\mathbf{k}) = 4$$

where $\mathbf{r} = x\mathbf{i} + y\mathbf{j} + z\mathbf{k}$ is the position vector for a point on the plane.

  a Let L be the line of intersection of the two planes P_1 and P_2.

    i Show that L is parallel to $3\mathbf{i} + 11\mathbf{j} + \mathbf{k}$.

    ii Show that the point $A(0, -1, -1)$ lies on the line L. Hence, or otherwise, find the equation of L.

  The equation of a third plane P_3 is given by

$$P_3: \mathbf{r} \cdot (-4\mathbf{i} + \mathbf{j} + \mathbf{k}) = c.$$

  b Determine the value of c for which the three planes, P_1, P_2 and P_3, intersect, and deduce whether this value of c gives a point of intersection or a line of intersection.

  c For $c = 5$,

    i show that the plane P_3 is parallel to the line L

    ii find the distance between the line L and the plane P_3. [IB May 98 P2 Q3]

 13 Using row reduction show that the following planes intersect in a line, and find the vector equation of the line of intersection.

$$x + 2y + z = 3$$
$$2x - y + 3z = 2$$
$$3x - 4y + 5z = 1$$

 14 **a** The line L_1 is parallel to the vector $\mathbf{v} = 3\mathbf{i} + \mathbf{j} + 3\mathbf{k}$ and passes through the point $(2, 3, 7)$. Find a vector equation of the line.

b The equation of a plane, E, is given by $2x + 3y - 4z + 21 = 0$. Find the point of intersection of the line L_1 and the plane E.

c Find an equation of a plane which passes through the point $(1, 2, 3)$ and is parallel to the plane E.

d The parametric equations of another line L_2 are $x = t$, $y = t$ and $z = -t$, $-\infty < t < \infty$.

Show that

  i L_1 is not parallel to L_2

  ii L_1 does not intersect L_2.

e Let O be the origin and P be the point. $(-1, 2, 4)$

  i Find a vector $\mathbf{w}$ that is parallel to the line L_2.

  ii Find the vector $\overrightarrow{PO}$.

  iii Find the shortest distance d between the lines L_1 and L_2 by using the formula $d = \left| \dfrac{\overrightarrow{PO} \cdot (\mathbf{v} \times \mathbf{w})}{|\mathbf{v} \times \mathbf{w}|} \right|$.
               [IB May 97 P2 Q3]

15 a Show that the line L whose vector equation is

    $\mathbf{r} = 3\mathbf{i} - 4\mathbf{j} + 2\mathbf{k} + \lambda(2\mathbf{i} + \mathbf{j} - 3\mathbf{k})$

is parallel to the plane P whose vector equation is $\mathbf{r} \cdot (\mathbf{i} + 4\mathbf{j} + 2\mathbf{k}) = 4$.

b What is the distance from the origin to the plane?

c Find, in the same form as the equation given above for P, the equation of the plane P_1 which contains L and is parallel to the plane P.

d Deduce that the plane P_1 is on the opposite side of the origin to the plane P. Hence, or otherwise, find the distance between the line L and the plane P.

e Show that the plane P_1 contains the line whose vector equation is

    $\mathbf{r} = -\mathbf{i} - 2\mathbf{j} + \gamma(2\mathbf{i} - \mathbf{j} + \mathbf{k})$.
           [IB May 93 P2 Q3]

16 a Show that the lines given by the parametric equations

    $x = 3 + 4m, y = 3 - 2m, z = 7 - 2m$ and

    $x = 7 + 2n, y = 1 - n, z = 8 + n$

intersect and find the coordinates of P, the point of intersection.

b Find the Cartesian equation of the plane π that contains these two lines.

c Find the coordinates of Q, the point of intersection of the plane and the line

$$\mathbf{r} = \begin{pmatrix} 2 \\ -1 \\ 0 \end{pmatrix} + \lambda \begin{pmatrix} 3 \\ 1 \\ 1 \end{pmatrix}.$$

d Find the coordinates of the point R if $|\overrightarrow{PR}| = |\overrightarrow{QR}| = 4$ and the plane of the triangle PQR is normal to the plane π.

14 Integration 1

Jakob Bernoulli was a Swiss mathematician born in Basel, Switzerland, on 27 December 1654. Along with his brother, Johann, he is considered to be one of the most important researchers of calculus after Newton and Leibniz. Jakob studied theology at university, but during this time he was studying mathematics and astronomy on the side, much against the wishes of his parents. After graduating in theology he travelled

Jakob Bernoulli

around Europe and worked with a number of the great mathematicians of the time. On return to Basel, it would have been natural for him to take an appointment in the church, but he followed his first love of mathematics and theoretical physics and took a job at the university. He was appointed professor of mathematics in 1687 and, along with his brother, Johann, started studying Leibniz's work on calculus. At this time Leibniz's theories were very new, and hence the work done by the two brothers was at the cutting edge. Jakob worked on a variety of mathematical ideas, but in 1690 he first used the term "integral" with the meaning it has today. Jakob held the chair of mathematics at the university in Basel until his death in 1705. Jakob had always been fascinated by the properties of the logarithmic spiral, and this was engraved on his tombstone along with the words "Eadem Mutata Resurgo" which translates as "I shall arise the same though changed."

14.1 Undoing differentiation

In Chapters 8–10 we studied differential calculus and saw that by using the techniques of differentiation the gradient of a function or the rate of change of a quantity can be found. If the rate of change is known and the original function needs to be found, it is necessary to "undo" differentiation. Integration is this "undoing", the reverse process to differentiation. Integration is also known as anti-differentiation, and this is often the best way of looking at it.

If $\frac{dy}{dx} = 2x$, what is the original function y?

This is asking what we started with in order to finish with a derived function of $2x$. Remembering that we differentiate by multiplying by the power and then subtracting one from the power, we must have started with x^2.

Similarly, if $\frac{dy}{dx} = 4$, then this must have started as $4x$.

Exercise 1

Find the original function.

1 $\dfrac{dy}{dx} = 5$ **2** $\dfrac{dy}{dx} = 10$ **3** $\dfrac{dy}{dx} = -2$ **4** $\dfrac{dy}{dx} = 4x$

5 $\dfrac{dy}{dx} = 12x$ **6** $\dfrac{dy}{dx} = 3x^2$ **7** $\dfrac{dy}{dx} = 4x^3$ **8** $\dfrac{dy}{dx} = 5x^4$

9 $\dfrac{dy}{dx} = 9x^2$ **10** $\dfrac{dy}{dx} = -4x^{-2}$

Looking at the answers to Exercise 1, we can form a rule for anti-differentiation.

If we describe the process of "undoing" differentiation for this type of function, we could say "add 1 to the power and divide by that new power".

In mathematical notation this is

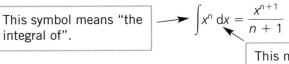

This symbol means "the integral of".

$$\int x^n \, dx = \frac{x^{n+1}}{n+1}$$

This means "with respect to x" – the variable we are concerned with.

14.2 Constant of integration

Again consider the situation of $\dfrac{dy}{dx} = 4$. Geometrically, this means that the gradient of the original function, y, is constant and equal to 4.

So $y = 4x + c$ (c is the y-intercept of the line from the general equation of a line $y = mx + c$).

Consider the lines $y = 4x - 3$, $y = 4x$ and $y = 4x + 5$.

For each line $\dfrac{dy}{dx} = 4$ (as the gradient is 4 each time). Remember that, when differentiating, a constant "disappears" because the gradient of a horizontal line is zero or alternatively the derivative of a constant is zero.

So $\int 4 \, dx = 4x + c$.

Unless more information is given (a point on the line), then the value of c remains unknown. So any of the lines below could be the original function with $\dfrac{dy}{dx} = 4$. In fact there are an infinite number of lines that could have been the original function.

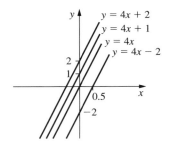

This "c" is called the **constant of integration**. It must be included in the answer of any integral, as we do not know what constant may have "disappeared" when the function was differentiated.

Example

Find y if $\dfrac{dy}{dx} = 8x$.

$y = \displaystyle\int 8x\,dx = 4x^2 + c$

Example

Find $\displaystyle\int 10x - 7\,dx$.

So $\displaystyle\int 10x - 7\,dx = \int 10x\,dx + \int -7\,dx$

So $\displaystyle\int 10x - 7\,dx = 5x^2 - 7x + c$

> We established in Chapter 8 that a function could be differentiated "term by term". In a similar way, this can be integrated term by term.

Example

Integrate $8x^{-2} + x^{\frac{3}{4}}$.

$\displaystyle\int 8x^{-2} + x^{\frac{3}{4}}\,dx = -8x^{-1} + \frac{4}{7}x^{\frac{7}{4}} + c$

The coefficient of a term has no effect on the process of integration, and so a constant can be "taken out" of the integral. This is demonstrated in the next example.

Example

Find the solution of $\dfrac{dy}{dx} = \dfrac{2}{9}x^{-\frac{1}{3}}$.

So

$y = \displaystyle\int \frac{2}{9}x^{-\frac{1}{3}}\,dx$

$y = \dfrac{2}{9}\displaystyle\int x^{-\frac{1}{3}}\,dx$

$y = \dfrac{2}{9}\cdot\dfrac{3}{2}x^{\frac{2}{3}} + c$

$y = \dfrac{1}{3}x^{\frac{2}{3}} + c$

> This is asking to find y by integrating.

As with the expressions differentiated in Chapters 8–10, it is sometimes necessary to simplify the function prior to integrating.

Example

Find $\int \dfrac{x^2 - x^5}{3\sqrt{x}} dx$.

$\int \dfrac{x^2 - x^5}{3\sqrt{x}} dx$

$= \dfrac{1}{3} \int x^{-\frac{1}{2}}(x^2 - x^5)\, dx$

$= \dfrac{1}{3} \int x^{\frac{3}{2}} - x^{\frac{9}{2}}\, dx$

$= \dfrac{1}{3}\left[\dfrac{2}{5}x^{\frac{5}{2}} - \dfrac{2}{11}x^{\frac{11}{2}} \right] + c$

$= \dfrac{2}{15}x^{\frac{5}{2}} - \dfrac{2}{33}x^{\frac{11}{2}} + c$

> The integral sign and the dx remain until the integration is performed.

> The c can remain outside all brackets as it is an arbitrary constant and so, when multiplied by another constant, it is still a constant.

Example

Find $\int \dfrac{4}{p^2} dp$.

$\int \dfrac{4}{p^2} dp$

$= \int 4p^{-2}\, dp$

$= -4p^{-1} + c$

$= -\dfrac{4}{p} + c$

> Here p is the variable and so the integral is with respect to p.

Exercise 2

Integrate these expressions.

1 $2x - 1$ **2** x^2 **3** x^3 **4** x^4

5 $6x^2 - 5$ **6** $8x^3 + 4x - 3$ **7** $5x^2 - 4$ **8** x^{-2}

9 $x^{\frac{1}{2}}$ **10** $x^{-\frac{2}{3}}$ **11** $7 - 4x^{-3}$

Find these integrals.

12 $\int x^{-\frac{1}{2}}\, dx$ **13** $\int 2x^6 - 5x^4\, dx$ **14** $\int 5x^{\frac{3}{2}} - 4x^{-3}\, dx$

15 $\int 4x^3 + 4x - 9\, dx$ **16** $\int 1 - 2x + 6x^2 - x^3\, dx$ **17** $\int \dfrac{6}{x^3} dx$

Find the solution of these.

18 $\dfrac{dy}{dx} = \dfrac{2}{x^5}$ **19** $\dfrac{dy}{dx} = \sqrt{x} - \dfrac{1}{\sqrt{x}}$ **20** $\dfrac{dy}{dx} = 8 - \dfrac{3}{x^{\frac{1}{3}}}$

21 $\dfrac{dy}{dx} = 8x(2x^2 - 3)$ **22** $\dfrac{dy}{dx} = (x - 9)(2x - 3)$ **23** $\dfrac{dy}{dx} = (3x - 4)^2$

24 $\dfrac{dy}{dx} = \dfrac{x^2 - 5}{x^5}$ **25** $\dfrac{dy}{dx} = \dfrac{4x^3 - 7x}{\sqrt{x}}$ **26** $\dfrac{dy}{dx} = \dfrac{7x^4 - 6x^{\frac{3}{4}}}{2x^{\frac{1}{4}}}$

Find y by integrating with respect to the relevant variable.

27 $\dfrac{dy}{dp} = \dfrac{12}{p^3}$ **28** $\dfrac{dy}{dk} = 8k^{\frac{5}{4}}$ **29** $\dfrac{dy}{dz} = z^3\left(z^2 - \dfrac{1}{z^2}\right)$

30 $\dfrac{dy}{dt} = \dfrac{(t + 3)^2}{t^4}$ **31** $\dfrac{dy}{dt} = \dfrac{\sqrt{t} - 4t^3}{3t}$

14.3 Initial conditions

In all of the integrals met so far, it was necessary to include the constant of integration, c. However, if more information is given (often known as initial conditions), then the value of c can be found.

Consider again the example of $\dfrac{dy}{dx} = 4$. If the line passes through the point (1, 3) then c can be evaluated and the specific line found.

So $\dfrac{dy}{dx} = 4$

$\Rightarrow y = \displaystyle\int 4 \, dx$

$\Rightarrow y = 4x + c$

Since we know that when $x = 1, y = 3$ these values can be substituted into the equation of the line.

So $3 = 4 \times 1 + c$

$\Rightarrow c = -1$

The equation of the line is $y = 4x - 1$.

When the value of c is unknown, this is called the **general solution**.

If the value of c can be found, this is known as the **particular solution**.

Example

Given that the curve passes through (1, 3) and $\dfrac{dy}{dx} = 2x - 1$, find the equation of the curve.

$\dfrac{dy}{dx} = 2x - 1$

$\Rightarrow y = \displaystyle\int 2x - 1 \, dx$

$\Rightarrow y = x^2 - x + c$

Using (1, 3) $\Rightarrow 3 = 1^2 - 1 + c$

$\Rightarrow c = 3$

So the equation of the curve is $y = x^2 - x + 3$.

Example

Find P given that $\dfrac{dP}{dt} = \dfrac{1}{\sqrt{t}}$ and $P = 7$ when $t = 100$

$$\frac{dP}{dt} = \frac{1}{\sqrt{t}}$$

$$\Rightarrow P = \int \frac{1}{\sqrt{t}} dt$$

$$\Rightarrow P = \int t^{-\frac{1}{2}} dt$$

$$\Rightarrow P = 2t^{\frac{1}{2}} + c$$

Since $P = 7$ when $t = 100$, $7 = 2 \cdot \sqrt{100} + c$

$$\Rightarrow c = -13$$

Hence $P = 2t^{\frac{1}{2}} - 13$

Exercise 3

Given the gradient of each curve, and a point on that curve, find the equation of the curve.

1 $\dfrac{dy}{dx} = 6$, $(2, 8)$

2 $\dfrac{dy}{dx} = 4x$, $(1, 5)$

3 $\dfrac{dy}{dx} = 8x - 3$, $(-2, 4)$

4 $\dfrac{dy}{dx} = -2x + 5$, $(4, 4)$

5 $\dfrac{dy}{dx} = 4x^3 - 6x^2 + 7$, $(1, 9)$

6 $\dfrac{dy}{dx} = 4x^2 + \dfrac{6}{x^2}$, $(4, -1)$

7 $\dfrac{dy}{dx} = \dfrac{8}{\sqrt{x}}$, $(9, 2)$

Find the particular solution, using the information given.

8 $\dfrac{dy}{dt} = t^2(t^4 - 3t^2 - 4)$, $y = 6$ when $t = 1$

9 $\dfrac{dQ}{dp} = \dfrac{p^3 - 4p^5}{3\sqrt{p}}$, $Q = 2$ when $p = 0$

14.4 Basic results

Considering the basic results from differentiation, standard results for integration can now be produced. For polynomials, the general rule is:

$$\int x^n \, dx = \frac{x^{n+1}}{n+1} + c$$

However, consider $\int \frac{1}{x} dx = \int x^{-1} \, dx$.

Using the above rule, we would obtain $\frac{x^0}{0}$, but this is not defined. However, it is known that $\frac{d}{dx}(\ln x) = \frac{1}{x}$.

This provides the result that $\int \frac{1}{x} dx = \ln|x| + c$

Remembering that $\ln x$ is defined only for positive values of x, we recognize that $\int \frac{1}{x} dx = \ln|x| + c$, taking the absolute value of x. As $\frac{1}{x}$ is defined for all $x \in \mathbb{R}, x \neq 0$ and $\ln x$ is defined only for $x > 0$, the absolute value sign is needed so that we can integrate $\frac{1}{x}$ for all values of x for which it is defined.

Similarly $\frac{d}{dx}(e^x) = e^x$ so $\int e^x \, dx = e^x + c$

When differentiating sine and cosine functions the following diagram was used and is now extended:

Differentiate — S / C / −S / −C — Integrate

The integrals of other trigonometric functions can be found by reversing the basic rules for differentiation, and will be discussed further in Chapter 15.

Standard results

Function	Integral + c
$f(x)$	$\int f(x) \, dx$
$x^n (n \neq -1)$	$\frac{x^{n+1}}{n+1}$
$\frac{1}{x}$	$\ln x$
e^x	e^x
$\sin x$	$-\cos x$
$\cos x$	$\sin x$

> **Example**
>
> Integrate $\sin x - e^x$.
>
> $$\int \sin x - e^x \, dx$$
>
> $$= -\cos x - e^x + c$$

> **Example**
>
> Integrate $\dfrac{3}{x} + 4 \cos x$.
>
> $$\int \dfrac{3}{x} + 4 \cos x \, dx$$
>
> $$= 3 \ln|x| + 4 \sin x + c$$

Exercise 4

Integrate these functions.

1 $x^3 - \dfrac{2}{x}$

2 $4e^x + \sin x$

3 $\dfrac{5}{x} - \cos x$

4 $6 \sin x - 6x^4$

5 $-8 \sin x + 7e^x$

6 $5e^x - 2 \sin x + \dfrac{3}{x}$

7 $\dfrac{e^x}{3} - \dfrac{5}{2x} + 7 \sin x$

8 $\dfrac{e^x}{15} - 15\sqrt{x} + \cos x$

14.5 Anti-chain rule

When functions of the type $(2x - 1)^5$, e^{8x} and $\sin\left(2x - \dfrac{\pi}{3}\right)$ are differentiated, the

chain rule is applied. The chain rule states that we multiply the derivative of the outside function by the derivative of the inside function. So to integrate functions of these types we consider what we started with to obtain that derivative.

> **Example**
>
> $\dfrac{dy}{dx} = (2x - 1)^3$. Find y.
>
> So $y = \displaystyle\int (2x - 1)^3 \, dx$
>
> When integrating, 1 is added to the power, so y must be connected to $(2x - 1)^4$. Since we multiply by the power and by the derivative of the inside function when differentiating, we need to balance this when finding y.
>
> So $y = \dfrac{1}{4} \cdot \dfrac{1}{2}(2x - 1)^4 + c$
>
> $\Rightarrow y = \dfrac{1}{8}(2x - 1)^4 + c$

Example

Find $\int \cos 3x \, dx$.

Using   Diff $\Big\downarrow \begin{array}{c} S \\ C \\ -S \\ -C \end{array} \Big\uparrow$ Int   this begins with sin $3x$.

Balancing to obtain cos $3x$ when differentiating

$\Rightarrow \int \cos 3x \, dx = \dfrac{1}{3}\sin 3x + c$

Example

Find $\int 6e^{4x} \, dx$.

This started with e^{4x} as $\dfrac{d}{dx}(e^{4x}) = 4e^{4x}$.

So $\int 6e^{4x} \, dx = 6\int e^{4x} \, dx$

$= 6 \cdot \dfrac{1}{4}e^{4x} + c$

$= \dfrac{3}{2}e^{4x} + c$

Example

Find y given that $\dfrac{dy}{dx} = \dfrac{1}{3x-4} - \sin\left(4x - \dfrac{\pi}{2}\right)$.

So $y = \int \dfrac{1}{3x-4} - \sin\left(4x - \dfrac{\pi}{2}\right) dx$

As $\dfrac{1}{3x-4} = (3x-4)^{-1}$

we recognize that this comes from $\ln|3x-4|$ as $\dfrac{d}{dx}(\ln|3x-4|) = \dfrac{3}{3x-4}$.

So $y = \dfrac{1}{3}\ln|3x-4| + \dfrac{1}{4}\cos\left(4x - \dfrac{\pi}{2}\right) + c$

For these simple cases of the "anti-chain rule", we divide by the derivative of the inside function each time. With more complicated integrals, which will be met in the next chapter, this is not always the case, and at that point the results will be formalized. This is why it is useful to consider these integrals as the reverse of differentiation.

Find these integrals.

1 $\displaystyle\int \sin 5x \, dx$ **2** $\displaystyle\int \cos 6x \, dx$ **3** $\displaystyle\int \sin 2x \, dx$ **4** $\displaystyle\int \sin\frac{1}{2}x \, dx$

5 $\displaystyle\int 8 \cos 4x \, dx$ **6** $\displaystyle\int -6 \sin 3x \, dx$ **7** $\displaystyle\int -5 \cos 2x \, dx$ **8** $\displaystyle\int e^{6x} \, dx$

9 $\displaystyle\int e^{5x} \, dx$ **10** $\displaystyle\int 4e^{4x} \, dx$ **11** $\displaystyle\int 8e^{6t} \, dt$ **12** $\displaystyle\int -5e^{6p} \, dp$

13 $\displaystyle\int 8x - e^{2x} \, dx$ **14** $\displaystyle\int 4e^{-2x} \, dx$

Find y if:

15 $\dfrac{dy}{dx} = \dfrac{1}{2x - 3}$ **16** $\dfrac{dy}{dx} = \dfrac{1}{8x + 7}$ **17** $\dfrac{dy}{dx} = \dfrac{4}{2x - 5}$

18 $\dfrac{dy}{dx} = (3x - 1)^5$ **19** $\dfrac{dy}{dx} = (4x - 7)^6$ **20** $\dfrac{dy}{dx} = (4x + 3)^{-3}$

21 $\dfrac{dy}{dx} = (3 - 2x)^4$ **22** $\dfrac{dy}{dx} = \dfrac{4}{(3x - 2)^3}$ **23** $\dfrac{dy}{dt} = \dfrac{3}{(2t - 1)^2}$

24 $\dfrac{dy}{dx} = \dfrac{4}{3x - 1}$ **25** $\dfrac{dy}{dx} = \dfrac{6}{3x - 5}$ **26** $\dfrac{dy}{dp} = \dfrac{8}{4 - p}$

27 $\dfrac{dy}{dt} = \dfrac{3}{6 - t}$

Integrate these functions.

28 $6e^{4x}$ **29** $\sin 3x - 4x$ **30** $4e^{-8x} - 4 \cos 2x$

31 $\dfrac{1}{2x - 1} + (3x + 4)^5$ **32** $6x^2 - \dfrac{2}{3x + 2}$

14.6 Definite integration

Definite integration is where the integration is performed between limits, and this produces a numerical answer.

A definite integral is of the form $\displaystyle\int_a^b f(x) \, dx$

Upper limit $(x = b)$

Lower limit $(x = a)$

When a definite integral is created, the lower limit is always smaller than the upper limit.

$\displaystyle\int_2^3 2x - 1 \, dx$

$= \left[x^2 - x \right]_2^3$

This notation means that the integration has taken place. The two values are now substituted into the function and subtracted.

$$= (3^2 - 3) - (2^2 - 2)$$
$$= 6 - 2$$
$$= 4$$

There is no constant of integration used here. This is because it cancels itself out and so does not need to be included.

$$\left[x^2 - x + c \right]_2^3$$
$$= (3^2 - 3 + c) - (2^2 - 2 + c)$$
$$= 6 + c - 2 - c$$
$$= 4$$

Example

$$\int_0^4 (2x - 3)^2 \, dx$$

$$= \left[\frac{1}{6}(2x - 3)^3 \right]_0^4$$

$$= \frac{1}{6}(8 - 3)^3 - \left(\frac{1}{6}(0 - 3)^3 \right)$$

$$= \frac{125}{6} + \frac{27}{6}$$

$$= \frac{152}{6}$$

$$= \frac{76}{3}$$

Example

$$\int_0^{\frac{\pi}{4}} \sin 2x + 1 \, dx$$

$$= \left[-\frac{1}{2}\cos 2x + x \right]_0^{\frac{\pi}{4}}$$

$$= \left(-\frac{1}{2}\cos\frac{\pi}{2} + \frac{\pi}{4} \right) - \left(-\frac{1}{2}\cos 0 + 0 \right)$$

$$= \frac{\pi}{4} + \frac{1}{2}$$

$$= \frac{\pi + 2}{4}$$

To differentiate trigonometric functions, we always use radians, and the same is true in integration.

Example

$$\int_{-2}^{2} \frac{2}{x-4}\,dx$$

$$= \left[2\ln|x-4| \right]_{-2}^{2}$$

$$= 2\ln|-2| - 2\ln|-6|$$

$$= 2\ln 2 - 2\ln 6$$

$$= \ln 4 - \ln 36$$

$$= \ln\frac{4}{36}$$

$$= \ln\frac{1}{9}$$

$$= -2.20$$

An answer could have been approximated earlier, but if an exact answer were required, this form would need to be given.

Example

$$\int_{2}^{a} e^{4x}\,dx \quad \text{where } a > 2$$

$$= \left[\frac{1}{4}e^{4x} \right]_{2}^{a}$$

$$= \frac{1}{4}e^{4a} - \frac{1}{4}e^{8}$$

$$= \frac{1}{4}(e^{4a} - e^{8})$$

Although there is no value for the upper limit, an answer can still be found that is an expression in a.

Exercise 6

Find the value of these definite integrals

1 $\int_{1}^{2} 2x\,dx$

2 $\int_{2}^{3} 6x^2\,dx$

3 $\int_{0}^{4} 5\,dx$

4 $\int_{0}^{2} 8x - 4x^3\,dx$

5 $\int_{2}^{3} \frac{4}{x^2}\,dx$

6 $\int_{-3}^{-1} \frac{6}{x^3}\,dx$

7 $\int_{0}^{3} e^{2x}\,dx$

8 $\int_{2}^{4} 4e^{3x}\,dx$

9 $\int_{-2}^{1} 2e^{-4x}\,dx$

10 $\int_{0}^{\frac{\pi}{3}} \sin 3x\,dx$

11 $\int_{0}^{\frac{\pi}{2}} \cos 2\theta\,d\theta$

12 $\int_{0}^{\frac{\pi}{6}} \cos 3t - 6\,dt$

13 $\displaystyle\int_{0}^{3} (2x - 1)^2 \, dx$

14 $\displaystyle\int_{-2}^{0} (3x - 1)^3 \, dx$

15 $\displaystyle\int_{-1}^{2} (3 - 2x)^4 \, dx$

16 $\displaystyle\int_{1}^{4} \frac{2}{(2x - 1)^3} \, dx$

17 $\displaystyle\int_{-2}^{-1} \frac{3}{(3x - 4)^2} \, dx$

18 $\displaystyle\int_{2}^{5} \frac{1}{2x - 1} \, dx$

19 $\displaystyle\int_{2}^{4} \frac{4}{3x - 4} \, dx$

20 $\displaystyle\int_{-4}^{-1} \frac{5}{t - 2} \, dt$

21 $\displaystyle\int_{1}^{2} 6p - \frac{3}{4p + 1} \, dp$

22 $\displaystyle\int_{-p}^{p} \sin 4x - 6x \, dx$

23 $\displaystyle\int_{0}^{k} \frac{4}{2x + 1} \, dx$

14.7 Geometric significance of integration

When we met differentiation, it was considered as a technique for finding the gradient of a function at any point. We now consider the geometric significance of integration.

Consider $\displaystyle\int_{1}^{5} 3 \, dx$.

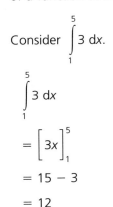

$$\int_{1}^{5} 3 \, dx$$

$$= \left[3x \right]_{1}^{5}$$

$$= 15 - 3$$

$$= 12$$

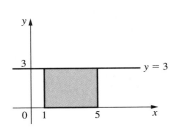

> Remember the limits are values of *x*.

This is the same as the area enclosed by the function, the *x*-axis and the vertical lines $x = 1$ and $x = 5$.

This suggests that the geometric interpretation of integration is **the area between the curve and the *x*-axis.**

Consider $y = 3x^2$.

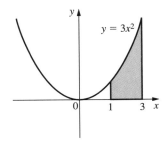

In the previous example where $y = 3$, it was easy to find the area enclosed by the function, the *x*-axis and the limits. However, to find the area under a curve is less obvious. This area could be approximated by splitting it into rectangles.

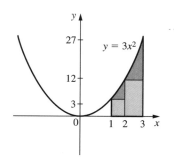

$$A \simeq 1 \times 3 + 1 \times 12$$

$$= 15 \text{ square units}$$

This is clearly not a very accurate approximation, so to make it more accurate thinner rectangles are used.

Four rectangles would make this more accurate.

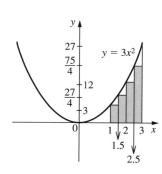

$$A \simeq \frac{1}{2} \times 3 + \frac{1}{2} \times \frac{27}{4} + \frac{1}{2} \times 12 + \frac{1}{2} \times \frac{75}{4}$$

$$A \simeq \frac{81}{4}$$

As the rectangles become thinner, the approximation becomes more accurate.

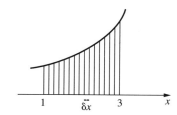

This can be considered in a more formal way. Each strip has width δx with height y and area δA.

So $\delta A \simeq y \cdot \delta x$

$$\Rightarrow A = \sum \delta A$$

$$\Rightarrow A \simeq \sum y \, \delta x$$

As δx gets smaller, the approximation improves

and so $A = \displaystyle\lim_{\delta x \to 0} \sum_{x=a}^{x=b} y \delta x.$

Now $y \simeq \dfrac{\delta A}{\delta x}$

$$\Rightarrow y = \lim_{\delta x \to 0} \frac{\delta A}{\delta x}$$

$$\Rightarrow y = \frac{dA}{dx}$$

Integrating gives $\displaystyle\int y \, dx = \int \frac{dA}{dx} dx = \int dA = A + c$

Limits are needed here to specify the boundaries of the area.

With the boundary conditions,

$$A = \int_a^b y \, dx$$

c can now be ignored.

This is the basic formula for finding the area between the curve and the x-axis.

Hence $\lim_{\delta x \to 0} \sum_{x=a}^{x=b} y \cdot \delta x = \int_a^b y \, dx$.

This now shows more formally that the geometric significance of integration is that it finds the area between the curve and the x-axis.

If the two notations for summation are compared, we find that sigma notation is used for a discrete variable and that integral notation is used for a continuous variable.

So the $\int$ sign actually means "sum of" (it is an elongated S).

$$\sum_1^3 3x^2 \quad \text{and} \quad \int_1^3 3x^2 \, dx$$

This is the sum of a continuous variable.

$$= \left[x^3 \right]_1^3$$

This is a sum of a discrete variable.

$$= 27 - 1$$

$$= 26$$

The area required is 26 square units.

Example

Find the area given by $\int_2^4 2x - 1 \, dx$.

$$\int_2^4 2x - 1 \, dx$$

$$= \left[x^2 - x \right]_2^4$$

$$= (16 - 4) - (4 - 2)$$

$$= 12 - 2$$

$$= 10 \text{ square units}$$

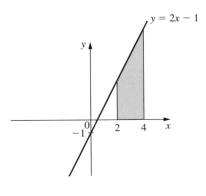

This integration can be performed on a calculator. Although a calculator cannot perform calculus algebraically, it can calculate definite integrals. This is shown in the diagram below, and we find the area is still 10 square units.

$\int f(x)dx=10$

Some definite integration questions require the use of a calculator. Where an exact value is required, one of the limits is a variable or the question is in a non-calculator paper, algebraic methods must be employed.

Example

Find the area enclosed by $y = \cos 2x$ and the x axis, between $x = \dfrac{\pi}{4}$ and $x = \dfrac{\pi}{2}$.

$$A = \int_{\frac{\pi}{4}}^{\frac{\pi}{2}} \cos 2x \ dx$$

$$= \left[\frac{1}{2}\sin 2x \right]_{\frac{\pi}{4}}^{\frac{\pi}{2}}$$

$$= \frac{1}{2}\sin \pi - \frac{1}{2}\sin \frac{\pi}{2}$$

$$= 0 - \frac{1}{2}$$

$$= -\frac{1}{2}$$

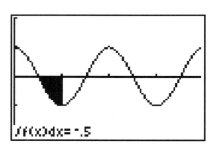

$\int f(x)dx= ^-.5$

The answer to this definite integral is negative. However, area is a scalar quantity (it has no direction, only magnitude) and so the area required is $\dfrac{1}{2}$ square unit.

As the negative sign has no effect on the area, the area can be considered to be the absolute value of the integral. So $A = \left| \displaystyle\int_{a}^{b} f(x) \ dx \right|$. Whether the calculation is done using the absolute value sign or whether we do the calculation and then ignore the negative sign at the end does not matter. As can be seen from the graph, the significance of the negative sign is that the area is contained below the x-axis.

Example

Find the area given by $\displaystyle\int_{-2}^{-1} \frac{1}{x}\,dx$.

$$A = \left| \int_{-2}^{-1} \frac{1}{x}\,dx \right|$$

$$= \left| \Big[\ln|x| \Big]_{-2}^{-1} \right|$$

$$= \big| \ln|-1| - \ln|-2| \big|$$

$$= |\ln 1 - \ln 2|$$

$$= |-\ln 2|$$

$$= 0.693$$

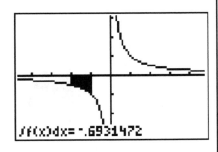

So the required area is 0.693 square units.

It should be noted that it is actually not possible to find the area given by $\displaystyle\int_{-2}^{1} \frac{1}{x}\,dx$ as

there is a vertical asymptote at $x = 0$. It is not possible to find the definite integral over an asymptote of any curve, as technically the area would be infinite.

This example also provides another explanation for the need for the modulus signs in

$\displaystyle\int \frac{1}{x}\,dx = \ln|x|$. Although logarithms are not defined for negative values of x, in order to

find the area under a hyperbola like $y = \dfrac{1}{x}$, which clearly exists, negative values need to

be substituted into a logarithm, and hence the absolute value is required. This was shown in the above example.

Example

Find the area enclosed by $y = -(2x - 1)^2 + 4$ and the x-axis.

First the limits need to be found.

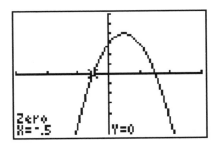

These are the roots of the graph (they can be found algebraically or by using a calculator)

$$-(2x - 1)^2 + 4 = 0$$

$$\Rightarrow 2x - 1 = \pm 2$$

$$\Rightarrow x = -\frac{1}{2} \text{ or } x = \frac{3}{2}$$

So the area is given by

$$\int_{-\frac{1}{2}}^{\frac{3}{2}} - (2x - 1)^2 + 4 \, dx$$

$$= \left[-\frac{1}{6}(2x - 1)^3 + 4x \right]_{-\frac{1}{2}}^{\frac{3}{2}}$$

$$= \frac{16}{3}$$

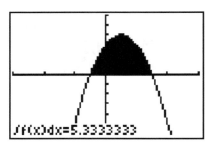

∫f(x)dx=5.3333333

Exercise 7

Find the area given by these definite integrals.

1 $\displaystyle\int_2^6 8 \, dx$

2 $\displaystyle\int_0^3 2x + 3 \, dx$

3 $\displaystyle\int_1^3 6x^2 - 1 \, dx$

4 $\displaystyle\int_0^2 (3x - 2)^4 \, dx$

5 $\displaystyle\int_0^\pi \sin x \, dx$

6 $\displaystyle\int_0^2 e^x \, dx$

7 $\displaystyle\int_1^4 \frac{4}{x} \, dx$

8 $\displaystyle\int_2^5 \frac{2}{2x + 3} \, dx$

9 $\displaystyle\int_{-4}^{-2} 3x - 2 \, dx$

10 $\displaystyle\int_{-4}^{-2} \frac{4}{2x + 1} \, dx$

11 $\displaystyle\int_0^1 e^{2x-1} - 4x \, dx$

12 $\displaystyle\int_0^{\frac{\pi}{3}} \cos 3x + 4x \, dx$

Find the area enclosed by the curve and the *x*-axis.

13 $y = x^2 + 2$ and the lines $x = 1$ and $x = 4$

14 $y = e^{6x}$ and the lines $x = -2$ and $x = 1$

15 $y = \sin x$ and the lines $x = 0$ and $x = \pi$

16 $y = \dfrac{4}{3x + 4}$ and the lines $x = -5$ and $x = -2$

17 $y = 1 - x^2$ and the lines $x = -1$ and $x = 1$

18 $y = -(4x + 1)^2 + 9$ and the lines $x = -1$ and $x = \dfrac{1}{2}$

19 $y = x^3 - 2x^2$ and the lines $x = 0$ and $x = 2$

20 $y = e^{2x} - \sin 2x$ and the line $x = -\dfrac{\pi}{2}$ and the *y*-axis

21 Find an expression in terms of p for this area.

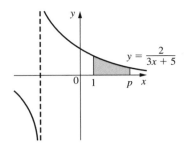

22 Find an expression in terms of p for this area.

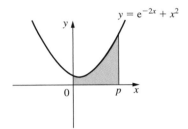

23 Find $k(k > 0)$ given that $\displaystyle\int_0^k 3x^{\frac{1}{2}}\, dx = 16$.

24 Find a given that $\displaystyle\int_{-a}^a \frac{25}{(9-x)^2}\, dx = \frac{5}{8}$.

14.8 Areas above and below the x-axis

In this case the formula needs to be applied carefully. Consider $y = x(x-1)(x-2)$ and the area enclosed by this curve and the x-axis.

$y = x^3 - 3x^2 + 2x$

If the definite integral $\displaystyle\int_0^2 x^3 - 3x^2 + 2x\, dx$ is calculated, we obtain an answer of 0 (remember the calculator uses a numerical process to calculate an integral) and so this result is interpreted as zero.

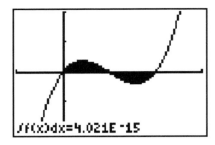

However, it is clear that the area is not zero.

We know that a definite integral for an area below the *x*-axis provides a negative answer. This explains the zero answer – the two (identical) areas have cancelled each other out. So although the answer to the definite integral is zero, the area is not zero.

To find an area that has parts above and below the *x*-axis, consider the parts separately.

So in the above example, area $= 2 \times \int_0^1 x^3 - 3x^2 + 2x \, dx$

$$= 2 \times \frac{1}{4}$$

$$= \frac{1}{2} \text{ unit}^2$$

This demonstrates an important point. The answer to finding an area and to finding the value of the definite integral may actually be different.

The method used in the example below shows how to avoid such problems.

Method

> **1** Sketch the curve to find the relevant roots of the graph.
> **2** Calculate the areas above and below the *x*-axis separately.
> **3** Add together the areas (ignoring the negative sign).

Example

Find the area enclosed by $y = x^2 - 4x + 3$, the *x*-axis and the *y*-axis.

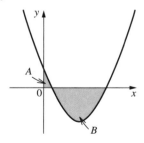

1. Sketch the curve and shade the areas required.

The roots of the graph are given by $x^2 - 4x + 3 = 0$

$$\Rightarrow (x - 1)(x - 3) = 0$$

$$\Rightarrow x = 1 \text{ or } x = 3$$

These results can be found using a calculator.

2. Work out the areas separately.

$$A = \left| \int_0^1 x^2 - 4x + 3 \, dx \right|$$

$$= \left| \left[\frac{1}{3}x^3 - 2x^2 + 3x \right]_0^1 \right|$$

$$= \left| \left(\frac{1}{3} - 2 + 3 \right) - (0) \right|$$

$$= \frac{4}{3}$$

$$B = \left| \int_1^3 x^2 - 4x + 3 \, dx \right|$$

$$= \left| \left[\frac{1}{3}x^3 - 2x^2 + 3x \right]_1^3 \right|$$

$$= \left| (9 - 18 + 9) - \left(\frac{4}{3} \right) \right|$$

$$= \left| -\frac{4}{3} \right| = \frac{4}{3}$$

3. These areas can be calculated on a calculator (separately) and then added.

So the total area $= \dfrac{8}{3}$ square units.

Example

Find the area bounded by $y = \sin x$, the y-axis and the line $x = \dfrac{5\pi}{3}$.

Graphing this on a calculator,

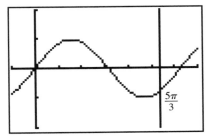

This area can be split into:

$$\int_0^\pi \sin x \, dx \qquad \text{and} \qquad \int_\pi^{\frac{5\pi}{3}} \sin x \, dx$$

$$= \left[-\cos x \right]_0^\pi$$

$$= (-\cos \pi) - (-\cos 0)$$

$$= 1 - (-1)$$

$$= 2$$

$$= \left[-\cos x \right]_\pi^{\frac{5\pi}{3}}$$

$$= \left(-\cos \frac{5\pi}{3} \right) - (-\cos \pi)$$

$$= \left(-\frac{1}{2} \right) - (1)$$

$$= -\frac{3}{2}$$

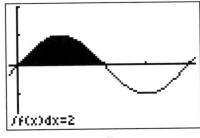

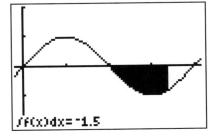

So the total area is $\dfrac{7}{2}$ square units.

Exercise 8

Find the shaded area on the following diagrams.

1 $y = x^2 - 8x + 12$

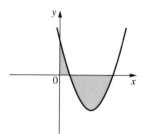

2 $y = -x^2 + 9x - 8$

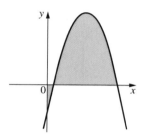

3 $y = 4 \sin x$

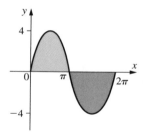

4 $y = 5x^3$

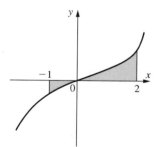

Find the area bounded by the curve and the *x*-axis in the following cases.

5 $y = x(x + 3)(x - 2)$
6 $y = -(x + 4)(2x + 1)(x - 3)$
7 $y = x^3 - x^2 - 16x + 16$
8 $y = 6x^3 - 5x^2 - 12x - 4$

Find the area bounded by the curve, the x-axis and the lines given.

9 $y = x^2 - x - 6$, $x = -2$ and $x = 4$

10 $y = 6x^3$, $x = -4$ and $x = 2$

11 $y = 3 \sin 2x$, $x = 0$ and $x = \dfrac{5\pi}{6}$

12 $y = \cos\left(2x - \dfrac{\pi}{6}\right)$, $x = 0$ and $x = \pi$

13 $y = 4 \cos 2x$, $x = \dfrac{\pi}{2}$ and $x = \dfrac{5\pi}{4}$

14 $y = x^3 - e^x$, $x = 0$ and $x = 4$

14.9 Area between two curves

The area contained between two curves can be found as follows.

The area under f(x) is given by $\displaystyle\int_a^b f(x)\, dx$ and under g(x) is given by $\displaystyle\int_a^b g(x)\, dx$.

So the shaded area is $\displaystyle\int_a^b f(x)\, dx - \int_a^b g(x)\, dx$.

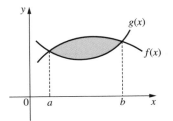

Combining these gives $\displaystyle\int_a^b f(x) - g(x)\, dx$.

This can be expressed as $\displaystyle\int_a^b$ upper curve – lower curve dx.

> As long as we always take upper – lower, the answer is positive and hence it is not necessary to worry about above and below the x-axis.

Example

Find the shaded area.

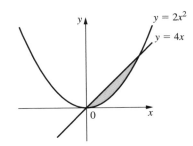

The functions intersect where $2x^2 = 4x$.
$$\Rightarrow 2x^2 - 4x = 0$$
$$\Rightarrow 2x(x - 2) = 0$$
$$\Rightarrow x = 0 \text{ or } x = 2$$

> These intersection points can also be found using a calculator.

So the area $= \displaystyle\int_0^2 4x - 2x^2 \, dx$

$\qquad\qquad = \left[2x^2 - \dfrac{2}{3}x^3 \right]_0^2$

$\qquad\qquad = \left(8 - \dfrac{16}{3} \right) - (0)$

$\qquad\qquad = \dfrac{8}{3}$

This function can be drawn using Y1-Y2 and then the area calculated.

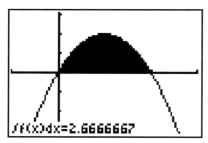

The two limits are roots of the resulting function.

Example

Find the area contained between $y = -e^x$ and $y = \sin x$ from $x = -\pi$ to $x = \pi$.

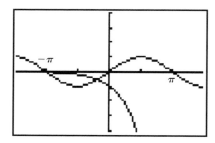

Note that these graphs cross twice within the given interval. So, we need to find the intersections and then treat each part separately (as the curve that is the upper one changes within the interval).

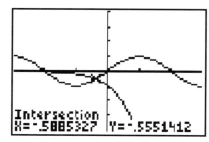

So we need to find

$$\int_{-\pi}^{-3.096} \sin x - (-e^x)\, dx \quad \int_{-3.096}^{-0.589} -e^x - \sin x\, dx \quad \text{and} \quad \int_{-0.589}^{\pi} \sin x - (-e^x)\, dx.$$

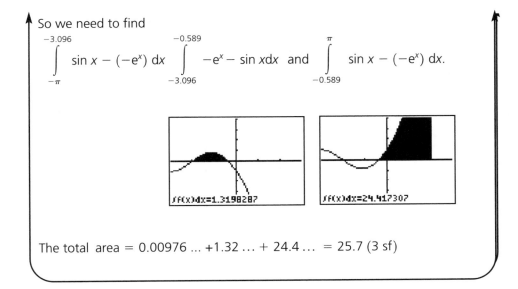

The total area = 0.00976 … + 1.32 … + 24.4 … = 25.7 (3 sf)

Area between a curve and the *y*-axis

Mostly we are concerned with the area bounded by a curve and the *x*-axis. However, for some functions it is more relevant to consider the area between the curve and the *y*-axis. This is particularly pertinent when volumes of revolution are considered in Chapter 16.

The area between a curve and the *x*-axis is $\int_a^b y\, dx$

To find the area between a curve and the *y*-axis we calculate

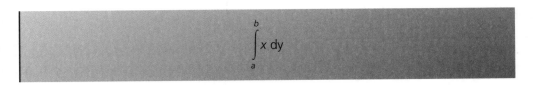

$$\int_a^b x\, dy$$

This is where *x* is a function of *y*.

This formula is proved in an identical way to the area between the curve and the *x*-axis, except that thin horizontal rectangles of length *x* and thickness δy are used.

Example

Consider $y^2 = 9 - x$.

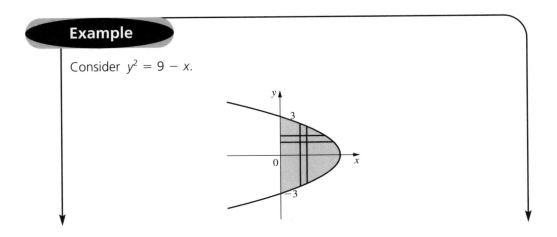

There is no choice here but to use horizontal strips as opposed to vertical strips, as vertical strips would have the curve at both ends, and hence the length of the rectangle would no longer be y and the formula would no longer work.

Area

$$= \int_{-3}^{3} x \, dy$$

$$= \int_{-3}^{3} 9 - y^2 \, dy$$

$$= \left[9y - \frac{1}{3}y^3 \right]_{-3}^{3}$$

$$= (27 - 9) - (-27 + 9)$$

$$= 18 + 18$$

$$= 36 \text{ square units}$$

Although the integration is performed with respect to y, a calculator can still be used to find the area (although of course it is not the correct graph).

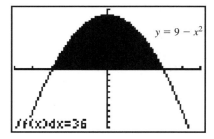

Exercise 9

Calculate the area enclosed by the two functions.

1 $y = x^2, y = x$

2 $y = 6x^2, y = 3x$

3 $y = x^3, y = x$

4 $y = x^2, y = \sqrt{x}$

5 $y = 8 - x^2, y = 2 - x$

6 $y = x^3 + 24, y = 3x^2 + 10x$

7 $y = 10 - x^2, y = 19 - 2x^2$

8 $y = e^x, y = 4 - x^2$

9 $y = -\frac{1}{2}x^2 + 6x - 10, \ y = 4x - \frac{1}{3}x^2$ and the x-axis. In this case draw the graphs and shade the area.

10 $y = 3 \cos 2x$ and $y = 1$ produce an infinite pattern as shown.

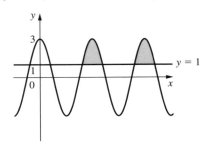

Find the area of each shaded part.

11 Find the area between $y = e^{2x}$, $y = 2 - x$, the x-axis and the y-axis as shown.

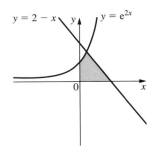

12 Find the area between $y = \dfrac{2}{4 - x}$, $y = 4 - x$, the x-axis and the y-axis as shown.

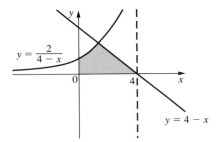

13 Find the shaded area.

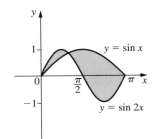

14 Find the shaded area.

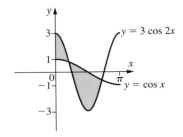

15 Find the area of the "curved triangle" shown below, the sides of which lie on the curves with equations $y = x(x + 3)$, $y = x - \frac{1}{4}x^2$ and $y = \frac{4}{x^2}$.

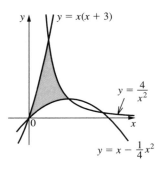

Find the area enclosed by the y-axis and the following curves.

16 $x = 4 - y^2$

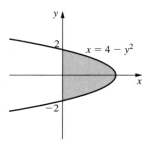

17 $y^2 = 16 - x$

18 $8y^2 = 18 - 2x$

19 Evaluate the shaded area **i** with respect to x and **ii** with respect to y.
$y = \sqrt{5 - x}$

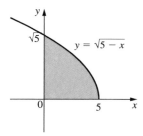

20 Find the shaded area.

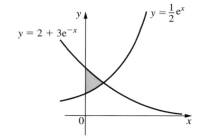

21 Find the shaded area.

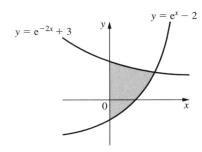

22 Find the shaded area.

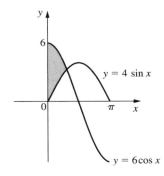

Review exercise

 1 Integrate these functions.

    a $4x^2 - 7$ **b** $9x^2 + 4x - 5$ **c** $\dfrac{8}{x^3}$ **d** $(3 - 2x)^2$

 2 Solve these equations.

    a $\dfrac{dy}{dx} = \dfrac{x^3 - 6}{x^5}$ **b** $\dfrac{dy}{dp} = p^2(3 - p^5)$ **c** $\dfrac{dy}{dt} = \dfrac{3t^2 - \sqrt{t}}{4t}$

 3 Given $\dfrac{dy}{dx} = -3x + 8$ and the curve passes through $(2, 8)$, find the

    equation of the curve.

 4 Find these integrals.

    a $\displaystyle\int 4e^x - \sin x \, dx$ **b** $\displaystyle\int 7 \cos x - \dfrac{4}{x} dx$ **c** $\displaystyle\int 2e^{6x} - \dfrac{5}{x} + 4 \sin x \, dx$

 5 Find these integrals.

    a $\displaystyle\int 6 \cos 2x \, dx$ **b** $\displaystyle\int 4e^{2x} \, dx$ **c** $\displaystyle\int \dfrac{2}{4x - 3} dx$

    d $\displaystyle\int (3x - 2)^6 \, dx$ **e** $\displaystyle\int 7e^{3x} - \dfrac{4}{(3x - 4)^5} \, dx$

 6 Let $f(x) = \sqrt{x}\left(2x - \dfrac{3}{x^{\frac{5}{2}}}\right)$. Find $\displaystyle\int f(x) \, dx$.

 7 Find these definite integrals.

    a $\displaystyle\int_1^3 4p - \dfrac{3}{(2p - 1)^3} \, dp$ **b** $\displaystyle\int_0^{\frac{\pi}{4}} \sin 4\theta + 1 \, d\theta$ **c** $\displaystyle\int_{-k}^{k} 3 \cos 2\theta - 4\theta \, d\theta$

8 Find the area given by these definite integrals.

a $\displaystyle\int_{2}^{5} e^x \, dx$ b $\displaystyle\int_{-3}^{1} 4x - 5 \, dx$ c $\displaystyle\int_{0}^{\frac{\pi}{6}} 2 \cos 3\theta \, d\theta$

 9 Find an expression in terms of p for this shaded area.

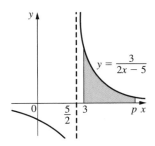

 10 Find the total area of the two regions enclosed by the curve
$y = x^3 - 3x^2 - 9x + 27$ and the line $y = x + 3$.   [IB Nov 04 P1 Q14]

 11 The figure below shows part of the curve $y = x^3 - 7x^2 + 14x - 7$.
The curve crosses the x-axis at the points A, B and C.

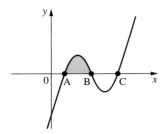

a Find the x-coordinate of A.
b Find the x-coordinate of B.
c Find the area of the shaded region.   [IB May 02 P1 Q13]

 12 Find the area bounded by the curve and the x-axis for:

a $y = 3 \cos 2\theta$, $\theta = 0$ and $\theta = \dfrac{3\pi}{4}$

b $y = x^4 - 2e^x$, $x = -1$ and $x = 3$

13 Find the area enclosed by:

a $y = e^x$ and $y = 6 - x^2$

b $y = \dfrac{3}{2 - x}$, $y = 2 - x$ and the x- and y-axes

14 Find the area between $x = 8 - y^2$ and the y-axis.

15 Integration 2 – Further Techniques

Johann Bernoulli, Jakob's brother, was born on 27 July 1667 and was the tenth child of Nicolaus and Margaretha Bernoulli. Johann's father wished him to enter the family business, but this did not suit Johann and in the end he entered the University of Basel to study medicine. However, he spent a lot of time studying mathematics with his brother, Jakob, as his teacher. He worked on Leibniz's papers on calculus and within two years he had become the equal of his brother in mathematical skill, and he moved to Paris where he worked with de l'Hôpital. He then returned to Basel and at this stage Johann and Jakob worked together and learned much from each other. However this was not to last and their friendly rivalry descended into open hostility over the coming years. Among Johann's many

Johann Bernoulli

mathematical achievements were work on the function $y = x^x$, and investigating series using the method of integration by parts. In this chapter we will use techniques that treat integration as the reverse of differentiation and this is exactly how Johann worked with it. His great success in mathematics was rewarded when in 1695 he accepted the offer of the chair of mathematics at the University of Groningen. This gave him equal status to his brother Jakob who was becoming increasingly jealous of Johann's progress. During the ten years he spent at Groningen the battle between the two brothers escalated. In 1705 he left Groningen to return to Basel unaware that his brother had died two days previously. Ironically, soon after his return to Basel he was offered his brother's position at the University of Basel, which he accepted. He stayed there until his death on 1 January 1748.

Function $f(x)$	Integral $+ c$ $\displaystyle\int f(x)\, dx$
x^n	$\dfrac{1}{n+1}x^{n+1}$
e^x	e^x
$\dfrac{1}{x}$	$\ln|x|$
$\sin x$	$-\cos x$
$\cos x$	$\sin x$
$\sec^2 x$	$\tan x$
$\operatorname{cosec}^2 x$	$-\cot x$
$\sec x \tan x$	$\sec x$
$\operatorname{cosec} x \cot x$	$-\operatorname{cosec} x$
$\dfrac{1}{\sqrt{1-x^2}}$	$\sin^{-1} x$
$\dfrac{1}{\sqrt{a^2-x^2}}$	$\sin^{-1}\dfrac{x}{a}$
$\dfrac{-1}{\sqrt{1-x^2}}$	$\cos^{-1} x$
$\dfrac{-1}{\sqrt{a^2-x^2}}$	$\cos^{-1}\dfrac{x}{a}$
$\dfrac{1}{1+x^2}$	$\tan^{-1} x$
$\dfrac{1}{a^2+x^2}$	$\dfrac{1}{a}\tan^{-1}\dfrac{x}{a}$
a^x	$\dfrac{a^x}{\ln|a|}$

Example

$$\int \sin 2x\, dx = -\frac{\cos 2x}{2} + c$$

Example

$$\int \frac{1}{2x-1}\, dx = \frac{1}{2}\ln|2x-1| + c$$

In this chapter we will look at the techniques of integrating more complicated functions and a wider range of functions. Above is the complete list of basic results.

These are as a result of inspection and were dealt with in Chapter 14.

15.1 Integration as a process of anti-differentiation – direct reverse

For more complex questions, the inspection method can still be used. We call this method **direct reverse**.

Method of direct reverse

1. Decide what was differentiated to get the function in the question and write it down ignoring the constants.
2. Differentiate this.
3. Divide or multiply by constants to find the required form.

Example

$$\int e^{2x}\, dx$$

1. We begin with $y = e^{2x}$.

2. $\dfrac{dy}{dx} = 2e^{2x}$

3. So $2\displaystyle\int e^{2x}\, dx = e^{2x}$, therefore $\displaystyle\int e^{2x}\, dx = \frac{1}{2}e^{2x} + c$

Example

$$\int 4^{2x}\, dx$$

1. We begin with $y = 4^{2x}$.

2. $\dfrac{dy}{dx} = \ln 4 \times 4^{2x} \times 2 = 2 \ln 4 \cdot 4^{2x}$

3. So $2 \ln 4 \displaystyle\int 4^{2x}\, dx = 4^{2x}$, therefore $\displaystyle\int 4^{2x}\, dx = \frac{4^{2x}}{2 \ln 4} + c$

Example

$$\int \cos\left(4\theta - \frac{\pi}{3}\right) d\theta$$

1. We begin with $y = \sin\left(4\theta - \dfrac{\pi}{3}\right)$.

2. $\dfrac{dy}{d\theta} = 4 \cos\left(4\theta - \dfrac{\pi}{3}\right)$

3. $4 \displaystyle\int \cos\left(4\theta - \frac{\pi}{3}\right) d\theta = \sin\left(4\theta - \frac{\pi}{3}\right)$,

   therefore $\displaystyle\int \cos\left(4\theta - \frac{\pi}{3}\right) d\theta = \frac{1}{4} \sin\left(4\theta - \frac{\pi}{3}\right) + c$

The method of direct reverse can also prove all the basic results given at the start of the chapter.

> **Example**
>
> $\int a^x \, dx$
>
> 1. We begin with $y = a^x$.
>
> 2. $\dfrac{dy}{dx} = a^x \ln|a|$
>
> 3. $\ln|a| \int a^x \, dx = a^x$, therefore $\int a^x \, dx = \dfrac{a^x}{\ln|a|} + c$

This technique allows us to do much more complicated examples including some products and quotients.

Integration of products and quotients using direct reverse

Unlike differentiation this is not quite so simple. Considering differentiation, an answer to a derivative may be a product or a quotient from more than one technique, e.g. chain rule, product rule or quotient rule. This means there has to be more than one way to integrate products and quotients. Here cases which can be done by direct reverse will be considered.

Products

This occurs when one part of the product is a constant multiplied by the derivative of the inside function. If we are integrating $f(x) \times gh(x)$, then $f(x) = ah'(x)$ where a is a constant for the technique to work.

This may seem complicated but it is easy to apply and is a quick way of doing some quite advanced integration.

> **Example**
>
> $\int x^2(x^3 + 3)^3 \, dx$
>
> Direct reverse works in this case, since if $h(x) = x^3 + 3$, $h'(x) = 3x^2$, then $a = \dfrac{1}{3}$.
>
> 1. This begins with $y = (x^3 + 3)^4$.
>
> 2. $\dfrac{dy}{dx} = 4(x^3 + 3)^3 \cdot 3x^2 = 12x^2(x^3 + 3)^3$
>
> 3. $12 \int x^2(x^3 + 3)^3 \, dx = (x^3 + 3)^4$,
>
>   therefore $\int x^2(x^3 + 3)^3 \, dx = \dfrac{1}{12}(x^3 + 3)^4 + c$

The value of a is not used in this method, but it is necessary to check that a constant exists for the method to work.

Now consider the example $\int x^2(2x - 1)^4 \, dx$. In this case $h(x) = 2x - 1$ and hence $h'(x) = 2$. Since $f(x) = x^2$, then $f(x) \neq ah'(x)$ and hence this technique cannot be used. Further techniques for integrating products and quotients which will deal with this will be discussed later in the chapter.

This technique can never be made to work by letting a be a function of x.

Example

$\int \sin x \sqrt{1 + \cos x} \, dx$

The method works in this case, since if $h(x) = 1 + \cos x$, $h'(x) = -\sin x$ then $a = -1$. We begin by writing the integral in the form $\int \sin x (1 + \cos x)^{\frac{1}{2}} \, dx$.

1. This begins with $y = (1 + \cos x)^{\frac{3}{2}}$.

2. $\dfrac{dy}{dx} = \dfrac{3}{2}(1 + \cos x)^{\frac{1}{2}}(-\sin x)$

3. $\dfrac{-3}{2} \int \sin x (1 + \cos x)^{\frac{1}{2}} \, dx = (1 + \cos x)^{\frac{3}{2}}$

therefore $\int \sin x (1 + \cos x)^{\frac{1}{2}} \, dx = -\dfrac{2}{3}(1 + \cos x)^{\frac{3}{2}} + c$

Sometimes the examples can be somewhat disguised.

Example

$\int (x + 1)(x^2 + 2x - 5)^6 \, dx$

It still works in this case, since if $h(x) = x^2 + 2x - 5$, $h'(x) = 2x + 2 = 2(x + 1)$, then $a = \dfrac{1}{2}$.

1. This begins with $y = (x^2 + 2x - 5)^7$.

2. $\dfrac{dy}{dx} = 7(x^2 + 2x - 5)^6(2x + 2) = 14(x + 1)(x^2 + 2x - 5)^6$

3. $14 \int (x + 1)(x^2 + 2x - 5)^6 \, dx = (x^2 + 2x - 5)^7$,

therefore $\int (x + 1)(x^2 + 2x - 5)^6 \, dx = \dfrac{1}{14}(x^2 + 2x - 5)^7 + c$

Example

$\int (2x + 1)e^{2x^2 + 2x} \, dx$

The method works here since if $h(x) = 2x^2 + 2x$, $h'(x) = 4x + 2 = 2(2x + 1)$, then $a = \dfrac{1}{2}$.

1. This begins with $y = e^{2x^2 + 2x}$.

2. $\dfrac{dy}{dx} = (4x + 2)e^{2x^2 + 2x} = 2(2x + 1)e^{2x^2 + 2x}$

3. $2 \int (2x + 1)e^{2x^2 + 2x} \, dx = e^{2x^2 + 2x}$,

therefore $\int (2x + 1)e^{2x^2 + 2x} \, dx = \dfrac{1}{2}e^{2x^2 + 2x} + c$

Quotients

This occurs when the numerator of the quotient is in the form of a constant multiplied by the derivative of the inside function.

If we are integrating $\dfrac{f(x)}{gh(x)}$ then $f(x) = ah'(x)$ for the technique to work. The quotients often come in the form $\dfrac{f'(x)}{f(x)}$ which was met in Chapter 9 when differentiating logarithmic functions. This came from the form $y = \ln(f(x))$.

Example

$$\int \frac{x^2}{1 + x^3}\,dx$$

Direct reverse works in this case, since if $h(x) = 1 + x^3$, $h'(x) = 3x^2$ then $a = \dfrac{1}{3}$.

1. This begins with $y = \ln|1 + x^3|$.

2. $\dfrac{dy}{dx} = \dfrac{3x^2}{1 + x^3}$

3. $3\displaystyle\int \frac{x^2}{1 + x^3}\,dx = \ln|1 + x^3|$, therefore $\displaystyle\int \frac{x^2}{1 + x^3}\,dx = \frac{1}{3}\ln|1 + x^3| + c$

Example

$$\int \frac{\sin x}{22 - \cos x}\,dx$$

The method works, since if $h(x) = 22 - \cos x$, $h'(x) = \sin x$ then $a = 1$.

1. This begins with $y = \ln|22 - \cos x|$.

2. $\dfrac{dy}{dx} = \dfrac{\sin x}{22 - \cos x}$

3. Therefore $\displaystyle\int \frac{\sin x}{22 - \cos x}\,dx = \ln|22 - \cos x| + k$

Example

$$\int \frac{\sin x}{\cos x}\,dx$$

The method works, since if $h(x) = \cos x$, $h'(x) = -\sin x$ then $a = -1$.

1. This begins with $y = \ln|\cos x|$.

2. $\dfrac{dy}{dx} = \dfrac{-\sin x}{\cos x}$

3. $-\displaystyle\int \frac{\sin x}{\cos x}\,dx = \ln|\cos x|$ therefore $\displaystyle\int \frac{\sin x}{\cos x}\,dx = -\ln|\cos x| + c$ ⋯⋯⋯⋯⋯ This is the method of integrating $\tan x$.

Example

$\int \dfrac{3e^x}{14 + e^x} dx$

Again this works, since if $h(x) = 14 + e^x$, $h'(x) = e^x$ then $a = 1$.

1. This begins with $y = \ln|14 + e^x|$.

2. $\dfrac{dy}{dx} = \dfrac{e^x}{14 + e^x}$

3. Therefore $\int \dfrac{e^x}{14 + e^x} dx = \ln|14 + e^x|$

   $\Rightarrow \int \dfrac{3e^x}{14 + e^x} dx = 3 \int \dfrac{e^x}{14 + e^x} dx = 3 \ln|14 + e^x| + c$

> This is a slightly different case since the constant is part of the question rather than being produced through the process of integration. However, it is dealt with in the same way.

There is a danger of assuming that all integrals of quotients become natural logarithms. Many are, but not all, so care needs to be taken. The following examples demonstrate this.

Example

$\int_2^p \dfrac{x^2}{(x^3 + 3)^4} dx$

The technique will still work in the case of definite integration, since if $h(x) = x^3 + 3$, $h'(x) = 3x^2$ then $a = \dfrac{1}{3}$. We begin by writing the integral in the form $\int_2^p x^2(x^3 + 3)^{-4} dx$.

1. This begins with $y = (x^3 + 3)^{-3}$.

2. $\dfrac{dy}{dx} = -3(x^3 + 3)^{-4} \times 3x^2 = -9x^2(x^3 + 3)^{-4}$

3. $-9 \int_2^p x^2(x^3+3)^{-4} dx = [(x^3+3)^{-3}]_2^p$ therefore $\int_2^p x^2(x^3+3)^{-4} dx$

$$= \left[-\dfrac{1}{9}(x^3 + 3)^{-3} \right]_2^p$$

$$= \left[-\dfrac{1}{9}(p^3 + 3)^{-3} + \dfrac{1}{9}(8 + 3)^{-3} \right]$$

$$= \left[-\dfrac{1}{9(p^3 + 3)^3} + \dfrac{1}{11\,979} \right]$$

If the question had asked for an exact answer to $\int_2^3 \dfrac{x^2}{(x^3 + 3)^4} dx$, then we would proceed as above, but the final lines would be:

> If the limits were both numbers, and an exact answer was not required, then a calculator could be used to evaluate the integral. However, if one or both of the limits are not known, an exact answer is required, or if the question appears on the non-calculator paper, then this technique must be used.

$$= \left[-\frac{1}{9}(27 + 3)^{-3} + \frac{1}{9}(8 + 3)^{-3} \right]$$

$$= \left[-\frac{1}{243\ 000} + \frac{1}{11\ 979} \right]$$

$$= \frac{231\ 021}{2\ 910\ 897\ 000}$$

Example

$$\int \frac{x}{\sqrt{x^2 - 1}}\, dx$$

The method works in this case, since if $h(x) = x^2 - 1$, $h'(x) = 2x$ then $a = \frac{1}{2}$.

$$\int x(x^2 - 1)^{-\frac{1}{2}}\, dx$$

1. This begins with $y = (x^2 - 1)^{\frac{1}{2}}$.

2. $\dfrac{dy}{dx} = \dfrac{1}{2}(x^2 - 1)^{-\frac{1}{2}}(2x) = x(x^2 - 1)^{-\frac{1}{2}}$

3. Therefore $\displaystyle\int x(x^2 - 1)^{-\frac{1}{2}}\, dx = (x^2 - 1)^{\frac{1}{2}} + c.$

Exercise 1

Without using a calculator, find the following integrals.

1 $\displaystyle\int \sin\left(\theta - \frac{3\pi}{4}\right) d\theta$

2 $\displaystyle\int \cos\left(3x + \frac{\pi}{4}\right) dx$

3 $\displaystyle\int e^{32x - 7}\, dx$

4 $\displaystyle\int_0^2 2\cos x e^{\sin x}\, dx$

5 $\displaystyle\int \frac{2}{8x - 9}\, dx$

6 $\displaystyle\int x^5(x^6 - 9)^8\, dx$

7 $\displaystyle\int_0^p x^3(2x^4 + 1)\, dx$

8 $\displaystyle\int x\sqrt{1 + x^2}\, dx$

9 $\displaystyle\int x\sqrt{4x^2 - 3}\, dx$

10 $\displaystyle\int \sec^2 x(3\tan x + 4)^3\, dx$

11 $\displaystyle\int \frac{x}{\sqrt{x^2 - 1}}\, dx$

12 $\displaystyle\int_0^{0.5} \sin x(2\cos x - 1)^4\, dx$

13 $\displaystyle\int_0^a e^{2x}(6e^{2x} - 7)\, dx$

14 $\displaystyle\int \sin 2x\sqrt{\cos 2x - 1}\, dx$

15 $\displaystyle\int (x + 1)(x^2 + 2x - 4)^5\, dx$

16 $\displaystyle\int \frac{2x - 3}{x^2 - 3x + 5}\, dx$

17 $\displaystyle\int \frac{3\sin x}{(\cos x + 8)^3}\, dx$

18 $\displaystyle\int \frac{e^x}{2e^x - 4}\, dx$

19 $\displaystyle\int \frac{\cos x}{3 \sin x - 12} \, dx$

20 $\displaystyle\int (3x^2 + 2)(3x^3 + 6x - 19)^{\frac{3}{2}} \, dx$

21 $\displaystyle\int \sin 2x(1 - 3 \cos 2x)^{\frac{5}{2}} \, dx$

22 $\displaystyle\int \frac{\sec^2 2x}{3 \tan 2x - 7} \, dx$

23 $\displaystyle\int_0^1 \frac{x}{(x^2 + 1)^5} \, dx$

24 $\displaystyle\int \frac{2x - 1}{3x^2 - 3x + 4} \, dx$

25 $\displaystyle\int \frac{2x - 1}{(3x^2 - 3x + 4)^4} \, dx$

26 $\displaystyle\int_1^p \frac{2 \ln x}{x} \, dx$

27 $\displaystyle\int \frac{2e^x}{e^x + e^{-x}} \, dx$

28 $\displaystyle\int_1^p \frac{2x + 1}{6x^2 + 6x - 15} \, dx$

15.2 Integration of functions to give inverse trigonometric functions

Questions on inverse trigonometric functions come in a variety of forms. Sometimes the given results can be used as they stand and in other cases some manipulation needs to be done first. In more difficult cases the method of direct reverse needs to be used.

Example

$\displaystyle\int \frac{1}{\sqrt{4 - x^2}} \, dx$

Using the result $\displaystyle\int \frac{1}{\sqrt{a^2 - x^2}} \, dx = \sin^{-1}\frac{x}{a} + c$

$\Rightarrow \displaystyle\int \frac{1}{\sqrt{4 - x^2}} \, dx = \sin^{-1}\frac{x}{2} + c$

Example

$\displaystyle\int \frac{1}{\sqrt{1 - \dfrac{x^2}{4}}} \, dx.$

This situation it is not given in the form $\dfrac{1}{\sqrt{a^2 - x^2}}$ and hence there are two options. It can either be rearranged into that form or alternatively we can use the method of direct reverse.

Rearranging gives:

$\displaystyle\int \frac{1}{\sqrt{1 - \dfrac{x^2}{4}}} \, dx = \int \frac{1}{\sqrt{\dfrac{1}{4}}\sqrt{4 - x^2}} \, dx$

$= \displaystyle\int \frac{2}{\sqrt{4 - x^2}} \, dx$

$= 2 \sin^{-1}\frac{x}{2} + c$

Using direct reverse we begin with $y = \sin^{-1}\dfrac{x}{2}$.

$$\frac{dy}{dx} = \frac{1}{\sqrt{1 - \dfrac{x^2}{4}}} \times \frac{1}{2}$$

Therefore $\displaystyle\int \frac{1}{2\sqrt{1 - \dfrac{x^2}{4}}}\, dx = \sin^{-1}\frac{x}{2}$

$$\Rightarrow \int \frac{1}{\sqrt{1 - \dfrac{x^2}{4}}}\, dx = 2\sin^{-1}\frac{x}{2} + c$$

The same thing happens with the inverse tan function.

Example

$$\int \frac{1}{1 + \dfrac{x^2}{9}}\, dx$$

Rearranging to the standard result of $\displaystyle\int \frac{1}{a^2 + x^2}\, dx = \frac{1}{a}\tan^{-1}\frac{x}{a} + c$ gives:

$$\int \frac{1}{1 + \dfrac{x^2}{9}}\, dx = \int \frac{1}{\dfrac{1}{9}(9 + x^2)}\, dx$$

$$= \int \frac{9}{9 + x^2}\, dx$$

$$= \frac{9}{3}\tan^{-1}\frac{x}{3} + c$$

$$= 3\tan^{-1}\frac{x}{3} + c$$

Using direct reverse we begin with $y = \tan^{-1}\dfrac{x}{3}$.

$$\frac{dy}{dx} = \frac{1}{1 + \dfrac{x^2}{9}} \times \frac{1}{3}$$

Therefore $\displaystyle\int \frac{1}{3\left(1 + \dfrac{x^2}{9}\right)}\, dx = \tan^{-1}\frac{x}{3}$

$$\Rightarrow \int \frac{1}{\left(1 + \dfrac{x^2}{9}\right)}\, dx = 3\tan^{-1}\frac{x}{3} + c$$

Now we need to look at more complicated examples. If it is a $\dfrac{number}{\sqrt{quadratic}}$ or a $\dfrac{number}{quadratic}$, then we need to complete the square and then use the method of direct reverse. It should be noted that this is not the case for every single example as they could integrate in different ways, but this is beyond the scope of this syllabus.

Example

$$\int \frac{1}{x^2 + 2x + 5}\,dx$$

Completing the square gives:

$$\int \frac{1}{x^2 + 2x + 5}\,dx = \int \frac{1}{(x + 1)^2 + 4}\,dx$$

$$= \int \frac{1}{4 + (x + 1)^2}\,dx$$

$$= \frac{1}{4}\int \frac{1}{1 + \left(\dfrac{x + 1}{2}\right)^2}\,dx$$

Using direct reverse, we begin with $y = \tan^{-1}\left(\dfrac{x + 1}{2}\right)$.

$$\frac{dy}{dx} = \frac{1}{1 + \left(\dfrac{x + 1}{2}\right)^2} \times \frac{1}{2}$$

Therefore $\dfrac{1}{2}\displaystyle\int \dfrac{1}{1 + \left(\dfrac{x + 1}{2}\right)^2}\,dx = \tan^{-1}\left(\dfrac{x + 1}{2}\right)$

$$\Rightarrow \frac{1}{4}\int \frac{1}{1 + \left(\dfrac{x + 1}{2}\right)^2}\,dx = \frac{1}{2}\tan^{-1}\left(\frac{x + 1}{2}\right) + c$$

Example

$$\int \frac{1}{\sqrt{-x^2 - 4x + 12}}\,dx$$

Completing the square gives:

$$\int \frac{1}{\sqrt{-x^2 - 4x + 12}}\,dx = \int \frac{1}{\sqrt{-(x^2 + 4x - 12)}}\,dx$$

$$= \int \frac{1}{\sqrt{-[(x + 2)^2 - 16]}}\,dx$$

$$= \int \frac{1}{\sqrt{16 - (x + 2)^2}}\,dx$$

$$= \frac{1}{\sqrt{16}}\int \frac{1}{\sqrt{1 - \left(\dfrac{x + 2}{4}\right)^2}}\,dx$$

$$= \frac{1}{4}\int \frac{1}{\sqrt{1 - \left(\dfrac{x + 2}{4}\right)^2}}\,dx$$

Using direct reverse, we begin with $y = \sin^{-1}\left(\dfrac{x + 2}{4}\right)$.

$$\frac{dy}{dx} = \frac{1}{\sqrt{1 - \left(\frac{x+2}{4}\right)^2}} \times \frac{1}{4}$$

Therefore $\frac{1}{4}\displaystyle\int \frac{1}{\sqrt{1 - \left(\frac{x+2}{4}\right)^2}}\, dx = \sin^{-1}\!\left(\frac{x+2}{4}\right) + c.$

Exercise 2

1 $\displaystyle\int \frac{1}{9 + x^2}\, dx$

2 $\displaystyle\int \frac{1}{\sqrt{25 - x^2}}\, dx$

3 $\displaystyle\int \frac{-1}{\sqrt{36 - x^2}}\, dx$

4 $\displaystyle\int \frac{1}{1 + \frac{x^2}{9}}\, dx$

5 $\displaystyle\int \frac{1}{\sqrt{2 - \frac{x^2}{4}}}\, dx$

6 $\displaystyle\int_1^p \frac{1}{3 + x^2}\, dx$

7 $\displaystyle\int_0^1 \frac{1}{49 + x^2}\, dx$

8 $\displaystyle\int \frac{1}{\sqrt{1 - 3x^2}}\, dx$

9 $\displaystyle\int \frac{1}{\sqrt{-x^2 + 2x}}\, dx$

10 $\displaystyle\int \frac{1}{4x^2 + 8x + 20}\, dx$

11 $\displaystyle\int \frac{1}{x^2 + 6x + 18}\, dx$

12 $\displaystyle\int \frac{-5}{\sqrt{-x^2 - 4x + 5}}\, dx$

13 $\displaystyle\int \frac{1}{9x^2 + 6x + 5}\, dx$

14 $\displaystyle\int \frac{1}{x^2 + 3x + 4}\, dx$

15 $\displaystyle\int \frac{4}{\sqrt{-x^2 + 3x + 10}}\, dx$

16 $\displaystyle\int \frac{1}{\sqrt{-9x^2 + 18x + 99}}\, dx$

17 $\displaystyle\int_0^{0.5} \frac{1}{x^2 + 4x + 7}\, dx$

18 $\displaystyle\int_{-2}^p \frac{1}{\sqrt{-x^2 - 6x - 6}}\, dx$

15.3 Integration of powers of trigonometric functions

To integrate powers of trigonometric functions the standard results and methods of direct reverse are again used, but trigonometric identities are also required.

Even powers of sine and cosine

For these we use the double angle identity $\cos 2x = \cos^2 x - \sin^2 x.$

Example

$$\int \sin^2 x \, dx$$

Knowing that $\cos 2x = \cos^2 x - \sin^2 x$

$\Rightarrow \cos 2x = 1 - 2 \sin^2 x$

$\Rightarrow \sin^2 x = \dfrac{1 - \cos 2x}{2}$

$\Rightarrow \displaystyle\int \sin^2 x \, dx = \int \left(\dfrac{1 - \cos 2x}{2} \right) dx = \dfrac{1}{2} \int (1 - \cos 2x) \, dx$

$\qquad = \dfrac{1}{2} \left(x - \dfrac{\sin 2x}{2} \right) = \dfrac{1}{4} (2x - \sin 2x) + c$

> We cannot use this idea with $\sin^3 x$. If $y = \sin^3 x$, then
>
> $\dfrac{dy}{dx} = 3 \sin^2 x \cos x$ and so
>
> there is a cosine term that creates a problem. This illustrates a major difference between differentiation and integration.

Example

$$\int \cos^4 x \, dx$$

Knowing that

$\qquad \cos 2x = \cos^2 x - \sin^2 x$

$\qquad \Rightarrow \cos 2x = 2 \cos^2 x - 1$

$\qquad \Rightarrow \cos^2 x = \dfrac{1 + \cos 2x}{2}$

$\Rightarrow \displaystyle\int \cos^4 x \, dx = \int \left(\dfrac{1 + \cos 2x}{2} \right)^2 dx$

$\qquad = \displaystyle\int \left(\dfrac{1 + 2 \cos 2x + \cos^2 2x}{4} \right) dx$

$\qquad = \dfrac{1}{4} \displaystyle\int (1 + 2 \cos 2x + \cos^2 2x) \, dx$

Using the double angle formula again on $\cos^2 2x$.

$\qquad \cos 4x = \cos^2 2x - \sin^2 2x$

$\qquad \Rightarrow \cos 4x = 2 \cos^2 2x - 1$

$\qquad \Rightarrow \cos^2 2x = \dfrac{1 + \cos 4x}{2}$

$\Rightarrow \displaystyle\int \cos^4 x \, dx = \dfrac{1}{4} \int \left(1 + 2 \cos 2x + \dfrac{1}{2} + \dfrac{\cos 4x}{2} \right) dx$

$\qquad = \dfrac{1}{8} \displaystyle\int (2 + 4 \cos 2x + 1 + \cos 4x) \, dx$

$\qquad = \dfrac{1}{8} \displaystyle\int (3 + 4 \cos 2x + \cos 4x) \, dx$

$\qquad = \dfrac{1}{8} \left(3x + \dfrac{4 \sin 2x}{2} + \dfrac{\sin 4x}{4} \right)$

$\qquad = \dfrac{1}{32} (12x + 8 \sin 2x + \sin 4x) + c$

> For higher even powers, it is a matter of repeating the process as many times as necessary. This can be made into a general formula, but it is beyond the scope of this curriculum.

If integration of even powers of multiple angles is required the same method can be used.

Example

$$\int \sin^2 8x \, dx$$

This time $\cos 16x = \cos^2 8x - \sin^2 8x$

$$\Rightarrow \cos 16x = 1 - 2\sin^2 8x$$

$$\Rightarrow \sin^2 8x = \frac{1 - \cos 16x}{2}$$

$$\int \sin^2 8x \, dx = \int \left(\frac{1 - \cos 16x}{2} \right) dx$$

$$= \frac{x}{2} - \frac{\sin 16x}{32} = \frac{1}{32}(16x - \sin 16x) + c$$

Odd powers of sine and cosine

For these use the Pythagorean identity $\cos^2 x + \sin^2 x = 1$ with the aim of leaving a single power of sine multiplied by a higher power of cosine or a single power of cosine multiplied by a higher power of sine.

Example

$$\int \sin^3 x \, dx$$

$$= \int \sin x \sin^2 x \, dx$$

$$= \int \sin x (1 - \cos^2 x) dx \text{ using the identity } \cos^2 x + \sin^2 x = 1$$

$$= \int (\sin x - \cos^2 x \sin x) dx$$

To find $\int \cos^2 x \sin x \, dx$ the method of direct reverse is used.

This begins with $y = \cos^3 x = (\cos x)^3$

$$\Rightarrow \frac{dy}{dx} = -3(\cos x)^2 \sin x = -3\cos^2 x \sin x$$

$$\Rightarrow -3 \int \cos^2 x \sin x \, dx = \cos^3 x$$

$$\Rightarrow \int \cos^2 x \sin x \, dx = -\frac{1}{3}\cos^3 x + k$$

$$\int (\sin x - \cos^2 x \sin x) dx = -\cos x + \frac{1}{3}\cos^3 x + c$$

Unlike even powers of cosine and sine, this is a one-stage process, no matter how high the powers become. This is demonstrated in the next example.

Example

$$\int \cos^7 x\,dx$$

$$= \int \cos x \cos^6 x\,dx$$

$$= \int \cos x(1 - \sin^2 x)^3\,dx$$

$$= \int \cos x(1 - 3\sin^2 x + 3\sin^4 x - \sin^6 x)\,dx$$

$$= \int (\cos x - 3\sin^2 x \cos x + 3\sin^4 x \cos x - \sin^6 x \cos x)\,dx$$

These can all be integrated using the method of direct reverse.

$$\int 3\sin^2 x \cos x\,dx \text{ begins with } y = \sin^3 x$$

$$\Rightarrow \frac{dy}{dx} = 3\sin^2 x \cos x$$

$$\int 3\sin^2 x \cos x\,dx = \sin^3 x + k_1$$

$$\int 3\sin^4 x \cos x\,dx \text{ begins with } y = \sin^5 x$$

$$\Rightarrow \frac{dy}{dx} = 5\sin^4 x \cos x$$

$$\Rightarrow 5\int \sin^4 x \cos x\,dx = \sin^5 x$$

$$\Rightarrow 3\int \sin^4 x \cos x\,dx = \frac{3}{5}\sin^5 x + k_2$$

$$\int \sin^6 x \cos x\,dx \text{ begins with } y = \sin^7 x$$

$$\Rightarrow \frac{dy}{dx} = 7\sin^6 x \cos x$$

$$\Rightarrow 7\int \sin^6 x \cos x\,dx = \sin^7 x$$

$$\Rightarrow \int \sin^6 x \cos x\,dx = \frac{1}{7}\sin^7 x + k_3$$

Hence $\int (\cos x - 3\sin^2 x \cos x + 3\sin^4 x \cos x - \sin^6 x \cos x)\,dx$

$$= \sin x - \sin^3 x + \frac{3}{5}\sin^5 x - \frac{1}{7}\sin^7 x + c, \text{ where } c = k_1 + k_2 + k_3.$$

Integrating odd powers of multiple angles works in the same way.

Example

$$\int \cos^3 4x\,dx$$

$$= \int \cos 4x \cos^2 4x\,dx$$

$$= \int \cos 4x(1 - \sin^2 4x)dx$$

$$= \int (\cos 4x - \sin^2 4x \cos 4x)dx$$

To find $\int \sin^2 4x \cos 4x\,dx$ the method of direct reverse is used.

This begins with $y = \sin^3 4x$

$$\Rightarrow \frac{dy}{dx} = 3 \sin^2 4x \cos 4x \times 4 = 12 \sin^2 4x \cos 4x$$

$$\Rightarrow 12 \int \sin^2 4x \cos 4x\,dx = \sin^3 4x$$

$$\Rightarrow \int \sin^2 4x \cos 4x\,dx = \frac{1}{12}\sin^3 4x + k$$

$$\Rightarrow \int (\cos 4x - \sin^2 4x \cos 4x)dx = \frac{\sin 4x}{4} - \frac{1}{12}\sin^3 4x + c$$

Often this technique will work with mixed powers of sine and cosine and the aim is still to leave a single power of sine multiplied by a higher power of cosine or a single power of cosine multiplied by a higher power of sine.

Example

$$\int \sin^3 x \cos^2 x\,dx$$

Since it is sine that has the odd power, this is the one that is split.

$$\int \sin x \sin^2 x \cos^2 x\,dx = \int \sin x(1 - \cos^2 x)\cos^2 x\,dx$$

$$= \int (\cos^2 x \sin x - \cos^4 x \sin x)dx$$

Now these can both be integrated using the method of direct reverse.

$$\int \cos^2 x \sin x\,dx \text{ begins with } y = \cos^3 x$$

$$\Rightarrow \frac{dy}{dx} = -3 \cos^2 x \sin x$$

$$\Rightarrow -3 \int \cos^2 x \sin x\,dx = \cos^3 x$$

$$\int \cos^2 x \sin x\,dx = -\frac{1}{3}\cos^3 x + k_1$$

$\int \cos^4 x \sin x dx$ begins with $y = \cos^5 x$

$$\Rightarrow \frac{dy}{dx} = -5\cos^4 x \sin x$$

$$\Rightarrow -5\int \cos^4 x \sin x dx = \cos^5 x$$

$$\Rightarrow \int \cos^4 x \sin x dx = -\frac{1}{5}\cos^5 x + k_2$$

Therefore $\int (\cos^2 x \sin x - \cos^4 x \sin x)dx = -\frac{1}{3}\cos^3 x + \frac{1}{5}\cos^5 x + c$, where $c = k_1 + k_2$.

Powers of tan x

In this case the identity $1 + \tan^2 x = \sec^2 x$ is used with the aim of getting $\tan x$, $\sec^2 x$ or a power of $\tan x$ multiplied by $\sec^2 x$. It should also be remembered that $\int \tan x dx = -\ln|\cos x| + c$ and $\int \sec^2 x dx = \tan x + c$.

Example

$\int \tan^3 x dx$

This is first turned into $\int \tan x \tan^2 x dx$.

Using the identity gives $\int \tan x(\sec^2 x - 1)dx = \int \tan x \sec^2 x - \tan x dx$

To find $\int \tan x \sec^2 x dx$

direct reverse is used. This happens because the derivative of $\tan x$ is $\sec^2 x$ and also explains why it is necessary to have $\sec^2 x$ with the power of $\tan x$.
To integrate this we begin with $y = \tan^2 x$.

So $\frac{dy}{dx} = 2\tan x \sec^2 x$

Hence $2\int \tan x \sec^2 x dx = \tan^2 x,$

Therefore $\int \tan x \sec^2 x dx = \frac{1}{2}\tan^2 x + k$

Thus $\int \tan^3 x dx = \int (\tan x \sec^2 x - \tan x)dx$

$$= \frac{1}{2}\tan^2 x + \ln|\cos x| + c$$

We need to extract $\tan^2 x$ out of the power of $\tan x$ in order to produce $\sec^2 x - 1$

Example

$$\int \tan^5 x dx$$

First change this to $\int \tan^3 x \tan^2 x dx$.

Using the identity: $\int \tan^3 x(\sec^2 x - 1)dx = \int (\tan^3 x \sec^2 x - \tan^3 x)dx$

The integral of $\tan^3 x$ was done in the example above and the result will just be quoted here. The integral of $\tan^3 x \sec^2 x$ is done by direct reverse.

This begins with $y = \tan^4 x$

$$\Rightarrow \frac{dy}{dx} = 4 \tan^3 x \sec^2 x$$

$$\Rightarrow 4 \int \tan^3 x \sec^2 x dx = \tan^4 x$$

$$\Rightarrow \int \tan^3 x \sec^2 x dx = \frac{1}{4} \tan^4 x + k$$

Hence $\int \tan^5 x dx = \frac{1}{4} \tan^4 x - \frac{1}{2} \tan^2 x - \ln \cos x + c$.

> As with even and odd powers of sine and cosine, as the powers get higher, we are just repeating earlier techniques and again this could be generalized.

If multiple angles are used, this does not change the method.

Example

$$\int \tan^2 2x dx$$

This time the identity $\tan^2 2x = \sec^2 2x - 1$ is used.

$$\int \tan^2 2x dx = \int (\sec^2 2x - 1)dx$$

$$= \frac{1}{2} \tan 2x - x + c$$

Exercise 3

1 $\int \cos^3 x dx$　　　　**2** $\int \sin^3 2x dx$　　　　**3** $\int \sin^5 x dx$

4 $\int_0^p \tan^2 2x dx$　　　　**5** $\int \cos^2 2x dx$　　　　**6** $\int \sin^2 2x dx$

7 $\int \sin^4 x dx$　　　　**8** $\int \sin^9 x dx$　　　　**9** $\int \tan^3 3x dx$

10 $\int \sin^2 x \cos^2 x dx$　　**11** $\int \tan^3 x \sec^4 x dx$　　**12** $\int \sin^3 2x \cos^2 2x dx$

13 $\int_0^p \frac{\sin^2 x}{\sec^3 x} dx$

15.4 Selecting the correct technique 1

The skill in integration is often to recognize which techniques to apply. Exercise 4 contains a mixture of questions.

Exercise 4

1 $\int (x + 2)^4 \, dx$

2 $\int (2 + 7x)^3 \, dx$

3 $\int \dfrac{1}{\sqrt{1 - 2x}} \, dx$

4 $\int \left(\dfrac{3}{(2x + 1)^3} + \sqrt{1 + 2x} \right) dx$

5 $\dfrac{1}{4} \int \sqrt{3 + 5x} \, dx$

6 $\int \left(\sqrt{1 - x} + \dfrac{1}{\sqrt{1 - x}} - \dfrac{1}{(1 - x)^2} \right) dx$

7 $3 \int \cos \left(4x - \dfrac{\pi}{2} \right) dx$

8 $\int \dfrac{2}{1 + 4x^2} \, dx$

9 $\int \dfrac{\sin x}{3 - 4 \cos x} \, dx$

10 $\int \sec^2 \left(\dfrac{\pi}{3} - 2x \right) dx$

11 $2 \int \sin(3x + \alpha) \, dx$

12 $\int e^{4x + 1} \, dx$

13 $\int 2^x \, dx$

14 $\int \dfrac{1}{3x + 1} \, dx$

15 $\int \dfrac{2x}{x^2 + 4} \, dx$

16 $\int \dfrac{x + 1}{x^2 + 2x + 3} \, dx$

17 $\int \dfrac{x^3}{x^4 + 3} \, dx$

18 $\int \dfrac{3}{\sqrt{1 - 9x^2}} \, dx$

19 $\int \dfrac{6}{4 + 16x^2} \, dx$

20 $\int (x + 3)(x^2 + 6x - 8)^6 \, dx$

21 $\int \cos 2x (\sin 2x + 3)^4 \, dx$

22 $\int \text{cosec}^2 \dfrac{x}{2} e^{1 - \cot \frac{x}{2}} \, dx$

23 $\int \sqrt{x}(1 + x^{\frac{3}{2}})^7 \, dx$

24 $\int \cos^3 x \sin x \, dx$

25 $\int \dfrac{\text{cosec}^2 x}{(\cot x - 3)^3} \, dx$

26 $\int \dfrac{e^x}{e^x + 2} \, dx$

27 $\int \dfrac{e^x}{(e^x + 2)^{\frac{1}{2}}} \, dx$

28 $\int \sin^4 2x \, dx$

29 $\int \dfrac{x + 1}{x^2 + 2x + 4} \, dx$

30 $\int \dfrac{2}{x^2 + 2x + 3} \, dx$

31 $\int \dfrac{2}{\sqrt{-x^2 + 4x + 5}} \, dx$

32 $\int \dfrac{-2x + 4}{\sqrt{-x^2 + 4x + 5}} \, dx$

33 $\int \sin 2x (\sin^2 x + 3)^4 \, dx$

15.5 Integration by substitution

The method we have called direct reverse is actually the same as substitution except that we do the substitution mentally. The questions we have met so far could all have been done using a method of substitution, but it is much more time consuming. However, certain more complicated questions require a substitution to be used. If a question requires substitution then this will often be indicated, as will the necessary substitution. Substitution is quite straightforward, apart from "dealing with the dx part". Below is a proof of the equivalence of operators, which will allow us to "deal with dx".

Proof

Consider a function of u, $f(u)$.

$$\frac{d}{dx}[f(u)] = \frac{du}{dx} \times f'(u)$$

Hence integrating both sides gives $\int \frac{du}{dx} f'(u) dx = f(u) + k$ (equation 1).

Also $f'(u) = \frac{d}{du} f(u)$.

Therefore $\int f'(u) du = f(u) + k$ (equation 2).

Combining equation 1 with equation 2 gives $\int \frac{du}{dx} f'(u) dx = \int f'(u) du$

Therefore $\int \ldots \frac{du}{dx} dx = \int \ldots du$ where $\ldots$ is the function being integrated.

This is known as the equivalence of operators.

The question is how to use it. There is a great temptation to treat "dx" as part of a fraction. In the strictest sense it is not, it is a piece of notation, but at this level of mathematics most people do treat it as a fraction and in the examples we will do so. The equivalence of operators shown above demonstrates that treating $\frac{dy}{dx}$ as a fraction will also work.

Example

Find $\int \cos x(1 + \sin x)^{\frac{1}{2}} dx$ using the substitution $u = 1 + \sin x$.

This example could also be done by direct reverse. It is possible that an examination could ask for a question to be done by substitution when direct reverse would also work.

$(1 + \sin x)^{\frac{1}{2}} = u^{\frac{1}{2}}$

$$\frac{du}{dx} = \cos x$$

$\Rightarrow \cos x dx = du$

This is the same as using the equivalence of operators, which would work as follows:

$$\int \ldots \frac{du}{dx} dx = \int \ldots du$$

$$\Rightarrow \int \ldots \cos x dx = \int \ldots du$$

Making the substitution gives $\int u^{\frac{1}{2}}\, du = \frac{2}{3}u^{\frac{3}{2}}$

The answer cannot be left in this form and we need to substitute for x.

So $\int \cos x(1 + \sin x)^{\frac{1}{2}}\, dx = \frac{2}{3}(1 + \sin x)^{\frac{3}{2}} + c$

Example

Find $\int 3x\sqrt{4x - 1}\, dx$ using the substitution $u = 4x - 1$.

$\sqrt{4x - 1} = u^{\frac{1}{2}}$

$3x = 3\left(\dfrac{u + 1}{4}\right)$

$\dfrac{du}{dx} = 4$

Hence $dx = \dfrac{du}{4}$

$\Rightarrow \int 3x\sqrt{4x - 1}\, dx = \int 3\left(\dfrac{u + 1}{4}\right)u^{\frac{1}{2}}\dfrac{du}{4}$

$= \dfrac{3}{16}\int u^{\frac{3}{2}} + u^{\frac{1}{2}}\, du$

$= \dfrac{3}{16}\left[\dfrac{2}{5}u^{\frac{5}{2}} + \dfrac{2}{3}u^{\frac{3}{2}}\right] = \dfrac{u^{\frac{3}{2}}}{40}[3u + 5]$

Hence $\int 3x\sqrt{4x - 1}\, dx = \dfrac{(4x - 1)^{\frac{3}{2}}}{40}[3(4x - 1) + 5]$

$= \dfrac{(4x - 1)^{\frac{3}{2}}}{40}(12x + 2)$

$= \dfrac{(4x - 1)^{\frac{3}{2}}}{20}(6x + 1) + c$

Definite integration works the same way as with other integration, but the limits in the substitution need to be changed.

Example

Find $\int_{0}^{p}(x + 1)(2x - 1)^9\, dx$ using the substitution $u = 2x - 1$.

$(2x - 1)^9 = u^9$

$x + 1 = \dfrac{u + 3}{2}$

If the question has two numerical limits and appears on a calculator paper, then perform the calculation directly on a calculator.

$$\frac{du}{dx} = 2$$

$$\Rightarrow dx = \frac{du}{2}$$

Because this is a question of definite integration, the limits must be changed. The reason for this is that the original limits are values of x and we now need values of u as we are integrating with respect to u.

When $x = 0$, $u = -1$

When $x = p$, $u = 2p - 1$

Hence the integral now becomes

$$\int_{-1}^{2p-1} \left(\frac{u+3}{2}\right) u^9 \frac{du}{2} = \frac{1}{4} \int_{-1}^{2p-1} (u^{10} + 3u^9)\,du$$

$$= \frac{1}{4}\left[\frac{u^{11}}{11} + \frac{3u^{10}}{10}\right]_{-1}^{2p-1}$$

$$= \frac{1}{4}\left[\left(\frac{(2p-1)^{11}}{11} + \frac{3(2p-1)^{10}}{10}\right) - \left(-\frac{1}{11} + \frac{3}{10}\right)\right]$$

$$= \frac{1}{440}[10(2p-1)^{11} + 33(2p-1)^{10} - 23]$$

When limits are changed using substitution, it is sometimes the case that the limits switch around and the lower limit is bigger than the upper limit.

Example

Evaluate $\displaystyle\int_{-1}^{1} \frac{(x+1)dx}{(2-x)^4}$.

On a calculator paper this would be done directly by calculator.

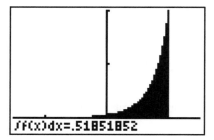

$$\int f(x)dx = .51851852$$

$$\Rightarrow \int_{-1}^{1} \frac{(x+1)dx}{(2-x)^4} = 0.519$$

On a non-calculator paper we would proceed as follows.

Let $u = 2 - x$

$(2 - x)^4 = u^4$

$x + 1 = 3 - u$

$$\frac{du}{dx} = -1$$

$\Rightarrow dx = -du$

When $x = -1$, $u = 3$

When $x = 1$, $u = 1$

Therefore $\displaystyle\int_{-1}^{1} \frac{(x + 1)dx}{(2 - x)^4}$ becomes

$$\int_{3}^{1} (u^{-3} - 3u^{-4})du = \left[\frac{u^{-2}}{-2} - \frac{3u^{-3}}{-3}\right]_{3}^{1}$$

$$= \left[\frac{-1}{2u^2} + \frac{1}{u^3}\right]_{3}^{1}$$

$$= \left(\frac{-1}{2} + 1\right) - \left(\frac{-1}{18} + \frac{1}{27}\right)$$

$$= \frac{14}{27}$$

The substitutions dealt with so far are fairly intuitive, but some of them are less obvious. In this case the question will sometimes state the substitution.

Example

Find $\displaystyle\int \sqrt{1 - x^2}\, dx$ using the substitution $x = \sin\theta$.

$\sqrt{1 - x^2} = \sqrt{1 - \sin^2\theta} = \sqrt{\cos^2\theta} = \cos\theta$

$\dfrac{dx}{d\theta} = \cos\theta \Rightarrow dx = \cos\theta\, d\theta$

Hence $\displaystyle\int \sqrt{1 - x^2}\, dx$ becomes $\displaystyle\int \cos^2\theta\, d\theta$.

Now $\qquad \cos 2\theta = \cos^2\theta - \sin^2\theta$

$\qquad \Rightarrow \cos 2\theta = 2\cos^2\theta - 1$

$\qquad \Rightarrow \cos^2\theta = \dfrac{1 + \cos 2\theta}{2}$

$\qquad \Rightarrow \displaystyle\int \cos^2\theta\, d\theta = \int \frac{1 + \cos 2\theta}{2}\, d\theta$

$\qquad\qquad\qquad\quad = \dfrac{\theta}{2} + \dfrac{\sin 2\theta}{4}$

We now substitute back for x.

Given that $x = \sin\theta \Rightarrow \theta = \sin^{-1}x$

Also $\sin 2\theta = 2\sin\theta\cos\theta$

From the triangle below, $\cos\theta = \sqrt{1 - x^2}$.

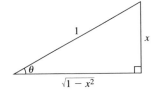

Hence $\displaystyle\int \sqrt{1 - x^2}\, dx = \frac{\sin^{-1} x}{2} + \frac{x\sqrt{1 - x^2}}{2} + c.$

Example

Find $\displaystyle\int \frac{2\tan x}{\cos 2x}\,dx$ using the substitution $t = \tan x$.

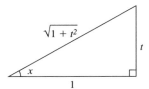

From the triangle above, $\sin x = \dfrac{t}{\sqrt{1 + t^2}}$ and $\cos x = \dfrac{1}{\sqrt{1 + t^2}}$.

Hence $\cos 2x = \cos^2 x - \sin^2 x = \dfrac{1}{1 + t^2} - \dfrac{t^2}{1 + t^2} = \dfrac{1 - t^2}{1 + t^2}$.

If $t = \tan x$, then $\dfrac{dt}{dx} = \sec^2 x = 1 + \tan^2 x = 1 + t^2$.

$$\Rightarrow dx = \frac{dt}{1 + t^2}$$

$$\int \frac{2\tan x}{\cos 2x}\,dx = \int \frac{2t}{\dfrac{1 - t^2}{1 + t^2}} \times \frac{dt}{1 + t^2} = \int \frac{2t}{1 - t^2}\,dt$$

This is now done by direct reverse and begins with $y = \ln|1 - t^2|$.

$$\Rightarrow \frac{dy}{dt} = \frac{-2t}{1 - t^2}$$

Therefore $\displaystyle\int \frac{2\tan x}{\cos 2x}\,dx = \int \frac{2t}{1 - t^2}\,dt = -\ln|1 - t^2| = -\ln|1 - \tan^2 x| + c.$

Example

Find $\displaystyle\int \frac{1}{4 + \sin x}\,dx$ using the substitution $t = \tan\dfrac{x}{2}$.

Since $\tan x = \dfrac{2\tan\dfrac{x}{2}}{1 - \tan^2\dfrac{x}{2}}$

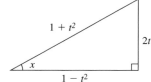

From the above diagram, $\sin x = \dfrac{2t}{1 + t^2}$.

$$\frac{dt}{dx} = \frac{1}{2}\sec^2\frac{x}{2} = \frac{1}{2}\left(1 + \tan^2\frac{x}{2}\right) = \frac{1}{2}(1 + t^2)$$

$$\Rightarrow dx = \frac{2dt}{1 + t^2}$$

Therefore $\displaystyle\int \frac{1}{4 + \sin x}\,dx = \int \frac{2\dfrac{dt}{1 + t^2}}{4 + \dfrac{2t}{1 + t^2}} = \int \frac{2}{4(1 + t^2) + 2t}\,dt$

$$= \int \frac{2}{4\left(t^2 + \dfrac{t}{2} + 1\right)}\, dt = \int \frac{2}{4\left[\left(t + \dfrac{1}{4}\right)^2 + \dfrac{15}{16}\right]}\, dt$$

$$= \frac{8}{15} \int \frac{1}{1 + \left[\dfrac{1}{\sqrt{15}}(4t + 1)\right]^2}\, dt$$

This is now integrated by direct reverse beginning with $y = \tan^{-1}\dfrac{1}{\sqrt{15}}(4t + 1)$.

$$\Rightarrow \frac{dy}{dt} = \frac{\dfrac{4}{\sqrt{15}}}{1 + \left[\dfrac{1}{\sqrt{15}}(4t + 1)\right]^2}$$

$$\Rightarrow \frac{4}{\sqrt{15}} \int \frac{1}{1 + \left[\dfrac{1}{\sqrt{15}}(4t + 1)\right]^2}\, dt = \tan^{-1}\frac{1}{\sqrt{15}}(4t + 1)$$

$$\Rightarrow \frac{8}{15} \int \frac{1}{1 + \left[\dfrac{1}{\sqrt{15}}(4t + 1)\right]^2}\, dt = \frac{2}{\sqrt{15}}\tan^{-1}\frac{1}{\sqrt{15}}(4t + 1)$$

$$\Rightarrow \int \frac{1}{4 + \sin x}\, dx = \frac{2}{\sqrt{15}}\tan^{-1}\frac{1}{\sqrt{15}}\left(4\tan\frac{x}{2} + 1\right) + c$$

Exercise 5

1 Find $\int x(x^2 + 3)^5\, dx$ using the substitution $u = x^2 + 3$.

2 Find $\int \dfrac{3x + 1}{6x^2 + 4x - 13}\, dx$ using the substitution $u = 6x^2 + 4x - 13$.

3 Find $\int \dfrac{\cos 2x}{\sqrt{1 - \sin 2x}}\, dx$.

4 Find $\int x\sqrt{x - 2}\, dx$ using the substitution $u = x - 2$.

5 Find $\int_1^p \dfrac{x}{\sqrt{2x - 1}}\, dx$.

6 Find $\int \dfrac{(x + 3)}{\sqrt{2x + 1}}\, dx$ using the substitution $u = 2x + 1$.

7 Find $\int \dfrac{2x}{1 + x^4}\, dx$ using the substitution $u = x^2$.

8 Find $\int (x + 2)\sqrt{3x - 4}\, dx$.

9 Find $\int_1^p (2x - 1)(x - 2)^3\, dx$ using the substitution $u = x - 2$.

10 Find $\int \dfrac{x}{(2x - 1)^4}\, dx$ using the substitution $u = 2x - 1$.

11 Find $\displaystyle\int \frac{2x+1}{(x-3)^6}\,dx.$ **12** Find $\displaystyle\int_9^p \frac{x}{\sqrt{x-2}}\,dx.$ **13** Find $\displaystyle\int \frac{x^3}{(x+5)^2}\,dx.$

14 Find $\displaystyle\int_2^p x\sqrt{5x+2}\,dx.$ **15** Find $\displaystyle\int \frac{x(x-4)}{(x-2)^2}\,dx.$

16 Find $\displaystyle\int \frac{1}{\cos^2 x + 4\sin^2 x}\,dx$ using the substitution $t = \tan x$.

17 Find $\displaystyle\int \sqrt{9 - 9x^2}\,dx$ using the substitution $x = \sin\theta$.

18 Find $\displaystyle\int \frac{1}{7 - 5\cos x}\,dx$ using the substitution $t = \tan\frac{x}{2}$.

19 Find $\displaystyle\int_{0.5}^p \sqrt{4 - x^2}\,dx$ using the substitution $x = 2\sin\theta$.

20 Find $\displaystyle\int \frac{1}{5\sin^2 x + \cos^2 x}\,dx$ using the substitution $t = \tan x$.

21 Find $\displaystyle\int_0^p \frac{6}{5 + 3\sin x}\,dx$ using the substitution $t = \tan\frac{x}{2}$.

22 Find $\displaystyle\int \frac{1}{8 + 8\cos 4x}\,dx$ using the substitution $t = \tan 2x$.

23 Find $\displaystyle\int \frac{4}{3x\sqrt{x^n - 1}}\,dx$ using the substitution $u^2 = x^n - 1$.

15.6 Integration by parts

As was mentioned earlier in the chapter, not all products can be integrated by the method of direct reverse. Integration by parts is another technique and tends to be used when one half of the product is not related to the other half. Direct reverse is basically undoing the chain rule and integration by parts is basically reversing the product rule. However, unlike direct reverse, this does not mean that it is used for those answers that came from the product rule.

We will begin by showing the formula.

We know that

$$\frac{d}{dx}(uv) = v\frac{du}{dx} + u\frac{dv}{dx} \quad \text{where } u \text{ and } v \text{ are both functions of } x.$$

$$\Rightarrow v\frac{du}{dx} = \frac{d}{dx}(uv) - u\frac{dv}{dx}$$

Integrating both sides with respect to x gives:

$$\int v\frac{du}{dx}\,dx = \int \frac{d}{dx}(uv)\,dx - \int u\frac{dv}{dx}\,dx$$

Now $\displaystyle\int \frac{d}{dx}(uv)\,dx$ is just uv, so:

$$\int v\frac{du}{dx}\,dx = uv - \int u\frac{dv}{dx}\,dx$$

This is the formula for integration by parts.

The basic method is as follows.

Let one part of the product be v and one part $\frac{du}{dx}$. Calculate u and $\frac{dv}{dx}$ and then use the formula.

Unlike the product rule in differentiation, in some cases it makes a difference which part is v and which part is $\frac{du}{dx}$ and in other cases it makes no difference. The choice depends on what can be integrated, and the aim is to make the problem easier. The table below will help.

One half of product	Other half of product	Which do you differentiate?
Power of x	Trigonometric ratio	Power of x
Power of x	Inverse trigonometric ratio	Power of x
Power of x	Power of e	Power of x
Power of x	ln f(x)	ln f(x)
Power of e	sin f(x), cos f(x)	Does not matter

This can also be summarized as a priority list.

Which part is v?
1. Choose ln f(x).
2. Choose the power of x.
3. Choose $e^{f(x)}$ or sin f(x), cos f(x).

Example

Find $\int xe^x\,dx$.

Using the formula $\int v\frac{du}{dx}\,dx = uv - \int u\frac{dv}{dx}\,dx$,

let $v = x$ and $\frac{du}{dx} = e^x$

$\Rightarrow \frac{dv}{dx} = 1$ and $u = \int e^x\,dx = e^x$

Now substitute the values in the formula.

$$\int xe^x\,dx = xe^x - \int 1e^x\,dx$$

$$\Rightarrow \int xe^x\,dx = xe^x - e^x + c$$

x is differentiated here since it will differentiate to 1 and allow the final integration to be carried out.

It is possible to leave out the mechanics of the question once you feel more confident about the technique.

Example

Find $\int x \sin x \, dx$.

$$\int x \sin x \, dx = -x \cos x - \int -\cos x \times 1 \, dx \ldots\ldots\ldots\ldots\ldots\ldots\ldots\ldots\ldots\ldots\ldots\ldots\ldots\ldots$$

$$= -x \cos x + \int \cos x \, dx$$

$$= -x \cos x + \sin x + c$$

It is not recommended to try to simplify the signs and constants at the same time as doing the integration!

Example

Find $\int x^3 \ln x \, dx$.

Using the formula $\int v \dfrac{du}{dx} \, dx = uv - \int u \dfrac{dv}{dx} \, dx$,

let $v = \ln x$ and $\dfrac{du}{dx} = x^3$

$\Rightarrow \dfrac{dv}{dx} = \dfrac{1}{x}$ and $u = \int x^3 \, dx = \dfrac{x^4}{4}$

Substituting the values in the formula gives:

$$\int x^3 \ln x \, dx = \frac{x^4}{4} \ln x - \int \frac{x^4}{4} \times \frac{1}{x} \, dx = \frac{x^4}{4} \ln x - \frac{1}{4} \int x^3 \, dx$$

Therefore $\int x^3 \ln x \, dx = \dfrac{x^4}{4} \ln x - \dfrac{x^4}{16} = \dfrac{x^4}{16}(4 \ln x - 1) + c$

There is no choice but to differentiate $\ln x$ since it cannot be integrated at this point.

Example

Find $\int_0^p 4x \cos x \, dx$.

Using the formula $\int v \dfrac{du}{dx} \, dx = uv - \int u \dfrac{dv}{dx} \, dx$,

let $v = 4x$ and $\dfrac{du}{dx} = \cos x$

$\Rightarrow \dfrac{dv}{dx} = 4$ and $u = \int \cos x \, dx = \sin x$

Substituting the values in the formula gives:

$$\int_0^p 4x \cos x \, dx = [4x \sin x]_0^p - \int_0^p 4 \sin x \, dx$$

$$= [4p \sin p - 0] - [-4 \cos x]_0^p$$

$$= [4p \sin p - 0] - [4 \cos p + 4]$$

$$= 4p \sin p + 4 \cos p - 4$$

Example

Find $\int x^2 e^x \, dx$.

Here the integration by parts formula will need to be applied twice.

Using the formula $\int v\dfrac{du}{dx}\,dx = uv - \int u\dfrac{dv}{dx}\,dx,$

let $v = x^2$ and $\dfrac{du}{dx} = e^x$

$\Rightarrow \dfrac{dv}{dx} = 2x$ and $u = \int e^x\,dx = e^x$

$\Rightarrow \int x^2 e^x\,dx = x^2 e^x - \int 2xe^x\,dx = x^2 e^x - 2\int xe^x\,dx$

> It is always a good idea to take the constants outside the integral sign.

We need to find $\int xe^x\,dx$. This is again done using the method of integration by parts.

Using the formula $\int v\dfrac{du}{dx}\,dx = uv - \int u\dfrac{dv}{dx}\,dx,$

let $v = x$ and $\dfrac{du}{dx} = e^x$

$\Rightarrow \dfrac{dv}{dx} = 1$ and $u = \int e^x\,dx = e^x$

$\Rightarrow \int x\,e^x\,dx = xe^x - \int 1e^x\,dx = xe^x - e^x$

Combining the two gives:

$\int x^2 e^x\,dx = x^2 e^x - 2(xe^x - e^x)$

$\qquad\qquad = x^2 e^x - 2xe^x + 2e^x + c$

Example

Find $\int e^{2x}\sin x\,dx$.

This is a slightly different case, since it makes no difference which part is integrated and which part is differentiated. With a little thought this should be obvious since, excluding constants, repeated integration or differentiation of these functions gives the same pattern of answers. Remember the aid

$$\text{Differentiate} \left.\begin{array}{c} S \\ C \\ -S \\ -C \end{array}\right\uparrow \text{Integrate}$$

and the fact that functions of e^x differentiate or integrate to themselves.

We begin by letting $\int e^{2x} \sin x\, dx = I$.

Using the formula $\int v\dfrac{du}{dx}\, dx = uv - \int u\dfrac{dv}{dx}\, dx$,

let $v = e^{2x}$ and $\dfrac{du}{dx} = \sin x$

$\Rightarrow \dfrac{dv}{dx} = 2e^{2x}$ and $u = \int \sin x\, dx = -\cos x$

$\Rightarrow I = \int e^{2x} \sin x\, dx = -e^{2x} \cos x - \int -\cos x \times 2e^{2x}\, dx$

$\qquad\qquad = -e^{2x} \cos x + 2\int e^{2x} \cos x\, dx$

Applying the formula again, being very careful to ensure that we continue to integrate the trigonometric function and differentiate the power of e gives:

$v = e^{2x}$ and $\dfrac{du}{dx} = \cos x$

$\Rightarrow \dfrac{dv}{dx} = 2e^{2x}$ and $u = \int \cos x\, dx = \sin x$

Hence $\int e^{2x} \cos x\, dx = e^{2x} \sin x - \int \sin x \times 2e^{2x}\, dx$

$\qquad\qquad\qquad = e^{2x} \sin x - 2\int e^{2x} \sin x\, dx$ .... This is the original integral I.

Hence $I = \int e^{2x} \sin x\, dx = -e^{2x} \cos x + 2\left(e^{2x} \sin x - 2\int e^{2x} \sin x\, dx\right)$

$\Rightarrow I = -e^{2x} \cos x + 2e^{2x} \sin x - 4I$

$\Rightarrow 5I = -e^{2x} \cos x + 2e^{2x} \sin x$ ·················· Calling the original integral I makes this rearrangement easier.

$\Rightarrow I = \dfrac{1}{5}(-e^{2x} \cos x + 2e^{2x} \sin x) = \dfrac{e^{2x}}{5}(-\cos x + 2\sin x) + c$

Example

Find $\int \ln x\, dx$.

This is done as a special case of integration by parts. However, it is not a product of two functions. To resolve this issue we let the other function be 1.

Hence this becomes $\int 1 \ln x\, dx$ and the integration by parts formula is applied as usual.

Using the formula $\int v\dfrac{du}{dx}\, dx = uv - \int u\dfrac{dv}{dx}\, dx$,

let $v = \ln x$ and $\dfrac{du}{dx} = 1$

$$\Rightarrow \frac{dv}{dx} = \frac{1}{x} \text{ and } u = \int 1 \, dx = x$$

Hence $\int 1 \ln x \, dx = x \ln x - \int x \times \frac{1}{x} \, dx = x \ln x - \int 1 \, dx = x \ln x - x + c$

To integrate inverse trigonometric functions an identical method is used, for example

$\int \cos^{-1} x \, dx = \int 1 \cdot \cos^{-1} x \, dx.$

Exercise 6

Find these integrals using the method of integration by parts.

1 $\int x \cos x \, dx$

2 $\int x \, e^{2x} \, dx$

3 $\int x^4 \ln x \, dx$

4 $\int x \sin 2x \, dx$

5 $\int_1^p x(x + 1)^9 \, dx$

6 $\int x^2 \sin x \, dx$

7 $\int x^2 \, e^{2x} \, dx$

8 $\int x^2 \ln 3x \, dx$

9 $\int 3x^2 \ln 8x \, dx$

10 $\int x^2 \, e^{-3x} \, dx$

11 $\int e^x \cos x \, dx$

12 $\int \sin^{-1} x \, dx$

13 $\int \tan^{-1} x \, dx$

14 $\int e^{2x}(2x - 1) \, dx$

15 $\int e^{3x} \cos x \, dx$

16 $\int e^{2x} \sin 3x \, dx$

17 $\int 2e^x \sin x \cos x \, dx$

18 $\int x^n \ln x \, dx$

19 $\int e^{ax} \sin bx \, dx$

20 $\int_0^p x(2x + 1)^n \, dx$

15.7 Miscellaneous techniques

There are two other techniques that need to be examined. These methods are normally only used when it is suggested by a question or when earlier techniques do not work.

Splitting the numerator

This is a trick that can really help when the numerator is made up of two terms. Often these questions cannot be tackled by a method of direct reverse as the derivative of the denominator does not give a factor of the numerator. Substitution is unlikely to simplify the situation and integration by parts does not produce an integral that is any simpler.

Example

Find $\int \frac{2x + 1}{x^2 + 1} \, dx.$

Splitting the numerator gives two integrals.

$\int \frac{2x + 1}{x^2 + 1} \, dx = 2 \int \frac{x}{x^2 + 1} \, dx + \int \frac{1}{x^2 + 1} \, dx$

The first integral can be done by direct reverse and the second one is a standard result. To integrate the first integral we begin with $y = \ln|x^2 + 1|$.

So $\dfrac{dy}{dx} = \dfrac{2x}{x^2 + 1}$

Hence $2\displaystyle\int \dfrac{x}{x^2 + 1}\, dx = \ln|x^2 + 1|$

Therefore $\displaystyle\int \dfrac{2x + 1}{x^2 + 1}\, dx = 2\int \dfrac{x}{x^2 + 1}\, dx + \int \dfrac{1}{x^2 + 1}\, dx$

$$= \ln|x^2 + 1| + \tan^{-1} x + c$$

Example

Find $\displaystyle\int \dfrac{2x}{x^2 + 2x + 26}\, dx$.

This is a slightly different case as we now make the numerator $2x + 2$ and then split the numerator. Hence

$\displaystyle\int \dfrac{2x}{x^2 + 2x + 26}\, dx$ becomes $\displaystyle\int \dfrac{2x + 2}{x^2 + 2x + 26}\, dx - \int \dfrac{2}{x^2 + 2x + 26}\, dx$.

The first integral is calculated by direct reverse and the second integral will become a function of inverse tan.

Consider the first integral.

$\displaystyle\int \dfrac{2x + 2}{x^2 + 2x + 26}\, dx$

To integrate this we know it began with something to do with

$\qquad y = \ln|x^2 + 2x + 26|$.

$\Rightarrow \dfrac{dy}{dx} = \dfrac{2x + 2}{x^2 + 2x + 26}$

Hence $\displaystyle\int \dfrac{2x + 2}{x^2 + 2x + 26}\, dx = \ln|x^2 + 2x + 26| + k_1$

Now look at the second integral.

$\displaystyle\int \dfrac{2}{x^2 + 2x + 26}\, dx = \int \dfrac{2}{(x + 1)^2 + 25}\, dx = \int \dfrac{2}{25 + (x + 1)^2}\, dx$ ....

$$= \dfrac{2}{25}\int \dfrac{1}{1 + \left(\dfrac{x + 1}{5}\right)^2}\, dx$$

Using direct reverse, we begin with $y = \tan^{-1}\left(\dfrac{x + 1}{5}\right)$.

> Here we complete the square on the denominator to produce an inverse tan result.

$$\frac{dy}{dx} = \frac{1}{1 + \left(\dfrac{x+1}{5}\right)^2} \times \frac{1}{5}$$

Therefore $\dfrac{1}{5}\displaystyle\int \dfrac{1}{1 + \left(\dfrac{x+1}{5}\right)^2}\,dx = \tan^{-1}\left(\dfrac{x+1}{5}\right)$

$$\Rightarrow \frac{2}{25}\int \frac{1}{1 + \left(\dfrac{x+1}{5}\right)^2}\,dx = \frac{2}{5}\tan^{-1}\left(\frac{x+1}{5}\right) + k_2$$

Putting the two integrals together gives:

$$\int \frac{2x}{x^2 + 2x + 26}\,dx = \int \frac{2x + 2}{x^2 + 2x + 26}\,dx - \int \frac{2}{x^2 + 2x + 26}\,dx$$

$$= \ln|x^2 + 2x + 26| - \frac{2}{5}\tan^{-1}\left(\frac{x+1}{5}\right) + c$$

Algebraic division

If the numerator is of higher or equal power to the denominator, then algebraic division may help. Again this only needs to be tried if other methods have failed. In Chapter 8, rational functions (functions of the form $\dfrac{P(x)}{Q(x)}$ where P(x) and Q(x) are both polynomials) were introduced when finding non-vertical asymptotes. Algebraic division was used if degree of P(x) $\geq$ degree of Q(x). In order to integrate these functions exactly the same thing is done.

Example

Find $\displaystyle\int \dfrac{x+3}{x+1}\,dx$.

Algebraically dividing the fraction:

$$\begin{array}{r} 1 \\ x + 1 \overline{)\, x + 3\,} \\ \underline{x + 1} \\ 2 \end{array}$$

So the question becomes $\displaystyle\int 1 + \dfrac{2}{x+1}\,dx$.

$$\int 1 + \frac{2}{x+1}\,dx = x + 2\ln|x+1| + c$$

Example

Find $\displaystyle\int \frac{x^2 + 1}{x + 1}\, dx$.

Algebraically dividing the fraction:

$$
\begin{array}{r}
x - 1 \\
x + 1\overline{)x^2 + 0x + 1} \\
\underline{x^2 + x} \\
-x + 1 \\
\underline{-x - 1} \\
2
\end{array}
$$

So the question becomes $\displaystyle\int (x - 1) + \frac{2}{x + 1}\, dx$.

$$\int (x - 1) + \frac{2}{x + 1}\, dx = \frac{x^2}{2} - x + 2\ln|x + 1| + c$$

Exercise 7

Evaluate these integrals.

1 $\displaystyle\int \frac{x + 1}{\sqrt{1 - x^2}}\, dx$ **2** $\displaystyle\int \frac{3x + 4}{x^2 + 4}\, dx$ **3** $\displaystyle\int \frac{x + 5}{x^2 + 3}\, dx$

4 $\displaystyle\int \frac{4x + 7}{x^2 + 4x + 8}\, dx$ **5** $\displaystyle\int \frac{2x + 3}{x^2 + 4x + 6}\, dx$ **6** $\displaystyle\int \frac{-2x - 5}{\sqrt{-x^2 - 6x - 4}}\, dx$

7 $\displaystyle\int \frac{x - 3}{x + 4}\, dx$ **8** $\displaystyle\int \frac{x^2 + 1}{x + 3}\, dx$

15.8 Further integration practice

All the techniques of integration for this curriculum have now been met. The following exercise examines all the techniques. It should be noted that there is often more than one technique that will work. For example, direct reverse questions can be done by substitution and some substitution questions can be done by parts.

Exercise 8

Find these integrals using the method of direct reverse.

1 $\displaystyle\int (\cos 3x + \sin 2x)\, dx$ **2** $\displaystyle\int (4\sqrt{x} + 4\sqrt{x + 1} - 4(1 - 3x)^3)\, dx$

3 $\displaystyle\int \frac{2x}{3x^2 + 1}\, dx$ **4** $\displaystyle\int \frac{4}{1 + 4x^2}\, dx$

5 $\displaystyle\int \sec^2\!\left(2x - \frac{\pi}{3}\right) dx$ **6** $\displaystyle\int \frac{\cos x}{\sqrt{1 + \sin x}}\, dx$

7 $\displaystyle\int (4x + 2)e^{x^2 + x - 5}\, dx$ **8** $\displaystyle\int \frac{\sin x}{\cos^n x}\, dx$

9 $\displaystyle\int \frac{2}{4x^2 + 8x + 5}\, dx$ **10** $\displaystyle\int \frac{2}{\sqrt{-x^2 - 8x + 9}}\, dx$

Find these by using a substitution.

11 $\displaystyle\int 2x(1-x)^7 \, dx$ using the substitution $u = 1 - x$

12 $\displaystyle\int \frac{x^2}{(x+5)^2} \, dx$ using the substitution $u = x + 5$

13 $\displaystyle\int 2x\sqrt{3x-4} \, dx$

14 $\displaystyle\int_1^p \frac{2x+1}{(x-3)^6} \, dx$

15 $\displaystyle\int \frac{1}{e^x + e^{-x}} \, dx$ using the substitution $u = e^x$

16 $\displaystyle\int \sqrt{4 - 4x^2} \, dx$ using the substitution $x = \sin\theta$

17 $\displaystyle\int \frac{1}{2 + \cos x} \, dx$ using the substitution $t = \tan\dfrac{x}{2}$

18 $\displaystyle\int \frac{1}{x^2\sqrt{4 - x^2}} \, dx$ using the substitution $x = 2\sin\theta$

19 $\displaystyle\int \frac{3x}{1 + x^4} \, dx$ using the substitution $x^2 = p$

20 $\displaystyle\int \frac{x+1}{x\sqrt{x-2}} \, dx$ using the substitution $x - 2 = p^2$

Find these by integrating by parts.

21 $\displaystyle\int \frac{1}{x^2}\ln x \, dx$

22 $\displaystyle\int x\,e^{3x} \, dx$

23 $\displaystyle\int x\cos\left(x + \frac{\pi}{6}\right) dx$

24 $\displaystyle\int e^{-2x}\cos 2x \, dx$

25 $\displaystyle\int x^2 \sin\frac{x}{2} \, dx$

26 $\displaystyle\int \ln(2x + 1) \, dx$

27 $\displaystyle\int \tan^{-1}\left(\frac{1}{x}\right) dx$

28 $\displaystyle\int e^{ax}\sin 2x \, dx$

Integrate these trigonometric powers.

29 $\displaystyle\int \cos x \sin^2 x \, dx$

30 $\displaystyle\int \frac{\cos^2 x}{\operatorname{cosec} x} \, dx$

31 $\displaystyle\int \frac{\tan^3 x}{\cos^2 x} \, dx$

32 $\displaystyle\int \cos^2\theta \, d\theta$

33 $\displaystyle\int \sin^2 3x \, dx$

34 $\displaystyle\int \cos^3 2x \, dx$

35 $\displaystyle\int \sin^5\frac{x}{4} \, dx$

36 $\displaystyle\int \tan^4\frac{x}{2} \, dx$

37 $\displaystyle\int 2\sin^2 ax \cos^2 ax \, dx$

38 $\displaystyle\int \tan^2 x \sec^4 x \, dx$

39 $\displaystyle\int \cos^4\frac{x}{6} \, dx$

Use either splitting the numerator or algebraic division to find these.

40 $\displaystyle\int \frac{2x-1}{\sqrt{1-x^2}} \, dx$

41 $\displaystyle\int \frac{3x-4}{\dfrac{3x^2}{2} - 2x + 3} \, dx$

42 $\displaystyle\int \frac{x^3+1}{x-1} \, dx$

43 $\displaystyle\int \frac{x^2+3}{2x-1} \, dx$

15.9 Selecting the correct technique 2

The different techniques of integration should now be familiar, but in many situations the technique will not be given. The examples below demonstrate how similar looking questions can require quite different techniques.

Example

Find $\displaystyle\int \frac{3x + 2}{3x^2 + 4x + 8}\, dx$.

This is direct reverse beginning with $y = \ln|3x^2 + 4x + 8|$.

So $\displaystyle\frac{dy}{dx} = \frac{6x + 4}{3x^2 + 4x + 8} = \frac{2(3x + 2)}{3x^2 + 4x + 8}$

$\displaystyle\Rightarrow 2\int \frac{3x + 2}{3x^2 + 4x + 8}\, dx = \ln|3x^2 + 4x + 8|$

$\displaystyle\Rightarrow \int \frac{3x + 2}{3x^2 + 4x + 8}\, dx = \frac{1}{2}\ln|3x^2 + 4x + 8| + c$

Example

Find $\displaystyle\int \frac{3x + 2}{(3x^2 + 4x + 8)^4}\, dx$.

Write the integral as $\displaystyle\int (3x + 2)(3x^2 + 4x + 8)^{-4}\, dx$.

This is direct reverse beginning with $y = (3x^2 + 4x + 8)^{-3}$.

So $\displaystyle\frac{dy}{dx} = -3(6x + 4)(3x^2 + 4x + 8)^{-4} = -6(3x + 2)(3x^2 + 4x + 8)^{-4}$

$\displaystyle\Rightarrow -6\int (3x + 2)(3x^2 + 4x + 8)^{-4}\, dx = (3x^2 + 4x + 8)^{-3}$

$\displaystyle\Rightarrow \int (3x + 2)(3x^2 + 4x + 8)^{-4}\, dx = -\frac{1}{6}(3x^2 + 4x + 8)^{-3} + c$

Example

Find $\displaystyle\int \frac{6x + 3}{3x^2 + 4x + 8}\, dx$.

This is a case of splitting the numerator.

Hence the integral becomes $\displaystyle\int \frac{6x + 4}{3x^2 + 4x + 8}\, dx - \int \frac{1}{3x^2 + 4x + 8}\, dx$.

The first integral is direct reverse of $y = \ln|3x^2 + 4x + 8|$.

So $\displaystyle\frac{dy}{dx} = \frac{6x + 4}{3x^2 + 4x + 8}$

$$\Rightarrow \int \frac{6x + 4}{3x^2 + 4x + 8}\, dx = \ln|3x^2 + 4x + 8| + k_1$$

The second integral requires completion of the square.

$$\int \frac{1}{3x^2 + 4x + 8}\, dx = \int \frac{1}{3\left(x^2 + \frac{4}{3}x + \frac{8}{3}\right)}\, dx$$

$$= \int \frac{1}{3\left[\left(x + \frac{2}{3}\right)^2 + \frac{20}{9}\right]}\, dx = \int \frac{1}{3\left(x + \frac{2}{3}\right)^2 + \frac{20}{3}}\, dx$$

$$= \frac{1}{20} \int \frac{1}{\frac{9}{3}\left(x + \frac{2}{3}\right)^2 + 1}\, dx = \frac{3}{20} \int \frac{1}{1 + \frac{9}{20}\left(x + \frac{2}{3}\right)^2}\, dx$$

Using direct reverse, this begins with $y = \tan^{-1} \frac{3}{\sqrt{20}}\left(x + \frac{2}{3}\right)$.

$$\frac{dy}{dx} = \frac{1}{1 + \frac{9}{20}\left(x + \frac{2}{3}\right)^2} \times \frac{3}{\sqrt{20}}$$

Therefore $\dfrac{3}{\sqrt{20}} \displaystyle\int \dfrac{1}{1 + \frac{9}{20}\left(x + \frac{2}{3}\right)^2}\, dx = \tan^{-1} \dfrac{3}{\sqrt{20}}\left(x + \dfrac{2}{3}\right)$

$$\Rightarrow \frac{3}{20} \int \frac{1}{1 + \frac{9}{20}\left(x + \frac{2}{3}\right)^2}\, dx = \frac{1}{\sqrt{20}} \tan^{-1} \frac{3}{\sqrt{20}}\left(x + \frac{2}{3}\right) + k$$

Hence

$$\int \frac{6x + 3}{3x^2 + 4x + 8}\, dx = \int \frac{6x + 4}{3x^2 + 4x + 8}\, dx - \int \frac{1}{3x^2 + 4x + 8}\, dx$$

$$= \ln|3x^2 + 4x + 8| - \frac{1}{\sqrt{20}} \tan^{-1} \frac{3}{\sqrt{20}}\left(x + \frac{2}{3}\right)$$

Even though all the techniques for this syllabus have been met, there are still a lot of functions that cannot be integrated. There are two reasons for this. First, there are other techniques which have not been covered and second there are some fairly simple looking functions which cannot be integrated by any direct method. An example of this is $\int e^{x^2}\, dx$. However, questions like these in the form of definite integrals can be asked as it is expected that these would be done on a calculator. If a definite integral is asked for on a calculator paper, then it should be done on a calculator unless there is a good reason (for example being asked for an exact answer).

Example

Work out $\displaystyle\int_{1}^{2} x\, e^{x^3}\, dx$.

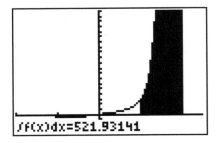

$\int f(x)dx = 521.93141$

This cannot be done by any direct method, so the only choice is to use a calculator which will give the answer of 522.

Exercise 9

Use a calculator where appropriate to find these.

1 $\displaystyle\int (x-3)^3\, dx$

2 $\displaystyle\int \sqrt{3x-5}\, dx$

3 $\displaystyle\int e^{4x-5}\, dx$

4 $\displaystyle\int (2x^{\frac{2}{3}} - 4x^{\frac{1}{4}})^2\, dx$

5 $\displaystyle\int \operatorname{cosec} 4x \cot 4x\, dx$

6 $\displaystyle\int_{\frac{\pi}{6}}^{\frac{\pi}{2}} \cos^6 x \sin 2x\, dx$

7 $\displaystyle\int x^2\, e^{-2x}\, dx$

8 $\displaystyle\int_{0}^{1} \frac{x^3}{(1+e^x)^{\frac{1}{3}}}\, dx$

9 $\displaystyle\int \frac{2x-3}{x^2+1}\, dx$

10 $\displaystyle\int \frac{1}{\sqrt{25-4x^2}}\, dx$

11 $\displaystyle\int \frac{e^x}{(2e^x+1)^3}\, dx$

12 $\displaystyle\int_{0}^{p} \frac{\sin x}{a+b\cos x}\, dx$

13 $\displaystyle\int \frac{x}{(2+5x)^3}\, dx$ using the substitution $u = 2 + 5x$

14 $\displaystyle\int \frac{3}{x^2-6x+25}\, dx$

15 $\displaystyle\int \cos^4 x\, dx$

16 $\displaystyle\int_{2}^{5} \frac{3}{5-7x^2}\, dx$

17 $\displaystyle\int_{0}^{2} e^{2x^2}\, dx$

18 $\displaystyle\int \frac{3x^2}{2}\sin 2x\, dx$

19 $\displaystyle\int \frac{1}{\sqrt{-x^2-4x+29}}\, dx$

20 $\displaystyle\int \log_4 x\, dx$

21 $\displaystyle\int \frac{1}{\sqrt{x(2-x)}}\, dx$

22 $\displaystyle\int x^4 \ln 2x\, dx$

23 $\displaystyle\int_{1}^{2} \frac{3x^4}{x^3+3}\, dx$

24 $\displaystyle\int_{0.1}^{0.5} \frac{\sqrt{3-5x}}{x}\, dx$

25 $\displaystyle\int e^{-3x} \cos x\, dx$

26 $\displaystyle\int \frac{x^2+7}{5\sqrt{x}}\, dx$

27 $\displaystyle\int \frac{\tan^4 x}{\cos^4 x}\, dx$

28 $\displaystyle\int \sqrt{\frac{1-2x}{1+2x}}\, dx$

29 $\displaystyle\int \frac{1}{3 - 2\cos x}\, dx$ using the substitution $t = \tan\dfrac{x}{2}$

30 $\displaystyle\int_{-1}^{0} \sqrt{4 - 3e^x}\, dx$ **31** $\displaystyle\int_{0}^{a} \frac{2x}{(x + 1)^4}\, dx$ using the substitution $u = x + 1$

32 $\displaystyle\int \cos^{-1} 2x\, dx$ **33** $\displaystyle\int_{0}^{\infty} x\, e^{-x}\, dx$

34 $\displaystyle\int_{0}^{a} x^2 \sqrt{a^2 - x^2}\, dx$ using the substitution $x = a\sin\theta$

35 $\displaystyle\int_{-1}^{0} \frac{3x^7}{2 - 13x}\, dx$ **36** $\displaystyle\int_{1}^{4} \sin^{-1}\frac{1}{x}\, dx$ **37** $\displaystyle\int \frac{\sin x \cos x}{\cos^2 x - \sin^2 x}\, dx$

38 $\displaystyle\int_{0}^{\frac{\pi}{3}} \cos 6x \cos 3x\, dx$

15.10 Finding the area under a curve

We will now look at finding areas under curves by using these techniques.

Example

Consider the curve $y = e^x \cos x$. Using a calculator, find the area bounded by the curve, the x-axis, the y-axis and the line $x = a$ where $2 \le a \le 4$.

Drawing the curve on a calculator gives:

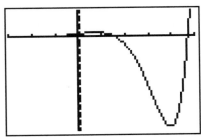

To do this question the first point of intersection of the curve with the x-axis needs to be found. Again this is done on a calculator.

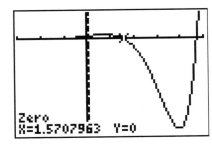

Hence the area is given by:

$$A = \left| \int_0^{1.57} e^x \cos x\,dx \right| + \left| \int_{1.57}^{a} e^x \cos x\,dx \right|$$

To find $\int e^x \cos x\,dx$ integration by parts is used.

Letting $\int e^x \cos x\,dx = I$

and using the formula $\int v\dfrac{du}{dx}\,dx = uv - \int u\dfrac{dv}{dx}\,dx,$

gives $v = e^x$ and $\dfrac{du}{dx} = \cos x$

$\Rightarrow \dfrac{dv}{dx} = e^x$ and $u = \int \cos x\,dx = \sin x$

Hence $\quad I = \int e^x \cos x\,dx = e^x \sin x - \int \sin x \times e^x\,dx = e^x \sin x - \int e^x \sin x\,dx$

Again using the formula $\int v\dfrac{du}{dx}\,dx = uv - \int u\dfrac{dv}{dx}\,dx,$

and letting $v = e^x$ and $\dfrac{du}{dx} = \sin x$

$\Rightarrow \dfrac{dv}{dx} = e^x$ and $u = \int \sin x\,dx = -\cos x$

Hence $\int e^x \sin x\,dx = -e^x \cos x - \int -\cos x \times e^x\,dx$

$$= -e^x \cos x + \int e^x \cos x\,dx$$

Putting it all together $\Rightarrow I = \int e^x \cos x\,dx = e^x \sin x + e^x \cos x - I$

$\Rightarrow I = \dfrac{1}{2}(e^x \sin x + e^x \cos x) + c$

$\Rightarrow A = \left| \left[\dfrac{1}{2}(e^x \sin x + e^x \cos x) \right]_0^{1.57} \right| + \left| \left[\dfrac{1}{2}(e^x \sin x + e^x \cos x) \right]_{1.57}^{a} \right|$

$\Rightarrow A = \left| \left[\dfrac{1}{2}(e^{1.57} \sin 1.57 + e^{1.57} \cos 1.57) - \dfrac{1}{2}(e^0 \sin 0 + e^0 \cos 0) \right] \right.$

$\left. + \left[\dfrac{1}{2}(e^a \sin a + e^a \cos a) - \dfrac{1}{2}(e^{1.57} \sin 1.57 + e^{1.57} \cos 1.57) \right] \right|$

$\Rightarrow A = \left| 2.41 - \dfrac{1}{2} \right| + \left| \dfrac{1}{2}(e^a \sin a + e^a \cos a) - 2.41 \right|$

Exercise 10

Use a calculator where appropriate.

1 Calculate the area bounded by the lines $y = 0$, $x = 1$ and the curve
$y = \dfrac{x^2}{x^2 + 1}$.

2 Find the area between the curves $y = \cos 2x$ and $y = x \sin x$ shaded in the diagram below.

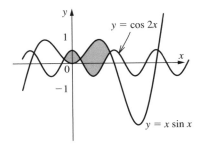

3 Find the area bounded by the curve $y = xe^{-2x}$, the x-axis and the lines $x = 0$ and $x = 1$.

4 Sketch the curve $y^2 = p^2(p - 2x)$, $p \in \mathbb{R}^+$ and show that the area bounded by the curve and the y-axis is $\dfrac{2}{3}p^{\frac{5}{2}}$.

5 Find the value of the shaded area in the diagram below which shows the curve.

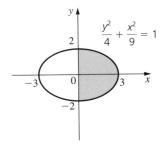

6 If $y = x\sqrt{4 - x^2} + 4 \sin^{-1}\dfrac{x}{2}$, find $\dfrac{dy}{dx}$.

Hence show that $\displaystyle\int_0^p \dfrac{1}{2}\sqrt{4 - x^2}\, dx = \dfrac{1}{4}p\sqrt{4 - p^2} + \sin^{-1}\left(\dfrac{p}{2}\right)$, $0 \le p \le 2$.

Draw a diagram showing the area this integral represents.

7 Consider the circle $y^2 + x^2 = a^2$ and the line $x = \dfrac{3a}{4}$. This line splits the circle into two segments. Using integration, find the area of the smaller segment.

8 Show that the exact ratio of $\displaystyle\int_0^{\pi} e^{-x} \cos x \, dx$ to $\displaystyle\int_{\pi}^{2\pi} e^{-x} \cos x \, dx$ is $e^{2\pi}$.

9 Show that $\displaystyle\int_2^3 \log_{10} x \, dx$ is $\dfrac{1}{\ln 10}(3 \ln 3 - 2 \ln 2 - 1)$.

10 The curve $y = (2x - 3)e^{2x}$ crosses the x-axis at P and the y-axis at Q. Find the area bounded by OP, OQ and the curve PQ in terms of e given that O is the origin.

11 Using the substitution $u^2 = 2x - 1$, find the area bounded by the curve $y = x\sqrt{2x - 1}$, the x-axis and the lines $x = 1$ and $x = a, a > 1$.

12 Find the area bounded by the curve $y = \dfrac{\sin x}{\sqrt{1 - \cos x}}$, the line $x = 1$ and the x-axis.

13 Find the area bounded by the curve $y = x^2 \sin x$ and the x-axis.

14 Find the area between the curves $y = x \sin x$ and $y = e^{3x}$.

Review exercise

1 Evaluate $\displaystyle\int_0^1 (e - ke^{kx}) \, dx$.

2 Using the substitution $u = \dfrac{1}{2}x + 1$, or otherwise, find the integral

$$\int x\sqrt{\dfrac{1}{2}x + 1} \, dx.$$

[IB May 99 P1 Q14]

3 Evaluate $\displaystyle\int_4^k \dfrac{x}{\sqrt{5}}\sqrt{x^2 - 4} \, dx$.

4 Find $\displaystyle\int \arctan x \, dx$. [IB May 98 P1 Q17]

5 Find the area bounded by the curve $y = \dfrac{1}{x^2 - 2x - 15}$, the lines $x = -2$, $x = 2$ and the x-axis.

6 Find the real numbers a and b such that $21 + 4x - x^2 = a - (x - b)^2$ for all values of x. Hence or otherwise find $\displaystyle\int \dfrac{dx}{\sqrt{21 + 4x - x^2}}$. [IB Nov 88 P1 Q15]

7 The area bounded by the curve $y = \dfrac{1}{1 + 4x^2}$, the x-axis and the lines $x = a$ and $x = a + 1$ is 0.1. Find the value of a given that $a > 0$.

8 Find the indefinite integral $\displaystyle\int x^2 e^{-2x} \, dx$. [IB May 97 P1 Q13]

9 a Find the equation of the tangent to the curve $y = \dfrac{1 + \ln x}{x}$ which passes through the origin.

   b Find the area bounded by the curve, the tangent and the x-axis.

 10 Find $\displaystyle\int \dfrac{dx}{x^2 + 6x + 13}$.   [IB Nov 96 P1 Q18]

 11 Find $\displaystyle\int \dfrac{a \cos x}{3 - b \sin x}\, dx$.

 12 Let $f: x \mapsto \dfrac{\sin x}{x}$, $\pi \leq x \leq 3\pi$. Find the area enclosed by the graph of f and the

   x-axis.   [IB May 01 P1 Q18]

 13 For the curve $y = \dfrac{1}{1 + x^2}$:

   a find the coordinates of any maximum of minimum points
   b find the equations of any asymptotes
   c sketch the curve

   d find the area bounded by the curve and the line $y = \dfrac{1}{2}$.

 14 Calculate the area bounded by the graph of $y = x \sin(x^2)$ and the x-axis, between $x = 0$ and the smallest positive x-intercept.   [IB Nov 00 P1 Q5]

 15 Let $f(x) = x \cos 3x$.

   a Use integration by parts to show that $\displaystyle\int f(x)\, dx = \dfrac{1}{3}x \sin 3x + \dfrac{1}{9}\cos 3x + c$.

   b Use your answer to part **a** to calculate the exact area enclosed by f(x) and the x-axis in each of the following cases. Give your answers in terms of π.

   i $\dfrac{\pi}{6} \leq x \leq \dfrac{3\pi}{6}$

   ii $\dfrac{3\pi}{6} \leq x \leq \dfrac{5\pi}{6}$

   iii $\dfrac{5\pi}{6} \leq x \leq \dfrac{7\pi}{6}$

   c Given that the above areas are the first three terms of an arithmetic sequence, find an expression for the total area enclosed by f(x) and the x-axis for

   $\dfrac{\pi}{6} \leq x \leq \dfrac{(2n + 1)\pi}{6}$, where $n \in \mathbb{Z}^+$. Give your answers in terms of n and π.
                   [IB May 01 P2 Q1]

16 Integration 3 – Applications

When students study integral calculus, the temptation is to see it as a theoretical subject. However, this is not the case. Pelageia Yakovlevna Polubarinova Kochina, who was born on 13 May 1899 in Astrakhan, Russia, spent much of her life working on practical applications of differential equations. Her field of study was fluid dynamics and *An application of the theory of linear differential equations to some problems of ground-water motion* is an example of her work. She graduated from the University of Petrograd in 1921 with a degree in pure mathematics. Following her marriage in 1925, Kochina had two daughters, Ira and Nina, and for this reason she resigned her position at the Main Geophysical Laboratory. However for the next ten years she continued to be active in her research and in 1934 she returned to full-time work after being given the position of professor at Leningrad University. In 1935 the family moved to Moscow and Kochina gave up her teaching position to concentrate on full-time research. She continued to publish until 1999, a remarkable achievement given that she was 100 years old!

16.1 Differential equations

An equation which relates two variables and contains a differential coefficient is called a differential equation. Differential coefficients are terms such as $\dfrac{dy}{dx}, \dfrac{d^2y}{dx^2}$ and $\dfrac{d^ny}{dx^n}$.

The order of a differential equation is the highest differential coefficient in the equation. Therefore, a **first order equation** contains $\dfrac{dy}{dx}$ only. For example $\dfrac{dy}{dx} + 5y = 0$. However, a **second order equation** contains $\dfrac{d^2y}{dx^2}$ and could also contain $\dfrac{dy}{dx}$. An example of this would be $\dfrac{d^2y}{dx^2} + 3\dfrac{dy}{dx} - 7y = 0$. Hence a differential equation of *n*th order would contain $\dfrac{d^ny}{dx^n}$ and possibly other lower orders.

A **linear differential equation** is one in which none of the differential coefficients are raised to a power other than one. Hence $x^2 + 5\left(\dfrac{dy}{dx}\right)^2 - 6y = 0$ is **not** a linear differential equation. Within the HL syllabus only questions on linear differential equations will be asked.

The solution to a differential equation has no differential coefficients within it. So to solve differential equations integration is needed. Now if $y = x^3 + k$, then $\dfrac{dy}{dx} = 3x^2$. $\dfrac{dy}{dx} = 3x^2$ is called the differential equation and $y = x^3 + k$ is called the solution.

Given that many things in the scientific world are dependent on rate of change it should come as no surprise that differential equations are very common and so the need to be able to solve them is very important. For example, one of the first researchers into population dynamics was Thomas Malthus, a religious minister at Cambridge University, who was born in 1766. His idea was that the rate at which a population grows is directly proportional to its current size. If t is used to represent the time that has passed since the beginning of the "experiment", then $t = 0$ would represent some reference time such as the year of the first census and p could be used to represent the population's size at time t. He found that $\dfrac{dp}{dt} = kp$ and this is the differential equation that was used as the starting point for his research.

A further example comes from physics. Simple harmonic motion refers to the periodic sinusoidal oscillation of an object or quantity. For example, a pendulum executes simple harmonic motion. Mathematically, simple harmonic motion is defined as the motion executed by any quantity obeying the differential equation $\dfrac{d^2x}{dt^2} = -\omega^2 x$.

Types of solution to differential equations

Consider the differential equation $\dfrac{dy}{dx} = e^{2x} - 4x$.

This can be solved using basic integration to give:

$$y = \frac{1}{2}e^{2x} - 2x^2 + k$$

There are two possible types of answer.

1. The answer above gives a family of curves, which vary according to the value which k takes. This is known as the **general solution**.

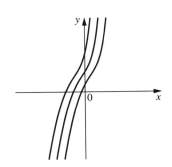

2. Finding the constant of integration, k, produces one specific curve, which is known as the **particular solution**. The information needed to find a particular solution is called the initial condition. For the example above, if we are told that $(0, 5)$ lies on the curve, then we could evaluate k and hence find the particular solution.

$$y = \frac{1}{2}e^{2x} - 2x^2 + k$$

$$\Rightarrow 5 = \frac{1}{2} - 0 + k$$

$$\Rightarrow k = \frac{9}{2} \quad\text{..}\quad \boxed{\text{This is not the final answer.}}$$

$$y = \frac{1}{2}e^{2x} - 2x^2 + \frac{9}{2} \quad\text{..}\quad \boxed{\text{The final answer should always be in this form.}}$$

Always give the answer to a differential equation in the form $y = f(x)$, if possible.

If the general solution is required, the answer will involve a constant.

If the initial condition is given, then the constant should be evaluated and the particular solution given.

16.2 Solving differential equations by direct integration

Differential equations of the form $\dfrac{d^n y}{dx^n} = f(x)$ can be solved by integrating both sides.

If we are asked to solve $\dfrac{dy}{dx} = \dfrac{x}{1 + x^2} + x \ln x$, then we can integrate to get

$$y = \int \frac{x}{1 + x^2}\,dx + \int x \ln x\,dx.$$

It was shown in Chapter 15 that the first integral could be found using direct reverse and the second can be solved using the technique of integration by parts.

For the first integral:

We begin with $y = \ln(1 + x^2)$

Hence $\dfrac{dy}{dx} = \dfrac{2x}{1 + x^2}$

And therefore $\displaystyle\int \frac{x}{1 + x^2}\,dx = \frac{1}{2}\ln(1 + x^2) + k_1$

For the second integral:

$$\int x \ln x\,dx = \frac{x^2}{2}\ln x - \int \frac{x^2}{2}\left(\frac{1}{x}\right)dx$$

using integration by parts

$$= \frac{x^2}{2}\ln x - \int \frac{x}{2}\,dx$$

$$= \frac{x^2}{2}\ln x - \frac{x^2}{4} + k_2$$

Hence the solution to $\dfrac{dy}{dx} = \dfrac{x}{1 + x^2} + x \ln x$ is $y = \dfrac{1}{2}\ln(1 + x^2) + \dfrac{x^2}{2}\ln x - \dfrac{x^2}{4} + k.$ $\quad$ $\boxed{\text{The two constants of integration } k_1 \text{ and } k_2 \text{ can be combined into one constant } k.}$

If the values of y and x are given then k can be calculated. Given the initial condition that $y = 0$ when $x = 1$ we find that:

$$0 = \frac{1}{2}\ln 2 + 0 - \frac{1}{4} + k$$

$$\Rightarrow k = \frac{1}{4} - \frac{1}{2}\ln 2$$

$$y = \frac{1}{2}\ln(1 + x^2) + \frac{x^2}{2}\ln x - \frac{x^2}{4} + \frac{1}{4} - \frac{1}{2}\ln 2$$

This is the particular solution.

Many questions involving differential equations are set in a real-world context as many natural situations can be modelled using differential equations.

Example

The rate of change of the volume (V) of a cone as it is filled with water is directly proportional to the natural logarithm of the time (t) it takes to fill. Given that $\frac{dV}{dt} = \frac{1}{2}$ cm³/s when $t = 5$ seconds and that $V = 25$ cm³ when $t = 8$ seconds, find the formula for the volume.

We start with $\frac{dV}{dt} \propto \ln t$.

To turn a proportion sign into an equals sign we include a constant of proportionality, say k, which then needs to be evaluated.

Hence $\frac{dV}{dt} = k \ln t$

Given that $\frac{dV}{dt} = \frac{1}{2}$ when $t = 5$ we get:

$$\frac{1}{2} = k \ln 5$$

$$\Rightarrow k = \frac{1}{2 \ln 5} = 0.310 \ldots$$

So $\frac{dV}{dt} = 0.311 \ln t$

$$\Rightarrow V = \int 0.311 \ln t \, dt$$

$$\Rightarrow V = 0.311 \int \ln t \, dt$$

Hence $V = 0.311\left[t \ln t - \int t\left(\frac{1}{t}\right) dt\right]$ using integration by parts

$$\Rightarrow V = 0.311[t \ln t - t + c]$$

The constant can be included within the brackets or it can be outside. It will evaluate to the same number finally.

Given that $V = 25$ when $t = 8$

$$25 = 0.311[8 \ln 8 - 8 + c]$$
$$\Rightarrow c = 71.8$$

Hence $V = 0.311[t \ln t - t + 71.8]$

or $V = 0.311t \ln t - 0.311t + 22.3$

Given that the question is dealing with volume and time, this formula is only valid for $t > 0$.

In certain situations we may be asked to solve differential equations other than first order.

Example

Solve the differential equation $\dfrac{d^4y}{dx^4} = \cos x$, giving the general solution.

From basic integration:

$$\frac{d^3y}{dx^3} = \int \cos x \, dx$$

$$\Rightarrow \frac{d^3y}{dx^3} = \sin x + k$$

Continuing to integrate:

$$\frac{d^2y}{dx^2} = -\cos x + kx + c$$

$$\frac{dy}{dx} = -\sin x + \frac{kx^2}{2} + cx + d$$

$$y = \cos x + \frac{kx^3}{6} + \frac{cx^2}{2} + dx + e$$

> If we were given the boundary conditions then the constants k, c, d, and e could be evaluated.

Example

Find the particular solution to the differential equation $\dfrac{d^3y}{dx^3} = 25e^{-5x} + 24x$

given that when $x = -1$, $\dfrac{d^2y}{dx^2} = -5e^5$, when $x = 1$, $\dfrac{dy}{dx} = -8$, and when $x = 0$, $y = 0$.

$$\frac{d^3y}{dx^3} = 25e^{-5x} + 24x$$

Hence $\dfrac{d^2y}{dx^2} = \displaystyle\int (25e^{-5x} + 24x) \, dx$

$$\Rightarrow \frac{d^2y}{dx^2} = -5e^{-5x} + 12x^2 + c$$

When $x = -1$, $\dfrac{d^2y}{dx^2} = -5e^5$

$$\Rightarrow -5e^5 = -5e^5 + 12 + c$$

$$\Rightarrow c = -12$$

$$\Rightarrow \frac{d^2y}{dx^2} = -5e^{-5x} + 12x^2 - 12$$

Integrating again gives:

$$\frac{dy}{dx} = \int (-5e^{-5x} + 12x^2 - 12) \, dx$$

$$\Rightarrow \frac{dy}{dx} = e^{-5x} + 4x^3 - 12x + d$$

Now when $x = 1, \dfrac{dy}{dx} = -8$

$\Rightarrow -8 = e^{-5} + 4 - 12 + d$

$\Rightarrow d = -e^{-5}$

So $\dfrac{dy}{dx} = e^{-5x} + 4x^3 - 12x - e^{-5}$

The final integration gives:

$$y = \int (e^{-5x} + 4x^3 - 12x - e^{-5})\, dx$$

$$\Rightarrow y = \dfrac{e^{-5x}}{-5} + x^4 - 6x^2 - e^{-5}x + f$$

When $x = 0, y = 0$ gives:

$$0 = -\dfrac{1}{5} + f$$

$$\Rightarrow f = \dfrac{1}{5}$$

Therefore $y = \dfrac{e^{-5x}}{-5} + x^4 - 6x^2 - e^{-5}x + \dfrac{1}{5}$

Exercise 1

Find the general solutions of these differential equations.

1 $\dfrac{dy}{dx} = x^2 + \sin x$ **2** $\dfrac{dy}{dx} = (3x - 7)^4$ **3** $\dfrac{dy}{dx} = 2x(1 - x^2)^{\frac{1}{2}}$

4 $\dfrac{dy}{dx} = x \sin x$ **5** $\dfrac{dy}{dx} = \dfrac{\cos x}{1 - \sin x}$ **6** $\dfrac{dy}{dx} = xe^{\frac{2kx}{3}}$

7 $\dfrac{dy}{dx} = \dfrac{5x}{\sqrt{1 - 15x^2}}$ **8** $\dfrac{dy}{dx} = \sin^2 2x$ **9** $\dfrac{d^2y}{dx^2} = (3x + 2)^{\frac{1}{2}}$

10 $\dfrac{d^2y}{dx^2} = \sec^2 x$ **11** $\dfrac{d^3y}{dx^3} = x \ln x$ **12** $\dfrac{d^4y}{dx^4} = x \cos x$

Find the particular solutions of these differential equations.

13 $\dfrac{dy}{dx} = \dfrac{4x}{4x^2 + 3}$ given that when $x = 2, y = 0$

14 $\dfrac{dy}{dx} = 3 \sin\left(4x - \dfrac{\pi}{2}\right)$ given that when $x = \dfrac{\pi}{4}, y = 2$

15 $\dfrac{d^2y}{dx^2} = (2x - 1)^4$ given that when $x = \dfrac{1}{2}, \dfrac{dy}{dx} = 2$ and that when

$x = 1, y = 4$

16 $\dfrac{d^2y}{dx^2} = \dfrac{2}{1 + x^2}$ given that when $x = \dfrac{\pi}{4}, \dfrac{dy}{dx} = 3$ and that when

$x = 0, y = 5$

16.3 Solving differential equations by separating variables

Equations in which the variables are separable can be written in the form $g(y)\dfrac{dy}{dx} = f(x)$. These can then be solved by integrating both sides with respect to x.

Method

> **1.** Put in the form $g(y)\dfrac{dy}{dx} = f(x)$
>
> This gives an equation in the form $\displaystyle\int g(y)\dfrac{dy}{dx}dx = \int f(x)\,dx$ which simplifies to
>
> $\displaystyle\int g(y)\,dy = \int f(x)\,dx$. The variables are now separated onto opposite sides of the equation.
>
> **2.** Integrate both sides with respect to x.
> **3.** Perform the integration.

> ### Example
>
> Find the general solution to the differential equation $\dfrac{dy}{dx} = \dfrac{4x - 1}{2y}$.
>
> Following step 1:
>
> $2y\dfrac{dy}{dx} = 4x - 1$
>
> Following step 2 and step 3:
>
> $\displaystyle\int 2y\dfrac{dy}{dx}dx = \int (4x - 1)\,dx$
>
> $\Rightarrow \displaystyle\int 2y\,dy = \int (4x - 1)\,dx$
>
> $\Rightarrow \dfrac{2y^2}{2} + k_1 = \dfrac{4x^2}{2} - x + k_2$
>
> $\Rightarrow y^2 = 2x^2 - x + k$
>
> $\Rightarrow y = \pm\sqrt{2x^2 - x + k}$

Since there is an integral on each side of the equation, a constant of integration is theoretically needed on each side. For simplicity, these are usually combined and written as one constant.

> ### Example
>
> Solve to find the general solution of the equation $e^x\dfrac{dy}{dx} = \dfrac{x}{y^2 + 1}$.
>
> Following step 1:
>
> $(y^2 + 1)\dfrac{dy}{dx} = xe^{-x}$
>
> Following step 2 and step 3:
>
> $\displaystyle\int (y^2 + 1)\dfrac{dy}{dx}dx = \int xe^{-x}\,dx$
>
> $\Rightarrow \displaystyle\int (y^2 + 1)\,dy = \int xe^{-x}\,dx$

$$\Rightarrow \frac{y^3}{3} + y = -xe^{-x} + \int 1e^{-x}\,dx$$

$$\Rightarrow \frac{y^3}{3} + y = -xe^{-x} - e^{-x} + k \cdots\cdots\cdots\cdots\cdots\cdots\cdots\cdots$$

> In this situation an explicit equation in y cannot be found.

As was shown earlier, k can be evaluated if the initial condition that fits the equation is given. In this situation, we could be told that when $x = 2$, $y = 1$ and we can evaluate k.

$$\frac{1}{3} + 1 = -2e^{-2} - e^{-2} + k$$

$$\Rightarrow k = \frac{4}{3} + 3e^{-2}$$

$$\Rightarrow k = 1.73\ldots$$

Therefore the final answer would be

$$\frac{y^3}{3} + y = -xe^{-x} - e^{-x} + 1.74$$

Example

The diagram below shows a tangent to a curve at a point P which cuts the x-axis at the point A. Given that OA is of length qx, show that the points on the curve all satisfy the equation

$$\frac{dy}{dx} = \frac{y}{x - qx}$$

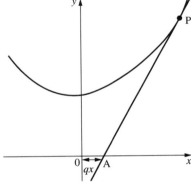

a Hence show that the equation is of the form $y = kx^{\frac{1}{1-q}}$ where k is a constant.

b Given that q is equal to 2, find the equation of the specific curve which passes through the point $(1, 1)$.

The gradient is $\dfrac{\Delta y}{\Delta x}$:

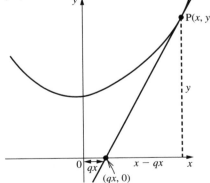

Therefore the gradient is $\dfrac{y - 0}{x - qx}$.

Hence $\dfrac{dy}{dx} = \dfrac{y}{x - qx}$

a Following the method of separating variables:

$$\frac{1}{y}\frac{dy}{dx} = \frac{1}{x - qx}$$

$$\Rightarrow \int \frac{1}{y}\frac{dy}{dx}dx = \int \frac{1}{x - qx}dx$$

$$\Rightarrow \int \frac{1}{y}dy = \int \frac{1}{x - qx}dx$$

$$\Rightarrow \int \frac{1}{y}dy = \int \frac{1}{x(1 - q)}dx$$

$$\Rightarrow \ln|y| = \frac{1}{1 - q}\ln|x| + c \quad\cdots\cdots\cdots\cdots\cdots\cdots\cdots\cdots\cdots\cdots\cdots\cdots\cdots\cdots\cdots\cdots\cdots$$

$$\Rightarrow \ln|y| = \frac{1}{1 - q}\ln|x| + \ln k$$

> To simplify equations of this type (i.e. where natural logarithms appear in all terms) it is often useful to let $c = \ln k$.

Now by using the laws of logarithms:

$$\ln|y| - \ln k = \frac{1}{1 - q}\ln|x|$$

$$\Rightarrow \ln\left|\frac{y}{k}\right| = \ln|x|^{\frac{1}{1-q}} \quad\cdots\cdots\cdots\cdots\cdots\cdots\cdots\cdots\cdots$$

$$\Rightarrow \frac{y}{k} = x^{\frac{1}{1-q}}$$

$$\Rightarrow y = kx^{\frac{1}{1-q}}$$

> Technically the absolute value signs should remain until the end, but in this situation they are usually ignored.

b The curve passes through the point (1, 1) and $q = 2$.

$$1 = k(1)^{\frac{1}{1-2}}$$

$$\Rightarrow 1 = k(1)^{-1}$$

$$\Rightarrow k = 1$$

$$\Rightarrow y = 1x^{-1}$$

$$\Rightarrow y = \frac{1}{x}$$

Another real-world application of differential equations comes from work done with kinematics.

Example

A body has an acceleration a, which is dependent on time t and velocity v and is linked by the equation

$a = v \sin kt$

Given that when $t = 0$ seconds, $v = 1\ ms^{-1}$ and when $t = 1$ second, $v = 2\ ms^{-1}$, and that k takes the smallest possible positive value, find the velocity of the body after 6 seconds.

From the work on kinematics, we know that acceleration is the rate of change of velocity with respect to time, i.e. $a = \frac{dv}{dt}$.

Therefore the equation can be rewritten as $\frac{dv}{dt} = v \sin kt$.

This can be solved by separating variables.

$$\frac{1}{v}\frac{dv}{dt} = \sin kt$$

$$\Rightarrow \int \frac{1}{v}\frac{dv}{dt}dt = \int \sin kt \, dt$$

$$\Rightarrow \int \frac{1}{v}dv = \int \sin kt \, dt$$

$$\Rightarrow \ln|v| = -\frac{1}{k}\cos kt + c$$

The problem here is that when we substitute values for v and t, there are still two unknown constants. This is why two conditions are given. Now when $t = 0$ seconds, $v = 1$ ms^{-1} gives:

$$\ln|1| = -\frac{1}{k}\cos 0 + c$$

$$\Rightarrow 0 = -\frac{1}{k} + c$$

$$\Rightarrow c = \frac{1}{k}$$

Hence $\ln|v| = -\frac{1}{k}\cos kt + \frac{1}{k}$

The other values can now be substituted to evaluate k:

$$\Rightarrow \ln 2 = -\frac{1}{k}\cos k + \frac{1}{k}$$

This equation cannot be solved by any direct means and so a graphing calculator needs to be used to find a value for k.

To do this, input the equation $y = \frac{1}{x}\cos x - \frac{1}{x} + \ln 2$ into a calculator and then solve it for $y = 0$. Since the question states that k should have the smallest possible positive value, then the value of k is the smallest positive root given by the calculator.

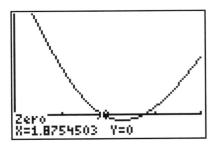

```
Zero
X=1.8754503   Y=0
```

This value is $k = 1.87 \ldots$ and hence $\frac{1}{k} = 0.533 \ldots$.

The equation now reads $\ln|v| = 0.533 \cos 1.88t + 0.533$

The value of v when $t = 6$ seconds, can now be found by substituting in the value $t = 6$.

$$\ln|v| = 0.533 \cos(1.88 \times 6) + 0.533$$

$$\Rightarrow \ln|v| = 0.683 \ldots$$

$$\Rightarrow v = e^{0.683}$$

$$\Rightarrow v = 1.98 \text{ ms}^{-1}$$

Exercise 2

Find the general solutions of these differential equations.

1 $y\dfrac{dy}{dx} = \tan x$

2 $2x\dfrac{dy}{dx} = y^2 + 1$

3 $\dfrac{dy}{dx} = \dfrac{3 + 2y}{4 - 3x}$

4 $(\sin x + \cos x)\dfrac{dy}{dx} = \cos x - \sin x$

5 $\dfrac{y^3}{x}\dfrac{dy}{dx} = \ln x$

6 $5x\dfrac{dy}{dx} = 6e^y$

7 $\dfrac{dy}{dx} = \dfrac{4y}{\sqrt{4 - x^2}}$

8 $s^2\dfrac{ds}{dt} = \sin^{-1} t$

9 $\dfrac{1}{x}\dfrac{dy}{dx} = \dfrac{(3x - 1)^9}{y^2}$ ....

> You will need to use the substitution $u = 3x - 1$ to perform the integration.

10 $v\dfrac{dv}{dt} = \cos^2 at$

11 $e^{2x+y}\dfrac{dy}{dx} = 1$

12 $3y(x + 1) = (x^2 + 2x)\dfrac{dy}{dx}$

Find the particular solutions of these differential equations.

13 $\dfrac{dy}{dx} = y(3 - x)^4$ given that when $x = 2, y = 4$

14 $e^{2x}\dfrac{dy}{dx} = \sqrt[3]{y}$ given that when $x = 1, y = 1$

15 $x\dfrac{dy}{dx} = \sin^2 y$ given that when $y = \dfrac{\pi}{4}, x = 4$

16 $\dfrac{2y}{3x}\dfrac{dy}{dx} = \dfrac{2y^2 + 3}{4x^2 - 1}$ given that when $x = 2, y = 8$

17 $\theta^2\dfrac{d\theta}{dt} = e^{2t} \sin t$ given that when $t = 0, \theta = \dfrac{\pi}{2}$

18 $\dfrac{ds}{dt} = \sqrt{t^2 - 9s^2t^2}$ given that when $t = 1, s = \dfrac{1}{3}$

The following exercise contains a mixture of questions on the material covered in this chapter so far.

Exercise 3

In questions 1 to 5, solve the differential equations.

1 $\dfrac{dy}{dx} = \tan x$

2 $\dfrac{d^2x}{dt^2} - \omega^2 t = 0$

3 $\dfrac{d^2x}{dt^2} = \omega \sin nt$

4 $\dfrac{1}{x^2}\dfrac{dy}{dx} = \dfrac{3}{y^2(1 + x^3)}$ given that when $x = 0, y = 3$

5 $\cos^2 x\dfrac{dy}{dx} = \cos^2 y$

6 Consider the expression $z = x + y$.

a Using differentiation, find an expression for $\dfrac{dz}{dx}$.

b Hence show that the differential equation $\dfrac{dy}{dx} = (x + y)^2$ can be changed

to $\dfrac{dz}{dx} = z^2 + 1$.

c Find the general solution of the differential equation $\dfrac{dz}{dx} = z^2 + 1$ and

hence write down the solution to the differential equation $\dfrac{dy}{dx} = (x + y)^2$.

7 Find the particular solution to the equation $A\dfrac{d^4y}{dx^4} = B$, where A and B are

constants that do not need to be evaluated, given that $y = 0$ and $\dfrac{d^2y}{dx^2} = 0$

for both $x = 0$ and $x = 1$.

8 A hollow cone is filled with water. The rate of increase of water with

respect to time is $\dfrac{dV}{dt} = 4\sin\left(t + \dfrac{\pi}{4}\right)$. Given that when $t = \dfrac{\pi}{12}$ seconds,

$V = 4$ cm³, find a general formula for the volume V at any time t.

9 Oil is dripping out of a hole in the engine of a car, forming a thin circular
film of the ground. The rate of increase of the radius of the circular film is

given by the formula $\dfrac{dr}{dt} = 2\ln t^2$. Given that when $t = 5$ seconds, the

radius of the film is 4 cm, find a general formula for the radius r at any time t.

10 The rate at which the height h of a tree increases is proportional to the
difference between its present height and its final height s. Show that its

present height is given by the formula $h = s - \dfrac{e^{-kt}}{B}$ where B and k are both

constants.

11 Show that the equation of the curve which satisfies the differential

equation $\dfrac{dy}{dx} = \dfrac{1 + y^2}{1 + x^2}$ and passes through the point $\left(\dfrac{\sqrt{3}}{3}, \sqrt{3}\right)$ is

$\dfrac{3x + \sqrt{3}}{3 - x\sqrt{3}}$.

16.4 Verifying that a particular solution fits a differential equation

The easiest way to tackle questions of this form is to differentiate the expression the
required number of times and then substitute into the differential equation to show that
it actually fits.

Example

Show that $y = 2e^{2x}$ is a solution to the differential equation

$\dfrac{d^2y}{dx^2} + 6\dfrac{dy}{dx} + 9y = 50e^{2x}$

We begin with the expression $y = 2e^{2x}$.

Differentiating using the chain rule: $\dfrac{dy}{dx} = 4e^{2x}$

Differentiating again: $\dfrac{d^2y}{dx^2} = 8e^{2x}$

Substituting back into the left-hand side of the original differential equation gives:

$8e^{2x} + 6 \cdot 4e^{2x} + 9 \cdot 2e^{2x} = 50e^{2x}$

Since this is the same as the right-hand side, this is verified.

Sometimes the question will involve constants and in this case, on substitution, they will cancel out.

Example

Show that $y = Ae^{-2x} + Be^{-3x}$ is a solution to $\dfrac{d^2y}{dx^2} + 5\dfrac{dy}{dx} + 6y = 0$.

We begin with the expression $y = Ae^{-2x} + Be^{-3x}$

Differentiating using the chain rule: $\dfrac{dy}{dx} = -2Ae^{-2x} - 3Be^{-3x}$

Differentiating again: $\dfrac{d^2y}{dx^2} = 4Ae^{-2x} + 9Be^{-3x}$

Substituting these back into the left-hand side of the original differential equation gives:

$$4Ae^{-2x} + 9Be^{-3x} + 5(-2Ae^{-2x} - 3Be^{-3x}) + 6(Ae^{-2x} + Be^{-3x})$$
$$= 4Ae^{-2x} + 9Be^{-3x} - 10Ae^{-2x} - 15Be^{-3x} + 6Ae^{-2x} + 6Be^{-3x}$$
$$= 10Ae^{-2x} + 15Be^{-3x} - 10Ae^{-2x} - 15Be^{-3x}$$
$$= 0$$

Since this is the same as the right-hand side, this is verified.

Exercise 4

Verify that these solutions fit the differential equations.

1 $y = 2x$, $\qquad\qquad\qquad\qquad\qquad \dfrac{d^2y}{dx^2} + y = 2x$

2 $y = -\dfrac{1}{4}e^{-2x}$, $\qquad\qquad\qquad \dfrac{d^2y}{dx^2} + 2\dfrac{dy}{dx} - 8y = 2e^{-2x}$

3 $y = e^x + \dfrac{1}{4}e^{2x} + \dfrac{1}{2}x + \dfrac{11}{4}$, $\qquad \dfrac{d^2y}{dx^2} - 3\dfrac{dy}{dx} + 2y = 4 + x$

4 $y = e^x(A + Bx) + e^{2x}$, $\qquad\qquad \dfrac{d^2y}{dx^2} - 2\dfrac{dy}{dx} + y = e^{2x}$

5 $y = e^x\left(A + Bx + \dfrac{1}{2}x^3\right)$, $\qquad \dfrac{d^2y}{dx^2} - 2\dfrac{dy}{dx} + y = 3xe^x$

6 $y = Ae^{-x} + Be^{-2x} + \dfrac{1}{10}(\sin x - 3\cos x)$, $\quad \dfrac{d^2y}{dx^2} + 3\dfrac{dy}{dx} + 2y = \sin x$

7 $y = Ae^{-\frac{x}{2}}\cos\dfrac{\sqrt{3}}{2}x + x$, $\qquad\qquad \dfrac{d^2y}{dx^2} + \dfrac{dy}{dx} + y = 1 + x$

16.5 Displacement, velocity and acceleration

This is one of the more important applications of integral calculus. In Chapter 10, velocity and acceleration were represented as derivatives.

Reminder: if s is displacement, v is velocity and t is time, then:

$$v = \frac{ds}{dt}$$

and $a = \dfrac{d^2s}{dt^2} = \dfrac{dv}{dt}$.

Since we can represent velocity and acceleration using differential coefficients, solving problems involving velocity and acceleration often involves solving a differential equation.

Example

Given that the velocity v of a particle at time t is given by the formula $v = (t - 1)^3$ and that when $t = 2, s = 6,$ find the formula for the displacement at any time t.

Beginning with the formula: $v = (t - 1)^3$

we know that $v = \dfrac{ds}{dt}$

So $\qquad \dfrac{ds}{dt} = (t - 1)^3$

Integrating both sides with respect to t gives:

$$\int \frac{ds}{dt}\,dt = \int (t - 1)^3\,dt$$

$$\Rightarrow \int ds = \int (t - 1)^3\,dt$$

$$\Rightarrow s = \frac{(t - 1)^4}{4} + k$$

Now when $t = 2, s = 6$

$$6 = \frac{(2 - 1)^4}{4} + k$$

$$\Rightarrow k = \frac{23}{4}$$

$$\Rightarrow s = \frac{(t - 1)^4}{4} + \frac{23}{4}$$

It is quite straightforward to find the displacement if there is a formula relating velocity and time and to find the velocity or the displacement if there is a formula relating acceleration and time. However, what happens when acceleration is related to displacement? The connection here was shown in Chapter 10 to be

$$a = v\frac{dv}{ds}.$$

Example

The acceleration of a particle is given by the formula $a = e^{2s}$. Given that when $s = 0$, $v = 2$, find the formula for the velocity in terms of the displacement s.

$$a = e^{2s}$$

$$\Rightarrow v\frac{dv}{ds} = e^{2s}$$

$$\Rightarrow \int v\frac{dv}{ds}ds = \int e^{2s}\,ds$$

$$\Rightarrow \int v\,dv = \int e^{2s}\,ds$$

$$\Rightarrow \frac{v^2}{2} = \frac{e^{2s}}{2} + k$$

We know that when $s = 0$, $v = 2$

$$\Rightarrow \frac{4}{2} = \frac{e^0}{2} + k$$

$$\Rightarrow k = \frac{3}{2}$$

Hence $\quad \dfrac{v^2}{2} = \dfrac{e^{2s}}{2} + \dfrac{3}{2}$

$$\Rightarrow v^2 = e^{2s} + 3$$

$$\Rightarrow v = \pm\sqrt{e^{2s} + 3}$$

Example

A particle moves in a straight line with velocity $v\ \text{ms}^{-1}$. Its initial velocity is $u\ \text{ms}^{-1}$. At any time t, the velocity v is given by the equation $\dfrac{dv}{dt} + 3 + v^2 = 2$.

Prove that the particle comes instantaneously to rest after $\tan^{-1} u$ seconds. Given that the particle moves s metres in t seconds, show that the particle first comes to rest after a displacement of $\dfrac{1}{2}\ln|1 + u^2|$ metres.

$$\frac{dv}{dt} + 3 + v^2 = 2$$

$$\Rightarrow \frac{dv}{dt} = -(1 + v^2)$$

We solve this using the method of variables separable.

$$\Rightarrow \frac{1}{1 + v^2}\frac{dv}{dt} = -1$$

$$\Rightarrow \int \frac{1}{1 + v^2}\frac{dv}{dt}dt = -\int 1\,dt$$

$$\Rightarrow \tan^{-1} v = -t + k$$

It is given that when $t = 0$, $v = u$.

$\Rightarrow \tan^{-1} u = -0 + k$

$\qquad \Rightarrow k = \tan^{-1} u$

So $\tan^{-1} v = -t + \tan^{-1} u$

The particle comes to instantaneous rest when $v = 0$.

$\Rightarrow \tan^{-1} 0 = -t + \tan^{-1} u$

$\qquad \Rightarrow t = \tan^{-1} u$

Hence the result is proved.

To find the displacement s we use the fact that $\dfrac{dv}{dt} = v\dfrac{dv}{ds}$.

The equation $\dfrac{dv}{dt} = -(1 + v^2)$ becomes $v\dfrac{dv}{ds} = -(1 + v^2)$.

Again we solve this using the method of variables separable.

$$\frac{v}{1 + v^2}\frac{dv}{ds} = -1$$

$$\Rightarrow \int \frac{v}{1 + v^2}\frac{dv}{ds}ds = \int -1\,ds$$

$$\Rightarrow \int \frac{v}{1 + v^2}dv = \int -1\,ds$$

To integrate the left-hand side, we use the method of direct reverse.

Letting $\qquad y = \ln(1 + v^2)$

$$\Rightarrow \frac{dy}{dv} = \frac{2v}{1 + v^2}$$

$$\Rightarrow \int \frac{v}{1 + v^2}dv = \frac{1}{2}\ln|1 + v^2| + k$$

So returning to the original equation:

$\dfrac{1}{2}\ln|1 + v^2| = -s + c$

When $v = u$, $s = 0$

$\Rightarrow \dfrac{1}{2}\ln|1 + u^2| = -0 + c$

$\qquad \Rightarrow c = \dfrac{1}{2}\ln|1 + u^2|$

So $\dfrac{1}{2}\ln|1 + v^2| = -s + \dfrac{1}{2}\ln|1 + u^2|$

We now find the displacement when the particle first comes to instantaneous rest, this is when $v = 0$.

$\dfrac{1}{2}\ln|1 + 0^2| = -s + \dfrac{1}{2}\ln|1 + u^2|$

$\qquad \Rightarrow 0 = -s + \dfrac{1}{2}\ln|1 + u^2|$

$\qquad \Rightarrow s = \dfrac{1}{2}\ln|1 + u^2|$

Hence the result is proved.

Exercise 5

1 The acceleration in ms^{-2} of a particle moving in a straight at time t is given by the formula $a = 4t^2 + 1$. When $t = 0$, $v = 0$ and $s = 0$. Find the velocity and displacement at any time t.

2 A particle starts to accelerate along a line AB with an initial velocity of $10\ ms^{-1}$ and an acceleration of $-6t^3$ at time t after leaving A.

    a Find a general formula for the velocity v of the particle.
    b Find the velocity after 8 seconds.
    c Find a general formula for the displacement s of the particle.
    d Find the displacement after 10 seconds.

3 The acceleration in ms^{-2} of a particle moving in a straight line at time t is given by the formula $a = 2t^3 + 3t - 4$. Given that the particle has an initial velocity of $6\ ms^{-1}$, find the distance travelled by the particle in the third second of its motion.

4 The acceleration in ms^{-2} of a particle moving in a straight line at time t is given by the formula $a = \cos 3t$. The particle is initially at rest when its displacement is 0.5 m from a fixed point O on the line.

    a Find the velocity and displacement of the particle from O at any time t.
    b Find the time that elapses before the particle comes to rest again.

5 The velocity of a particle is given by the formula $v = \dfrac{t^2 + 1}{1 - t}$ and is valid for all $t > 1$. Given that when $t = 2$, $s = 10$, find the displacement at any time $t > 1$.

6 The acceleration of a particle is given by the formula $a = \sin\left(s + \dfrac{\pi}{4}\right)$. Given that when $s = \dfrac{\pi}{4}$, $v = 2$, find the formula for the velocity as a function of displacement for any s.

7 Consider a particle moving with acceleration $a = se^{2s}$. Given that when $s = 0$, $v = 2$, find the formula for the velocity as a function of displacement for any s.

8 A particle moves along a line AB. Given that A is 2 metres from O and it starts at A with a velocity of $2\ ms^{-1}$, find the formula for the velocity of the particle as a function of s, given that it has acceleration $a = (2s - 1)^4$.

9 The acceleration in ms^{-2} of a particle moving in a straight line at time t is given by the formula $a = -\dfrac{1}{t^2}$. When $t = 2$ seconds, $v = 8\ ms^{-1}$.

    a Find the velocity when $t = 4$.
    b Show that the particle has a terminal velocity of $7\dfrac{1}{2}\ ms^{-1}$.

10 Consider a particle with acceleration $e^{2t} - 4$. The particle moves along a straight line, PQ, starting from rest at P.

    a Show that the greatest speed of the particle in its motion along PQ is $\left(\dfrac{3}{2} - \ln 16\right) ms^{-1}$.

b Find the distance covered by the particle in the first four seconds of its motion.

11 A bullet is decelerating at a rate of kv ms^{-2} when its velocity is v ms^{-1}.

During the first $\dfrac{1}{2}$ second the bullet's velocity is reduced from 220 ms^{-1} to 60 ms^{-1}.

 a Find the value of k.
 b Deduce a formula for the velocity at any time t.
 c Find the distance travelled during the time it takes for the velocity to reduce from 220 ms^{-1} to 60 ms^{-1}.

12 Consider a particle with acceleration $\sin \omega t$ ms^{-2}. The particle starts from rest and moves in a straight line.

 a Find the maximum velocity of the particle.
 b The particle's motion is periodic. Give the time period of the particle. (This is the time taken between it achieving its maximum speeds.)

13 A particle is moving vertically downwards. Gravity is pulling it downwards, but there is a force of kv acting against gravity. Hence the acceleration experienced by the particle at time t is $g - kv$. If the particle starts from rest, find the velocity at any time t. Does this velocity have a limiting case?

16.6 Volumes of solids of revolution

Consider the problem of finding the volume of this tree trunk.

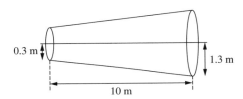

There are a number of ways that this can be done.

First, we could assume that the tree trunk is a cylinder of uniform radius and calculate the volume of the cylinder.

To do this, we need to calculate an average radius. This would be $\dfrac{1.3 + 0.3}{2} = 0.8$ m.

Hence the volume of the tree trunk is $\pi r^2 h = \pi \times 0.8^2 \times 10 = 6.4\pi$ m^3 = 20.1 m^3.

This is an inaccurate method.

A better way would be to divide the tree trunk into 10 equal portions as shown below and then calculate the volume of each portion.

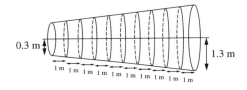

If we take an average radius for each section and assume that each portion is approximately a cylinder with height 1 m, then the volume will be:

$(\pi \cdot 0.35^2 \cdot 1) + (\pi \cdot 0.45^2 \cdot 1) + (\pi \cdot 0.55^2 \cdot 1) + (\pi \cdot 0.65^2 \cdot 1) + (\pi \cdot 0.75^2 \cdot 1) +$

$\quad (\pi \cdot 0.85^2 \cdot 1) + (\pi \cdot 0.95^2 \cdot 1) + (\pi \cdot 1.05^2 \cdot 1) + (\pi \cdot 1.15^2 \cdot 1) + (\pi \cdot 1.25^2 \cdot 1)$

$\quad = \pi \cdot 1(0.35^2 + 0.45^2 + 0.55^2 + 0.65^2 + 0.75^2 + 0.85^2 + 0.95^2 + 1.05^2 +$

$\quad 1.15^2 + 1.25^2)$

$\quad = 22.7 \text{ m}^3$

This is still an approximation to the actual answer, but it is a better approximation than the first attempt. As we increase the number of portions that the tree trunk is split into, the better the accuracy becomes.

Consider the case of the curve $y = x^2$. If the part of the curve between $x = 1$ and $x = 3$ is rotated around the x-axis, then a volume is formed as shown below.

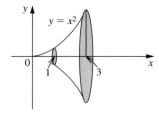

This is known as a volume of solid of revolution. The question now is how to calculate this volume. In Chapter 14, to find the area under the curve, the curve was split into infinitesimally thin rectangles and then summed using integration. Exactly the same principle is used here except rather than summing infinitesimally thin rectangles, we sum infinitesimally thin cylinders.

Effectively, this is what we did when we found the volume of the tree trunk.

Consider the diagram below.

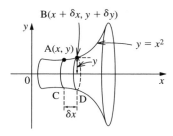

Look at the element ABCD where A is on the curve and has coordinates (x, y). Since in this case $y = x^2$, then the coordinates of A are (x, x^2). ABCD is approximately a cylinder with radius y and whose "height" is δx.

Therefore the volume of ABCD $\approx \pi y^2 \, \delta x$ and hence the volume V of the entire solid $\approx \displaystyle\sum_{x=a}^{x=b} \pi y^2 \, \delta x.$ ....

> a and b are the boundary conditions which ensure that the volume is finite.

The smaller δx becomes, the closer this approximation is to V,

i.e. $V = \lim_{\delta x \to 0} \displaystyle\sum_{x=a}^{x=b} \pi y^2 \, \delta x.$

$$V = \pi \int_a^b y^2 \, dx$$

This is the formula for a full revolution about the x-axis.

In this case $y = x^2$,

$$V = \pi \int_1^3 y^2 \, dx$$

$$\Rightarrow V = \pi \int_1^3 x^4 \, dx$$

If this question appeared on a calculator paper then the integration can be done on a calculator.

$$\Rightarrow V = \pi \left[\frac{x^5}{5} \right]_1^3$$

$$\Rightarrow V = \pi \left[\frac{243}{5} - \frac{1}{5} \right]$$

$$\Rightarrow V = \frac{242\pi}{5}$$

Example

Find the volume generated when one complete wavelength of the curve $y = \sin 2x$ is rotated through 2π radians about the x-axis.

By drawing the curve on a graphing calculator it is evident that there are an infinite number of complete wavelengths. In this case the one which lies between 0 and π will be chosen.

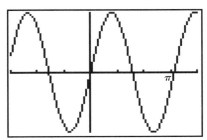

The volume of the solid formed is given by the formula $V = \pi \int_0^\pi y^2 \, dx$.

Hence $V = \pi \int_0^\pi \sin^2 2x \, dx$.

At this stage the decision on whether to use a graphing calculator or not will be based on whether the question appears on the calculator or non-calculator paper. The calculator display for this is shown below. In this case an answer of 4.93 units³ is found.

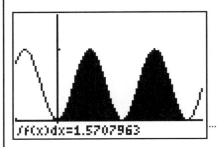

∫f(x)dx=1.5707963

This answer will need to be multiplied by π.

On a non-calculator paper we would proceed as follows.
From the trigonometrical identities we have:

$\cos 4x = \cos^2 2x - \sin^2 2x$

and $\cos^2 2x + \sin^2 2x = 1$,

giving $\cos^2 2x = 1 - \sin^2 2x$

So $\cos 4x = 1 - \sin^2 2x - \sin^2 2x$

$\Rightarrow \cos 4x = 1 - 2\sin^2 2x$

$\Rightarrow \sin^2 2x = \dfrac{1 - \cos 4x}{2}$

Hence

$$V = \pi \int_0^\pi \left(\dfrac{1 - \cos 4x}{2}\right) dx$$

$$\Rightarrow V = \dfrac{\pi}{2} \int_0^\pi (1 - \cos 4x)\, dx$$

$$\Rightarrow V = \dfrac{\pi}{2}\left[x - \dfrac{\sin 4x}{4}\right]_0^\pi$$

$$\Rightarrow V = \dfrac{\pi}{2}\left[\left(\pi - \dfrac{\sin 4\pi}{4}\right) - \left(0 - \dfrac{\sin 0}{4}\right)\right]$$

$$\Rightarrow V = \dfrac{\pi}{2}\left[(\pi - 0) - (0 - 0)\right]$$

$$\Rightarrow V = \dfrac{\pi^2}{2} \text{ units}^3$$

It is possible to rotate curves around a variety of different lines, but for the purposes of this syllabus it is only necessary to know how to find the volumes of solids of revolution formed when rotated about the x- or the y-axes.

Volumes of solids of revolution when rotated about the y-axis

The method is identical to finding the volume of the solid formed when rotating about the x-axis. Consider the curve $y = x^2$.

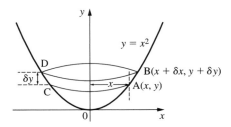

Look at the element ABCD where A is on the curve and has coordinates (x, y). If $y = x^2$, then $x = y^{\frac{1}{2}}$. Hence the coordinates of A are $(y^{\frac{1}{2}}, y)$. ABCD is approximately a cylinder with radius x and whose "height" is δy.

Therefore the volume of ABCD $\approx \pi x^2\,\delta y$

and the volume V of the entire solid $\approx \displaystyle\sum_{y=a}^{y=b} \pi x^2\,\delta y.$

The smaller δy becomes, the closer this approximation is to V,

i.e. $V = \lim_{\delta y \to 0} \displaystyle\sum_{y=a}^{y=b} \pi x^2\,\delta y$

$$V = \pi\int_a^b x^2\,dy$$

This is the formula for a full revolution about the y-axis.

In this case $x = y^{\frac{1}{2}}$. To find the volume of the solid formed when the part of the curve between $y = 0$ and $y = 2$ is rotated about the y-axis, we proceed as follows.

$$V = \pi\int_0^2 x^2\,dy$$

$$\Rightarrow V = \pi\int_0^2 y\,dy$$

$$\Rightarrow V = \pi\left[\frac{y^2}{2}\right]_0^2$$

$$\Rightarrow V = \pi\left[\frac{4}{2} - 0\right]$$

$$\Rightarrow V = 2\pi$$

Example

Find the volume of the solid of revolution formed when the area bounded by the curve $xy = 2$ and the lines $x = 0, y = 3, y = 6$ is rotated about the y-axis.

As before, plotting the curve on a calculator or drawing a diagram first is a good idea.

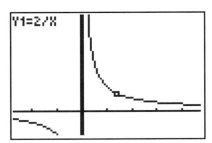

As the rotation is taking place about the y-axis, the required formula is:

$$V = \pi\int_3^6 x^2\,dy$$

Since $xy = 2$, then $x = \dfrac{2}{y}$

So we have:

$$V = \pi \int_{3}^{6} \left(\frac{2}{y}\right)^2 \, dy$$

$$\Rightarrow V = \pi \int_{3}^{6} \frac{4}{y^2} \, dy$$

$$\Rightarrow V = \pi \int_{3}^{6} 4y^{-2} \, dy$$

$$\Rightarrow V = \pi \left[\frac{4y^{-1}}{-1}\right]_{3}^{6}$$

$$\Rightarrow V = \pi \left[\left(\frac{-4}{6}\right) - \left(\frac{-4}{3}\right)\right]$$

$$\Rightarrow V = \frac{2\pi}{3}$$

The calculator display is shown below.

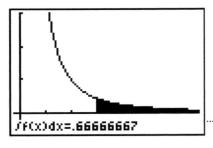

∫f(x)dx=.6666667

Remember that this answer will need to be multiplied by π.

Hence $V = 2.09$ units^3.

Up until now it appears that volumes of solids of revolution are a theoretical application of integration. However, this is not the case. In the field of computer-aided design, volumes of solids of revolution are important. If, for example, we wanted to design a wine glass, then we could rotate the curve $y = x^2$ around the y-axis to give a possible shape. If we wanted a thinner wine glass, then we could rotate the curve $y = x^4$. Because this can be all modelled on a computer, and a three-dimensional graphic produced, designers can work out the shape that they want.

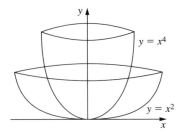

Exercise 6

1 Find the volumes generated when the following areas are rotated through 2π radians about the x-axis.

a The area bounded by the curve $y = 3x + 2$, the x-axis, the y-axis and the line $x = 2$.

b The area bounded by the curve $y = 4x - x^2$ and the x-axis.

c The area bounded by the curve $y = x^3$, the x-axis and the line $x = 2$.

d The area bounded by the curve $y = 1 + \sqrt{x}$, the x-axis, the y-axis and the line $x = 1$.

e The area bounded by the curve $y = x^2 - 1$, the x-axis and the line $x = 3$.

f The area bounded by the curve $y = \frac{1}{3}(2x - 1)^2$, the x-axis and the line $x = 5$.

g The area bounded by the curve $y = 9x - x^2 - 14$ and the x-axis.

h The area bounded by the curve $y = \sin 4x$, the x-axis and the lines $x = 0$ and $x = \frac{\pi}{4}$.

i The area bounded by the curve $y = \tan 2x$, the x-axis and line $x = \frac{\pi}{6}$.

j The area bounded by the curve $y = e^{2x} \sin x$, the x-axis and the lines $x = 0.5$ and $x = 1.5$.

2 Find the volumes generated when the following areas are rotated through 2π radians about the y-axis.

a The area bounded by the curve $y = 4 - x^2$ and the x-axis.

b The area bounded by the curve $y = x^3$, the y-axis and the line $y = 2$.

c The area bounded by the curve $y = e^x$, the y-axis and the line $y = 2$.

d The area bounded by the curve $y = \sin x$, the y-axis and the lines $y = 0.2$ and $y = 0.8$.

e The area bounded by the curve $y = x^2 - 4x$, the y-axis, the x-axis and the line $y = 2$.

f The area bounded by the curve $y = \sin^{-1} x$, the y-axis and the line $y = \frac{\pi}{3}$.

g The area bounded by the curve $y = \ln(x + 1)$, the y-axis and the line $y = 1.5$.

3 Find a general formula for the volume generated when the area bounded by the curve $y = x^2$, the x axis and the line $x = a$ is rotated through 2π radians about the x-axis.

4 Find the volume obtained when the region bounded by the curve $y = 3 + \frac{4}{x}$, the x-axis and the lines $x = 3$ and $x = 6$ is rotated through $360°$ about the x-axis.

5 Consider the curve $y = x^{\frac{1}{2}}$. The part of the curve between $y = 2$ and $y = 5$ is rotated through 2π radians about the y-axis. Find the volume of the solid of revolution formed.

6 Find the volume generated when the area bounded by the curve $y = \ln x$, the x-axis, the y-axis and the line $y = 2$ is rotated through $360°$ about the y-axis.

7 Consider the curve $y = \frac{1}{5}x^2$. A volume is formed by revolving this curve through $360°$ about the y-axis. The radius of the rim of this volume is 5 cm.

Find the depth of the shape and its volume.

8 Sketch the curve $y = |x^2 - 1|$ and shade the area that is bounded by the curve and the x-axis. This area is rotated through 2π radians about the x-axis. Find the volume generated. What is the volume when it is rotated through 2π radians about the y-axis?

9 The parabola $y = 8x^2$ is rotated through $360°$ about its axis of symmetry, thus forming a volume of solid of revolution. Calculate the volume enclosed between this surface and a plane perpendicular to the y-axis. This plane is a distance of 7 units from the origin.

10 Find the volume of the solid of revolution formed when the area included between the x-axis and one wavelength of the curve $y = b \sin\frac{x}{a}$ is rotated through $360°$ about the x-axis.

11 Consider the part curve $y^2 = x^2 \sin x$ which lies between $x = (n - 1)\pi$ and $x = (n + 1)\pi$ where n is an integer. Find the volume generated when this area is rotated through 2π radians about the x-axis.

12 a Using the substitution $x = \sin\theta$, evaluate $\int \sqrt{1 - x^2}\, dx$.

b Hence or otherwise, find the volume generated when the area bounded by the curve $y^2 = 1 - x^4$, the y-axis and the line $y = a$, $0 < a < 1$, is rotated through 2π radians about the y-axis.

Review exercise

1 The region A is bounded by the curve $y = \sin\left(2x + \frac{\pi}{3}\right)$ and by the lines $x = 0$ and $x = \frac{\pi}{6}$. Find the exact value of the volume formed when the area A is rotated fully about the x-axis.

2 Solve the differential equation $xy\frac{dy}{dx} = 1 + y^2$, given that $y = 0$ when $x = 2$.

[IB Nov 00 P1 Q17]

3 A particle moves in a straight line with velocity, in metres per second, at time t seconds, given by $v(t) = 6t^2 - 6t$, $t \geq 0$.

Calculate the total distance travelled by the particle in the first two seconds of motion.

[IB Nov 02 P1 Q11]

4 Consider the region bounded by the curve $y = e^{-2x}$, the x-axis and the lines $x = \pm a$. Find in terms of a, the volume of the solid generated when this region is rotated through 2π radians about the x-axis.

5 Solve the differential equation $\dfrac{dy}{dx} = 5xy$ and sketch one of the solution curves which does not pass through $y = 0$.

6 The acceleration of a body is given in terms of the displacement s metres as $a = \dfrac{3s}{s^2 + 1}$. Determine a formula for the velocity as a function of the displacement given that when $s = 1$ m, $v = 2$ ms^{-1}. Hence find the exact velocity when the body has travelled 5 m.

7 The temperature $T°C$ of an object in a room, after t minutes, satisfies the differential equation $\dfrac{dT}{dt} = k(T - 22)$ where k is a constant.

  a Solve this equation to show that $T = Ae^{kt} + 22$ where A is a constant.

  b When $t = 0$, $T = 100$ and when $t = 15$, $T = 70$.

    i Use this information to find the value of A and of k.

    ii Hence find the value of t when $T = 40$.   [IB May 04 P1 Q4]

8 The velocity of a particle is given by the formula $v = \dfrac{1}{t^2\sqrt{t - 1}}$ for $t > 1$.

Using the substitution $t = \sec^2\theta$, find the displacement travelled between $t = 2$ seconds and $t = T$ seconds.

9 Consider the curve $y = -\dfrac{r}{h}x + r$. The triangular region of this curve which occupies the first quadrant is rotated fully about the x-axis. Show that the volume of the cone formed is $\dfrac{1}{3}\pi r^2 h$.

10 A sample of radioactive material decays at a rate which is proportional to the amount of material present in the sample. Find the half-life of the material if 50 grams decay to 48 grams in 10 years.   [IB Nov 01 P1 Q19]

11 The acceleration in ms^{-2} of a particle moving in a straight line at time t, is given by the formula $a = \sin\dfrac{2\pi}{3}t$. The particle starts from rest from a point where its displacement is 0.5 m from a fixed point O on the line.

  a Find the velocity and displacement of the particle from O at any time t.

  b Find the time that elapses before the particle comes to rest again.

12 a Let $y = \sin(kx) - kx\cos(kx)$, where k is a constant.

    Show that $\dfrac{dy}{dx} = k^2 x \sin(kx)$.

A particle is moving along a straight line so that t seconds after passing through a fixed point O on the line, its velocity $v(t)$ ms^{-1} is given by

$$v(t) = t\sin\left(\frac{\pi}{3}t\right).$$

b Find the values of t for which $v(t) = 0$, given that $0 \leq t \leq 6$.

c **i** Write down a mathematical expression for the **total** distance travelled by the particle in the first six seconds after passing through O.

    ii Find this distance. [IB Nov 01 P2 Q2]

 13 When air is released from an inflated balloon it is found that the rate of decrease of the volume of the balloon is proportional to the volume of the balloon. This can be represented by the differential equation $\dfrac{dv}{dt} = -kv$, where v is the volume, t is the time and k is the constant of proportionality.

a If the initial volume of the balloon is v_0, find an expression, in terms of k, for the volume of the balloon at time t.

b Find an expression, in terms of k, for the time when the volume is $\dfrac{v_0}{2}$.

[IB May 99 P1 Q19]

 14 Show by means of the substitution $x = \tan\theta$ that

$$\int_0^1 \frac{1}{(x^2 + 1)^2}\,dx = \int_0^{\frac{\pi}{4}} \cos^2\theta\,d\theta.$$ Hence find the exact value of the volume

formed when the curve $y = \dfrac{1}{x^2 + 1}$ bounded by the lines $x = 0$ and $x = 1$

is rotated fully about the x-axis.

 15 Consider the curve $y^2 = 9a(4a - x)$.

a Sketch the part of the curve that lies in the first quadrant.

b Find the exact value of the volume V_x when this part of the curve is rotated through $360°$ about the x-axis.

c Show that $\dfrac{V_y}{V_x} = \dfrac{p}{q}$ where V_y is the volume generated when the curve is

rotated fully about the y-axis and p and q are integers.

17 Complex Numbers

Abraham de Moivre was born in Vitry-le-François in France on 26 May 1667. It was not until his late teenage years that de Moivre had any formal mathematics training. In 1685 religious persecution of Protestants became very serious in France and de Moivre, as a practising Protestant, was imprisoned for his religious beliefs. The length of time for which he was imprisoned is unclear, but by 1688 he had moved to England and was a private tutor of mathematics, and was also teaching in the coffee houses of London. In the last decade of the 15th century he met Newton and his first mathematics paper arose from his study of fluxions in

Abraham de Moivre

Newton's *Principia*. This first paper was accepted by the Royal Society in 1695 and in 1697 de Moivre was elected as a Fellow of the Royal Society. He researched mortality statistics and probability and during the first decade of the 16th century he published his theory of probability. In 1710 he was asked to evaluate the claims of Newton and Leibniz to be the discoverers of calculus. This was a major and important undertaking at the time and it is interesting that it was given to de Moivre despite the fact he had found it impossible to gain a university post in England. In many ways de Moivre is best known for his work with the formula $(\cos x + i \sin x)^n$. The theorem that comes from this bears his name and will be introduced in this chapter.

De Moivre was also famed for predicting the day of his own death. He noted that each night he was sleeping 15 minutes longer and by treating this as an arithmetic progression and summing it, he calculated that he would die on the day that he slept for 24 hours. This was 27 November 1754 and he was right!

17.1 Imaginary numbers

Up until now we have worked with any number k that belongs to the real numbers and has the property $k^2 \geq 0$. Hence we have not been able to find $\sqrt{\text{negative number}}$ and have not been able to solve equations such as $x^2 = -1$. In this chapter we begin by defining a new set of numbers called imaginary numbers and state that $i = \sqrt{-1}$.

An **imaginary number** is any number of the form

$$\sqrt{-n^2} = \sqrt{n^2 \times -1}$$
$$= \sqrt{n^2} \times \sqrt{-1}$$
$$= ni$$

Adding and subtracting imaginary numbers

Imaginary numbers are added in the usual way and hence $3i + 7i = 10i$.

They are also subtracted in the usual way and hence $3i - 7i = -4i$.

Multiplying imaginary numbers

When we multiply two imaginary numbers we need to consider the fact that powers of i can be simplified as follows:

$$i^2 = i \times i = \sqrt{-1} \times \sqrt{-1} = -1$$
$$i^3 = i^2 \times i = -1 \times i = -i$$
$$i^4 = i^2 \times i^2 = -1 \times -1 = 1$$
$$i^5 = i^4 \times i = 1 \times i = i$$

This pattern now continues and is shown in the diagram:

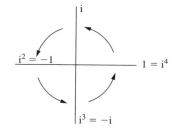

This reminds us that every fourth multiple comes full circle.

Example

Simplify i^{34}.

Since $i^4 = 1$ we simplify this to the form $i^{4n} \times i^x$.

Hence $i^{34} = i^{32} \times i^2$
$$= (i^4)^8 \times i^2$$
$$= 1^8 \times -1$$
$$= -1$$

> **Example**
>
> Simplify $15i^7 \times 3i^{18}$.
> $$15i^7 \times 3i^{18} = 45i^{25}$$
> $$= 45i^{24} \times i$$
> $$= 45(i^4)^6 \times i$$
> $$= 45(1)^6 \times i$$
> $$= 45i$$

Dividing imaginary numbers

This is done in the same way as multiplication.

> **Example**
>
> Simplify $60i^{27} \div 25i^{18}$.
> $$60i^{27} \div 25i^{18} = \frac{60i^{27}}{25i^{18}}$$
> $$= \frac{12}{5}i^9$$
> $$= \frac{12}{5}(i^4)^2 i$$
> $$= \frac{12}{5}(1)^2 i$$
> $$= \frac{12}{5}i$$

If the power of i in the numerator is lower than the power of i in the denominator then we need to use the fact that $i^4 = 1$.

> **Example**
>
> Simplify $27i^{20} \div 18i^{25}$.
> $$27i^{20} \div 18i^{25} = \frac{27i^{20}}{18i^{25}}$$
> $$= \frac{3}{2i^5}$$
> $$= \frac{3i^8}{2i^5} \text{ since } i^8 = (i^4)^2 = 1$$
> $$= \frac{3}{2}i^3$$
> $$= -\frac{3}{2}i$$

Hence when performing these operations the answers should not involve powers of i.

1 Add the following imaginary numbers.

    a $3i + 15i$ **b** $20i + 18i$

    c $5i + 70i + 35i + 2i$ **d** $15i + 45i$

2 Subtract the following imaginary numbers.

    a $20i - 8i$

    b $38i - 23i$

    c $56i - 80i$

    d $25i - 31i + 16i - 62i$

3 Multiply the following imaginary numbers giving the answer in the form n or ni where $n \in \mathbb{R}$.

    a $16 \times 15i$ **b** $4i \times 8i$

    c $15i^2 \times 3i^3$ **d** $8i \times 12i^4$

    e $9i^2 \times 8i^5$ **f** $7i^7 \times 5i^5$

    g $3i^2 \times 5i^4 \times 6i^5$

4 Divide the following imaginary numbers giving the answer in the form n or ni where $n \in \mathbb{R}$.

    a $15i^3 \div 2i$ **b** $6i^7 \div 3i^3$

    c $\dfrac{15i^3}{6i^2}$ **d** $16 \div i$

5 Find x if:

    a $xi + 3i^3 - 4i^5 = 2i$ **b** $\dfrac{3 + 2i^2}{i} = xi$

6 Simplify these.

    a $3i^3 + 6i^5 - 8i^7 - 2i^9$ **b** $\dfrac{2i^3 + 3i^3 - 7i^4}{3i}$

    c $\dfrac{3i^4}{2i} + \dfrac{2i^5}{i^2} - \dfrac{3i}{i^2}$ **d** $3i^5 \times \dfrac{2i^6}{6i^3}$

    e $\dfrac{6i - 3i^2 + 2i^3}{4i}$

17.2 Complex numbers

A **complex number** is defined as one that has a real and an imaginary part. Examples of these would be $2 + 3i$ or $6 - 5i$.

They are generally written in the form $z = x + iy$ where x and y can have any real value including zero.

Hence 6 is a complex number since it can be written in the form $6 + 0i$ and $5i$ is a complex number since it can be written as $0 + 5i$.

Hence both real numbers and imaginary numbers are actually subsets of complex numbers and the notation for this set is $\mathbb{C}$.

Thus we can say $3 + 5i \in \mathbb{C}$.

Adding and subtracting complex numbers

This is done by adding or subtracting the real parts and the imaginary parts in separate groups.

Example

Simplify $(5 + 7i) + (2 - 3i)$.
$$(5 + 7i) + (2 - 3i) = (5 + 2) + (7i - 3i)$$
$$= 7 + 4i$$

Example

Simplify $(9 - 2i) - (4 - 7i)$.
$$(9 - 2i) - (4 - 7i) = (9 - 4) + (-2i - -7i)$$
$$= 5 + 5i$$

Multiplication of complex numbers

This is done by applying the distributive law to two brackets and remembering that $i^2 = -1$. It is similar to expanding two brackets to form a quadratic expression.

Example

Simplify $(2i + 3)(3i - 2)$.
$$(2i + 3)(3i - 2) = 6i^2 + 9i - 4i - 6$$
$$= 6(-1) + 9i - 4i - 6$$
$$= -12 + 5i$$

Example

Simplify $(6 + i)(6 - i)$.
$$(6 + i)(6 - i) = 36 + 6i - 6i - i^2$$
$$= 36 - (-1)$$
$$= 37$$

We can also use the binomial theorem to simplify complex numbers.

Example

Express $(3 - 2i)^5$ in the form $x + iy$.
$$(3 - 2i)^5 = {}^5C_0(3)^5(-2i)^0 + {}^5C_1(3)^4(-2i)^1 + {}^5C_2(3)^3(-2i)^2$$
$$+ {}^5C_3(3)^2(-2i)^3 + {}^5C_4(3)^1(-2i)^4 + {}^5C_5(3)^0(-2i)^5$$
$$= 243 + 405(-2i) + 270(4i^2) + 90(-8i^3) + 15(16i^4) + (-32i^5)$$
$$= 243 + 405(-2i) + 270(-4) + 90(8i) + 15(16) + (-32i)$$
$$= -597 - 122i$$

Division of complex numbers

Before we do this, we have to introduce the concept of a conjugate complex number. Any pair of complex numbers of the form $x + iy$ and $x - iy$ are said to be **conjugate** and $x - iy$ is said to be the conjugate of $x + iy$.

If $x + iy$ is denoted by z, then its conjugate $x - iy$ is denoted by $\bar{z}$ or $z*$.

Conjugate complex numbers have the property that when multiplied the result is real. This was demonstrated in the example on the previous page and the result in general is

$$
\begin{aligned}
(x + iy)(x - iy) &= x^2 + ixy - ixy - i^2y^2 \\
&= x^2 - (-1)y^2 \\
&= x^2 + y^2
\end{aligned}
$$

> Note the similarity to evaluating the difference of two squares.

To divide two complex numbers we use the property that if we multiply the numerator and denominator of a fraction by the same number, then the fraction remains unchanged in size. The aim is to make the denominator real, and hence we multiply numerator and denominator by the conjugate of the denominator. This process is called realizing the denominator. This is very similar to rationalizing the denominator of a fraction involving surds.

Example

Write $\dfrac{2 + 3i}{2 - i}$ in the form $a + ib$.

$$
\begin{aligned}
\frac{2 + 3i}{2 - i} &= \frac{(2 + 3i)}{(2 - i)} \times \frac{(2 + i)}{(2 + i)} \\
&= \frac{4 + 6i + 2i + 3i^2}{4 - 2i + 2i - i^2} \\
&= \frac{4 + 8i - 3}{4 - (-1)} \\
&= \frac{1 + 8i}{5} \\
&= \frac{1}{5} + \frac{8}{5}i
\end{aligned}
$$

Zero complex number

A complex number is only zero if both the real and imaginary parts are zero, i.e. $0 + 0i$.

Equal complex numbers

Complex numbers are only equal if both the real and imaginary parts are separately equal. This allows us to solve equations involving complex numbers.

Example

Solve $x + iy = (3 + i)(2 - 3i)$.

$$x + iy = (3 + i)(2 - 3i)$$

$$\Rightarrow x + iy = 6 + 2i - 9i - 3i^2$$

$$\Rightarrow x + iy = 6 - 7i - 3(-1)$$

$$\Rightarrow x + iy = 9 - 7i$$

Equating the real parts of the complex number gives $x = 9$.
Equating the imaginary parts of the complex number gives $y = -7$.

This idea also allows us to find the square root of a complex number.

Example

Find the values of $\sqrt{3 + 4i}$ in the form $a + ib$.

Let $\sqrt{3 + 4i} = a + ib$ ···

$$\Rightarrow (\sqrt{3 + 4i})^2 = (a + ib)^2$$

$$\Rightarrow 3 + 4i = a^2 + 2iab + i^2b^2$$

$$\Rightarrow 3 + 4i = a^2 - b^2 + 2iab$$

We now use the idea of equal complex numbers and equate the real and imaginary parts.

Equating real parts $\Rightarrow a^2 - b^2 = 3$
Equating imaginary parts $\Rightarrow 2ab = 4 \Rightarrow ab = 2$
These equations can be solved simultaneously to find a and b.

If we substitute $b = \dfrac{2}{a}$ into $a^2 - b^2 = 3$ we find

$$a^2 - \left(\frac{2}{a}\right)^2 = 3$$

$$\Rightarrow a^4 - 3a^2 - 4 = 0$$

$$\Rightarrow (a^2 - 4)(a^2 + 1) = 0$$

Ignoring the imaginary roots

$$\Rightarrow a = 2 \text{ or } a = -2$$

$$\Rightarrow b = 1 \text{ or } b = -1$$

Therefore $\sqrt{3 + 4i} = 2 + i$ or $-2 - i$ ·········

If we had used the imaginary values for a then·······

$$a = i \quad \text{or} \quad a = -i$$

$$\Rightarrow b = \frac{2}{i} \quad \text{or} \quad b = -\frac{2}{i}$$

> This can also be done in a different way that will be dealt with later in the chapter.

> As with the square root of a real number, there are two answers and one is the negative of the other.

> It is usually assumed that a and b are real numbers and we ignore imaginary values for a and b, but if we assume they are imaginary the same answers result.

$$\Rightarrow b = \frac{2i^4}{i} \text{ or } b = -\frac{2i^4}{i}$$

$$\Rightarrow b = 2i^3 \text{ or } b = -2i^3$$

$$\Rightarrow b = -2i \text{ or } b = 2i$$

So $\sqrt{3 + 4i} = i + i(-2i) \text{ or } -i + i(2i)$

$$= i - 2i^2 \text{ or } -i + 2i^2$$

$$= 2 + i \text{ or } -2 - i \text{ as before}$$

Complex roots of a quadratic equation

In Chapter 2 we referred to the fact that when a quadratic equation has the property $b^2 - 4ac < 0$, then it has no real roots. We can now see that there are two complex conjugate roots.

Example

Solve the equation $x^2 - 2x + 4 = 0$.

Using the quadratic formula $x = \dfrac{2 \pm \sqrt{4 - 16}}{2}$

$$\Rightarrow x = \frac{2 \pm \sqrt{-12}}{2} = \frac{2 \pm \sqrt{-1}\sqrt{12}}{2}$$

$$\Rightarrow x = \frac{2 \pm 2i\sqrt{3}}{2}$$

$$\Rightarrow x = 1 + i\sqrt{3} \text{ or } x = 1 - i\sqrt{3}$$

The $\pm$ sign in the formula ensures that the complex roots of quadratic equations are always conjugate.

Example

Form a quadratic equation which has a complex root of $2 + i$.
Since one complex root is $2 + i$ the other root must be its complex conjugate.
Hence the other root is $2 - i$.
Thus the quadratic equation is

$$[x - (2 + i)][x - (2 - i)] = 0$$

$$\Rightarrow x^2 - (2 + i)x - (2 - i)x + (2 + i)(2 - i) = 0$$

$$\Rightarrow x^2 - 4x + (4 + 2i - 2i - i^2) = 0$$

$$\Rightarrow x^2 - 4x + 5 = 0$$

Complex roots of a polynomial equation

We know from Chapter 4 that solving any polynomial equation with real coefficients always involves factoring out the roots. Hence if any polynomial has complex roots these will always occur in conjugate pairs. For a polynomial equation it is possible that some of the roots will be complex and some will be real. However, the number of complex roots is always even. Hence a polynomial of degree five could have:

- five real roots;
- three real roots and two complex roots; or
- one real root and four complex roots.

Having two real roots and three complex roots is not possible.

To find the roots we need to use long division.

Example

Given that $z = 2 + i$ is a solution to the equation $z^3 - 3z^2 + z + 5 = 0$, find the other two roots.

Since one complex root is $2 + i$ another complex root must be its conjugate. Hence $2 - i$ is a root.

Thus a quadratic factor of the equation is

$$[z - (2 + i)][z - (2 - i)]$$

$$\Rightarrow z^2 - (2 + i)z - (2 - i)z + (2 + i)(2 - i)$$

$$\Rightarrow z^2 - 4z + (4 + 2i - 2i - i^2)$$

$$\Rightarrow z^2 - 4z + 5$$

Using long division:

$$
\begin{array}{r}
z + 1 \\
(z^2 - 4z + 5)\overline{)z^3 - 3z^2 + z + 5} \\
-\underline{z^3 - 4z^2 + 5z} \\
z^2 - 4z + 5 \\
-\underline{z^2 - 4z + 5} \\
0
\end{array}
$$

Hence $z^3 - 3z^2 + z + 5 = (z + 1)(z^2 - 4z + 5) = 0$

$$\Rightarrow z = -1, 2 + i, 2 - i$$

Many of the operations we have covered so far could be done on a calculator.

Example

Find $(6 + 6i) + (7 - 2i)$.

```
(6+6i)+(7-2i)
              13+4i
■
```

$(6 + 6i) + (7 - 2i) = 13 + 4i$

Example

Find $\sqrt{5 - 12i}$.

```
√(5-12i)
            3-2i
```

As with real numbers the calculator only gives one value for the square root, unless the negative square root is specified. To find the second square root, we find the negative of the first.

$\sqrt{5 - 12i} = 3 - 2i$ or $-3 + 2i$

Example

Express $(5 - 4i)^7$ in the form $x + iy$.

```
(5-4i)^7
        4765+441284i
■
```

$(5 - 4i)^7 = 4765 + 441\,284i$

Exercise 2

1 Add these pairs of complex numbers.

  a $2 + 7i$ and $6 + 9i$

  b $5 + 12i$ and $13 + 16i$

  c $4 + 8i$ and $3 - 7i$

  d $8 - 7i$ and $6 - 14i$

  e $-4 - 18i$ and $-3 + 29i$

2 Find $u - v$.

  a $u = 5 + 8i$ and $v = 2 + 13i$

  b $u = 16 + 7i$ and $v = 3 - 15i$

  c $u = -3 + 6i$ and $v = 17 - 4i$

  d $u = -5 - 4i$ and $v = -12 - 17i$

3 Simplify these.

  a $(3 + 2i)(2 + 5i)$ **b** $(5 + 3i)(10 + i)$ **c** $i(2 - 7i)$

  d $(9 - 5i)(15 - 4i)$ **e** $(15 - 9i)(15 + 9i)$ **f** $(a + bi)(a - bi)$

  g $i(5 + 2i)(12 - 5i)$ **h** $(11 + 2i)^2$ **i** $(x + iy)^2$

  j $(m + in)^3$ **k** $i(m + 2i)^3(3 - i)$

4 Realize the denominator of each of the following fractions and hence express each in the form $a + bi$.

  a $\dfrac{2}{3 - i}$ **b** $\dfrac{5i}{2 - 3i}$ **c** $\dfrac{3 + 4i}{5 - 2i}$ **d** $\dfrac{-5 + 4i}{-2 - 5i}$

  e $\dfrac{4 + 12i}{2i}$ **f** $\dfrac{x + iy}{2x - iy}$ **g** $\dfrac{2 + i}{2 - i}$ **h** $i\left(\dfrac{3 + 4i}{3 - i}\right)$

  i $\dfrac{2 + 5i}{x - iy}$ **j** $ix\left(\dfrac{3x - iy}{y + ix}\right)$

5 Express these in the form $x + iy$.

  a $(1 - i)^4$ **b** $(1 + 2i)^3$ **c** $(3 + 4i)^5$ **d** $(-1 - 3i)^8$

6 Solve these equations for x and y.

  a $x + iy = 15 - 7i$ **b** $x + iy = 8$ **c** $x + iy = -3i$

  d $x + iy = (3 + 2i)^2$ **e** $x + iy = 6i^2 - 3i$ **f** $x + iy = \dfrac{3i + 2}{i - 4}$

  g $x + iy = (2 + 5i)^2$ **h** $(x + iy)^2 = 15i$ **i** $x + iy = \dfrac{2 + 7i}{3 - 4i} - 2$

  j $x + iy = 6i^2 - 3i + (2 + i)^2$ **k** $x + iy = \left(\dfrac{2 + i}{3 - 2i}\right)^2 + 15i$

7 Find the real and imaginary parts of these.

  a $(6 + 5i)(2 - 3i)$ **b** $\dfrac{3 - 7i}{8 + 11i}$ **c** $\dfrac{3}{2 + 5i} + \dfrac{5}{3 - 4i}$

  d $\dfrac{3 - i}{2 + i} + \dfrac{1}{7 - 4i}$ **e** $\dfrac{a}{2 + ib} - \dfrac{3}{4 - ia}$ **f** $\dfrac{x}{x + iy} - \dfrac{x}{x - iy}$

  g $(3 + 2i)^5$ **h** $\left(\cos\dfrac{\pi}{3} + i\sin\dfrac{\pi}{3}\right)^2$ **i** $[2(1 + i\sqrt{3})]^4$

  j $\dfrac{x}{1 + iy} + \dfrac{3}{4 - 3xi}$

8 Find the square roots of these.

    a $-15 + 8i$ **b** $1 + i$ **c** $4 - 3i$ **d** $12 + 13i$

    e $2 + 5i$ **f** $\dfrac{2 + i}{3i}$ **g** $\dfrac{3 - i}{4 + 3i}$ **h** $2i + \dfrac{3 + 2i}{1 - i}$

    i $i\left(\dfrac{2 - i}{4 + i}\right)^2$

9 Solve these equations.

    a $x^2 + 6x + 10 = 0$ **b** $x^2 + x + 1 = 0$ **c** $x^2 + 3x + 15 = 0$

    d $3x^2 + 6x + 5 = 0$ **e** $2(x + 4)(x - 1) = 3(x - 7)$

10 Form an equation with these roots.

    a $2 + 3i, 2 - 3i$ **b** $3 + i, 3 - i$ **c** $4 + 3i, 4 - 3i$

    d $1 + 2i, 1 - 2i, 1$ **e** $3 + 2i, 3 - 2i, 2$ **f** $5 + 2i, 5 - 2i, 3, -3$

11 Find a quadratic equation with the given root.

    a $2 + 7i$ **b** $4 - 3i$ **c** $7 - 6i$ **d** $a + ib$

12 Find a quartic equation given that two of its roots are $2 + i$ and $3 - 2i$.

13 Given that $z = 1 + i$ is a root of the equation $z^3 - 5z^2 + 8z - 6 = 0$, find the other two roots.

14 Find, in the form $a + ib$, all the solutions of these equations.

    a $z^3 - z^2 - z = 15$ **b** $z^3 + 6z = 20$

15 Given that $z = -2 + 7i$, express $2z + \dfrac{1}{z}$ in the form $a + ib$, where a and b are real numbers.

16 Given that the complex number z and its conjugate z^* satisfy the equation $zz^* + iz = 66 - 8i$, find the possible values of z.

17 If $z = 12 - 5i$, express $2z + \dfrac{39}{z}$ in its simplest form.

18 Let z_1 and z_2 be complex numbers. Solve the simultaneous equations

$z_1 + 2z_2 = 4$
$2z_1 + iz_2 = 3 + i$

Give your answer in the form $x + iy$ where $x, y \in \mathbb{Q}$.

19 If $z = 1 + \dfrac{2}{1 + i\sqrt{3}}$, find z in the form $x + iy$ where $x, y \in \mathbb{R}$.

20 Consider the equation $4(p - iq) = 2q - 3ip - 3(2 + 3i)$ where p and q are real numbers. Find the values of p and q.

21 If $\sqrt{z} = \dfrac{3}{1 + 2i} + 4 - 3i$, find z in the form $x + iy$ where $x, y \in \mathbb{R}$.

17.3 Argand diagrams

We now need a way of representing complex numbers in two-dimensional space and this is done on an Argand diagram, named after the mathematician Jean-Robert Argand. It looks like a standard two-dimensional Cartesian plane, except that real numbers are represented on the x-axis and imaginary numbers on the y-axis.

Hence on an Argand diagram the complex number $2 + 5i$ is represented as the vector $\begin{pmatrix} 2 \\ 5 \end{pmatrix}$.

For this reason it is known as the **Cartesian form** of a complex number.

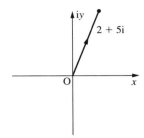

As with a vector the complex number will usually have an arrow on it to signify the direction and is often denoted by z.

On an Argand diagram the complex number $2 + 9i$ can be represented by the vector $\overrightarrow{OA}$ where A has coordinates $(2, 9)$. However, since it is the line that represents the complex number, the vectors $\overrightarrow{BC}$ and $\overrightarrow{DE}$ also represent the complex number $2 + 9i$.

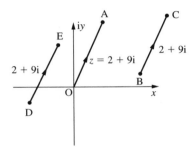

This is similar to the idea of position vectors and tied vectors introduced in Chapter 12.

Example

The complex numbers $z_1 = \dfrac{m}{1-i}$ and $z_2 = \dfrac{n}{3+4i}$, where m and n are real numbers, have the property $z_1 + z_2 = 2$.

a Find the values of m and n.

b Using these values of m and n, find the distance between the points which represent z_1 and z_2 in the Argand diagram.

a
$$\frac{m}{1-i} + \frac{n}{3+4i} = 2$$

$$\Rightarrow \frac{m(1+i)}{(1-i)(1+i)} + \frac{n(3-4i)}{(3+4i)(3-4i)} = 2$$

$$\Rightarrow \frac{m(1+i)}{2} + \frac{n(3-4i)}{25} = 2$$

$$\Rightarrow \frac{m}{2} + \frac{3n}{25} + i\left(\frac{m}{2} - \frac{4n}{25}\right) = 2$$

Equating real parts:

$$\frac{m}{2} + \frac{3n}{25} = 2$$

$$\Rightarrow 25m + 6n = 100 \quad \text{equation (i)}$$

Equating imaginary parts:

$$\frac{m}{2} - \frac{4n}{25} = 0$$

$$\Rightarrow 25m - 8n = 0 \quad \text{equation (ii)}$$

Subtracting equation (ii) from equation (i): $14n = 100 \Rightarrow n = \dfrac{50}{7}$

Substituting in equation (i): $m = \dfrac{16}{7}$

b From part **a** $z_1 = \dfrac{m(1 + i)}{2} = \dfrac{8(1 + i)}{7}$ and $z_2 = \dfrac{n(3 - 4i)}{25} = \dfrac{2(3 - 4i)}{7}$

These are shown on the Argand diagram.

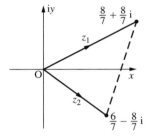

Hence the distance between the points is $\sqrt{\left(\dfrac{6}{7} - \dfrac{8}{7}\right)^2 + \left(-\dfrac{8}{7} - \dfrac{8}{7}\right)^2} = \dfrac{\sqrt{260}}{7}$.

Addition and subtraction on the Argand diagram

This is similar to the parallelogram law for vectors which was explained in Chapter 12.

Consider two complex numbers z_1 and z_2 represented by the vectors $\overrightarrow{OA}$ and $\overrightarrow{OB}$.

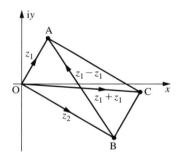

If AC is drawn parallel to OB, then $\overrightarrow{AC}$ also represents z_2. We know from vectors that $\overrightarrow{OA} + \overrightarrow{AC} = \overrightarrow{OC}$.

Hence $z_1 + z_2$ is represented by the diagonal $\overrightarrow{OC}$.

Similarly $\overrightarrow{OB} + \overrightarrow{BA} = \overrightarrow{OA}$

$$\Rightarrow \overrightarrow{BA} = \overrightarrow{OA} - \overrightarrow{OB}$$

Hence $z_1 - z_2$ is represented by the diagonal $\overrightarrow{BA}$.

Example

Show $(5 + 3i) + (2 - 5i)$ and $(5 + 3i) - (2 - 5i)$ on an Argand diagram.

Let $z_1 = 5 + 3i$ and $z_2 = 2 - 5i$ and represent them on the diagram by the vectors $\overrightarrow{OA}$ and $\overrightarrow{OB}$ respectively. Let C be the point which makes OACB a parallelogram.

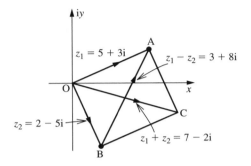

From the diagram it is clear that $\overrightarrow{OC} = \begin{pmatrix} 7 \\ -2 \end{pmatrix}$ and $\overrightarrow{BA} = \begin{pmatrix} 5 \\ 3 \end{pmatrix} - \begin{pmatrix} 2 \\ -5 \end{pmatrix} = \begin{pmatrix} 3 \\ 8 \end{pmatrix}$

and these two diagonals represent $z_1 + z_2$ and $z_1 - z_2$ respectively. This is confirmed by the fact that $z_1 + z_2 = (5 + 3i) + (2 - 5i) = 7 - 2i$ and

$z_1 - z_2 = (5 + 3i) - (2 - 5i) = 3 + 8i$.

Multiplication by i on the Argand diagram

Consider the complex number $z = x + iy$.

Hence

$$iz = ix + i^2y$$

$$\Rightarrow iz = -y + ix$$

These are shown on the Argand diagram.

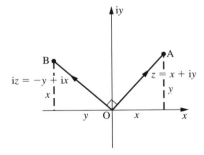

Considering this diagram the gradient of OA is $\dfrac{y}{x}$ and the gradient of OB is $-\dfrac{x}{y}$. Since the product of gradients is -1, these two lines are perpendicular. Hence if we multiply a complex number by i, the effect on the Argand diagram is to rotate the vector representing it by $90°$ anticlockwise.

Example

If $z = 3 + 4i$, draw z and iz on an Argand diagram and state iz in the form $a + ib$.

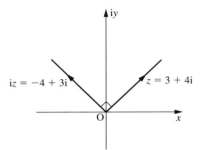

$iz = 3i + 4i^2$

$\qquad = -4 + 3i$

Notation for complex numbers

So far we have only seen the representation of a complex number in Cartesian form, that is $x + iy$. However, there are two other forms which are very important.

Polar coordinate form

This is more commonly called the **modulus-argument form** or the mod-arg form. It defines the complex number by a distance r from a given point and an angle θ radians from a given line. Consider the diagram below.

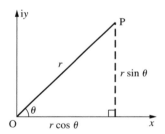

$\overrightarrow{OP}$ represents the complex number $x + iy$. $\overrightarrow{OP}$ has magnitude r and is inclined at an angle of θ radians.

From the diagram $x = r\cos\theta$ and $y = r\sin\theta$.

Thus

$$x + iy = r(\cos\theta + i\sin\theta).$$

This is the modulus-argument form of a complex number where the modulus is r and the angle, known as the argument, is θ. It is usual to give θ in radians. We sometimes express this as (r, θ).

If we are asked to express a complex number given in modulus-argument form in Cartesian form, then we use the fact that $x = r\cos\theta$ and $y = r\sin\theta$.

Example

Express the complex number $\left(2, \dfrac{\pi}{6}\right)$ in Cartesian form.

$x = r \cos \theta$

$\Rightarrow x = 2 \cos \dfrac{\pi}{6} = \sqrt{3}$

$y = r \sin \theta$

$\Rightarrow y = 2 \sin \dfrac{\pi}{6} = 1$

Hence in Cartesian form the complex number is $\sqrt{3} + i$.

If we are asked to express a complex number given in Cartesian form in modulus-argument form, then we proceed as follows.

If $x = r \cos \theta$ and $y = r \sin \theta$

Then

$x^2 + y^2 = r^2 \cos^2 \theta + r^2 \sin^2 \theta$

$\qquad\qquad = r^2(\cos^2 \theta + \sin^2 \theta)$

$\cos^2 \theta + \sin^2 \theta = 1 \Rightarrow r = \sqrt{x^2 + y^2}$

The modulus of a complex number is assumed positive and hence we can ignore the negative square root.

Also $\quad \dfrac{y}{x} = \dfrac{r \sin \theta}{r \cos \theta}$

$\Rightarrow \tan \theta = \dfrac{y}{x}$

$\Rightarrow \theta = \arctan\left(\dfrac{y}{x}\right)$

Example

Express $3 + 4i$ in polar form.

$r = \sqrt{x^2 + y^2}$

$\Rightarrow r = \sqrt{3^2 + 4^2} = \sqrt{25} = 5$

$\Rightarrow \theta = \arctan\left(\dfrac{y}{x}\right)$

$\Rightarrow \theta = \arctan\left(\dfrac{4}{3}\right)$

$\Rightarrow \theta = 0.927\ldots$

Hence in polar form the complex number is $5(\cos 0.927 + i \sin 0.927)$.

This leads us on to the problem of which quadrant the complex number lies in. From the work done in Chapter 1 we know that $\theta = \arctan\left(\dfrac{y}{x}\right)$ has infinite solutions. To resolve this problem, when calculating the argument in questions like this, it is essential to draw a sketch. Also, by convention, the argument always lies in the range $-\pi < x \leq \pi$. This is slightly different to the method used in Chapter 1 for finding angles in a given range. We will demonstrate this in the example below.

Example

Express the following in modulus-argument form.

a $12 - 5i$

b $-12 + 5i$

c $-12 - 5i$

In all cases the modulus is the same since the negative signs do not have an effect.

$$r = \sqrt{x^2 + y^2}$$

$$\Rightarrow r = \sqrt{12^2 + 5^2} = \sqrt{169} = 13$$

In terms of the argument we will examine each case in turn.

a

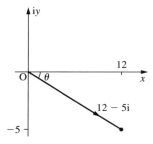

From the diagram it is clear that the complex number lies in the fourth quadrant and hence the argument must be a negative acute angle.

$$\theta = \arctan\left(\frac{y}{x}\right)$$

$$\Rightarrow \theta = \arctan\left(\frac{-5}{12}\right)$$

$$\Rightarrow \theta = -0.395\ldots \quad \cdots\cdots\cdots\cdots\cdots\cdots\cdots\cdots\cdots\cdots\cdots\cdots\cdots\cdots\cdots\cdots$$

This comes directly from a calculator.

Hence in modulus-argument form the complex number is $13[\cos(-0.395) + i\sin(-0.395)]$.

b

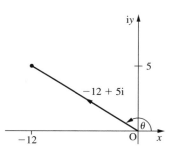

From the diagram it is clear that the complex number lies in the second quadrant and hence the argument must be a positive obtuse angle.

$$\theta = \arctan\left(\frac{y}{x}\right)$$

$$\Rightarrow \theta = \arctan\left(\frac{5}{-12}\right)$$

$$\Rightarrow \theta = 0.395\ldots$$

However, this is clearly in the wrong quadrant and hence to find the required angle we need to add π to this giving $\theta = 2.75\ldots$

Hence in modulus-argument form the complex number is $13[\cos(2.75) + \text{i}\sin(2.75)]$.

c

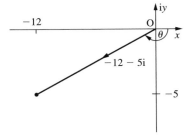

From the diagram it is clear that the complex number lies in the third quadrant and hence the argument must be a negative obtuse angle.

$$\theta = \arctan\left(\frac{y}{x}\right)$$

$$\Rightarrow \theta = \arctan\left(\frac{-5}{-12}\right)$$

$$\Rightarrow \theta = 0.395\ldots$$

However, this is clearly in the wrong quadrant and hence to find the required angle we need to subtract π from this giving $\theta = -2.75\ldots$

Hence in modulus-argument form the complex number is $13[\cos(-2.75) + \text{i}\sin(-2.75)]$.

Example

Express $\dfrac{3}{1+2\text{i}}$ in polar form.

To do this we begin by expressing the complex number in Cartesian form, by realizing the denominator.

$$\frac{3}{1+2\text{i}} = \frac{3(1-2\text{i})}{(1+2\text{i})(1-2\text{i})}$$

$$= \frac{3-6\text{i}}{5}$$

Hence $r\cos\theta = \dfrac{3}{5}$ and $r\sin\theta = -\dfrac{6}{5}$

$$\Rightarrow r^2 = \left(\frac{3}{5}\right)^2 + \left(-\frac{6}{5}\right)^2$$

$$\Rightarrow r = \sqrt{\frac{9}{5}} = 1.34\ldots$$

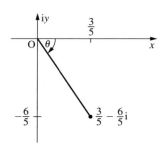

From the diagram the complex number lies in the fourth quadrant and hence the argument is a negative acute angle.

$$\frac{r\sin\theta}{r\cos\theta} = \frac{\dfrac{-6}{5}}{\dfrac{3}{5}}$$

$$\Rightarrow \tan\theta = -2$$

$$\Rightarrow \theta = -1.10\ldots$$

Hence $\dfrac{3}{1+2i} = 1.34[\cos(-1.11) + i\sin(-1.11)]$.

Exponential form

This is similar to the mod-arg form and is sometimes called the Euler form. A complex number in this form is expressed as $re^{i\theta}$ where r is the modulus and θ is the argument.

Hence $5\left(\cos\dfrac{4\pi}{3} + i\sin\dfrac{4\pi}{3}\right)$ becomes $5e^{i\frac{4\pi}{3}}$ in exponential form.

A calculator will also give complex numbers in exponential form if required.

To express Cartesian form in exponential form or vice versa, we proceed in exactly the same way as changing between polar form and Cartesian form.

We will now show that polar form and exponential form are equivalent.

Let $z = r(\cos\theta + i\sin\theta)$

$$\Rightarrow \frac{dz}{d\theta} = r(-\sin\theta + i\cos\theta)$$

$$\Rightarrow \frac{dz}{d\theta} = ir(\cos\theta + i\sin\theta)$$

$$\Rightarrow \frac{dz}{d\theta} = iz$$

We now treat this as a variables separable differential equation.

$$\Rightarrow \frac{1}{z}\frac{dz}{d\theta} = i$$

$$\Rightarrow \int \frac{1}{z}\frac{dz}{d\theta}\,d\theta = \int i\,d\theta$$

$$\Rightarrow \int \frac{1}{z}\,dz = \int i\,d\theta$$

$$\Rightarrow \ln z = i\theta + \ln c$$

When $\theta = 0$, $z = r \Rightarrow \ln r = \ln c$

$$\Rightarrow \ln z - \ln r = i\theta$$

$$\Rightarrow \ln \frac{z}{r} = i\theta$$

$$\Rightarrow \frac{z}{r} = e^{i\theta} \Rightarrow z = re^{i\theta}$$

Example

Express $-2 + 5i$ in exponential form.

$r \cos \theta = -2$ and $r \sin \theta = 5$

$$\Rightarrow r^2 = (-2)^2 + (5)^2$$
$$\Rightarrow r = \sqrt{29}$$

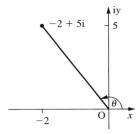

From the diagram above the complex number lies in the second quadrant and hence the argument is a positive obtuse angle.

$$\frac{r \sin \theta}{r \cos \theta} = \frac{5}{-2}$$

$$\Rightarrow \tan \theta = -2.5$$

From the calculator $\theta = -1.19\ldots$ However, this is clearly in the wrong quadrant and hence to find the required angle we need to add π to this giving $\theta = 1.95\ldots$

$$\Rightarrow -2 + 5i = \sqrt{29}e^{1.95i}$$

Products and quotients in polar form

If $z_1 = a(\cos \alpha + i \sin \alpha)$ and $z_2 = b(\cos \beta + i \sin \beta)$

Then $z_1 z_2 = ab(\cos \alpha + i \sin \alpha)(\cos \beta + i \sin \beta)$

$$\Rightarrow z_1 z_2 = ab(\cos \alpha \cos \beta + i \sin \alpha \cos \beta + i \cos \alpha \sin \beta + i^2 \sin \alpha \sin \beta)$$

$$\Rightarrow z_1 z_2 = ab[(\cos \alpha \cos \beta - \sin \alpha \sin \beta) + i(\sin \alpha \cos \beta + \cos \alpha \sin \beta)]$$

Remembering the compound angle formulae from Chapter 7

$$\Rightarrow z_1 z_2 = ab[\cos(\alpha + \beta) + i \sin(\alpha + \beta)]$$

Hence if we multiply two complex numbers in polar form, then we multiply the moduli and add the arguments.

$$|z_1 z_2| = |z_1| \times |z_2| \text{ and } \arg(z_1 z_2) = \arg z_1 + \arg z_2$$

Similarly $\dfrac{z_1}{z_2} = \dfrac{a}{b}[\cos(\alpha - \beta) + i \sin(\alpha - \beta)]$.

The standard notation for the modulus of a complex number z is $|z|$ and the standard notation for the argument of a complex number z is arg (z).

Hence if we divide two complex numbers in polar form, then we divide the moduli and subtract the arguments.

$$\left|\frac{z_1}{z_2}\right| = \frac{|z_1|}{|z_2|} \text{ and } \arg\left(\frac{z_1}{z_2}\right) = \arg z_1 - \arg z_2$$

Example

Let $z_1 = 2 - i$ and $z_2 = 3 - i$.

a Find the product $z_1 z_2$ in the form $x + iy$.

b Find z_1, z_2 and $z_1 z_2$ in exponential form.

c Hence show that $-\dfrac{\pi}{4} = \arctan\left(-\dfrac{1}{2}\right) + \arctan\left(-\dfrac{1}{3}\right)$.

a $z_1 z_2 = (2 - i)(3 - i)$

$\quad\quad\quad = 6 - 2i - 3i + i^2$

$\quad\quad\quad = 5 - 5i$

b For z_1, $r \cos\theta = 2$ and $r \sin\theta = -1$

$\Rightarrow r^2 = (2)^2 + (-1)^2$

$\quad\Rightarrow r = \sqrt{5}$

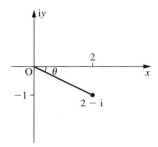

The diagram shows the complex number lies in the fourth quadrant and hence the argument is a negative acute angle.

$$\frac{r \sin\theta}{r \cos\theta} = \frac{-1}{2}$$

$$\Rightarrow \tan\theta = -\frac{1}{2}$$

$$\Rightarrow \theta = \arctan\left(-\frac{1}{2}\right)$$

$$\Rightarrow z_1 = \sqrt{5}e^{i(\arctan(-\frac{1}{2}))}$$

For z_2, $r \cos\theta = 3$ and $r \sin\theta = -1$

$\Rightarrow r^2 = (3)^2 + (-1)^2$

$\quad\Rightarrow r = \sqrt{10}$

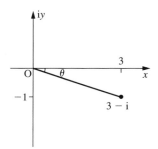

The diagram shows the complex number lies in the fourth quadrant and hence the argument is a negative acute angle.

$$\frac{r \sin \theta}{r \cos \theta} = \frac{-1}{3}$$

$$\Rightarrow \tan \theta = -\frac{1}{3}$$

$$\Rightarrow \theta = \arctan\left(-\frac{1}{3}\right)$$

$$\Rightarrow z_2 = \sqrt{10}\, e^{i(\arctan(-\frac{1}{3}))}$$

For $z_1 z_2$, $r \cos \theta = 5$ and $r \sin \theta = -5$

$$\Rightarrow r^2 = (5)^2 + (-5)^2$$

$$\Rightarrow r = 5\sqrt{2}$$

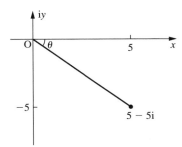

The diagram shows the complex number lies in the fourth quadrant and hence the argument is a negative acute angle.

$$\frac{r \sin \theta}{r \cos \theta} = \frac{-5}{5}$$

$$\Rightarrow \tan \theta = -1$$

$$\Rightarrow \theta = -\frac{\pi}{4}$$

$$\Rightarrow z_1 z_2 = 5\sqrt{2}\, e^{-i\left(\frac{\pi}{4}\right)}$$

c Since $\arg(z_1 z_2) = \arg z_1 + \arg z_2$

$$-\frac{\pi}{4} = \arctan\left(-\frac{1}{2}\right) + \arctan\left(-\frac{1}{3}\right)$$

Exercise 3

1 If $z_1 = 1 - 2i$, $z_2 = 2 + 4i$, $z_3 = -4 + 3i$ and $z_4 = -5 - i$, using the parallelogram law, represent these lines on an Argand diagram, showing the direction of each line by an arrow.

  a $z_1 + z_3$ **b** $z_2 + z_3$ **c** $z_1 - z_4$ **d** $z_4 + z_1$ **e** $z_3 - z_4$

2 Express these complex numbers in the form $r(\cos \theta + i \sin \theta)$.

  a $1 + i\sqrt{3}$ **b** $-2 - 2i$ **c** $-5 + i$ **d** $4 - 5i$ **e** 10 **f** $6i$

3 Express these complex numbers in the form $r e^{i\theta}$.

  a $4 + 4i$ **b** $-3 - 4i$ **c** $-2 + 7i$ **d** $1 - 9i$ **e** 8 **f** $2i$

4 Express these in the form $a + ib$.

a $2\left(\cos\dfrac{\pi}{3} + i\sin\dfrac{\pi}{3}\right)$

b $\sqrt{5}\left(\cos\dfrac{5\pi}{6} + i\sin\dfrac{5\pi}{6}\right)$

c $10\left(\cos\left(-\dfrac{\pi}{4}\right) + i\sin\left(-\dfrac{\pi}{4}\right)\right)$

d $\sqrt{15}\left(\cos\left(-\dfrac{\pi}{12}\right) + i\sin\left(-\dfrac{\pi}{12}\right)\right)$

e $3e^{i\frac{3\pi}{4}}$ **f** $\sqrt{5}e^{i\frac{2\pi}{3}}$ **g** $15e^{-i\frac{\pi}{6}}$ **h** $\sqrt{19}e^{-i\frac{3\pi}{8}}$

5 If $m = 5 + 7i$ and $n = 2 - i$, find the modulus and argument of:

a $2m + n$ **b** $3m - 5n$ **c** $2mn$ **d** $\dfrac{4m}{n}$

6 Find the modulus and argument of each root of these equations.

a $z^2 + 3z + 7 = 0$ **b** $z^2 - 2z + 5 = 0$ **c** $z^2 - 4z + 7 = 0$

7 a Express these complex numbers in exponential form.

 i $z_1 = 5 - 12i$ **ii** $z_2 = -3 + 4i$

 iii $z_3 = 24 - 7i$ **iv** $z_4 = 1 + i\sqrt{3}$

b Hence find the modulus and argument of:

 i $z_1 z_2$ **ii** $z_3 z_1$ **iii** $z_4 z_1$ **iv** $\dfrac{z_2}{z_4}$ **v** $\dfrac{z_3}{z_2}$

 vi $\dfrac{z_1}{z_4}$ **vii** $2\dfrac{z_3}{z_4}$ **viii** $3z_3 z_4$

8 a If $z_1 = 3 - 5i$ and $z_2 = 2 - 3i$, draw z_1 and z_2 on an Argand diagram.

b If $z_3 = -iz_1$ and $z_4 = -iz_2$, draw z_3 and z_4 on an Argand diagram.

c Write down the transformation which maps the line segment $z_1 z_2$ onto the line segment $z_3 z_4$.

9 a If $z_1 = 2 - i\sqrt{3}$ and $z_2 = -1 - i$, express z_1 and z_2 in polar form.

b Hence find the modulus and argument of $z_1 z_2$ and $\dfrac{z_1}{z_2}$.

10 If $z_1 = \dfrac{3 + i}{2 - 7i}$ and $z_2 = \dfrac{2 + i}{3 - 2i}$, express z_1 and z_2 in the form $x + iy$.

Sketch an Argand diagram showing the points P representing the complex number $106z_1 + 39z_2$ and Q representing the complex number

$106z_1 - 39z_2$.

11 a Show that $z^3 - 1 = (z - 1)(z^2 + z + 1)$.

b Hence find the roots to the equation $z^3 = 1$ in the form $x + iy$.

c Let the two complex roots be denoted by z_1 and z_2. Verify that $z_1 = z_2^2$ and $z_2 = z_1^2$.

12 a The two complex numbers z_1 and z_2 are represented on an Argand diagram. Show that $|z_1 + z_2| \le |z_1| + |z_2|$.

b If $|z_1| = 3$ and $z_2 = 12 - 5i$, find:

 i the greatest possible value of $|z_1 + z_2|$

 ii the least possible value of $|z_1 + z_2|$.

13 If $z = \cos\theta + i\sin\theta$ where θ is real, show that $\dfrac{1}{1 - z} = \dfrac{1}{2} + \dfrac{1}{2}i\cot\dfrac{\theta}{2}$.

14 a Find the solutions to the equation $3z^2 - 4z + 3 = 0$ in modulus-argument form.

b On the Argand diagram, the roots of this equation are represented by the points P and Q. Find the angle POQ.

15 a Find the modulus and argument of the complex number
$$z = \frac{(\sqrt{2} + i)(1 - i\sqrt{2})}{(1 - i)^2}.$$

b Shade the region in the Argand plane such that $\frac{\pi}{2} < \arg \omega < \frac{3\pi}{4}$ and $\frac{1}{2} < |\omega| < 3$ for any complex number ω.

c Determine if z lies in this region.

17.4 de Moivre's theorem

We showed earlier that $z = re^{i\theta}$.

Hence $z^n = (re^{i\theta})^n$
$$\Rightarrow z^n = r^n e^{in\theta}$$

This is more often stated in polar form.

If $z = r(\cos \theta + i \sin \theta)$ then $z^n = r^n(\cos \theta + i \sin \theta)^n = r^n(\cos n\theta + i \sin n\theta)$.

This is de Moivre's theorem. ......

An alternative proof of de Moivre's theorem, using the method of proof by induction, will be shown in Chapter 18.

Remember: The argument of a complex number lies in the range $-\pi < \theta \leq \pi$.

Example

Write $\left(\cos\frac{\pi}{3} + i \sin\frac{\pi}{3}\right)^{15}$ in the form $\cos n\theta + i \sin n\theta$. ......

Using de Moivre's theorem
$$\left(\cos\frac{\pi}{3} + i \sin\frac{\pi}{3}\right)^{15} = \cos\frac{15\pi}{3} + i \sin\frac{15\pi}{3}$$
$$= \cos 5\pi + i \sin 5\pi$$
$$= \cos \pi + i \sin \pi$$

Example

Write $\left[2\left(\cos\frac{\pi}{3} + i \sin\frac{\pi}{3}\right)\right]^8$ in the form $r(\cos n\theta + i \sin n\theta)$.

$$\left[2\left(\cos\frac{\pi}{3} + i \sin\frac{\pi}{3}\right)\right]^8 = 2^8\left(\cos\frac{\pi}{3} + i \sin\frac{\pi}{3}\right)^8$$
$$= 256\left(\cos\frac{8\pi}{3} + i \sin\frac{8\pi}{3}\right)$$
$$= 256\left(\cos\frac{2\pi}{3} + i \sin\frac{2\pi}{3}\right)$$

Example

Write $\cos\dfrac{\theta}{5} - \text{i}\sin\dfrac{\theta}{5}$ in the form $(\cos\theta + \text{i}\sin\theta)^n$.

We know that $\cos(-\theta) = \cos\theta$ and $\sin(-\theta) = -\sin\theta$

Hence

$$\cos\frac{\theta}{5} - \text{i}\sin\frac{\theta}{5} = \cos\left(-\frac{\theta}{5}\right) + \text{i}\sin\left(-\frac{\theta}{5}\right)$$

$$= (\cos\theta + \text{i}\sin\theta)^{-\frac{1}{5}}$$

Example

Simplify $\dfrac{(4\cos 4\theta + 4\text{i}\sin 4\theta)(\cos 2\theta - \text{i}\sin 2\theta)}{(\cos 3\theta + \text{i}\sin 3\theta)}$.

Since de Moivre's theorem is used on expressions of the form $r(\cos n\theta + \text{i}\sin n\theta)$ we need to put all expressions in this form:

$$\frac{(4\cos 4\theta + 4\text{i}\sin 4\theta)(\cos(-2\theta) + \text{i}\sin(-2\theta))}{(\cos 3\theta + \text{i}\sin 3\theta)}$$

We now apply de Moivre's theorem:

$$\frac{4(\cos\theta + \text{i}\sin\theta)^4(\cos\theta + \text{i}\sin\theta)^{-2}}{(\cos\theta + \text{i}\sin\theta)^3}$$

$$= 4(\cos\theta + \text{i}\sin\theta)^{-1}$$

$$= 4(\cos(-\theta) + \text{i}\sin(-\theta))$$

Since $\cos(-\theta) = \cos\theta$ and $\sin(-\theta) = -\sin\theta$

$$= 4(\cos\theta - \text{i}\sin\theta)$$

Example

Use de Moivre's theorem to derive expressions for $\cos 4\theta$ and $\sin 4\theta$ in terms of $\cos\theta$ and $\sin\theta$.

From de Moivre's theorem we know that

$$(\cos\theta + \text{i}\sin\theta)^4 = \cos 4\theta + \text{i}\sin 4\theta$$

Using Pascal's triangle or the binomial theorem, we find

$$\cos 4\theta + \text{i}\sin 4\theta = \cos^4\theta + 4(\cos^3\theta)(\text{i}\sin\theta) + 6(\cos^2\theta)(\text{i}\sin\theta)^2$$

$$+ 4\cos\theta(\text{i}\sin\theta)^3 + (\text{i}\sin\theta)^4$$

$$= \cos^4\theta + 4\text{i}\cos^3\theta\sin\theta - 6\cos^2\theta\sin^2\theta - 4\text{i}\cos\theta\sin^3\theta + \sin^4\theta$$

By equating real parts we find $\cos 4\theta = \cos^4\theta - 6\cos^2\theta\sin^2\theta + \sin^4\theta$.

And by equating imaginary parts we find $\sin 4\theta = 4\cos^3\theta\sin\theta - 4\cos\theta\sin^3\theta$.

> Since $\text{i}^2 = -1$, $\text{i}^3 = -\text{i}$ and $\text{i}^4 = 1$.

This is an alternative way of finding multiple angles of sin and cos in terms of powers of sin and cos.

Example

Using de Moivre's theorem, show that $\tan 3\theta = \dfrac{3t - t^3}{1 - 3t^2}$ where $t = \tan \theta$ and use the equation to solve $t^3 - 3t^2 - 3t + 1 = 0$.

Since we want $\tan 3\theta$ we need expressions for $\sin 3\theta$ and $\cos 3\theta$.
From de Moivre's theorem we know that
$(\cos \theta + i \sin \theta)^3 = \cos 3\theta + i \sin 3\theta$
Using Pascal's triangle we find
$\cos 3\theta + i\sin 3\theta = \cos^3\theta + 3(\cos^2\theta)(i\sin\theta) + 3(\cos\theta)(i\sin\theta)^2 + (i\sin\theta)^3$
$\quad = \cos^3 \theta + 3i \cos^2 \theta \sin \theta - 3 \cos \theta \sin^2 \theta - i \sin^3 \theta$ · · · · · · · · · · · · · · · · · ·

> Since $i^2 = -1$ and $i^3 = -i$

By equating real parts we find $\cos 3\theta = \cos^3 \theta - 3 \cos \theta \sin^2 \theta$.
And by equating imaginary parts we find $\sin 3\theta = 3 \cos^2 \theta \sin \theta - \sin^3 \theta$.

Hence $\tan 3\theta = \dfrac{\sin 3\theta}{\cos 3\theta} = \dfrac{3 \cos^2 \theta \sin \theta - \sin^3 \theta}{\cos^3 \theta - 3 \cos \theta \sin^2 \theta}$

$\Rightarrow \tan 3\theta = \dfrac{3\dfrac{\sin \theta}{\cos \theta} - \dfrac{\sin^3 \theta}{\cos^3 \theta}}{\dfrac{\cos^3 \theta}{\cos^3 \theta} - 3\dfrac{\sin^2 \theta}{\cos^2 \theta}}$ · · · · · · · · · · · · · · · · · ·

> Dividing numerator and denominator by $\cos^3 \theta$

$\Rightarrow \tan 3\theta = \dfrac{3t - t^3}{1 - 3t^2}$ · · · · · · · · · · · · · · · · · ·

> Letting $t = \tan \theta$

If we now let $\tan 3\theta = 1$

$\dfrac{3t - t^3}{1 - 3t^2} = 1$

$\Rightarrow t^3 - 3t^2 - 3t + 1 = 0$

Hence this equation can be solved using $\tan 3\theta = 1$

$\Rightarrow 3\theta = -\dfrac{3\pi}{4}, \dfrac{\pi}{4}, \dfrac{5\pi}{4}$

$\Rightarrow \theta = -\dfrac{\pi}{4}, \dfrac{\pi}{12}, \dfrac{5\pi}{12}$

Hence the solutions to the equation are $\tan\left(-\dfrac{\pi}{4}\right), \tan\dfrac{\pi}{12}, \tan\dfrac{5\pi}{12}$

Example

If $z = \cos\theta + i\sin\theta$, using de Moivre's theorem, show that $\dfrac{1}{z} = \cos \theta - i \sin \theta$.

$\dfrac{1}{z} = z^{-1} = (\cos \theta + i \sin \theta)^{-1}$

$\qquad = \cos(-\theta) + i \sin(-\theta)$

Since $\cos(-\theta) = \cos \theta$ and $\sin(-\theta) = -\sin \theta$

$\dfrac{1}{z} = \cos \theta - i \sin \theta$

This now leads to four useful results.

If $z = \cos \theta + i \sin \theta$ and $\dfrac{1}{z} = \cos \theta - i \sin \theta$, by adding the two equations together we find

$$z + \frac{1}{z} = 2\cos\theta$$

If we subtract the two equations we find

$$z - \frac{1}{z} = 2i\sin\theta$$

This can be generalized for any power of z.

If $z = \cos\theta + i\sin\theta$, then $z^n = (\cos\theta + i\sin\theta)^n = \cos n\theta + i\sin n\theta$

and $z^{-n} = (\cos\theta + i\sin\theta)^{-n} = \cos(-n\theta) + i\sin(-n\theta) = \cos n\theta - i\sin n\theta$.

Once again by adding and subtracting the equations we find

$$z^n + \frac{1}{z^n} = 2\cos n\theta$$

$$\text{and} \quad z^n - \frac{1}{z^n} = 2i\sin n\theta$$

Example

Using the result $z^n - \frac{1}{z^n} = 2i\sin n\theta$, show that $\sin^3\theta = \dfrac{3\sin\theta - \sin 3\theta}{4}$.

We know that $z - \dfrac{1}{z} = 2i\sin\theta$

Hence $\left(z - \dfrac{1}{z}\right)^3 = (2i\sin\theta)^3 = -8i\sin^3\theta$

$\Rightarrow -8i\sin^3\theta = z^3 - 3z + \dfrac{3}{z} - \dfrac{1}{z^3}$

$\qquad\qquad = z^3 - \dfrac{1}{z^3} - 3\left(z - \dfrac{1}{z}\right)$

Hence $-8i\sin^3\theta = 2i\sin 3\theta - 6i\sin\theta$

$\Rightarrow \sin^3\theta = \dfrac{\sin 3\theta - 3\sin\theta}{-4} = \dfrac{3\sin\theta - \sin 3\theta}{4}$

Roots of complex numbers

Earlier in the chapter we found the square root of a complex number. We can also do this using de Moivre's theorem, which is a much more powerful technique as it will allow us to find any root.

Method

1. Write the complex number in polar form.
2. Add $2n\pi$ to the argument then put it to the necessary power. This will allow us to find multiple solutions.
3. Apply de Moivre's theorem.
4. Work out the required number of roots, ensuring that the arguments lie in the range $-\pi < \theta \leq \pi$. Remember the number of roots is the same as the denominator of the power.

Important points to note

1. The roots are equally spaced around the Argand diagram. Thus for the square root they are π apart. Generally for the nth root they are $\dfrac{2\pi}{n}$ apart.
2. All the roots have the same moduli.

Example

Find the cube roots of $2 + 2i$.

Step 1. Let $2 + 2i = r(\cos\theta + i\sin\theta)$

Equating real and imaginary parts

$\Rightarrow r\cos\theta = 2$

$\Rightarrow r\sin\theta = 2$

$\qquad\Rightarrow r = \sqrt{2^2 + 2^2} = \sqrt{8}$

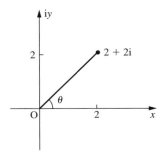

The diagram shows the complex number lies in the first quadrant and hence the argument is a positive acute angle.

$\dfrac{r\sin\theta}{r\cos\theta} = \dfrac{2}{2}$

$\Rightarrow \tan\theta = 1$

$\qquad\Rightarrow \theta = \dfrac{\pi}{4}$

$\Rightarrow 2 + 2i = \sqrt{8}\left(\cos\dfrac{\pi}{4} + i\sin\dfrac{\pi}{4}\right)$

Step 2. $(2 + 2i)^{\frac{1}{3}} = 8^{\frac{1}{6}}\left\{\cos\left(\dfrac{\pi}{4} + 2n\pi\right) + i\sin\left(\dfrac{\pi}{4} + 2n\pi\right)\right\}^{\frac{1}{3}}$

Step 3. $(2 + 2i)^{\frac{1}{3}} = 8^{\frac{1}{6}}\left\{\cos\left(\dfrac{\pi}{12} + \dfrac{2n\pi}{3}\right) + i\sin\left(\dfrac{\pi}{12} + \dfrac{2n\pi}{3}\right)\right\}$

Step 4. If we now let $n = -1, 0, 1$ we will find the three solutions.

Hence $(2 + 2i)^{\frac{1}{3}} = 8^{\frac{1}{6}}\left(\cos\left(-\dfrac{7\pi}{12}\right) + i\sin\left(-\dfrac{7\pi}{12}\right)\right),$

$8^{\frac{1}{6}}\left(\cos\left(\dfrac{3\pi}{4}\right) + i\sin\left(\dfrac{3\pi}{4}\right)\right), \; 8^{\frac{1}{6}}\left(\cos\left(\dfrac{\pi}{12}\right) + i\sin\left(\dfrac{\pi}{12}\right)\right)$

These can be converted to the form $x + iy$.

$(2 + 2i)^{\frac{1}{3}} = -0.366 - 1.37i, \; 1.37 + 0.366i, \; -1 + i$

This calculation can also be done directly on the calculator.

We can also use the exponential form to evaluate roots of a complex number.

Example

Find $(1 - i)^{\frac{1}{4}}$.

Step 1. Let $1 - i = re^{i\theta}$

Equating real and imaginary parts

$\Rightarrow r \cos \theta = 1$

$\Rightarrow r \sin \theta = -1$

$\Rightarrow r = \sqrt{1^2 + (-1)^2} = \sqrt{2}$

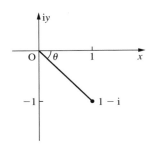

The diagram shows the complex number lies in the fourth quadrant and hence the argument is a negative acute angle.

$\dfrac{r \sin \theta}{r \cos \theta} = \dfrac{1}{-1}$

$\Rightarrow \tan \theta = -1$

$\Rightarrow \theta = -\dfrac{\pi}{4}$

$\Rightarrow 1 - i = \sqrt{2}e^{-i\frac{\pi}{4}}$

Step 2. $(1 - i)^{\frac{1}{4}} = (\sqrt{2}e^{i(-\frac{\pi}{4} + 2n\pi)})^{\frac{1}{4}}$

Step 3. $(1 - i)^{\frac{1}{4}} = 2^{\frac{1}{8}}e^{i(-\frac{\pi}{16} + \frac{n\pi}{2})}$

Step 4. Clearly, if we let $n = -1, 0, 1$ we will find three solutions, but does $n = 2$ or $n = -2$ give the fourth solution? Since $\dfrac{\pi}{16}$ is negative, then using $n = -2$ takes the argument out of the range $-\pi < x \leq \pi$. Hence we use $n = 2$.

Thus $(1 - i)^{\frac{1}{4}} = 2^{\frac{1}{8}}e^{-i\frac{9\pi}{16}}, 2^{\frac{1}{8}}e^{-i\frac{\pi}{16}}, 2^{\frac{1}{8}}e^{i\frac{7\pi}{16}}, 2^{\frac{1}{8}}e^{i\frac{15\pi}{16}}$

These can be converted to the form $x + iy$.

$(1 - i)^{\frac{1}{4}} = -0.213 - 1.07i, \ 1.07 - 0.213i, \ 0.213 + 1.07i, \ -1.07 + 0.213i$

Again, this calculation can also be done directly on the calculator.

Roots of unity

We can find the complex roots of 1 and these have certain properties.

1. Since the modulus of 1 is 1, then the modulus of all roots of 1 is 1.
2. We know that the roots are equally spaced around an Argand diagram. Since one root of unity will always be 1, we can measure the arguments relative to the real axis.

Hence the cube roots of unity on an Argand diagram will be:

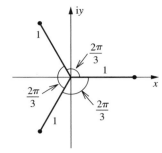

The fourth roots of unity will be:

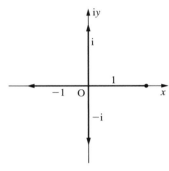

3. Since the roots of unity are equally spaced and all have modulus 1, if we call one complex root b, say, then for the cube roots of unity the other roots will be 1 and b^2. Similarly for the fifth roots, if one complex root is b, then the other roots will be 1, b^2, b^3 and b^4.

Example

a Simplify $(\omega - 1)(1 + \omega + \omega^2)$.

b Hence factorise $z^3 = 1$.

c If ω is a complex root of this equation, simplify:

  i ω^3

  ii $1 + \omega + \omega^2$

  iii ω^4

  iv $(\omega - 1)(\omega^2 + \omega)$

a $(\omega - 1)(1 + \omega + \omega^2) = \omega + \omega^2 + \omega^3 - 1 - \omega - \omega^2$

$$= \omega^3 - 1$$

b $z^3 = 1 \Rightarrow z^3 - 1 = 0$

$$\Rightarrow (z - 1)(1 + z + z^2) = 0$$

c

  i Since $z = \omega$, $\omega^3 = 1$

  ii Since $z = \omega$ and from part **b** $1 + z + z^2 = 0$, $1 + \omega + \omega^2 = 0$

  iii $\omega^4 = \omega^3 \times \omega$

      Since $\omega^3 = 1$, $\omega^4 = 1 \times \omega = \omega$

  iv $(\omega - 1)(\omega^2 + \omega) = \omega^3 + \omega^2 - \omega^2 - \omega$

$$= 1 - \omega$$

Exercise 4

1 Use de Moivre's theorem to express each of these complex numbers in the form $r(\cos n\theta + i \sin n\theta)$.

a $[2(\cos \theta + i \sin \theta)]^{10}$ **b** $(\cos \theta + i \sin \theta)^{25}$ **c** $[3(\cos \theta + i \sin \theta)]^{-5}$

d $(\cos \theta + i \sin \theta)^{-9}$ **e** $(\cos \theta + i \sin \theta)^{\frac{1}{2}}$ **f** $[4(\cos \theta + i \sin \theta)]^{-\frac{1}{3}}$

g $\left(\cos\frac{\pi}{3} + i \sin\frac{\pi}{3}\right)^6$ **h** $\left(\cos\frac{\pi}{6} + i \sin\frac{\pi}{6}\right)^9$ **i** $\left(\cos\frac{\pi}{4} + i \sin\frac{\pi}{4}\right)^{-5}$

j $\left(\cos\frac{\pi}{5} + i \sin\frac{\pi}{5}\right)^{\frac{1}{2}}$

2 Express each of these in the form $r(\cos \theta + i \sin \theta)^n$.

a $\cos 7\theta + i \sin 7\theta$ **b** $4 \cos\frac{1}{2}\theta + 4i \sin\frac{1}{2}\theta$

c $6 \cos(-3\theta) + 6i \sin(-3\theta)$ **d** $\cos\left(-\frac{1}{4}\theta\right) + i \sin\left(-\frac{1}{4}\theta\right)$

e $\cos 2\theta - i \sin 2\theta$ **f** $\cos\frac{1}{8}\theta - i \sin\frac{1}{8}\theta$

3 Simplify these expressions.

a $(\cos 3\theta + i \sin 3\theta)(\cos 5\theta + i \sin 5\theta)$

b $(\cos 2\theta + i \sin 2\theta)\left(\cos\frac{1}{2}\theta + i \sin\frac{1}{2}\theta\right)$

c $\dfrac{(\cos 8\theta + i \sin 8\theta)}{(\cos 5\theta + i \sin 5\theta)}$

d $\dfrac{(\cos 4\theta + i \sin 4\theta)}{(\cos 5\theta + i \sin 5\theta)}$

e $\dfrac{(\cos 10\theta + i \sin 10\theta)(\cos 2\theta + i \sin 2\theta)}{(\cos \theta + i \sin \theta)}$

f $(\cos 4\theta + i \sin 4\theta)(\cos 7\theta - i \sin 7\theta)$

g $\left(\cos\frac{1}{3}\theta + i \sin\frac{1}{3}\theta\right)\left(\cos\frac{1}{2}\theta - i \sin\frac{1}{2}\theta\right)$

h $\dfrac{\left(\cos\frac{\pi}{4} + i \sin\frac{\pi}{4}\right)^5\left(\cos\frac{\pi}{3} + i \sin\frac{\pi}{3}\right)^2}{\left(\cos\frac{\pi}{6} + i \sin\frac{\pi}{6}\right)^4}$

i $\dfrac{\left(\cos\frac{\pi}{8} - i \sin\frac{\pi}{8}\right)^5\left(\cos\frac{\pi}{16} + i \sin\frac{\pi}{16}\right)^{-2}}{\left(\cos\frac{\pi}{8} + i \sin\frac{\pi}{8}\right)^4}$

j $\sqrt[4]{\left(\cos\frac{\pi}{3} + i \sin\frac{\pi}{3}\right)}$

4 Use de Moivre's theorem to find these roots.

a the square root of $-5 + 12i$ **b** the square root of $-2 - 2i$

c the cube roots of $1 - i$ **d** the cube root of $3 - 5i$

e the fourth roots of $3 + 4i$ **f** the fifth roots of $-5 - 12i$

g the sixth roots of $\sqrt{3} + i$

5 Without first calculating them, illustrate the nth roots of unity on an Argand diagram where n is:

   a 3   **b** 6   **c** 8   **d** 9

6 a Express the complex number $16i$ in polar form.

   b Find the fourth roots of $16i$ in both polar form and Cartesian form.

7 a Write $1 + i\sqrt{3}$ in polar form.

   b Hence find the real and imaginary parts of $(1 + i\sqrt{3})^{16}$.

8 Prove these trigonometric identities using methods based on de Moivre's theorem.

   a $\sin 3\theta \equiv 3\cos^2\theta \sin\theta - \sin^3\theta$

   b $\tan 6\theta \equiv 2\left(\dfrac{3 - 10t^2 + 3t^4}{1 - 15t^2 + 15t^4 - t^6}\right)$ where $t = \tan\theta$

9 a Use de Moivre's theorem to show that $\tan 4\theta = \dfrac{4t - 4t^3}{1 - 6t^2 + t^4}$ where $t = \tan\theta$.

   b Use your result to solve the equation $t^4 + 4t^3 - 6t^2 - 4t + 1 = 0$.

10 Let $z_1 = m\left(\cos\dfrac{\pi}{6} + i\sin\dfrac{\pi}{6}\right)$ and $z_2 = m\left(\cos\dfrac{\pi}{3} + i\sin\dfrac{\pi}{3}\right)$. Express $\left(\dfrac{z_1}{z_2}\right)^4$ in the form $x + iy$.

11 Let $z_1 = r\left(\cos\dfrac{\pi}{3} + i\sin\dfrac{\pi}{3}\right)$ and $z_2 = 3 - 4i$.

   a Write z_2 in modulus-argument form.

   b Find r if $|z_1 z_2^2| = 4$.

12 Given that ω is a complex cube root of unity, $\omega^3 = 1$ and $1 + \omega + \omega^2 = 0$, simplify each of the expressions $(1 + 3\omega + \omega^2)$ and $(1 + \omega + 3\omega^2)$ and find the product and the sum of these two expressions.

13 By considering the ninth roots of unity, show that:

$$\cos\frac{2\pi}{9} + \cos\frac{4\pi}{9} + \cos\frac{6\pi}{9} + \cos\frac{8\pi}{9} = -\frac{1}{2}$$

14 a If $z = \cos\theta + i\sin\theta$, show that $z^n + \dfrac{1}{z^n} = 2\cos n\theta$ and

$$z^n - \frac{1}{z^n} = 2i\sin n\theta.$$

   b Hence show that:

     i $\cos^4\theta + \sin^4\theta = \dfrac{1}{4}(\cos 4\theta + 3)$

     ii $\sin^6\theta = \dfrac{1}{32}(-\cos 6\theta + 6\cos 4\theta - 15\cos 2\theta + 10)$

15 Consider $z^7 = 128$.

   a Find the root to this equation in the form $r(\cos\theta + i\sin\theta)$ which has the smallest positive argument. Call this root z_1.

   b Find $z_1^2, z_1^3, z_1^4, z_1^5, z_1^6, z_1^7$ in modulus-argument form.

   c Plot the points that represent $z_1, z_1^2, z_1^3, z_1^4, z_1^5, z_1^6, z_1^7$ on an Argand diagram.

   d The point z_1^n is mapped to z_1^{n+1} by a composition of two linear transformations. Describe these transformations.

16 a Show that $-i$ satisfies the equation $z^3 = i$.

   b Knowing that the three roots of the equation $z^3 = i$ are equally spaced around the Argand diagram and have equal modulii, write down the other two roots, z_1 and z_2, of the equation in modulus-argument form. (z_1 lies in the second quadrant.)

c Find the complex number ω such that $\omega z_1 = z_2$ and $\omega z_2 = -i$.

17 The complex number z is defined by $z = \cos\theta + i\sin\theta$.

   a Show that $\dfrac{1}{z} = \cos(-\theta) + i\sin(-\theta)$.

   b Deduce that $z^n + \dfrac{1}{z^n} = 2\cos n\theta$.

   c Using the binomial theorem, expand $(z + z^{-1})^6$.

   d Hence show that $\cos^6\theta = a\cos 6\theta + b\cos 4\theta + c\cos 2\theta + d$ giving the values of a, b, c and d.

Review exercise

 1 Find the modulus and argument of the complex number $\dfrac{5 - 7i}{1 + 2i}$.

 2 Find the real number k for which $1 + ki$, $(i = \sqrt{-1})$, is a zero of the polynomial $z^2 + kz + 5$.   [IB Nov 00 P1 Q10]

 3 If $z = 1 + 2i$ is a root of the equation $z^2 + az + b$, find the values of a and b.

 4 If z is a complex number and $|z + 16| = 4|z + 1|$, find the value of $|z|$.   [IB Nov 00 P1 Q18]

 5 a Show that $(1 + i)^4 = -4$.

   b Hence or otherwise, find $(1 + i)^{64}$.

 6 Solve the equation $\dfrac{-i}{x - iy} = \dfrac{4 + 7i}{5 - 3i}$ for x and y, leaving your answers as rational numbers.   [IB May 94 P1 Q15]

 7 Find a cubic equation with real coefficients, given that two of its roots are 3 and $1 - i\sqrt{3}$.

 8 If $z = x + iy$, find the real part and the imaginary part of $z + \dfrac{1}{z}$.

 9 Given that $z = (b + i)^2$, where b is real and positive, find the exact value of b when $\arg z = 60°$.   [IB May 01 P1 Q14]

 10 a If $z = 1 + i\sqrt{3}$, find the modulus and argument of z.

   b Hence find the modulus and argument of z^2.

   c i On an Argand diagram, point A represents the complex number $0 + i$, B represents the complex number z and C the complex number z^2. Draw these on an Argand diagram.

   ii Calculate the area of triangle OBC where O is the origin.

   iii Calculate the area of triangle ABC.

 11 a Verify that $(z - 1)(1 + z + z^2) = z^3 - 1$.

   b Hence or otherwise, find the cube roots of unity in the form $a + ib$.

   c Find the cube roots of unity in polar form and draw them on an Argand diagram.

   d These three roots form the vertices of a triangle. State the length of each side of the triangle and find the area of the triangle.

12 Given that $(2 - 3i)a + 3b = 2 + 5i$, find the values of a and b if

    a a and b are real

    b a and b are conjugate complex numbers.

 13 Let $y = \cos \theta + i \sin \theta$.

    a Show that $\dfrac{dy}{d\theta} = iy$.

    [You may assume that for the purposes of differentiation and integration, i may be treated in the same way as a real constant.]

    b Hence show, using integration, that $y = e^{i\theta}$.

    c Use this result to deduce de Moivre's theorem.

    d i Given that $\dfrac{\sin 6\theta}{\sin \theta} = a \cos^5 \theta + b \cos^3 \theta + c \cos \theta$, where $\sin \theta \neq 0$, use

       de Moivre's theorem with $n = 6$ to find the values of the constants a, b and c.

      ii Hence deduce the value of $\lim\limits_{\theta \to 0} \dfrac{\sin 6\theta}{\sin \theta}$.   [IB Nov 06 P2 Q5]

 14 Given that z and ω are complex numbers, solve the simultaneous equations

    $z + \omega = 11$

    $iz + 5\omega = 29$

    expressing your solution in the form $a + bi$ where a and b are real. [IB Nov 89 P1 Q20]

 15 Let $z = \cos \theta + i \sin \theta$ for $-\dfrac{\pi}{4} < \theta < \dfrac{\pi}{4}$.

    a i Find z^3 using the binomial theorem.

      ii Use de Moivre's theorem to show that $\cos 3\theta = 4 \cos^3 \theta - 3 \cos \theta$ and

    $\sin 3\theta = 3 \sin \theta - 4 \sin^3 \theta$.

    b Hence prove that $\dfrac{\sin 3\theta - \sin \theta}{\cos 3\theta + \cos \theta} = \tan \theta$.

    c Given that $\sin \theta = \dfrac{1}{3}$, find the exact value of $\tan 3\theta$.   [IB May 06 P2 Q2]

 16 Consider the complex number $z = \dfrac{\left(\cos\dfrac{\pi}{4} - i \sin\dfrac{\pi}{4}\right)^2 \left(\cos\dfrac{\pi}{3} + i \sin\dfrac{\pi}{3}\right)^3}{\left(\cos\dfrac{\pi}{24} - i \sin\dfrac{\pi}{24}\right)^4}$.

    a i Find the modulus of z.

      ii Find the argument of z, giving your answer in radians.

    b Using de Moivre's theorem, show that z is a cube root of one, i.e. $z = \sqrt[3]{1}$.

    c Simplify $(1 + 2z)(2 + z^2)$, expressing your answer in the form $a + bi$, where a and b are exact real numbers.   [IB Nov 02 P2 Q2]

 17 In this Argand diagram, a circle has centre the origin and radius 5, $\theta = \dfrac{\pi}{3}$ and the line which is parallel to the imaginary axis has equation $x = -2$. The complex number z corresponds to a point inside, or on, the boundary of the shaded region. Write down inequalities which $|z|$, $\text{Arg } z$ and $\text{Re } z$ must satisfy. ($\text{Re } z$ means the real part of z.)

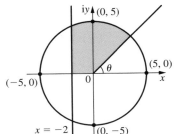

18 Let $z = 3 + ik$ and $\omega = k + 7i$ where $k \in \mathbb{R}$ and $i = \sqrt{-1}$.

    a Express $\dfrac{z}{\omega}$ in the form $a + ib$ where $a, b \in \mathbb{R}$.

    b For what values of k is $\dfrac{z}{\omega}$ a real number?

 19 a Find all three solutions of the equation $z^3 = 1$ where z is a complex number.

    b If $z = \omega$ is the solution of the equation $z^3 = 1$ which has the smallest positive root, show that $1 + \omega + \omega^2 = 0$.

    c Find the matrix product $\begin{pmatrix} 1 & 1 & 1 \\ 1 & \omega & \omega^2 \\ 1 & \omega^2 & \omega \end{pmatrix} \begin{pmatrix} 1 & 1 & 1 \\ 1 & \omega^2 & \omega \\ 1 & \omega & \omega^2 \end{pmatrix}$ giving your answer in its simplest form (that is, not in terms of ω).

    d Solve the system of simultaneous equations

       $x + y + z = 3$

       $x + \omega y + \omega^2 z = -3$

       $x + \omega^2 y + \omega z = -3$

    giving your answer in numerical form (that is, not in terms of ω).   [IB Nov 98 P2 Q4]

 20 z_1 and z_2 are complex numbers on the Argand diagram relative to the origin. If $|z_1 + z_2| = |z_1 - z_2|$, show that $\arg z_1$ and $\arg z_2$ differ by $\dfrac{\pi}{2}$.

 21 a Find the two square roots of $3 - 4i$ in the form $x + iy$ where x and y are real.

    b Draw these on the Argand diagram, labelling the points A and B.

    c Find the two possible points C_1 and C_2 such that triangles ABC_1 and ABC_2 are equilateral.

18 Mathematical Induction

Abu Bekr ibn Muhammad ibn al-Husayn Al-Karaji was born on 13 April 953 in Baghdad, Iraq and died in about 1029. His importance in the field of mathematics is debated by historians and mathematicians. Some consider that he only reworked previous ideas, while others see him as the first person to use arithmetic style operations with algebra as opposed to geometrical operations.

In his work, *Al-Fakhri*, Al-Karaji succeeded in defining x, x^2, x^3, ... and $\dfrac{1}{x}$, $\dfrac{1}{x^2}$, $\dfrac{1}{x^3}$, ... and gave rules for finding the products of any pair without reference to geometry. He was close to giving the rule $x^n x^m = x^{m+n}$ for all integers n and m

but just failed because he did not define $x^0 = 1$.

In his discussion and demonstration of this work Al-Karaji used a form of mathematical induction where he proved a result using the previous result and noted that this process could continue indefinitely. As we will see in this chapter, this is not a full proof by induction, but it does highlight one of the major principles.

Al-Karaji used this form of induction in his work on the binomial theorem, binomial coefficients and Pascal's triangle. The table shown is one that Al-Karaji used, and is actually Pascal's triangle in its side.

He also worked on the sums of the first n natural numbers, the squares of the first n natural numbers and the cubes of these numbers, which we introduced in Chapter 6.

col 1	col 2	col 3	col 4	col 5	· · ·
1	1	1	1	1	· · ·
1	2	3	4	5	· · ·
	1	3	6	10	· · ·
		1	4	10	· · ·
			1	5	· · ·
				1	· · ·

18.1 Introduction to mathematical induction

Mathematical induction is a method of mathematical proof. Most proofs presented in this book are direct proofs – that is proofs where one step leads directly from another to the required result. However, there are a number of methods of indirect proof including proof by contradiction, proof by contrapositive and proof by mathematical induction. In this curriculum, we only consider proof by induction for positive integers.

Mathematical induction is based on the idea of proving the next step to be true if the previous one is true. If the result is true for an initial value, then it is true for all values. This is demonstrated by the following metaphor.

Consider a ladder that is infinite in one direction. We want to prove that the ladder is completely safe, that is each rung on the ladder is sound.

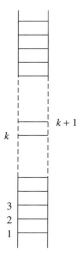

First, test the bottom rung on the ladder and check that it is sound. Then assume that a rung on the ladder, somewhere further up, is also sound. Call this the kth rung. Using this assumption, show that the next rung up, the $(k + 1)$th rung, is also sound **if** the assumption is true. Since we know the first rung is sound, we can now say the second one is sound. As the second one is sound, the third one is sound and so on. So the whole ladder is safe.

Method for mathematical induction

1. Prove the result is true for an initial value (normally $n = 1$).
2. Assume the result to be true for another value, $n = k, k > 1$, stating this result.
3. Consider the case for $n = k + 1$, writing down the goal – the required form.
4. Using the assumption, show that the result is then true for $n = k + 1$.
5. Communicate why this proves the result using mathematical induction.

Example

Prove $\displaystyle\sum_{r=1}^{n} r = \frac{n(n + 1)}{2}$ $\forall n \in \mathbb{Z}^+$ by mathematical induction.

Remembering the meaning of this notation, we know that $\forall n \in \mathbb{Z}^+$ means that we need to prove it is true for all positive integers, i.e. $n \geq 1, n \in \mathbb{Z}$.

1 Prove the result is true for $n = 1$.

It is important to show this very clearly (even though it is often obvious).

$$\text{LHS} = \sum_{r=1}^{1} r \qquad\qquad \text{RHS} = \frac{1(2)}{2}$$
$$= 1 \qquad\qquad\qquad\qquad = 1$$

Since LHS = RHS, the result is true for $n = 1$.

2 Assume the result is true for $n = k$, $k > 1$, $k \in \mathbb{Z}$,

i.e. $\displaystyle\sum_{r=1}^{k} r = \frac{k(k + 1)}{2}$

3 Now consider the result for $n = k + 1$. We want to show that

$$\sum_{r=1}^{k+1} r = \frac{(k + 1)(k + 1 + 1)}{2} = \frac{(k + 1)(k + 2)}{2}$$

4 For $n = k + 1$,

$$\sum_{r=1}^{k+1} r = \sum_{r=1}^{k} r + (k + 1) \quad\cdots\cdots\cdots\cdots$$

> Adding on the $(k + 1)$th term.

$$= \frac{k(k + 1)}{2} + (k + 1) \quad\cdots\cdots\cdots$$

> We are using the assumption here.

$$= \frac{k(k + 1)}{2} + \frac{2(k + 1)}{2}$$

$$= \frac{k(k + 1) + 2(k + 1)}{2}$$

$$= \frac{(k + 1)(k + 2)}{2}$$

which is the required form.

> This communication is identical for virtually all induction proofs. It is worth learning its form.

5 So the result is true for $n = k + 1$ when true for $n = k$. Since the result is true for $n = 1$, it is true $\forall n \in \mathbb{Z}^{+}$ by mathematical induction.

Example

Prove that $\displaystyle\sum_{r=1}^{n} 3r^2 - 5r = n(n + 1)(n - 2) \; \forall n \in \mathbb{Z}^{+}$ by mathematical induction.

We cannot use the standard results for $\displaystyle\sum_{r=1}^{n} r$ and $\displaystyle\sum_{r=1}^{n} r^2$ here as we are being asked to prove it by mathematical induction.

1 For $n = 1$,

$$\text{LHS} = \sum_{r=1}^{1} 3r^2 - 5r \qquad\qquad \text{RHS} = 1(1 + 1)(1 - 2)$$
$$= 3(1)^2 - 5(1) \qquad\qquad\qquad = 1(2)(-1)$$
$$= 3 - 5 \qquad\qquad\qquad\qquad = -2$$
$$= -2$$

Since LHS = RHS, the result is true for $n = 1$.

2 Assume the result to be true for $n = k$,

i.e. $\displaystyle\sum_{r=1}^{k} 3r^2 - 5r = k(k + 1)(k - 2)$

3 Consider $n = k + 1$. We want to show that

$$\sum_{r=1}^{k+1} 3r^2 - 5r = (k + 1)(k + 1 + 1)(k + 1 - 2) = (k + 1)(k + 2)(k - 1)$$

4 For $n = k + 1$,

$$\sum_{r=1}^{k+1} 3r^2 - 5r$$

$$= \sum_{r=1}^{k} (3r^2 - 5r) + 3(k + 1)^2 - 5(k + 1) \cdots\cdots\cdots$$ | Adding on the $(k + 1)$th term. |

$$= k(k + 1)(k - 2) + 3(k + 1)^2 - 5(k + 1) \cdots\cdots$$ | We are using the assumption here. |

$$= (k + 1)[k(k - 2) + 3(k + 1) - 5]$$

$$= (k + 1)[k^2 + k - 2]$$

$$= (k + 1)(k + 2)(k - 1)$$

which is the required form.

5 So the result is true for $n = k + 1$ when true for $n = k$. Since the result is true for $n = 1$, it is true $\forall n \in \mathbb{Z}^+$ by mathematical induction.

Example

Prove that $\displaystyle\sum_{r=1}^{n} \frac{1}{r(r + 1)} = \frac{n}{n + 1}$ $\forall n \in \mathbb{Z}^+$ by mathematical induction.

1 For $n = 1$,

$$\text{LHS} = \sum_{r=1}^{1} \frac{1}{r(r + 1)} \qquad\qquad \text{RHS} = \frac{1}{1 + 1}$$

$$= \frac{1}{1(2)} \qquad\qquad\qquad\qquad\quad = \frac{1}{2}$$

$$= \frac{1}{2}$$

Since LHS = RHS, the result is true for $n = 1$.

2 Assume the result to be true for $n = k$,

i.e. $\displaystyle\sum_{r=1}^{k} \frac{1}{r(r + 1)} = \frac{k}{k + 1}$

3 Consider $n = k + 1$. We want to show that

$$\sum_{r=1}^{k+1} \frac{1}{r(r + 1)} = \frac{k + 1}{k + 1 + 1} = \frac{k + 1}{k + 2}$$

4 For $n = k + 1$,

$$\sum_{r=1}^{k+1} \frac{1}{r(r+1)}$$

$$= \sum_{r=1}^{k} \frac{1}{r(r+1)} + \frac{1}{(k+1)(k+2)}$$ .... Adding on the $(k+1)$th term.

$$= \frac{k}{k+1} + \frac{1}{(k+1)(k+2)}$$ .... We are using the assumption here.

$$= \frac{k(k+2)}{(k+1)(k+2)} + \frac{1}{(k+1)(k+2)}$$

$$= \frac{k^2 + 2k + 1}{(k+1)(k+2)}$$

$$= \frac{(k+1)^2}{(k+1)(k+2)}$$

$$= \frac{k+1}{k+2}$$

which is the required form.

5 So the result is true for $n = k + 1$ when true for $n = k$. Since the result is true for $n = 1$, it is true $\forall n \in \mathbb{Z}^+$ by mathematical induction.

Example

Prove that $\sum_{r=1}^{n} 3^r = \frac{3}{2}(3^n - 1) \forall n \in \mathbb{Z}^+$ by mathematical induction.

1 For $n = 1$,

$$\text{LHS} = \sum_{r=1}^{1} 3^r \qquad\qquad \text{RHS} = \frac{3}{2}(3^1 - 1)$$

$$= 3^1 \qquad\qquad\qquad = \frac{3}{2}(2)$$

$$= 3 \qquad\qquad\qquad\quad = 3$$

Since LHS = RHS, the result is true for $n = 1$.

2 Assume the result to be true for $n = k$,

i.e. $\sum_{r=1}^{k} 3^r = \frac{3}{2}(3^k - 1)$

3 Consider $n = k + 1$. We want to show that $\sum_{r=1}^{k+1} 3^r = \frac{3}{2}(3^{k+1} - 1)$

4 For $n = k + 1$,

$$\sum_{r=1}^{k+1} 3^r$$

$$= \sum_{r=1}^{k} 3^r + 3^{k+1}$$ .... Adding on the $(k+1)$th term.

$$= \frac{3}{2}(3^k - 1) + 3^{k+1}$$ .... Using the assumption.

$$= \frac{3}{2} \cdot 3^k - \frac{3}{2} + 3.3^k$$

$$= 3^k \left(\frac{3}{2} + 3 \right) - \frac{3}{2}$$

$$= 3^k \left(\frac{9}{2} \right) - \frac{3}{2}$$

$$= \frac{1}{2}(9.3^k - 3)$$

$$= \frac{3}{2}(3.3^k - 1)$$

$$= \frac{3}{2}(3^{k+1} - 1)$$

which is the required form.

5 So the result is true for $n = k + 1$ when true for $n = k$. Since the result is true for $n = 1$, it is true $\forall n \in \mathbb{Z}^+$ by mathematical induction.

It can be seen from these examples that sigma notation is very useful when proving a result by induction.

Example

Prove that $1.4 + 2.5 + 3.6 + \ldots + n(n + 3) = \frac{1}{3}n(n + 1)(n + 5) \; \forall n \in \mathbb{Z}^+$.

It is simpler to express the LHS using sigma notation. Hence the result becomes
$$\sum_{r=1}^{n} r(r + 3) = \frac{1}{3}n(n + 1)(n + 5).$$

1 For $n = 1$,

$$\text{LHS} = \sum_{r=1}^{1} r(r + 3) \qquad\qquad \text{RHS} = \frac{1}{3}(1)(1 + 1)(1 + 5)$$

$$= (1)(1 + 3) \qquad\qquad\qquad\qquad = \frac{1}{3}(2)(6)$$

$$= 4 \qquad\qquad\qquad\qquad\qquad\qquad = 4$$

Since LHS = RHS, the result is true for $n = 1$.

2 Assume the result to be true for $n = k$,

i.e. $\displaystyle\sum_{r=1}^{k} r(r + 3) = \frac{1}{3}k(k + 1)(k + 5)$

3 Consider $n = k + 1$. We want to show that

$$\sum_{r=1}^{k+1} r(r + 3) = \frac{1}{3}(k + 1)(k + 1 + 1)(k + 1 + 5)$$

$$= \frac{1}{3}(k + 1)(k + 2)(k + 6)$$

4 For $n = k + 1$,

$$\sum_{r=1}^{k+1} r(r + 3)$$

$$= \sum_{r=1}^{k} r(r + 3) + (k + 1)(k + 4) \quad\text{........................}$$

Adding on the $(k + 1)$th term.

$$= \frac{1}{3}k(k + 1)(k + 5) + (k + 1)(k + 4) \quad\text{.....................}$$

Using the assumption.

$$= \frac{1}{3}(k + 1)[k(k + 5) + 3(k + 4)]$$

$$= \frac{1}{3}(k + 1)[k^2 + 8k + 12]$$

$$= \frac{1}{3}(k + 1)(k + 2)(k + 6)$$

which is the required form.

5 So the result is true for $n = k + 1$ when true for $n = k$. Since the result is true for $n = 1$, it is true $\forall n \in \mathbb{Z}^+$ by mathematical induction.

Exercise 1

Prove these results $\forall n \in \mathbb{Z}^+$ by mathematical induction.

1 $\displaystyle\sum_{r=1}^{n} r^2 = \frac{1}{6}n(n + 1)(2n + 1)$

2 $\displaystyle\sum_{r=1}^{n} 2r - 1 = n^2$

3 $\displaystyle\sum_{r=1}^{n} 3r + 4 = \frac{11}{2}n(3n + 1)$

4 $\displaystyle\sum_{r=1}^{n} 5r - 2 = \frac{1}{2}n(5n + 1)$

5 $\displaystyle\sum_{r=1}^{n} 8 - 3r = \frac{1}{2}n(1 - 3n)$

6 $\displaystyle\sum_{r=1}^{n} 4r^2 - 3 = \frac{4}{3}n(4n^2 + 6n - 9)$

7 $\displaystyle\sum_{r=1}^{n} 6 + 2r - r^2 = 6n(41 + 3n - n^2)$

8 $\displaystyle\sum_{r=1}^{n} r^3 = \frac{1}{4}n^2(n + 1)^2$

9 $\displaystyle\sum_{r=1}^{n} (2r - 1)^3 = n^2(2n^2 - 1)$

10 $\displaystyle\sum_{r=1}^{n} r(r + 1) = \frac{1}{3}(n + 1)(n + 2)$

11 $\displaystyle\sum_{r=1}^{n} r(r + 1)(r + 2) = \frac{1}{4}n(n + 1)(n + 2)(n + 3)$

12 $\displaystyle\sum_{r=1}^{n} (2r)^2 = \frac{2}{3}n(n + 1)(2n + 1)$

13 $\displaystyle\sum_{r=1}^{n} 4^r = \frac{4}{3}(4^n - 1)$

14 $\displaystyle\sum_{r=1}^{n} \frac{1}{r(r + 1)(r + 2)} = \frac{n(n + 3)}{4(n + 1)(n + 2)}$

15 $\displaystyle\sum_{r=1}^{n} \frac{r}{2^r} = 2 - \left(\frac{1}{2}\right)^n (n + 2)$

16 $4 + 5 + 6 + \ldots + (n + 3) = \frac{1}{2}n(n + 7)$

17 $5 + 3 + 1 + \ldots + (7 - 2n) = n(6 - n)$

18 $3 + 6 + 11 + \ldots + (n^2 + 2) = \frac{1}{6}n(2n^2 + 3n + 13)$

19 $1.2 + 2.3 + 3.4 + \ldots + n(n + 1) = \frac{1}{3}n(n + 1)(n + 2)$

20 $-4 + 0 + 6 + \ldots + (n - 2)(n + 3) = \frac{1}{3}n(n^2 + 3n - 16)$

18.2 Proving some well-known results

So far we have concentrated on proving results that involve sigma notation. However, mathematical induction can be used to prove results from a variety of mathematical spheres. These include results from calculus, complex numbers and matrices as well as algebra.

In earlier chapters, it was stated that proofs would be provided using mathematical induction for the binomial theorem and de Moivre's theorem. In this syllabus, knowledge of the proof of de Moivre's theorem is expected but not for the binomial theorem.

Proof of de Moivre's theorem using mathematical induction

This was proved in Chapter 17 using calculus, but this method must also be known.

Prove de Moivre's theorem for all positive integers, i.e.
$(\cos\theta + i\sin\theta)^n = \cos n\theta + i\sin n\theta$

1 For $n = 1$,

LHS $= (\cos\theta + i\sin\theta)^1$ RHS $= \cos(1\theta) + i\sin(1\theta)$
   $= \cos\theta + i\sin\theta$ $= \cos\theta + i\sin\theta$

Since LHS $=$ RHS, the result is true for $n = 1$.

2 Assume the result to be true for $n = k$, i.e. $(\cos\theta + i\sin\theta)^k = \cos k\theta + i\sin k\theta$

> Substituting $(k + 1)$ for k in the result.

3 Consider $n = k + 1$. We want to show that
$(\cos\theta + i\sin\theta)^{k+1} = \cos(k + 1)\theta + i\sin(k + 1)\theta$

> We are multiplying the result for $n = k$ by $(\cos\theta + i\sin\theta)$ to get the result for $n = k + 1$.

4 For $n = k + 1$,
$(\cos\theta + i\sin\theta)^{k+1}$
$= (\cos\theta + i\sin\theta)(\cos\theta + i\sin\theta)^k$
$= (\cos\theta + i\sin\theta)(\cos k\theta + i\sin k\theta)$

> Using the assumption.

$= \cos\theta\cos k\theta + i\sin k\theta\cos\theta + i\sin\theta\cos k\theta + i^2\sin\theta\sin k\theta$
$= \cos\theta\cos k\theta - \sin\theta\sin k\theta + i(\sin k\theta\cos\theta + \sin\theta\cos k\theta)$
$= \cos(\theta + k\theta) + i(\sin(k\theta + \theta))$
$= \cos(k + 1)\theta + i\sin(k + 1)\theta$
which is the required form.

5 So the result is true for $n = k + 1$ when true for $n = k$. Since the result is true for $n = 1$, it is true $\forall n \in \mathbb{Z}^+$ by mathematical induction.

This can be extended to negative integers by considering $n = -m$ where m is a positive integer.

Proof of the binomial theorem for positive integer powers

Prove the binomial theorem, i.e. $(x + y)^n = \sum_{r=0}^{n} \binom{n}{r} x^{n-r} y^r$.

1 For $n = 1$,

$\text{LHS} = (x + y)^1$

$\hspace{4em} = x + y$

$\text{RHS} = \sum_{r=0}^{1} \binom{1}{r} x^{1-r} y^r$

$\hspace{4em} = \binom{1}{0} x^1 y^0 + \binom{1}{1} x^0 y^1$

$\hspace{4em} = x + y$

Since $\text{LHS} = \text{RHS}$, the result is true for $n = 1$.

2 Assume the result to be true for $n = k$, i.e.

$(x + y)^k$

$= \sum_{r=0}^{k} \binom{k}{r} x^{k-r} y^r = \binom{k}{0} x^k y^0 + \binom{k}{1} x^{k-1} y^1 + \binom{k}{2} x^{k-2} y^2 + \ldots + \binom{k}{r} x^{k-r} y^r + \ldots + \binom{k}{k} x^0 y^k$

3 Consider $n = k + 1$. We want to show that $(x + y)^{k+1} = \sum_{r=0}^{k+1} \binom{k+1}{r} x^{k+1-r} y^r$

4 For $n = k + 1$,

$(x + y)^{k+1}$

$= (x + y)(x + y)^k$

$= (x + y) \sum_{r=0}^{k} \binom{k}{r} x^{k-r} y^r$

$= (x + y)\left[\binom{k}{0} x^k y^0 + \binom{k}{1} x^{k-1} y^1 + \binom{k}{2} x^{k-2} y^2 + \ldots + \binom{k}{r-1} x^{k-r+1} y^{r-1} \right.$

$\hspace{2em} \left. + \binom{k}{r} x^{k-r} y^r + \ldots + \binom{k}{k} x^0 y^k \right]$

$= \binom{k}{0} x^{k+1} y^0 + \binom{k}{1} x^k y^1 + \binom{k}{2} x^{k-1} y^2 + \ldots + \binom{k}{r-1} x^{k-r+2} y^{r-1} + \binom{k}{r} x^{k+1-r} y^r$

$\hspace{2em} + \ldots + \binom{k}{k} x^1 y^k$

$\hspace{2em} + \binom{k}{0} x^k y^1 + \binom{k}{1} x^{k-1} y^2 + \binom{k}{2} x^{k-2} y^3 + \ldots + \binom{k}{r-1} x^{k-r+1} y^r + \binom{k}{r} x^{k-r} y^{r+1}$

$\hspace{2em} + \ldots + \binom{k}{k} x^0 y^{k+1}$

$= \binom{k}{0} x^{k+1} + \left[\binom{k}{1} + \binom{k}{0} \right] x^k y^1 + \left[\binom{k}{2} + \binom{k}{1} \right] x^{k-1} y^2 + \ldots$

$\hspace{2em} + \left[\binom{k}{r} + \binom{k}{r-1} \right] (x^{k+1-r} y^r) + \ldots + \left[\binom{k}{k} + \binom{k}{k-1} \right] x y^k + \binom{k}{k} y^{k+1}$

We can now use the result that $\binom{n}{r} + \binom{n}{r-1} = \binom{n+1}{r}$ (which was proved in Chapter 6).

So the general term becomes $\left[\binom{k}{r} + \binom{k}{r-1}\right](x^{k+1-r}y^r)$

$$= \binom{k+1}{r}x^{k+1-r}y^r$$

The expansion is therefore

$$\binom{k+1}{0}x^{k+1} + \binom{k+1}{1}x^k y^1 + \binom{k+1}{2}x^{k-1}y^2 + \ldots + \binom{k+1}{r}(x^{k+1-r}y^r) + \ldots$$

$$+ \binom{k+1}{k}xy^k + \binom{k+1}{k+1}y^{k+1}$$

$$= \sum_{r=0}^{k+1}\binom{k+1}{r}x^{k+1-r}y^r$$

which is the required form.

5 So the result is true for $n = k + 1$ when true for $n = k$. Since the result is true for $n = 1$, it is true $\forall n \in \mathbb{Z}^+$ by mathematical induction.

We also use mathematical induction to prove divisibility. This is demonstrated in the example below.

Example

Prove that $3^{2n} + 7$ is divisible by 8 for $n \in \mathbb{Z}^+$.

This can be restated as $3^{2n} + 7 = 8p, p \in \mathbb{N}$.

1 For $n = 1$,

$$3^{2n} + 7 = 3^2 + 7$$
$$= 16$$

As 8 is a factor of 16, or $16 = 8 \times 2$, the result is true for $n = 1$.

2 Assume the result to be true for $n = k$, i.e. $3^{2k} + 7 = 8p, p \in \mathbb{N}$.

3 Consider $n = k + 1$. We want to show that $3^{2(k+1)} + 7 = 8t, t \in \mathbb{N}$.

4 For $n = k + 1$,

$$3^{2(k+1)} + 7 = 3^{2k+2} + 7$$
$$= 3^2 3^{2k} + 7$$
$$= 9.3^{2k} + 7$$
$$= 9(3^{2k} + 7 - 7) + 7 \quad \cdots\cdots\cdots\cdots\cdots \text{This allows us to use the assumption.}$$
$$= 9(8p - 7) + 7$$
$$= 9.8p - 63 + 7$$
$$= 9.8p - 56$$
$$= 8(9p - 7)$$

Since $(9p - 7) \in \mathbb{N}$, we can say that $t = 9p - 7$ which is the required form.

5 So the result is true for $n = k + 1$ when true for $n = k$. Since the result is true for $n = 1$, it is true $\forall n \in \mathbb{Z}^+$ by mathematical induction.

There are other algebraic results that we can prove using mathematical induction, as exemplified below.

Example

Prove that $2^n > 2n + 1$ for all $n \geq 3$, $n \in \mathbb{N}$ using mathematical induction.

1 Notice here that the initial value is not $n = 1$.

For $n = 3$,

LHS $= 2^3$ RHS $= 2(3) + 1$
$\quad = 8$ $\quad = 7$

Since LHS $>$ RHS, the result is true for $n = 3$.

2 Assume the result to be true for $n = k$, $k > 3$, i.e. $2^k > 2k + 1$.

3 Consider $n = k + 1$. We want to show that $2^{k+1} > 2(k + 1) + 1$

$$\Rightarrow 2^{k+1} > 2k + 3$$

4 For $n = k + 1$,

$2^{k+1} = 2.2^k$

$\qquad > 2(2k + 1)$ ·························· Using the assumption.

$\qquad = 2k + 2k + 2$

We know that $2k + 2 > 3$, $\forall k \geq 3$ and so $2k + 2k + 2 > 2k + 3$

Hence $2^{k+1} > 2k + 3$ which is the required form.

5 So the result is true for $n = k + 1$ when true for $n = k$. Since the result is true for $n = 3$, it is true $\forall n \geq 3$, $n \in \mathbb{N}$ by mathematical induction.

Induction can also be used to prove results from other spheres of mathematics such as calculus and matrices.

Example

Prove that $\dfrac{d}{dx}(x^n) = nx^{n-1}$, $\forall n \in \mathbb{Z}^+$.

1 For $n = 1$,

LHS $= \dfrac{d}{dx}(x^1)$ RHS $= (1)x^{1-1}$
$\qquad = 1$ $\qquad = x^0$
$\qquad\qquad\qquad\qquad\qquad\qquad\qquad = 1$

We know the LHS is equal to 1 as the gradient of $y = x$ is 1. However, we should prove this by differentiation by first principles as part of a proof.

Let $f(x) = x$

$\dfrac{f(x + h) - f(x)}{h} = \dfrac{x + h - x}{h}$

$\qquad\qquad\qquad = \dfrac{h}{h}$

$\qquad\qquad\qquad = 1$

Hence $\lim\limits_{h \to 0} \dfrac{f(x + h) - f(x)}{h} = 1$

Since LHS = RHS, the result is true for $n = 1$.

2 Assume the result to be true for $n = k$, i.e.

$$\frac{d}{dx}(x^k) = kx^{k-1}$$

3 Consider $n = k + 1$. We want to show that $\dfrac{d}{dx}(x^{k+1}) = (k + 1)x^k$.

4 For $n = k + 1$,

$$\frac{d}{dx}(x^{k+1}) = \frac{d}{dx}(x.x^k)$$

$$= 1.x^k + x.kx^{k-1} \dotfill$$ Using the product rule and the assumption.

$$= x^k + kx^k$$

$$= x^k(1 + k)$$

which is the required form.

5 So the result is true for $n = k + 1$ when true for $n = k$. Since the result is true for $n = 1$, it is true $\forall n \in \mathbb{Z}^+$ by mathematical induction.

Example

Prove that $\begin{pmatrix} 1 & -1 \\ 0 & 1 \end{pmatrix}^n = \begin{pmatrix} 1 & -n \\ 0 & 1 \end{pmatrix} \forall n \in \mathbb{Z}^+$ using mathematical induction.

1 For $n = 1$,

$$\text{LHS} = \begin{pmatrix} 1 & -1 \\ 0 & 1 \end{pmatrix}^1 \qquad\qquad \text{RHS} = \begin{pmatrix} 1 & -1 \\ 0 & 1 \end{pmatrix}$$

$$= \begin{pmatrix} 1 & -1 \\ 0 & 1 \end{pmatrix}$$

Since LHS = RHS, the result is true for $n = 1$.

2 Assume the result to be true for $n = k$,

i.e. $\begin{pmatrix} 1 & -1 \\ 0 & 1 \end{pmatrix}^k = \begin{pmatrix} 1 & -k \\ 0 & 1 \end{pmatrix}$

3 Consider $n = k + 1$. We want to show that

$$\begin{pmatrix} 1 & -1 \\ 0 & 1 \end{pmatrix}^{k+1} = \begin{pmatrix} 1 & -(k + 1) \\ 0 & 1 \end{pmatrix}.$$

4 For $n = k + 1$,

$$\begin{pmatrix} 1 & -1 \\ 0 & 1 \end{pmatrix}^{k+1}$$

$$= \begin{pmatrix} 1 & -1 \\ 0 & 1 \end{pmatrix}\begin{pmatrix} 1 & -1 \\ 0 & 1 \end{pmatrix}^k$$

$$= \begin{pmatrix} 1 & -1 \\ 0 & 1 \end{pmatrix}\begin{pmatrix} 1 & -k \\ 0 & 1 \end{pmatrix} \quad \cdots\cdots\cdots\cdots\cdots\cdots\cdots \boxed{\text{Using the assumption.}}$$

$$= \begin{pmatrix} 1 + 0 & -k - 1 \\ 0 + 0 & 0 + 1 \end{pmatrix}$$

$$= \begin{pmatrix} 1 & -(k + 1) \\ 0 & 1 \end{pmatrix}$$

which is the required form.

5 So the result is true for $n = k + 1$ when true for $n = k$. Since the result is true for $n = 1$, it is true $\forall n \in \mathbb{Z}^+$ by mathematical induction.

Exercise 2

Prove these results using mathematical induction.

1 Prove that $2^{3n} - 1$ is divisible by 7, $\forall n \in \mathbb{Z}^+$.

2 Prove that $3^{2n} - 5$ is divisible by 4, $\forall n \in \mathbb{Z}^+$.

3 Prove that $5^n + 3$ is divisible by 4, $\forall n \in \mathbb{Z}^+$.

4 Prove that $9^n - 8n - 1$ is divisible by 64, $\forall n \in \mathbb{Z}^+$.

5 Prove that $6^n + 4$ is divisible by 10, $\forall n \in \mathbb{Z}^+$.

6 Prove that $n(n^2 - 1)(3n + 2)$ is divisible by 24, $\forall n \in \mathbb{Z}^+$.

7 Prove that $n! > 2^n$, for $n \geq 4, n \in \mathbb{Z}^+$.

8 Prove that $n! > n^2$, for $n \geq 4, n \in \mathbb{Z}^+$.

9 Prove that $2^n > n^3$, for $n \geq 10, n \in \mathbb{Z}^+$.

10 Find the smallest integer t for which $n! > 3^n$.
Hence prove by induction that $n! > 3^n$ for all $n > t$.

11 For $A = \begin{pmatrix} 2 & 0 \\ 1 & 1 \end{pmatrix}$, prove that $A^n = \begin{pmatrix} 2^n & 0 \\ 2^n - 1 & 1 \end{pmatrix} \forall n \in \mathbb{Z}^+$.

12 Prove that $\dfrac{d^n}{dx^n}(e^{px}) = p^n e^{px}$, $\forall n \in \mathbb{Z}^+$.

13 Prove that $\dfrac{d^n}{dx^n}(\sin 2x) = 2^{n-1} \sin\left(2x + \dfrac{(n-1)\pi}{2}\right)$, for all positive integer values of n.

14 For $M = \begin{pmatrix} 1 & 2 & 0 \\ 0 & p & 0 \\ 0 & 0 & 3 \end{pmatrix}$, prove that $M^n = \begin{pmatrix} 1 & \dfrac{2(1 - p^n)}{1 - p} & 0 \\ 0 & p^n & 0 \\ 0 & 0 & 3^n \end{pmatrix}$, $\forall n \in \mathbb{Z}^+$.

15 For $T = \begin{pmatrix} 4 & t \\ 0 & 1 \end{pmatrix}$, prove that $T^n = \begin{pmatrix} 4^n & \sum\limits_{r=1}^{n} 4^{r-1} \\ 0 & 1 \end{pmatrix}$, $\forall n \in \mathbb{Z}^+$.

18.3 Forming and proving conjectures

For all of the examples met so far, the result to be proved was given in the question. This is not always the case; sometimes, it is necessary to form a conjecture which can then be proved using mathematical induction.

Example

Form a conjecture for the sum $\dfrac{1}{1 \times 2} + \dfrac{1}{2 \times 3} + \dfrac{1}{3 \times 4} + \ldots + \dfrac{1}{n(n+1)}$.

In order to form the conjecture, consider the results for the first few values of n.

n	1	2	3	4	5
sum	$\dfrac{1}{2}$	$\dfrac{1}{2} + \dfrac{1}{6} = \dfrac{2}{3}$	$\dfrac{2}{3} + \dfrac{1}{12} = \dfrac{3}{4}$	$\dfrac{3}{4} + \dfrac{1}{20} = \dfrac{4}{5}$	$\dfrac{4}{5} + \dfrac{1}{30} = \dfrac{5}{6}$

Looking at the pattern, we can make a conjecture that

$$\frac{1}{1 \times 2} + \frac{1}{2 \times 3} + \frac{1}{3 \times 4} + \ldots + \frac{1}{n(n+1)} = \frac{n}{n+1}$$

We now try to prove this conjecture using mathematical induction.

This can be expressed as $\displaystyle\sum_{r=1}^{n} \frac{1}{r(r+1)}$.

1 For $n = 1$,

$$\text{LHS} = \sum_{r=1}^{1} \frac{1}{r(r+1)} \qquad\qquad \text{RHS} = \frac{1}{1+1}$$

$$= \frac{1}{1 \times 2} \qquad\qquad\qquad\quad = \frac{1}{2}$$

$$= \frac{1}{2}$$

Since LHS = RHS, the conjecture is true for $n = 1$.

2 Assume the result to be true for $n = k$,

i.e. $\displaystyle\sum_{r=1}^{k} \frac{1}{r(r+1)} = \frac{k}{k+1}$

3 Consider $n = k + 1$. We want to show that

$$\sum_{r=1}^{k+1} \frac{1}{r(r+1)} = \frac{k+1}{k+1+1} = \frac{k+1}{k+2}.$$

4 For $n = k + 1$,

$$\sum_{r=1}^{k+1} \frac{1}{r(r+1)}$$

$$= \sum_{r=1}^{k} \frac{1}{r(r+1)} + \frac{1}{(k+1)(k+2)}$$

$$= \frac{k}{k+1} + \frac{1}{(k+1)(k+2)} \quad\cdots\cdots\cdots\cdots\cdots\cdots\cdots\cdots \boxed{\text{Using the assumption.}}$$

$$= \frac{k(k+2)}{(k+1)(k+2)} + \frac{1}{(k+1)(k+2)}$$

$$= \frac{k(k + 2) + 1}{(k + 1)(k + 2)}$$

$$= \frac{k^2 + 2k + 1}{(k + 1)(k + 2)}$$

$$= \frac{(k + 1)^2}{(k + 1)(k + 2)}$$

$$= \frac{k + 1}{k + 2}$$

which is the required form.

5 So the conjecture is true for $n = k + 1$ when true for $n = k$. Since it is true for $n = 1$, it is true $\forall n \in \mathbb{Z}^+$ by mathematical induction.

Example

Form a conjecture for the pentagonal numbers as shown below. Prove your conjecture by mathematical nduction.

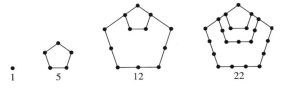

So the sequence of pentagonal numbers begins 1, 5, 12, 22, 35, ...

Remembering that these are formed by adding "a new layer" each time, we can consider this as a sum,

$$1 + 4 + 7 + 10 + \ldots + (3n - 2)$$

We are trying to find a formula for this. This is an arithmetic progression and so we can apply the formula for the sum to n terms with $a = 1$, $d = 3$.

Hence $S_n = \dfrac{n}{2}(2 + 3(n - 1))$

$$= \frac{n}{2}(3n - 1)$$

$$= \frac{3n^2 - n}{2}$$

Again, our conjecture can be expressed using sigma notation:

$$\sum_{r=1}^{n} 3r - 2 = \frac{3n^2 - n}{2}$$

1 For $n = 1$,

LHS $= \displaystyle\sum_{r=1}^{1} 3r - 2$ RHS $= \dfrac{3(1^2) - 1}{2}$

$$= 3 - 2 \qquad\qquad\qquad = \frac{3 - 1}{2}$$

$$= 1 \qquad\qquad\qquad\qquad = 1$$

Since LHS $=$ RHS, the conjecture is true for $n = 1$.

2 Assume the result to be true for $n = k$,

i.e. $\displaystyle\sum_{r=1}^{k} 3r - 2 = \frac{3k^2 - k}{2}$

3 Consider $n = k + 1$. We want to show that

$$\sum_{r=1}^{k+1} 3r - 2 = \frac{3(k + 1)^2 - (k + 1)}{2}$$

$$= \frac{3k^2 + 6k + 3 - k - 1}{2}$$

$$= \frac{3k^2 + 5k + 2}{2}$$

4 For $n = k + 1$,

$$\sum_{r=1}^{k+1} 3r - 2$$

$$= \sum_{r=1}^{k} 3r - 2 + 3(k + 1) - 2$$

$$= \frac{3k^2 - k}{2} + 3k + 1 \quad\text{·····························}\quad \boxed{\text{Using the assumption.}}$$

$$= \frac{3k^2 - k}{2} + \frac{6k + 2}{2}$$

$$= \frac{3k^2 + 5k + 2}{2}$$

which is the required form.

5 So the conjecture is true for $n = k + 1$ when true for $n = k$. Since it is true for $n = 1$, it is true $\forall n \in \mathbb{Z}^+$ by mathematical induction.

Example

For the matrix $A = \begin{pmatrix} 3 & 0 \\ 0 & 2 \end{pmatrix}$, form a conjecture for $A^n, n \in \mathbb{Z}^+$. Prove your conjecture by mathematical induction.

To form the conjecture, find the results for the first few values of n.

$$A = \begin{pmatrix} 3 & 0 \\ 0 & 2 \end{pmatrix}$$

$$A^2 = \begin{pmatrix} 9 & 0 \\ 0 & 4 \end{pmatrix}$$

$$A^3 = \begin{pmatrix} 27 & 0 \\ 0 & 8 \end{pmatrix}$$

$$A^4 = \begin{pmatrix} 81 & 0 \\ 0 & 16 \end{pmatrix}$$

From this we can make a conjecture that $A^n = \begin{pmatrix} 3^n & 0 \\ 0 & 2^n \end{pmatrix}$.

We can now prove this using mathematical induction.

1 For $n = 1$,

$$\text{LHS} = \begin{pmatrix} 3 & 0 \\ 0 & 2 \end{pmatrix}^1 \qquad\qquad \text{RHS} = \begin{pmatrix} 3^1 & 0 \\ 0 & 2^1 \end{pmatrix}$$

$$= \begin{pmatrix} 3 & 0 \\ 0 & 2 \end{pmatrix} \qquad\qquad = \begin{pmatrix} 3 & 0 \\ 0 & 2 \end{pmatrix}$$

Since LHS = RHS, the conjecture is true for $n = 1$.

2 Assume the result to be true for $n = k$,

i.e. $\begin{pmatrix} 3 & 0 \\ 0 & 2 \end{pmatrix}^k = \begin{pmatrix} 3^k & 0 \\ 0 & 2^k \end{pmatrix}$

3 Consider $n = k + 1$. We want to show that $\begin{pmatrix} 3 & 0 \\ 0 & 2 \end{pmatrix}^{k+1} = \begin{pmatrix} 3^{k+1} & 0 \\ 0 & 2^{k+1} \end{pmatrix}$.

4 For $n = k + 1$,

$$\begin{pmatrix} 3 & 0 \\ 0 & 2 \end{pmatrix}^{k+1} = \begin{pmatrix} 3 & 0 \\ 0 & 2 \end{pmatrix}\begin{pmatrix} 3 & 0 \\ 0 & 2 \end{pmatrix}^k$$

$$= \begin{pmatrix} 3 & 0 \\ 0 & 2 \end{pmatrix}\begin{pmatrix} 3^k & 0 \\ 0 & 2^k \end{pmatrix} \quad\cdots\cdots\cdots\cdots\cdots\cdots\boxed{\text{Using the assumption.}}$$

$$= \begin{pmatrix} 3.3^k + 0 & 0 + 0 \\ 0 + 0 & 0 + 2.2^k \end{pmatrix}$$

$$= \begin{pmatrix} 3^{k+1} & 0 \\ 0 & 2^{k+1} \end{pmatrix}$$

which is the required form.

5 So the conjecture is true for $n = k + 1$ when true for $n = k$. Since the conjecture is true for $n = 1$, it is true $\forall n \in \mathbb{Z}^+$ by mathematical induction.

Exercise 3

1 For the matrix $D = \begin{pmatrix} 1 & 1 \\ 0 & 2 \end{pmatrix}$, form a conjecture for D^n. Prove your conjecture using mathematical induction.

2 Form a conjecture for the sum $5 + 8 + 11 + 14 + \ldots + (3n + 2)$. Prove your conjecture using mathematical induction.

3 Form a conjecture for the series suggested by the initial values of $-3 + 1 + 5 + 9 + 13 + 17 + \ldots$. Prove your conjecture by mathematical induction.

4 With an unlimited supply of 4p and 7p stamps, make a conjecture about what values >17 of postage it is possible to create. Prove your conjecture using mathematical induction.

5 Make a conjecture about the sum of the first n odd numbers. Prove your conjecture using mathematical induction.

6 Find an expression for the *n*th term of the sequence 5, 10, 17, 26, 37, ... Prove this result to be true using mathematical induction.

7 For the sequence below where each new pattern is made by adding a new "layer", make a conjecture for the number of dots in the *n*th term of the pattern. Prove this result to be true using mathematical induction.

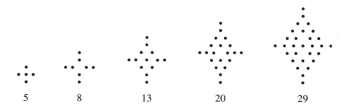

| 5 | 8 | 13 | 20 | 29 |

Review exercise

 1 Prove that $\sum_{r=1}^{n} r^4 = \frac{1}{30}n(n + 1)(2n + 1)(3n^2 + 3n - 1)$, $\forall n \in \mathbb{Z}^+$.

 2 Prove that $n^5 - n$ is divisible by 5, $\forall n \in \mathbb{Z}^+$.

 3 Prove that $\forall n \in \mathbb{Z}^+$, $11^{n+1} + 12^{2n-1}$ is divisible by 133.

 4 Prove, using mathematical induction, that
$$\frac{d^n}{dx^n}(xe^{px}) = p^{n-1}e^{px}(px + 1), \quad \forall n \in \mathbb{Z}^+.$$

 5 Prove, using mathematical induction, that for $T = \begin{pmatrix} 2 & 0 \\ p & 1 \end{pmatrix}$,

$$T^n = \begin{pmatrix} 2^n & 0 \\ p(2^n - 1) & 1 \end{pmatrix}, \forall n \in \mathbb{Z}^+.$$

6 Prove, using mathematical induction, that
$$2 \cdot 6 \cdot 10 \cdot 14 \cdot \ldots \cdot (4n - 2) = \frac{(2n)!}{n!}, \forall n \in \mathbb{Z}^+.$$

7 Prove that $\sin \theta + \sin 3\theta + \ldots + \sin(2n - 1)\theta = \frac{\sin^2 n\theta}{\sin \theta}$, $\forall n \in \mathbb{Z}^+$.

8 Form a conjecture for the sum $1 \times 1! + 2 \times 2! + 3 \times 3! + \ldots + n \times n!$. Prove your conjecture by mathematical induction.

 9 Using mathematical induction, prove that $\frac{d^n}{dx^n}(\cos x) = \cos\left(x + \frac{n\pi}{2}\right)$, for all

positive integer values of *n*. [IB May 01 P2 Q4]

10 a Prove using mathematical induction that $\begin{pmatrix} 2 & 1 \\ 0 & 1 \end{pmatrix}^n = \begin{pmatrix} 2^n & 2^n - 1 \\ 0 & 1 \end{pmatrix}$ for all positive integer values of *n*.
b Determine whether or not this result is true for $n = -1$.

[IB May 02 P2 Q3]

 11 Prove, using mathematical induction, that for a positive integer *n*,
$(\cos \theta + i \sin \theta)^n = \cos n\theta + i \sin n\theta$ where $i^2 = -1$. [IB May 03 P2 Q3]

12 Using mathematical induction, prove that $\sum_{r=1}^{n} (r + 1)2^{r-1} = n(2^n)$ for all

positive integers. [IB May 05 P2 Q4]

 13 The function f is defined by $f(x) = e^{px}(x + 1)$, where $p \in \mathbb{R}$.

a Show that $f'(x) = e^{px}(p(x - 1) + 1)$.

b Let $f^{(n)}(x)$ denote the result of differentiating $f(x)$ with respect to x, n times. Use mathematical induction to prove that

$f^{(n)}(x) = p^{n-1}e^{px}(p(x + 1) + n), n \in \mathbb{Z}^+$. [IB May 05 P2 Q2]

 14 For $T = \begin{pmatrix} -1 & 3 & 0 \\ 0 & 2 & 0 \\ 0 & 0 & s \end{pmatrix}$, prove that $T^n = \begin{pmatrix} (-1)^n & 2^n - (-1)^n & 0 \\ 0 & 2^n & 0 \\ 0 & 0 & s^n \end{pmatrix}, n \in \mathbb{Z}^+$

using mathematical induction. [IB May 06 P2 Q5]

 15 Consider the sequence $\{a_n\}$, $\{1, 1, 2, 3, 5, 8, 13, \dots\}$ where $a_1 = a_2 = 1$

and $a_{n+1} = a_n + a_{n-1}$ for all integers $n \geq 2$.

Given the matrix, $Q = \begin{pmatrix} 1 & 1 \\ 1 & 0 \end{pmatrix}$ use the principle of mathematical induction

to prove that $Q^n = \begin{pmatrix} a_{n+1} & a_n \\ a_n & a_{n-1} \end{pmatrix}$ for all integers $n \geq 2$.

[IB Nov 01 P2 Q4]

 16 The matrix M is defined as $M = \begin{pmatrix} 2 & -1 \\ 1 & 0 \end{pmatrix}$.

a Find M^2, M^3 and M^4.

b **i** State a conjecture for M^n, i.e. express M^n in terms of n, where $n \in \mathbb{Z}^+$.

ii Prove this conjecture using mathematical induction.

[IB Nov 02 P2 Q1]

19 Statistics

One of the most famous quotes about statistics, of disputed origin, is "Lies, damned lies and statistics". This joke demonstrates the problem quite succinctly:

Did you hear about the statistician who drowned while crossing a stream that was, on average, 6 inches deep?

Statistics is concerned with displaying and analysing data. Two early forms of display are shown here. The first pie chart was used in 1801 by William Playfair. The pie chart shown was used in 1805.

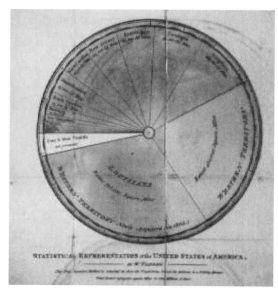

The first cumulative frequency curve, a graph that we will use in this chapter, was used by Jean Baptiste Joseph Fourier in 1821 and is shown below.

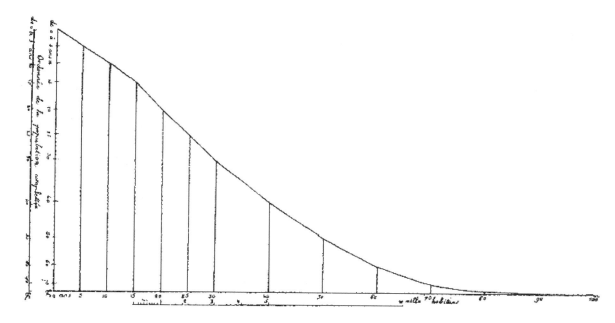

19.1 Frequency tables

Introduction

Statistics involves the collection, display and interpretation of data. This syllabus concentrates on the interpretation of data. One of the most common tools used to interpret data is the calculation of measures of central tendency. There are three measures of central tendency (or **averages**) which are presumed knowledge for this syllabus, the mean, median and mode.

The **mean** is the arithmetic average and is defined as $\bar{x} = \dfrac{\sum x}{n}$, where n is the number of pieces of data.

The **median** is in the middle of the data when the items are written in an ordered list. For an odd number of data items in the data set, this will be a data item. For an even number of data items, this will be the mean of the two middle data items. The median is said to be the $\dfrac{n+1}{2}$th data item.

The **mode** is the most commonly occurring data item.

Definitions

When interpreting data, we are often interested in a particular group of people or objects. This group is known as the **population**. If data are collected about all of these people or objects, then we can make comments about the population. However, it is not always possible to collect data about every object or person in the population.

A **sample** is part of a population. In statistical enquiry, data are collected about a sample and often then used to make informed comment about that sample and the population. For the comment to be valid about a population, the sample must be representative of that population. This is why most samples that are used in statistics are random samples. Most statistics quoted in the media, for example, are based on samples.

Types of data

Data can be categorized into two basic types: discrete and continuous. The distinction between these two types can be thought of as countables and uncountables.

Discrete data are data that can only take on exact values, for example shoe size, number of cars, number of people.

Continuous data do not take on exact values but are measured to a degree of accuracy. Examples of this type of data are height of children, weight of sugar.

The distinction between these two types of data is often also made in language. For example, in English the distinction is made by using "fewer" or "less". The sentence "there are fewer trees in my garden than in David's garden" is based on discrete data, and the sentence "there is less grass in David's garden than in my garden" is based on continuous data.

It is important to understand and be aware of the distinction as it is not always immediately obvious which type of data is being considered. For example, the weight of bread is continuous data but the number of loaves of bread is discrete data.

One way of organizing and summarizing data is to use a **frequency table**. Frequency tables take slightly different forms for discrete and continuous data. For discrete data, a frequency table consists of the various data points and the frequency with which they occur. For continuous data, the data points are grouped into intervals or "classes".

Frequency tables for discrete data

The three examples below demonstrate the different ways that frequency tables are used with discrete data.

Example

Ewan notes the colour of the first 20 cars passing him on a street corner. Organize this data into a frequency table, stating the modal colour.

Blue	Black	Silver	Red	Green
Silver	Blue	Blue	Silver	Black
Red	Black	Blue	Silver	Blue
Yellow	Blue	Silver	Silver	Black

The colour of cars noted by Ewan

Colour of car	Tally	Frequency
Black	‖‖	4
Blue	⧸⧸⧸⧸ ‖	6
Green	‖	1
Red	‖	2
Silver	⧸⧸⧸⧸ ‖	6
Yellow	‖	1
	Total	**20**

From this frequency table, we can see that there are two modes: blue and silver.

> We use tallies to help us enter data into a frequency table.

> As these data are not numerical it is not possible to calculate the mean and median.

Example

Laura works in a men's clothing shop and records the waist size (in inches) of jeans sold one Saturday. Organize this data into a frequency table, giving the mean, median and modal waist size.

30	28	34	36	38	36	34	32	32	34
34	32	40	32	28	34	30	32	38	34
30	28	30	38	34	36	32	32	34	34

These data are discrete and the frequency table is shown below.

Waist size (inches)	Tally	Frequency
28	‖‖	3
30	‖‖‖	4
32	⧸⧸⧸⧸ ‖	7
34	⧸⧸⧸⧸ ‖‖‖	9
36	‖‖	3
38	‖‖	3
40	‖	1
	Total	**30**

It is immediately obvious that the data item with the highest frequency is 34 and so the modal waist size is 34 inches.

In order to find the median, we must consider its position. In 30 data items, the median will be the mean of the 15th and 16th data items. In order to find this, it is useful to add a cumulative frequency column to the table. Cumulative frequency is another name for a running total.

Waist size (inches)	Tally	Frequency	Cumulative frequency
28					3	3
30						4	7
32	ℍ ||	7	14
34	ℍ ||||	9	23
36					3	26
38					3	29
40			1	30
	Total	**30**	

From the cumulative frequency column, it can be seen that the 15th and 16th data items are both 34 and so the median waist size is 34 inches.

In order to find the mean, it is useful to add a column of data × frequency to save repeated calculation.

Waist size (inches)	Tally	Frequency	Size × frequency
28					3	84
30						4	120
32	ℍ ||	7	224
34	ℍ ||||	9	306
36					3	108
38					3	114
40			1	40
	Total	**30**	**996**

The mean is given by $\bar{x} = \dfrac{\sum x}{n} = \dfrac{996}{30} = 33.2$. So the mean waist size is 33.2 inches.

Discrete frequency tables can also make use of groupings as shown in the next example.

The groups are known as **class intervals** and the range of each class is known as its **class width**. It is common for class widths for a particular distribution to be all the same but this is not always the case.

The upper interval boundary and lower interval boundary are like the boundaries used in sigma notation. So, for a class interval of 31–40, the lower interval boundary is 31 and the upper interval boundary is 40.

Example

Alastair records the marks of a group of students in a test scored out of 80, as shown in the table. What are the class widths? What is the modal class interval?

Mark	Frequency
21–30	5
31–40	12
41–50	17
51–60	31
61–70	29
71–80	16

The class widths are all 10 marks. The modal class interval is the one with the highest frequency and so is 51–60.

Finding averages from a grouped frequency table

The modal class interval is the one with the highest frequency. This does not determine the mode exactly, but for large distributions it is really only the interval that is important.

> The modal class interval only makes sense if the class widths are all the same.

Similarly, it is not possible to find an exact value for the median from a grouped frequency table. However, it is possible to find the class interval in which the median lies. In the above example, the total number of students was 110 and so the median lies between the 55th and 56th data items. Adding a cumulative frequency column helps to find these:

Mark	Frequency	Cumulative frequency
21–30	5	5
31–40	12	17
41–50	17	34
51–60	31	65
61–70	29	94
71–80	16	110

From the cumulative frequency column, we can see that the median lies in the interval of 51–60. The exact value can be estimated by assuming that the data are equally distributed throughout each class.

The median is the 55.5th data item which is the 21.5th data item in the 51–60 interval. Dividing this by the frequency $\frac{21.5}{31} = 0.693 \ldots$ provides an estimate of how far through the class the median would lie (if the data were equally distributed). Multiplying this fraction by 10 (the class width) gives $6.93 \ldots$, therefore an estimate for the median is $50 + 6.93 \ldots = 56.9$ (to 1 decimal place).

> It is often sufficient just to know which interval contains the median.

Finding the mean from a grouped frequency table also involves assuming the data is equally distributed. To perform the calculation, the mid-interval values are used. The **mid-interval value** is the median of each interval.

So for our example:

Mark	Mid-interval value	Frequency	Mid-value × frequency
21–30	25.5	5	127.5
31–40	35.5	12	426
41–50	45.5	17	773.5
51–60	55.5	31	1720.5
61–70	65.5	29	1899.5
71–80	75.5	16	1208
	Totals	110	6155

So the mean is $\bar{x} = \dfrac{6155}{110} = 56.0$ (to 1 decimal place).

Again, this value for the mean is only an estimate.

Frequency tables for continuous data

Frequency tables for continuous data are nearly always presented as grouped tables. It is possible to round the data so much that it effectively becomes a discrete distribution, but most continuous data are grouped.

The main difference for frequency tables for continuous data is in the way that the class intervals are constructed. It is important to recognize the level of accuracy to which the data have been given and the intervals should reflect this level of accuracy. The upper class boundary of one interval will be the lower class boundary of the next interval. This means that class intervals for continuous data are normally given as inequalities such as $19.5 \leq x < 24.5$, $24.5 \leq x < 29.5$ etc.

Example

A police speed camera records the speeds of cars passing in km/h, as shown in the table. What was the mean speed? Should the police be happy with these speeds in a 50 km/h zone?

Speed (km/h)	Frequency
$39.5 \leq x < 44.5$	5
$44.5 \leq x < 49.5$	65
$49.5 \leq x < 54.5$	89
$54.5 \leq x < 59.5$	54
$59.5 \leq x < 64.5$	12
$64.5 \leq x < 79.5$	3

The interval widths are 5, 5, 5, 5, 5, 15. However, to find the mean, the method is the same: we use the mid-interval value.

Speed	Mid-interval value	Frequency	Mid-value × frequency
$39.5 \leq x < 44.5$	42	5	210
$44.5 \leq x < 49.5$	47	65	3055
$49.5 \leq x < 54.5$	52	89	4628
$54.5 \leq x < 59.5$	57	54	3078
$59.5 \leq x < 64.5$	62	12	744
$64.5 \leq x < 79.5$	72	3	216
	Totals	228	11931

By choosing these class intervals with decimal values, an integral mid-interval value is created.

So the estimated mean speed is $\bar{x} = \dfrac{11\,931}{228} = 52.3$ km/h (to 1 decimal place).

We will discuss how we work with this mathematically later in the chapter.

Using this figure alone does not say much about the speeds of the cars. Although most of the cars were driving at acceptable speeds, the police would be very concerned about the three cars driving at a speed in the range $64.5 \leq x < 79.5$ km/h.

Frequency distributions

Frequency distributions are very similar to frequency tables but tend to be presented horizontally. The formula for the mean from a frequency distribution is written as $\bar{x} = \dfrac{\sum fx}{\sum f}$ but has the same meaning as $\bar{x} = \dfrac{\sum x}{n}$.

Example

Students at an international school were asked how many languages they could speak fluently and the results are set out in a frequency distribution. Calculate the mean number of languages spoken.

Number of languages, x	1	2	3	4
Frequency	31	57	42	19

So the mean for this distribution is given by

$$\bar{x} = \frac{1 \times 31 + 2 \times 57 + 3 \times 42 + 4 \times 19}{31 + 57 + 42 + 19} = \frac{347}{149} = 2.33 \text{ (to 2 d.p.)}$$

Example

The time taken (in seconds) by students running 100 m was recorded and grouped as shown.

What is the mean time?

Time, t	Frequency
$10.5 \le t < 11$	5
$11 \ \le t < 11.5$	11
$11.5 \le t < 12$	12
$12 \ \le t < 12.5$	15
$12.5 \le t < 13$	8
$13 \ \le t < 13.5$	10

As the data are grouped, we use the mid-interval values to calculate the mean.

$$\bar{t} = \frac{10.75 \times 5 + 11.25 \times 11 + 11.75 \times 12 + 12.25 \times 15 + 12.75 \times 8 + 13.25 \times 10}{5 + 11 + 12 + 15 + 8 + 10}$$

$$= \frac{736.75}{61}$$

$$= 12.1 \ (\text{to 1 d.p.})$$

Exercise 1

1 State whether the data are discrete or continuous.

  a Height of tomato plants   **b** Number of girls with blue eyes

  c Temperature at a weather station   **d** Volume of helium in balloons

2 Mr Coffey collected the following information about the number of people in his students' households:

4	2	6	7	3	3	2	4	4	4
5	5	4	5	4	3	4	3	5	6

Organize these data into a frequency table. Find the mean, median and modal number of people in this class's households.

3 Fiona did a survey of the colour of eyes of the students in her class and found the following information:

Blue	Blue	Green	Brown	Brown	Hazel	Brown	Green	Blue	Blue
Green	Blue	Blue	Green	Hazel	Blue	Brown	Blue	Brown	Brown
Blue	Brown	Blue	Brown	Green	Brown	Blue	Brown	Blue	Green

Construct a frequency table for this information and state the modal colour of eyes for this class.

4 The IBO recorded the marks out of 120 for HL Mathematics and organized the data into a frequency table as shown below:

Mark	Frequency
0–20	104
21–40	230
41–50	506
51–60	602
61–70	749
71–80	1396
81–90	2067
91–100	1083
101–120	870

a What are the class widths?

b Using a cumulative frequency column, determine the median interval.

c What is the mean mark?

5 Ganesan is recording the lengths of earthworms for his Group 4 project. His data are shown below.

Length of earthworm (cm)	Frequency
$4.5 \leq l < 8.5$	3
$8.5 \leq l < 12.5$	12
$12.5 \leq l < 16.5$	26
$16.5 \leq l < 20.5$	45
$20.5 \leq l < 24.5$	11
$24.5 \leq l < 28.5$	2

What is the mean length of earthworms in Ganesan's sample?

6 The heights of a group of students are recorded in the following frequency table.

Height (m)	Frequency
$1.35 \leq h < 1.40$	5
$1.40 \leq h < 1.45$	13
$1.45 \leq h < 1.50$	10
$1.50 \leq h < 1.55$	23
$1.55 \leq h < 1.60$	19
$1.60 \leq h < 1.65$	33
$1.65 \leq h < 1.70$	10
$1.70 \leq h < 1.75$	6
$1.75 \leq h < 1.80$	9
$1.80 \leq h < 2.10$	2

a Find the mean height of these students.

b Although these data are fairly detailed, why is the mean not a particularly useful figure to draw conclusions from in this case?

7 Rosemary records how many musical instruments each child in the school plays in a frequency distribution. Find the mean number of instruments played.

Number of instruments, x	0	1	2	3	4
Frequency	55	49	23	8	2

8 A rollercoaster operator records the heights (in metres) of people who go on his ride in a frequency distribution.

Height, *h*	Frequency
$1.30 \leq h < 1.60$	0
$1.60 \leq h < 1.72$	101
$1.72 \leq h < 1.84$	237
$1.84 \leq h < 1.96$	91
$1.96 \leq h < 2.08$	15

a Why do you think the frequency for $1.30 \leq h < 1.60$ is zero?

b Find the mean height.

19.2 Frequency diagrams

A frequency table is a useful way of organizing data and allows for calculations to be performed in an easier form. However, we sometimes want to display data in a readily understandable form and this is where diagrams or graphs are used.

One of the most simple diagrams used to display data is a pie chart. This tends to be used when there are only a few (2–8) distinct data items (or class intervals) with the relative area of the sectors (or length of the arcs) signifying the frequencies. Pie charts provide an immediate visual impact and so are often used in the media and in business applications. However, they have been criticized in the scientific community as area is more difficult to compare visually than length and so pie charts are not as easy to interpret as some diagrams.

Histograms

A histogram is another commonly used frequency diagram. It is very similar to a bar chart but with some crucial distinctions:

1 The bars must be adjacent with no spaces between the bars.
2 What is important about the bars is their area, not their height. In this curriculum, we have equal class widths and so the height can be used to signify the frequency but it should be remembered that it is the area of each bar that is proportional to the frequency.

A histogram is a good visual representation of data that gives the reader a sense of the central tendency and the spread of the data.

Example

Draw a bar chart to represent the information contained in the frequency table.

The colour of cars noted by Ewan

Colour of car	Frequency
Black	4
Blue	6
Green	1
Red	2
Silver	6
Yellow	1
Total	**20**

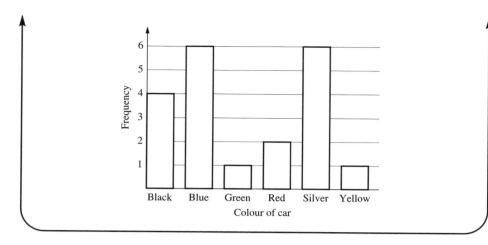

Example

The distances thrown in a javelin competition were recorded in the frequency table below. Draw a histogram to represent this information.

Distances thrown in a javelin competition (metres)

Distance	Frequency
$44.5 \leq d < 49.5$	2
$49.5 \leq d < 54.5$	2
$54.5 \leq d < 59.5$	4
$59.5 \leq d < 64.5$	5
$64.5 \leq d < 69.5$	12
$69.5 \leq d < 74.5$	15
$74.5 \leq d < 79.5$	4
$79.5 \leq d < 84.5$	3
Total	**47**

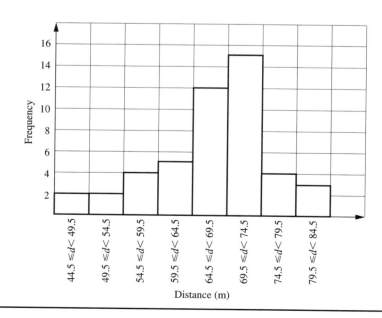

Box and whisker plots

A box and whisker plot is another commonly used diagram that provides a quick and accurate representation of a data set. A box and whisker plot notes five major features of a data set: the maximum and minimum values and the quartiles.

The **quartiles** of a data set are the values that divide the data set into four equal parts. So the lower quartile (denoted Q_1) is the value that cuts off 25% of the data.

The second quartile, normally known as the median but also denoted Q_2, cuts the data in half.

The third or upper quartile (Q_3) cuts off the highest 25% of the data.

These quartiles are also known as the 25th, 50th and 75th percentiles respectively.

A simple way of viewing quartiles is that Q_1 is the median of the lower half of the data, and Q_3 is the median of the upper half. Therefore the method for finding quartiles is the same as for finding the median.

Example

Find the quartiles of this data set.

Age	Frequency	Cumulative frequency
14	3	3
15	4	7
16	8	15
17	5	20
18	6	26
19	3	29
20	1	30
Total	**30**	

Here the median is the 15.5th piece of data (between the 15th and 16th) which is 16.5.

Each half of the data set has 15 data items. The median of the lower half will be the data item in the 8th position, which is 16. The median of the upper half will be the data item in the $15 + 8 = 23$rd position. This is 18.

So for this data set,

$Q_1 = 16$
$Q_2 = 16.5$
$Q_3 = 18$

There are a number of methods for determining the positions of the quartiles. As well as the method above, the lower quartile is sometimes calculated to be the $\dfrac{n + 1}{4}$ th data item, and the upper quartile calculated to be the $\dfrac{3(n + 1)}{4}$ th data item.

A box and whisker plot is a representation of the three quartiles plus the maximum and minimum values. The box represents the "middle" 50% of the data, that is the data between Q_1 and Q_3. The whiskers are the lowest 25% and the highest 25% of the data. It is very important to remember that this is a graph and so a box and whisker plot should be drawn with a scale.

For the above example, the box and whisker plot would be:

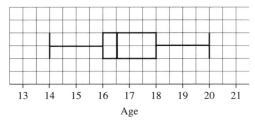

13 14 15 16 17 18 19 20 21
Age

This is the simplest form of a box and whisker plot. Some statisticians calculate what are known as outliers before drawing the plot but this is not part of the syllabus. Box and whisker plots are often used for discrete data but can be used for grouped and continuous data too. Box and whisker plots are particularly useful for comparing two distributions, as shown in the next example.

Example

Thomas and Catherine compare the performance of two classes on a French test, scored out of 90 (with only whole number marks available). Draw box and whisker plots (on the same scale) to display this information. Comment on what the plots show about the performance of the two classes.

Thomas' class

Score out of 90	Frequency	Cumulative frequency
$0 \leq x \leq 10$	1	1
$11 \leq x \leq 20$	2	3
$21 \leq x \leq 30$	4	7
$31 \leq x \leq 40$	0	7
$41 \leq x \leq 50$	6	13
$51 \leq x \leq 60$	4	17
$61 \leq x \leq 70$	3	20
$71 \leq x \leq 80$	2	22
$81 \leq x \leq 90$	1	23
Total	**23**	

Catherine's class

Score out of 90	Frequency	Cumulative frequency
$0 \leq x \leq 10$	0	0
$11 \leq x \leq 20$	0	0
$21 \leq x \leq 30$	3	3
$31 \leq x \leq 40$	5	8
$41 \leq x \leq 50$	8	16
$51 \leq x \leq 60$	6	22
$61 \leq x \leq 70$	1	23
$71 \leq x \leq 80$	0	23
$81 \leq x \leq 90$	0	23
Total	**23**	

As the data are grouped, we use the mid-interval values to represent the classes for calculations. For $n = 23$, the quartiles will be the 6th, 12th and 18th data items.

The five-figure summaries for the two classes are:

Thomas
min = 5
$Q_1 = 25$
$Q_2 = 45$
$Q_3 = 65$
max = 85

Catherine
min = 25
$Q_1 = 35$
$Q_2 = 45$
$Q_3 = 55$
max = 65

The box and whisker plots for the two classes are:

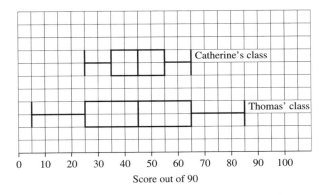

It can be seen that although the median mark is the same for both classes, there is a much greater spread of marks in Thomas' class than in Catherine's class.

Cumulative frequency diagrams

A cumulative frequency diagram, or ogive, is another diagram used to display frequency data. Cumulative frequency goes on the y-axis and the data values go on the x-axis. The points can be joined by straight lines or a smooth curve. The graph is always rising (as cumulative frequency is always rising) and often has an S-shape.

Example

Draw a cumulative frequency diagram for these data:

Age	Frequency	Cumulative frequency
14	3	3
15	4	7
16	8	15
17	5	20
18	6	26
19	3	29
20	1	30
Total	**30**	

By plotting age on the *x*-axis and cumulative frequency on the *y*-axis, plotting the points and then drawing lines between them, we obtain this diagram:

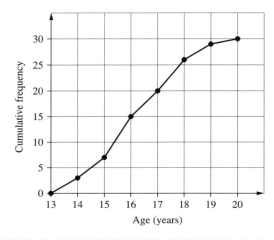

These diagrams are particularly useful for large samples (or populations).

Example

The IBO recorded the marks out of 120 for HL Mathematics and organized the data into a frequency table:

Mark	Frequency	Cumulative frequency
0–20	104	104
21–40	230	334
41–50	506	840
51–60	602	1442
61–70	749	2191
71–80	1396	3587
81–90	2067	5654
91–100	1083	6737
101–120	870	7607

Draw a cumulative frequency diagram for the data.

For grouped data like this, the upper class limit is plotted against the cumulative frequency to create the cumulative frequency diagram:

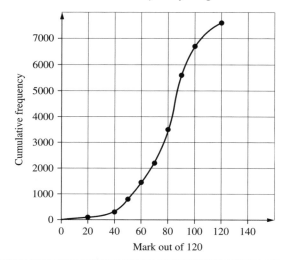

Estimating quartiles and percentiles from a cumulative frequency diagram

We know that the median is a measure of central tendency that divides the data set in half. So the median can be considered to be the data item that is at half of the total frequency. As previously seen, cumulative frequency helps to find this and for large data sets, the median can be considered to be at 50% of the total cumulative frequency, the lower quartile at 25% and the upper quartile at 75%.

These can be found easily from a cumulative frequency diagram by drawing a horizontal line at the desired level of cumulative frequency (*y*-axis) to the curve and then finding the relevant data item by drawing a vertical line to the *x*-axis.

> When the quartiles are being estimated for large data sets, it is easier to use these percentages than to use $\dfrac{n + 1}{4}$ etc.

Example

The cumulative frequency diagram illustrates the data set obtained when the numbers of paper clips in 80 boxes were counted. Estimate the quartiles from the cumulative frequency diagram.

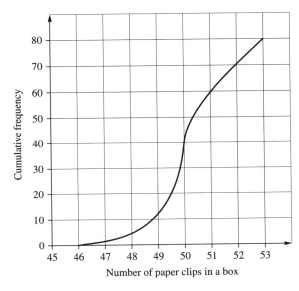

So for this data set,
$Q_1 = 49.5$
$Q_2 = 50$
$Q_3 = 51$

This can be extended to find any percentile. A percentile is the data item that is given by that percentage of the cumulative frequency.

Example

The weights of babies born in December in a hospital were recorded in the table. Draw a cumulative frequency diagram for this information and hence find the median and the 10th and 90th percentiles.

Weight (kg)	Frequency	Cumulative frequency
$2.0 \leq x < 2.5$	1	1
$2.5 \leq x < 3.0$	4	5
$3.0 \leq x < 3.5$	15	20
$3.5 \leq x < 4.0$	38	58
$4.0 \leq x < 4.5$	45	103
$4.5 \leq x < 5.0$	15	118
$5.0 \leq x < 5.5$	2	120

This is the cumulative frequency diagram:

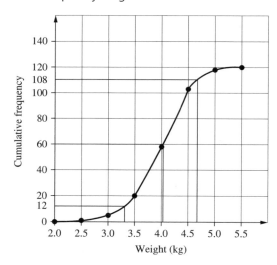

The 10th percentile is given by a cumulative frequency of 10% of $120 = 12$. The median is given by a cumulative frequency of 60 and the 90th percentile is given by a cumulative frequency of 108.

Drawing the lines from these cumulative frequency levels as shown above gives:

90th percentile = 4.7

Median = 4.1

10th percentile = 3.3

Exercise 2

1 The nationalities of students at an international school were recorded and summarized in the frequency table. Draw a bar chart of the data.

Nationality	Frequency
Swedish	85
British	43
American	58
Norwegian	18
Danish	11
Chinese	9
Polish	27
Other	32

2 The ages of members of a golf club are recorded in the table below. Draw a histogram of this data set.

Age	Frequency
$10 < x \le 18$	36
$18 < x \le 26$	24
$26 < x \le 34$	37
$34 < x \le 42$	27
$42 < x \le 50$	20
$50 < x \le 58$	17
$58 < x \le 66$	30
$66 < x \le 74$	15
$74 < x \le 82$	7

3 The contents of 40 bags of nuts were weighed and the results in grams are shown below. Group the data using class intervals $26.5 \le x < 27.5$ etc. and draw a histogram.

28.4	29.2	28.7	29.0	27.1	28.6	30.8	29.9
30.3	30.7	27.6	28.8	29.0	28.1	27.7	30.1
29.4	29.9	31.4	28.9	30.9	29.1	27.8	29.3
28.5	27.9	30.0	29.1	31.2	30.8	29.2	31.1
29.0	29.8	30.9	29.2	29.4	28.7	29.7	30.2

4 The salaries in US$ of teachers in an international school are shown in the table below. Draw a box and whisker plot of the data.

Salary	Frequency
25 000	8
32 000	12
40 000	26
45 000	14
58 000	6
65 000	1

5 The stem and leaf diagram below shows the weights of a sample of eggs. Draw a box and whisker plot of the data.

```
4 | 4  4  6  7  8  9
5 | 0  1  2  4  4  7  8
6 | 1  1  3  6  8
7 | 0  0  2  2  3  4
```

$n = 24$ key: $6\,|\,1$ means 61 grams

6 The Spanish marks of a class in a test out of 30 are shown below.

16	14	12	27	29	21	19	19
15	22	26	29	22	11	12	30
19	20	30	8	25	30	23	21
18	23	27					

a Draw a box and whisker plot of the data.

b Find the mean mark.

7 The heights of boys in a basketball club were recorded. Draw a box and whisker plot of the data.

Height (cm)	Frequency
$140 \leq x < 148$	3
$148 \leq x < 156$	3
$156 \leq x < 164$	9
$164 \leq x < 172$	16
$172 \leq x < 180$	12
$180 \leq x < 188$	7
$188 \leq x < 196$	2

8 The heights of girls in grade 7 and grade 8 were recorded in the table. Draw box and whisker plots of the data and comment on your findings.

Height (cm)	Grade 7 frequency	Grade 8 frequency
$130 \leq x < 136$	5	2
$136 \leq x < 142$	6	8
$142 \leq x < 148$	10	12
$148 \leq x < 154$	12	13
$154 \leq x < 160$	8	6
$160 \leq x < 166$	5	3
$166 \leq x < 172$	1	0

9 The ages of children attending a drama workshop were recorded. Draw a cumulative frequency diagram of the data. Find the median age.

Age	Frequency	Cumulative frequency
11	8	8
12	7	15
13	15	30
14	14	44
15	6	50
16	4	54
17	1	55
Total	**55**	

10 The ages of mothers giving birth in a hospital in one month were recorded. Draw a cumulative frequency diagram of the data. Estimate the median age from your diagram.

Age	Frequency
$14 \leq x < 18$	7
$18 \leq x < 22$	26
$22 \leq x < 26$	54
$26 \leq x < 30$	38
$30 \leq x < 34$	21
$34 \leq x < 38$	12
$38 \leq x < 42$	3

11 A survey was conducted among girls in a school to find the number of pairs of shoes they owned. A cumulative frequency diagram of the data is shown. From this diagram, estimate the quartiles of this data set.

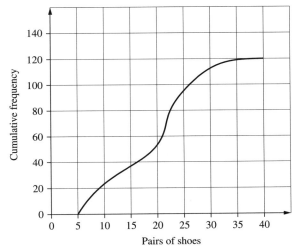

12 The numbers of sweets in a particular brand's packets are counted. The information is illustrated in the cumulative frequency diagram. Estimate the quartiles and the 10th percentile.

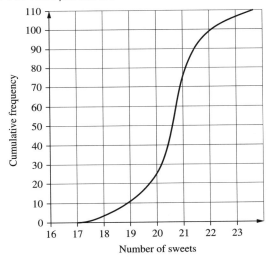

13 There was a competition to see how far girls could throw a tennis ball. The results are illustrated in the cumulative frequency diagram. From the diagram, estimate the quartiles and the 95th and 35th percentiles.

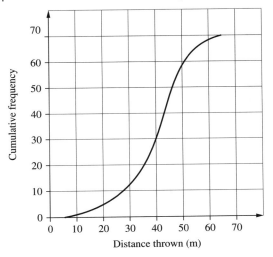

19.3 Measures of dispersion

Consider the two sets of data below, presented as dot plots.

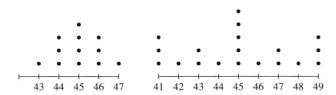

It is quickly obvious that both sets of data have a mean, median and mode of 45 but the two sets are not the same. One of them is much more spread out than the other. This brings us back to the joke at the start of the chapter: it is not only the average that is important about a distribution. We also want to measure the spread of a distribution, and there are a number of measures of spread used in this syllabus.

Diagrams can be useful for obtaining a sense of the spread of a distribution, for example the dot plots above or a box and whisker plot.

There are three measures of dispersion that are associated with the data contained in a box and whisker plot.

The **range** is the difference between the highest and lowest values in a distribution.

$$\text{Range} = \text{maximum value} - \text{minimum value}$$

The **interquartile range** is the difference between the upper and lower quartiles.

$$\text{IQ range} = Q_3 - Q_1$$

The **semi-interquartile range** is half of the interquartile range.

$$\text{Semi-IQ range} = \frac{Q_3 - Q_1}{2}$$

These measures of spread are associated with the median as the measure of central tendency.

Example

Donald and his son, Andrew, played golf together every Saturday for 20 weeks and recorded their scores.

Donald									
81	78	77	78	82	79	80	80	78	79
77	79	79	80	81	78	80	79	78	78
Andrew									
80	73	83	74	72	75	73	77	79	78
84	73	71	75	79	75	73	84	72	74

Draw box and whisker plots of their golf scores, and calculate the interquartile range for each player.

Comment on their scores.

By ordering their scores, we can find the necessary information for the box and whisker plots.

Donald
77 77 78 78 78 78 78 78 79 79 79 79 79 80 80 80 80 81 81 82
↑   ↑   ↑   ↑   ↑

min Q_1 Q_2 Q_3 max

Andrew
71 72 72 73 73 73 73 74 74 75 75 75 77 78 79 79 80 83 84 84
↑   ↑   ↑   ↑   ↑

min Q_1 Q_2 Q_3 max

The box and whisker plots are presented below:

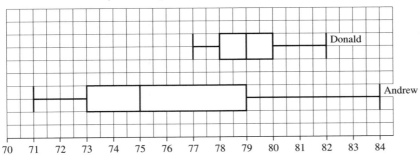

Donald   IQ range $= 80 - 78 = 2$

Andrew   IQ range $= 79 - 73 = 6$

From these statistics, we can conclude that Andrew is, on average, a better player than Donald as his median score is 4 lower than Donald's. However, Donald is a more consistent player as his interquartile range is lower than Andrew's.

Standard deviation

The measures of spread met so far (range, interquartile range and semi-interquartile range) are all connected to the median as the measure of central tendency. The measure of dispersion connected with the mean is known as **standard deviation**.

Here we return to the concepts of population and sample which were discussed at the beginning of this chapter. Most statistical calculations are based on a sample as data about the whole population is not available.

There are different notations for measures related to population and sample.

> The population mean is denoted μ and the sample mean is denoted $\bar{x}$.

Commonly, the sample mean is used to estimate the population mean. This is known as statistical inference. It is important that the sample size is reasonably large and representative of the population. We say that when the estimate is unbiased, $\bar{x}$ is equal to μ.

The standard deviation of a sample is defined to be $s = \sqrt{\dfrac{\sum(x - \bar{x})^2}{n}}$, where n is the sample size.

Standard deviation provides a measure of the spread of the data and comparing standard deviations for two sets of similar data is useful. For most sets of data, the majority of the distribution lies within two standard deviations of the mean. For normal distributions, covered in Chapter 22, approximately 95% of the data lies within two standard deviations of the mean.

The units of standard deviation are the same as the units of the original data.

Example

For the following sample, calculate the standard deviation.

5, 8, 11, 12, 12, 14, 15

It is useful to present this as a table to perform the calculation:

This is the deviation from the mean.

The deviation is then squared so it is positive.

x	$x - \bar{x}$	$(x - \bar{x})^2$
5	−6	36
8	−3	9
11	0	0
12	1	1
12	1	1
14	3	9
15	4	16
Total = 77		Total = 72

$\bar{x} = \dfrac{77}{7} = 11$

From the table, $\sum(x - \bar{x})^2 = 72$

So $s = \sqrt{\dfrac{\sum(x - \bar{x})^2}{n}} = \sqrt{\dfrac{72}{7}} = 3.21$ (to 2 d.p.)

Although the formula above for sample standard deviation is the one most commonly used, there are other forms including this one:

$$s = \sqrt{\dfrac{\sum x^2}{n} - (\bar{x})^2}$$

Example

For the following sample, find the standard deviation.

6, 8, 9, 11, 13, 15, 17

x	x^2
6	36
8	64
9	81
11	121
13	169
15	225
17	289
$\sum x = 79$	$\sum x^2 = 985$

So $s = \sqrt{\dfrac{\sum x^2}{n} - (\overline{x})^2} = \sqrt{\dfrac{985}{7} - \left(\dfrac{79}{7}\right)^2} = 3.65$ (to 2 d.p.)

It is clear that the first method is simpler for calculations without the aid of a calculator.

These formulae for standard deviation are normally applied to a sample. The standard deviation of a population is generally not known and so the sample standard deviation is used to find an estimate.

The notation for the standard deviation of a population is σ.

The standard deviation of a population can be estimated using this formula:

$$\sigma = \sqrt{\frac{n}{n-1}} \times s$$

Variance

Variance is another measure of spread and is defined to be the square of the standard deviation.

So the variance of a sample is s^2 and of a population is σ^2. The formula connecting the standard deviation of a sample and a population provides a similar result for variance:

$$\sigma^2 = \frac{n}{n-1} s^2$$

Example

For the following sample, find the standard deviation. Hence estimate the variance for the population.

8, 10, 12, 13, 13, 16

x	$x - \bar{x}$	$(x - \bar{x})^2$
8	−4	16
10	−2	4
12	0	0
13	1	1
13	1	1
16	4	16
Total = 72		Total = 38

$\bar{x} = \dfrac{72}{6} = 12$

So $s = \sqrt{\dfrac{\sum(x - \bar{x})^2}{n}} = \sqrt{\dfrac{38}{6}} = 2.52$ (to 2 d.p.)

The variance of the sample is $\dfrac{38}{6}$ and so the estimate of the variance of the population is $\dfrac{6}{5} \times \dfrac{38}{6} = \dfrac{38}{5} = 7.6$.

For large samples, with repeated values, it is useful to calculate standard deviation by considering the formula as $s = \sqrt{\dfrac{\sum\limits_{i=1}^{k} f_i(x_i - \bar{x})^2}{n}}$.

Example

Find the standard deviation for this sample and find an estimate for the population from which it comes.

Age	Frequency
16	12
17	18
18	26
19	32
20	17
21	13

Here $\bar{x} = 18.5$

We can still use the table by adding columns.

Age, x	Frequency, f	$x - \bar{x}$	$(x - \bar{x})^2$	$f \times (x - \bar{x})^2$
16	12	−2.5	6.25	75
17	18	−1.5	2.25	40.5
18	26	−0.5	0.25	6.5
19	32	0.5	0.25	8
20	17	1.5	2.25	38.25
21	13	2.5	6.25	81.25
Totals	**118**			**249.5**

$$\sum_{i=1}^{k} f_i(x_i - \bar{x})^2 = 249.5 \quad \text{and} \quad n = \sum f = 118$$

$$\text{So } s = \sqrt{\dfrac{\sum_{i=1}^{k} f_i(x_i - \bar{x})^2}{n}} = \sqrt{\dfrac{249.5}{118}} = 1.45\ldots$$

$$\sigma = \sqrt{\dfrac{118}{117}} \times 1.45\ldots = 1.46$$

Exercise 3

1 For these sets of data, calculate the median and interquartile range.

 a 5, 7, 9, 10, 13, 15, 17

 b 54, 55, 58, 59, 60, 62, 64, 69

 c 23, 34, 45, 56, 66, 68, 78, 84, 92, 94

 d 103, 107, 123, 134, 176, 181, 201, 207, 252

 e

Shoe size	Frequency
37	8
38	14
39	19
40	12
41	24
42	9

2 Compare these two sets of data by calculating the medians and interquartile ranges.

Age	Set A: Frequency	Set B: Frequency
16	0	36
17	0	25
18	37	28
19	34	17
20	23	16
21	17	12
22	12	3
23	9	2
24	6	1

3 University students were asked to rate the quality of lecturing on a scale ranging from 1 (very good) to 5 (very poor). Compare the results for medicine and law students, by drawing box and whisker plots and calculating the interquartile range for each set of students.

Rating	Medicine	Law
1	21	25
2	67	70
3	56	119
4	20	98
5	6	45

4 For these samples, calculate the standard deviation.

   a 5, 6, 8, 10, 11

   b 12, 15, 16, 16, 19, 24

   c 120, 142, 156, 170, 184, 203, 209, 224

   d 15, 17, 22, 25, 28, 29, 30

   e 16, 16, 16, 18, 19, 23, 37, 40

5 Calculate the mean and standard deviation for this sample of ages of the audience at a concert. Estimate the standard deviation of the audience.

Age	Frequency
14	6
15	14
16	18
17	22
18	12
19	8
20	4
21	6
36	3
37	3
38	4

6 The contents of milk containers labelled as 500 ml were measured.
Find the mean and variance of the sample.

Volume (ml)	Frequency
498	4
499	6
500	28
501	25
502	16
503	12
504	8
505	3

7 The lengths of all films (in minutes) shown at a cinema over the period of a year were recorded in the table below. For this data, find:

   a the median and interquartile range

   b the mean and standard deviation.

115	120	118	93	160	117	116	125	98	93
156	114	112	123	100	99	105	119	100	102
134	101	96	92	88	102	114	112	122	100
104	107	109	110	96	91	90	106	111	100
112	103	100	95	92	105	112	126	104	149
125	103	105	100	96	105	177	130	102	100
103	99	123	116	109	114	113	97	104	112

19.4 Using a calculator to perform statistical calculations

Calculators can perform statistical calculations and draw statistical diagrams, normally by entering the data as a list. Be aware of the notation that is used to ensure the correct standard deviation (population or sample) is being calculated.

Example

Draw a box and whisker plot of the following data set, and state the median.

16.4	15.3	19.1	18.7	20.4
15.7	19.1	14.5	17.2	12.6
15.9	19.4	18.5	17.3	13.9

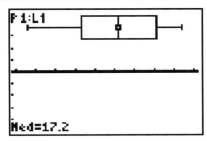

Median = 17.2

Example

Find the mean and standard deviation for this sample of best times (in seconds) for the 200 m at an athletics event. Estimate the standard deviation of the population.

20.51	22.45	23.63	21.91	24.03	23.80	21.98
19.98	20.97	24.19	22.54	22.98	21.84	22.96
20.46	23.86	21.76	23.01	22.74	23.51	20.02

```
1-Var Stats
x̄=22.33952381
Σx=469.13
Σx²=10515.9625
Sx=1.338314149
σx=1.306060876
↓n=21
```

It is important to be careful when using a calculator for standard deviation as the notation used is different to that used in this curriculum. The standard deviation that is given by the formula $s = \sqrt{\dfrac{\sum(x - \bar{x})^2}{n}}$ is σ on the calculator and so $\bar{x} = 22.3$ seconds and $s = 1.31$. An estimate for the population standard deviation is given by Sx on the calculator and hence $\sigma = 1.34$.

Transformations of statistical data

We need to consider the effect of these transformations:

- Adding on a constant c to each data item
- Multiplying each data item by a constant k.

Adding on a constant c to each data item

The mean is the original mean $+ c$.

The standard deviation is unaltered.

Multiplying each data item by a constant k

The mean is multiplied by k.

The standard deviation is multiplied by k.

Example

The salaries of a sample group of oil workers (in US $) are given below:

42 000	55 120	48 650	67 400	63 000
54 000	89 000	76 000	63 000	72 750
71 500	49 500	98 650	74 000	52 500

a What is the mean salary and the standard deviation?

The workers are offered a $2500 salary rise or a rise of 4%.

b What would be the effect of each rise on the mean salary and the standard deviation?

c Which would you advise them to accept?

```
1-Var Stats
x̄=65138
Σx=977070
Σx²=6.70819E10
Sx=15669.68465
σx=15138.35359
↓n=15
■
```

a So the mean salary is $65 100 and the standard deviation is $15 100.

b For a $2500 rise, the mean salary would become $67 600 and the standard deviation would remain at $15 100.

For a 4% rise, this is equivalent to each salary being multiplied by 1.04. So the mean salary would be $67 700 and the standard deviation would be $15 700.

c The $2500 rise would benefit those with salaries below $62 500 (6 out of 15 workers) while the 4% rise would benefit those with higher salaries. The percentage rise would increase the gap between the salaries of these workers. As more workers would benefit from the 4% rise, this one should be recommended.

Exercise 4

1 For these samples, find
   i the quartiles   **ii** the mean and standard deviation.

   a 9.9, 6.7, 10.5, 11.9, 12.1, 9.2, 8.3
   b 183, 129, 312, 298, 267, 204, 301, 200, 169, 294, 263
   c 29 000, 43 000, 63 000, 19 500, 52 000, 48 000, 39 000, 62 500
   d 0.98, 0.54, 0.76, 0.81, 0.62, 0.75, 0.85, 0.75, 0.24, 0.84, 0.98, 0.84, 0.62, 0.52, 0.39, 0.91, 0.63, 0.81, 0.92, 0.72

2 Using a calculator, draw a box and whisker plot of this data set and calculate the interquartile range.

x	Frequency
17	8
18	19
19	26
21	15
30	7

3 Daniel and Paul regularly play ten-pin bowling and record their scores.

Using a calculator, draw box and whisker plots to compare their scores, and calculate the median and range of each.

Daniel

185	202	186	254	253	212	109	186	276	164
112	243	200	165	172	199	166	231	210	175
163	189	182	120	204	225	185	174	144	122

Paul

240	176	187	199	169	201	205	210	195	190
210	213	226	223	218	205	187	182	181	169
172	174	200	198	183	192	190	201	200	211

4 Karthik has recorded the scores this season for his innings for the local cricket team.

   a Calculate his mean score and his standard deviation.

64	0	102	8	83	52
1	44	64	0	73	26
50	24	40	44	36	12

   b Karthik is considering buying a new bat which claims to improve batting scores by 15%. What would his new mean and standard deviation be?

5 Mhairi records the ages of the members of her chess club in a frequency table.

Age	Frequency
12	8
13	15
14	17
15	22
16	19
17	8

If the membership remains the same, what will be the mean age and standard deviation in two years' time?

Review exercise

 1 State whether the data is discrete or continuous.

    a Height of girls   **b** Number of boys playing different sports

    c Sizes of shoes stocked in a store   **d** Mass of bicycles

 2 Jenni did a survey of the colours of cars owned by the students in her class and found the following information:

Blue	Black	Silver	Red	Red	Silver	Black	White	White	Black
Green	Red	Blue	Red	Silver	Yellow	Black	White	Blue	Red
Blue	Silver	Blue	Red	Silver	Black	Red	White	Red	Silver

Construct a frequency table for this information and state the modal colour of car for this class.

 3 Katie has recorded the lengths of snakes for her Group 4 project.

Length of snake (cm)	Frequency
$30 \le l < 45$	2
$45 \le l < 60$	8
$60 \le l < 75$	22
$75 \le l < 90$	24
$90 \le l < 105$	10
$105 \le l < 120$	3

What is the mean length of snakes in Katie's sample? What is the standard deviation?

 4 Nancy records how many clubs each child in the school attends in a frequency distribution. Find the mean number of clubs attended.

Number of clubs, x	0	1	2	3	4
Frequency	40	64	36	28	12

 5 The heights of students at an international school are shown in the frequency table. Draw a histogram of this data.

Height	Frequency
$1.20 \le h < 1.30$	18
$1.30 \le h < 1.40$	45
$1.40 \le h < 1.50$	62
$1.50 \le h < 1.60$	86
$1.60 \le h < 1.70$	37
$1.70 \le h < 1.80$	19

6 A class's marks out of 60 in a history test are shown below.

  a Draw a box plot of this data.

  b Calculate the interquartile range.

  c Find the mean mark.

58	34	60	21	45	44	29	55
34	48	41	40	36	38	39	29
59	36	37	45	49	51	27	12
57	51	52	32	37	51	33	30

7 A survey was conducted among students in a school to find the number of hours they spent on the internet each week. A cumulative frequency diagram of the data is shown. From this diagram, estimate the quartiles of the data set.

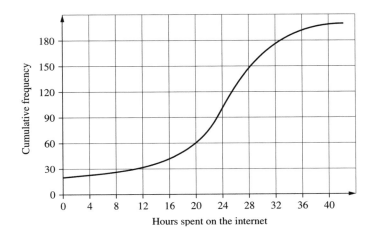

Hours spent on the internet

8 The number of goals scored by a football team in each match is shown below. For this data, find

  a the median and interquartile range

  b the mean and standard deviation.

0	3	2	1	1	0	3	4	2	2
0	2	1	1	0	1	3	1	2	0
7	2	1	0	5	1	1	0	4	3
1	2	1	0	0	1	2	3	1	1

9 The weekly wages of a group of employees in a factory (in £) are shown below.

208	220	220	265	208	284	312	296	284
220	364	300	285	240	220	290	275	264

  a Find the mean wage, and the standard deviation.

The following week, they all receive a 12% bonus for meeting their target.

  b What is the mean wage and standard deviation as a result?

10 A machine produces packets of sugar. The weights in grams of 30 packets chosen at random are shown below.

Weight (g)	29.6	29.7	29.8	29.9	30.0	30.1	30.2	30.3
Frequency	2	3	4	5	7	5	3	1

Find unbiased estimates of
a the mean of the population from which this sample is taken
b the standard deviation of the population from which this sample is taken.

[IB May 01 P1 Q6]

11 The 80 applicants for a sports science course were required to run 800 metres and their times were recorded. The results were used to produce the following cumulative frequency graph.

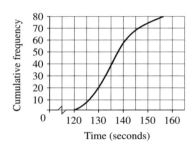

Estimate
a the median
b the interquartile range. [IB May 02 P1 Q14]

12 A teacher drives to school. She records the time taken on each of 20 randomly chosen days. She finds that,

$$\sum_{i=1}^{20} x_i = 626 \text{ and } \sum_{i=1}^{20} x_i^2 = 19780.8$$

where x_i denotes the time, in minutes, taken on the ith day.
Calculate an unbiased estimate of

a the mean time taken to drive to school
b the variance of the time taken to drive to school. [IB May 03 P1 Q19]

13 The cumulative frequency curve below indicates the amount of time 250 students spend eating lunch.

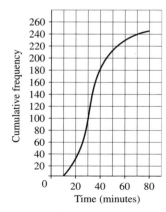

a Estimate the number of students who spend between 20 and 40 minutes eating lunch.
b If 20% of the students spend more than x minutes eating lunch, estimate the value of x. [IB Nov 03 P1 Q2]

20 Probability

Have you ever asked around a group of your classmates and been surprised to find that two of them share the same birthday? At first thought it seems likely that for two people in the room to share a birthday there will need to be a lot of people in the room. However, this is not actually the case. If we want to have more than a 50% chance of two people in the room having the same birthday then we can calculate, using probability, that the number required is 23 or more. However, if we change the problem to having 50% chance of finding someone in the room with the same birthday as you, then the number required increases dramatically – approximately 254 people are needed. Hence it should come as no surprise that two presidents of the United States of America have shared the same birthday and three presidents have died on the same day!

The way in which we can investigate problems like this is by using probability, but we will begin with much simpler problems.

20.1 Introduction to probability

In the presumed knowledge section (Chapter 0 – see accompanying CD) we considered the idea that when we look at an experimental situation we find answers that indicate that a theoretical application is appropriate. This theoretical approach is called probability and is what we will explore in this chapter. Consider a number of equally likely outcomes of an event. What is the probability of one specific outcome of that event? For example, if we have a cubical die what is the probability of throwing a six? Since there are six equally likely outcomes and only one of them is throwing a six, then the probability of throwing a six is 1 in 6. We would normally write this as a fraction $\frac{1}{6}$ or as a decimal or a percentage. Since probability is a theoretical concept, it does not mean that if we throw a die six times we will definitely get a six on one of the throws. However, as the number of trials increases, the number of sixes becomes closer to $\frac{1}{6}$ of the total.

Generally, if the probability space S consists of a finite number of equally likely outcomes, then the probability of an event E, written $P(E)$ is defined as:

$$P(E) = \frac{n(E)}{n(S)}$$ where $n(E)$ is the number of occurrences of the event E and $n(S)$ is the total number of possible outcomes.

Hence in a room of fifteen people, if seven of them have blue eyes, then the probability that a person picked at random will have blue eyes is $\frac{7}{15}$.

Important results

$$0 \leq P(A) \leq 1$$

If an event A can never happen, the probability is 0, and if it will certainly happen, the probability is 1. This can be seen from the fact that $P(A) = \dfrac{n(A)}{n(S)}$ since if the event can never happen then $n(A) = 0$ and if the event is certain to happen then $n(A) = n(S)$. Since $n(S)$ is always greater than or equal to $n(A)$, a probability can never be greater than 1.

A' is known as the complement of A.

$P(A) + P(A') = 1$ where $P(A')$ is the probability that the event A does not occur.

Example

A bag contains a large number of tiles. The probability that a tile drawn from the bag shows the letter A is $\dfrac{3}{10}$, the probability that a tile drawn from the bag shows the letter B is $\dfrac{5}{10}$, and the probability that a tile drawn from the bag shows the letter C is $\dfrac{2}{10}$.

What is the probability that a tile drawn at random from the bag

  a shows the letter A or B

  b does not show the letter C?

  a Since the probability of showing the letter A is $\dfrac{3}{10}$ and the probability of showing the letter B is $\dfrac{5}{10}$, the probability of showing the letter A or B is

$$\dfrac{3}{10} + \dfrac{5}{10} = \dfrac{8}{10} = \dfrac{4}{5}.$$

  b The probability of not showing the letter C can be done in two ways. We can use the complement and state that

$$P(\text{not } C) = P(C') = 1 - P(C) = 1 - \dfrac{2}{10} = \dfrac{4}{5}.$$

Alternatively, we can see that this is the same as showing a letter A or B and hence is the same as the answer to part **a**.

$$P(A \text{ or } B) = P(A) + P(B) - P(A \text{ and } B)$$
or using set notation: $P(A \cup B) = P(A) + P(B) - P(A \cap B)$

In this case the two intersecting sets A and B represent the events A and B and the universal set represents the sample space S. This is shown in the Venn diagram. Venn diagrams and set notation are introduced in the presumed knowledge chapter (Chapter 0 – see accompanying CD).

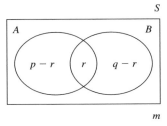

Proof

If $n(S) = m$, $n(A) = p$, $n(B) = q$ and $n(A \cap B) = r$ then

$$P(A \cup B) = \frac{n(A \cup B)}{n(S)}$$

$$= \frac{(p - r) + r + (q - r)}{m}$$

$$= \frac{p}{m} + \frac{q}{m} - \frac{r}{m}$$

$$= P(A) + P(B) - P(A \cap B)$$

Example

A tetrahedral die and a cubical die are thrown. What is the probability of throwing a five on the cubical die or a four on the tetrahedral die?

We could use the formula above directly, but we will demonstrate what is happening here by using a possibility space diagram that shows all the possible outcomes of the event "throwing a cubical die and a tetrahedral die".

```
4 │(4,1)  (4,2)  (4,3)  (4,4)  (4,5)  (4,6)

3 │(3,1)  (3,2)  (3,3)  (3,4)  (3,5)  (3,6)

2 │(2,1)  (2,2)  (2,3)  (2,4)  (2,5)  (2,6)

1 │(1,1)  (1,2)  (1,3)  (1,4)  (1,5)  (1,6)
  └──────────────────────────────────────
    1      2      3      4      5      6
```

From the diagram we see that the probability of throwing a five on the cubical die is $\frac{4}{24}$.

The probability of throwing a four on the tetrahedral die is $\frac{6}{24}$.

Hence at first it may appear that the probability of throwing a five on the cubical die or a four on the tetrahedral die is $\frac{6}{24} + \frac{4}{24}$. However, from the diagram we notice that the occurrence four on the tetrahedral die and five on the cubical die appears in both calculations. Hence we need to subtract this probability. Therefore the probability of throwing a five on the cubical die or a four on the tetrahedral die is $\frac{6}{24} + \frac{4}{24} - \frac{1}{24} = \frac{9}{24} = \frac{3}{8}$.

Example

In a group of 20 students, there are 12 girls and 8 boys. Two of the boys and three of the girls wear red shirts. What is the probability that a person chosen randomly from the group is either a boy or someone who wears a red shirt?

Let A be the event "being a boy" and B be the event "wearing a red shirt".

Hence $P(A) = \frac{8}{20}$, $P(B) = \frac{5}{20}$ and $P(A \cap B) = \frac{2}{20}$.

Now

$$P(A \cup B) = P(A) + P(B) - P(A \cap B)$$
$$\Rightarrow P(A \cup B) = \frac{8}{20} + \frac{5}{20} - \frac{2}{20} = \frac{11}{20}$$

If an event A can occur or an event B can occur but A and B cannot both occur, then the two events A and B are said to be mutually exclusive.
In this case $P(A \text{ and } B) = P(A \cap B) = 0$.

We can see this from the Venn diagram, where there is no overlap between the sets.

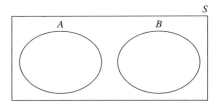

For mutually exclusive events $P(A \cup B) = P(A) + P(B)$.

Example

A bag contains 3 red balls, 4 black balls and 3 yellow balls. What is the probability of drawing either a red ball or a black ball from the bag?

Let the event "drawing a red ball" be A and the event "drawing a black ball" be B.

$P(A) = \frac{3}{10}$ and $P(B) = \frac{4}{10}$

Since these are mutually exclusive events

$$P(A \cup B) = P(A) + P(B)$$
$$\Rightarrow P(A \cup B) = \frac{3}{10} + \frac{4}{10} = \frac{7}{10}$$

If two events A and B are such that $A \cup B = S$, where S is the total probability space, then $P(A \cup B) = 1$ and the events A and B are said to be exhaustive.

Example

The events A and B are exhaustive. If $P(A) = 0.65$ and $P(B) = 0.44$, find $P(A \cap B)$.

We know that $P(A \cup B) = P(A) + P(B) - P(A \cap B)$.
Since the events are exhaustive $P(A \cup B) = 1$
$\Rightarrow 1 = 0.65 + 0.44 - P(A \cap B)$
$\Rightarrow P(A \cap B) = 0.09$

Exercise 1

1 An unbiased tetrahedral die is thrown. What is the probability of throwing
   a a three
   b an even number
   c a prime number?

2 A spinner has the numbers 1 to 10 written on it. When spun, it is equally likely to stop on any of the ten numbers. What is the probability that it will stop on
   a a three
   b an odd number
   c a multiple of 3
   d a prime number?

3 A bag contains 5 black balls, 6 white balls, 7 pink balls and 2 blue balls. What is the probability that a ball drawn randomly from the bag will
   a be a black ball
   b be either a white or a pink ball
   c not be a blue ball
   d be either a black, white or blue ball
   e be a red ball?

4 Sheila is picking books off her shelf. The shelf only contains mathematics books and novels. The probability that she picks a novel is 0.48 and the probability that she picks a mathematics book is 0.52.
   a What is the probability that she does not pick a mathematics book?
   b Explain why these events are exhaustive.

5 If $P(A) = \frac{1}{5}$, $P(B) = \frac{2}{3}$ and $P(A \cap B) = \frac{4}{15}$, are A and B exhaustive events?

6 The probability that Hanine goes to the local shop is $\frac{3}{7}$. The probability that she does not cycle is $\frac{4}{11}$. The probability that she goes to the shop and cycles is $\frac{4}{15}$.
   a What is the probability that she cycles?
   b What is the probability that she cycles or goes to the shop?

7 It is given that for two events A and B, $P(A) = \dfrac{3}{8}$, $P(A \cup B) = \dfrac{11}{16}$ and $P(A \cap B) = \dfrac{3}{16}$. Find $P(B)$.

8 In a class, 6 students have brown eyes, 3 students have blue eyes, 4 students have grey eyes and 2 students have hazel eyes. A student is chosen at random. Find the probability that
  a a student with blue eyes is chosen
  b a student with either blue or brown eyes is chosen
  c a student who does not have hazel eyes is chosen
  d a student with blue, brown or grey eyes is chosen
  e a student with grey or brown eyes is chosen.

9 Two tetrahedral dice are thrown. What is the probability that
  a the sum of the two scores is 5
  b the sum of the two scores is greater than 4
  c the difference between the two scores is 3
  d the difference between the two scores is less than 4
  e the product of the two scores is an even number
  f the product of the two scores is greater than or equal to 6
  g one die shows a 3 and the other die shows a number greater than 4?

10 Two cubical dice are thrown. What is the probability that
  a the sum of the two scores is 9
  b the sum of the two scores is greater than 4
  c the difference between the two scores is 3
  d the difference between the two scores is at least 4
  e the product of the two scores is 12
  f the product of the two scores is an odd number
  g the first die shows an even number or the second die shows a multiple of 3?

11 The probability that John passes his mathematics examination is 0.9, and the probability that he passes his history examination is 0.6. These events are exhaustive. What is the probability that
  a he does not pass his mathematics examination
  b he passes his history examination or his mathematics examination
  c he passes his mathematics examination and his history examination?

12 In a school's IB diploma programme, 30 students take at least one of physics or chemistry. If 15 students take physics and 18 students take chemistry, find the probability that a student chosen at random studies both physics and chemistry.

13 There are 20 students in a class. In a class survey on pets, it is found that 12 students have a dog, 5 students have a dog and a rabbit and 3 students do not have a dog or a rabbit. Find the probability that a student chosen at random will have a rabbit.

14 In a survey of people living in a village, all respondents either shop at supermarket A, supermarket B or both. It is found that the probability that a person will shop at supermarket A is 0.65 and the probability that he/she will shop at supermarket B is 0.63. If the probability that a person shops at both supermarkets is 0.28, find the probability that a person from the village chosen at random will shop at supermarket A or supermarket B, but not both.

15 Two cubical dice are thrown. What is the probability that the sum of the two scores is
  a a multiple of 3
  b greater than 5
  c a multiple of 3 and greater than 5

d a multiple of 3 or greater than 5

e less than 4 or one die shows a 5?

f Explain why the events in part **e** are mutually exclusive.

16 A class contains 15 boys and 17 girls. Of these 10 boys and 8 girls have blonde hair. Find the probability that a student chosen at random is a boy or has blonde hair.

17 Two tetrahedral dice are thrown. What is the probability that the sum of the scores is

a even

b prime

c even or prime?

18 When David goes fishing the probability of him catching a fish of type A is 0.45, catching a fish of type B is 0.75 and catching a fish of type C is 0.2. David catches four fish.

If the event X is David catching two fish of type A and two other fish, the event Y is David catching two fish of type A and two of type B and the event Z is David catching at least one fish of type C, for each of the pairs of X, Y and Z state whether the two events are mutually exclusive, giving a reason.

19 If A and B are exhaustive events, and $P(A) = 0.78$ and $P(B) = 0.37$, find $P(A \cap B)$.

20 A whole number is chosen from the numbers 1 to 500. Find the probability that the whole number is

a a multiple of 6

b a multiple of both 6 and 8.

20.2 Conditional probability

If A and B are two events, then the probability of A given that B has already occurred is written as $P(A|B)$. This is known as **conditional probability**.

$$P(A|B) = \frac{P(A \cap B)}{P(B)}$$

On the Venn diagram below the possibility space is the set B as this has already occurred.

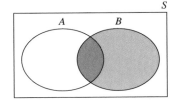

Hence $P(A|B) = \dfrac{n(A \cap B)}{n(B)}$

If all the possible events are represented by the universal set S, then

$$P(A|B) = \frac{\dfrac{n(A \cap B)}{n(S)}}{\dfrac{n(B)}{n(S)}}$$

$$\Rightarrow P(A|B) = \frac{P(A \cap B)}{P(B)}$$

Example

A card is picked at random from a pack of 20 cards numbered 1, 2, 3, ..., 20. Given that the card shows an even number, find the probability that it is a multiple of 4.

Let the event A be "picking a card showing a multiple of 4" and let the event B be "picking a card showing an even number".

Hence we require $P(A|B) = \dfrac{P(A \cap B)}{P(B)}$

In this case $P(A \cap B) = \dfrac{5}{20}$ and $P(B) = \dfrac{10}{20}$

$\Rightarrow P(A|B) = \dfrac{\frac{5}{20}}{\frac{10}{20}} = \dfrac{1}{2}$

Alternatively, we could write the result as $P(A \cap B) = P(A|B) \times P(B)$.

Example

Two tetrahedral dice are thrown; one is red and the other is blue. The faces are marked 1, 2, 3, 4. Given that the red die lands on an odd number, the probability that the sum of the scores on the dice is 6 is $\dfrac{1}{8}$. Find the probability that the sum of the scores on the dice is 6 and the red die lands on an odd number, assuming the red die is fair.

Let A be the event "the sum of the scores is 6" and let B be the event "the red die lands on an odd number".

We know $P(A|B) = \dfrac{1}{8}$ and $P(B) = \dfrac{1}{2}$

Hence $P(A \cap B) = P(A|B) \times P(B) = \dfrac{1}{8} \times \dfrac{1}{2} = \dfrac{1}{16}$

If A and B are mutually exclusive events then since $P(A \cap B) = 0$ and $P(B) \neq 0$, it follows that $P(A|B) = 0$.

Exercise 2

1 For two events A and B it is given that $P(A) = \dfrac{5}{18}$, $P(B) = \dfrac{5}{9}$ and $P(A|B) = \dfrac{3}{14}$. Find

    a $P(A \cap B)$ **b** $P(B|A)$

2 A bag contains 6 balls, each with a number between 4 and 9 written on it. Each ball has a different number written on it. Find the probability that if two balls are drawn

    a the sum of the scores is greater than 12

    b the second ball shows a 7, given that the sum of the scores is greater than 12

    c the first ball is even, given that the difference between the numbers is 3.

3 Two tetrahedral dice are thrown. Find the probability that

    a at least one of the dice shows a 3

    b the difference between the scores on the two dice is 2

c given that at least one of the dice shows a 3, the difference between the scores on the dice is 2

d given that the difference between the scores on the dice is 2, the product of the scores on the dice is 8.

4 In a game of Scrabble, Dalene has the seven letters A, D, E, K, O, Q and S. She picks two of these letters at random.

   a What is the probability that one is a vowel and the other is the letter D?

   b If the first letter she picks is a consonant, what is the probability that the second letter is the E?

   c Given that she picks the letter Q first, what is the probability that she picks the letter D or the letter K second?

5 There are ten discs in a bag. Each disc has a number on it between 0 and 9. Each number only appears once. Hamish picks two discs at random. Given that the first disc drawn shows a multiple of 4, what is the probability that

   a the sum of the numbers on the two discs is less than 10

   b the sum of the numbers on the two discs is even

   c the difference between the two numbers on the discs is less than 3?

6 On any given day in June the probability of it raining is 0.24. The probability of Suzanne cycling to work given that it is raining is 0.32. Find the probability that Suzanne cycles to work and it is raining.

7 Events A and B are such that $P(A) = \dfrac{4}{13}$ and $P(B) = \dfrac{9}{16}$. The conditional probability $P(A|B) = 0$.

   a Find $P(A \cup B)$.

   b Are A and B exhaustive events? Give a reason for the answer.

8 The probability of Nick gaining a first class degree at university given that he does 25 hours revision per week is 0.85. The probability that he gains a first class degree and does 25 hours revision per week is 0.7. Find the probability that he does 25 hours revision.

9 A team of two is to be picked from Alan, Bruce, Charlie and Danni.

   a Draw a possibility space diagram to show the possible teams of two.

   b What is the probability that if Danni is chosen, either Alan or Bruce will be her partner?

   c Given that exactly one of Alan or Bruce are chosen, what is the probability that Danni will be the other person?

20.3 Independent events

If the occurrence or non-occurrence of an event A does not influence in any way the probability of an event B then the event B is said to be **independent** of event A.

In this case $P(B|A) = P(B)$.

Now we know that $P(A \cap B) = P(B|A)P(A)$

$\Rightarrow P(A \cap B) = P(B) \times P(A) = P(A) \times P(B)$

> This is only true if A and B are independent events.

For independent events, $P(A \text{ and } B) = P(A) \times P(B)$.

To tackle questions on independent events we sometimes use a possibility space diagram but a more powerful tool is a **tree diagram**.

Example

A man visits his local supermarket twice in a week. The probability that he pays by credit card is 0.4 and the probability that he pays with cash is 0.6. Find the probability that

a he pays cash on both visits

b he pays cash on the first visit and by credit card on the second visit.

We will use A to mean paying by cash and B to mean paying with a credit card. The tree diagram is shown below.

First visit	Second visit	Outcome	Probability
	A *	AA	0.36
	B ▵	AB	0.24
	A	BA	0.24
	B	BB	0.16

The probabilities of his method of payment are written on the branches.
By multiplying the probabilities along one branch we find the probability of one outcome. Hence in this situation there are four possible outcomes. If we add all the probabilities of the outcomes together, the answer will be 1.

a In this case we need the branch marked *.

$P(AA) = 0.6 \times 0.6 = 0.36$ ....

b In this case we need the branch marked ▵.

$P(AB) = 0.6 \times 0.4 = 0.24$

> $P(AA)$ means paying by cash on the first visit and on the second visit.

Example

A bag contains 3 red sticks, 5 white sticks and 2 blue sticks. A stick is taken from the bag, the colour noted then replaced in the bag. Another stick is then taken. Find the probability that

a both sticks are red

b a blue stick is drawn first and then a white stick

c one blue stick and one white stick are taken

d at least one stick is blue.

We will use R to mean taking a red stick, W to mean taking a white stick and B to mean taking a blue stick. The tree diagram is shown below.

First selection	Second selection	Outcome	Probability
	R * □	RR	$\frac{9}{100}$
	W □	RW	$\frac{15}{100}$
	B	RB	$\frac{6}{100}$
	R □	WR	$\frac{15}{100}$
	W □	WW	$\frac{25}{100}$
	B ○	WB	$\frac{10}{100}$
	R	BR	$\frac{6}{100}$
	W ▵	BW	$\frac{10}{100}$
	B	BB	$\frac{4}{100}$

The probability of taking a certain colour of stick is written on the appropriate branch.

By multiplying the probabilities along one branch we find the probability of one outcome. Hence there are nine possible outcomes. If we add all the probabilities of the outcomes together, the answer will be 1.

a We need the branch marked *.

$$P(RR) = \frac{3}{10} \times \frac{3}{10} = \frac{9}{100}$$

P(RR) means the probability of taking a red stick and then taking another red stick.

b We need the branch marked △.

$$P(BW) = \frac{2}{10} \times \frac{5}{10} = \frac{10}{100} = \frac{1}{10}$$

c The order in which we take the blue stick and the white stick does not matter. Thus we require the probability of taking a blue stick followed by a white stick and the probability of taking a white stick followed by a blue stick. Hence we use two separate branches and then add the answers. The two branches required are marked △ and ○.

$$P(BW) + P(WB) = \left(\frac{2}{10} \times \frac{5}{10}\right) + \left(\frac{5}{10} \times \frac{2}{10}\right) = \frac{20}{100} = \frac{1}{5}$$

d We could add all the branches that contain an event B. Since we know the total probability is 1, it is actually easier to subtract the probabilities of those branches that do not contain an event B from 1. These are marked ☐.

$$P(\text{at least one blue}) = 1 - \{P(RR) + P(WW) + P(RW) + P(WR)\}$$

$$\Rightarrow P(\text{at least one blue}) = 1 - \left\{\left(\frac{3}{10} \times \frac{3}{10}\right) + \left(\frac{5}{10} \times \frac{5}{10}\right) + \left(\frac{3}{10} \times \frac{5}{10}\right)\right.$$

$$\left. + \left(\frac{5}{10} \times \frac{3}{10}\right)\right\}$$

$$\Rightarrow P(\text{at least one blue}) = \frac{36}{100} = \frac{9}{25}$$

This is an example of sampling with replacement, which means that the probabilities do not change from one event to the subsequent event. Below is an example of sampling with replacement where they do change.

Example

A box contains 3 red balls, 5 blue balls and 4 yellow balls. Keith draws a ball from the box, notes its colour and discards it. He then draws another ball from the box and again notes the colour. Find the probability that
a both balls are yellow
b Keith draws a blue ball the first time and a yellow ball the second time
c Keith draws a red ball and a blue ball.

We will use R to mean drawing a red ball, B to mean drawing a blue ball and Y to mean drawing a yellow ball. The tree diagram is shown below.

First selection	Second selection	Outcome	Probability

Tree diagram:

First selection branches from start: $\frac{3}{12}$ to R, $\frac{5}{12}$ to B, $\frac{4}{12}$ to Y.

From R: $\frac{2}{11}$ to R (Outcome RR, Probability $\frac{6}{132}$); $\frac{5}{11}$ to B □ (Outcome RB, Probability $\frac{15}{132}$); $\frac{4}{11}$ to Y (Outcome RY, Probability $\frac{12}{132}$).

From B: $\frac{3}{11}$ to R ○ (Outcome BR, Probability $\frac{15}{132}$); $\frac{4}{11}$ to B (Outcome BB, Probability $\frac{20}{132}$); $\frac{4}{11}$ to Y Δ (Outcome BY, Probability $\frac{20}{132}$).

From Y: $\frac{3}{11}$ to R (Outcome YR, Probability $\frac{12}{132}$); $\frac{5}{11}$ to B (Outcome YB, Probability $\frac{20}{132}$); $\frac{3}{11}$ to Y * (Outcome YY, Probability $\frac{12}{132}$).

a We need the branch marked *.

$$P(YY) = \frac{4}{12} \times \frac{3}{11} = \frac{1}{11}$$

b We need the branch marked Δ.

$$P(BY) = \frac{5}{12} \times \frac{4}{11} = \frac{20}{132} = \frac{5}{33}$$

c The order in which Keith takes the red ball and the blue ball does not matter. We require the probability of taking a red ball followed by a blue ball and the probability of taking a blue ball followed by a red. Hence we use two separate branches and then add the answers. The two branches required are marked □ and ○.

$$P(RB) + P(BR) = \left(\frac{3}{12} \times \frac{5}{11}\right) + \left(\frac{5}{12} \times \frac{3}{11}\right) = \frac{30}{132} = \frac{5}{22}$$

We can also use tree diagrams to help us with conditional probability.

Example

In the school canteen 60% of students have salad as a starter. Of the students who have salad as a starter, 30% will have cheesecake as dessert. Of those who do not have salad as a starter, 70% will have cheesecake as dessert.
   a Show this information on a tree diagram.
   b Given that Levi chooses to have salad as a starter, what is the probability that he will choose cheesecake as dessert?
   c Given that Levi does not have cheesecake as dessert, what is the probability that he chose salad as a starter?

   a We will use S to mean having salad and C to mean having cheesecake.

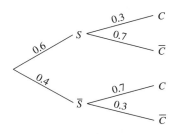

b We want to find $P(C|S)$. We could use the formula here, but it can be seen directly from the tree diagram that $P(C|S) = 0.3$.

By the formula

$$P(C|S) = \frac{P(C \cap S)}{P(S)} = \frac{P(S \cap C)}{P(S)}$$

$$= \frac{0.6 \times 0.3}{0.6} = 0.3$$

c Here we want $P(S|\bar{C})$ and in this situation it is easier to use the formula.

$$P(S|\bar{C}) = \frac{P(S \cap \bar{C})}{P(\bar{C})}$$

$$= \frac{0.6 \times 0.7}{(0.6 \times 0.7) + (0.4 \times 0.3)} = \frac{7}{9}$$

Exercise 3

1 In a mathematics test the probability that Aly scores more than 70% is 0.6. In a physics test the probability that he scores more than 70% is 0.5. What is the probability that
  a he scores more than 70% in both tests
  b he scores more than 70% in only one test?

2 A cubical die is thrown twice.
  a Draw a tree diagram to show the outcomes "throwing a three" and "not throwing a three".
  b What is the probability that both dice show a three?
  c What is the probability that neither dice shows a three?
  d Draw a tree diagram to show the outcomes "throwing a number less than four" and "throwing a number greater than or equal to four".
  e What is the probability that only one die shows a number less than four?
  f What is the probability that at least one die shows a number less than four?

3 On any particular day, the probability that it rains is 0.2. The probability that a soccer team will win is 0.6 if it is raining and 0.7 if it is not raining. The team plays once in a week.
  a Draw a tree diagram to show these events and their outcomes.
  b What is the probability that it will rain and the team will win?
  c What is the probability that the team will not win?
  d Given that it is not raining, what is the probability that they will not win?
  e Given that they win, what is the probability that it was raining?

4 Three fair coins are tossed. Each coin can either land on a head or a tail. What is the probability of gaining
  a three heads
  b two heads and a tail

   c a tail on the first toss followed by a head or a tail in either order on the second and third tosses

   d at least one tail?

5 Two fair coins are tossed. Each coin can either land on a head or a tail.

   a Show the possible outcomes on a tree diagram.

   b What is the probability of getting at least one head?

   A third coin is now tossed which is twice as likely to show heads as tails.

   c Add an extra set of branches to the tree diagram to show the possible outcomes.

   d What is the probability of getting two heads and a tail?

   e What is the probability of getting two tails and a head, given that the third coin lands on tails?

6 The letters of the word PROBABILITY are placed in a bag. A letter is selected, it is noted whether it is a vowel or a consonant, and returned to the bag. A second letter is then selected and the same distinction is noted.

   a Draw a tree diagram to show the possible outcomes.

   b What is the probability of noting two consonants?

   c What is the probability of noting a vowel and a consonant?

7 A box contains 4 blue balls, 3 red balls and 5 green balls. Three balls are drawn from the box without replacement. What is the probability that

   a all three balls are green

   b one ball of each colour is drawn

   c at least one blue ball is drawn

   d a pink ball is drawn

   e no red balls are drawn?

   f Given that the second ball is blue, what is the probability that the other two are either both red, both green, or one each of red or green?

8 Bag A contains 6 blue counters and 4 green counters. Bag B contains 9 blue counters and 5 green counters. A counter is drawn at random from bag A and two counters are drawn at random from bag B. The counters are not replaced.

   a Find the probability that the counters are all blue.

   b Find the probability that the counters are all the same colour.

   c Given that there are two blue counters and one green counter, what is the probability that the green counter was drawn from bag B?

9 Events A and B are such that $P(B) = \dfrac{2}{5}$, $P(A|B) = \dfrac{1}{3}$ and $P(A \cup B) = \dfrac{4}{5}$.

   a Find $P(A \cap B)$.

   b Find $P(A)$.

   c Show that A and B are not independent.

10 Six cards a placed face down on a table. Each card has a single letter on it. The six letters on the cards are B, H, K, O, T and U. Cards are taken from the table and not replaced. Given that the first card drawn shows a vowel, what is the probability that the second card shows

   a the letter B

   b one of the first ten letters of the alphabet

   c the letter T or the letter K?

11 Zahra catches the train to school every day from Monday to Friday. The probability that the train is late on a Monday is 0.35. The probability that it is late on any other day is 0.42. A day is chosen at random. Given that the train is late that day, what is the probability that the day is Monday?

12 Jane and John are playing a game with a biased cubical die. The probability that the die lands on any even number is twice that of the die landing on any odd number. The probability that the die lands on an even number is $\frac{2}{9}$. If the die shows a 1, 2, 3 or 4 the player who threw the die wins the game. If the die shows a 5 or a 6 the other player has the next throw. Jane plays first and the game continues until there is a winner.

   a What is the probability that Jane wins on her first throw?

   b What is the probability that John wins on his first throw?

   c Calculate the probability that Jane wins the game.

13 The events A and B are independent. If $P(A) = 0.4$ and $P(B|A) = 0.2$, find

   a $P(B)$

   b the probability that A occurs or B occurs, but not both A and B.

14 Janet has gone shopping to buy a new dress. To keep herself entertained whilst shopping she is listening to her iPod, which she takes off when she tries on a new dress. The probability that she leaves her iPod in the shop is 0.08. After visiting two shops in succession she finds she has left her iPod in one of them. What is the probability that she left her iPod in the first shop?

15 A school selects three students at random from a shortlist of ten students to be prefects. There are six boys and four girls.

   a What is the probability that no girl is selected?

   b Find the probability that two boys and one girl are selected.

16 A and B are two independent events. $P(A) = 0.25$ and $P(B) = 0.12$. Find

   a $P(A \cap B)$ **b** $P(A \cup B)$ **c** $P(A|B)$ **d** $P(B|A)$

17 A soccer player finds that when the weather is calm, the probability of him striking his target is 0.95. When the weather is windy, the probability of him striking his target is 0.65. According to the local weather forecast, the probability that any particular day is windy is 0.45.

   a Find the probability of him hitting the target on any randomly chosen day.

   b Given that he fails to hit the target, what is the probability that the day is calm?

20.4 Bayes' theorem

We begin with the result

$$P(B|A) = \frac{P(A|B) \times P(B)}{P(A)}$$

Proof

We know $P(A|B) = \dfrac{P(A \cap B)}{P(B)} \Rightarrow P(A \cap B) = P(A|B)P(B)$

and $P(B|A) = \dfrac{P(B \cap A)}{P(A)} = \dfrac{P(A \cap B)}{P(A)} \Rightarrow P(A \cap B) = P(B|A)P(A)$

So $P(A|B)P(B) = P(B|A)P(A)$

$\Rightarrow P(B|A) = \dfrac{P(A|B) \times P(B)}{P(A)}$

This result can be written in a different form.

From the Venn diagram we see that

$P(A) = P(A \cap B) + P(A \cap B')$

$\qquad = P(B)\dfrac{P(A \cap B)}{P(B)} + P(B')\dfrac{P(A \cap B')}{P(B')}$

$\qquad = P(B)P(A|B) + P(B')P(A|B')$

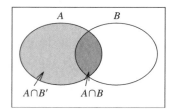

Substituting into the formula for $P(B|A)$ gives:

$$P(B|A) = \frac{P(A|B) \times P(B)}{P(B)P(A|B) + P(B')P(A|B')}$$

This is also a useful result to remember.

This is known as **Bayes' theorem for two events** and is used if we are given $P(A|B)$ and need $P(B|A)$.

Example

A town has only two bus routes. Route A has twice as many buses as route B. The probability of a bus running late on route A is $\dfrac{1}{8}$ and the probability of it running late on route B is $\dfrac{1}{10}$. At a certain point on the route the buses run down the same road. If a passenger standing at the bus stop sees a bus running late, use Bayes' theorem to find the probability that it is a route B bus.

Let the probability of a route B bus be $P(B)$ and the probability of a bus being late be $P(L)$.

Since there are only two bus routes the probability of a route A bus is $P(B')$.

Bayes' theorem states $P(B|L) = \dfrac{P(L|B) \times P(B)}{P(B)P(L|B) + P(B')P(L|B')}$.

We are given that $P(B) = \dfrac{1}{3}$, $P(B') = \dfrac{2}{3}$, $P(L|B) = \dfrac{1}{10}$ and $P(L|B') = \dfrac{1}{8}$.

$\Rightarrow P(B|L) = \dfrac{\dfrac{1}{10} \times \dfrac{1}{3}}{\left(\dfrac{1}{3} \times \dfrac{1}{10}\right) + \left(\dfrac{2}{3} \times \dfrac{1}{8}\right)} = \dfrac{2}{7}$

This question could have been done using a tree diagram and conditional probability.

"At least" problems

We have already met the idea of the total probability being one and that sometimes it is easier to find a probability by subtracting the answer from one. In some situations we do not have a choice in this, as shown in the example below.

Example

A student is practising her goal scoring for soccer. The probability that the ball hits the net on any particular attempt is 0.7 and she does not improve with practice.

   a Find how many balls should be kicked so that the probability that she hits the net at least once is greater than 0.995.

b Find how many balls should be kicked so that the probability that she does not hit the net is less than 0.001.

We begin by drawing part of the tree diagram. Because we do not know how many times she kicks the ball, the diagram potentially has an infinite number of branches.

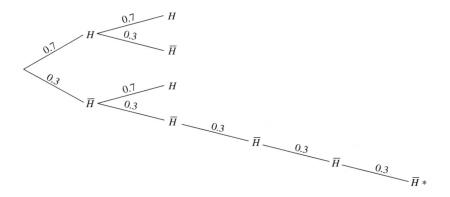

a From the tree diagram we can see that the only branch of the tree where she never hits the net is the one marked *.
Thus

$$1 - (0.3)^n > 0.995$$

$$\Rightarrow 0.3^n < 0.005$$

$$\Rightarrow \log 0.3^n < \log 0.005$$

$$\Rightarrow n \log 0.3 < \log 0.005$$

$$\Rightarrow n > \frac{\log 0.005}{\log 0.3}$$

$$\Rightarrow n > 4.40 \ldots$$

$$\Rightarrow n = 5 \text{ since } n \in \mathbb{N}$$

> Taking logs of both sides

> Using the laws of logs from Chapter 5

> The inequality changes because log 0.3 is negative.

b In this case the branch of the tree diagram that we are interested in is again the one marked *.
We want

$$0.3^n < 0.001$$

$$\Rightarrow \log 0.3^n < \log 0.001$$

$$\Rightarrow n \log 0.3 < \log 0.001$$

$$\Rightarrow n > \frac{\log 0.001}{\log 0.3}$$

$$\Rightarrow n > 5.73 \ldots$$

$$\Rightarrow n = 6$$

We will now look at two examples that bring together a number of these results.

Example

The results of a traffic survey on cars are shown below.

	Less than 3 years old	Between 3 and 6 years old	More than 6 years old
Grey	30	45	20
Black	40	37	17
White	50	30	31

a What is the probability that a car is less than 3 years old?

b What is the probability that a car is grey or black?

c Are a and b independent events?

d Given that a car is grey, what is the probability that it is less than 3 years old?

e Given that a car is more than 6 years old, what is the probability that it is white?

a Since there are 300 cars in the survey, the probability that a car is less than 3 years old is

$$\frac{30 + 40 + 50}{300} = \frac{2}{5}$$

b Since these are mutually exclusive events, the probability that it is grey or black is the probability that it is grey + the probability that it is black

$$= \frac{30 + 45 + 20}{300} + \frac{40 + 37 + 17}{300} = \frac{189}{300} = \frac{63}{100}$$

c We define the probability that a car is grey as $P(G)$, the probability that a car is black as $P(B)$ and the probability that it is less than 3 years old as $P(X)$. If these events are independent then $P[(B \cup G)|X] = P(B \cup G)$.

$$P[(B \cup G)|X] = \frac{P[(B \cup G) \cap X]}{P(X)}$$

$$= \frac{\frac{70}{300}}{\frac{120}{300}} = \frac{7}{12}$$

Since this is not the same as $\frac{63}{100}$ the events are not independent.

d We require $P(X|G) = \frac{P(X \cap G)}{P(G)}$

$$= \frac{\frac{30}{300}}{\frac{95}{300}} = \frac{30}{95} = \frac{6}{19}$$

e We begin by defining the probability that a car is white as $P(W)$ and the probability that it is more than 6 years old as $P(Y)$.

We require $P(W|Y) = \frac{P(W \cap Y)}{P(Y)}$

$$= \frac{\frac{31}{300}}{\frac{68}{300}} = \frac{31}{68}$$

Example

Mike and Belinda are both keen cyclists and want to see who is the best cyclist. They do this by competing in a series of independent races, and decide that the best cyclist will be the one to win three races. These races only have the two of them as contestants. However, the probability of either of them winning a race is dependent on the weather. In the rain the probability that Mike will win is 0.8, but when it is dry the probability that Mike will win is 0.3. In every race the weather is either defined as rainy or dry. The probability that on the day of a race the weather is rainy is 0.3.

   a Find the probability that Mike wins the first race.

   b Given that Mike wins the first race, what is the probability that the weather is rainy?

   c Given that Mike wins the first race, what is the probability that Mike is the best cyclist?

Let the probability that Mike wins a race be $P(M)$, the probability that Belinda wins a race be $P(B)$, and the probability that it rains be $P(R)$.

   a We begin by drawing a tree diagram.

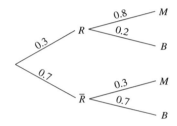

$$P(M) = 0.3 \times 0.8 + 0.7 \times 0.3 = 0.45$$

   b We require $\dfrac{P(R \cap M)}{P(M)} = \dfrac{0.3 \times 0.8}{0.45} = \dfrac{8}{15}$

   c We now draw a tree diagram showing the different ways that Mike and Belinda can win three races. We now know that $P(M) = 0.45$ and $P(B) = 0.55$ from part **a**.

Since it is given that Mike has won the first race we can ignore this part of the tree diagram.

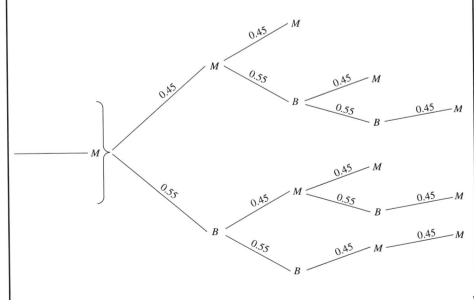

Hence the probability that Mike is the best cyclist

$$= (0.45 \times 0.45) + (0.45 \times 0.55 \times 0.45) + (0.45 \times 0.55 \times 0.55$$
$$\times 0.45) + (0.55 \times 0.45 \times 0.45) + (0.55 \times 0.45 \times 0.55 \times 0.45)$$
$$+ (0.55 \times 0.55 \times 0.45 \times 0.45)$$
$$= 0.609$$

Exercise 4

1 Shonil is practising playing darts. The probability that he hits the dartboard is 0.6.
   a Find the probability that he hits the dartboard at least once in four throws.
   b How many darts must he throw in order that the probability that he hits the dartboard at least once is greater that 0.99?
   c How many darts must Shonil throw in order that the probability that the dartboard is not hit is less than 0.01?

2 A die is biased such that the probability of throwing a six is 0.2.
   a Find the probability of throwing at least one six in the first eight throws.
   b How many times must the die be thrown in order that the probability of getting at least one six is greater than 0.995?

3 To proceed to the second round in a mathematics competition, Katie must get at least one of the four questions in the first round completely correct. The probability that she gets a question completely correct is 0.65. What is the probability that she proceeds to the second round?

4 Terri and Robyn are playing a game with a tetrahedral die. Terri goes first and they take turns at throwing the die. The first person to throw a one is the winner. What is the probability that Robyn wins?

5 Ayesha cycles to school. She has a choice of two routes, route A and route B. She is three times more likely to travel by route A than route B. If she travels by route A the probability that she will be late is 0.1 and if she travels by route B the probability that she will be late is 0.15. On a particular Monday Ayesha is late for school. Use Bayes' theorem to find the probability that she travelled by route A.

6 Ali should take two examinations, one in mathematics and one in English. On the day of the examinations, he takes only one. He is five times more likely to take mathematics than English. If he takes mathematics the probability that he passes is 0.9 and if he takes English the probability that he passes is 0.8. Ali passes the examination. Use Bayes' theorem to find the probability that he took mathematics.

7 Nicolle goes shopping to buy a present for her partner Ian. She has the choice of buying him a book or a DVD. She is four times more likely to buy him a book than a DVD. If she buys a book the probability that she pays with cash is 0.4 and if she buys a DVD the probability that she pays cash is 0.65. She pays cash for the present. Use Bayes' theorem to find the probability that Nicolle bought Ian a book.

8 In class 12A there are 30 students. 12 of the students hope to go to university A, 8 students hope to go to university B, 4 students hope to go to university C and 6 students hope to go to university D.
   a Four students are picked at random.
      i What is the probability that all four hope to go to university A?
      ii What is the probability that all four hope to go to the same university?
      iii Given that the first person picked hopes to go to university A, what is the probability that the other three hope to go to university B?
   b Find the probability that exactly four other students will be selected before a student who hopes to go to university C is selected.

9 Bill and David decide to go out for the day. They will either go to the beach or go to the mountains. The probability that they will go to the beach is 0.4. If they go to the beach the probability that they will forget the sunscreen is 0.1 and if they go to the mountains the probability they will forget the sunscreen is 0.35. They forget the sunscreen. Use Bayes' theorem to determine the probability that they go to the mountains.

10 Jerry and William play squash every week. In a certain week the probability that Jerry will win is 0.6. In subsequent weeks the probabilities change depending on the score the week before. For the winner, the probability of winning the following week increases by a factor of 1.05.

    a What is the probability that Jerry wins the first week and loses the second week?

    b What is the probability that William wins for three consecutive weeks?

    c Given that Jerry wins the first week, what is the probability that William wins for the following two weeks?

    d How many games must William play to have less than a 1% chance of always winning?

11 In a class of students, there are five students with blue eyes, seven students with brown eyes, four students with hazel eyes and four students with green eyes. It is found in this class that it is only boys who have either blue or green eyes and only girls who have brown or hazel eyes. Three students are chosen at random.

    a What is the probability that all three have blue eyes?

    b What is the probability that exactly one student with brown eyes is chosen?

    c What is the probability that two girls are chosen given that exactly one blue-eyed boy is chosen?

    d What is the probability that the group contains exactly one hazel-eyed girl or exactly one green-eyed boy or both?

12 Arnie, Ben and Carl are going out for the night and decide to meet in town. However, they cannot remember where they decided to meet. Arnie cannot re-member whether they were meeting in the square or outside the cinema. To make a decision he flips an unbiased coin. Ben cannot remember whether they were meeting outside the cinema or outside the theatre. He also flips a coin, but the coin is biased. The probability that it will land on a head is 0.6. If it lands on a head he will go to the cinema, but if it lands on a tail he will go to the theatre. Carl knows they are meeting either in the square, outside the cinema or outside the theatre. He flips an unbiased coin. If it lands on heads he goes to the the-atre, but if it lands on tails he flips again. On the second flip, if it lands on heads he goes to the square and if it lands on tails he goes to the theatre.

    a What is the probability that Arnie and Ben meet?

    b What is the probability that Ben and Carl meet?

    c What is the probability that all three meet?

    d Given that Carl goes to the cinema, what is the probability that all three will meet?

13 To promote the sale of biscuits, a manufacturer puts cards showing pictures of celebrities in the packets. There are four different celebrities on the cards. Equal numbers of cards showing each celebrity are randomly distributed in the packets and each packet has one card.

    a If Ellen buys three packets of biscuits, what is the probability that she gets a picture of the same celebrity in each packet?

    b If she buys four packets and the first packet she opens has a card with a picture of celebrity A, what is the probability that the following three packets will contain cards with celebrity B on them?

c Ellen's favourite celebrity is celebrity B. How many packets must she buy to have at least a 99.5% chance of having at least one picture of celebrity B?

14 At the ninth hole on Sam's local golf course he has to tee off over a small lake. If he uses a three wood, the probability that the ball lands in the lake is 0.15. If he uses a five wood, the probability that the ball lands in the lake is 0.20. If he uses a three iron, the probability that the ball lands in the lake is 0.18. If he tees off and the shot lands in the lake, he has to tee off again.

   a What is the probability that if he tees off with a three wood, he needs three shots to get over the lake?

   b Sam decides that if his shot lands in the lake on the first tee off, on the next tee off he will use a different club. He uses the three wood for the first shot, the five wood for the second, the three iron for the third and then returns to the three wood for the fourth and continues in that order.

     i What is the probability that he successfully tees off over the lake on his second use of the three wood?

     ii Given that he uses a three wood twice, what is the probability that he successfully hits over the lake on his sixth shot?

15 An author is writing a new textbook. The probability that there will be a mistake on a page is $\frac{1}{20}$ if he is writing in the evening and $\frac{1}{30}$ if he is writing in the morning.

   a What is the probability that if he is writing in the morning there is one mistake on each of three consecutive pages?

   b How many pages must he write in the evening for there to be a greater than 99% chance of at least one error?

   c He writes page 200 of the book in the morning, page 201 in the evening and page 202 in the morning. Given that page 201 has no mistakes on it, what is the probability that both pages 200 and 202 have a mistake on them?

20.5 Permutations and combinations

We met the idea of permutations and combinations in Chapter 6. A combination of a given number of articles is a set or group of articles selected from those given where the order of the articles in the set or group **is not** taken into account. A permutation of a given number of articles is a set or group of articles selected from those given where the order of the articles in the set or group **is** taken into account. At that point we looked at straightforward questions and often used the formulae to calculate the number of permutations or combinations. We will now look at some more complicated examples where the formulae do not work directly.

Permutations

Example

How many arrangements can be made of three letters chosen from the word PLANTER if the first letter is a vowel and each arrangement contains three different letters?

We split this into two separate calculations.

Assume we begin with the letter A. Hence the other two letters can be chosen in 6×5 ways = 30 ways.

If we begin with the letter E, then the other two letters can also be chosen in 30 ways.

Since these are the only two possibilities for beginning with a vowel, there are $30 + 30 = 60$ possible permutations.

Example

How many three-digit numbers can be made from the set of integers
{1, 2, 3, 4, 5, 6, 7, 8, 9} if
  a the three digits are all different
  b the three digits are all the same
  c the number is greater than 600
  d the number is even and each digit can only be used once?

a The first digit can be chosen in nine ways.
The second digit can be chosen in eight ways.
The third digit can be chosen in seven ways.
$\Rightarrow$ Total number of three-digit numbers that are all different
$= 9 \times 8 \times 7 = 504$

b If the three digits are all the same then there are nine possible three-digit numbers, since the only possibilities are 111, 222, 333, 444, 555, 666, 777, 888 and 999.

c If the number is greater than 600 then there are only four choices for the first digit: 6, 7, 8 or 9.
The second and third digits can each be chosen in nine ways.
$\Rightarrow$ Total number of three-digit numbers $= 4 \times 9 \times 9 = 324$

d In this case we start with the last digit as this is the one with the restriction. The last digit can be chosen in four ways.
The other two digits can be chosen in eight ways and seven ways respectively.
$\Rightarrow$ Total number of three-digit numbers $= 4 \times 8 \times 7 = 224$

Example

Find the number of two- and three-digit numbers greater than 20 that can be made from 1, 2 and 3, assuming each digit is only picked once.

We split this into the two separate problems of finding the two-digit numbers and the three-digit numbers.
For the two-digit numbers, the first digit must be either a 2 or a 3. Hence there are two ways of picking the first digit. The second digit can also be picked in two ways as we can choose either of the two left.
$\Rightarrow$ Total number of two-digit numbers $= 2 \times 2 = 4$
For the three-digit number, this is just the number of permutations, which is $3! = 6$.
$\Rightarrow$ Total number of two- and three-digit numbers greater than $20 = 4 + 6 = 10$

Example

In how many ways can six people be sat around a circular dining table?

At first this appears to be a simple permutation, where the answer is 6! However, if we look at the two situations below, where the chairs are labelled from 1 to 6 and the people from A to F, we can see that they appear as different permutations, but are actually the same.

Because of the actual situation, every person has the same people on either side in both cases. If the chairs had been distinguishable, then this would no longer be the case.

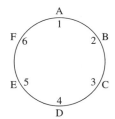 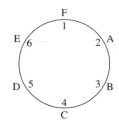

Hence for every permutation of people sat around the table, there are five more permutations which are the same and hence the answer is six times too big. These are shown in the diagram below.

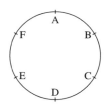 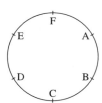 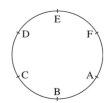

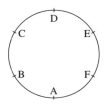 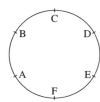 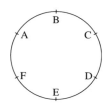

Therefore the number of ways that six people can be sat around a circular dining table is $\dfrac{6!}{6} = 5! = 120$.

For any situation like this the answer can be generalised to $(n-1)!$

Example

Jenny is making a necklace. In how many ways can 4 beads chosen from 12 beads be threaded on a string?

This is similar to the example above. As with the example above the answer $12 \times 11 \times 10 \times 9$ needs to be divided by 4 because of the repetitions caused by the fact that it is on a circle. However in this situation there is another constraint because the necklace can be turned over giving an equivalent answer. In the diagrams below, these two situations are actually the same, but appear as two separate permutations of the answer.

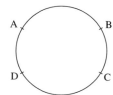 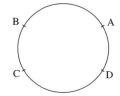

Hence we need to divide the answer by 2.

Therefore the number of permutations is $\dfrac{12 \times 11 \times 10 \times 9}{4 \times 2} = 1485$

Example

a Find the number of arrangements of the letters of the word LITTER.
b Find the number of arrangements where the T's are together.
c Find the number of arrangements where the T's are separated.

a In this question we treat it as a simple permutation and hence the answer would appear to be 6! However, the two T's are indistinguishable and hence LIT_1T_2LE and LIT_2T_1LE are actually the same arrangement but appear as two separate permutations. As this happens in every single case the number of arrangements is $\frac{6!}{2!} = 360$.

b With the T's together we treat the two T's as one letter. If we give TT the symbol Θ, then we are finding the permutations of LI Θ ER, which are $5! = 120$ arrangements.

c For the T's separated, we remove the T's initially and find the number of permutations of LIER which is 4!

If we now consider the specific permutation REIL, then the two T's can be placed in two of five positions. This is shown in the diagram below.

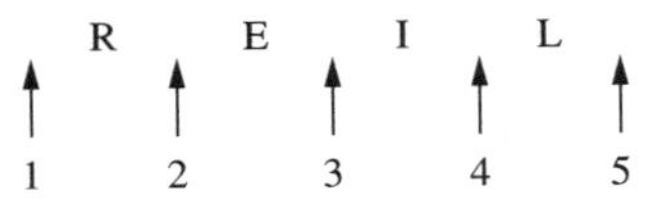

Hence for the permutation REIL there are $^5C_2 = 10$ ways of positioning the T's. As this can happen with each of the 4! permutations of the four letters, then the total number of permutations is $4! \times 10 = 240$ arrangements.
We could also think about this another way. As we know the total number of arrangements is 360 and the T's either have to be together or separated then the number of arrangements where they are separated is 360 minus the number of arrangements where they are together.
This gives $360 - 120 = 240$ arrangements.

Example

Find the number of arrangements of the letters in the word REFERENCE where the E's are separated.

We begin by considering RFRNC. The number of arrangements of these letters is $\frac{5!}{2!}$ since the two R's are indistinguishable.

Again considering one possible arrangement of the letters, say RFNRC, the positions that the three E's can take are shown below.

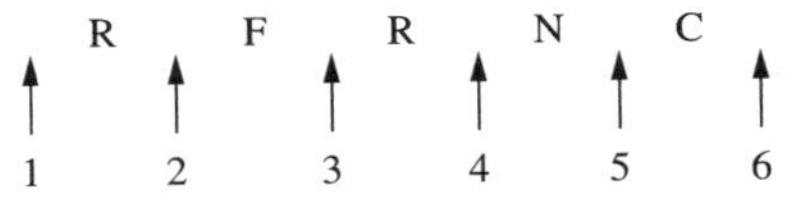

We need to find the number of combinations of four from six positions, which is $^6C_4 = 15$.

Hence the number of arrangements where the E's are separated is

$\frac{5!}{2!} \times 15 = 900$.

Combinations

> ### Example
>
> A team of 4 children is to be selected from a class of 20 children, to compete in a quiz game. In how many ways can the team be chosen if
> **a** any four can be chosen
> **b** the four chosen must include the oldest in the class?
>
> **a** This is a straightforward combination with an answer of $^{20}C_4 = 4845$.
>
> **b** In this situation we remove the oldest in the class since this child has to be part of every group. Hence the problem is actually to find how many teams of 3 children can be found from 19. Therefore the number of teams is $^{19}C_3 = 969$.

> ### Example
>
> Ten students in a class are divided into two groups of five to play in a five-a-side soccer tournament. In how many ways can the two teams of five be selected?
>
> This appears to be very similar to the example above, but there is a subtle difference. The number of ways of selecting a team of five is $^{10}C_5 = 252$. Let us imagine that the chosen team is ABCDE. Hence the other team would automatically be FGHIJ. However another possible combination of a team of five would be FGHIJ and this would then automatically select the other team as ABCDE. In other words the calculation picks each pair of teams twice. Hence the actual number of teams is $\dfrac{^{10}C_5}{2} = 126$.

> ### Example
>
> Anisa goes into her local supermarket and finds that there are 20 different types of chocolate on offer and 15 different types of soft drink. She wants to buy seven different bars of chocolate and four different cans of soft drink for herself and her friends. Find the number of different ways in which she can do this.
>
> The number of ways she can choose seven bars of chocolate is $^{20}C_7 = 77\,520$. The number of ways she can choose four cans of soft drink is $^{15}C_4 = 1365$. Since with any particular combination of chocolate bars she can put all the particular combinations of cans of soft drink, the total number of choices is $77\,520 \times 1365 = 105\,814\,800$.

> ### Example
>
> A box contains four red, two blue, one yellow and one pink ball. How many different selections of three balls may be made?
>
> All three the same: The only possibility here is three red balls and hence there is only one way of doing this.
> Two the same, one different: There are two possibilities for two the same, red and blue. The third ball can then be chosen from any of the others, so there are

three possibilities because we cannot choose the same colour again. Therefore the total number of ways is $2 \times 3 = 6$.

All three different: Since there are four different colours of ball this is $^4C_3 = 4$.

Hence the number of different selections that can be made is $1 + 6 + 4 = 11$.

Unlike the previous example, we added the combinations here as opposed to multiplying them.

Exercise 5

1 In how many ways can six different files be arranged in a row on a desk?

2 In how many ways can two boys and two girls be chosen from a group of 15 boys and 18 girls?

3 In how many ways can four different letters be put in four different envelopes?

4 In how many ways can three different coats be arranged on five hooks in a row?

5 Giulia has ten different mathematics books and four different chemistry books. In how many ways can she arrange seven of the mathematics books and one chemistry book on a shelf if the chemistry book must always be at one end?

6 In how many ways can the letters of the word PHOTOGRAPH be arranged?

7 Two sets of books contain seven different novels and four different autobiographies. In how many ways can the books be arranged on a shelf if the novels and the autobiographies are not mixed up?

8 In how many ways can ten different examinations be arranged so that the two mathematics examinations are not consecutive and the two French examinations are not consecutive?

9 Given that each digit can be used more than once, how many two-digit numbers can be made from the set $\{2, 4, 6, 7, 8, 9\}$ if
a any two digits can be used
b the two digits must be the same
c the number must be odd
d the number must be greater than 60?

10 A quiz team of five students is to be chosen from nine students. The two oldest students cannot both be chosen. In how many ways can the quiz team be chosen?

11 Consider the letters of the word DIFFICULT.
a How many different arrangements of the letters can be found?
b How many of these arrangements have the two I's together and the two F's together?
c How many of the arrangements begin and end with the letter F?

12 Margaret wants to put eight new plants in her garden. They are all different.
a She first of all decides to plant them in a row. In how many ways can she do this?
b She then decides that they would look better in a circle. In how many ways can she do this?
c She now realizes that two of the plants are identical. How many arrangements are there for planting them in a row and for planting them in a circle?

13 a How many different arrangements of the word ARRANGEMENT can be made?
b How many different arrangements are there which start with a consonant and end with a vowel?

14 Jim is having a dinner party for four couples.

    a In how many ways can the eight people be seated at Jim's circular dining table?

    b John and Robin are a couple, but do not want to sit next to each other at the dinner party. In how many ways can the eight people now be seated?

    c Jim decides that the two oldest guests should sit next to each other. In how many ways can the eight people now be seated?

15 **a** How many numbers greater than 300 can be made from the set {1, 2, 5, 7} if each integer can be used only once?

    b How many of these numbers are even?

16 **a** A local telephone number has seven digits and cannot start with zero. How many local numbers are there?

    b The telephone company realizes that they do not have enough numbers. It decides to add an eighth digit to each number, but insists that all the eight-digit numbers start with an odd number and end with an even number. The number still cannot start with a zero. Does this increase or decrease the number of possible telephone numbers and if so by how many?

17 **a** How many different arrangements are there of the letters of the word INQUISITION?

    b How many arrangements are there where the four I's are separate?

    c How many arrangements are there where the S and the T are together?

18 Five different letters are written and five different envelopes are addressed. In how many ways can at least one letter be placed in the wrong envelope?

19 On an examination paper of 20 questions a student obtained either 6 or 7 marks for each question. If his total mark is 126, in how many different ways could he have obtained this total?

20 Four boxes each contain six identical coloured counters. In the first box the counters are red, in the second box the counters are orange, in the third box the counters are green and in the fourth box the counters are purple. In how many ways can four counters be arranged in a row if

    a they are all the same

    b three are the same and one is different

    c they are all different

    d there is no restriction on the colours of the counters?

21 **a** In how many ways can six different coloured beads be arranged on a ring?

    b If two beads are the same colour, how many ways are there now?

22 **a** How many different combinations of six numbers can be chosen from the digits 1, 2, 3, 4, 5, 6, 7, 8 if each digit is only chosen once?

    b In how many ways can the digits be divided into a group of six digits and a group of two digits?

    c In how many ways can the digits be divided into two groups of four digits?

23 A shop stocks ten different types of shampoo. In how many ways can a shopper buy three types of shampoo if

    a each bottle is a different type

    b two bottles are the same type and the third is different?

24 A mixed team of 10 players is chosen from a class of 25 students. 15 students are boys and 10 students are girls. In how many ways can this be done if the team has five boys and five girls?

25 Find the number of ways in which ten people playing five-a-side football can be divided into two teams of five if Alex and Bjorn must be in different teams.

26 A tennis team of four is chosen from seven married couples to represent a club at a match. If the team must consist of two men and two women and a husband and wife cannot both be in the team, in how many ways can the team be formed?

27 Nick goes to the shop to buy seven different packets of snacks and four bottles of drink. At the shop he find he has to choose from 15 different packets of snacks and 12 different bottles of drink. In how many different ways can he make his selection?

28 In how many ways can three letters from the word BOOKS be arranged in a row if at least one of the letters is O?

20.6 Probability involving permutations and combinations

Sometimes we can use the idea of permutations and combinations when solving questions about probability.

Example

From a group of 12 people, 8 are chosen to serve on a committee.
  a In how many different ways can the committee be chosen?
  b One of the 12 people is called Sameer. What is the probability that he will be on the committee?
  c Among the 12 people there is one married couple. Find the probability that both partners will be chosen.
  d Find the probability that the three oldest people will be chosen.

a This is the combination $^{12}C_8 = 495$, which acts as the total possibility space.

b Since Sameer must be on the committee, we need to choose 7 people from 11. This can be done in $^{11}C_7 = 330$ ways. Hence the probability that Sameer will be on the committee is $\frac{330}{495} = \frac{2}{3}$.

c As in part **b**, we know that the married couple will be on the committee and need to choose six people from ten. This can be done in $^{10}C_6 = 210$ ways. Hence the probability that the married couple will be on the committee is $\frac{210}{495} = \frac{14}{33}$.

d Since the three oldest people have been chosen, we now choose the other five people from the nine remaining. This can be done in $^9C_5 = 126$ ways. Hence the probability that the three oldest people will be on the committee is $\frac{126}{495} = \frac{14}{55}$.

Example

Four letters are picked from the word EXAMPLES.
  a How many different arrangements are there of four letters?
  b What is the probability that the arrangement of four letters will not contain a letter E?
  c What is the probability that the arrangement will contain both of the letter E's?

d Given that the arrangement contains both of the letter E's, what is the probability that the two letter E's will be separated?

a Because of the repetition of the letter E we need to do this in groups.

Consider XAMPLS. The number of arrangements of four letters is $^6P_4 = 360$.

Consider XAMPLSE with the condition that the arrangement must contain one E. To do this we find the number of permutations of three letters from XAMPLS $= {}^6P_3 = 120$. Within each of these permutations there are four positions that the E can take. This is shown below, using XAM as an example.

Hence the number of permutations that contain one E is $4 \times 120 = 480$.

Consider XAMPLSEE with the condition that the arrangement must contain two E's. To do this we find the number of permutations of two letters from XAMPLS $= {}^6P_2 = 30$. Within each of these permutations there are six positions that the two E's can take. These are shown below, using XA as an example.

EEXA	EXEA	EXAE
XAEE	XEEA	XEAE

The number of permutations that contain two E's is $6 \times 30 = 180$. Hence the total possible number of arrangements is $240 + 480 + 180 = 1020$.

b Since the total number of arrangements is 1020 and the number that do not contain an E is 360, the probability is $\dfrac{360}{1020} = \dfrac{6}{17}$.

c Since the total number of arrangements is 1020 and the number that contain both E's is 180, the probability is $\dfrac{180}{1020} = \dfrac{3}{17}$.

d Because this is a conditional probability question we are limiting the sample space to only containing two E's. Consider the permutation PL. The two E's can be positioned as shown below.

EEPL	EPEL	EPLE
PLEE	PEEL	PELE

Hence for any permutation of two letters, 3 out of the 6 cases will have the letter E's separated. This is true for every permutation of two letters. Hence the probability is $\dfrac{3}{6} = \dfrac{1}{2}$.

Exercise 6

1 A group of three boys and one girl is chosen from six boys and five girls.
   a How many different groups can be formed?
   b What is the probability that the group contains the oldest boy?
   c What is the probability that it contains the youngest girl?
   d Within the six boys there are two brothers. What is the probability that the group contains both brothers?

2 a In how many ways can ten people be sat around a circular table?
   b Within the ten people there are two sisters.
      i What is the probability that the sisters will sit together?
      ii What is the probability that the sisters will not sit together?

c The ten people consist of five men and five women. What is the probability that no two men and no two women are sat next to each other?

3 Allen is predicting the results of six soccer matches.

   a In how many ways can he predict exactly four correct results?

   b His favourite team is A and they are playing in one of the six matches. Given that he predicts exactly four correct results, what is the probability that he will predict correctly the result of the match in which team A are playing?

4 Consider the letters of the word EATING.

   a How many different arrangements of four letters can be formed?

   b What is the probability that the four-letter arrangement contains the letter A?

   c What is the probability that the four-letter arrangement contains either the letter T or the letter G, but not both?

5 At a local squash club there are 40 members. League A consists of six people.

   a If league A is made up randomly from the 40 members, in how many different ways can league A be made?

   b What is the probability that league A will contain the oldest member of the club?

   c Given that league A contains the oldest member of the club, what is the probability that it also contains the youngest?

6 Six letters are picked from the word CULTURES.

   a How many different arrangements of six letters can be formed?

   b What is the probability that an arrangement contains exactly one U?

   c Given that the arrangement contains both U's, what is the probability that both U's are together at the start of the arrangement?

7 a How many even numbers less than 500 can be formed from the digits 2, 4, 5, 7 and 9 using each digit only once?

   b An even number is picked at random.

      i What is the probability that it is a two-digit number?

      ii What is the probability that it is greater than 200?

      iii What is the probability that it is a three-digit number beginning with 4?

8 Laura has ten plants to put in a row along the fence of her garden. There are four identical roses, four identical clematis and two identical honeysuckle.

   a In how many different ways can she plant them along the fence?

   b What is the probability that the four roses are all together?

   c What is the probability that there is a honeysuckle at each end of the row?

   d What is the probability that no clematis is next to another clematis?

9 A committee meeting takes place around a rectangular table. There are six members of the committee and six chairs. Each position at the table has different papers at that position.

   a How many different arrangements are there of the six committee members sitting at the table?

   b Two friends Nikita and Fatima are part of the committee. What is the probability that either of them sit at the table with papers A and B in front of them?

   c The only two men on the committee are Steve and Martin. What is the probability that they sit in positions C and F respectively?

10 Jim is trying to arrange his DVD collection on the shelf. He has ten DVDs, three titles starting with the letter A, four titles starting with the letter C and three titles starting with the letter S. Even though they start with the same letter the DVDs are distinguishable from each other.

   a In how many ways can Jim arrange the DVDs on the shelf?

   b What is the probability that the four starting with the letter C will be together?

   c What is the probability that no DVD starting with the letter A will be next to another DVD starting with the letter A?

d Given that the three DVDs starting with the letter S are together, what is the probability that the three DVDs starting with the letter A will be together?

Review exercise

 A calculator may be used in all questions in this exercise where necessary.

1 In a group of 30 boys, they all have black hair or brown eyes or both. 20 of the boys have black hair and 15 have brown eyes. A boy is chosen at random.
  a What is the probability that he has black hair and brown eyes?
  b Are these two events independent, mutually exclusive or exhaustive? Give reasons.

2 Roy drives the same route to school every day. Every morning he goes through one set of traffic lights. The probability that he has to stop at the traffic lights is 0.35.
  a In a five-day week, what is the probability that he will have to stop on at least one day?
  b How many times does Roy have to go through the traffic lights to be able to say that there is a 95% chance that he will have to stop?

3 There are ten seats in a waiting room. There are six people in the room.
  a In how many different ways can they be seated?
  b In the group of six people, there are three sisters who must sit next to each other. In how many different ways can the group be seated? [IB May 06 P1 Q19]

4 Tushar and Ali play a game in which they take turns to throw an unbiased cubical die. The first one to throw a one is the winner. Tushar throws first.
  a What is the probability that Tushar wins on his first throw?
  b What is the probability that Ali wins on his third throw?
  c What is the probability that Tushar wins on his nth throw?

5 A bag contains numbers. It is twice as likely that an even number will be drawn than an odd number. If an odd number is drawn, the probability that Chris wins a prize is $\frac{3}{16}$. If an even number is drawn, the probability that Chris wins a prize is $\frac{5}{16}$. Chris wins a prize. Use Bayes' theorem to find the probability that Chris drew an even number.

6 In how many ways can six different coins be divided between two students so that each student receives at least one coin? [IB Nov 00 P1 Q19]

7 a In how many ways can the letters of the word PHOTOGRAPH be arranged?
  b In how many of these arrangements are the two O's together?
  c In how many of these arrangements are the O's separated?

8 The local football league consists of ten teams. Team A has a 40% chance of winning any game against a higher-ranked team, and a 75% chance of winning any game against a lower-ranked team. If A is currently in fourth position, find the probability that A wins its next game. [IB Nov 99 P1 Q13]

9 Kunal wants to invite some or all of his four closest friends for dinner.
  a In how many different ways can Kunal invite one or more of his friends to dinner?
  b Mujtaba is his oldest friend. What is the probability that he will be invited?
  c Two of his closest friends are Anna and Meera. What is the probability that they will both be invited?

10 The probability of Prateek gaining a grade 7 in Mathematics HL given that he revises is 0.92. The probability of him gaining a grade 7 in Mathematics HL given that he does not revise is 0.78. The probability of Prateek revising is 0.87. Prateek gains a grade 7 in Mathematics HL. Use Bayes' theorem to find the probability that he revised.

11 A committee of five people is to be chosen from five married couples. In how many ways can the committee be chosen if
  a there are no restrictions on who can be on the committee
  b the committee must contain at least one man and at least one woman
  c the committee must contain the youngest man
  d both husband and wife cannot be on the committee?

12 In a bilingual school there is a class of 21 pupils. In this class, 15 of the pupils speak Spanish as their first language and 12 of these 15 pupils are Argentine. The other 6 pupils in the class speak English as their first language and 3 of these 6 pupils are Argentine. A pupil is selected at random from the class and is found to be Argentine. Find the probability that the pupil speaks Spanish as his/her first language.   [IB May 99 P1 Q8]

13 How many different arrangements, each consisting of five different digits, can be formed from the digits 1, 2, 3, 4, 5, 6, 7, if
  a each arrangement begins and ends with an even digit
  b in each arrangement odd and even digits alternate?   [IB Nov 96 P1 Q12]

14 Three suppliers A, B and C produce respectively 45%, 30% and 25% of the total number of a certain component that is required by a car manufacturer. The percentages of faulty components in each supplier's output are, again respectively, 4%, 5% and 6%. What is the probability that a component selected at random is faulty?   [IB May 96 P1 Q4]

15 a How many different arrangements of the letters of the word DISASTER are there?
  b What is the probability that if one arrangement is picked at random, the two S's are together?
  c What is the probability that if one arrangement is picked at random, it will start with the letter D and finish with the letter T?

16 Note: In this question all answers must be given exactly as rational numbers.
  a A man can invest in at most one of two companies, A and B. The probability that he invests in A is $\frac{3}{7}$ and the probability that he invests in B is $\frac{2}{7}$, otherwise he makes no investment. The probability that an investment yields a dividend is $\frac{1}{2}$ for company A and $\frac{2}{3}$ for company B. The performances of the two companies are totally unrelated. Draw a probability tree to illustrate the various outcomes and their probabilities. What is the probability that the investor receives a dividend and, given that he does, what is the probability that it was from his investment in company A?
  b Suppose that a woman must decide whether or not to invest in each company. The decisions she makes for each company are independent and the probability of her investing in company A is $\frac{3}{10}$ while the probability of her investing in company B is $\frac{6}{10}$. Assume that there are the same probabilities of the investments yielding a dividend as in part **a**.
    i Draw a probability tree to illustrate the investment choices and whether or not a dividend is received. Include the probabilities for the various outcomes on your tree.
    ii If she decides to invest in both companies, what is the probability that she receives a dividend from at least one of her investments?
    iii What is the probability that she decides not to invest in either company?
    iv If she does not receive a dividend at all, what is the probability that she made no investment?   [IB May 96 P2 Q4]

17 An advanced mathematics class consists of six girls and four boys.

   a How many different committees of five students can be chosen from this class?

   b How many such committees can be chosen if class members Jack and Jill cannot both be on the committee?

   c How many such committees can be chosen if there must be more girls than boys on the committee? [IB Nov 95 P1 Q12]

18 Each odd number from 1 to $3n$, where $n \in \mathbb{N}$ and n is odd, is written on a disc and the discs are placed in a box.

   a How many discs are there in the box?

   b What is the probability, in terms of n, that a disc drawn at random from the box has a number that is divisible by 3? [IB May 95 P1 Q19]

19 A box contains 20 red balls and 10 white balls. Three balls are taken from the box without replacement. Find the probability of obtaining three white balls. Let p_k be the probability that k white balls are obtained. Show by evaluating p_0, p_1, p_2 and p_3 that $\displaystyle\sum_{k=0}^{k=3} p_k = 1$. [IB Nov 87 P1 Q19]

20 Pat is playing computer games. The probability that he succeeds at level 1 is 0.7. If he succeeds at level 1 the next time he plays he goes to level 2 and

the probability of him succeeding is $\dfrac{2}{3}$ the probability of him succeeding on

level 1. If he does not succeed he stays on level 1. If he succeeds on level 2 he

goes to level 3 where the probability of him succeeding is $\dfrac{2}{3}$ the probability of

him succeeding on level 2. If he fails level 2 he goes back to level 1

with the initial probability of success. If he succeeds on level 3 he goes to

level 4 where the probability of success is again $\dfrac{2}{3}$ the probability of him

succeeding on level 3. If he fails on level 3 he goes back to level 2 and the probability of success is again the same as it was before. This continues in all games that Pat plays.

   a What is the probability that after the third game he is on level 2?

   b What is the probability that after the fourth game he is on level 1?

   c Given that he wins the first game, what is the probability that after the fourth game he is on level 3?

21 In Kenya, at a certain doctor's surgery, in one week the doctor is consulted by 90 people all of whom think they have malaria. 50 people test positive for the disease. However, the probability that the test is positive when the patient does not have malaria is 0.05 and the probability that the test is negative when the patient has malaria is 0.12.

   a Find the probability that a patient who tested positive in the surgery has malaria.

   b Given that a patient has malaria, what is the probability that the patient tested negative?

   c Given that a patient does not have malaria, what is the probability that the patient tested positive?

21 Discrete Probability Distributions

In this chapter we will meet the concept of a discrete probability distribution and one of these is called the Poisson distribution. This was named after Siméon-Denis Poisson who was born in Pithivier in France on 21 June 1781. His father was a great influence on him and it was he who decided that a secure future for his son would be the medical profession. However, Siméon-Denis was not suited to being a surgeon, due to his lack of interest and also his lack of coordination. In 1796 Poisson went to Fontainebleau to study at the École Centrale, where he

Siméon Denis Poisson

showed a great academic talent, especially in mathematics. Following his success there, he was encouraged to sit the entrance examinations for the École Polytechnique in Paris, where he gained the highest mark despite having had much less formal education than most of the other entrants. He continued to excel at the École Polytechnique and his only weakness was the lack of coordination which made drawing mathematical diagrams virtually impossible. In his final year at the École Polytechnique he wrote a paper on the theory of equations that was of such a high quality that he was allowed to graduate without sitting the final examinations. On graduation in 1800, he became répétiteur at the École Polytechnique, which was rapidly followed by promotion to deputy professor in 1802, and professor in 1806. During this time Poisson worked on differential calculus and later that decade published papers with the Academy of Sciences, which included work on astronomy and confirmed the belief that the Earth was flattened at the poles.

Poisson was a tireless worker and was dedicated to both his research and his teaching. He played an ever increasingly important role in the organization of mathematics in France and even though he married in 1817, he still managed to take on further duties. He continued to research widely in a range of topics based on applied mathematics. In *Recherches sur la probabilité des jugements en matière criminelle et matière civile*, published in 1837, the idea of the Poisson distribution first appears. This describes the probability that a random event will occur when the event is evenly spaced, on average, over an infinite space. We will learn about this distribution in this chapter. Overall, Poisson published between 300 and 400 mathematical works and his name is attached to a wide variety of ideas, including Poisson's integral, Poisson brackets in differential equations, Poisson's ratio in elasticity, and Poisson's constant in electricity. Poisson died on 25 April 1840.

21.1 Introduction to discrete random variables

In Chapter 20 we met the idea of calculating probability given a specific situation and found probabilities using tree diagrams and sample spaces. Once we have obtained these values, we can write them in the form of a table and further work can be done with them. Also, we can sometimes find patterns that allow us to work more easily in terms of finding the initial distribution of probabilities. In this chapter we will work with discrete random variables. A **discrete random variable** has the following properties.

- It is a discrete (exact) variable.
- It can only assume certain values, $x_1, x_2, \ldots, x_n$.
- Each value has an associated probability, $P(X = x_1) = p_1$, $P(X = x_2) = p_2$ etc.
- The probabilities add up to 1, that is $\sum_{i=1}^{i=n} P(X = x_i) = 1$.
- A discrete variable is only random if the probabilities add up to 1.

A discrete random variable is normally denoted by an upper case letter, e.g. X, and the particular value it takes by a lower case letter, e.g. x.

Example

Write out the probability distribution for the number of threes obtained when two tetrahedral dice are thrown. Confirm that it is a discrete random variable.

$P(\text{no threes}) = \dfrac{3}{4} \times \dfrac{3}{4} = \dfrac{9}{16}$

$P(\text{1 three}) = P(3\bar{3}) + P(\bar{3}3)$

$\qquad\qquad = \dfrac{1}{4} \times \dfrac{3}{4} + \dfrac{3}{4} \times \dfrac{1}{4} = \dfrac{6}{16}$

$P(\text{2 threes}) = \dfrac{1}{4} \times \dfrac{1}{4} = \dfrac{1}{16}$

Hence the probability distribution is:

Number of threes	0	1	2
Probability	$\dfrac{9}{16}$	$\dfrac{6}{16}$	$\dfrac{1}{16}$

The distribution is discrete because we can only find whole-number values for the number of threes obtained. Finding a value for 2.5 threes does not make sense. It is also random because $\dfrac{9}{16} + \dfrac{6}{16} + \dfrac{1}{16} = \dfrac{16}{16} = 1$.

We can now use the notation introduced above and state that if X is the number of threes obtained when two tetrahedral dice are thrown, then X is a discrete random variable, where $P(X = 0) = \dfrac{9}{16}$, $P(X = 1) = \dfrac{6}{16}$, $P(X = 2) = \dfrac{1}{16}$. In a table:

X	0	1	2
$P(X = x)$	$\dfrac{9}{16}$	$\dfrac{6}{16}$	$\dfrac{1}{16}$

These probability distributions can be found in three ways. Firstly they can be found from tree diagrams or probability space diagrams, secondly they can be found from a specific formula and thirdly they can be found because they follow a set pattern and hence form a special probability distribution.

Example

A bag contains 3 red balls and 4 black balls. Write down the probability distribution for R, where R is the number of red balls chosen when 3 balls are picked without replacement, and show that it is random.
The tree diagram for this is shown below.

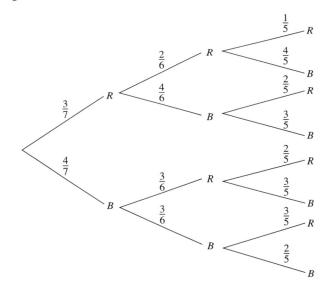

Hence $P(R = 0) = \dfrac{24}{210}$

$P(R = 1) = \dfrac{108}{210}$

$P(R = 2) = \dfrac{72}{210}$

$P(R = 3) = \dfrac{6}{210}$

In tabular form this can be represented as:

r	0	1	2	3
$P(R = r)$	$\dfrac{24}{210}$	$\dfrac{108}{210}$	$\dfrac{72}{210}$	$\dfrac{6}{210}$

To check that it is random, we add together the probabilities.

$$\frac{24}{210} + \frac{108}{210} + \frac{72}{210} + \frac{6}{210} = \frac{210}{210} = 1$$

Thus we conclude that the number of red balls obtained is a random variable.

Alternatively the probabilities may be assigned using a function which is known as the probability density function (p.d.f) of X.

Example

The probability density function of a discrete random variable Y is given by

$P(Y = y) = \dfrac{y^2}{14}$ for $y = 0, 1, 2$ and 3. Find $P(Y = y)$ for $y = 0, 1, 2$ and 3,

verify that Y is a random variable and state the mode.

y	0	1	2	3
$P(Y = y)$	0	$\dfrac{1}{14}$	$\dfrac{4}{14}$	$\dfrac{9}{14}$

To check that it is random, we add together the probabilities.

$$0 + \frac{1}{14} + \frac{4}{14} + \frac{9}{14} = \frac{14}{14} = 1$$

Thus we conclude that Y is a random variable.
The mode is 3 since this is the value with the highest probability.

Example

The probability density function of a discrete random variable X is given by
$P(X = x) = kx$ for $x = 9, 10, 11$ and 12. Find the value of the constant k.

In this case we are told that the variable is random and hence $\sum P(X = x) = 1$.

Therefore $9k + 10k + 11k + 12k = 1$

$$\Rightarrow k = \frac{1}{42}$$

Exercise 1

1 A discrete random variable X has this probability distribution:

x	0	1	2	3	4	5	6
$P(X = x)$	0.05	0.1	0.3	b	0.15	0.15	0.05

Find
a the value of b **b** $P(1 \leq X \leq 3)$ **c** $P(X < 4)$ **d** $P(1 < X \leq 5)$
e the mode.

2 A discrete random variable X has this probability distribution:

x	4	5	6	7	8	9	10
$P(X = x)$	0.02	0.15	0.25	a	0.12	0.1	0.03

Find
a the value of a **b** $P(4 \leq X \leq 8)$ **c** $P(X < 8)$

d $P(4 < X < 8)$ **e** $P(5 < X < 7)$ **f** the mode.

3 Find the discrete probability distribution for X in the following cases and verify that the variable is random. X is defined as

  a the number of tails obtained when three fair coins are tossed

  b the number of black balls drawn with replacement from a bag of 4 black balls and 3 white balls, when 3 balls are picked

  c the number of sixes obtained on a die when it is rolled three times

  d the sum of the numbers when two dice are thrown

  e the number of times David visits his local restaurant in three consecutive days, given that the probability of him visiting on any specific day is 0.2 and is an independent event.

4 Write down the discrete probability distributions given the following probability density functions:

  a $P(X = x) = \dfrac{x^2}{55}$ for $0 \leq x \leq 5,\, x \in \mathbb{N}$

  b $P(X = x) = \dfrac{x}{21}$ for $x = 1, 2, 3, 4, 5, 6$

  c $P(Y = y) = \dfrac{y - 1}{30}$ for $y = 7, 8, 9, 10$

  d $P(S = s) = \dfrac{s - 3}{42}$ for $s = 12, 13, 14, 15$

5 Find the value of k in each of the probability density functions shown below, such that the variable is random. In each case write out the probability distribution.

  a $P(X = x) = k(x - 1)$ for $x = 3, 4, 5$

  b $P(X = x) = k(x^2 - 1)$ for $x = 4, 5, 6$

  c $P(Y = y) = ky^3$ for $y = 1, 2, 3, 4, 5$

  d $P(B = b) = \dfrac{b + 2}{k}$ for $b = 3, 4, 5, 6, 7$

6 A man has six blue shirts and three grey shirts that he wears to work. Once a shirt is worn, it cannot be worn again in that week. If X is the discrete random variable "the number of blue shirts worn in the first three days of the week", find

  a the probability distribution for X

  b the probability that he wears at least one blue shirt during the first three days.

7 In a game a player throws three unbiased tetrahedral dice. If X is the discrete random variable "the number of fours obtained", find

  a the probability distribution for X

  b $P(X \geq 2)$.

8 Five women and four men are going on holiday. They are travelling by car and the first car holds four people including the driver. If Y is the discrete random variable "the number of women travelling in the first car", write down

  a the probability distribution for Y

  b the probability that there is at least one woman in the first car.

21.2 Expectation and variance

The expectation, E(X)

In a statistical experiment:

• A practical approach results in a frequency distribution and a mean value.

• A theoretical approach results in a probability distribution and an expected value.

The **expected value** is what we would expect the mean to be if a large number of terms were averaged.

The expected value is found by multiplying each score by its corresponding probability and summing.

$$E(X) = \sum_{\text{all } x} x \cdot P(X = x)$$

> If the probability distribution is symmetrical about a mid-value, then E(X) will be this mid-value.

Example

The probability distribution of a discrete random variable X is as shown in the table.

x	1	2	3	4	5
$P(X = x)$	0.2	0.4	a	0.1	0.05

a Find the value of a.

b Find E(X).

a Since it is random

$$0.2 + 0.4 + a + 0.1 + 0.05 = 1$$
$$\Rightarrow a = 0.25$$

b $E(X) = 1 \times 0.2 + 2 \times 0.4 + 3 \times 0.25 + 4 \times 0.1 + 5 \times 0.05 = 2.4$

Example

The probability distribution of a discrete random variable Y is shown below.

y	5	6	7	8	9
$P(Y = y)$	0.05	0.2	b	0.2	0.05

a Find the value of b.

b Find E(Y).

a Since the variable is random $0.05 + 0.2 + b + 0.2 + 0.05 = 1$
$$\Rightarrow b = 0.5$$

b In this case we could use the formula $E(Y) = \sum_{\text{all } y} y \cdot P(Y = y)$ to find the expectation, but because the distribution is symmetrical about $y = 7$ we can state immediately that $E(Y) = 7$.

Example

A discrete random variable has probability density function $P(X = x) = \dfrac{kx^3}{2}$ for $x = 1, 2, 3$ and 4.

a Find the value of k.
b Find E(X).

a $\dfrac{k}{2} + 4k + \dfrac{27k}{2} + 32k = 1$

$$\Rightarrow k = \dfrac{1}{50}$$

b $E(X) = 1 \times \dfrac{1}{100} + 2 \times \dfrac{8}{100} + 3 \times \dfrac{27}{100} + 4 \times \dfrac{64}{100}$

$\Rightarrow E(X) = 3.54$

Example

A discrete random variable X can only take the values 1, 2 and 3. If $P(X = 1) = 0.15$ and $E(X) = 2.4$, find the probability distribution for X.

The probability distribution for X is shown below:

x	1	2	3
$P(X = x)$	0.15	p	q

Since the variable is random $0.15 + p + q = 1$

$$\Rightarrow p + q = 0.85$$

If $E(X) = 2.4$ then $0.15 + 2p + 3q = 2.4$

$$\Rightarrow 2p + 3q = 2.25$$

Solving these two equations simultaneously gives $p = 0.3$ and $q = 0.55$. Hence the probability distribution function for X is:

x	1	2	3
$P(X = x)$	0.15	0.3	0.55

Example

Alan and Bob play a game in which each throws an unbiased die. The table below shows the amount in cents that Alan receives from Bob for each possible outcome of the game. For example, if both players throw a number greater than 3, Alan receives 50 cents from Bob while if both throw a number less than or equal to 3, Alan pays Bob 60 cents.

		B	
		≤ 3	> 3
A	≤ 3	-60	x
	> 3	40	50

Find

a **i** the expected value of Alan's gain in one game in terms of x

ii the value of x which makes the game fair to both players

iii the expected value of Alan's gain in 20 games if $x = 40$.

b Alan now discovers that the dice are biased and that the dice are three times more likely to show a number greater than 3 than a number less than or equal to 3. How much would Alan expect to win if $x = 30$?

a **i** On throwing a die, if X is the number thrown, then $P(X \leq 3) = P(X > 3) = \frac{1}{2}$.

The probability of each combination of results for Alan and Bob is

$\frac{1}{2} \times \frac{1}{2} = \frac{1}{4}$.

Hence the probability distribution table is shown below where the discrete random variable X is Alan's gain.

x	-60	x	40	50
$P(X = x)$	$\dfrac{1}{4}$	$\dfrac{1}{4}$	$\dfrac{1}{4}$	$\dfrac{1}{4}$

$\Rightarrow E(X) = \frac{1}{4} \times -60 + \frac{1}{4}x + \frac{1}{4} \times 40 + \frac{1}{4} \times 50 = \frac{30 + x}{4}$

ii If the game is fair, then neither player should gain or lose anything and hence $E(X) = 0$

$\Rightarrow \dfrac{30 + x}{4} = 0$

$\Rightarrow x = -30$

iii His expected gain in one game when $x = 40$ is $\dfrac{30 + 40}{4} = 17.5$ cents.

Hence his expected gain in 20 games is $20 \times 17.5 = 350$ cents.

b In this case the probability of die showing a number greater than 3 is $\frac{3}{4}$ and the probability of it showing a number less than or equal to 3 is $\frac{1}{4}$.

Hence the probability distribution table is now:

x	-60	30	40	50
$P(X = x)$	$\dfrac{1}{16}$	$\dfrac{3}{16}$	$\dfrac{3}{16}$	$\dfrac{9}{16}$

$\Rightarrow E(X) = \frac{1}{16} \times -60 + \frac{3}{16} \times 30 + \frac{3}{16} \times 40 + \frac{9}{16} \times 50 = 37.5$ cents

The expectation of any function f(x)

If $E(X) = \sum_{\text{all } x} x \cdot P(X = x)$, then $E(X^2) = \sum_{\text{all } x} x^2 \cdot P(X = x)$, $E(X^3) = \sum_{\text{all } x} x^3 \cdot P(X = x)$ etc.

In general, $E(f(x)) = \sum_{\text{all } x} f(x) \cdot P(X = x)$.

Example

For the probability distribution shown below, find:

x	0	1	2	3	4	5
P(X = x)	0.08	0.1	0.2	0.4	0.15	0.07

a $E(X)$ **b** $E(X^2)$ **c** $E(2X)$ **d** $E(2X - 1)$

a $E(X) = 0 \times 0.08 + 1 \times 0.1 + 2 \times 0.2 + 3 \times 0.4 + 4 \times 0.15$
$+ 5 \times 0.07 = 2.65$

b In this case the probability distribution is shown below:

x^2	0	1	4	9	16	25
$P(X = x^2)$	0.08	0.1	0.2	0.4	0.15	0.07

$\Rightarrow E(X^2) = 0 \times 0.08 + 1 \times 0.1 + 4 \times 0.2 + 9 \times 0.4 + 16 \times 0.15$
$+ 25 \times 0.07 = 8.65$

c The probability distribution for this is:

2x	0	2	4	6	8	10
$P(X = 2x)$	0.08	0.1	0.2	0.4	0.15	0.07

$\Rightarrow E(2X) = 0 \times 0.08 + 2 \times 0.1 + 4 \times 0.2 + 6 \times 0.4 + 8 \times 0.15$
$+ 10 \times 0.07 = 5.3$

d The probability distribution for this is:

2x − 1	−1	1	3	5	7	9
$P(X = 2x - 1)$	0.08	0.1	0.2	0.4	0.15	0.07

$\Rightarrow E(2X - 1) = -1 \times 0.08 + 1 \times 0.1 + 3 \times 0.2 + 5 \times 0.4$
$+ 7 \times 0.15 + 9 \times 0.07 = 4.3$

This idea becomes important when we need to find the variance.

The variance, Var(X)

From Chapter 19, we know that for a frequency distribution with mean $\bar{x}$, the variance is given by

$$s^2 = \frac{\sum f(x - \bar{x})^2}{\sum f} \quad \text{or} \quad s^2 = \frac{\sum fx^2}{\sum f} - \bar{x}^2$$

Using the first formula we can see that the variance is the mean of the squares of the deviations from the mean. If we now take a theoretical approach using a probability distribution from a discrete random variable, where we define $E(X) = \mu$ and apply the same idea, we find

$$\text{Var}(X) = E(X - \mu)^2.$$

However, we do not normally use this form and the alternative form we usually use is shown below.

$$\text{Var}(X) = E(X - \mu)^2$$

$$= E[X^2 - 2\mu X + \mu^2]$$

$$= E(X^2) - 2\mu E(X) + \mu^2$$

$$= E(X^2) - 2\mu^2 + \mu^2$$

$$= E(X^2) - \mu^2$$

$$\boxed{\text{Var}(X) = E(X^2) - E^2(X)}$$

> The variance can never be negative. If it is, then a mistake has been made in the calculation.

Example

For the probability distribution shown below for a discrete random variable X, find:

x	-2	-1	0	1	2
$P(X = x)$	0.1	0.25	0.3	0.25	0.1

a $E(X)$ **b** $E(X^2)$ **c** $\text{Var}(X)$

a $E(X) = 0$ since the distribution is symmetrical.

b $E(X^2) = 4 \times 0.1 + 1 \times 0.25 + 0 \times 0.3 + 1 \times 0.25 + 4 \times 0.1 = 1.3$

c $\text{Var}(X) = E(X^2) - E^2(X)$

$\qquad = 1.3 - 0^2 = 1.3$

Example

A cubical die and a tetrahedral die are thrown together.
   a If X is the discrete random variable "total scored", write down the probability distribution for X.
   b Find $E(X)$.
   c Find $\text{Var}(X)$.

A game is now played with the two dice. Anna has the cubical die and Beth has the tetrahedral die. They each gain points according to the following rules:

- If the number on both dice is greater than 3, then Beth gets 6 points.
- If the tetrahedral die shows 3 and the cubical die less than or equal to 3, then Beth gets 4 points.
- If the tetrahedral die shows 4 and the cubical die less than or equal to 3, then Beth gets 2 points.
- If the tetrahedral die shows a number less than 3 and the cubical die shows a 3, then Anna gets 5 points.
- If the tetrahedral die shows a number less than 3 and the cubical die shows a 1 or a 2, then Anna gets 3 points.
- If the tetrahedral die shows 3 and the cubical die greater than 3, then Anna gets 2 points.
- If the tetrahedral die shows a number less than 3 and the cubical die shows a number greater than 3, then Anna gets 1 point.

    d Write out the probability distribution for Y, "the number of points gained by Anna".

    e Calculate $E(Y)$ and $Var(Y)$.

    f The game is now to be made fair by changing the number of points Anna gets when the tetrahedral die shows a number less than 3 and the cubical die shows a 1 or a 2. What is this number of points to the nearest whole number?

    a A probability space diagram is the easiest way to show the possible outcomes.

6	7	8	9	10
5	6	7	8	9
4	5	6	7	8
3	4	5	6	7
2	3	4	5	6
1	2	3	4	5
	1	2	3	4

Hence the probability distribution for X is:

x	2	3	4	5	6	7	8	9	10
$P(X=x)$	$\frac{1}{24}$	$\frac{2}{24}$	$\frac{3}{24}$	$\frac{4}{24}$	$\frac{4}{24}$	$\frac{4}{24}$	$\frac{3}{24}$	$\frac{2}{24}$	$\frac{1}{24}$

    b Since the probability distribution is symmetrical, $E(X) = 6$.

    c $E(X^2) = 4 \times \frac{1}{24} + 9 \times \frac{2}{24} + 16 \times \frac{3}{24} + 25 \times \frac{4}{24} + 36 \times \frac{4}{24}$

$$+ 49 \times \frac{4}{24} + 64 \times \frac{3}{24} + 81 \times \frac{2}{24} + 100 \times \frac{1}{24}$$

$$= \frac{964}{24} = \frac{241}{6}$$

$$Var(X) = E(X^2) - E^2(X)$$
$$= \frac{241}{6} - 6^2 = \frac{25}{6}$$

d By considering the possibility space diagram again the probability distribution is:

y	-6	-4	-2	5	3	2	1
$P(Y = y)$	$\dfrac{1}{8}$	$\dfrac{1}{8}$	$\dfrac{1}{8}$	$\dfrac{1}{12}$	$\dfrac{1}{6}$	$\dfrac{1}{8}$	$\dfrac{1}{4}$

e $E(Y) = -6 \times \dfrac{1}{8} - 4 \times \dfrac{1}{8} - 2 \times \dfrac{1}{8} + 5 \times \dfrac{1}{12} + 3 \times \dfrac{1}{6} + 2 \times \dfrac{1}{8} + 1 \times \dfrac{1}{4} = -\dfrac{1}{12}$

$E(Y^2) = 36 \times \dfrac{1}{8} + 16 \times \dfrac{1}{8} + 4 \times \dfrac{1}{8} + 25 \times \dfrac{1}{12} + 9 \times \dfrac{1}{6} + 4 \times \dfrac{1}{8} + 1 \times \dfrac{1}{4} = \dfrac{34}{3}$

$Var(Y) = E(Y^2) - E^2(Y)$

$= \dfrac{34}{3} - \dfrac{1}{144} = \dfrac{1631}{144}$

f Let the number of points Anna gains when the tetrahedral die shows a number less than 3 and the cubical die shows a 1 or a 2 be y.

In this case the probability distribution is now:

y	-6	-4	-2	5	y	2	1
$P(Y = y)$	$\dfrac{1}{8}$	$\dfrac{1}{8}$	$\dfrac{1}{8}$	$\dfrac{1}{12}$	$\dfrac{1}{6}$	$\dfrac{1}{8}$	$\dfrac{1}{4}$

$E(Y) = -6 \times \dfrac{1}{8} - 4 \times \dfrac{1}{8} - 2 \times \dfrac{1}{8} + 5 \times \dfrac{1}{12} + y \times \dfrac{1}{6} + 2 \times \dfrac{1}{8} + 1 \times \dfrac{1}{4}$

Since the game is now fair, $E(Y) = 0$

$\Rightarrow -\dfrac{7}{12} + \dfrac{y}{6} = 0$

$\Rightarrow y = 3.5$

Exercise 2

1 Find the value of b and $E(X)$ in these distributions.

a

x	3	4	5	6
$P(X = x)$	0.1	b	0.4	0.2

b

x	0	1	2	3	4
$P(X = x)$	0.05	0.15	0.6	b	0.05

c

x	-3	0	3	6	9
$P(X = x)$	0.15	0.2	b	0.22	0.15

d

x	-4	1	3	5	6
$P(X = x)$	0.1	0.25	0.3	0.25	b

2 If three unbiased cubical dice are thrown, what is the expected number of threes that will occur?

3 A discrete random variable has a probability distribution function given by

$f(x) = \dfrac{cx^2}{12}$ for $x = 1, 2, 3, 4, 5, 6$.

a Find the value of c.   **b** Find E(X).

4 Caroline and Lisa play a game that involves each of them tossing a fair coin. The rules are as follows:
- If Caroline and Lisa both get heads, Lisa gains 6 points.
- If Caroline and Lisa both get tails, Caroline gains 6 points.
- If Caroline gets a tail and Lisa gets a head, Lisa gains 3 points.
- If Caroline gets a head and Lisa gets a tail, Caroline gains x points.

a If X is the discrete random variable "Caroline's gain", find E(X) in terms of x.

b What value of x makes the game fair?

5 A discrete random variable X can only take the values -1 and 1. If E(X) = 0.4, find the probability distribution for X.

6 A discrete random variable X can only take the values -1, 1 and 3. If P($X = 1$) = 0.25 and E(X) = 1.9, find the probability distribution for X.

7 A discrete random variable Y can only take the values 0, 2, 4 and 6. If P($Y \le 4$) = 0.6, P($Y \le 2$) = 0.5, P($Y = 2$) = P($Y = 4$) and E(Y) = 2.4, find the probability distribution for Y.

8 A five-a-side soccer team is to be chosen from four boys and five girls. If the team members are chosen at random, what is the expected number of girls on the team?

9 In a chemistry examination each question is a multiple choice with four possible answers. Given that Kevin randomly guesses the answers to the first four questions, how many of the first four questions can he expect to get right?

10 A discrete random variable X has probability distribution:

x	0	1	2	3	4
P($X = x$)	0.1	0.2	0.35	0.25	0.1

Find:

a E(X)   **b** E(X^2)   **c** E($2X - 1$)   **d** E($3X + 2$)

11 A discrete random variable X has probability distribution:

x	-2	0	2	4	6
P($X = x$)	0.05	0.15	0.25	0.35	0.2

Find:

a E(X)   **b** E(X^2)   **c** E($2X + 1$)   **d** E($3X - 1$)

12 Find Var(X) for each of these probability distributions.

a

x	0	1	2	3	4
P($X = x$)	0.2	0.2	0.3	0.15	0.15

b

x	1	3	5	7	9
P($X = x$)	0.05	0.2	0.2	0.3	0.25

c

x	−4	−2	0	2	4
$P(X = x)$	0.1	0.3	p	0.2	0.15

d

x	−2	−1	0	1	2
$P(X = x)$	0.03	0.2	p	0.35	0.1

13 If X is the sum of the numbers shown when two unbiased dice are thrown, find:

a $E(X)$ **b** $E(X^2)$ **c** $Var(X)$

14 Three members of a school committee are to be chosen from three boys and four girls. If Y is the random variable "number of boys chosen", find:

a $E(Y)$ **b** $E(Y^2)$ **c** $Var(Y)$

15 A discrete random variable X has probability distribution:

x	0	1	2	3	4
$P(X = x)$	k	0.2	$2k$	0.3	$4k$

a Find the value of k. **b** Calculate $E(X)$. **c** Calculate $Var(X)$.

16 A teacher randomly selects 4 students from a class of 15 to attend a careers talk. In the class there are 7 girls and 8 boys. If Y is the number of girls selected and each selection is independent of the others, find

a the probability distribution for Y **b** $E(Y)$ **c** $Var(Y)$

17 A discrete random variable X takes the values $x = 1, 3, 5$, with probabilities $\frac{1}{7}, \frac{5}{14}$ and k respectively. Find

a k

b the mean of X

c the standard deviation of X.

18 One of the following expressions can be used as a probability density function for a discrete random variable X. Identify which one and calculate its mean and standard deviation.

a $f(x) = \dfrac{x^2 + 1}{35}, x = 0, 1, 2, 3, 4$

b $g(x) = \dfrac{x - 1}{7}, x = 0, 1, 2, 4, 5$

19 Jim has been writing letters. He has written four letters and has four envelopes addressed. Unfortunately he drops the letters on the floor and he has no way of distinguishing which letters go in which envelopes so he puts each letter in each envelope randomly. Let X be the number of letters in their correct envelopes.

a State the values which X can take.

b Find the probabilities for these values of X.

c Calculate the mean and variance for X.

20 A box contains ten numbered discs. Three of the discs have the number 5 on them, four of the discs have the number 6 on them, and three of the discs have the number 7 on them. Two discs are drawn without replacement and the score is the sum of the numbers shown on the discs. This is denoted by X.

a Write down the values that X can take.

b Find the probabilities of these values of X.

c Calculate the expectation and variance of X.

d Two children, Ahmed and Belinda, do this. Find the probability that Ahmed gains a higher score than Belinda.

21 Pushkar buys a large box of fireworks. The probability of there being X fireworks that fail is shown in the table below.

x	0	1	2	3	≥ 4
$P(X = x)$	$9k$	$3k$	k	k	0

a Find the value of k.

b Find $E(X)$ and $Var(X)$.

c His friend, Priya, also buys a box. They put their fireworks together and the total number of fireworks that fail, Y, is determined. What values can Y take?

d Write down the distribution for Y.

e Find the expectation and variance of Y.

21.3 Binomial distribution

This is a distribution that deals with events that either occur or do not occur, so there are two complementary outcomes. We are usually told the number of times an event occurs and we are given the probability of the event happening or not happening.

Consider the three pieces of Mathematics Higher Level homework done by Jay. The probability of him seeking help from his teacher is 0.8.

The tree diagram to represent this is shown below.

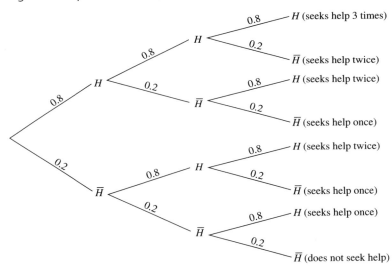

If X is the number of times he seeks help, then from the tree diagram:

$P(X = 0) = 0.2 \times 0.2 \times 0.2 = 0.008$

By using the different branches of the tree diagram we can calculate the values for $x = 1, 2, 3$.

Without using the tree diagram we can see that the probability of him never seeking help is $0.2 \times 0.2 \times 0.2$ and this can happen in $^3C_0$ ways, giving $P(X = 0) = 0.008$.

The probability of him seeking help once is $0.8 \times 0.2 \times 0.2$ and this can happen in $^3C_1$ ways, giving $P(X = 1) = 0.096$.

The probability of him seeking help twice is $0.8 \times 0.8 \times 0.2$ and this can happen in $^3C_2$ ways, giving $P(X = 2) = 0.384$.

The probability of him seeking help three times is $0.8 \times 0.8 \times 0.8$ and this can happen in $^3C_3$ ways, giving $P(X = 3) = 0.512$.

Without using the tree diagram we can see that for 20 homeworks, say, the probability of him seeking help once would be $^{20}C_1 \times 0.8 \times 0.2^{19}$.

If we were asked to do this calculation using a tree diagram it would be very time consuming!

Generalizing this leads to a formula for a binomial distribution.

If a random variable X follows a **binomial distribution** we say $X \sim \text{Bin}(n, p)$ where
$n = $ number of times an event occurs and $p = $ probability of success.
The probability of failure $= q = 1 - p$.
n and p are called the parameters of the distribution.

If $X \sim \text{Bin}(n, p)$ then $P(X = x) = {}^nC_x p^x q^{n-x}$.

Example

If $X \sim \text{Bin}\left(7, \dfrac{1}{4}\right)$, find:

  a $P(X = 6)$

  b $P(X \leq 2)$

  a In this case $n = 7, p = \dfrac{1}{4}$ and $q = 1 - \dfrac{1}{4} = \dfrac{3}{4}$.

      Hence $P(X = 6) = {}^7C_6 \left(\dfrac{1}{4}\right)^6 \left(\dfrac{3}{4}\right)^1 = 0.00128$

  b $P(X \leq 2) = P(X = 0) + P(X = 1) + P(X = 2)$

$$= {}^7C_0 \left(\dfrac{1}{4}\right)^0 \left(\dfrac{3}{4}\right)^7 + {}^7C_1 \left(\dfrac{1}{4}\right)^1 \left(\dfrac{3}{4}\right)^6 + {}^7C_2 \left(\dfrac{1}{4}\right)^2 \left(\dfrac{3}{4}\right)^5 = 0.756$$

It is usual to do these calculations on a calculator. The screen shots for these are shown below.

  a

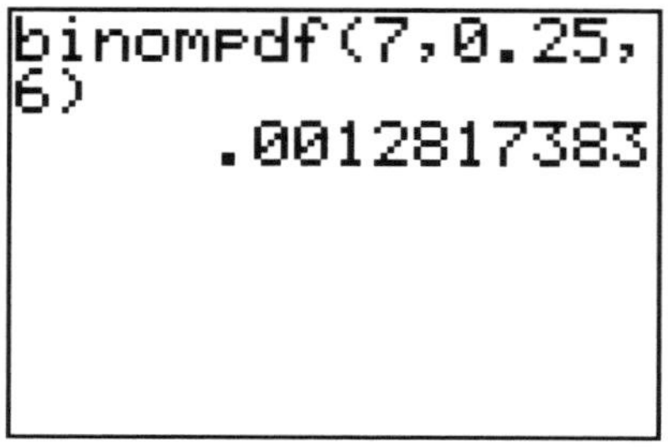

```
binompdf(7,0.25,
6)
           .0012817383
```

  b

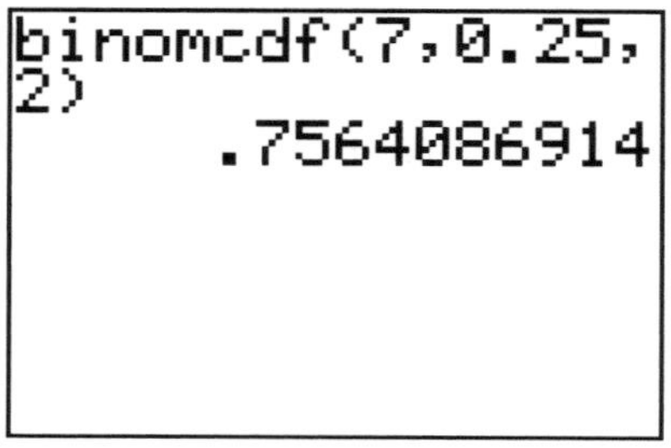

```
binomcdf(7,0.25,
2)
           .7564086914
```

So how do we recognize a binomial distribution? For a distribution to be binomial there must be an event that happens a finite number of times and the probability of that event happening must not change and must be independent of what happened before. Hence if we have 8 red balls and 6 black balls in a bag, and we draw 7 balls from the bag one after the other with replacement, X = "the number of red balls drawn" follows a binomial distribution. Here the number of events is 7 and the probability of success (drawing a red ball) is constant. If the problem were changed to the balls not being replaced, then the probability of drawing a red ball would no longer be constant and the distribution would no longer follow a binomial distribution.

Example

Market research is carried out at a supermarket, looking at customers buying cans of soup. If a customer buys one can of soup, the probability that it is tomato soup is 0.75. If ten shoppers buy one can of soup each, what is the probability that

   a exactly three buy tomato soup

   b less than six buy tomato soup

   c more than four buy tomato soup?

The distribution for this is $X \sim \text{Bin}(10, 0.75)$.

a $P(X = 3) = {}^{10}C_3 \left(\dfrac{3}{4}\right)^3 \left(\dfrac{1}{4}\right)^7 = 0.00309$

Or from the calculator:

```
binompdf(10,0.75
,3)
        .0030899048
■
```

b $P(X < 6) = P(X = 0) + P(X = 1) + P(X = 2) + P(X = 3)$
$$+ P(X = 4) + P(X = 5)$$

$$= {}^{10}C_0 \left(\frac{3}{4}\right)^0 \left(\frac{1}{4}\right)^{10} + {}^{10}C_1 \left(\frac{3}{4}\right)^1 \left(\frac{1}{4}\right)^9 + {}^{10}C_2 \left(\frac{3}{4}\right)^2 \left(\frac{1}{4}\right)^8$$

$$+ {}^{10}C_3 \left(\frac{3}{4}\right)^3 \left(\frac{1}{4}\right)^7 + {}^{10}C_4 \left(\frac{3}{4}\right)^4 \left(\frac{1}{4}\right)^6 + {}^{10}C_5 \left(\frac{3}{4}\right)^5 \left(\frac{1}{4}\right)^5$$

$$= 0.0781$$

Or from the calculator:

```
binomcdf(10,0.75
,5)
        .0781269074
```

c In a binomial distribution, the sum of the probabilities is one and hence it is sometimes easier to subtract the answer from one.

In this case $P(X > 4) = 1 - P(X \le 4)$

$$= 1 - \{P(X = 0) + P(X = 1) + P(X = 2) + P(X = 3) + P(X = 4)\}$$

$$= 1 - \left\{ {}^{10}C_0\left(\frac{3}{4}\right)^0\left(\frac{1}{4}\right)^{10} + {}^{10}C_1\left(\frac{3}{4}\right)^1\left(\frac{1}{4}\right)^9 + {}^{10}C_2\left(\frac{3}{4}\right)^2\left(\frac{1}{4}\right)^8 \right.$$

$$\left. + {}^{10}C_3\left(\frac{3}{4}\right)^3\left(\frac{1}{4}\right)^7 + {}^{10}C_4\left(\frac{3}{4}\right)^4\left(\frac{1}{4}\right)^6 \right\}$$

$$= 1 - 0.0197\ldots = 0.980$$

On the calculator we also subtract the answer from 1.

```
binomcdf(10,0.75
,4)
         .0197277069
1-Ans
         .9802722931
```

> In questions involving discrete distributions, ensure you read the question. If a question asks for more than 2, this is different from asking for at least 2. This also affects what is inputted into the calculator.

Example

Scientists have stated that in a certain town it is equally likely that a woman will give birth to a boy or a girl. In a family of seven children, what is the probability that there will be at least one girl?

"At least" problems, i.e. finding $P(X \ge x)$, can be dealt with in two ways. Depending on the number, we can either calculate the answer directly or we can work out $P(X < x)$, and subtract the answer from 1.

In this case, $X \sim \text{Bin}(7, 0.5)$ and we want $P(X \ge 1)$.

$P(X \ge 1) = 1 - P(X = 0)$

$$= 1 - {}^{7}C_0(0.5)^0(0.5)^7 = 0.992$$

Or from the calculator:

```
binompdf(7,0.5,0
)
            .0078125
1-Ans
            .9921875
```

Example

The probability of rain on any particular day in June is 0.45. In any given week in June, what is the most likely number of days of rain?

If we are asked to find the most likely value, then we should work through all the probabilities and then state the value with the highest probability. In this case the calculator is very helpful.

If we let X be the random variable "the number of rainy days in a week in June", then the distribution is $X \sim \text{Bin}(7, 0.45)$.

From the calculator, the results are:

x	P(X = x)
0	0.0152 …
1	0.0871 …
2	0.214 …
3	0.291 …
4	0.238 …
5	0.117 …
6	0.0319 …
7	0.00373 …

Hence we can state that the most likely number of days is 3.

If asked to do a question of this sort, it is not usual to write out the whole table. It is enough to write down the highest value and one either side and state the conclusion from there. This is because in the binomial distribution the probabilities increase to a highest value and then decrease again and hence once we have found where the highest value occurs we know it will not increase beyond this value elsewhere.

Expectation and variance of a binomial distribution

$$\text{If } X \sim \text{Bin}(n, p)$$
$$E(X) = np$$
$$\text{Var}(X) = npq$$

The proofs for these are shown below, but they will not be asked for in examination questions.

Proof that $E(X) = np$

Let $X \sim \text{Bin}(n, p)$

Hence $P(X = x) = {}^nC_x p^x q^{n-x}$

Therefore the probability distribution for this is:

x	0	1	2	…	n
P(X = x)	q^n	$nq^{n-1}p$	$\dfrac{n(n-1)}{2!}q^{n-2}p^2$		p^n

Now $E(X) = \displaystyle\sum_{\text{all } x} x \cdot P(X = x)$

$$= 0 \cdot q^n + 1 \cdot nq^{n-1}p + 2 \cdot \frac{n(n-1)}{2!}q^{n-2}p^2 + \ldots + n \cdot p^n$$
$$= np[q^{n-1} + (n-1)q^{n-2}p + \ldots + p^{n-1}]$$
$$= np(q + p)^{n-1}$$

Since $q + p = 1$, $E(X) = np$.

Proof that $\text{Var}(X) = npq$

$\text{Var}(X) = E(X^2) - E^2(X)$

Now $E(X^2) = \sum_{\text{all } x} x^2 \cdot P(X = x)$

$= 0 \cdot q^n + 1 \cdot n q^{n-1} p + 4 \cdot \dfrac{n(n-1)}{2!} q^{n-2} p^2 + 9 \cdot \dfrac{n(n-1)(n-2)}{3!} q^{n-3} p^3 + \dots$

$\qquad + n^2 \cdot p^n$

$= np \left[q^{n-1} + 2(n-1)q^{n-2}p + \dfrac{3(n-1)(n-2)}{2!} q^{n-3} p^2 + \dots + np^{n-1} \right]$

This can be split into two series:

$= np \Bigg\{ \left[q^{n-1} + (n-1)q^{n-2}p + \dfrac{(n-1)(n-2)}{2!} q^{n-3} p^2 + \dots + p^{n-1} \right]$

$\qquad + \left[(n-1)q^{n-2}p + \dfrac{2(n-1)(n-2)}{2!} q^{n-3} p^2 + \dots + (n-1)p^{n-1} \right] \Bigg\}$

$= np(q+p)^{n-1} + np \Bigg[(n-1)q^{n-2}p + \dfrac{2(n-1)(n-2)}{2!} q^{n-3} p^2 + \dots$

$\qquad + (n-1)p^{n-1} \Bigg]$

> Since the first series is the same as the one in the proof of $E(X)$.

$= np\{1 + (n-1)p[q^{n-2} + (n-2)q^{n-3}p + \dots + p^{n-2}]\}$

> Since $p + q = 1$.

$= np\{1 + (n-1)p(q+p)^{n-2}\}$

$= np\{1 + (n-1)p\}$

> Again since $p + q = 1$.

Hence $\text{Var}(X) = np\{1 + (n-1)p\} - (np)^2$

$\qquad\qquad\quad = np + n^2p^2 - np^2 - n^2p^2$

$\qquad\qquad\quad = np(1-p) = npq$

Example

X is a random variable such that $X \sim \text{Bin}(n, p)$. Given that $E(X) = 3.6$ and $p = 0.4$, find n and the standard deviation of X.

Since $p = 0.4$ then $q = 1 - 0.4 = 0.6$.

Using the formula $\quad E(X) = np$

$\qquad\qquad\qquad \Rightarrow 3.6 = 0.4n$

$\qquad\qquad\qquad \Rightarrow n = 9$

$\text{Var}(X) = npq = 9 \times 0.4 \times 0.6 = 2.16$

Hence the standard deviation is $\sqrt{2.16} = 1.47$.

Example

In a class mathematics test, the probability of a girl passing the test is 0.62 and the probability of a boy passing the test is 0.65. The class contains 15 boys and 17 girls.

a What is the expected number of boys to pass?

b What is the most likely number of girls to pass?

c What is the probability that more than eight boys fail?

If X is the random variable "the number of boys who pass" and Y is the random variable "the number of girls who pass", then $X \sim Bin(15, 0.65)$ and $Y \sim Bin(17, 0.62)$.

a $E(X) = np = 15 \times 0.65 = 9.75$

b From the calculator the results are:

y	10	11	12
$P(Y = y)$	0.186 …	0.193 …	0.158 …

Hence we can state that the most likely number of girls passing is 11.

c The probability of more than eight boys failing is the same as the probability of no more than six boys passing, hence we require $P(X \le 6)$.

```
binomcdf(15,0.65
,6)
        .042193838
```

The expectation is the theoretical equivalent of the mean, whereas the most likely is the equivalent of the mode.

Therefore the probability that more than eight boys fail is 0.0422.

Example

Annabel always takes a puzzle book on holiday with her and she attempts a puzzle every day. The probability of her successfully solving a puzzle is 0.7. She goes on holiday for four weeks.

a Find the expected value and the standard deviation of the number of successfully solved puzzles in a given week.

b Find the probability that she successfully solves at least four puzzles in a given week.

c She successfully solves a puzzle on the first day of the holiday. What is the probability that she successfully solves at least another three during the rest of that week?

d Find the probability that she successfully solves four or less puzzles in only one of the four weeks of her holiday.

Let X be the random variable "the number of puzzles successfully completed by Annabel". Hence $X \sim Bin(7, 0.7)$.

a $E(X) = np = 7 \times 0.7 = 4.9$

$Var(X) = npq = 7 \times 0.7 \times 0.3 = 1.47$

Hence standard deviation $= \sqrt{1.47} = 1.21$

b $P(X \geq 4) = 1 - [P(X = 0) + P(X = 1) + P(X = 2) + P(X = 3)]$

$$= 1 - \left\{ {}^7C_0(0.7)^0(0.3)^7 + {}^7C_1(0.7)^1(0.3)^6 \right.$$

$$\left. + {}^7C_2(0.7)^2(0.3)^5 + {}^7C_3(0.7)^3(0.3)^4 \right\}$$

$$= 1 - 0.126\ldots = 0.874$$

```
binomcdf(7,0.7,3
)
             .126036
1-Ans
             .873964
```

c This changes the distribution and we now want $P(Y \geq 3)$ where $Y \sim \text{Bin}(6, 0.7)$.

$P(Y \geq 3) = 1 - P(Y \leq 2)$

$$= 1 - \left\{ {}^6C_0(0.7)^0(0.3)^6 + {}^6C_1(0.7)^1(0.3)^5 + {}^6C_2(0.7)^2(0.3)^4 \right\}$$

$$= 1 - 0.0704\ldots = 0.930$$

```
binomcdf(6,0.7,2
)
              .07047
1-Ans
              .92953
```

d We first calculate the probability that she successfully completes four or less in a week, $P(X \leq 4)$.

$P(X \leq 4) = P(X = 0) + P(X = 1) + P(X = 2) + P(X = 3) + P(X = 4)$

$$= {}^7C_0(0.7)^0(0.3)^7 + {}^7C_1(0.7)^1(0.3)^6 + {}^7C_2(0.7)^2(0.3)^5$$

$$+ {}^7C_3(0.7)^3(0.3)^4 + {}^7C_4(0.7)^4(0.3)^3$$

$$= 0.353$$

```
binomcdf(7,0.7,4
)
            .3529305
```

We now want $P(A = 1)$ where $A \sim \text{Bin}(4, 0.353)$.

$P(A = 1) = {}^4C_1(0.353)^1(0.647)^3 = 0.382$

Exercise 3

1 If $X \sim \text{Bin}(7, 0.35)$, find:
   a $P(X = 3)$ **b** $P(X \leq 2)$ **c** $P(X > 4)$

2 If $X \sim \text{Bin}(10, 0.4)$, find:
   a $P(X = 5)$ **b** $P(X \geq 3)$ **c** $P(X \leq 5)$

3 If $X \sim \text{Bin}(8, 0.25)$, find:
   a $P(X = 3)$ **b** $P(X \geq 5)$ **c** $P(X \leq 4)$ **d** $P(X = 0 \text{ or } 1)$

4 A biased coin is tossed ten times. On each toss, the probability that it will land on a head is 0.65. Find the probability that it will land on a head at least six times.

5 Given that $X \sim \text{Bin}(6, 0.4)$, find
   a $E(X)$ **b** $\text{Var}(X)$ **c** the most likely value for X.

6 In a bag of ten discs, three of them are numbered 5 and seven of them are numbered 6. A disc is drawn at random, the number noted, and then it is replaced. This happens eight times. Find
   a the expected number of 5's
   b the variance of the number of 5's drawn
   c the most likely number of 5's drawn.

7 A random variable Y follows a binomial distribution with mean 1.75 and variance 1.3125.
   a Find the values of n, p and q.
   b What is the probability that Y is less than 2?
   c Find the most likely value(s) of Y.

8 An advert claims that 80% of dog owners, prefer Supafood dog food. In a sample of 15 dog owners, find the probability that
   a exactly seven buy Supafood
   b more than eight buy Supafood
   c ten or more buy Supafood.

9 The probability that it will snow on any given day in January in New York is given as 0.45. In any given week in January, find the probability that it will snow on
   a exactly one day   **b** more than two days
   c at least three days   **d** no more than four days.

10 A student in a mathematics class has a probability of 0.68 of gaining full marks in a test. She takes nine tests in a year. What is the probability that she will
   a never gain full marks
   b gain full marks three times in a year
   c gain full marks in more than half the tests
   d gain full marks at least eight times?

11 Alice plays a game that involves kicking a small ball at a target. The probability that she hits the target is 0.72. She kicks the ball eight times.
   a Find the probability that she hits the target exactly five times.
   b Find the probability that she hits the target for the first time on her fourth kick.

12 In a school, 19% of students fail the IB Diploma. Find the probability that in a class of 15 students
   a exactly two will fail   **b** less than five will fail
   c at least eight will pass.

13 A factory makes light bulbs that it distributes to stores in boxes of 20. The probability of a light bulb being defective is 0.05.

  a Find the probability that there are exactly three defective bulbs in a box of light bulbs.

  b Find the probability that there are more than four defective light bulbs in a box.

  c If a certain store buys 25 boxes, what is the probability that at least two of them have more than four defective light bulbs?

  The quality control department in the company decides that if a randomly selected box has no defective light bulbs in it, then all bulbs made that day will pass and if it has two or more defective light bulbs in it, then all light bulbs made that day will be scrapped. If it has one defective light bulb in it, then another box will be tested, and if that has no defective light bulbs in it, all light bulbs made that day will pass. Otherwise all light bulbs made that day will be scrapped.

  d What is the probability that the first box fails but the second box passes?

  e What is the probability that all light bulbs made that day will be scrapped?

14 A multiple choice test in biology consists of 40 questions, each with four possible answers, only one of which is correct. A student chooses the answers to the questions at random.

  a What is the expected number of correct answers?

  b What is the standard deviation of the number of correct answers?

  c What is the probability that the student gains more than the expected number of correct answers?

15 In a chemistry class a particular experiment is performed with a probability of success p. The outcomes of successive experiments are independent.

  a Find the value of p if probability of gaining three successes in six experiments is the same as gaining four successes in seven experiments.

  b If p is now given as 0.25, find the number of times the experiment must be performed in order that the probability of gaining at least one success is greater than 0.99.

16 The probability of the London to Glasgow train being delayed on a weekday is $\frac{1}{15}$. Assuming that the delays occur independently, find

  a the probability that the train experiences exactly three delays in a five-day week

  b the most likely number of delays in a five-day week

  c the expected number of delays in a five-day week

  d the number of days such that there is a 20% probability of the train having been delayed at least once

  e the probability of being delayed at least twice in a five-day week

  f the probability of being delayed at least twice in each of two weeks out of a four-week period (assume each week has five days in it).

17 It is known that 14% of a large batch of light bulbs is defective. From this batch of light bulbs, 15 are selected at random.

  a Write down the distribution and state its mean and variance.

  b Calculate the most likely number of defective light bulbs.

  c What is the probability of exactly three defective light bulbs?

  d What is the probability of at least four defective light bulbs?

  e If six batches of 15 light bulbs are selected randomly, what is the probability that at least three of them have at least four defective light bulbs?

18 In the game scissors, paper, rock, a girl never chooses paper, and is twice as likely to choose scissors as rock. She plays the game eight times.

    a Write down the distribution for X, the number of times she chooses rock.

    b Find $P(X = 1)$.

    c Find $E(X)$.

    d Find the probability that X is at least one.

19 On a statistics course at a certain university, students complete 12 quizzes. The probability that a student passes a quiz is $\frac{2}{3}$.

    a What is the expected number of quizzes a student will pass?

    b What is the probability that the student will pass more than half the quizzes?

    c What is the most likely number of quizzes that the student will pass?

    d At the end of the course, the student takes an examination. The probability of passing the examination is $\frac{n}{55}$, given that n is the number of quizzes passed. What is the probability that the student passes four quizzes and passes the examination?

21.4 Poisson distribution

Consider an observer counting the number of cars passing a specific point on a road during 100 time intervals of 30 seconds. He finds that in these 100 time intervals a total of 550 cars pass.

Now if we assume from the beginning that 550 cars will pass in these time intervals, that a car passing is independent of another car passing, and that it is equally likely that they will pass in any of the time intervals, then the probability that a car passes in any specific time interval is $\frac{1}{100}$. The probability that a second car arrives in this time interval is also $\frac{1}{100}$ as the events are independent, and so on. Hence the number of cars passing this point in this time period follows a binomial distribution $X \sim \text{Bin}\left(550, \frac{1}{100}\right)$.

Unfortunately, this is not really the case as we do not know exactly how many cars will pass in any interval. What we do know from experience is the mean number of cars that will pass. Also, as n gets larger, p must become smaller. That is, the more cars we observe, the less likely it is that a specific car will pass in a given interval. Hence the distribution we want is one where n increases as p decreases and where the mean np stays constant. This is called a Poisson distribution and occurs when an event is evenly spaced, on average, over an infinite space.

If a random variable X follows a **Poisson distribution**, we say $X \sim \text{Po}(\lambda)$ where λ is the parameter of the distribution and is equal to the mean of the distribution.

If $X \sim \text{Po}(\lambda)$ then $P(X = x) = \dfrac{e^{-\lambda}\lambda^{x}}{x!}$.

Example

If $X \sim Po(2)$, find:

a $P(X = 3)$

b $P(X \leq 4)$

a $P(X = 3) = \dfrac{e^{-2}2^3}{3!} = 0.180$

b $P(X \leq 4) = P(X = 0) + P(X = 1) + P(X = 2) + P(X = 3) + P(X = 4)$

$$= \frac{e^{-2}2^0}{0!} + \frac{e^{-2}2^1}{1!} + \frac{e^{-2}2^2}{2!} + \frac{e^{-2}2^3}{3!} + \frac{e^{-2}2^4}{4!} = 0.947$$

As with the binomial distribution, it is usual to do these calculations on the calculator.

a

b

To recognize a Poisson distribution we normally have an event that is randomly scattered in time or space and has a mean number of occurrences in a given interval of time or space.

Unlike the binomial distribution, X can take any positive integer value up to infinity and hence if we want $P(X \geq x)$ we must always subtract the answer from 1. As x becomes very large, the probability becomes very small.

Example

The mean number of zebra per square kilometre in a game park is found to be 800. Given that the number of zebra follows a Poisson distribution, find the probability that in one square kilometre of game park there are

a 750 zebra

b less than 780 zebra

c more than 820 zebra.

Let X be the number of zebra in one square kilometre.

Hence $X \sim Po(800)$.

a We require $P(X = 750) = e^{-800} \cdot \dfrac{800^{750}}{750!}$.

Because of the numbers involved, we have to use the Poisson function on a calculator.

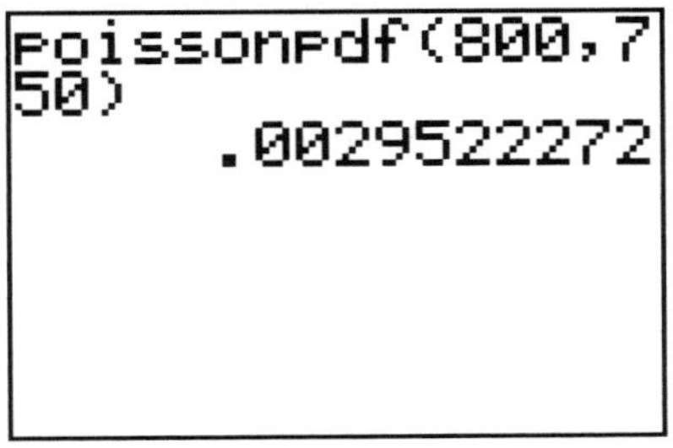

```
Poissonpdf(800,7
50)
          .0029522272
```

$P(X = 750) = 0.00295$

b In this case we have to use a calculator. We want less than 780, which is the same as less than or equal to 779.

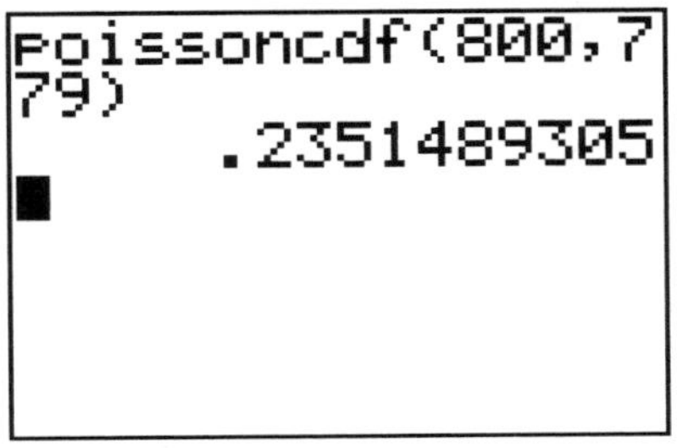

```
Poissoncdf(800,7
79)
          .2351489305
■
```

$P(X < 780) = 0.235$

c We calculate $P(X > 820)$ using $1 - P(X \le 820)$ on a calculator.

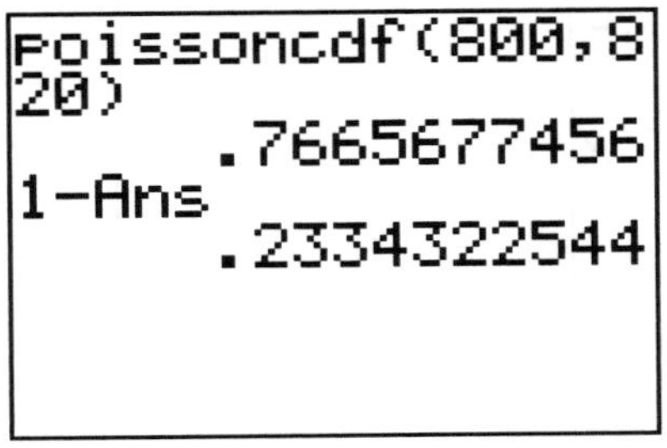

```
Poissoncdf(800,8
20)
          .7665677456
1-Ans
          .2334322544
```

$P(X > 820) = 0.233$

With a Poisson distribution we are sometimes given the mean over a certain interval. We can sometimes assume that this can then be recalculated for a different interval.

Example

The mean number of telephone calls arriving at a company's reception is five per minute and follows a Poisson distribution. Find the probability that there are

a exactly six phone calls in a given minute

b more than three phone calls in a given minute

c more than 20 phone calls in a given 5-minute period

d less than ten phone calls in a 3-minute period.

Let X be the "number of telephone calls in a minute". Hence $X \sim \text{Po}(5)$.

a $P(X = 6) = \dfrac{e^{-5}5^6}{6!} = 0.146$

b $P(X > 3) = 1 - [P(X = 0) + P(X = 1) + P(X = 2) + P(X = 3)]$

$$= 1 - \left\{ \frac{e^{-5}5^0}{0!} + \frac{e^{-5}5^1}{1!} + \frac{e^{-5}5^2}{2!} + \frac{e^{-5}5^3}{3!} \right\} = 0.735$$

On a calculator:

```
Poissoncdf(5,3)
         .2650259153
1-Ans
         .7349740847
■
```

c If there are five calls in a minute period, then in a 5-minute period there are, on average, 25 calls. Hence if Y is "the number of telephone calls in a 5-minute period", then $Y \sim \text{Po}(25)$. We require $P(Y > 20)$. Because of the numbers involved we need to solve this on a calculator.

```
Poissoncdf(25,20
)
         .1854923028
1-Ans
         .8145076972
■
```

$P(Y > 20) = 0.815$

d If A is "the number of telephone calls in a 3-minute period", then $A \sim \text{Po}(15)$. We require $P(Y < 10)$. Because of the numbers involved, again we solve this on a calculator.

```
Poissoncdf(15,9)
         .0698536607
```

$P(Y < 10) = 0.0699$

Example

Passengers arrive at the check-in desk of an airport at an average rate of seven per minute.

Assuming that the passengers arriving at the check-in desk follow a Poisson distribution, find

a the probability that exactly five passengers will arrive in a given minute

b the most likely number of passengers to arrive in a given minute

c the probability of at least three passengers arriving in a given minute

d the probability of more than 30 passengers arriving in a given 5-minute period.

If X is "the number of passengers checking-in in a minute", then $X \sim \text{Po}(7)$.

a $P(X = 5) = \dfrac{e^{-7}7^5}{5!} = 0.128$

b As with the binomial distribution, we find the probabilities on a calculator and look for the highest. This time we select a range of values around the mean. As before, it is only necessary to write down the ones either side as the distribution rises to a maximum probability and then decreases again. Written as a table, the results are:

x	5	6	7	8
$P(X = x)$	0.127...	0.149...	0.149...	0.130...

Since there are two identical probabilities in this case, the most likely value is either 6 or 7.

c $P(X \geq 3) = 1 - P(X \leq 2)$

$$= 1 - \left\{ \frac{e^{-7}7^0}{0!} + \frac{e^{-7}7^1}{1!} + \frac{e^{-7}7^2}{2!} \right\} = 0.970$$

On the calculator:

```
Poissoncdf(7,2)
          .0296361639
1-Ans
          .9703638361
```

d If seven people check-in in a minute period, then on average 35 people will check-in in a 5-minute period. Hence if Y is "the number of people checking-in in a 5-minute period", then $Y \sim \text{Po}(35)$. We require $P(Y > 30)$. Because of the numbers involved we need to solve this on a calculator.

```
Poissoncdf(35,30
)
          .2269424471
1-Ans
          .7730575529
```

$P(Y > 30) = 0.773$

Expectation and variance of a Poisson distribution

If $X \sim \text{Po}(\lambda)$
$E(X) = \lambda$
$\text{Var}(X) = \lambda$

The proofs for these are shown below, but they will not be asked for in examination questions.

Proof that $E(x) = \lambda$

The probability distribution for $X \sim \text{Po}(\lambda)$ is:

x	0	1	2	3	...
$P(X = x)$	$e^{-\lambda}$	$\lambda e^{-\lambda}$	$\dfrac{\lambda^2}{2!}e^{-\lambda}$	$\dfrac{\lambda^3}{3!}e^{-\lambda}$	

Now $E(X) = \displaystyle\sum_{\text{all } x} x \cdot P(X = x)$

$$= 0 \cdot e^{-\lambda} + 1 \cdot \lambda e^{-\lambda} + 2 \cdot \frac{\lambda^2}{2!}e^{-\lambda} + 3 \cdot \frac{\lambda^3}{3!}e^{-\lambda} + \cdots$$

$$= \lambda e^{-\lambda}\left(1 + \lambda + \frac{\lambda^2}{2!} + \frac{\lambda^3}{3!} \cdots\right)$$

The series in the bracket has a sum of e^λ (the proof of this is beyond the scope of this curriculum).

Hence $E(X) = \lambda$.

Proof that $\text{Var}(X) = \lambda$

$\text{Var}(X) = E(X^2) - E^2(X)$

Now $E(X^2) = \displaystyle\sum_{\text{all } x} x^2 \cdot P(X = x)$

$$= 0 \cdot e^{-\lambda} + 1 \cdot \lambda e^{-\lambda} + 4 \cdot \frac{\lambda^2}{2!}e^{-\lambda} + 9 \cdot \frac{\lambda^3}{3!}e^{-\lambda} + 16 \cdot \frac{\lambda^4}{4!}e^{-\lambda} + \cdots$$

$$= \lambda e^{-\lambda}\left(1 + 2\lambda + \frac{3\lambda^2}{2!} + \frac{4\lambda^3}{3!} + \cdots\right)$$

We now split this into two series.

$$= \lambda e^{-\lambda}\left(1 + \lambda + \frac{\lambda^2}{2!} + \frac{\lambda^3}{3!} + \cdots + \lambda + \frac{2\lambda^2}{2!} + \frac{3\lambda^3}{3!} + \cdots\right)$$

The first of these series is the same as in the proof for $E(X)$ and has a sum of e^λ.

$$= \lambda e^{-\lambda}\left\{e^\lambda + \lambda\left(1 + \lambda + \frac{\lambda^2}{2!} + \cdots\right)\right\}$$

$$= \lambda e^{-\lambda} - {}^\lambda(e^\lambda + \lambda e^\lambda)$$

$$= \lambda + \lambda^2$$

Hence $\text{Var}(X) = E(X^2) - E^2(X)$

$$= \lambda + \lambda^2 - \lambda^2$$

$$= \lambda$$

Example

In a given Poisson distribution it is found that $P(X > 1) = 0.25$. Find the variance of the distribution.

Let the distribution be $X \sim Po(m)$.

If $P(X > 1) = 0.25$ then

$$1 - P(X = 0) - P(X = 1) = 0.25$$

$$\Rightarrow 1 - e^{-m} - me^{-m} = 0.25$$

$$\Rightarrow e^{-m} + me^{-m} - 0.75 = 0$$

This can be solved on a calculator.

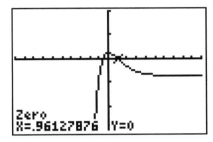

```
Zero
X=.96127876  Y=0
```

Since m is not negative, $m = 0.961$ and this is also $Var(X)$.

Example

In a fireworks factory, the number of defective fireworks follows a Poisson distribution with an average of three defective fireworks in any given box.

a Find the probability that there are exactly three defective fireworks in a given box.

b Find the most likely number of defective fireworks in a box.

c Find the probability that there are more than five defective fireworks in a box.

d Find the probability that in a sample of 15 boxes, at least three boxes have more than five defective fireworks in them.

Let X be the number of defective fireworks.
Hence $X \sim Po(3)$.

a $P(X = 3) = \dfrac{e^{-3}3^3}{3!} = 0.224$

b We select a range of values around the mean to find the most likely value. In this case we choose 1, 2, 3, 4, 5 and use a calculator.
Written as a table, the results are:

x	1	2	3	4	5
$P(X = x)$	0.149 …	0.224 …	0.224 …	0.168 …	0.100 …

Since there are two identical probabilities in this case, the most likely value is either 2 or 3.

c $P(X > 5) = 1 - [P(X = 0) + P(X = 1) + P(X = 2) + P(X = 3)$
$$+ P(X = 4) + P(X = 5)]$$

$$= 1 - \left\{ \frac{e^{-3}3^0}{0!} + \frac{e^{-3}3^1}{1!} + \frac{e^{-3}3^2}{2!} + \frac{e^{-3}3^3}{3!} + \frac{e^{-3}3^4}{4!} + \frac{e^{-3}3^5}{5!} \right\}$$

$$= 0.0839$$

```
poissoncdf(3,5)
         .9160820581
1-Ans
         .0839179419
```

d This is an example of where the question now becomes a binomial distribution Y, which is the number of boxes with more than five defective fireworks in them. Hence $Y \sim \text{Bin}(15, 0.0839)$.

$$P(Y \geq 3) = 1 - [P(Y = 0) + P(Y = 1) + P(Y = 2)]$$

$$= 1 - \left\{ {}^{15}C_0(0.0839)^0(0.916)^{15} + {}^{15}C_1(0.0839)^1(0.916)^{14} \right.$$

$$\left. + {}^{15}C_2(0.839)^2(0.916)^{13} \right\}$$

$$= 1 - 0.874\ldots = 0.126$$

```
binomcdf(15,0.08
39,2)
         .8742221686
1-Ans
         .1257778314
```

Exercise 4

1 If $X \sim \text{Po}(3)$, find:
  a $P(X = 2)$ **b** $P(X \leq 2)$ **c** $P(X > 3)$ **d** $E(X)$

2 If $X \sim \text{Po}(6)$, find:
  a $P(X = 4)$ **b** $P(X \leq 3)$ **c** $P(X > 5)$ **d** $E(X)$

3 If $X \sim \text{Po}(10)$, find:
  a $P(X = 9)$ **b** $P(X \leq 8)$ **c** $P(X \geq 6)$ **d** $\text{Var}(X)$

4 If $X \sim \text{Po}(m)$ and $E(X^2) = 4.5$, find:
  a m **b** $P(X = 4)$ **c** $P(X \leq 3)$

5 If $X \sim \text{Po}(\lambda)$ and $P(X \leq 2) = 0.55$, find:
  a $E(X)$ **b** $P(X = 3)$ **c** $P(X < 4)$ **d** $P(X > 5)$

6 If $X \sim \text{Po}(m)$ and $P(X > 1) = 0.75$, find
  a $\text{Var}(X)$

  b $P(X = 5)$

  c the probability that X is greater than 4

    d the probability that X is at least 3

    e the probability that X is less than or equal to 5.

7 If $X \sim \text{Po}(n)$ and $E(X^2) = 6.5$, find

    a n

    b the probability that X equals 4

    c the probability that X is greater than 5

    d the probability that X is at least 3.

8 A Poisson distribution is such that $X \sim \text{Po}(n)$.

    a Given that $P(X = 5) = P(X = 3) + P(X = 4)$, find the value of n.

    b Find the probability that X is at least 2.

9 On a given road, during a specific period in the morning, the number of drivers who break the speed limit, X, follows a Poisson distribution with mean m. It is calculated that $P(X = 1)$ is twice $P(X = 2)$. Find

    a the value of m **b** $P(X \leq 3)$.

10 At a given road junction, the occurrence of an accident happening on a given day follows a Poisson distribution with mean 0.1. Find the probability of

    a no accidents on a given day

    b at least two accidents on a given day

    c exactly three accidents on a given day.

11 Alexander is typing out a mathematics examination paper. On average he makes 3.6 mistakes per examination paper. His colleague, Roy, makes 3.2 mistakes per examination paper, on average. Given that the number of mistakes made by each author follows a Poisson distribution, calculate the probability that

    a Alexander makes at least two mistakes

    b Alexander makes exactly four mistakes

    c Roy makes exactly three mistakes

    d Alexander makes exactly four mistakes and Roy makes exactly three mistakes.

12 A machine produces carpets and occasionally minor faults are produced. The number of faults in a square metre of carpet follows a Poisson distribution with mean 2.7. Calculate

    a the probability of there being exactly five faults in a square metre of carpet

    b the probability of there being at least two faults in a square metre of carpet

    c the most likely number of faults in a square metre of carpet

    d the probability of less than six faults in 3 m^2 of carpet

    e the probability of more than five faults in 2 m^2 of carpet.

13 At a local airport the number of planes that arrive between 10.00 and 12.00 in the morning is 6, on average. Given that these arrivals follow a Poisson distribution, find the probability that

    a only one plane lands between 10.00 and 12.00 next Saturday morning

    b either three or four planes will land next Monday between 10.00 and 12.00.

14 X is the number of Annie dolls sold by a shop per day. X has a Poisson distribution with mean 4.

    a Find the probability that no Annie dolls are sold on a particular Monday.

    b Find the probability that more than five are sold on a particular Saturday.

    c Find the probability that more than 20 are sold in a particular week, assuming the shop is open seven days a week.

   d If each Annie doll sells for 20 euros, find the mean and variance of the sales for a particular day.

   Y is the number of Bobby dolls sold by the same shop per day. Y has a Poisson distribution with mean 6.

   e Find the probability that the shop sells at least four Bobby dolls on a particular Tuesday.

   f Find the probability that on a certain day, the shop sells three Annie dolls and four Bobby dolls.

15 A school office receives, on average, 15 calls every 10 minutes. Assuming this follows a Poisson distribution, find the probability that the office receives

   a exactly nine calls in a 10-minute period

   b at least seven calls in a 10-minute period

   c exactly two calls in a 3-minute period

   d more than four calls in a 5-minute period

   e more than four calls in three consecutive 5-minute periods.

16 The misprints in the answers of a mathematics textbook are distributed following a Poisson distribution. If a book of 700 pages contains exactly 500 misprints, find

   a **i** the probability that a particular page has exactly one misprint

   ii the mean and variance of the number of misprints in a 30-page chapter

   iii the most likely number of misprints in a 30-page chapter.

   b If Chapters 12, 13 and 14 each have 40 pages, what is the probability that exactly one of them will have exactly 50 misprints?

17 A garage sells Super Run car tyres. The monthly demand for these tyres has a Poisson distribution with mean 4.

   a Find the probability that they sell exactly three tyres in a given month.

   b Find the probability that they sell no more than five tyres in a month.

   A month consists of 22 days when the garage is open.

   c What is the probability that exactly one tyre is bought on a given day?

   d What is the probability that at least one tyre is bought on a given day?

   e How many tyres should the garage have at the beginning of the month in order that the probability that they run out is less than 0.05?

18 Between 09.00 and 09.30 on a Sunday morning, 15 children and 35 adults enter the local zoo, on average. Find the probability that on a given Sunday between 09.00 and 09.30

   a exactly ten children enter the zoo

   b at least 30 adults enter the zoo

   c exactly 14 children and 28 adults enter the zoo

   d exactly 25 adults and 5 children enter the zoo.

Review exercise

 All questions in this exercise will require a calculator.

1 The volumes (V) of four bottles of drink are 1 litre, 2 litres, 3 litres and 4 litres. The probability that a child selects a bottle of drink of volume V is cV.

   a Find the value of c.

   b Find E(X) where X is the volume of the selected drink.

   c Find Var(X).

2 The random variable X follows a Poisson distribution. Given that
$P(X \leq 1) = 0.2$, find:
a the mean of the distribution
b $P(X \leq 2)$ [IB Nov 06 P1 Q7]

3 The probability that a boy in a class has his birthday on a Monday or a
Tuesday during a school year is $\dfrac{1}{4}$. There are 15 boys in the class.

a What is the probability that exactly three of them have birthdays on a
Monday or a Tuesday?

b What is the most likely number of boys to have a birthday on a Monday or
a Tuesday?

c In a particular year group, there are 70 boys. The probability of one of these
boys having a birthday on a Monday or a Tuesday is also $\dfrac{1}{4}$. What is the
expected number of boys having a birthday on a Monday or Tuesday?

4 In a game a player rolls a ball down a chute. The ball can land in one of six
slots which are numbered 2, 4, 6, 8, 10 and x. The probability that it lands in
a slot is the number of the slot divided by 50.
a If this is a random variable, calculate the value of x.
b Find $E(X)$.
c Find $Var(X)$.

5 The number of car accidents occurring per day on a highway follows a Poisson
distribution with mean 1.5.
a Find the probability that more than two accidents will occur on a given
Monday.
b Given that at least one accident occurs on another day, find the probability
that more than two accidents occur on that day.   [IB May 06 P1 Q16]

6 The most popular newspaper according to a recent survey is the Daily Enquirer,
which claims that 65% of people read the newspaper on a certain bus route.
Consider the people sitting in the first ten seats of a bus.
a What is the probability that exactly eight people will be reading the Daily
Enquirer?
b What is the probability that more than four people will be reading the Daily
Enquirer?
c What is the most likely number of people to be reading the Daily Enquirer?
d What is the expected number of people to be reading the Daily Enquirer?
e On a certain bus route, there are ten buses between the hours of 09.00 and
10.00. What is the probability that on exactly four of these buses at least six
people in the first ten seats are reading the Daily Enquirer?

7 The discrete random variable X has the following probability distribution.

$$P(X = x) = \frac{k}{x}, x = 1, 2, 3, 4$$

$$= 0 \text{ otherwise}$$

Calculate:
a the value of the constant k
b $E(X)$ [IB May 04 P1 Q13]

8 An office worker, Alan, knows that the number of packages delivered in a
day to his office follows a Poisson distribution with mean 5.
a On the first Monday in June, what is the probability that the courier company
delivers four packages?

b On another day, Alan sees the courier van draw up to the office and hence knows that he will receive a delivery. What is the probability that he will receive three packages on that day?

9 The number of cats found in a particular locality follows a Poisson distribution with mean 4.1.

   a Find the probability that the number of cats found will be exactly 5.

   b What is the most likely number of cats to be found in the locality?

   c A researcher checks half the area. What is the probability that he will find exactly two cats?

   d Another area is found to have exactly the same Poisson distribution. What is the probability of finding four cats in the first area and more than three in the second?

10 The probability of the 16:55 train being delayed on a weekday is $\frac{1}{10}$.

   Assume that delays occur independently.

   a What is the probability, correct to three decimal places, that a traveller experiences 2 delays in a given 5-day week?

   b How many days must a commuter travel before having a 90% probability of having been delayed at least once? [IB Nov 90 P1 Q20]

11 Two children, Alan and Belle, each throw two fair cubical dice simultaneously. The score for each child is the sum of the two numbers shown on their respective dice.

   a **i** Calculate the probability that Alan obtains a score of 9.

   ii Calculate the probability that Alan and Belle both obtain a score of 9.

   b **i** Calculate the probability that Alan and Belle obtain the same score.

   ii Deduce the probability that Alan's score exceeds Belle's score.

   c Let X denote the largest number shown on the four dice.

   i Show that $P(X \leq x) = \left(\dfrac{x}{6}\right)^4$ for $x = 1, 2, \ldots 6$.

   ii Copy and complete the following probability distribution table.

x	1	2	3	4	5	6
$P(X = x)$	$\dfrac{1}{1296}$	$\dfrac{15}{1296}$				$\dfrac{671}{1296}$

   iii Calculate E(X). [IB May 02 P2 Q4]

12 The probability of finding the letter z on a page in a book is 0.05.

   a In the first ten pages of a book, what is the probability that exactly three pages contain the letter z?

   b In the first five pages of the book, what is the probability that at least two pages contain the letter z?

   c What is the most likely number of pages to contain the letter z in a chapter of 20 pages?

   d What would be the expected number of pages containing the letter z in a book of 200 pages?

   e Given that the first page of a book does not contain the letter z, what is the probability that it occurs on more than two of the following five pages?

13 A biased die with four faces is used in a game. A player pays 10 counters to roll the die. The table below shows the possible scores on the die, the probability of each score and the number of counters the player receives for each score.

Score	1	2	3	4
Probability	$\frac{1}{2}$	$\frac{1}{5}$	$\frac{1}{5}$	$\frac{1}{10}$
Number of counters player receives	4	5	15	n

Find the value of n in order for the player to get an expected return of 9 counters per roll. [IB May 99 P1 Q17]

14 The number of accidents in factory A in a week follows a Poisson distribution X, where $\text{Var}(X) = 2.8$.

a Find the probability that there are exactly three accidents in a week.

b Find the probability that there is at least one accident in a week.

c Find the probability of more than 15 accidents in a four-week period.

d Find the probability that during the first two weeks of a year, the factory will have no accidents.

e In a neighbouring factory B, the probability of one accident in a week is the same as the probability of two accidents in a week in factory A. Assuming that this follows a Poisson distribution with mean n, $0 < n < 1$, find the value of n.

f What is the probability that in the first week of July, factory A has no accidents and factory B has one accident?

g Given that in the first week of September factory A has two accidents, what is the probability that in the same week factory B has no more than two accidents?

15 a Give the definition of the conditional probability that an event A occurs given that an event B (with $P(B) > 0$) is known to have occurred.

b If A_1 and A_2 are mutually exclusive events, express $P(A_1 \cup A_2)$ in terms of $P(A_1)$ and $P(A_2)$.

c State the multiplication rule for two independent events E_1 and E_2.

d Give the conditions that are required for a random variable to have a binomial distribution.

e A freight train is pulled by four locomotives. The probability that any locomotive works is θ and the working of a locomotive is independent of the other locomotives.

i Write down an expression for the probability that k of the four locomotives are working.

ii Write down the mean and variance of the number of locomotives working.

iii In order that the train may move, at least two of the locomotives must be working. Write down an expression, in terms of θ, for P, the probability that the train can move. (Simplification of this expression is not required.)

iv Calculate P for the cases when $\theta = 0.5$ and when $\theta = 0.9$.

v If the train is moving, obtain a general expression for the conditional probability that j locomotives are working. (Again, simplification of the expression is not required.) Verify that the sum of the possible conditional probabilities is unity.

vi Evaluate the above conditional probability when $j = 2$, for the cases when $\theta = 0.5$ and when $\theta = 0.9$.

vii For calculate the probability that at least one of the three trains is able to move, assuming that they all have four locomotives and that different trains work independently. [IB May 94 P2 Q15]

16 In a game, a player pays 10 euros to flip six biased coins, which are twice as likely to show heads as tails. Depending on the number of heads he obtains, he receives a sum of money. This is shown in the table below:

Number of heads	0	1	2	3	4	5	6
Amount received in euros	30	25	15	12	18	25	40

  a Calculate the probability distribution for this.

  b Find the player's expected gain in one game.

  c What is the variance?

  d What would be his expected gain, to the nearest euro, in 15 games?

17 The table below shows the probability distribution for a random variable X. Find α and E(X).   [IB May 93 P1 Q20]

x	1	2	3	4
$P(X = x)$	2α	$4\alpha^2$	$2\alpha^2 + 3\alpha$	$\alpha^2 + \alpha$

18 An unbiased coin is tossed n times and X is the number of heads obtained. Write down an expression for the probability that $X = r$.

State the mean and standard deviation of X.

Two players, A and B, take part in the game. A has three coins and B has two coins. They each toss their coins and count the number of heads which they obtain.

  a If A obtains more heads than B, she wins 5 cents from B. If B obtains more heads than A, she wins 10 cents from A. If they obtain an equal number of heads then B wins 1 cent from A. Show that, in a series of 100 such games, the expectation of A's winnings is approximately 31 cents.

  b On another occasion they decide that the winner shall be the player obtaining the greater number of heads. If they obtain an equal number of heads, they toss the coins again, until a definite result is achieved. Calculate the probability that

    i no result has been achieved after two tosses

    ii A wins the game.   [IB Nov 89 P2 Q8]

22 Continuous Probability Distributions

In Chapter 21, we saw that the binomial distribution could be used to solve problems such as "If an unbiased cubical die is thrown 50 times, what is the probability of throwing a six more than 25 times?" To solve this problem, we compute the probability of throwing a six 25 times then the probability of throwing a six 26 times, 27 times, etc., which before the introduction of calculators would have taken a very long time to compute. Abraham de Moivre, who we met in Chapter 17, noted that when the number of events (throwing a die in this case) increased to a large enough number, then the shape of the binomial distribution approached a very smooth curve.

De Moivre realised that if he was able to find a mathematical expression for this curve, he would be able to find probabilities much more easily. This curve is what we now call

Binomial distributions for 2, 4 and 12 throws

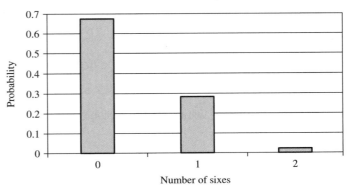

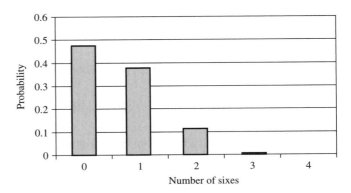

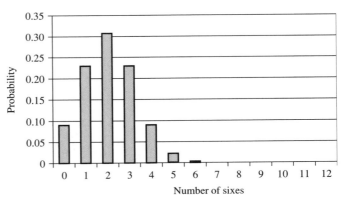

a normal curve and the distribution associated with it is introduced in this chapter. It is shown here approximating the binomial distribution for 12 throws of an unbiased die.

The normal distribution is of great importance because many natural phenomena are at least approximately normally distributed. One of the earliest applications of the normal distribution was connected to error analysis in astronomical observations. Galileo in the 17th century hypothesized several distributions for these errors, but it was not until two centuries later that it was discovered that they followed a normal distribution. The normal distribution had also been discovered by Laplace in 1778 when he derived the extremely important central limit theorem. Laplace showed that for any distribution, provided that the sample size is large, the distribution of the means of repeated samples from the distribution would be approximately normal, and that the larger the sample size, the closer the distribution would be to a normal distribution.

22.1 Continuous random variables

In Chapter 19 we discussed the difference between discrete and continuous data and in Chapter 21 we met discrete data where $\sum_{\text{all } x} P(X = x) = 1$. In this chapter we consider continuous data. To find the probability that the height of a man is 1.85 metres, correct to 3 significant figures, we need to find $P(1.845 \le H < 1.855)$. Hence for continuous data we construct ranges of values for the variable and find the probabilities for these different ranges.

For a discrete random variable, a table of probabilities is normally given. For a continuous random variable, a probability density function is normally used instead. In Chapter 21, we met probability density functions where the variable was discrete. When the variable is continuous the function f(x) can be integrated over a particular range of values to give the probability that the random variable X lies in that particular range.

Hence for a continuous random variable valid over the range $a \le x \le b$ we can say that $\int_a^b f(x) \, dx = 1$. This is analogous to $\sum_{x=a}^{x=b} P(X = x) = 1$ for discrete data and also the idea of replacing sigma notation with integral notation when finding the area under the curve, as seen in Chapter 14.

> The area under the curve represents the probability.

Thus, if $a \le x_1 \le x_2 \le b$ then $P(x_1 \le X \le x_2) = \int_{x_1}^{x_2} f(x) \, dx$ as shown in the diagram.

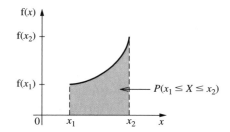

> Since many of the calculations involve definite integration, if the questions were to appear on a calculator paper, the calculations could be performed on a calculator.

Example

Consider the function $f(x) = \dfrac{3}{4}$, $\dfrac{1}{3} \le x \le \dfrac{5}{3}$, which is being used as a probability density function for a continuous random variable X.

   a Show that $f(x)$ is a valid probability density function.

   b Find the probability that X lies in the range $\dfrac{3}{4}$ to $\dfrac{5}{4}$.

   c Show this result graphically.

   a $f(x)$ is a valid probability density function if $\displaystyle\int_{\frac{1}{3}}^{\frac{5}{3}} \dfrac{3}{4}\,dx = 1$.

$$\int_{\frac{1}{3}}^{\frac{5}{3}} \frac{3}{4}\,dx = \left[\frac{3x}{4}\right]_{\frac{1}{3}}^{\frac{5}{3}}$$

$$= \frac{5}{4} - \frac{1}{4} = 1$$

Hence $f(x)$ can be used as a probability density function for a continuous random variable.

   b $P\left(\dfrac{3}{4} \le X \le \dfrac{5}{4}\right) = \displaystyle\int_{\frac{3}{4}}^{\frac{5}{4}} \dfrac{3}{4}\,dx$

$$= \left[\frac{3x}{4}\right]_{\frac{3}{4}}^{\frac{5}{4}} = \frac{15}{16} - \frac{9}{16} = \frac{6}{16} = \frac{3}{8}$$

   c

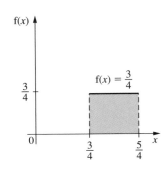

> In this case we did not have to use integration as the area under the curve is given by the area of a rectangle.

Example

The continuous random variable X has probability density function $f(x)$ where $f(x) = \dfrac{3}{26}(x - 1)^2$, $2 \le x \le k$.

   a Find the value of the constant k.

   b Sketch $y = f(x)$.

   c Find $P(2.5 \le X \le 3.5)$ and show this on a diagram.

   d Find $P(X \ge 2.5)$.

a $\displaystyle\int_{2}^{k} \frac{3}{26}(x-1)^2 \, dx = 1$

$\Rightarrow \left[\dfrac{(x-1)^3}{26}\right]_{2}^{k} = 1$

$\Rightarrow \dfrac{(k-1)^3}{26} - \dfrac{1}{26} = 1$

$\Rightarrow \dfrac{(k-1)^3}{26} = \dfrac{27}{26}$

$\Rightarrow k - 1 = 3 \Rightarrow k = 4$

b

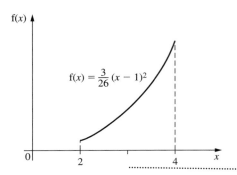

$f(x) = \frac{3}{26}(x-1)^2$

> The graph has a domain of $2 \le x \le 4$.

c $\displaystyle P(2.5 \le X \le 3.5) = \int_{2.5}^{3.5} \frac{3}{26}(x-1)^2 \, dx$

$= \left[\dfrac{(x-1)^3}{26}\right]_{2.5}^{3.5} = \dfrac{125}{208} - \dfrac{27}{208} = \dfrac{98}{208} = \dfrac{49}{104}$

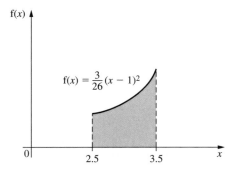

$f(x) = \frac{3}{26}(x-1)^2$

d $\displaystyle P(X \ge 2.5) = \int_{2.5}^{4} \frac{3}{26}(x-1)^2 \, dx$

$= \left[\dfrac{(x-1)^3}{26}\right]_{2.5}^{4} = \dfrac{27}{26} - \dfrac{27}{208} = \dfrac{189}{208}$

Sometimes the probability density function for a continuous random variable can use two or more different functions.

However, it works in exactly the same way.

> This is similar to the piecewise functions met in Chapter 3.

Example

The continuous random variable X has probability density function

$$f(x) = \begin{cases} k(4-x)^2 & 0 \le x \le 2 \\ 4k & 2 < x \le \dfrac{8}{3} \\ 0 & \text{otherwise} \end{cases}$$

where k is a constant.

a Find the value of the constant k.

b Sketch $y = f(x)$.

c Find $P(1 \le X \le 2.5)$.

d Find $P(X \ge 1)$.

a $\displaystyle\int_0^2 k(4-x)^2 \, dx + \int_2^{\frac{8}{3}} 4k \, dx = 1$

$\Rightarrow \left[\dfrac{k(4-x)^3}{-3}\right]_0^2 + \left[4kx\right]_2^{\frac{8}{3}} = 1$

$\Rightarrow -\dfrac{8k}{3} + \dfrac{64k}{3} + \dfrac{32k}{3} - 8k = 0$

$\Rightarrow \dfrac{64k}{3} = 1$

$\Rightarrow k = \dfrac{3}{64}$

b

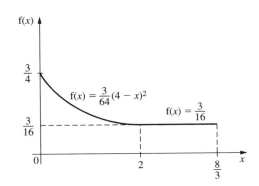

$f(x) = \dfrac{3}{64}(4-x)^2$

$f(x) = \dfrac{3}{16}$

c As the area spans the two distributions, we integrate over the relevant domains.

$$P(1 \le X \le 2.5) = \int_1^2 \dfrac{3}{64}(4-x)^2 \, dx + \int_2^{2.5} \dfrac{3}{16} \, dx$$

$$= \left[-\dfrac{1}{64}(4-x)^3\right]_1^2 + \left[\dfrac{3x}{16}\right]_2^{2.5}$$

$$= -\dfrac{1}{8} + \dfrac{27}{64} + \dfrac{15}{32} - \dfrac{3}{8} = \dfrac{25}{64}$$

d $\displaystyle P(X \ge 1) = \int_1^2 \dfrac{3}{64}(4-x)^2 \, dx + \int_2^{\frac{8}{3}} \dfrac{3}{16} \, dx$

$$= \left[-\frac{1}{64}(4-x)^3\right]_1^2 + \left[\frac{3x}{16}\right]_2^{\frac{8}{3}}$$

$$= -\frac{1}{8} + \frac{27}{64} + \frac{1}{2} - \frac{3}{8} = \frac{27}{64}$$

These integrals could be done directly on a calculator.

Exercise 1

1 A continuous probability density function is defined as

$$f(x) = \begin{cases} k - \dfrac{x}{4} & 1 \leq x \leq 3 \\ 0 & \text{otherwise} \end{cases}$$

where k is a constant.

   a Find the value of k.
   b Sketch $y = f(x)$.
   c Find $P(1.5 \leq X \leq 2.5)$ and show this on a sketch.
   d Find $P(X \leq 2.5)$.

2 Let X be a continuous random variable with probability density function

$$f(x) = \begin{cases} \dfrac{x}{2} - 1 & 2 \leq x \leq c \\ 0 & \text{otherwise} \end{cases}$$

where c is a constant.

   a Find the value of c.
   b Sketch $y = f(x)$.
   c Find $P(2.5 \leq X \leq 3)$ and show this on a sketch.
   d Find $P(4.5 \leq X \leq 5.2)$.

3 A continuous random variable X has probability density function

$$f(x) = \begin{cases} k \cos x & 0 \leq x \leq \dfrac{\pi}{4} \\ 0 & \text{otherwise} \end{cases}$$

where k is a constant.

   a Find the value of k.
   b Sketch $y = f(x)$.
   c Find $P\left(0 \leq X \leq \dfrac{\pi}{6}\right)$ and show this on a sketch.
   d Find $P\left(X \geq \dfrac{\pi}{12}\right)$.

4 The probability density function $f(x)$ of a continuous random variable X is defined by

$$f(x) = \begin{cases} \dfrac{1}{2}x(4 - x^2) & k \leq x \leq 2 \\ 0 & \text{otherwise} \end{cases}$$

where k is a constant.

   a Find the value of k.
   b Sketch $y = f(x)$.
   c Find $P(1.1 \leq X \leq 1.3)$ and show this on a sketch.
   d Find $P(X \leq 1.5)$.

5 The time taken for a worker to perform a particular task, t minutes, has probability density function

$$f(t) = \begin{cases} kt^2 & 0 \leq t \leq 5 \\ 0.4k(2 + t) & 5 < t \leq 15 \\ 0 & \text{otherwise} \end{cases}$$

where k is a constant.

a Find the value of k.

b Sketch $y = f(t)$.

c Find $P(4 \leq X \leq 11)$ and show this on a sketch.

d Find $P(X \geq 9)$.

22.2 Using continuous probability density functions

Expectation

For a discrete random variable $E(X) = \sum_{\text{all } x} x \cdot P(X = x)$.

Hence $E(X) = \int_a^b x\, f(x)\, dx$ for a continuous random variable valid over the range $a \leq x \leq b$.

If the probability density function is symmetrical then $E(X)$ is the value of the line of symmetry.

> This is similar to the result for discrete data.

> For continuous data we are often dealing with a population, so $E(X)$ is denoted as μ. For discrete data we are often dealing with a sample, so $E(X)$ is denoted as $\bar{x}$. In both cases $E(X)$ is referred to as the mean of X.

Example

If X is a continuous random variable with probability density function $f(x) = \frac{1}{9}x^2$, $0 \leq x \leq 3$, find $E(X)$.

$$E(X) = \int_0^3 x\left(\frac{1}{9}x^2\right)dx$$

$$= \frac{1}{9}\int_0^3 x^3\, dx$$

$$= \left[\frac{x^4}{36}\right]_0^3$$

$$= \frac{81}{36} = \frac{9}{4}$$

Example

The continuous random variable X has probability density function
$f(x) = k(1 - x)(x - 5)$, $1 \le x \le 5$.

a Find the value of the constant k.

b Sketch $y = f(x)$.

c Find $E(X)$.

d Find $P(1.5 \le X \le 3.5)$.

a
$$\int_1^5 k(1 - x)(x - 5)\, dx = 1$$

$$\Rightarrow k \int_1^5 (-x^2 + 6x - 5)\, dx = 1$$

$$\Rightarrow k \left[-\frac{x^3}{3} + 3x^2 - 5x \right]_1^5 = 1$$

$$\Rightarrow k \left[\left(-\frac{125}{3} + 75 - 25 \right) - \left(-\frac{1}{3} + 3 - 5 \right) \right] = 1$$

$$\Rightarrow \frac{32}{3} k = 1$$

$$\Rightarrow k = \frac{3}{32}$$

b

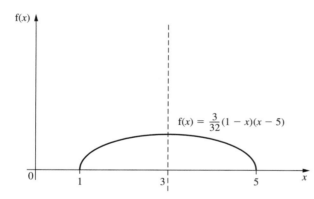

$f(x) = \frac{3}{32}(1 - x)(x - 5)$

c Since the distribution is symmetrical, $E(X) = 3$ from the above diagram.

d $P(1.5 \le X \le 3.5) = \displaystyle\int_{1.5}^{3.5} \frac{3}{32}(1 - x)(x - 5)\, dx$

$$= \frac{3}{32} \int_{1.5}^{3.5} (-x^2 + 6x - 5)\, dx$$

$$= \frac{3}{32} \left[-\frac{x^3}{3} + 3x^2 - 5x \right]_{1.5}^{3.5}$$

$$= \frac{3}{32} \left[\left(-\frac{343}{24} + \frac{147}{4} - \frac{35}{2} \right) - \left(-\frac{9}{8} + \frac{27}{4} - \frac{15}{2} \right) \right]$$

$$= \frac{41}{64}$$

Example

The time taken in hours for a particular insect to digest food is a continuous random variable whose probability density function is given by

$$f(x) = \begin{cases} k(x-1)^2 & 1 \le x \le 2 \\ k(8-x) & 2 < x \le 4 \\ 0 & \text{otherwise} \end{cases}$$

Find

a the value of the constant k

b the mean time taken

c the probability that it takes an insect between 1.5 and 3 hours to digest its food

d the probability that two randomly chosen insects each take between 1.5 and 3 hours to digest their food.

a $\displaystyle\int_1^2 k(x-1)^2 \, dx + \int_2^4 k(8-x) \, dx = 1$

$$\Rightarrow k\left[\frac{(x-1)^3}{3}\right]_1^2 + k\left[8x - \frac{x^2}{2}\right]_2^4 = 1$$

$$\Rightarrow k\left[\left(\frac{1}{3} - 0\right) + (24 - 14)\right] = 1$$

$$\Rightarrow k = \frac{3}{31}$$

b $\displaystyle E(X) = \int_1^2 kx(x-1)^2 \, dx + \int_2^4 kx(8-x) \, dx$

$$= \frac{3}{31}\left[\int_1^2 (x^3 - 2x^2 + x) \, dx + \int_2^4 (8x - x^2) \, dx\right]$$

$$= \frac{3}{31}\left\{\left[\frac{x^4}{4} - \frac{2x^3}{3} + \frac{x^2}{2}\right]_1^2 + \left[4x^2 - \frac{x^3}{3}\right]_2^4\right\}$$

$$= \frac{3}{31}\left[\left(4 - \frac{16}{3} + 2\right) - \left(\frac{1}{4} - \frac{2}{3} + \frac{1}{2}\right) + \left(64 - \frac{64}{3}\right) - \left(16 - \frac{8}{3}\right)\right]$$

$$= 2.90 \text{ hours}$$

c $\displaystyle P(1.5 \le X \le 3) = \int_{1.5}^2 k(x-1)^2 \, dx + \int_2^3 k(8-x) \, dx$

$$= \frac{3}{31}\left\{\left[\frac{(x-1)^3}{3}\right]_{1.5}^2 + \left[8x - \frac{x^2}{2}\right]_2^3\right\}$$

$$= \frac{3}{31}\left[\left(\frac{1}{3} - \frac{1}{24}\right) + \left(\frac{39}{2} - 14\right)\right]$$

$$= 0.560$$

d P(two randomly chosen insects each take between 1.5 and 3 hours to digest their food) $= 0.560^2 = 0.314$

For a continuous random variable valid over the range $a \leq x \leq b$

$$E[g(X)] = \int_a^b g(x)\, f(x)\, dx$$

This is similar to the result for discrete data.

where $g(x)$ is any function of the continuous random variable X and $f(x)$ is the probability density function.

Hence we have the result

$$E(X^2) = \int_a^b x^2\, f(x)\, dx$$

Example

The continuous random variable X has probability density function $f(x)$ where $f(x) = \dfrac{1}{18}(6 - x),\ 0 \leq x \leq 6$. Find:

a $E(X)$

b $E(2X - 1)$

c $E(X^2)$

a $E(X) = \displaystyle\int_0^6 \dfrac{1}{18} x\,(6 - x)\, dx$

$= \dfrac{1}{18} \displaystyle\int_0^6 (6x - x^2)\, dx$

$= \dfrac{1}{18} \left[3x^2 - \dfrac{x^3}{3} \right]_0^6$

$= \dfrac{1}{18} \left(108 - \dfrac{216}{3} \right) = 2$

b $E(2X - 1) = \displaystyle\int_0^6 \dfrac{1}{18} (2x - 1)(6 - x)\, dx$

$= \dfrac{1}{18} \displaystyle\int_0^6 (-2x^2 + 13x - 6)\, dx$

$= \dfrac{1}{18} \left[\dfrac{-2x^3}{3} + \dfrac{13x^2}{2} - 6x \right]_0^6$

$= \dfrac{1}{18} (-144 + 234 - 36) = 3$

c $E(X^2) = \displaystyle\int_0^6 \dfrac{1}{18} x^2 (6 - x)\, dx$

$$= \frac{1}{18} \int_0^6 (6x^2 - x^3)\, dx$$

$$= \frac{1}{18} \left[2x^3 - \frac{x^4}{4} \right]_0^6$$

$$= \frac{1}{18}(432 - 324) = 6$$

Variance

We are now in a position to calculate the variance. As with discrete data

$$\text{Var}(X) = E(X - \mu)^2$$
$$= E(X^2) - E^2(X)$$

Therefore for a continuous random variable with probability density function valid over the domain $a \le x \le b$

$$\text{Var}(X) = \int_a^b x^2\, f(x)\, dx - \left(\int_a^b x\, f(x)\, dx \right)^2$$

The standard deviation of X is $\sigma = \sqrt{\text{Var}(X)}$.

Example

The continuous random variable X has probability density function f(x) where
$f(x) = \frac{2}{63}(1 - 2x)^2, 2 \le x \le \frac{7}{2}$. Find:

a $E(X)$
b $E(X^2)$
c $\text{Var}(X)$
d σ

a $E(X) = \int_2^{\frac{7}{2}} \frac{2}{63} x (1 - 2x)^2\, dx$

$$= \frac{2}{63} \int_2^{\frac{7}{2}} (x - 4x^2 + 4x^3)\, dx$$

$$= \frac{2}{63} \left[\frac{x^2}{2} - \frac{4x^3}{3} + x^4 \right]_2^{\frac{7}{2}}$$

$$= \frac{2}{63} \left[\left(\frac{49}{8} - \frac{343}{6} + \frac{2401}{16} \right) - \left(2 - \frac{32}{3} + 16 \right) \right]$$

$$= \frac{163}{56}$$

b $E(X^2) = \displaystyle\int_{2}^{\frac{7}{2}} \frac{2}{63} x^2 (1 - 2x)^2 \, dx$

$\qquad = \dfrac{2}{63} \displaystyle\int_{2}^{\frac{7}{2}} (x^2 - 4x^3 + 4x^4) \, dx$

$\qquad = \dfrac{2}{63} \left[\dfrac{x^3}{3} - x^4 + \dfrac{4x^5}{5} \right]_{2}^{\frac{7}{2}}$

$\qquad = \dfrac{2}{63} \left[\left(\dfrac{343}{24} - \dfrac{2401}{16} + \dfrac{16\,807}{40} \right) - \left(\dfrac{8}{3} - 16 + \dfrac{128}{5} \right) \right]$

$\qquad = \dfrac{2419}{280}$

c $\text{Var}(X) = E(X^2) - E^2(X)$

$\qquad = \dfrac{2419}{280} - \left(\dfrac{163}{56} \right)^2 = 0.167$

d $\sigma = \sqrt{\text{Var}(X)} = \sqrt{0.167} = 0.409$

Example

A particular road has been altered so that the traffic has to keep to a lower speed and at one point in the road traffic can only go through one way at a time. At this point traffic in one direction will have to wait. The time in minutes that vehicles have to wait has probability density function

$$f(x) = \begin{cases} \dfrac{1}{2}\left(1 - \dfrac{x}{4}\right) & 0 \le x \le 4 \\ 0 & \text{otherwise} \end{cases}$$

a Find the mean waiting time.
b Find the standard deviation of the waiting time.
c Find the probability that three cars out of the first six to arrive after 8.00 am in the morning have to wait more than 2 minutes.

a The mean waiting time is given by $E(X)$.

$$E(X) = \int_{0}^{4} \frac{1}{2} x \left(1 - \frac{x}{4} \right) dx$$

$$= \int_{0}^{4} \frac{1}{2} \left(x - \frac{x^2}{4} \right) dx$$

$$= \frac{1}{2} \left[\frac{x^2}{2} - \frac{x^3}{12} \right]_{0}^{4}$$

$$= \frac{1}{2} \left(8 - \frac{16}{3} \right) = \frac{4}{3}$$

b $E(X^2) = \displaystyle\int_0^4 \frac{1}{2} x^2 \left(1 - \frac{x}{4}\right) dx$

$= \displaystyle\int_0^4 \frac{1}{2}\left(x^2 - \frac{x^3}{4}\right) dx$

$= \dfrac{1}{2}\left[\dfrac{x^3}{3} - \dfrac{x^4}{16}\right]_0^4$

$= \dfrac{1}{2}\left(\dfrac{64}{3} - 16\right) = \dfrac{8}{3}$

$\mathrm{Var}(X) = E(X^2) - E^2(X)$

$= \dfrac{8}{3} - \left(\dfrac{4}{3}\right)^2 = \dfrac{8}{9}$

Hence $\sigma = \sqrt{\mathrm{Var}(X)} = \sqrt{\dfrac{8}{9}} = 0.943$

c First we need to calculate the probability that a car has to wait more than 2 minutes.

$P(X > 2) = \displaystyle\int_2^4 \frac{1}{2}\left(1 - \frac{x}{4}\right) dx$

$= \dfrac{1}{2}\left[x - \dfrac{x^2}{8}\right]_2^4$

$= \dfrac{1}{2}\left[(4 - 2) - \left(2 - \dfrac{1}{2}\right)\right]$

$= \dfrac{1}{4}$

Since we are now considering six cars, this follows the binomial distribution $Y \sim \mathrm{Bin}\left(6, \dfrac{1}{4}\right)$.

We want $P(Y = 3) = {}^6C_3 \left(\dfrac{1}{4}\right)^3 \left(\dfrac{3}{4}\right)^3 = 0.132$.

Example

A continuous random variable has probability density function f(x) where

$f(x) = \begin{cases} \dfrac{3}{128} x^2 & 0 \le x \le 4 \\ \dfrac{1}{4} & 4 < x \le 6 \\ 0 & \text{otherwise} \end{cases}$

Calculate:

a $E(X)$

b $\mathrm{Var}(X)$

c σ

d $P(|X - \mu| < \sigma)$

a $E(X) = \int\limits_{0}^{4} \frac{3}{128}x^3\,dx + \int\limits_{4}^{6} \frac{1}{4}x\,dx$

$= \left[\frac{3x^4}{512}\right]_{0}^{4} + \left[\frac{x^2}{8}\right]_{4}^{6}$

$= \frac{3}{2} + \frac{9}{2} - 2 = 4$

b $E(X^2) = \int\limits_{0}^{4} \frac{3}{128}x^4\,dx + \int\limits_{4}^{6} \frac{1}{4}x^2\,dx$

$= \left[\frac{3x^5}{640}\right]_{0}^{4} + \left[\frac{x^3}{12}\right]_{4}^{6}$

$= \frac{24}{5} + 18 - \frac{16}{3} = \frac{262}{15}$

$Var(X) = E(X^2) - E^2(X)$

$= \frac{262}{15} - (4)^2 = \frac{22}{15}$

c $\sigma = \sqrt{Var(X)} = \sqrt{\frac{22}{15}} = 1.21$

d $P(|X - \mu| < \sigma) = P(|X - 4| < 1.21)$

$= P(-1.21 < X - 4 < 1.21)$

$= P(2.79 < X < 5.21)$

$= \int\limits_{2.79}^{4} \frac{3}{128}x^2\,dx + \int\limits_{4}^{5.21} \frac{1}{4}\,dx$

$= \left[\frac{x^3}{128}\right]_{2.79}^{4} + \left[\frac{x}{4}\right]_{4}^{5.21}$

$= 0.5 - 0.169\ldots + 1.3025 - 1 = 0.633$

The mode

Since the mode is the most likely value for X, it is found at the value of X for which $f(x)$ is greatest, in the given range of X. Provided the probability density function has a maximum point, it is possible to determine the mode by finding this point.

> To find the mode we differentiate, but when we find the mean and the median we integrate.

Example

The continuous random variable X has probability density function $f(x)$ where

$f(x) = \frac{3}{38}(3 + 2x)(3 - x)$, $1 \le x \le 3$. Find the mode.

To find the mode we need to find the value of X for which $f(x)$ is greatest. This function does not have a local maximum between $x = 1$ and $x = 3$. (There is a local maximum at $x = 0.75$ but this is not in the given domain.) The mode is the value of X which gives the maximum point on the graph and since this is a decreasing function between $x = 1$ and $x = 3$ the mode is 1.

The median

Since the probability is given by the area under the curve, the median splits the area under the curve $y = f(x)$, $a \le x \le b$, into two halves. So if the median is m, then

$$\int_a^m f(x)\ dx = 0.5$$

Example

The continuous random variable X has probability density function $f(x)$ where

$f(x) = \dfrac{3}{10}(x - 4)(1 - x)$, $1 \le x \le 3$. Find the median.

Let the median be m.

$$\int_1^m \frac{3}{10}(x - 4)(1 - x)\ dx = 0.5$$

$$\Rightarrow \frac{3}{10}\int_1^m (-x^2 + 5x - 4)\ dx = 0.5$$

$$\Rightarrow \frac{3}{10}\left[-\frac{x^3}{3} + \frac{5x^2}{2} - 4x \right]_1^m = 0.5$$

$$\Rightarrow \frac{3}{10}\left[\left(-\frac{m^3}{3} + \frac{5m^2}{2} - 4m \right) - \left(-\frac{1}{3} + \frac{5}{2} - 4 \right) \right] = 0.5$$

$$\Rightarrow -2m^3 + 15m^2 - 24m + 1 = 0$$

We now solve this on a calculator.

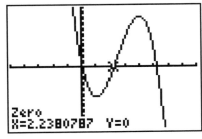

```
Zero
X=2.2380787  Y=0
```

There are three solutions to this equation, but only one lies in the domain and hence $m = 2.24$.

This becomes a little more complicated if the probability density function is made up of more than one function, as we have to calculate in which domain the median lies.

Example

A continuous random variable X has a probability density function

$$f(x) = \begin{cases} 4x & 0 \le x \le \frac{1}{4} \\ 1 & \frac{1}{4} \le x \le \frac{1}{2} \\ -\frac{4}{5}x + \frac{7}{5} & \frac{1}{2} \le x \le \frac{7}{4} \\ 0 & \text{otherwise} \end{cases}$$

a Sketch $y = f(x)$.

b Find the median m.

c Find $P\left(|X - m| < \frac{1}{2}\right)$.

a

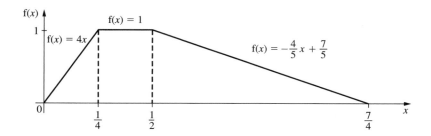

b We now have to determine in which section of the function the median occurs. We will do this by integration, but it can be done using the areas of triangles and rectangles.

$$P\left(X \le \frac{1}{4}\right) = \int_0^{\frac{1}{4}} 4x \, dx$$

$$= \left[2x^2\right]_0^{\frac{1}{4}} = \frac{1}{8}$$

Since $\frac{1}{8} < 0.5$ the median does not lie in this region.

$$P\left(X \le \frac{1}{2}\right) = \int_0^{\frac{1}{4}} 4x \, dx + \int_{\frac{1}{4}}^{\frac{1}{2}} 1 \, dx$$

$$= \left[2x^2\right]_0^{\frac{1}{4}} + \left[x\right]_{\frac{1}{4}}^{\frac{1}{2}} = \frac{1}{8} + \frac{1}{2} - \frac{1}{4} = \frac{3}{8}$$

Since $\frac{3}{8} < 0.5$ the median does not lie in this region so it must be in the third region.

Hence $\displaystyle\int_0^{\frac{1}{4}} 4x \, dx + \int_{\frac{1}{4}}^{\frac{1}{2}} 1 \, dx + \int_{\frac{1}{2}}^{m} \left(-\frac{4}{5}x + \frac{7}{5}\right) dx = 0.5$

$$\Rightarrow \frac{3}{8} + \frac{1}{5}\left[-2x^2 + 7x\right]_{\frac{1}{2}}^{m} = 0.5$$

$$\Rightarrow \frac{3}{8} + \frac{1}{5}\left(-2m^2 + 7m + \frac{1}{2} - \frac{7}{2}\right) = 0.5$$

$$\Rightarrow 15 - 16m^2 + 56m - 24 = 20$$

$$\Rightarrow 16m^2 - 56m + 29 = 0$$

$$\Rightarrow m = 0.632 \text{ or } 2.86$$

Since $m = 2.86$ is not defined for the probability density function, $m = 0.632$.

c $P\left(|X - m| < \frac{1}{2}\right) = P\left(|X - 0.632| < \frac{1}{2}\right)$

$$= P\left(-\frac{1}{2} < X - 0.632 < \frac{1}{2}\right)$$

$$= P(0.132 < X < 1.132)$$

$$= \int_{0.132}^{\frac{1}{4}} 4x \, dx + \int_{\frac{1}{4}}^{\frac{1}{2}} 1 \, dx + \int_{\frac{1}{2}}^{1.132} \left(-\frac{4}{5}x + \frac{7}{5}\right) dx$$

$$= [2x^2]_{0.132}^{\frac{1}{4}} + [x]_{\frac{1}{4}}^{\frac{1}{2}} + \frac{1}{5}[-2x^2 + 7x]_{\frac{1}{2}}^{1.132}$$

$$= 0.125 - 0.0348 + 0.5 - 0.25 + 1.07 - 0.6 = 0.812$$

Exercise 2

1 A continuous random variable has probability density function

$$f(x) = \begin{cases} kx & 0 \le x \le 2 \\ 0 & \text{otherwise} \end{cases}$$

where k is a constant.

  a Find the value of k.   **b** Find $E(X)$.   **c** Find $Var(X)$.

2 A continuous random variable has probability density function

$$f(x) = \begin{cases} \dfrac{k}{x} & 1 \le x \le 3 \\ 0 & \text{otherwise} \end{cases}$$

where k is a constant.

Without using a calculator, find:

  a k **b** $E(X)$ **c** $Var(X)$

3 The probability density function of a continuous random variable Y is given by

$$f(y) = \begin{cases} y(2 + 3y) & 0 < y < c \\ 0 & \text{otherwise} \end{cases}$$

where c is a constant.

  a Find the value of c.   **b** Find the mean of Y.

4 A continuous random variable X has probability density function

$$p(x) = \begin{cases} kx & 0 \le x \le 4 \\ 4k & 4 \le x \le 6 \\ 0 & \text{otherwise} \end{cases}$$

where k is a constant.

Find:

a k **b** $E(X)$ **c** $Var(X)$ **d** the median of x **e** $P(3 \le X \le 5)$

5 A continuous random variable X has a probability density function

$$f(x) = \begin{cases} kx^2 & 0 \le x \le 3 \\ 0 & \text{otherwise} \end{cases}$$

where k is a constant.

Find:

a k **b** $E(X)$ **c** $Var(X)$ **d** the median of X.

6 A continuous probability density function is described as

$$f(x) = \begin{cases} ce^x & 0 \le x \le 1 \\ 0 & \text{otherwise} \end{cases}$$

where c is a constant.

a Find the value of c.   **b** Find the mean of the distribution.

7 A continuous random variable X has a probability density function

$$f(x) = k \cos x, \ 0 \le x \le \frac{\pi}{2}.$$

Find:

a k **b** $E(X)$ **c** $Var(X)$ **d** the median of X **e** $P\left(|X - m| > \frac{1}{2}\right)$

8 A continuous probability distribution is defined as

$$p(x) = \begin{cases} \dfrac{1}{1 + x^2} & 0 \le x \le k \\ 0 & \text{otherwise} \end{cases}$$

where k is a constant.

Find:

a k
b the mean, μ
c the standard deviation, σ
d the median, m
e $P\left(|X - m| > \dfrac{1}{4}\right)$

9 a If $x \ge 0$, what is the largest domain of the function $f(x) = \dfrac{1}{\sqrt{1 - 4x^2}}$?

The function $f(x) = \dfrac{1}{\sqrt{1 - 4x^2}}$ is now to be used as a probability density function for a continuous random variable X.

b For it to be a probability density function for a continuous random variable X, what is the domain, given that the lower bound of the domain is 0?

c Find the mean of X.

d Find the standard deviation of X.

10 A continuous random variable has a probability density function given by

$$f(x) = \begin{cases} \dfrac{2}{\pi(1 + x^2)} & -1 \le x \le 1 \\ 0 & \text{otherwise} \end{cases}$$

a Without using a calculator, find $P\left(|X| \le \tan\dfrac{\pi}{4}\right)$.

b Find the mean of X.

c Find the standard deviation of X.

11 A continuous random variable X has probability density function

$$f(x) = \begin{cases} kx^2 e^{-cx} & 0 \le x \le 2 \\ 0 & \text{otherwise} \end{cases}$$

where k and c are positive constants. Show that $k = \dfrac{-c^3 e^{2c}}{2(2c^2 + c + 1)}$.

12 The time taken in minutes for a carpenter in a factory to make a wooden shelf follows the probability density function

$$f(t) = \begin{cases} \dfrac{3}{56}(15t - t^2 - 50) & 6 \le t \le 10 \\ 0 & \text{otherwise} \end{cases}$$

a Find:

i μ

ii σ^2

b A carpenter is chosen at random. Find the probability that the time taken for him to complete the shelf lies in the interval $[\mu - \sigma, \mu]$.

13 The lifetime of Superlife batteries is X years where X is a continuous random variable with probability density function

$$f(x) = \begin{cases} 0 & x < 0 \\ ke^{\frac{-x}{3}} & 0 \le x \le 6 \end{cases}$$

where k is a constant.

a Find the exact value of k.

b Find the probability that a battery fails after 4 months.

c A computer keyboard takes six batteries, but needs a minimum of four batteries to operate. Find the probability that the keyboard will continue to work after 9 months.

14 The probability that an express train is delayed by more than X minutes is modelled by the probability density function $f(x) = \dfrac{1}{72\,000}(x - 60)^2$, $0 \le x \le 60$. It is assumed that no train is delayed by more than 60 minutes.

a Sketch the curve.

b Find the standard deviation of X.

c Find the median, m, of X.

15 A continuous random variable X has probability density function

$$f(x) = \begin{cases} kx^2 + c & 0 \le x \le 2 \\ 0 & \text{otherwise} \end{cases}$$

where k and c are constants.

The mean of X is $\dfrac{3}{2}$.

a Find the values of k and c.

b Find the variance of X.

c Find the median, m, of X.

d Find $P(|X - 1| < \sigma)$ where σ is the standard deviation of X.

22.3 Normal distributions

We found in Chapter 21 that there were special discrete distributions, which modelled certain types of data. The same is true for continuous distributions and the normal distribution is probably the most important continuous distribution in statistics since it models data from natural situations quite effectively. This includes heights and weights of human beings. The probability density function for this curve is quite complex and contains two parameters, μ the mean and σ^2 the variance.

The probability density function for a normal distribution is $f(x) = \dfrac{e^{\frac{-(x-\mu)^2}{2\sigma^2}}}{\sigma\sqrt{2\pi}}$.

If a random variable X follows a **normal distribution**, we say $X \sim N(\mu, \sigma^2)$.

When we draw the curve it is a bell-shaped distribution as shown below.

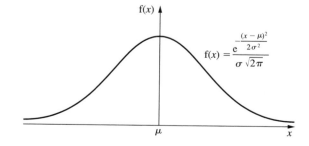

The exact shape of the curve is dependent on the values of μ and σ and four examples are shown below.

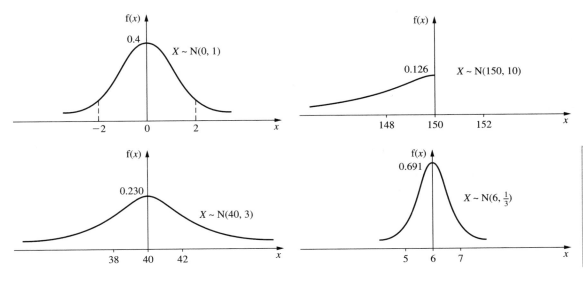

We normally make μ the axis of symmetry, but we could draw them as translations of the normal curve centred on $\mu = 0$.

Important results

1. The area under the curve is 1, meaning that f(x) is a probability density function.

2. The curve is symmetrical about μ, that is the part of curve to the left of $x = \mu$ is the mirror image of the part to the right. Hence $P(-a \le X \le a) = 2P(0 \le X \le a)$ and $P(X \ge \mu) = P(X \le \mu) = 0.5$.

3. We can find the probability for any value of x since the probability density function is valid for all values between $\pm\infty$. The further away the value of x is from the mean, the smaller the probability becomes.

4. Approximately 95% of the distribution lies within two standard deviations of the mean.

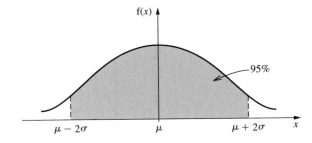

5. Approximately 99.8% of the distribution lies within three standard deviations of the mean.

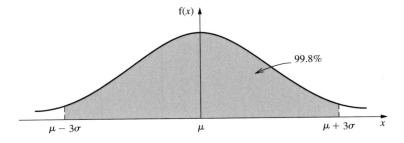

6. The maximum value of f(x) occurs when $x = \mu$ and is given by $f(x) = \dfrac{1}{\sigma\sqrt{2\pi}}$. Hence in the case of a normal distribution, the mean and the mode are the same.

7. $E(X) = \mu$. The proof of this involves mathematics beyond the scope of this syllabus.

8. $Var(X) = \sigma^2$. Again the proof of this involves mathematics beyond the scope of this syllabus.

9. The curve has points of inflexion at $x = \mu - \sigma$ and $x = \mu + \sigma$.

Finding probabilities from the normal distribution

Theoretically, this works in exactly the same way as for any continuous random variable and hence if we have a normal distribution with $\mu = 0$ and $\sigma^2 = 1$, that is $X \sim N(0, 1)$,

and we want to find $P(-0.5 \leq X \leq 0.5)$ the calculation we do is $\displaystyle\int_{-0.5}^{0.5} \dfrac{e^{\frac{-x^2}{2}}}{\sqrt{2\pi}}\, dx$. This could be done on a calculator, but would be very difficult to do manually. In fact there is no direct way of integrating this function manually and approximate methods would need to be used. In the past this problem was resolved by looking up values for the different probabilities in tables of values, but now graphing calculators will do the calculation directly. Within this syllabus you will not be required to use tables of values and it is unlikely that a question on normal distributions would appear on a non-calculator paper.

Since there are infinite values of μ and σ there are an infinite number of possible distributions. Hence we designate what we call a standard normal variable Z and these are the values that appear in tables and are the default values on a calculator. The standard normal distribution is one that has mean 0 and variance 1, that is $Z \sim N(0, 1)$.

Example

Find $P(Z \leq 1.5)$.

The diagram for this is shown below.

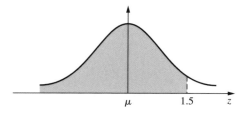

We do the calculation directly on a calculator.

```
normalcdf(-1ᴇ99,
1.5)
         .9331927713
```

$P(Z \leq 1.5) = 0.933$

It is often a good idea to draw a sketch showing what you need.

The value -1×10^{99} was chosen as the lower bound because the number is so small that the area under the curve to the left of that bound is negligible.

If we need to find a probability where Z is greater than a certain value or between two values, this works in the same way.

Example

Find $P(-1.8 \leq Z \leq 0.8)$.

The diagram for this is shown below.

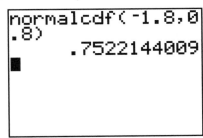

Again, we do the calculation directly on a calculator.

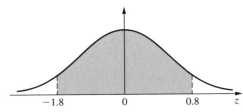

$P(-1.8 \leq Z \leq 0.8) = 0.752$

More often than not we will be using normal distributions other than the standard normal distribution. This works in the same way, except we need to tell the calculator the distribution from which we are working.

Example

If $X \sim N(2, 1.5^2)$, find $P(X \geq 3)$.

The diagram for this is shown below.

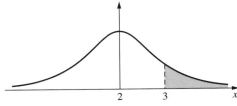

Again, we do the calculation directly on a calculator.

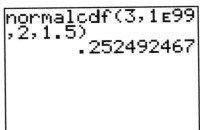

$P(X \geq 3) = 0.252$

Since we are dealing with continuous distributions, if we are asked to find the probability of X being a specific value, then we need to turn this into a range.

Example

If $X \sim N(20, 1.2^2)$, find $P(X = 25)$, given that 25 is correct to 2 significant figures.

In terms of continuous data $P(X = 25) = P(24.5 \leq X < 25.5)$ and hence this is what we calculate.
This is shown in the diagram.

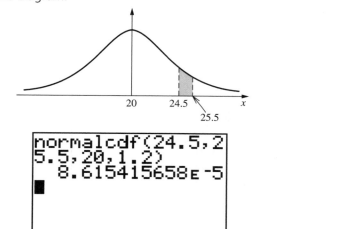

$P(X = 25) = 8.62 \times 10^{-5}$

> For the normal distribution, calculating the probability of X "less than" or the probability of X "less than or equal to" amounts to exactly the same calculation.

Up until now we have been calculating probabilities. Now we also need to be able to find the values that give a defined probability using a calculator.

Example

Find a if $P(Z \leq a) = 0.73$.

In this case we are using the standard normal distribution and we are told the area is 0.73, that is the probability is 0.73, and we want the value. This is shown in the diagram.

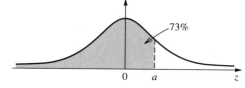

We do the calculation directly on a calculator.

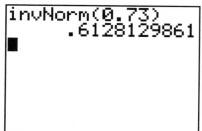

$a = 0.613$

The calculator will only calculate the value that provides what we call the **lower tail** of the graph and if we wanted $P(Z \geq a) = 0.73$, which we call the **upper tail**, we would need to undertake a different calculation. An upper tail is an area greater than a certain value and a lower tail is an area less than a certain value. This is why it can be very useful to draw a sketch first to see what is required.

Example

Find a if $P(Z \geq a) = 0.73$.

This is shown in the diagram.

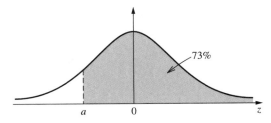

In this case $P(Z \leq a) = 1 - 0.73 = 0.27$.
We do this calculation directly on a calculator.

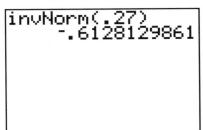

$a = -0.613$

> Because of the symmetry of the curve, the answer is the negative of the answer in the previous example. You can use this property, but it is probably easier to always use the lower tail of the distribution. This negative property only appears on certain distributions, including the standard normal distribution, since the values to the left of the mean depend on the value of the mean.

If we do not have a standard normal distribution, we can still do these questions on a calculator, but this time we need to specify μ and σ.

Example

If $X \sim N(20, 3.2^2)$, find a where $P(X \geq a) = 0.6$.

This is shown in the diagram.

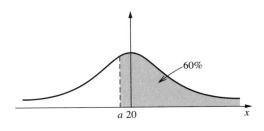

In this case $P(X \leq a) = 1 - 0.6 = 0.4$.
We do this calculation directly on a calculator.

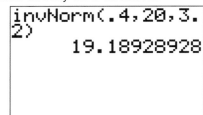

```
invNorm(.4,20,3.
2)
           19.18928928
```

$a = 19.2$

Example

If $X \sim N(15, 0.8^2)$, find a where $P(|X - \mu| \leq a) = 0.75$.
This is the same as finding $P(-a \leq X - 15 \leq a) = 0.75$
or $P(15 - a \leq X \leq 15 + a) = 0.75$.
This is shown in the diagram.

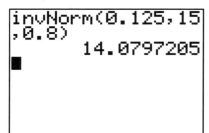

In this case $P(X \leq 15 - a) = \dfrac{1 - 0.75}{2} = 0.125$.

We do this calculation directly on a calculator.

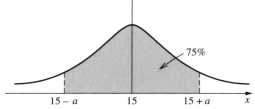

```
invNorm(0.125,15
,0.8)
           14.0797205
```

$15 - a = 14.0797\ldots$
$a = 0.920$

Exercise 3

1 If $Z \sim N(0, 1)$, find:

   a $P(Z \leq 0.756)$ **b** $P(Z \leq 0.224)$

   c $P(Z \geq -0.341)$ **d** $P(Z \leq -1.76)$

   e $P(Z \leq 1.43)$ **f** $P(0.831 < Z < 1.25)$

   g $P(-0.561 < Z < -0.0232)$ **h** $P(-1.28 < Z < 0.419)$

   i $P(|Z| < 1.41)$ **j** $P(|Z| > 0.614)$

2 If $Z \sim N(0, 1)$, find a where

a $P(Z < a) = 0.548$ **b** $P(Z < a) = 0.937$

c $P(Z < a) = 0.346$ **d** $P(Z < a) = 0.249$

e $P(Z > a) = 0.0456$ **f** $P(Z > a) = 0.686$

g $P(Z > a) = 0.159$ **h** $P(Z > a) = 0.0598$

i $P(|Z| > a) = 0.611$ **j** $P(|Z| < a) = 0.416$

3 If $X \sim N(250, 49)$, find:

a $P(X > 269)$ **b** $P(X > 241)$

c $P(X < 231)$ **d** $P(X < 263)$

4 If $X \sim N(63, 9)$, find:

a $P(X > 67)$ **b** $P(X > 54.5)$ **c** $P(X < 68)$

d $P(X < 59.5)$ **e** $P(X = 62)$

5 If $X \sim N(-15, 16)$, find:

a $P(X > -10)$ **b** $P(X > -18.5)$ **c** $P(X < -3.55)$

d $P(X < -20.1)$ **e** $P(X = -14)$

6 If $X \sim N(125, 70)$, find:

a $P(85 < X < 120)$ **b** $P(90 < X < 100)$

c $P(|X - 125| < \sqrt{70})$ **d** $P(|X - 100| < 9)$

7 If $X \sim N(80, 22)$, find:

a $P(75 < X < 90)$ **b** $P(60 < X < 73)$

c $P(|X - 80| < \sqrt{22})$ **d** $P(|X - 80| < 3\sqrt{22})$

8 If $X \sim N(40, 4)$, find a where

a $P(X < a) = 0.617$ **b** $P(X < a) = 0.293$

c $P(X > a) = 0.173$ **d** $P(X > a) = 0.651$

9 If $X \sim N(85, 15)$, find a where

a $P(X < a) = 0.989$ **b** $P(X < a) = 0.459$

c $P(X > a) = 0.336$ **d** $P(X > a) = 0.764$

10 If $X \sim N(300, 49)$, find a where

a $P(|X - 300| < a) = 0.6$ **b** $P(|X - 300| < a) = 0.95$

c $P(|X - 300| < a) = 0.99$ **d** $P(|X - 300| < a) = 0.45$

11 Z is a standardized normal random variable with mean 0 and variance 1. Find the upper quartile and the lower quartile of the distribution.

12 Z is a standardized normal random variable with mean 0 and variance 1. Find the value of a such that $P(|Z| \le a) = 0.65$.

13 A random variable X is normally distributed with mean -2 and standard deviation 1.5. Find the probability that an item chosen from this distribution will have a positive value.

14 The diagram below shows the probability density function for a random variable X which follows a normal distribution with mean 300 and standard deviation 60.

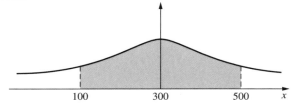

Find the probability represented by the shaded region.

15 The random variable Y is distributed normally with mean 26 and standard deviation 1.8. Find $P(23 \leq Y \leq 30)$.

16 A random variable X is normally distributed with mean zero and standard deviation 8. Find the probability that $|X| > 12$.

22.4 Problems involving finding μ and σ

To do this, we need to know how to convert any normal distribution to the standard normal distribution. Since the standard normal distribution is N(0,1) we standardize X which is N(μ, σ^2) using $Z = \dfrac{X - \mu}{\sigma}$. This enables us to find the area under the curve by finding the equivalent area on a standard curve. Hence if $X \sim$ N$(2, 0.5^2)$ and we want $P(X \leq 2.5)$, this is the same as finding $P\left(Z \leq \dfrac{2.5 - 2}{0.5}\right) = P(Z \leq 1)$ on the standard curve. The way to find μ and/or σ if they are unknown is best demonstrated by example.

Example

If $X \sim$ N$(\mu, 7^2)$ and $P(X \geq 22) = 0.729$, find the value of μ.

We begin by drawing a sketch.

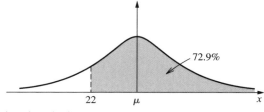

It is clear from the sketch that $\mu > 22$.
Since the question gives an upper tail, we want the value of Z associated with a probability of $1 - 0.729 = 0.271$ which can be found on a calculator to be -0.610.

Since $Z = \dfrac{X - \mu}{\sigma}$ we have $-0.610 = \dfrac{22 - \mu}{7}$

$$\Rightarrow \mu = 26.3$$

Example

If $X \sim N(221, \sigma^2)$ and $P(X \leq 215) = 0.218$, find the value of σ.

Again, we begin by drawing a sketch.

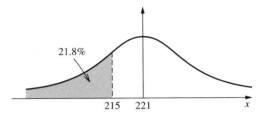

The question gives a lower tail and hence we want the value of Z associated with a probability of 0.218, which can be found on a calculator to be -0.779.

Since $Z = \dfrac{X - \mu}{\sigma}$ we have $-0.779 = \dfrac{215 - 221}{\sigma}$

$$\Rightarrow \sigma = 7.70$$

Example

If $X \sim N(\mu, \sigma^2)$, $P(X \leq 30) = 0.197$ and $P(X \geq 65) = 0.246$, find the values of μ and σ.

In this case we will have two equations and we will need to solve them simultaneously. Again we begin by drawing a sketch.

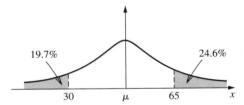

We first want the value of Z associated with a probability of 0.197, which can be found on a calculator to be -0.852.

$$-0.852 = \frac{30 - \mu}{\sigma}$$

$$\Rightarrow -0.852\sigma = 30 - \mu \quad \text{equation (i)}$$

To find the value of Z associated with 0.246 we need to use $1 - 0.246 = 0.754$ as it is an upper tail that is given. From the calculator we find the required value is 0.687.

$$0.687 = \frac{65 - \mu}{\sigma}$$

$$\Rightarrow 0.687\sigma = 65 - \mu \quad \text{equation (ii)}$$

We now subtract equation (i) from equation (ii) to find $\sigma = 22.7$.

Substituting back in equation (i) allows us to find $\mu = 49.4$.

Exercise 4

1 If $X \sim N(\mu, 1.5)$ and $P(X > 15.5) = 0.372$, find the value of μ.

2 If $X \sim N(\mu, 18)$ and $P(X > 72.5) = 0.769$, find the value of μ.

3 If $X \sim N(\mu, 7)$ and $P(X < 28.5) = 0.225$, find the value of μ.

4 If $X \sim N(\mu, 3.5)$ and $P(X < 41) = 0.852$, find the value of μ.

5 If $X \sim N(56, \sigma^2)$ and $P(X \leq 49) = 0.152$, find the value of σ.

6 If $X \sim N(15, \sigma^2)$ and $P(X \leq 18.5) = 0.673$, find the value of σ.

7 If $X \sim N(535, \sigma^2)$ and $P(X \geq 520) = 0.856$, find the value of σ.

8 If $X \sim N(125, \sigma^2)$ and $P(X \geq 135) = 0.185$, find the value of σ.

9 If $X \sim N(\mu, \sigma^2)$, $P(X \leq 8.5) = 0.247$ and $P(X \geq 14.5) = 0.261$, find the values of μ and σ.

10 If $X \sim N(\mu, \sigma^2)$, $P(X \leq 45) = 0.384$ and $P(X \geq 42.5) = 0.811$, find the values of μ and σ.

11 If $X \sim N(\mu, \sigma^2)$, $P(X \leq 268) = 0.0237$ and $P(X \geq 300) = 0.187$, find the values of μ and σ.

12 A random variable X is normally distributed with mean μ and standard deviation σ such that $P(X > 30.1) = 0.145$ and $P(X < 18.7) = 0.211$.

    a Find the values of μ and σ.   **b** Hence find $P(|X - \mu| < 3.5)$.

13 The random variable X is normally distributed and $P(X \leq 14.1) = 0.715$, $P(X \leq 18.7) = 0.953$. Find $E(X)$.

22.5 Applications of normal distributions

Normal distributions have many applications and are used as mathematical models within science, commerce etc. Hence problems with normal distributions are often given in context, but the mathematical manipulation is the same.

Example

The life of a certain make of battery is known to be normally distributed with a mean life of 150 hours and a standard deviation of 15 hours. Estimate the probability that the life of such a battery will be
    a greater than 170 hours
    b less than 120 hours
    c within the range 135 hours to 155 hours.
Six batteries are chosen at random. What is the probability that
    d exactly three of them have a life of between 135 hours and 155 hours
    e at least one of them has a life of between 135 hours and 155 hours?

$X \sim N(150, 15^2)$

a We require $P(X > 170)$.

This is shown below.

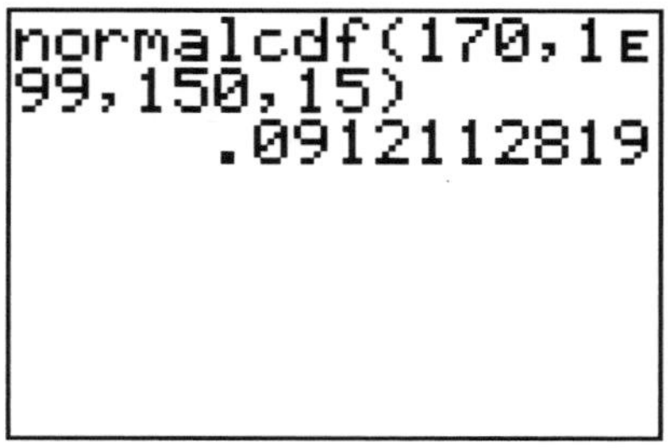

```
normalcdf(170,1E
99,150,15)
          .0912112819
```

$P(X > 170) = 0.0912$

b We require $P(X < 120)$.

This is shown below.

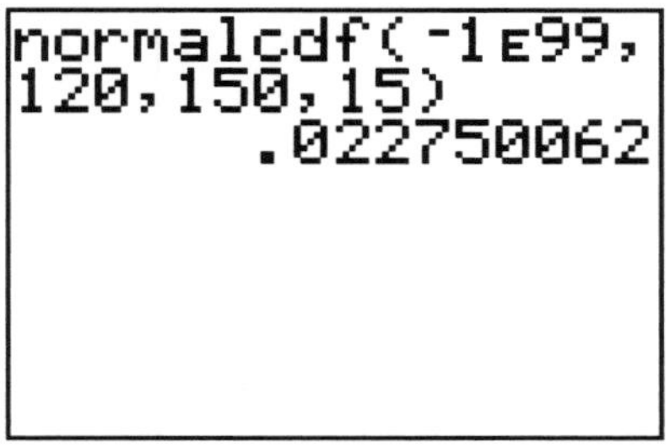

```
normalcdf(-1E99,
120,150,15)
          .022750062
```

$P(X < 120) = 0.0228$

c We require $P(135 < X < 155)$.

This is shown below.

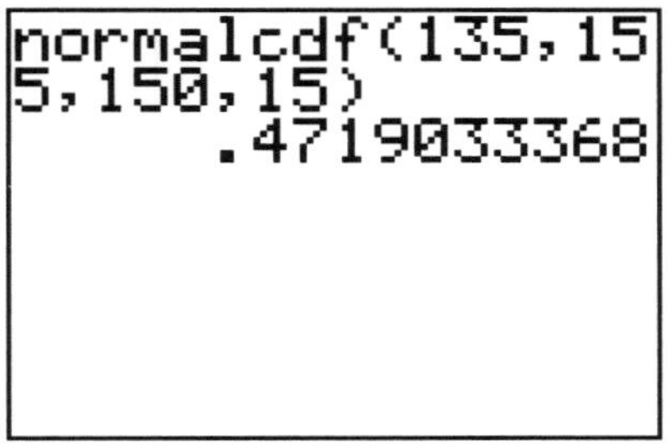

```
normalcdf(135,15
5,150,15)
          .4719033368
```

$P(135 < X < 155) = 0.472$

d We can now model this using a binomial distribution, $Y \sim \text{Bin}(6, 0.472)$.

$P(Y = 3) = {}^6C_3(0.472)^3(0.528)^3$

$\qquad\qquad = 0.310$

e Again, we use the binomial distribution and in this case we need

$P(Y \geq 1) = 1 - P(Y = 0)$

$\qquad\qquad = 0.978$

It is quite common for questions to involve finding a probability using the normal distribution and then taking a set number of these events, which leads to setting up a binomial distribution.

Example

The weight of chocolate bars produced by a particular machine follows a normal distribution with mean 80 grams and standard deviation 4.5 grams. A chocolate bar is rejected if its weight is less than 75 grams or more than 83 grams.

a Find the percentage of chocolate bars which are accepted.

The setting of the machine is altered so that both the mean weight and the standard deviation change. With the new setting, 2% of the chocolate bars are rejected because they are too heavy and 3% are rejected because they are too light.

b Find the new mean and the new standard deviation.

c Find the range of values of the weight such that 95% of chocolate bars are equally distributed about the mean.

$X \sim N(80, 4.5^2)$

a We require $P(75 < X < 83)$.

This is shown below.

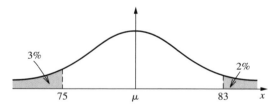

```
normalcdf(75,83,
80,4.5)
        .6142472266
```

$P(75 < X < 83) = 0.614$

$\Rightarrow 61.4\%$ are accepted.

b In this case we need to solve two equations simultaneously.
We begin by drawing a sketch.

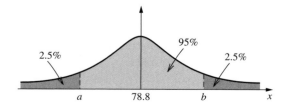

We first want the value of Z associated with a probability of 0.03. Since this is a lower tail, it can be found directly on a calculator to be -1.88.

$$-1.88 = \frac{75 - \mu}{\sigma}$$

$$\Rightarrow -1.88\sigma = 75 - \mu \text{ equation (i)}$$

To find the value of Z associated with 0.02 we need to use $1 - 0.02 = 0.98$ as it is an upper tail that is given. From the calculator we find the required value is 2.05.

$$2.05 = \frac{83 - \mu}{\sigma}$$

$$\Rightarrow 2.05\sigma = 83 - \mu \text{ equation (ii)}$$

If we now subtract equation (i) from equation (ii) we find $\sigma = 2.04$. Substituting back in equation (i) allows us to find $\mu = 78.8$.

c Again we begin by drawing a sketch.

In this case $X \sim N(78.8, 2.04^2)$ and we require $P(X < a) = 0.025$.
This is shown below.

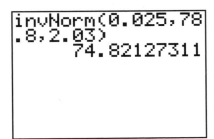

Hence the lower bound of the range is 74.8.

The upper bound is given by $78.8 + (78.8 - 74.8) = 82.8$.

The range of values required is $74.8 < X < 82.8$.

Exercise 5

1 The weights of a certain breed of otter are normally distributed with mean 2.5 kg and standard deviation 0.55 kg.

   a Find the probability that the weight of a randomly chosen otter lies between 2.25 kg and 2.92 kg.

   b What is the weight of less than 35% of this breed of otter?

2 Jars of jam are produced by Jim's Jam Company. The weight of a jar of jam is normally distributed with a mean of 595 grams and a standard deviation of 8 grams.

   a What percentage of jars has a weight of less than 585 grams?

   b Given that 50% of the jars of jam have weights between m grams and n grams, where m and n are symmetrical about 595 grams and $m < n$, find the values of m and n.

3 The temperature T on the first day of July in England is normally distributed with mean 18°C and standard deviation 4°C. Find the probability that the temperature will be

   a more than 20°C

   b less than 15°C

   c between 17°C and 22°C.

4 The heights of boys in grade 11 follow a normal distribution with mean 170 cm and standard deviation 8 cm. Find the probability that a randomly chosen boy from this grade has height

   a less than 160 cm   b less than 175 cm

   c more than 168 cm   d more than 178 cm

   e between 156 cm and 173 cm   f between 167 cm and 173 cm.

5 The mean weight of 600 male students in a college is 85 kg with a standard deviation of 9 kg. The weights are normally distributed.

   a Find the number of students whose weight lies in the range 75 kg to 95 kg.

   b 62% of students weigh more than a kg. Find the value of a.

6 The standard normal variable has probability density function $f(x) = \dfrac{e^{\frac{-(x-\mu)^2}{2\sigma^2}}}{\sigma\sqrt{2\pi}}$.
Find the coordinates of the two points of inflexion.

7 A manufacturer makes ring bearings for cars. Bearings below 7.5 cm in diameter are too small while those above 8.5 cm are too large. If the diameter of bearings produced is normally distributed with mean 7.9 cm and standard deviation 0.3 cm, what is the probability that a bearing chosen at random will fit?

8 The mean score for a mathematics quiz is 70 with a standard deviation of 15. The test scores are normally distributed.

    a Find the number of students in a class of 35 who score more than 85 in the quiz.

    b What score should more than 80% of students gain?

9 For the delivery of a package to be charged at a standard rate by a courier company, the mean weight of all the packages must be 1.5 kg with a standard deviation of 100 g. The packages are assumed to be normally distributed. A company sends 50 packages, hoping they will all be charged at standard rate. Find the number of packages that should have a weight

    a of less than 1.4 kg

    b of more than 1.3 kg

    c of between 1.2 kg and 1.45 kg.

10 At Sandy Hollow on a highway, the speeds of cars have been found to be normally distributed. 80% of cars have speeds greater than 55 kilometres per hour and 10% have speeds less than 50 kilometres per hour. Calculate the mean speed and its standard deviation.

11 Packets of biscuits are produced such that the weight of the packet is normally distributed with a mean of 500 g and a standard deviation of 50 g.

    a If a packet of biscuits is chosen at random, find the probability that the weight lies between 490 g and 520 g.

    b Find the weight exceeded by 10% of the packets.

    c If a supermarket sells 150 packets in a day, how many will have a weight less than 535 g?

12 Bags of carrots are sold in a supermarket with a mean weight of 0.5 kg and standard deviation 0.05 kg. The weights are normally distributed. If there are 120 bags in the supermarket, how many will have a weight

    a less than 0.45 kg

    b more than 0.4 kg

    c between 0.45 kg and 0.6 kg?

13 The examination scores in an end of year test are normally distributed with a mean of 70 marks and a standard deviation of 15 marks.

    a If the pass mark is 50 marks, find the percentage of candidates who pass the examination.

    b If 5% of students gain a prize for scoring above y marks, find the value of y.

14 The time taken to get to the desk in order to check in on a flight operated by Surefly Airlines follows a normal distribution with mean 40 minutes and standard deviation 12 minutes. The latest time that David can get to the desk for a flight is 1400. If he arrives at the airport at 1315, what is the probability that he will miss the flight?

15 Loaves of bread made in a particular bakery are found to follow a normal distribution X with mean 250 g and standard deviation 30 g.

    a 3% of loaves are rejected for being underweight and 4% of loaves are rejected for being overweight. What is the range of weights of a loaf of bread such that it should be accepted?

b If three loaves of bread are chosen at random, what is the probability that exactly one of them has a weight of more than 270 g?

16 Students' times to run a 200 metre race are measured at a school sports day. There are ten races and five students take part in each race. The results are shown in the table below.

Time to nearest second	26	27	28	29	30	31
Number of students	3	7	15	14	9	2

a Find the mean and the standard deviation of these times.

b Assuming that the distribution is approximately normal, find the percentage of students who would gain a time between 27.5 seconds and 29.5 seconds.

17 Apples are sold on a market stall and have a normal distribution with mean 300 grams and standard deviation 30 grams.

a If there are 500 apples on the stall, what is the expected number with a weight of more than 320 grams?

b Given that 25% of the apples have a weight less than m grams, find the value of m.

18 The lengths of screws produced in a factory are normally distributed with mean μ and standard deviation 0.055 cm. It is found that 8% of screws have a length less than 1.35 cm.

a Find μ.

b Find the probability that a screw chosen at random will be between 1.55 cm and 1.70 cm.

19 In a zoo, it is found that the height of giraffes is normally distributed with mean height H metres and standard deviation 0.35 metres. If 15% of giraffes are taller than 4.5 metres, find the value of H.

20 The weights of cakes sold by a baker are normally distributed with a mean of 280 grams. The weights of 18% of the cakes are more than 310 grams.

a Find the standard deviation.

b If three cakes are chosen at random, what is the probability that exactly two of them have weights of less than 260 grams?

21 A machine in a factory is designed to produce boxes of chocolates which weigh 0.5 kg. It is found that the average weight of a box of chocolates is 0.57 kg. Assuming that the weights of the boxes of chocolate are normally distributed, find the variance if 2.3% of the boxes weigh below 0.5 kg.

22 The marks in an examination are normally distributed with mean μ and standard deviation σ. 5% of candidates scored more than 90 and 15% of candidates scored less than 40. Find the mean μ and the standard deviation σ.

23 The number of hours, T, that a team of secretaries works in a week is normally distributed with a mean of 37 hours. However, 15% of the team work more than 42 hours in a week.

a Find the standard deviation of T.

b Andrew and Balvinder work on the team. Find the probability that both secretaries work more than 40 hours in a week.

Review exercise

 A calculator may be used in all questions unless exact answers are required.

1 A man is arranging flowers in a vase. The lengths of the flowers in the vase are normally distributed with a mean of μ cm and a standard deviation of σ cm. When he checks, he finds that 5% of the flowers are longer than 41 cm and 8% of flowers are shorter than 29 cm.

  a Find the mean μ and the standard deviation σ of the distribution.

  b Find the probability that a flower chosen at random is less than 45 cm long.

2 In a school, the heights of all 14-year-old students are measured. The heights of the girls are normally distributed with mean 155 cm and standard deviation 10 cm. The heights of the boys are normally distributed with mean 160 cm and standard deviation 12 cm.

  a Find the probability that a girl is taller than 170 cm.

  b Given that 10% of the girls are shorter than x cm, find x.

  c Given that 90% of the boys have heights between q cm and r cm where q and r are symmetrical about 160 cm, and $q < r$, find the values of q and r.

  In a group of 14-year-old students, 60% are girls and 40% are boys. The probability that a girl is taller than 170 cm was found in part **a**. The probability that a boy is taller than 170 cm is 0.202. A 14-year-old student is selected at random.

  d Calculate the probability that the student is taller than 170 cm.

  e Given that the student is taller than 170 cm, what is the probability that the student is a girl?   [IB May 06 P2 Q4]

3 In a certain college the weight of men is normally distributed with mean 80 kg and standard deviation 6 kg. Find the probability that a man selected at random will have a weight which is

  a between 65 kg and 90 kg

  b more than 75 kg.

  Three men are chosen at random from the college. Find the probability that

  c none of them weigh more than 70 kg, giving your answer to 5 decimal places.

  d at least one of them will weigh more than 70 kg.

4 A random variable X has probability density function f(x) where

$$f(x) = \begin{cases} \dfrac{1}{4}x & 0 \le x < 1 \\[2mm] \dfrac{1}{4} & 1 \le x < 3 \\[2mm] \dfrac{1}{12}(6 - x) & 3 \le x \le 6 \\[2mm] 0 & \text{otherwise} \end{cases}$$

Find the median value of X.   [IB Nov 97 P1 Q15]

5 A factory makes hooks which have one hole in them to attach them to a surface. The diameter of the hole produced on the hooks follows a normal distribution with mean diameter 11.5 mm and a standard deviation of 0.15 mm. A hook is rejected if the hole on the hook is less than 10.5 mm or more than 12.2 mm.

  a Find the percentage of hooks that are accepted.

The settings on the machine are altered so that the mean diameter changes but the standard deviation remains unchanged. With the new settings 5% of hooks are rejected because the hole is too large.

b Find the new mean diameter of the hole produced on the hooks.

c Find the percentage of hooks rejected because the hole is too small in diameter.

d Six hooks are chosen at random. What is the probability that exactly three of them will have a hole in them that is too small in diameter?

6 a A machine is producing components whose lengths are normally distributed with a mean of 8.00 cm. An upper tolerance limit of 8.05 cm is set and on one particular day it is found that one in sixteen components is rejected. Estimate the standard deviation.

b The next day, due to production difficulties, it is found that one in twelve components is rejected. Assuming that the standard deviation has not changed, estimate the mean of the day's production.

c If 3000 components are produced during each day, how many would be expected to have lengths in the range 7.95 cm to 8.05 cm on each of the two days? [IB May 93 P2 Q8]

7 A continuous random variable X has probability density function defined by

$$f(x) = \begin{cases} \dfrac{k}{1 + x^2} & \text{for } -\dfrac{1}{\sqrt{3}} \le x \le \sqrt{3} \\ 0 & \text{otherwise} \end{cases}$$

a Show that $k = \dfrac{2}{\pi}$.

b Sketch the graph of f(x) and state the mode of X.

c Find the median of X.

d Find the expected value of X.

e Find the variance of X. [IB Nov 93 P2 Q8]

8 A continuous random variable X has probability density function defined by

$$f(x) = \begin{cases} k|\sin x| & 0 \le x \le 2\pi \\ 0 & \text{otherwise} \end{cases}$$

a Find the exact value of k.

b Calculate the mean and the variance of X.

c Find $P\left(\dfrac{\pi}{2} \le X \le \dfrac{5\pi}{4}\right)$.

9 A company buys 44% of its stock of bolts from manufacturer A and the rest from manufacturer B. The diameters of the bolts produced by each manufacturer follow a normal distribution with a standard deviation of 0.16 mm.

The mean diameter of the bolts produced by manufacturer A is 1.56 mm. 24.2% of the bolts produced by manufacturer B have a diameter less than 1.52 mm.

a Find the mean diameter of the bolts produced by manufacturer B.

A bolt is chosen at random from the company's stock.

b Show that the probability that the diameter is less than 1.52 mm is 0.312 to 3 significant figures.

   c The diameter of the bolt is found to be less than 1.52 mm. Find the probability that the bolt was produced by manufacturer B.

   d Manufacturer B makes 8000 bolts in one day. It makes a profit of $1.50 on each bolt sold, on condition that its diameter measures between 1.52 mm and 1.83 mm. Bolts whose diameters measure less than 1.52 mm must be discarded at a loss of $0.85 per bolt. Bolts whose diameters measure over 1.83 mm are sold at a reduced profit of $0.50 per bolt. Find the expected profit for manufacturer B.   [IB May 05 P2 Q4]

10 The ages of people in a certain country with a large population are presently normally distributed. 40% of the people in this country are less than 25 years old.

   a If the mean age is twice the standard deviation, find, in years, correct to 1 decimal place, the mean and the standard deviation.

   b What percentage of the people in this country are more than 45 years old?

   c According to the normal distribution, 2.28% of the people in this country are less than x years old. Find x and comment on your answer.

   d If three people are chosen at random from this population, find the probability that

      i all three are less than 25

      ii two of the three are less than 25

      iii at least one is less than 25.

   e 40% of the people on a bus are less than 25 years old. If three people on this bus are chosen at random, what is the probability that all three are less than 25 years old?

   f Explain carefully why there is a difference between your answers to **d i** and **e**.   [IB Nov 91 P2 Q8]

11 A business man spends X hours on the telephone during the day. The probability density function of X is given by

$$f(x) = \begin{cases} \dfrac{1}{12}(8x - x^3) & \text{for } 0 \le x \le 2 \\ 0 & \text{otherwise} \end{cases}$$

   a **i** Write down an integral whose value is E(X).

      ii Hence evaluate E(X).

   b **i** Show that the median, m, of X satisfies the equation
        $m^4 - 16m^2 + 24 = 0$.

      ii Hence evaluate m.

   c Evaluate the mode of X.   [IB May 03 P2 Q4]

12 A machine is set to produce bags of salt, whose weights are distributed normally with a mean of 110 g and standard deviation 1.142 g. If the weight of a bag of salt is less than 108 g, the bag is rejected. With these settings, 4% of the bags are rejected.
The settings of the machine are altered and it is found that 7% of the bags are rejected.

   a **i** If the mean has not changed, find the new standard deviation, correct to 3 decimal places.

     The machine is adjusted to operate with this new value of the standard deviation.

     ii Find the value, correct to 2 decimal places, at which the mean should be set so that only 4% of the bags are rejected.

b With the new settings from part **a**, it is found that 80% of the bags of salt have a weight which lies between A g and B g, where A and B are symmetric about the mean. Find the values of A and B, giving your answers correct to 2 decimal places. [IB May 00 P2 Q4]

13 A farmer's field yields a crop of potatoes. The number of thousands of kilograms of potatoes the farmer collects is a continuous random variable X with probability density function $f(x) = k(3 - x)^3, 0 \leq x \leq 3$ and $f(x) = 0$ otherwise, where k is a constant.

a Find the value of k.

b Find the mean of X.

c Find the variance of X.

d Potatoes are sold at 30 cents per kilogram, but cost the farmer 15 cents per kilogram to dig up. What is the expected profit?

14 The difference of two independent normally distributed variables is itself normally distributed. The mean is the difference between the means of the two variables, but the variance is the sum of the two variances.

Two brothers, Oliver and John, cycle home from school every day. The times taken for them to travel home from school are normally distributed and are independent. Oliver's times have a mean of 25 minutes and a standard deviation of 4 minutes. John's times have a mean of 20 minutes and a standard deviation of 5 minutes. What is the probability that on a given day, John arrives home before Oliver?

15 The continuous random variable X has probability density function $f(x)$ where

$$f(x) = \begin{cases} e - ke^{kx}, & 0 \leq x \leq 1 \\ 0 & \text{otherwise} \end{cases}$$

a Show that $k = 1$.

b What is the probability that the random variable X has a value that lies between $\dfrac{1}{4}$ and $\dfrac{1}{2}$? Give your answer in terms of e.

c Find the mean and variance of the distribution. Give your answer exactly in terms of e.

The random variable X above represents the lifetime, in years, of a certain type of battery.

d Find the probability that a battery lasts more than six months.

A calculator is fitted with three of these batteries. Each battery fails independently of the other two. Finds the probability that at the end of six months

e none of the batteries has failed

f exactly one of the batteries has failed. [IB Nov 99 P2 Q4]

Answers

1 a 135°   **b** 20°   **c** 72°   **d** 150°   **e** 105°   **f** 22.5°   **g** 110°   **h** 114.6°   **i** 85.9°   **j** 229.2°   **k** 206.3°   **l** 22.9°

2 a $\frac{\pi}{6}$ **b** $\frac{7\pi}{6}$ **c** $\frac{3\pi}{4}$ **d** $\frac{7\pi}{4}$ **e** $\frac{4\pi}{3}$ **f** $\frac{7\pi}{18}$ **g** $\frac{2\pi}{5}$ **h** $\frac{3\pi}{10}$ **3 a** 0.611   **b** 1.75   **c** 5.24   **d** 1.40   **e** 2.30   **f** 4.85

4 a 30.7cm^2 **b** 4.71m^2 **c** 489cm^2 **d** 2430cm^2 **5 a** 6.28cm **b** 8.55m **c** 295cm **d** 113mm

6 a 66.8cm **b** 293cm **c** 177mm **7 a** 126cm **b** 2670cm^2 **8** 52.2cm^2 **9** 46.9m^2 **10** 200°

11 9.30cm **12** 30cm **13** 99.5cm **14** 7:2   **15 a** 17.9cm **b** 99.1cm^2

1 a $\frac{1}{2}$ **b** 0.174   **c** $-\frac{1}{\sqrt{2}}$ **d** −0.996   **e** $\frac{\sqrt{3}}{2}$ **f** $\frac{1}{\sqrt{2}}$ **g** $-\frac{\sqrt{3}}{2}$ **h** −0.737

2 a $\frac{1}{2}$ **b** $\frac{1}{2}$ **c** 1   **d** $\frac{\sqrt{3}}{2}$ **e** $\frac{1}{2}$ **f** $\frac{1}{\sqrt{2}}$ **g** $\frac{1}{\sqrt{2}}$ **h** 0   **i** −1   **j** 0

3 a $x° = 30°, 150°$ **b** $x° = 70.5°, 289.5°$ **c** $x° = 41.8°, 138.2°$ **d** $x° = 80.4°, 279.6°$ **e** $x° = 22.0°, 158.0°$ **f** $x° = 55.2°, 304.8°$

4 a $\theta = \frac{\pi}{3}, \frac{5\pi}{3}$ **b** $\theta = \frac{\pi}{3}, \frac{2\pi}{3}$ **c** $\theta = \frac{\pi}{4}, \frac{3\pi}{4}$ **d** No Solution   **e** $\theta = \frac{\pi}{6}, \frac{11\pi}{6}$ **f** $\theta = 0.290, 2.85$ **g** $\theta = 1.20, 5.08$ **h** $\theta = 0.775, 2.37$

1 a 24cm^2 **b** 20.7cm^2 **c** 160cm^2 **d** 26.0m^2 **2** 8220m^2 **3 a** 84.9cm^2 **b** 120cm^2 **4** 63.7m^3

5 311m^2 **6 a** $x = 5.82cm$ **b** $x = 6.37cm$ **c** $x = 7.64m$ **d** $n = 7.35m$ **e** $a = 14.2mm$

7 a $x° = 32.4°$ **b** $x° = 127.7°$ **8** 44.7° or 135.3°   **9** 10.75° or 169.25°

10 a $x = 6.07cm$ **b** $x = 22.7m$ **c** $p = 381mm$ **d** $t = 10.7cm$ **11** 2.63m **12 a** 59.6°   **b** 143.2°   **13** 68.7°

14 $A = 39.4°, B = 46.5°, C = 94.1°$ **15 a** $x = 7.64m$ **b** $x° = 50.9°$ **c** $x° = 83.8°$ **d** $x = 15.8cm$ **e** $x = 16.9m$

16 312km **17** 21.6m **18** 57.9°   **19** 204 cm^2

1 a 180°   **b** 120°   **c** $\frac{\pi}{2}$ **d** 90°   **e** 360°   **f** 120°   **g** 180°   **h** $\frac{\pi}{2}$ **i** 36°   **j** 3°

2 a
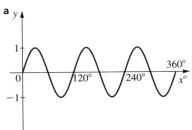

b
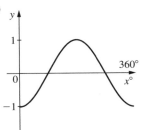

c
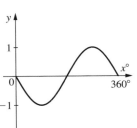

d
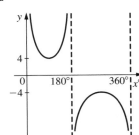

2 e
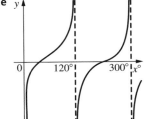

f
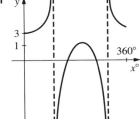

673

Answers

3 a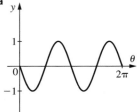

b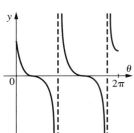

c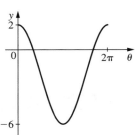

3 d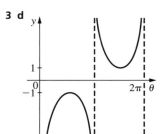

e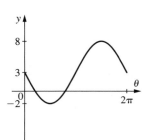

4 a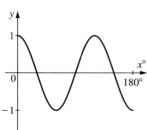

b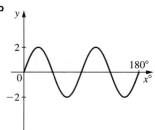

c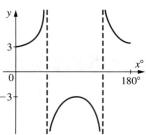

4 d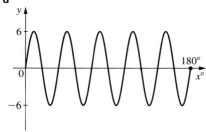

e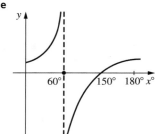

f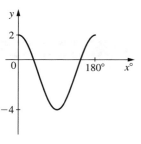

5 a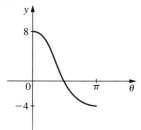

b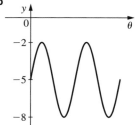

c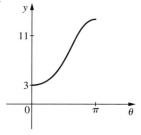

5 d 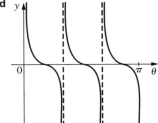 **e**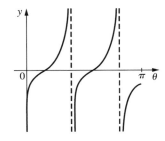

6 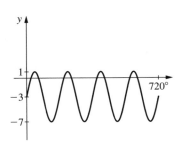 **7** 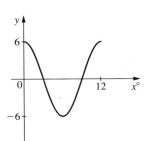 **8**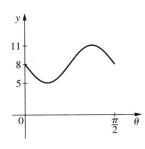

9 a $y = 4\cos 2\theta$ **b** $y = \sin x° + 1$ **c** $y = \tan 2\theta$ **d** $y = -4\cos x° + 2$ **e** $y = -4\sin 2\theta + 4$ **f** $y = -\dfrac{7}{2}\cos 4x° - \dfrac{7}{2}$

9 g $y = -\cot(15x° - 45°)$ **h** $y = 3\sin 24\theta + 8$ **i** $y = 2\csc 3x°$ **j** $y = 3\sec\dfrac{3}{2}\theta$

Chapter 1 — Exercise 5

1 a $-\dfrac{1}{2}$ **b** -1 **c** $\dfrac{1}{2}$ **d** $\dfrac{1}{2}$ **e** 1 **f** $-\dfrac{\sqrt{3}}{2}$ **g** $-\sqrt{3}$ **h** $-\dfrac{\sqrt{3}}{2}$ **i** $\dfrac{\sqrt{3}}{2}$ **j** $-\dfrac{\sqrt{3}}{2}$

2 a $\dfrac{1}{\sqrt{3}}$ **b** $\dfrac{1}{\sqrt{2}}$ **c** $\dfrac{\sqrt{3}}{2}$ **d** $-\sqrt{3}$ **e** $-\dfrac{1}{\sqrt{2}}$ **f** $-\dfrac{1}{\sqrt{3}}$ **g** 0 **h** $-\dfrac{\sqrt{3}}{2}$ **i** 1 **j** $4\sqrt{3}$

3 a $\sin 43°$ **b** $\cos 50°$ **c** $\tan 20°$ **d** $-\sin 50°$ **e** $-\cos 23°$ **f** $-\tan 34°$ **g** $-\cos 15°$ **h** $-\sin 20°$ **i** $-\tan 46°$

4 a $x° = 30°, 150°$ **b** $x° = 30°, 330°$ **c** $x° = 60°, 240°$ **d** $x° = 135°, 315°$

5 a $\theta = \dfrac{\pi}{3}, \dfrac{2\pi}{3}$ **b** $\theta = \dfrac{\pi}{6}, \dfrac{7\pi}{6}$ **c** $\theta = \dfrac{\pi}{3}, \dfrac{5\pi}{3}$ **d** $\theta = \dfrac{5\pi}{6}, \dfrac{7\pi}{6}$

Chapter 1 — Exercise 6

1 a $60°, 240°$ **b** $60°, 300°$ **c** $60°, 120°$ **d** $210°, 330°$ **e** $150°, 210°$ **f** $180°$ **g** $90°$ **h** $30°, 150°$ **i** $45°, 225°$

2 a $\dfrac{\pi}{6}, \dfrac{11\pi}{6}$ **b** $\dfrac{7\pi}{6}, \dfrac{11\pi}{6}$ **c** $\dfrac{5\pi}{6}, \dfrac{11\pi}{6}$ **d** $\dfrac{\pi}{4}, \dfrac{5\pi}{4}$ **e** $\dfrac{\pi}{6}, \dfrac{5\pi}{6}$ **f** $\dfrac{\pi}{6}, \dfrac{7\pi}{6}$ **g** $\dfrac{\pi}{2}, \dfrac{5\pi}{6}$ **h** $\dfrac{\pi}{6}, \dfrac{11\pi}{6}$

3 a $15°, 75°, 195°, 255°$ **b** $10°, 110°, 130°, 230°, 250°, 350°$ **c** $11.25°, 56.25°, 101.25°, 146.25°, 191.25°, 236.25°, 281.25°, 326.25°$

3 d $60°, 120°, 240°, 300°$ **e** $25°, 45°, 145°, 165°, 265°, 285°$ **f** $40°, 80°, 160°, 200°, 280°, 320°$

4 a $\dfrac{\pi}{12}, \dfrac{5\pi}{12}, \dfrac{7\pi}{12}, \dfrac{11\pi}{12}, \dfrac{13\pi}{12}, \dfrac{17\pi}{12}, \dfrac{19\pi}{12}, \dfrac{23\pi}{12}$ **b** $\dfrac{\pi}{6}, \dfrac{2\pi}{3}, \dfrac{7\pi}{6}, \dfrac{5\pi}{3}$ **c** $\dfrac{\pi}{30}, \dfrac{5\pi}{30}, \dfrac{13\pi}{30}, \dfrac{17\pi}{30}, \dfrac{25\pi}{30}, \dfrac{29\pi}{30}, \dfrac{37\pi}{30}, \dfrac{41\pi}{30}, \dfrac{49\pi}{30}, \dfrac{53\pi}{30}$

4 d $\dfrac{5\pi}{12}, \dfrac{7\pi}{12}, \dfrac{17\pi}{12}, \dfrac{19\pi}{12}$ **5** $15°, 105°$ **6** $\dfrac{7\pi}{24}, \dfrac{11\pi}{24}, \dfrac{19\pi}{24}, \dfrac{23\pi}{24}$ **7** $1°, 5°, 13°, 17°$ **8** $-120°, 60°$ **9** $-\dfrac{8\pi}{9}, -\dfrac{4\pi}{9}, -\dfrac{2\pi}{9}, \dfrac{2\pi}{9}, \dfrac{4\pi}{9}, \dfrac{8\pi}{9}$

10 a $19.5°, 160.5°$ **b** $41.4°, 318.6°$ **c** $58.0°, 238°$ **d** $48.2°, 311.8°$ **e** $48.6°, 131.4°$

10 f $31.3°, 288.7°$ **g** $75.5°, 284.5°$ **h** $45°, 135°$ **i** $228.7°, 341.3°$ **j** $36.9°, 143.1°$

10 k $20.9°, 69.1°, 200.9°, 249.1°$ **l** $25.2°, 94.8°, 145.2°, 214.8°, 265.2°, 334.8°$

10 m $24.9°, 60.9°, 96.9°, 132.9°, 168.9°, 204.9°, 240.9°, 276.9°, 312.9°, 348.9°$, **n** $70.5°, 289.5°$

10 o $8.2°, 171.8°$ **p** $48.2°, 311.8°$ **q** $33.7°, 213.7°$ **r** $15°, 75°, 105°, 165°, 195°, 255°, 285°, 345°$ **s** $1.4°, 4.6°, 13.4°, 16.6°,\ldots$

11 a $0.253, 2.89$ **b** $2.03, 4.25$ **c** $1.17, 4.31$ **d** $1.11, 5.18$ **e** $2.01, 0.08$ **f** $3.34, 6.08$ **g** $1.23, 4.37$ **h** $0.41, 1.68, 2.50, 3.78, 4.60, 5.87$

11 i $1.66, 3.06, 4.80, 6.20$ **j** $0.47, 1.26, 2.04, 2.83, 3.61, 4.40, 5.18, 5.97$ **k** $0.28, 0.76, 2.38, 2.86, 4.47, 4.95$ **l** $1.14, 5.15$ **m** $3.45, 5.98$

11 n $0.14, 3.28$ **o** $0.11, 0.67, 1.68, 2.25, 3.25, 3.82, 4.82, 5.39$ **12** $70.5°, 289.5°, 430.5°, 649.5°$ **13** $-\dfrac{35\pi}{36}, -\dfrac{19\pi}{36}, -\dfrac{11\pi}{36}, \dfrac{5\pi}{36}, \dfrac{13\pi}{36}, \dfrac{29\pi}{36}$

Answers

14 a 3 minutes **b**

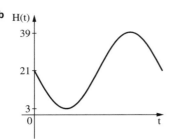

14 c i 2 mins 15 secs **ii** 45 secs **15 a** 7500 **b** 7060 **c** 12 years, 4500 fish

Chapter 1 — Review Exercise

1 a 30° **b** 75° **2 a** $\dfrac{2\pi}{3}$ **b** $\dfrac{13\pi}{12}$ **3** 7.01 cm^2 **4** 14.0 m **5** 20.6 cm^2

6 a 15.1 cm **b** 44.8° **c** 48.9° **d** 8.63 mm **7** $2.35 < BC < 5$

8 a $-\dfrac{1}{2}$ **b** $\dfrac{1}{\sqrt{2}}$ **c** $-\dfrac{1}{\sqrt{3}}$ **d** $-\dfrac{1}{2}$ **e** $\dfrac{1}{\sqrt{2}}$ **f** $-\dfrac{\sqrt{3}}{2}$ **g** $\sqrt{3}$ **h** $-\dfrac{1}{\sqrt{2}}$ **i** $-\dfrac{1}{\sqrt{3}}$ **j** 2 **k** $-\dfrac{2}{\sqrt{3}}$

9 a

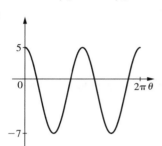

b

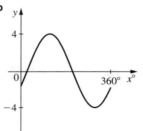

c

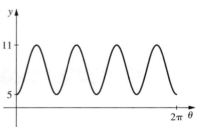

9 d

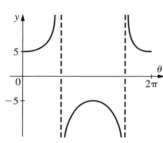

e

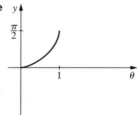

10 a $y = -3\cos 2\theta - 1$ **b** $y = \dfrac{5}{2}\cos(x - 20)° + \dfrac{5}{2}$ **11 a** $\theta = \dfrac{\pi}{6}, \dfrac{5\pi}{6}$ **b** $\theta = \dfrac{5\pi}{6}, \dfrac{7\pi}{6}$ **c** $\theta = \dfrac{\pi}{4}, \dfrac{5\pi}{4}$

11 d $\theta = \dfrac{\pi}{12}, \dfrac{\pi}{6}, \dfrac{7\pi}{12}, \dfrac{2\pi}{3}, \dfrac{13\pi}{12}, \dfrac{7\pi}{6}, \dfrac{19\pi}{12}, \dfrac{5\pi}{3}$ **e** $\theta = \dfrac{5\pi}{12}, \dfrac{11\pi}{12}, \dfrac{17\pi}{12}, \dfrac{23\pi}{12}$ **12 a** $x° = 135°, 315°$ **b** $x° = 90°$ **c** $x° = 210°, 330°$

12 d $x° = 60°, 240°$ **e** $x° = 15°, 165°, 195°, 345°$ **f** $x° = 10°, 50°, 130°, 170°, 250°, 290°$

13 a $x° = 64.6°, 295.4°$ **b** $x° = 109.3°, 160.7°, 289.3°, 340.7°$ **c** 20.7°, 80.7°, 140.7°, 200.7°, 260.7°, 320.7° **d** 64.6°, 295.4°

14 $-\dfrac{\pi}{3}, 0\,\dfrac{\pi}{3}$ **15** 20.9°, 69.1° **16 a** 14.5cm **b** 0.169s

17 a

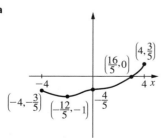

b

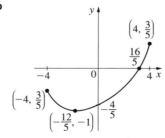

c $x = 1.57$

676

Chapter 2 — Exercise 1

1 a $x = 4, x = 1$ **b** $x = 3, x = -2$ **c** $x = -\frac{1}{2}, x = -8$ **d** $x = -1, x = 3$

2 a $x = 2, x = 5$ **b** $x = -3, x = 8$ **c** $x = -4, x = \frac{3}{2}$ **d** $x = -\frac{3}{2}, x = \frac{2}{3}$ **e** $x = \frac{2}{3}, x = 3$

3 a $x = 4.73, x = 1.27$ **b** $x = -0.854, x = 5.85$ **c** $x = -0.314, x = -3.19$ **d** $x = -0.260, x = -1.54$ **e** $x = -2.14, x = 0.468$

4 a $x = -5.45, x = -0.551$ **b** $x = -1.61, x = 5.61$ **c** $x = -1.77, -0.566$ **d** $x = -3.91, x = 1.41$ **e** $x = -1.78, x = 0.281$

Chapter 2 — Exercise 2

1 a $(x + 1)^2 + 4$ **b** $\left(x - \frac{3}{2}\right)^2 + \frac{3}{4}$ **c** $-\left(x - \frac{3}{2}\right)^2 - \frac{11}{4}$ **d** $3(x + 1)^2 - 11$ **e** $5\left(x + \frac{7}{10}\right)^2 - \frac{349}{100}$

2 a $(x + 3)^2 - 5$ Minimum $(-3, -5)$ y intercept: 4 x intercepts: $-0.764, -5.24$

2 b $(x - 2)^2 - 1$ Minimum $(2, -1)$ y intercept: 3 x intercepts: 3,1

2 c $\left(x + \frac{5}{2}\right)^2 - \frac{17}{4}$ Minimum $\left(-\frac{5}{2}, -\frac{17}{4}\right)$ y intercept: 2 x intercepts: 4.56,0.438

2 d $-(x + 2)^2 + 7$ Maximum $(-2, 7)$ y intercept: 3 x intercepts: $-4.65, 0.646$

2 e $-(x - 4)^2 + 19$ Maximum $(4, 19)$ y intercept: 3 x intercepts: $-0.359, 8.36$

2 f $2\left(x + \frac{5}{2}\right)^2 - \frac{47}{2}$ Minimum $\left(-\frac{5}{2}, -\frac{47}{2}\right)$ y intercept: -11 x intercepts: $0.928, -5.93$

2 g $4\left(x - \frac{3}{8}\right)^2 + \frac{7}{16}$ Minimum $\left(\frac{3}{8}, \frac{7}{16}\right)$ y intercept: 1 x intercepts: none

2 h $3\left(x + \frac{5}{6}\right)^2 - \frac{1}{2}$ Minimum $\left(-\frac{5}{6}, -\frac{1}{12}\right)$ y intercept: 2 x intercepts: $-1, -\frac{2}{3}$

2 i $-2\left(x - \frac{3}{4}\right)^2 + \frac{23}{8}$ Minimum $\left(\frac{3}{4}, \frac{23}{8}\right)$ y intercept: 4 x intercepts: $-0.851, 2.35$

3 a Minimum $(-1.25, -10.125)$ y intercept: -7 x intercepts: $-3.5, 1$

3 b Maximum $(2.5, -0.75)$ y intercept: -7 x intercepts: none   **c** Minimum $(-0.6, 14.2)$ y intercept: 16 x intercepts: none

3 d Maximum $(0.833, 11.1)$ y intercept: 9 x intercepts: $-1.09, 2.76$

Chapter 2 — Exercise 3

1 a
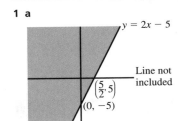

b
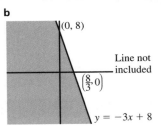

c
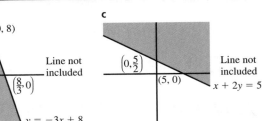

d
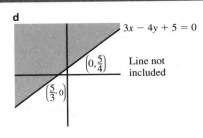

2 a $x > -\frac{1}{4}$ **b** $x < 2$ **c** $x \geq -\frac{1}{5}$ **d** $x > 3, x < 5$ **e** $5 < x < 6$ **f** $-\frac{1}{2} \leq x \leq \frac{3}{4}$ **g** $3 \leq x \leq \frac{7}{2}$ **h** $x < -4, x > -3$

2 i $-2 < x < -1$ **j** $-3 \leq x \leq 2$ **k** $-8 \leq x \leq -\frac{1}{2}$ **l** $-\frac{4}{3} < x < -\frac{1}{2}$

3 a $x < 0.764, x > 5.24$ **b** $x < -0.854, x > 5.85$ **c** $x < 0.157, x > 1.59$ **d** $-2.14 \leq x \leq 0.468$ **e** $-0.207 \leq x \leq 1.21$

Chapter 2 — Exercise 4

1 a Imaginary roots   **b** Real distinct roots   **c** Imaginary roots   **d** Real distinct roots   **e** Real equal roots   **f** Imaginary roots

2 $p = \pm 28$ **3** $q = \frac{9}{8}$ **5** $a - b - 8ab^3 + 16a^2b = 0$ **7** $p = \frac{1}{2}$ **9** $m = \frac{32}{9}$ **11 a** (ii)   **b** (iii)   **c** (i)   **d** (iii)

Chapter 2 — Review Exercise

1 $4 - (x - 2)^2$ **2** $-1 < x < 3$ **3** $k = 4.08, k = -2.58$ **4 i.** $x = \frac{1}{3}, -2, y = \frac{7}{3}, 0$ **ii** $x = \frac{9}{2}, 6, y = \frac{9}{2}, 3$ **5** $-2.46 < k < 0.458$

6 $m = \pm 2$ **8** $x = 16, x = 1$ **9** $k \leq 0, k \geq 8$ **10**

$a = \frac{9}{4}, b = \frac{7}{2}$. Maximum point is $\left(\frac{7}{2}, \frac{9}{4}\right)$. Line of symmetry is $x = \frac{7}{2}$

11 6 years old **12** $\cos C = \frac{c^2 + 3}{4c}$. True for all real values of c.

Chapter 3 — Exercise 1

1 a 11 **b** 1 **c** 130 **d** 6 **2 a** -17 **b** 12 **c** -27 **d** $\frac{19}{4}$

3 $\{-7, 1, 17\}$

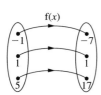

4 $\left\{-2, \frac{1}{2}, 2, 4\right\}$

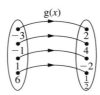

5 $\{y: -5 \leq y \leq 2\}$

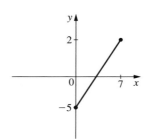

6 $\{y: 5 \leq y \leq 15\}$

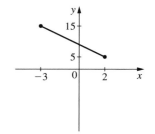

7 $\left\{y: -\frac{25}{4} \leq y \leq 24\right\}$

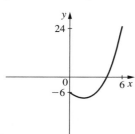

8 $\{y: 0.959 \leq y \leq 221\}$

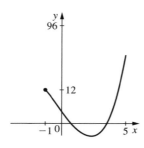

9 $\{y: 5 \leq y < \infty\}$

10 a Domain $\{x: -3 \leq x \leq 2\}$, Range $\{y: -9 \leq y \leq 11\}$

10 b Domain $\{x: 1 \leq x \leq 7\}$, Range $\{y: 2 \leq y \leq 18\}$

10 c Domain $\{x: -2 \leq x \leq 1\}$, Range $\{y: 0 \leq y \leq 4\}$ **11 a** $6x - 2$ **b** $-3x - 2$ **c** $\frac{3}{x} - 2$ **12 a** $4x^2 - 6x$ **b** $x^2 + 5x + 4$

12 c $36x^2 - 18x$ **d** $4x^2 - 10x + 4$ **13 a** $\frac{-x}{x + 2}$ **b** $\frac{2x}{1 - 2x}$ **c** $\frac{1}{2x - 1}$ **d** $\frac{-x - 2}{x}$ **14** x

Chapter 3 — Exercise 2

1 a 14 **b** 24 **c** 2 **d** $4x + 24$ **2 a** 61 **b** 5 **c** 46 **d** $3x^2 - 7$ **3 a** $\frac{\pi + 3}{3}$ **b** $\frac{\sqrt{3}}{2}$ **c** $\frac{\sqrt{3}}{2}$ **d** $\sin\left(\theta + \frac{\pi}{3}\right)$

4 a i $2x^2 - 1$ **ii** $(2x - 1)^2$ **b i** $9x^2 + 30x + 21$ **ii** $3x^2 - 7$ **c i** $(x^2 - 6)^3$ **ii** $x^6 - 6$ **d i** $\cos(3x^2)$ **ii** $3\cos^2 x$

4 e i $8x^3 + 36x^2 + 52x + 31$ **ii** $2x^3 - 2x + 17$ **f i** $(x^2 + 4)^6 - 2(x^2 + 4) + 3$ **ii** $(x^6 - 2x + 3)^2 + 4$ **g i** $\sin(2x^2 - 14)$ **ii** $\sin^2 2x - 7$

5 a $6x - 3p + 4$ **b** $6x + 8 - p$ **c** $p = -2$ **6 a** $12x^2 - 3$ **b** $6(2x - 3)^2$ **c** $216x^4$ **d** $4x - 9$ **7 a** $\cos\left(x + \frac{\pi}{2}\right)$ **b** $\cos x + \frac{\pi}{2}$

7 c $x + \pi$ **d** $\cos(\cos(x))$ **8 a i** $\frac{2}{3x - 2}$ **ii** $\frac{x + 3}{x - 3}$ **b i** $\frac{9 - 9x}{x^2}$ **ii** $\frac{3}{x^2 - 3x}$ **c i** $\frac{2 - 3x}{x + 1}$ **ii** $\frac{2 - 5x}{3 - 5x}$

8 d i $\frac{2x}{3 - x}$ **ii** $\frac{3x - 1}{2}$ **e i** $\frac{1}{x + 1}$ **ii** $\frac{2x + 4}{x}$ **9** x

Chapter 3   Exercise 3

1 a yes   **b** no   **c** yes   **d** no   **2 a** $f^{-1}(x) = \dfrac{x}{4}$ **b** $f^{-1}(x) = x + 5$ **c** $f^{-1}(x) = x - 6$ **d** $f^{-1}(x) = \dfrac{3}{2}x$ **e** $f^{-1}(x) = 7 - x$

2 f $f^{-1}(x) = \dfrac{9 - x}{4}$ **g** $f^{-1}(x) = \dfrac{x - 9}{2}$ **h** $f^{-1}(x) = (x + 6)^{\frac{1}{3}}$ **i** $f^{-1}(x) = \dfrac{x^{\frac{1}{3}}}{2}$ **3 a** $x \neq 3, x \in \mathbb{R}$ **b** $x \neq -4, x \in \mathbb{R}$ **c** $x \neq \dfrac{1}{2}, x \in \mathbb{R}$

3 d $x \geq -5, x \in \mathbb{R}$ **e** $x \leq 9, x \in \mathbb{R}$ **f** $x \geq \dfrac{1}{2}, x \in \mathbb{R}$ **g** $0 \leq x \leq \pi$ **4 a i** $x \neq 6$ **ii** $f^{-1}(x) = \dfrac{6x + 1}{x}$ **b i** $x \neq -7$ **ii** $f^{-1}(x) = \dfrac{3 - 7x}{x}$

4 c i $x \neq \dfrac{2}{3}$ **ii** $f^{-1}(x) = \dfrac{2x + 5}{3x}$ **d i** $x \neq 2$ **ii** $f^{-1}(x) = \dfrac{2x - 7}{x}$ **e i** $x \neq \dfrac{9}{4}$ **ii** $f^{-1}(x) = \dfrac{9x + 8}{4x}$ **f i** $x \neq -\dfrac{6}{5}$ **ii** $f^{-1}(x) = \dfrac{4 - 6x}{5x}$

4 g i $x > 0$ **ii** $f^{-1}(x) = \sqrt{\dfrac{x}{6}}$ **h i** $x > 0$ **ii** $f^{-1}(x) = \sqrt{x + 4}$ **i i** $x \geq 0$ **ii** $f^{-1}(x) = \sqrt{\dfrac{x - 3}{2}}$ **j i** $x \leq 16, x \in \mathbb{R}$ **ii** $f^{-1}(x) = \sqrt{\dfrac{16 - x}{9}}$

4 k i $x \geq 0$ **ii** $f^{-1}(x) = x^{\frac{1}{4}}$ **l i** $\mathbb{R}$ **ii** $f^{-1}(x) = \left(\dfrac{x + 5}{2}\right)^{\frac{1}{3}}$ **5 a** $h(x) = 3x - 6$ **b** $h^{-1}(x) = \dfrac{(x + 6)}{3}$

Chapter 3   Exercise 4

1 a $f^{-1}(x) = \dfrac{x}{2}$

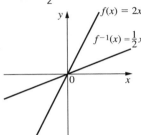

b $f^{-1}(x) = x - 2$

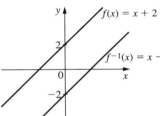

c $f^{-1}(x) = x + 3$

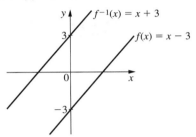

1 d $f^{-1}(x) = \dfrac{x - 1}{3}$

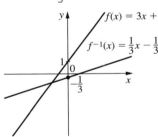

e $f^{-1}(x) = \dfrac{x + 4}{2}$

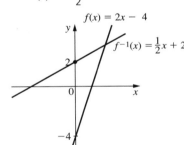

2 a $f^{-1}(x) = \sqrt{x}$

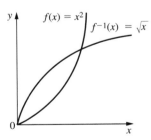

b $f^{-1}(x) = \sqrt{\dfrac{x}{3}}$

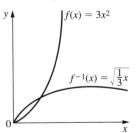

c $f^{-1}(x) = \sqrt{x - 4}$

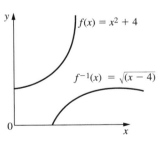

d $f^{-1}(x) = \sqrt{5 - x}$

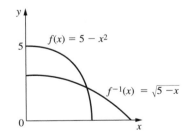

3 a $f^{-1}(x) = \dfrac{1 - 2x}{x}$

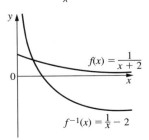

b $f^{-1}(x) = \dfrac{1 - 5x}{x}$

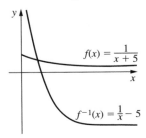

c $f^{-1}(x) = \dfrac{2 - x}{x}$

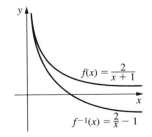

4 a

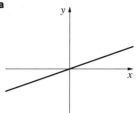

b

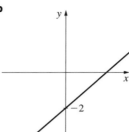

c

No inverse

d

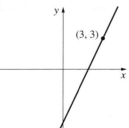
(3, 3)

4 e

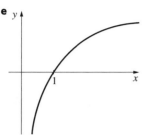

f

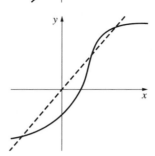

Chapter 3 — Exercise 5

1 $f(x) = \begin{cases} 2 - x, & x < 2 \\ x - 2, & x \geq 2 \end{cases}$ **2** $f(x) = \begin{cases} -2x - 1, & x < \frac{1}{2} \\ 2x + 1, & x \geq \frac{1}{2} \end{cases}$ **3** $f(x) = \begin{cases} 12 + x - x^2, & -3 < x < 4 \\ x^2 - x - 12, & x \geq 4 \\ 2x^2 - 5x - 3, & x \leq -\frac{1}{2} \end{cases}$ **4** $f(x) = \begin{cases} 3 + 5x - 2x^2, & -\frac{1}{2} < x < 3 \\ 2x^2 - 5x - 3, & x \geq 3 \end{cases}$

5

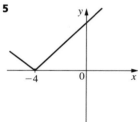

6

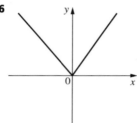

7

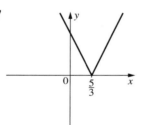

8

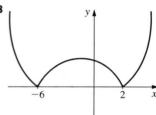

9

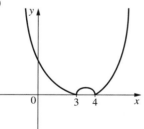

10

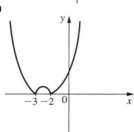

11

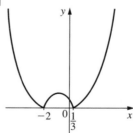

12 $x = -5, x = 1$ **13** $x = 4, x = 6$

14 $x = -4, x = -1$ **15** $x = 2, x = 5$ **16** $x = -3.37, x = -2.56, x = 1.56, x = 2.37$ **17** $x = -2.91, x = -2, x = \frac{3}{2}, x = 2.41$

18 $-7 < x < 3$ **19** $-4 \leq x \leq 5$ **20** $2 < x < \frac{5}{2}$ **21** $-6.80 \leq x \leq -5, 1 \leq x \leq 2.80$ **22** $-4.72 < x < -3, \frac{1}{2} < x < 2.22$

Chapter 3 — Exercise 6

1 a

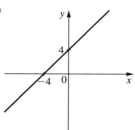

b

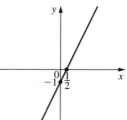

c

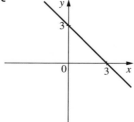

d

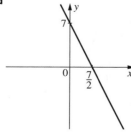

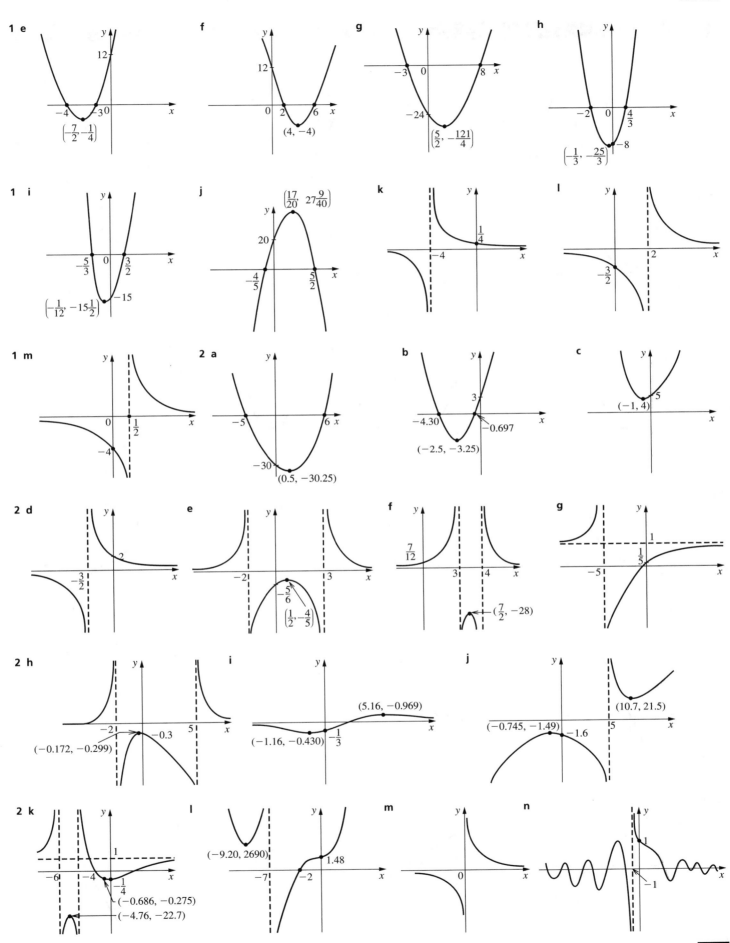

Chapter 3   Exercise 7

1 a 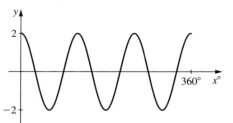   **b** 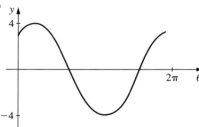   **c**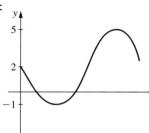

2 a 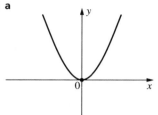   **b** 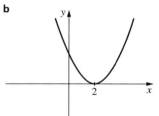   **c**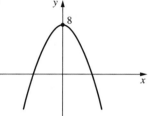

3 a i 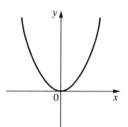 **ii** 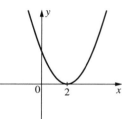 **iii** 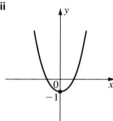 **iv**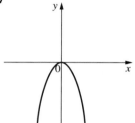

3 b i 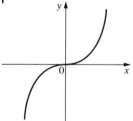 **ii** 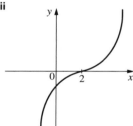 **iii** 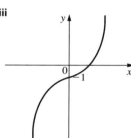 **iv**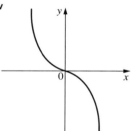

3 c i 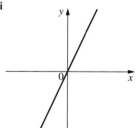 **ii** 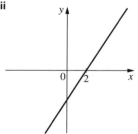 **iii** 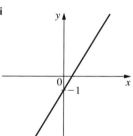 **iv**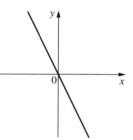

3 d i 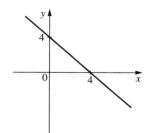 **ii** 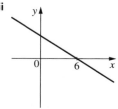 **iii** 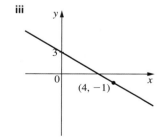 **iv**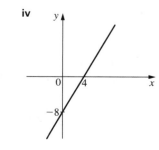

3 e i 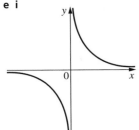 **ii** 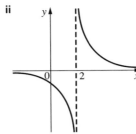 **iii** 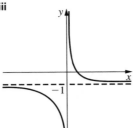 **iv**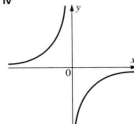

3 f i 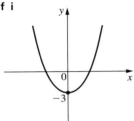 **ii** 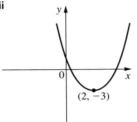 **iii** 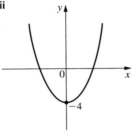 **iv**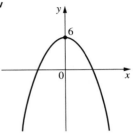

3 g i 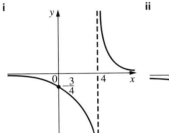 **ii** 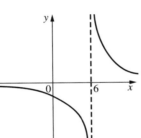 **iii** 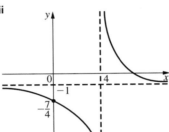 **iv**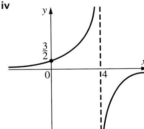

3 h i 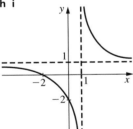 **ii** 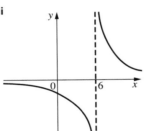 **iii** 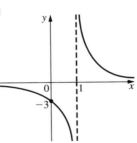 **iv**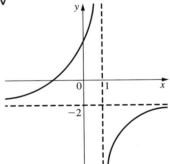

4 a $1 - 2x - x^2$ **b** $1 - 3x - 3x^2 - x^3$ **c** $-3x - 1$ **d** $x - 1$ **e** $\dfrac{2x + 1}{x + 1}$ **f** $4 - 2x - x^2$ **g** $\dfrac{2x - 9}{x - 3}$ **h** $\dfrac{x - 3}{x}$

5 a i 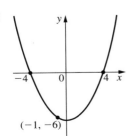 **ii** 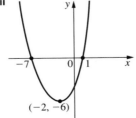 **iii**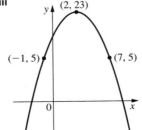

5 b i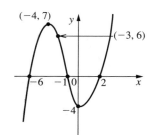

ii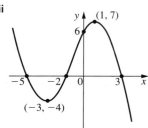

iii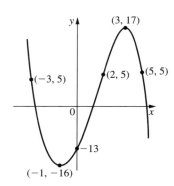

5 c i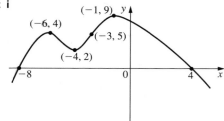

ii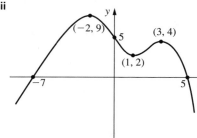

iii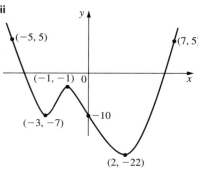

6 a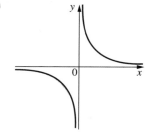

b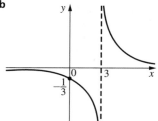

c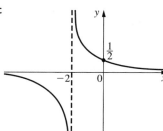

d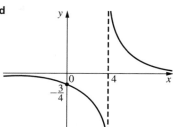

6 e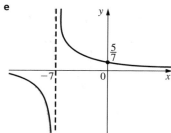

f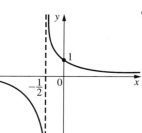

g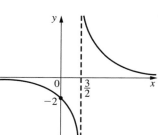

h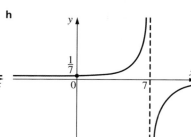

6 i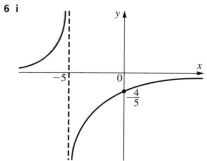

j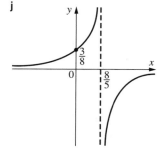

7 a

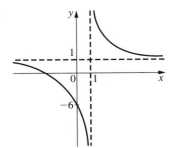

b

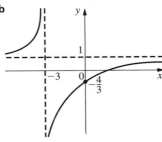

c

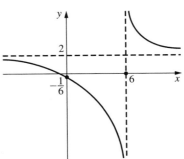

7 d

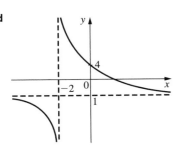

e

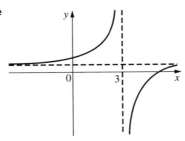

f

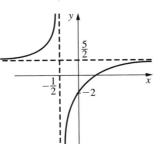

7 g

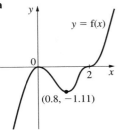

h

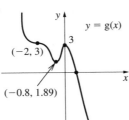

8 a $x = \dfrac{1}{2}$ **b** $x = \dfrac{10}{3}$ **c** $x = -\dfrac{1}{5}$ **d** $x = \dfrac{2}{3}$ **e** $x = \dfrac{37}{8}$

9 a $x > 0$ **b** $x \geq \dfrac{13}{4}$ **c** $x > 0$ **d** $x > \dfrac{9}{2}$

9 e $x > \dfrac{4}{3}$ **f** $x \geq 4$ **g** $x > \dfrac{7}{4}$ **h** $x > \dfrac{3}{2}$

10 a

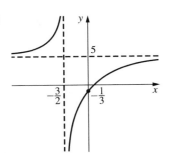

b

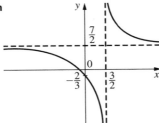

c

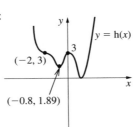

11 a

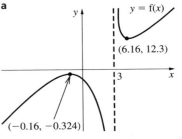

11 b

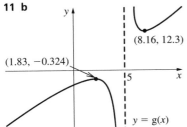

c

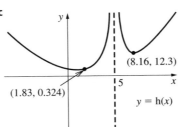

12 a

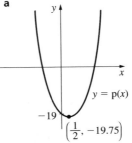

b

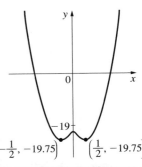

Chapter 3   Review Exercise

1 $\dfrac{7}{2}$ **2** $\left\{ \dfrac{7}{6}, -\dfrac{1}{2}, -3, \dfrac{11}{3} \right\}$

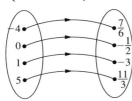

3 a $21x - 4$ **b** $14x - 11$ **c** $\dfrac{7 - 4x}{x}$ **4 a** $\dfrac{5x - 8}{x - 1}$ **b** $\dfrac{8 - 3x}{7 - 3x}$ **c** $9x - 16$ **d** x

5 a $x > 0, f^{-1}(x) = \sqrt{x + 6}$ **b** $x \neq -5, f^{-1}(x) = \dfrac{1 - 5x}{x}$ **c** $x \neq -\dfrac{3}{2}, f^{-1}(x) = \dfrac{7 - 3x}{2x}$

Answers

6 a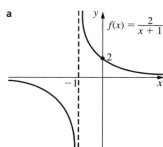
$f(x) = \dfrac{2}{x+1}$

b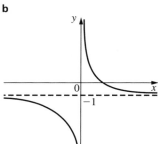

7 a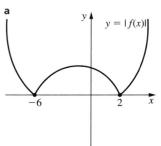
$y = |f(x)|$

b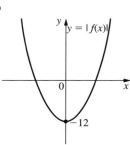
$y = |f(x)|$

8 $x = -8, x = -1$ **9** $\dfrac{4}{5} < x < 2$

10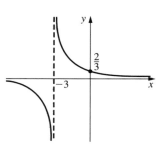

11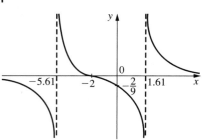

12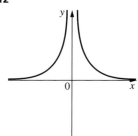

13 a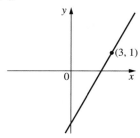

13 b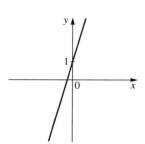

c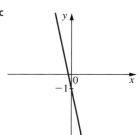

14 a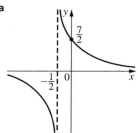

b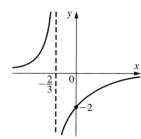

14 c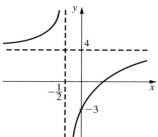

15 $x = \dfrac{19}{15}$ **16** $x \geq \dfrac{5}{12}$

17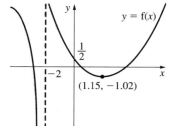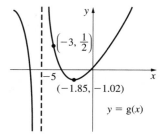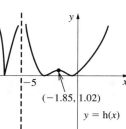
$y = f(x)$
$y = g(x)$
$y = h(x)$

18 a $-\dfrac{1}{2} < y < 2$ **b** $f^{-1}(x) = \dfrac{-2x-1}{x-2}$ **19** $-3 \leq x \leq \dfrac{1}{3}$ **20** $f^{-1}(x) = \sqrt{\dfrac{-x-1}{x-1}}$

21 a $x < \dfrac{1}{\sqrt{2}}$ **b** $0 \leq y < \infty$ **22** $x < -1$ and $x > 4$

Chapter 4 — Exercise 1

1 14 **2** −47 **3** 21 **4 a** 51 **b** 1224 **c** 674 **d** −70 **5** $\dfrac{9}{4}$ **6 a** $Q(x) = x + 8, R = 13$ **b** $Q(x) = x^2 - 3x + 2, R = 1$

6 c $Q(x) = 2x^2 + 13x + 70, R = 427$ **d** $Q(x) = x^2 + x - 5, R = 11$ **e** $Q(x) = x^3 - x^2 - 4x + 7, R = 0$

6 f $Q(x) = \dfrac{x^4}{2} + \dfrac{x^3}{4} + \dfrac{x^2}{8} - \dfrac{7x}{16} - \dfrac{87}{32}, R = -\dfrac{439}{32}$ **g** $Q(x) = \dfrac{t^2}{2} - \dfrac{t}{4} - \dfrac{27}{8}, R = \dfrac{99}{8}$ **h** $Q(x) = -\dfrac{3x^3}{4} - \dfrac{7x^2}{16} - \dfrac{59x}{64} + \dfrac{177}{256}, R = \dfrac{2797}{256}$

7 a $f(x) = (3x - 1)(x - 2)$ **b** $f(x) = (x^2 + 11x + 47)(x - 5) + 242$ **c** $f(x) = (4x^2 - 5x + 6)(x + 3) - 35$

7 d $f(x) = (5x^4 - 20x^3 + 76x^2 - 304x + 1219)(x + 4) - 4878$ **e** $f(x) = (2x^5 - 2x^4 - 3x^3 + 3x^2 - 3x + 3)(x + 1) + 6$

7 f $f(x) = \left(\dfrac{x^2}{2} - \dfrac{13x}{4} + \dfrac{3}{8}\right)(2x - 1) - \dfrac{13}{8}$ **g** $f(x) = \left(x^3 - \dfrac{x^2}{2} - \dfrac{7x}{4} + \dfrac{7}{8}\right)(2x + 1) + \dfrac{81}{8}$

Chapter 4 — Exercise 2

7 (c) (d) and (f) **8 a** $(x - 1)^2(x + 1)$ **b** $(x - 1)(x - 2)(x + 3)$ **c** $(x + 2)(x - 1)(x - 5)$ **d** $(x + 1)(x - 1)(x^2 + 1)$

8 e $(2x - 1)(x + 3)(x - 4)$ **f** $(2x + 1)(x + 4)(x + 6)$ **g** $(2x + 3)(2x + 1)(3x - 4)$ **h** $(x - 3)(x + 3)(x^2 + 2)$ **i** $(x + 3)(x^2 + 1)(2x^2 + 5)$

8 j $(6x - 1)(3x + 4)(2x + 5)(x^2 + 4)$

Chapter 4 — Exercise 3

1 17 **2** $\dfrac{133}{16}$ **3** $p = 24$ **4** $k = 2$ **5** $k = 10, (x - 3)(x - 1)(2x - 1)$ **6** $a = 2, (x + 2)(x - 3)(x + 3)$ **7** $p = 0, q = -1$

8 $k = -46, (2x + 1)(x - 4)(x + 6)$ **9** $p = 11, q = -21$ **10** $k = 5$

Chapter 4 — Exercise 4

1 $x = -1, x = -3$ **2** $x = -\dfrac{1}{2}, x = 6$ **3** $x = -2, x = 5, x = \dfrac{1}{2}, x = 4$ **4 a** $x = -1, x = 3, x = 4$ **b** $x = -7, x = -2, x = 2$

4 c $x = -11, x = -3$ **d** $x = -3, x = -2, x = 4, x = 5$ **e** $x = -2, x = 4$ **f** $x = -2, x = \dfrac{1}{2}, x = 8$ **g** $x = -\dfrac{2}{3}, x = \dfrac{1}{2}, x = \dfrac{3}{2}$

5 $x = -2, x = 1, x = 9$ **7** $x = 3$ **8 a** $p = 30$ **b** $x = 2, x = 3, x = 5$ **9 a** $k = 83$ **b** $x = -6, x = \dfrac{1}{2}, x = 7$

10 a $1 \le x \le 7$ **b** $2 \le x \le 5$ **c** $0 \le x \le 1$ **d** $3 \le x \le 7$ **e** $\dfrac{1}{2} \le x \le 35$ **11** $6 < t < 10$

Chapter 4 — Exercise 5

1 $y = x^2 + 3x - 4$ **2** $y = -x^2 + 6x$ **3** $y = 10 + 3x - x^2$ **4** $y = 2x^2 - 4x - 6$ **5** $y = 3x^2 - 12x + 12$ **6** $y = x^3 + 2x^2 - 11x - 12$

7 $y = \dfrac{1}{2}x^3 - \dfrac{3}{2}x^2 - 5x + 12$ **8** $y = -2x^3 + 2x^2 + 28x - 48$ **9** $y = x^4 - 2x^3 - 13x^2 + 14x + 24$

10 $y = 20x^4 - 180x^3 + 420x^2 + 20x - 600$ **11** $y = -2x^4 + 6x^3 + 18x^2 - 46x + 24$

Chapter 4 — Exercise 6

3 $x + 3$ **4** $2x + 1$ **5** $3x - 11, 15$ **7** $x + 5 + \dfrac{-14x - 21}{x^2 + 5}$ **8** $x - \dfrac{9}{2} + \dfrac{-30x + 85}{2(x^2 + 7)}$

9 $3x^3 - 6x^2 + 10x - 20 + \dfrac{47}{x + 2}$ **10** $2x^3 + 6x + \dfrac{18x + 13}{x^2 - 3}$ **11** $-\dfrac{x^3}{2} + \dfrac{x^2}{4} + \dfrac{27x}{8} - \dfrac{67}{16} + \dfrac{211}{16(2x + 1)}$ **12** $-x^4 + x^2 + 5x - 1 + \dfrac{12 - 5x}{x^2 + 1}$

13 $x = -4, x = -3, x = 1, x = 2$ **14** $x = -2, x = 2, x = -6$

Chapter 4 — Exercise 7

1 $x = -1, x = \dfrac{1}{2}, x = 3$ **2** $x = -7, x = -1, x = -\dfrac{1}{2}, x = 4$ **3** $x = -2, x = -\dfrac{1}{2}, x = 3$ **4** $x = 3$ **5** $x = -4.29, x = -0.428, x = 2.72$

6 $x = 5.50$ **7** $x = 0.388, x = 1$ **8** $x = 4.92, x = 1.02, x = -2.19, x = -3.74$ **9** $x = 10.2, x = 0.203, x = -1.44$

10 $x = -0.953$ **11** $f(x) = -4x^3 + 10x^2 + 28x - 16, x = -2, x = 0.388, x = 3.86$ **12** $(x + 5)(x - 1)(x - 4)$

13 $(3x + 4)(3x - 4)(2x - 1)(2x + 1)$ **14 a** $x < -6.964$ and $-0.239 < x < 1.203$ **b** $-3.877 < a < 59.877$

15 a $-1.575 < x < 2.130$ and $x < -6.556$ **b** $7.89 < a < 29.1$

Answers

Chapter 4   Review Exercise

1 -35 **2** $3x^3 + 4x^2 + 8x + 22 + \dfrac{45}{x-2}$ **3** $f(x) = \left(x^4 + \dfrac{x^3}{2} + \dfrac{x^2}{4} + \dfrac{17x}{8} + \dfrac{17}{16}\right)(2x-1) - \dfrac{95}{16}$ **4** $(x-3), (x-9)$

5 $f(x) = (2x+3)(x-4)(x+1)$ **6** $g(x) = (2x+1)(x-5)(x+2)(3x-2)$ **7** $k(x) = (x-4)(x+1)(x^2+5)$ **8** $x = -7, x = 3$

9 $x = -4, x = \dfrac{1}{2}, x = 1$ **10** $f(x) = 2(x+3)(2x-1)(x-4)^2$ **11** $x = -1.47$ **12** $x = 6.59, x = 2.38, x = -1.97$

13 $\dfrac{x^2}{2} - \dfrac{11x}{4} + \dfrac{35}{8} - \dfrac{67}{8(2x-1)}$ **14** $x^2 - 5 + \dfrac{3x}{x^2+1}$ **15** $a = -\dfrac{4}{5}, b = -\dfrac{6}{5}$ **16** $a = -6$ **17** $a = 4$ **18** $k = 6$

Chapter 5   Exercise 1

1 a p^9 **b** p^5 **c** x^{15} **d** $21y^5$ **e** $16x^{12}$ **f** t^2 **g** $2p^2$ **h** $6p^7$ **2 a** 4 **b** 3 **c** $\dfrac{1}{10}$ **d** 1 **e** 125 **f** $\dfrac{1}{3}$ **g** $\dfrac{1}{4}$ **h** $\dfrac{1}{8}$ **i** 9

3 a x^6 **b** $\dfrac{4}{3}y^4$ **c** $10p^{-6}$ **d** t^2 **e** $6m^{-\frac{2}{3}}$ **f** $12x^5 + 15x$ **g** $2x + 1$ **h** $x + 2 + x^{-1}$

4 a
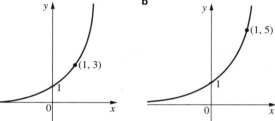

b
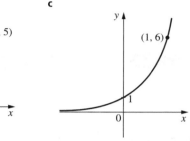

c

d
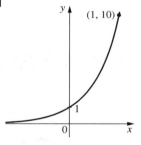

e
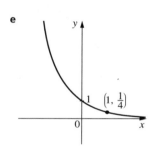

f
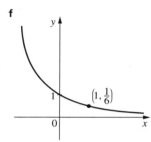

g
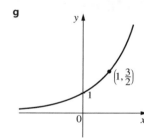

h
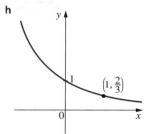

Chapter 5   Exercise 2

1 a
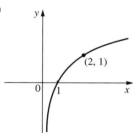

b
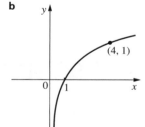

c
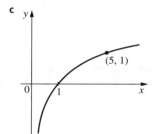

d
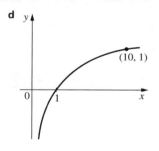

2 a
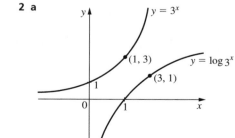

b
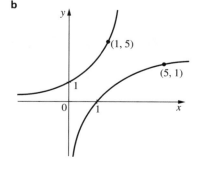

3 a 2 **b** 2 **c** 2 **d** 3 **e** 3 **f** 5

3 g 3 **h** 6 **i** 2 **j** 2 **k** 3 **l** 1 **m** 1 **n** 0

4 a $\dfrac{1}{3}$ **b** $\dfrac{1}{2}$ **c** $\dfrac{1}{2}$ **d** $\dfrac{1}{2}$ **e** $\dfrac{1}{2}$ **f** $\dfrac{1}{6}$

4 g $\dfrac{2}{3}$ **h** $\dfrac{3}{4}$ **5 a** -4 **b** -3 **c** -2

5 d -3 **e** -4 **f** $-\dfrac{1}{3}$ **g** $-\dfrac{1}{2}$

6 a 1 **b** 2 **c** $\dfrac{1}{2}$ **d** -1 **e** -1

Chapter 5 — Exercise 3

1 a $\log_a 18$ **b** $\log_a 15$ **c** $\log_a 5$ **d** $\log_a 64$ **e** $\log_a 24$ **f** $\log_a 4$ **g** $\log_a 32$ **h** $\log_a 6$ **i** $\log_a \frac{1}{8}$ **j** $\log_a 72$ **k** $\log_a 12$

2 a $\log_3 15$ **b** $\log_2\left(\frac{5}{2}\right)$ **c** $\log_2\left(\frac{8}{9}\right)$ **d** $\log_a\left(\frac{xy^2}{t^3}\right)$ **3 a** 2.699 **b** 2 **c** 1 **d** $\frac{1}{2}$ **e** 2 **f** 2 **g** 1 **h** 0 **i** -3 **j** 4 **k** $-\frac{7}{2}$

4 a $\log_a 3x^3$ **b** $\log_a\left(\frac{2}{x}\right)$ **c** $\log_a(x-1)$ **d** $\log_a\left(\frac{(x+2)^2}{3}\right)$ **5 a** 2 **b** $\frac{1}{2}$ **c** $\frac{3}{2}$ **d** $\frac{2}{3}$ **e** 2 **f** $\frac{1}{2}$ $\log_x y = \frac{1}{\log_y x}$ **6** $y = 4x^3$

7 $y = 9x^4$ **8** $y = px^5$ **10 a** $x = 7$ **b** $x = 57$ **c** $x = 6$ **d** $x = 9$ **e** $x = 8$ **f** $x = \frac{1}{6}$ **g** $x = 25$ **h** $x = 9$

11 a $x = 2$ **b** $x = 4$ **c** $x = 3$ **d** $x = 63$ **e** $x = \frac{26}{3}$ **f** $x = 10$ **12** $S_1 = 400$

Chapter 5 — Exercise 4

1 a 3 **b** 0.903 **c** 1.415 **d** 0.477 **e** -0.301 **2 a** 2.30 **b** 2.20 **c** 3.43 **d** -0.693 **e** 2.01 **f** -1.11

3 a 2 **b** 3.17 **c** 2.18 **d** 2.04 **e** 2.72 **f** 1.39 **g** 1.13 **h** 1.85 **i** $-\frac{1}{2}$ **j** -1.11 **k** -1.07 **l** $-\frac{2}{3}$ **m** -0.226

3 n -0.209 **o** 1.07 **p** 0.824 **q** -1.49 **r** -0.513

4 a
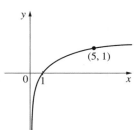

b
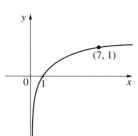

c
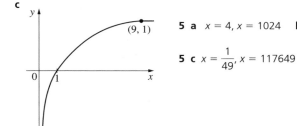

5 a $x = 4$, $x = 1024$ **b** $x = \frac{1}{4}$, $x = \frac{1}{8}$

5 c $x = \frac{1}{49}$, $x = 117649$

Chapter 5 — Exercise 5

1 a $x = 8$ **b** $x = 3.36$ **c** $x = 1.86$ **d** $x = 3.53$ **e** $x = 0.862$ **f** $x = 0.416$ **2 a** $x = 2.48$ **b** $x = 3.40$ **c** $x = 5.60$

2 d $x = 1.50$ **e** $x = -0.981$ **3 a** $x = 8100$ **b** $x = 7.39$ **c** $x = 22000$ **d** $x = 8890000$ **e** $x = 1.22$

4 a $x = 9$ **b** $x = 6$ **c** $x = 4$ **d** $x = 6$ **5 a** 40 **b** 132 **c** 3.84 days **6** 37.8 months **7 a** 80° C **b** 10.8 mins

8 a 2100g **b** 1650g **c** 57.8 years **9 a** 20100 km **b** 22.7 years **10 a** 220 **b** 57 **c** 2027

11 a $k = 0.0133$ **b** 67.2 hours **c** 9.6 hours longer **12 a** $k = 0.0114$ **b** 60.6 years **13** $\dfrac{\ln\frac{4}{3}}{\ln 6}$ **14** $\dfrac{\ln 2}{\ln 6}$ **15** $\dfrac{\ln\frac{2}{5}}{\ln 5}$ **16** $\dfrac{\ln\frac{64}{3}}{\ln\frac{9}{2}}$

17 $x = 2 - \log_2 5$ **18** $x = -1 + \log_4 3$ **19** $x = -\frac{3}{2} + \log_4 5$ **20** $x = \log_6 4$ **21** $x = 0$ **22** $x = 0.631$

Chapter 5 — Exercise 6

1 a
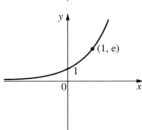

b
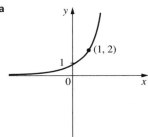

c
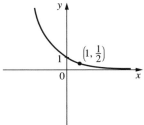

d
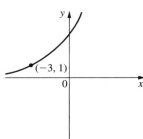

2 a
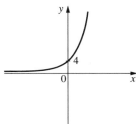

b
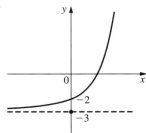

c
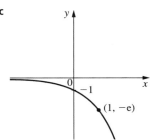

d
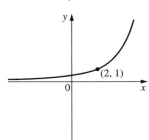

Answers

3 a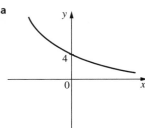
b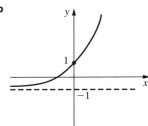
c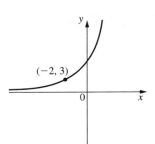
d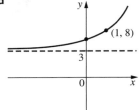

4 a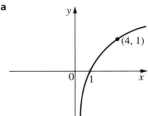
b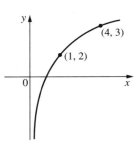
c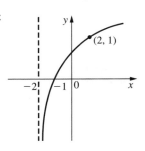
d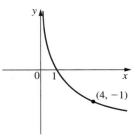

5 a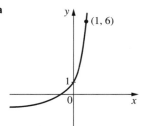
b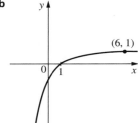
c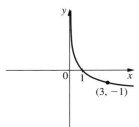
d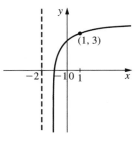

6 $k = 5$ **7** $k = 2, p = 3$ **8** $a = 3$ **9** $a = 4, p = 3$ **10** $p = 3, a = 6$

Chapter 5   Review Exercise

1 **a** x^6 **b** $15p^{-\frac{1}{4}}$ **c** $2 + 4x^{-2}$

2 a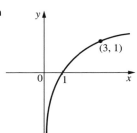
b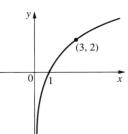

3 a 5   **b** 3   **c** $\frac{1}{3}$ **d** -2 **4 a** $\log_a 48$ **b** $\log_p 2$ **c** $\log_a 25a$

5 a $\log_3 8x$ **b** 1   **6 a** $x = 9$ **b** $x = 81$ **c** $x = 3$ **d** $\frac{11}{8}$

7 a 1.40   **b** 0.631   **7 c** 0.289   **d** 2.32   **8 a** 8100   **b** 24200000

8 c 44.7   **9 a** $x = 5.25$ **b** $x = 0.356$ **c** $x = 2.08$ **d** $x = 1.34$

10 a 5   **b** 5   **c** 6

11 a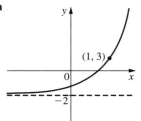
b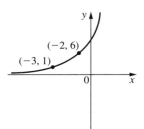
c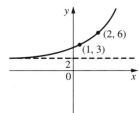
d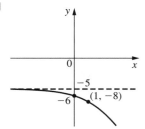

12 a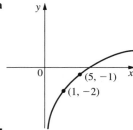
b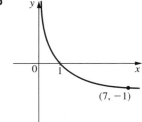
c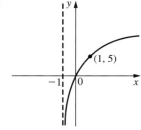

13 $k = 7$ **14** $p = 3, q = 5$ **15 a** $k = 0.0633$ **b** Yes as it will be 34 units at the end of the trip. The pressure will be ok for a total of 8.07 hours.

16 $x = \pm 6$ **17** $\dfrac{\ln 9}{\ln 8}$ **18** $-1 + \log_5 3$ **19** $x = 4, y = 4$ **20** $x = -5, y = 6; x = 1, y = 0$ **21** $x = \dfrac{15}{7}, y = \dfrac{10}{7}$ **22 a** $x < -\dfrac{2}{3}, x > \dfrac{3}{2}, x \in \mathbb{R}$

b $y \in \mathbb{R}$ **23** $x = \pm 8$ **24 a** $k = \dfrac{3}{2}, m = 1$ **b** $x = \sqrt{8}$

Chapter 6 — Exercise 1

1 $u_n = 2n + 3$ **2** $u_n = 5n - 4$ **3** $u_n = 6n + 2$ **4** $u_n = -9n + 69$ **5** $u_n = -4n + 8$ **6** $u_n = 11n - 4, u_{20} = 216$

7 $u_n = 110n + 90, u_{13} = 1520$ **8** $u_n = -7n + 24, u_{19} = -109$ **9** $u_n = \dfrac{1}{2}n + \dfrac{1}{2}, u_{15} = 8$ **10** $n = 143$ **11** $n = 23$ **12** $n = 27$

13 $u_n = 4n - 3$ **14** $u_n = 9n + 7$ **15** $u_n = -6n + 50$ **16** $u_n = -\dfrac{1}{2}n - 4$ **17** $k = 2$ **18** $k = 8$ **19** $k = 3$ **20** $k = -9, k = 6$

Chapter 6 — Exercise 2

1 $\dfrac{n}{2} + \dfrac{3}{2}n^2$ **2** $n^2 + 7n$ **3** $-\dfrac{3}{2}n^2 + \dfrac{163n}{2}$ **4** $2014n - 6n^2$ **5** $\dfrac{n}{3} + \dfrac{n^2}{6}$ **6** 203 **7** 354 **8** 1050 **9** 5586 **10** $\dfrac{n(n + 1)}{2}$

11 n^2 **12 a** $d = 2$ **b** 348

Chapter 6 — Exercise 3

1 $u_6 = \dfrac{1}{4}, u_n = 2^{4-n}$ **2** $u_6 = \dfrac{5}{64}, u_n = 80\left(\dfrac{1}{4}\right)^{n-1}$ **3** $u_6 = 486, u_n = 2(3)^{n-1}$ **4** $u_6 = -160, u_n = 5(-2)^{n-1}$ **5** $u_6 = \dfrac{-25}{8}, u_n = 100\left(-\dfrac{1}{2}\right)^{n-1}$

6 $u_6 = 384, u_n = 12(2)^{n-1}$ **7** $u_6 = 18750, u_n = 6(5)^{n-1}$ **8 a** $\dfrac{255}{16}$ **b** 107 **c** 6560 **d** -425 **e** 66.4 **f** 3060 **g** 585936

9 $S_n = \dfrac{x(1 - x^n)}{1 - x}$ **10** $S_n = \dfrac{1 - (-x)^n}{1 + x}$ **11** $S_n = \dfrac{(1 - (-3x)^n)}{1 + 3x}$ **12** $u_n = 5(2)^{n-1}$ **13** $u_n = 270\left(\dfrac{1}{3}\right)^{n-1}$ **14** $u_n = 4(-3)^{n-1}$

15 $u_n = \dfrac{1}{8}(-4)^{n-1}$ **16** $k = -2$ or $k = 10$ **17** $k = 1$ or $k = 3$ **18** $k = 8$ **19** $n = 9$ **20** $n = 5$ **21** $n = 8$ **22** $a = \dfrac{4}{3}, r = -4$

Chapter 6 — Exercise 4

1 $S_\infty = 40$ **2** $S_\infty = \dfrac{243}{2}$ **3** Does not converge **4** $S_\infty = -\dfrac{512}{13}$ **5** Does not converge **6** $S_\infty = 12$ **7** $S_\infty = 300$ **8** $S_\infty = 50$

9 $S_\infty = \dfrac{36}{7}$ **10** $-1 < x < 1$ **11** $x < -1$ or $x > 1$ **12** $\dfrac{58}{9}$ **13** $\dfrac{214}{99}$ **14** $\dfrac{727}{99}$ **15 a** 1950 **b** $\dfrac{58025}{32}$ **c** $\dfrac{32}{9}$

Chapter 6 — Exercise 5

1 28 days **2** $2720.98 **3** 98691 dkk **4** 22 years **5** 4.3% **6** £8820.36 **7 a** 17623 rats **b** 14.2 months **8 a** 187 leopards **b** 2012

9 a 3.07m **b** 10

Chapter 6 — Exercise 6

1 a 35 **b** 252 **c** 896 **2 a** $\displaystyle\sum_{r=1}^{5} 4r$ **b** $\displaystyle\sum_{r=1}^{n+2} 5r - 7$ **c** $\displaystyle\sum_{r=1}^{\infty} 4r + 5$

3 a $3n^2 + n$ **b** $\dfrac{2}{3}n^3 + \dfrac{3}{2}n^2 + \dfrac{17}{6}n$ **c** $-\dfrac{8}{3}n^3 - 2n^2 + \dfrac{53}{3}n$ **d** $\dfrac{1}{2}(k + 1)(7k + 8)$

Chapter 6 — Exercise 7

1 a 30 **b** 336 **c** 56 **d** 126 **e** 70 **2** 360 **3** 2002 **4 a** 39916800 **b** 990 **c** 330 **5** 1712304 **6** 76275360 **7** 12870

8 a 216 **b** 120 **9 a** $n = 6$ **b** $n = 5$ **c** $n = 4$ **d** $n = 6$ **10 a** $n = 4$ **b** $n = 10$ **c** $n = 9$ **11 a** $n = 7$ **b** $n = 3$ **c** $n = 3$

Chapter 6 — Exercise 8

1 a $a^4 + 4a^3b + 6a^2b^2 + 4ab^3 + b^4$ **b** $729x^6 + 2916x^5 + 4860x^4 + 4320x^3 + 2160x^2 + 576x + 64$ **c** $1 - 4x + 6x^2 - 4x^3 + x^4$

1 d $32p^5 - 240p^4q + 720p^3q^2 - 1080p^2q^3 + 810pq^4 - 243q^5$ **2 a** $x^3 + 3x + 3x^{-1} + x^{-3}$ **b** $x^5 + 10x^3 + 40x + \dfrac{80}{x} + \dfrac{80}{x^3} + \dfrac{32}{x^5}$

2 c $x^6 - 6x^4 + 15x^2 - 20 + \dfrac{15}{x^2} - \dfrac{6}{x^4} + \dfrac{1}{x^6}$ **d** $16t^4 - 8t^2 + \dfrac{3}{2} - \dfrac{1}{8t^2} + \dfrac{1}{256t^4}$ **3** $x^6 + 9x^5 + 30x^4 + 45x^3 + 30x^2 + 9x + 1$

4 a 40 **b** 7000 **c** 3840 **d** 6480 **e** −150994944 **f** 21 **g** −8 **h** 90720 **5 a** 242 **b** −54 **c** −243 **d** −54

6 a $x^7 - x^6 - 69x^5 + 109x^4 + 1616x^3 - 3360x^2 - 12800x + 32000$ **b** $x^6 - 6x^5 + 15x^4 - 26x^3 + 39x^2 - 42x + 37 - \dfrac{30}{x} + \dfrac{12}{x^2} - \dfrac{8}{x^3}$

6 c $x^7 - 2x^5 - 6x^3 + 8x + \dfrac{17}{x} - \dfrac{6}{x^3} - \dfrac{20}{x^5} - \dfrac{8}{x^7}$ **7 a** 22 **b** 952 **c** −5888 **8** 1360 **9 a** 1.04 **b** 0.210 **c** 15800000

10 $-22688x^2 + 12480x - 3200$ **11** Proof: $p^2 = -\dfrac{1}{5}$

Chapter 6 — Review Exercise

1 −7 **2 a** $r = 4$ **b** $S_n = 16(4^n - 1)$ **3 a** $S_n = \dfrac{3n^2}{2} + \dfrac{n}{2}$ **b** $n = 30$ **4 a** $r = \dfrac{2}{3}$ **b** $a = 9$

5 a 1,5,9 **b** $u_n = 4n - 3$ **6 a** $|x| < \dfrac{3}{2}$ **b** 5 **7 a** $8n - 3$ **b** 50 **8** $a = \pm 3$ **9** $a = 9$ **10 a** $n = 6$ **b** $A = 15, r = 4$

11 $a = 2, b = -3$ **12 a** $x^5 + 10x^4 + 40x^3 + 80x^2 + 80x + 32$ **b** 32.8080401 **13 a** $r = \dfrac{1}{2}$ **b** $d = -\dfrac{9}{20}$ **14 a** 59 **b** $n = 12, d = 0.25$

14 c i 99 **ii** 100 **15** 280 and 84 **16** $k = \dfrac{54}{5p^3}$ **17** 34642080 **18** 4455 **19** $6n - \dfrac{5}{6}n(n + 1)(2n + 1),\ -972$ **20** $n = 10$

Chapter 7 — Exercise 1

1 a $\pm\dfrac{\sqrt{3}}{2}$ **b** $\pm\dfrac{1}{\sqrt{2}}$ **c** $\pm\dfrac{2\sqrt{6}}{7}$ **d** No possible value **2 a** $\pm\dfrac{1}{2}$ **b** $\pm\dfrac{3}{5}$ **c** No possible value **d** $\pm\dfrac{2\sqrt{6}}{5}$ **4** $\pm\dfrac{1}{3}$

5 a $3\sin\theta$ **b** $\tan\theta$ **c** $-\sin\theta$ **d** $7\cot^2\theta$ **7 a** 0.464, 2.68, 3.61, 5.82 **b** $\dfrac{\pi}{2}, \dfrac{3\pi}{2}$ **c** 0.785, 2.90, 3.93, 6.04

7 d $\dfrac{\pi}{4}, \dfrac{5\pi}{4}$ **e** 3.53, 5.90 **f** 1.05, π, 5.24

Chapter 7 — Exercise 2

1 a $\dfrac{1 + \sqrt{3}}{2\sqrt{2}}$ **b** $\dfrac{\sqrt{3} - 1}{2\sqrt{2}}$ **c** $\dfrac{1 + \sqrt{3}}{\sqrt{3} - 1}$ **2 a and b** $\dfrac{\sqrt{3} - 1}{2\sqrt{2}}$ **3** $\dfrac{1 - \sqrt{3}}{2\sqrt{2}}$ **4** $-\dfrac{(1 + \sqrt{3})}{2\sqrt{2}}$ **8 a** 0 **b** $\dfrac{\sqrt{3}}{2}$

9 a $\dfrac{4}{5}$ **b** $\dfrac{117}{125}$ **c** $-\dfrac{44}{117}$ **10** $-\dfrac{16}{65}$ **11** 0.951 **12** $-\dfrac{2}{7}$ **13** 0.2829, −0.3877 **14** 52.5°, 232.5° **16** $\dfrac{6}{10}$ **17 a** $\dfrac{24}{25}$ **b** $-\dfrac{7}{25}$

17 c $\dfrac{336}{625}$ **d** $\dfrac{527}{625}$ **18 a** 40.9°, 220.9° **b** 0°, 180° **c** 7.6°, 187.6° **19 a** 2.88, 6.02 **b** 2.36, 5.50 **c** 1.70, 4.84

Chapter 7 — Exercise 3

1 a $\dfrac{\sqrt{3}}{2}$ **b** $\dfrac{1}{2}$ **c** $-\sqrt{3}$ **2** $\dfrac{120}{169}$ **3** $\dfrac{161}{289}$ **4 a** $\dfrac{1}{\sqrt{5}}$ **b** $\dfrac{2}{\sqrt{5}}$ **5** $\cos\theta(1 - 8\sin^2\theta\cos^2\theta) - 4\sin\theta\cos\theta(1 - 2\sin^2\theta)$

6 $2\cos\theta[16\sin^5\theta - 16\sin^3\theta + 3\sin\theta]$ **7** $-\dfrac{119}{169}$

Chapter 7 — Exercise 4

3 $\dfrac{\sqrt{2 + \sqrt{3}}}{2}$ **4** $\dfrac{\sqrt{2 + \sqrt{2}}}{2}$ **5 a** $\dfrac{\pi}{6}, \dfrac{\pi}{3}, \dfrac{7\pi}{6}, \dfrac{4\pi}{3}$ **b** $\dfrac{\pi}{6}, \dfrac{5\pi}{6}, \dfrac{7\pi}{6}, \dfrac{11\pi}{6}$ **6 a** 30°, 90°, 150°, 270° **b** 0°, 180°

6 c 60°, 90°, 270°, 300° **d** 90° **e** 60°, 300° **f** 60°, 300° **g** 41.41°, 180°, 318.59° **h** 210°, 330°

7 a $\dfrac{\pi}{3}, \pi, \dfrac{5\pi}{3}$ **b** $0, \dfrac{2\pi}{3}, \pi, \dfrac{4\pi}{3}$ **c** $\dfrac{\pi}{3}, \dfrac{2\pi}{3}, \dfrac{4\pi}{3}, \dfrac{5\pi}{3}$ **d** $\dfrac{\pi}{6}, \dfrac{\pi}{2}, \dfrac{5\pi}{6}, \dfrac{3\pi}{2}$ **e** $0, \dfrac{2\pi}{3}, \dfrac{4\pi}{3}$ **f** π **g** $0, \pi, \dfrac{7\pi}{6}, \dfrac{11\pi}{6}$

8 a $-\pi, 0$ **b** $-\dfrac{\pi}{3}, \dfrac{\pi}{3}$ **c** No solution **d** $-\pi, -\dfrac{2\pi}{3}, \dfrac{2\pi}{3}$ **9 a** 0.841, 1.82, 4.46, 5.44 **b** 0.605, 2.54

Chapter 7   Exercise 5

1 a $10\cos(x - 53.1)°$ **b** $13\cos(x - 67.4)°$ **c** $\sqrt{10}\cos(x - 288.4)°$ **d** $\sqrt{5}\cos(x - 153.4)°$ **2 a** $2\cos\left(\theta - \dfrac{11\pi}{6}\right)$ **b** $\sqrt{2}\cos\left(\theta - \dfrac{7\pi}{4}\right)$

2 c $\sqrt{5}\cos(\theta - 4.25)$ **d** $2\cos\left(\theta - \dfrac{2\pi}{3}\right)$ **3 a** $17\cos(x + 28.1)°$ **b** $\sqrt{\dfrac{37}{2}}\cos(x + 54.5)°$ **4 a** $2\sin\left(\theta + \dfrac{2\pi}{3}\right)$ **b** $\sqrt{2}\sin\left(\theta + \dfrac{3\pi}{4}\right)$

5 a $\sqrt{10}\sin(x - 108.4)°$ **b** $2\sin(x - 240)°$ **6 a** $\sqrt{2}\cos(2x - 45)°$ **b** $2\sin(3x + 60)°$ **c** $6\cos\left(\theta - \dfrac{11\pi}{6}\right)$ **d** $\sqrt{2}\cos\left(30\theta - \dfrac{3\pi}{4}\right)$

7 a Minimum $(135°, -5\sqrt{2})$ Maximum $(315°, 5\sqrt{2})$ **b** Minimum $(150°, 3)$ Maximum $(330°, 7)$

7 c Minimum $(157.5°, -8.90)$, $(337.5°, -8.90)$ Maximum $(67.5°, 10.9)$, $(247.5°, 10.9)$

8 a Minimum $(5.33, -\sqrt{12})$ Maximum $(2.19, \sqrt{12})$ **b** Minimum $\left(\dfrac{\pi}{18}, -2\right)$, $\left(\dfrac{13\pi}{18}, -2\right)$, $\left(\dfrac{25\pi}{18}, -2\right)$ Maximum $\left(\dfrac{7\pi}{18}, 2\right)$, $\left(\dfrac{19\pi}{18}, 2\right)$, $\left(\dfrac{31\pi}{18}, 2\right)$

8 c Minimum $\left(\dfrac{5\pi}{24}, -5.83\right)$, $\left(\dfrac{11\pi}{24}, -5.83\right)$, $\left(\dfrac{17\pi}{24}, -5.83\right)$, $\left(\dfrac{23\pi}{24}, -5.83\right)$ $\left(\dfrac{29\pi}{24}, -5.83\right)$, $\left(\dfrac{35\pi}{24}, -5.83\right)$, $\left(\dfrac{41\pi}{24}, -5.83\right)$, $\left(\dfrac{47\pi}{24}, -5.83\right)$

8 c Maximum $\left(\dfrac{\pi}{12}, -0.172\right)$, $\left(\dfrac{\pi}{3}, -0.172\right)$, $\left(\dfrac{7\pi}{12}, -0.172\right)$, $\left(\dfrac{5\pi}{6}, -0.172\right)$ $\left(\dfrac{13\pi}{12}, -0.172\right)$, $\left(\dfrac{4\pi}{3}, -0.172\right)$, $\left(\dfrac{19\pi}{12}, -0.172\right)$, $\left(\dfrac{11\pi}{6}, -0.172\right)$

9 a $\dfrac{\pi}{2}, \pi$ **b** $0, \dfrac{5\pi}{3}$ **c** $0, \dfrac{3\pi}{2}$ **d** $0, \dfrac{\pi}{3}, \dfrac{\pi}{2}, \dfrac{5\pi}{6}, \pi, \dfrac{4\pi}{3}, \dfrac{3\pi}{2}, \dfrac{11\pi}{6}$ **10 a** $102.4°, 195.7°$ **b** $7.3°, 34.0°, 127.3°, 154.0°, 247.3°, 274.0°$

Chapter 7   Review Exercise

1 $\pm\dfrac{\sqrt{3}}{2}$ **2 b i** $\dfrac{120}{169}$ **ii** $-\dfrac{119}{169}$ **6** $\dfrac{\pi}{6}, \dfrac{\pi}{3}, \dfrac{\pi}{2}, \dfrac{2\pi}{3}, \dfrac{5\pi}{6}$ **7 a** $5\cos(\theta - 0.644)$ **b** $\theta = 0.644$ **8** 0.905 **9** $\dfrac{1 - \sqrt{3}}{2\sqrt{2}}$ **10** $-\dfrac{13}{85}$ **11** $\dfrac{2}{\sqrt{2 - \sqrt{2}}}$

12 $-\dfrac{35}{12}$ **13** $\sqrt{112}\sin(\theta - 3.86)$ **14** $11.25°, 22.5°, 56.25°, 67.5°$ **17 b** $\alpha = 0.464$, $k = -0.559$, $p = 0.5$ **c** $A = \dfrac{\sqrt{5} + 2}{4}$, $\theta = 1.34$

Chapter 8   Exercise 1

1 $f'(x) = 5$ **2** $f'(x) = 8$ **3** $f'(x) = -2$ **4** $f'(x) = 2x$ **5** $f'(x) = 3x^2$ **6** $f'(x) = 4x^3$ **7** $f'(x) = 4x$ **8** $f'(x) = 10x$ **9** $f'(x) = 12x^2$

10 $f'(x) = 0$ **11** $f'(x) = \dfrac{-3}{x^2}$ **12** $f'(x) = 2x$ **13** $f'(x) = -3$ **14** $f'(x) = 2x - 4$ **15** $f'(x) = 2 + \dfrac{1}{x^2}$

Chapter 8   Exercise 2

1 $f'(x) = 18x$ **2** $f'(x) = 30x^2$ **3** $f'(x) = 24x^3$ **4** $f'(x) = -15x^4$ **5** $f'(x) = 0$ **6** $f'(x) = 7$ **7** $f'(x) = 11$ **8** $f'(x) = 8$ **9** $f'(x) = -\dfrac{8}{x^3}$

10 $f'(x) = \dfrac{5}{2\sqrt{x}}$ **11** $f'(x) = 2x + 5$ **12** $f'(x) = -\dfrac{25}{2\sqrt{x^7}}$ **13** $\dfrac{dy}{dx} = 3x^2 + 10x - 7$ **14** $\dfrac{dy}{dx} = 12x + \dfrac{2}{x^2}$ **15** $\dfrac{dy}{dx} = \dfrac{1}{4\sqrt[4]{x^3}}$ **16** $\dfrac{dy}{dx} = \dfrac{5\sqrt[3]{x^2}}{3}$

17 $\dfrac{dy}{dx} = \dfrac{4}{3} + \dfrac{4}{x^2}$ **18** $\dfrac{dy}{dx} = \dfrac{27}{10}(\sqrt{x^7} - \sqrt{x})$ **19** $f'(3) = 2$ **20** $g'(6) = -\dfrac{10}{9}$ **21** $\dfrac{dy}{dx} = 6$ **22** $\dfrac{dy}{dx} = \dfrac{3081}{128}$ **23** $(5, 7)$

24 $\left(-\dfrac{1}{2}, \dfrac{199}{24}\right)$ and $\left(5, -\dfrac{217}{6}\right)$

Chapter 8   Exercise 3

1 a Tangent $y = 6x - 3$, Normal $y = -\dfrac{1}{6}x + \dfrac{19}{6}$ **b** Tangent $y = x - 4$, Normal $y = -x$ **c** Tangent $y = -4x - 3$, Normal $y = \dfrac{1}{4}x + \dfrac{5}{4}$

1 d Tangent $y = \dfrac{1}{6}x + \dfrac{3}{2}$, Normal $y = -6x + 57$ **e** Tangent $y = -8x + 12$, Normal $y = \dfrac{1}{8}x + \dfrac{31}{8}$ **f** Tangent $y = 18x + 47$, Normal $y = -\dfrac{1}{18}x - \dfrac{43}{6}$

2 At A, $y = 4x - 4$, At B, $y = 3x - 3$ **3** $y = \dfrac{1}{3}x + \dfrac{4}{3}$ **4** $Q(-3, -20)$ **5** $y = 2x + 11$ **6** $\left(\dfrac{1}{4}, -\dfrac{9}{4}\right)$ **7** $y = -7x - \dfrac{28}{3}$ and $y = 19x - 57$

8 $y = 3x - \dfrac{38}{9}$ **9** $y = 5x - 15$ and $y = -5x - 10$, $y = -5x + 15$ and $y = 5x + 10$ **10** $y = 2x - 8$, $y = -\dfrac{x}{7} + \dfrac{4}{7}$, $17\dfrac{1}{7}$

Chapter 8   Exercise 4

1 a $(4, -13)$ Minimum Turning Point   **b** $(-2, 23)$ Maximum Turning Point and $(2, -9)$ Minimum Turning Point   **c** $(0, 0)$ Minimum Turning Point

1 d $\left(-\dfrac{1}{3}, -\dfrac{25}{3}\right)$ Minimum Turning Point   **e** $\left(-\dfrac{1}{2}, -4\right)$ Maximum Turning Point and $\left(\dfrac{1}{2}, 4\right)$ Minimum Turning Point

2 a $(-2, -13)$ Minimum Turning Point **b** $(-1, 25)$ Maximum Turning Point **c** $\left(\dfrac{4}{3}, \dfrac{256}{27}\right)$ Maximum Turning Point and $(4, 0)$ Minimum Turning Point

2 d $(1, 10)$ Maximum Turning Point and $(2, 9)$ Minimum Turning Point **e** $(0, 0)$ Rising Point of Inflexion

3 a $\left(1, -\dfrac{8}{3}\right)$ Maximum Turning Point and $(3, -4)$ Minimum Turning Point **b** $\left(\dfrac{5}{2}, 0\right)$ Minimum Turning Point **c** $\left(-\dfrac{1}{2}, -12\right)$ Maximum Turning Point

3 d $(0, 0)$ Minimum Turning Point **e** $(-1.57, 12.5)$ Maximum Turning Point and $(0, 2)$ Minimum Turning Point **4** 2

Chapter 8 — Exercise 5

1 a $\left(-2, \dfrac{224}{3}\right)$ Non-Stationary, $(0,0)$ Stationary, $\left(2, -\dfrac{224}{3}\right)$ Non-Stationary **b** $(-1, 15)$ Non-Stationary **c i** None **ii** None

iii None Quadratic functions have no points of inflexion **2 a** $(0,0)$ Stationary **b** $(1, -3)$ Non-Stationary **c** $(-2, -35)$ Non-Stationary

2 d $x = -\dfrac{b}{3a}$, $y = \dfrac{8b^3}{27a^2} - \dfrac{bc}{3a} + d$ There is always one point of inflexion for cubic functions

3 a $(-1, 3)$ Non-Stationary and $(1, 3)$ Non-Stationary **b** $\left(\dfrac{1}{6}, 4.11\right)$ Non-Stationary and $(-1, -9)$ Non-Stationary

3 c $(0,0)$ Stationary **d** No points of inflexion, $(0, -3)$ is a minimum turning point. **4** $y = -21x + 36$ **5** $\sqrt{1312} \approx 36.2$

Chapter 8 — Exercise 6

1 Vertical: $x = 0$, Horizontal $y = 0$ **2** Vertical: $x = 3$, Horizontal $y = 1$ **3** Vertical: $x = 0$, Oblique $y = x$

4 Vertical: $x = 2$, Horizontal $y = 1$ **5** Vertical: $x = -2$, Oblique $y = x - 2$ **6** Vertical: $x = 3$, Oblique $y = 2x + 9$

7 Vertical: None, Oblique $y = x - 2$ **8** Vertical: None, Oblique $y = x$ **9** Vertical: $x = -1$ and $x = 4$, Horizontal $y = 0$

10 Vertical: $x = -1$ and $x = 1$, Horizontal $y = 1$ **11** Vertical: $x = -3$ and $x = 3$, Horizontal $y = 3$

12 Vertical: $x = -2$ and $x = 3$, Oblique $y = 4x + 4$

Graphs of the following functions, including asymptotes, stationary points and intercepts:

13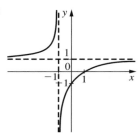
$y = \dfrac{x - 1}{x + 1}$

14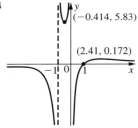
$y = \dfrac{x - 1}{x(x + 1)}$

15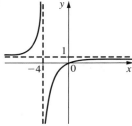
$y = \dfrac{x}{x + 4}$

16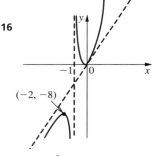
$y = \dfrac{2x^2}{x + 1}$

17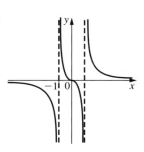
$y = \dfrac{x}{x^2 - 1}$

18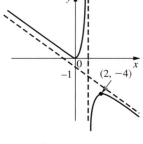
$y = \dfrac{x^2}{1 - x}$

19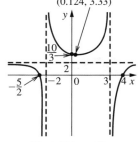
$y = \dfrac{(2x + 5)(x - 4)}{(x + 2)(x - 3)}$

20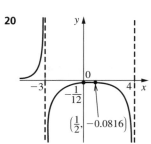
$y = \dfrac{1}{x^2 - x - 12}$

21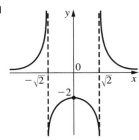

$$y = \frac{4}{x^2 - 2}$$

22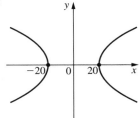

$$\frac{x^2}{16} - \frac{y^2}{9} = 25$$

Chapter 8 — Exercise 7

1 a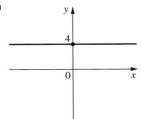

b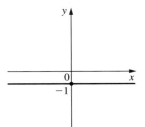

c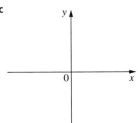

d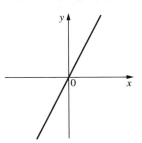

1 e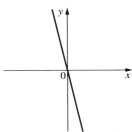

f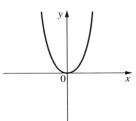

g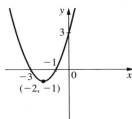

h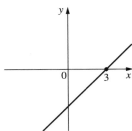

1 i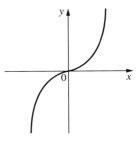

2 a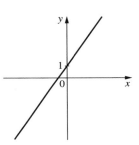

b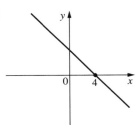

c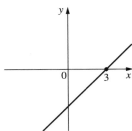

2 d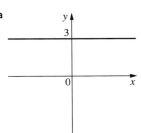

e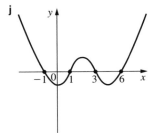

f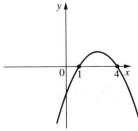

g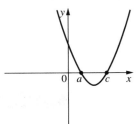

2 h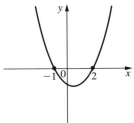

i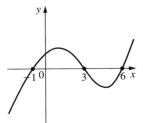

j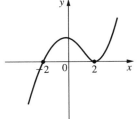

Answers

3 a 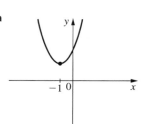 **b** 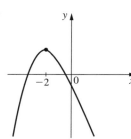 **c** 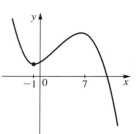 **d**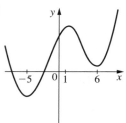

4 a 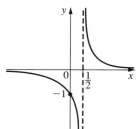 **b** 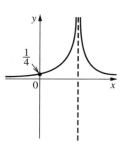 **c** 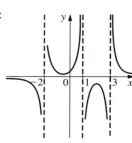 **d**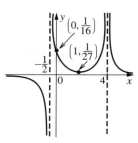

4 e 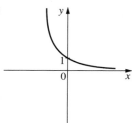 **f** 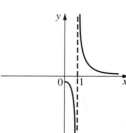 **g** 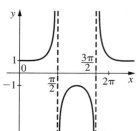 **h**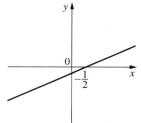

5 a 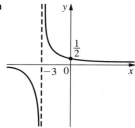 **b** 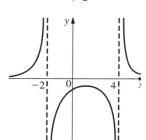 **c** 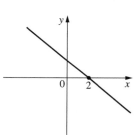 **d**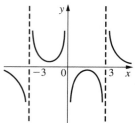

e 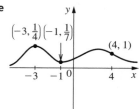 **f**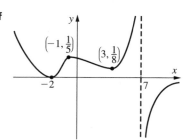

Chapter 8   Review Exercise

1 $f'(x) = 3x^2 - 4$ **2** $\dfrac{3}{16}$ **3** $\dfrac{dy}{dx} = \dfrac{3}{4}x^{-\frac{1}{2}} - \dfrac{17}{4}x^{\frac{15}{2}}$ **4** $x < 0,\ x > 4$ **5** $h(x) = 45x^2 + 60x + 19,\ h'(x) = 90x + 60$ **6** $x = 2,\ y = -6x + 16$

7 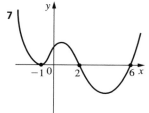   **8** $\left(-\dfrac{1}{2}, \dfrac{9}{2}\right)$ max TP, $\left(\dfrac{2}{3}, -\dfrac{50}{27}\right)$ min TP   **9** $y = 5x + 1,\ y = -\dfrac{1}{2}x + \dfrac{15}{2}, \left(\dfrac{13}{11}, \dfrac{76}{11}\right)$

10

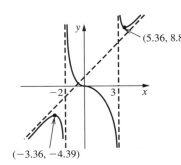

(5.36, 8.87)

(−3.36, −4.39)

11 $y = 1, x = 1, x = 4$ **12** $a = -4, b = 18$

13 a

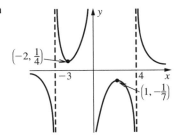

$\left(-2, \frac{1}{4}\right)$

$\left(1, -\frac{1}{7}\right)$

b

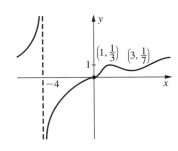

$\left(1, \frac{1}{3}\right)$ $\left(3, \frac{1}{7}\right)$

Chapter 9 — Exercise 1

1 $\frac{dy}{dx} = \sec^2 x$ **2** $\frac{dy}{dx} = \cos x + \operatorname{cosec} x \cot x$ **3** $\frac{dy}{dx} = \cos x + 12x$ **4** $\frac{dy}{dx} = -5 \sin x$ **5** $\frac{dy}{dx} = -7 \operatorname{cosec}^2 x$ **6** $\frac{dy}{dx} = -3 \sec x \tan x$

7 $\frac{dy}{dx} = 18x + 4 \sin x$ **8** $\frac{dy}{dx} = 7 - 5 \cos - \sec x \tan x$

Chapter 9 — Exercise 2

1 $f'(x) = 2(x + 4)$ **2** $f'(x) = 4(2x + 3)$ **3** $f'(x) = 6(3x - 4)$ **4** $f'(x) = 20(5x - 4)^3$ **5** $f'(x) = -3(5 - x)^2$ **6** $f'(x) = -8(7 - 2x)^3$

7 $\frac{dy}{dx} = -20(9 - 4x)^4$ **8** $\frac{dy}{dx} = 48(2x + 3)^5$ **9** $\frac{dy}{dx} = \frac{3}{2}(3x + 8)^{-\frac{1}{2}}$ **10** $\frac{dy}{dx} = \frac{10}{3}(2x - 9)^{\frac{2}{3}}$ **11** $\frac{dy}{dx} = 2(6x - 5)^{-\frac{2}{3}}$

12 $\frac{dy}{dx} = -\frac{3}{2}(3x - 2)^{-\frac{3}{2}}$ **13** $f'(x) = -20(5x - 4)^{-2}$ **14** $f'(x) = 56(3 - 8x)^{-2}$ **15** $\frac{dP}{dk} = 18(4 - 3k)^{-3}$ **16** $\frac{dN}{dp} = \frac{75}{2}(8 - 5p)^{-\frac{5}{2}}$

17 $\frac{dy}{dx} = 4 \cos 4x$ **18** $\frac{dy}{dx} = -3 \sin 3x$ **19** $\frac{dy}{dx} = -\frac{1}{2} \cos \frac{1}{2}x$ **20** $\frac{dy}{dx} = 6 \sec^2 6x$ **21** $\frac{dy}{dx} = 9 \sec 9x \tan 9x$ **22** $\frac{dy}{dx} = 6 - 3 \operatorname{cosec}^2 3x$

23 $\frac{dy}{dx} = -2 \operatorname{cosec} 2x \cot 2x + 12(3x + 2)^3$ **24** $\frac{dy}{dx} = 5 \cos 5x + 30(3x + 4)^{-\frac{7}{2}}$ **25** $\frac{dy}{dx} = 3 \sin^2 x \cos x$ **26** $\frac{dy}{dx} = 8 \tan 4x \sec^2 4x$

27 $\frac{dy}{dx} = 12x^3 + 3 \cos^2 x \sin x$ **28** $\frac{dy}{dx} = -30(3x - 4)^{-6} - 2 \sec^2 2x \tan 2x$ **29** $\frac{dy}{dx} = -3 \sin\left(3x - \frac{\pi}{4}\right)$ **30** $\frac{dy}{dx} = \frac{1}{2(x + 1)^{\frac{1}{2}}} \sec^2(\sqrt{x + 1})$

Chapter 9 — Exercise 3

1 $f'(x) = 3e^{3x}$ **2** $f'(x) = 7e^{7x}$ **3** $f'(x) = -4e^{4x}$ **4** $f'(x) = -10e^{-5x}$ **5** $f'(x) = 54e^{-9x}$ **6** $f'(x) = 2xe^{x^2}$ **7** $f'(x) = 2e^{2x+3}$ **8** $f'(x) = \frac{1}{x}$

9 $f'(x) = \frac{1}{x}$ **10** $f'(x) = -\frac{2}{x}$ **11** $f'(x) = \frac{2x}{x^2 + 2}$ **12** $\frac{dy}{dx} = \ln 4 \cdot 4^x$ **13** $\frac{dy}{dx} = \ln 10 \cdot 10^x$ **14** $\frac{dy}{dx} = 6 \ln 5 \cdot 5^x$ **15** $\frac{dy}{dx} = 3e^{3x} - \ln 3 \cdot 3^x$

16 $\frac{dy}{dx} = \frac{1}{x} - \ln 2 \cdot 2^x$ **17** $\frac{dy}{dx} = \frac{1}{x \ln 2}$ **18** $\frac{dy}{dx} = \frac{1}{x \ln 8}$ **19** $\frac{dy}{dx} = \ln 4 \cdot 4^x - \frac{1}{x \ln 5}$ **20** $\frac{dy}{dx} = 4e^{4x} - 2 \cos 2x + \frac{1}{x}$ **21** $\frac{dy}{dx} = -\tan x$

22 $\frac{dy}{dx} = \operatorname{cosec} x \sec x$ **23** $\frac{dy}{dx} = \frac{1}{x} \sec^2(\ln x)$

Chapter 9 — Exercise 4

1 $\frac{dy}{dx} = x(2 \sin x + x \cos x)$ **2** $\frac{dy}{dx} = x^2(3 \cos x - x \sin x)$ **3** $\frac{dy}{dx} = 3xe^x(2 + x)$ **4** $\frac{dy}{dx} = e^{3x}(3 \sin x + \cos x)$ **5** $\frac{dy}{dx} = \frac{\sin x}{x} + \ln x \cos x$

6 $\frac{dy}{dx} = \cos^2 x - \sin^2 x$ **7** $\frac{dy}{dx} = 3 \cos 3x \cos 2x - 2 \sin 3x \sin 2x$ **8** $\frac{dy}{dx} = 2x(x - 1)(2x - 1)$ **9** $\frac{dy}{dx} = x^2(x - 2)^3(7x - 6)$

10 $\frac{dy}{dx} = 6x^2(3x + 2)(5x + 2)$ **11** $\frac{dy}{dx} = (2x + 1)^2(8x - 11)$ **12** $\frac{dy}{dx} = 2(x + 5)(3x - 2)^3(9x + 28)$ **13** $\frac{dy}{dx} = 6(5 - 2x)^2(3x + 4)(1 - 5x)$

14 $\frac{dy}{dx} = (3x + 4)^2[9 \sin x + (3x + 4) \cos x]$ **15** $\frac{dy}{dx} = 5^x(\ln 5 \cdot \cos x - \sin x)$ **16** $\frac{dy}{dx} = x^2\left(3 \log_6 x + \frac{1}{\ln 6}\right)$ **17** $\frac{dy}{dx} = e^{4x} \sec 3x(4 + 3 \tan 3x)$

18 $\frac{dy}{dx} = 3(2x + 1)^2 \operatorname{cosec} 3x[2 - (2x - 1) \cot 3x]$ **19** $\frac{dy}{dx} = 4^x\left(\ln 4 \log_8 x + \frac{1}{x \ln 8}\right)$ **20** $\frac{dy}{dx} = \ln(2x + 3) + \frac{2x}{2x + 3}$

21 $\frac{dy}{dx} = 8x\left[\ln(x^2 + 2x + 5) + \frac{x(x + 1)}{x^2 + 2x + 5}\right]$ **22** $\frac{dy}{dx} = e^{3x} \sec\left(2x - \frac{\pi}{4}\right)\left[3 + 2 \tan\left(2x - \frac{\pi}{4}\right)\right]$

23 $\frac{dy}{dx} = -3x^{-5}\left(4 \tan\left(3x + \frac{\pi}{2}\right) - 3x \sec^2\left(3x + \frac{\pi}{2}\right)\right)$ **24** $\frac{dy}{dx} = x((2 \ln x + 1)\sin x + x \ln x \cos x)$ **25** $\frac{dy}{dx} = e^{3x}(x + 2)[(3x + 8) \tan x + (x + 2) \sec^2 x]$

Answers

Chapter 9 — Exercise 5

1 $f'(x) = \dfrac{e^x(\cos x + \sin x)}{\cos^2 x}$ **2** $f'(x) = \dfrac{6x(x+6)}{(x+3)^2}$ **3** $f'(x) = \dfrac{7\tan x - 7x\sec^2 x}{\tan^2 x}$ **4** $f'(x) = \dfrac{1 - \ln x}{4x^2}$ **5** $f'(x) = \dfrac{e^x(x-5)}{(x-4)^2}$

6 $f'(x) = \dfrac{-6}{(x-3)^2}$ **7** $f'(x) = \dfrac{-3(x+12)}{2x^3\sqrt{x+9}}$ **8** $f'(x) = \dfrac{4^x(2x\ln 4 - 1)}{2x^{\frac{3}{2}}}$ **9** $f'(x) = \dfrac{x-2}{(x-1)^{\frac{3}{2}}}$ **10** $\dfrac{dy}{dx} = \dfrac{e^{3x}(3x-2)}{9x^3}$

11 $\dfrac{dy}{dx} = \dfrac{\frac{1}{x\ln 6}(x+6) - \log_6 x}{(x+6)^2}$ **12** $\dfrac{dy}{dx} = \dfrac{\frac{1}{x}\ln(x-4) - \frac{1}{x-4}\ln x}{(\ln(x-4))^2}$ **13** $\dfrac{dy}{dx} = \dfrac{-2}{(e^x - e^{-x})^2}$ **14** $\dfrac{dy}{dx} = \dfrac{2(\cos 2x - 3\sin 2x)}{e^{6x}}$

15 $\dfrac{dy}{dx} = \dfrac{12(3x-2)^4(4x+19)}{(2x+3)^4}$ **16** $\dfrac{dy}{dx} = \dfrac{(1-x)\sin x + x\cos x}{e^x}$ **17** $\dfrac{dy}{dx} = \dfrac{xe^{3x}(3x^2 + 15x + 10)}{(x+5)^3}$ **18** $\dfrac{dy}{dx} = \dfrac{\tan\left(x + \frac{\pi}{4}\right) - 2}{e^{2x}\cos\left(x + \frac{\pi}{4}\right)}$

19 $\dfrac{dy}{dx} = \dfrac{-2(3x+1)\operatorname{cosec}^2\left(2x - \frac{\pi}{3}\right)\ln(3x+1) - 3\cot\left(2x - \frac{\pi}{3}\right)}{e^{2x}}$

Chapter 9 — Exercise 6

1 $\dfrac{dy}{dx} = \dfrac{-3x^2 - y}{x}$ **2 a** $\dfrac{x}{2+3y}$, $\dfrac{(2+3y)^2 - 3x^2}{(2+3y)^3}$ **b** $\dfrac{\cos x - 4y}{4x+1}$, $\dfrac{-(4x+1)\sin x - 8\cos x - 32y}{(4x+1)^2}$ **c** $\dfrac{-1}{x}$, $\dfrac{1}{x^2}$

3 $\dfrac{dy}{dx} = \dfrac{1}{6\sqrt{xy^2}}$

Chapter 9 — Exercise 7

1 $\dfrac{dy}{dx} = \dfrac{1}{\sqrt{25 - x^2}}$ **2** $\dfrac{dy}{dx} = \dfrac{-1}{\sqrt{64 - x^2}}$ **3** $\dfrac{dy}{dx} = \dfrac{10}{100 + x^2}$ **4** $\dfrac{dy}{dx} = \dfrac{2}{\sqrt{9 - 4x^2}}$ **5** $\dfrac{dy}{dx} = \dfrac{-3}{\sqrt{1 - 9x^2}}$ **6** $\dfrac{dy}{dx} = \dfrac{2e^x}{4 + e^{2x}}$

7 $\dfrac{dy}{dx} = \dfrac{-1}{2\sqrt{-(x+4)(x+3)}}$ **8** $\dfrac{dy}{dx} = \dfrac{1}{2x^2 - 2x + 1}$ **9** $\dfrac{dy}{dx} = \dfrac{1}{x\sqrt{1 - (\ln 5x)^2}}$

Chapter 9 — Exercise 8

1 $f'(x) = 2x - 5$ **2** $\dfrac{dy}{dx} = 6(2x - 7)^2$ **3** $f'(x) = -8\sin 8x - \dfrac{3}{2}x^{-\frac{1}{2}}$ **4** $\dfrac{dy}{dx} = \sec x \tan x - 5e^{5x}$ **5** $f'(x) = x^2 e^{-4x}(3 - 4x)$

6 $\dfrac{dy}{dx} = x(2\ln x + 1)$ **7** $f'(x) = \dfrac{3\cos 3x - \sin 3x}{e^x}$ **8** $\dfrac{dy}{dx} = \dfrac{\sin x - 4x\sin x + x\cos x}{e^{4x}}$ **9** $f'(x) = 3^x(\ln 3 \cdot \sin x + \cos x)$

10 $\dfrac{dy}{dx} = \dfrac{\frac{x-4}{x\ln 2} - 3\log_2 x}{(x-4)^4}$ **11** $\dfrac{dy}{dx} = \dfrac{x\ln x(x+18) + x(x+9)}{(x+9)^2}$ **12** $\dfrac{dy}{dx} = -6(\sin 2x \sin 4x - 2\cos 2x \cos 4x)$ **13** $\dfrac{dy}{dx} = \dfrac{12}{\sqrt{1 - 4x^2}}$

14 $f'(x) = \dfrac{-\left(\frac{x}{\sqrt{1-x^2}} + 2\cos^{-1} x\right)}{3x^3}$ **15** $\dfrac{dy}{dx} = \sin x(\ln x + 1) + x\ln x \cos x$ **16** $\dfrac{dy}{dx} = \dfrac{-(\sec x \operatorname{cosec} x + \ln(\cot x))}{e^x}$ **17** $f'(2) = -\dfrac{20}{3}$

18 $f'(4) = \dfrac{-\pi + 2}{64}$ **19** $\dfrac{dy}{dx} = \dfrac{-y(2x + e^x y)}{x^2 + 2e^x y}$ **20** $\dfrac{dy}{dx} = \dfrac{y(4x^3 y^2 - \cos x)}{(\sin x - 3x^4 y^2)}$

Chapter 9 — Exercise 9

1 $\dfrac{3}{4}$ **2** 1 **3** $\dfrac{4\ln 4 - 3}{e^2}$ **4** $y = -\dfrac{4}{5}x + \dfrac{14}{5}$, $y = -\dfrac{6}{5}x - \dfrac{9}{5}$ **5** $\dfrac{4}{3}$ **6** $\dfrac{4 - 4\pi}{e^\pi}$ **7** $(0,0)$ and $\left(2, \dfrac{4}{e^2}\right)$ **8** $\left(e^{-\frac{1}{2}}, \dfrac{-2}{e}\right)$

9 $(0, 0)$ rising point of inflexion and $\left(3, \dfrac{27}{e^3}\right)$ max TP

Chapter 9 — Review Exercise

1 a $60(3x - 2)^3$ **b** $14x(3 - 2x^2)^{-\frac{3}{2}}$ **c** $6 - 3\tan 3t \sec 3t$ **d** $48e^{8x}$ **e** $\dfrac{1}{x} - 3^x \ln 3$ **2 a** $4e^{4x}\sin 3x + 3e^{4x}\cos 3x$ **b** $\dfrac{1}{x} + \cot x$

2 c $\dfrac{e^{5x}}{\sqrt{x+4}}\left(5 - \dfrac{1}{2(x+4)}\right)$ **d** $\dfrac{3}{3x+4} - \dfrac{2}{2x-1}$ **e** $\dfrac{1}{\ln 10}\left(2 - 3\tan 3x - \dfrac{2}{x+4}\right)$ **3 a** $\dfrac{6xy}{8y - 3x^2}$ **b** $\dfrac{3x^2}{\ln x} - \dfrac{y}{x\ln x}$

4 $\dfrac{(2\sin x + 4x(\cos x - \sin x) - 2x^{2\cos x - 7})}{e^x} - \dfrac{2x\sin x - x^2\cos x}{e^{2x}} + y$ **5** $\dfrac{-2(1 + \csc x \cot x + \cot^2 x)}{1 + \left(\dfrac{1 + \cos x}{\sin x}\right)^2}$, which reduces to -1

6 $-\dfrac{16\sqrt{x} - 4\pi\sqrt{x}}{\pi^2}$ **7** $-\dfrac{3}{4}$ **8** $-\dfrac{1}{3}$ **9** $y = \dfrac{4}{3}x - \dfrac{5}{3}$ **10** $(0, 0)$ **11** 2.533 **12** $y = x - 4$ **13** $\cdot\dfrac{-4}{(2x - 1)^2}$ **14 a** $6\sec 2t \tan 2t + 5$

b i $3 + 5\pi$ **ii** 5 **15 a** -1 **b** $\dfrac{4}{5}$

Chapter 10　　Exercise 1

1 $v = \sqrt{500}\,ms^{-1}$ **2** All three sides are $\sqrt{50}\,cm$ **3** $14.65\,cm$ **4** $5.62\,cm, 4.22\,cm, 16.9\,cm$ **5** $7\,cm, 3.5\,cm$ **6** $3.414\,cm^2$ **7** 78.125 **8 b** $\dfrac{4r}{3}$

9 $x = y = 75\,cm$. It is a maximum value.   **10 b** $V = 60\left(\dfrac{450x - 60x^3}{13}\right)$ **10 c** $x = 1.58\,cm$. It is a maximum value.

11 $t = 6.29\ seconds, t = 12.6\ seconds$ **12** 11100 **13** 50 Yen, $14.14\ Kmh^{-1}$ **14** 20 **15 a** $\dfrac{7 - x}{7} = \dfrac{y}{24}$ **b** $A = \dfrac{168x - 24x^2}{7}$ **c** $42\,cm^2$

Chapter 10　　Exercise 2

1 $128\pi cm^3 s^{-1}$ **2** $\dfrac{15}{32\pi}$ **3** $\dfrac{2}{h}$ **4** $611N_0$ **5** $216\,cm^3 s^{-1}$ **6** $0.954\,ms^{-1}$ **7** $2.988\,cm^2 s^{-1}$ **8** $0.011\,cms^{-1}$ **9** $-\dfrac{5}{3}$

10 a $(r\theta - r\sin\theta, r - r\cos\theta)$ **11** $100\,cm^3 s^{-1}$ **13 a** $\dfrac{2}{9}\,cms^{-1}$ **b** $\dfrac{4\pi}{3}\,cm^2 s^{-1}$

Chapter 10　　Exercise 3

1 $v = 3\dfrac{1}{4}\,ms^{-1}$ $a = -3\,ms^{-2}$ $t = \dfrac{2\sqrt{3}}{3}$ seconds   **2 a** $8\,ms^{-1}$ **b** $16 - 12t$ **c** $-20\,ms^{-2}$ **3 a** $-4(1 - 2t)$ **b** $t = \dfrac{1}{2}$ second   **c** $0\,ms^{-2}$

4 a $v = \dfrac{t^2\cos t - t\cos t - \sin t}{(t - 1)^2}$ **b** $-1.74\,ms^{-1}$ **c** $a = \dfrac{t^3\sin t + 2t^2\sin t - 2t\cos t + 2\cos t - t\sin t + 2\sin t}{(t - 1)^3}$ **7** Period $= \pi$

8 a $s = \dfrac{e^{2t}(t^2 - 2t - 1)}{(t^2 - 1)^2}$ **b** $v = \dfrac{-e^4}{9}$ **c** $a = \dfrac{4e^{2t}}{t^2 - 1} - \dfrac{(8t + 2)e^{2t}}{(t^2 - 1)^2} + \dfrac{4t^2 e^{2t}}{(t^2 - 1)^3}$ **d** $a = \dfrac{-2}{27e^4}$ **10 a** $v = 3k\cos(kt + c)$

10 b $t = \dfrac{\pi - c}{k}$ **c** $a = 0.0213\,ms^{-2}$ **11 a** $k = \dfrac{1}{100}\ln 300$ **b.** $v = e^{\frac{\ln 300}{100}t^2}\left(1 + \dfrac{\ln 300}{50}t^2\right)$ **c** $a = \dfrac{\ln 300}{50}e^{\frac{\ln 300}{100}t^2}\left(3t + \dfrac{\ln 300}{50}t^3\right)$ **d** $1.927s$

12 a Min is $(2, 1.39)$ **b** $v = \dfrac{2}{t} - \dfrac{1}{t - 1}$ $t \geq 2$ **c** $a = -\dfrac{2}{t^2} - \dfrac{1}{(t - 1)^2}$ $t \geq 2$ **d** $v = \dfrac{4}{45}\,ms^{-1}$ $a = -\dfrac{31}{4050}\,ms^{-2}$

13 a $v = \dfrac{2t - s}{t + s}$ **b** $a = \dfrac{3(s^2 + 2st - 2t^2)}{(t + s)^3}$ **c** $s = t(-1 \mp \sqrt{3})$

Chapter 10　　Review Exercise

1 $120\ kmh^{-1}$ **2 a** $13\ ms^{-2}$ **b** $t = 4.70$ seconds and $t = 1.05$ seconds

2 c Maximum velocity is $0.541\ ms^{-1}$ Minimum velocity is $-1.35\ ms^{-1}$ **3** $\dfrac{1}{2\pi}\,cms^{-1}$ **4** $a = b = 24.42$ cm   **5** -1.6

6 a $A = 2\pi rh + \pi r^2$ **b** 4.30 cm   **c** 1120 cm^2 **7 a** $6t\sin 5t + 15t^2\cos 5t$ **b** 0.457 hours   **c** 0.304 hours   **8 a** $(\pi - 2x)\sin x$

8 b 1.12 units^2 **9** $\dfrac{\pi}{3}\,cm^2 s^{-1}$ **10** $-\dfrac{56}{27}\,ms^{-2}$ **11 a** 6.28 hours.   **b** 0.524 hours.   **12** 256 cm^3 s^{-1} **13 a** $BC = 2(x - h\csc\theta)$

13 b $2xh + h^2(\cot\theta - 2\csc\theta)$ **c** $2h\left(x + h\left(2 - \dfrac{1}{\sqrt{3}}\right)\right)$ **14 a** $t = 2$ minutes   **b** $\dfrac{112t^3}{(8 + t^4)^2}$ cm min^{-1} **c** 19.55 cm^2 min^{-1} **d** 1.48 minutes

Chapter 11　　Exercise 1

1 a 1×3 **b** 2×3 **c** 3×3 **d** 4×1 **2** Week 1 $\begin{pmatrix} 3 & 1 \\ 2 & 2 \\ 4 & 4 \end{pmatrix}$ Week 2 $\begin{pmatrix} 1 & 2 \\ 4 & 1 \\ 0 & 1 \end{pmatrix}$ Week 3 $\begin{pmatrix} 4 & 2 \\ 1 & 0 \\ 1 & 1 \end{pmatrix}$

The operation is addition.

	Magazines	Newspapers
Alan	8	5
Bill	7	3
Colin	5	6

Answers

3 a $\begin{pmatrix} 8 & 4 & 12 \\ 20 & -8 & 12 \\ 28 & -16 & 4 \end{pmatrix}$ **b** $\begin{pmatrix} -18 & -24 \\ 6 & -12 \\ 18 & 24 \end{pmatrix}$ **c** $\begin{pmatrix} 3k & 6k \\ -4k & -k \\ 12k & 4k \end{pmatrix}$ **d** $\begin{pmatrix} 3(k-1) & 2(k-1) \\ 1-k & 0 \end{pmatrix}$ **4 a** $k = 6$ **b** $k = 0, 1$ **c** $k = 1$ **d** $k = 0, 3$

4 e $k = \dfrac{1}{2}$ **f** $k = 5$ **g** $x = 0, y = 11$ **5 a** not possible **b** $\begin{pmatrix} 10 & -2 \\ 0 & 4 \end{pmatrix}$ **c** not possible **d** $\begin{pmatrix} -11 & 5 \\ 6 & 0 \\ 8 & 1 \end{pmatrix}$ **e** $\begin{pmatrix} 7 & 1 \\ -1 & 3 \end{pmatrix}$ **f** $\begin{pmatrix} 9 & -2 \\ 0 & 3 \end{pmatrix}$

5 g $\begin{pmatrix} 18 & 5 \\ -8 & -15 \\ -4 & 37 \end{pmatrix}$ **h** $\begin{pmatrix} -15 & 1 \\ 1 & -5 \end{pmatrix}$ **6 a** $\begin{pmatrix} -24 & 11 \\ -6 & 29 \end{pmatrix}$ **b** (-26) **c** $\begin{pmatrix} 11 & 29 \\ 6 & 22 \\ 12 & -7 \end{pmatrix}$ **d** $\begin{pmatrix} 29 & 39 & -11 \\ 11 & 41 & -9 \\ 2 & -29 & 2 \end{pmatrix}$ **e** $\begin{pmatrix} 6+k^2 & 2+2k \\ k^2-3 & 2k-1 \end{pmatrix}$

6 f $\begin{pmatrix} 7+3k^2-4k \\ -5 \\ 4-k \\ 1+2k-2k^2 \end{pmatrix}$ **7 a** $\begin{pmatrix} 45 & 4 \\ 39 & -18 \end{pmatrix}$ **b** $\begin{pmatrix} 45 & 45k+4 \\ 39 & 39k-18 \end{pmatrix}$ **c** $\begin{pmatrix} 45 & 45k+4 \\ 39 & 39k-18 \end{pmatrix}$ **d** $\begin{pmatrix} 45+39k & 4-18k \\ 39 & -18 \end{pmatrix}$ **e** $\begin{pmatrix} 9 & 9k+12 \\ 27 & 27k-6 \end{pmatrix}$

7 f $\begin{pmatrix} 0 & 0 \\ -11 & 7-11k \end{pmatrix}$ **g** $\begin{pmatrix} -6 & 84-9k \\ -6 & 52-5k \end{pmatrix}$ **h** $\begin{pmatrix} -264 & 168 \\ -99 & 63 \end{pmatrix}$ **8 a** $x = 2, y = 1$ **b** $x = 0, y = 1$ **c** $x = 10, y = 1$ **d** $x = 4, y = -3$

9 $A^2 = \begin{pmatrix} 13 & -4 \\ 12 & -3 \end{pmatrix}$ $A^3 = \begin{pmatrix} 40 & -13 \\ 39 & -12 \end{pmatrix}$ **10 a** $A = \begin{pmatrix} 3 & 2 & 4 \\ 5 & 7 & 2 \end{pmatrix}$ **b** $B = \begin{pmatrix} 1900 \\ 1300 \\ 1100 \end{pmatrix}$ **c** $AB = \begin{pmatrix} 12700 \\ 20800 \end{pmatrix}$

Minimum total number of calories consumed on day 1 and day 2. **d i** (33500). Minimum total calories consumed.

ii $\begin{pmatrix} 9 \\ 14 \end{pmatrix}$ Total number of people dieting on each day.

iii (4300) Total number of calories consumed by one man, one women and one child. **11 a** $P = \begin{pmatrix} 3 & 4 & 7 \\ 6 & 2 & 6 \\ 10 & 1 & 3 \\ 3 & 9 & 2 \\ 8 & 3 & 3 \end{pmatrix}$ **b** $Q = \begin{pmatrix} 3 \\ 1 \\ 0 \end{pmatrix}$ **c** $\begin{pmatrix} 13 \\ 20 \\ 31 \\ 18 \\ 27 \end{pmatrix}$

12 a $\begin{pmatrix} -1 & 6 \\ -4 & 1 \end{pmatrix}$ **b** $\begin{pmatrix} -7 & 9 \\ -6 & -4 \end{pmatrix}$ **c** $\begin{pmatrix} -6 & 0 \\ 0 & -6 \end{pmatrix}$ **13** $m = 1, n = -6$ **14** $k = 10$ **16** $c = 5$ **17** $c = -2$

18 $PQ = \begin{pmatrix} 4 & 1-3c \\ 5c+4 & c \end{pmatrix}$ $QP = \begin{pmatrix} 5+c & 22 & -15 \\ 2 & 8 & -6 \\ 3+c^2 & 12+2c & -9 \end{pmatrix}$ **19 a** $x = -\dfrac{8}{5}, y = 10$ **b** $x = -5, y = 0$ **c** $x = \dfrac{13}{3}, y = 0$ **20** $\begin{pmatrix} a & b \\ 0 & a \end{pmatrix}$

Chapter 11 · Exercise 2

1 a $\dfrac{1}{11}\begin{pmatrix} 5 & 2 \\ -3 & 1 \end{pmatrix}$ **b** $\dfrac{1}{41}\begin{pmatrix} 10 & -7 \\ 3 & 2 \end{pmatrix}$ **c** $\dfrac{1}{76}\begin{pmatrix} 1 & -8 \\ 9 & 4 \end{pmatrix}$ **d** $\begin{pmatrix} \dfrac{11}{137} & -\dfrac{6}{137} & \dfrac{29}{274} \\ \dfrac{10}{137} & \dfrac{7}{137} & -\dfrac{11}{274} \\ -\dfrac{24}{137} & \dfrac{38}{137} & \dfrac{1}{274} \end{pmatrix}$ **e** $\begin{pmatrix} \dfrac{2}{37} & -\dfrac{25}{333} & -\dfrac{8}{333} \\ \dfrac{3}{74} & \dfrac{1}{37} & \dfrac{11}{74} \\ \dfrac{5}{74} & \dfrac{5}{111} & \dfrac{19}{222} \end{pmatrix}$ **f** $\begin{pmatrix} \dfrac{37}{304} & -\dfrac{33}{152} & \dfrac{9}{76} \\ \dfrac{1}{38} & \dfrac{5}{19} & -\dfrac{1}{19} \\ \dfrac{17}{152} & \dfrac{9}{76} & \dfrac{1}{38} \end{pmatrix}$

1 g $-\dfrac{1}{14k}\begin{pmatrix} 1 & -5 \\ -3k & k \end{pmatrix}$ **h** $\dfrac{1}{k^2+7k}\begin{pmatrix} k+2 & -k \\ 1-2k & 3k \end{pmatrix}$ **2** $X = \begin{pmatrix} 1 & -\dfrac{3}{14} \\ -1 & \dfrac{5}{14} \end{pmatrix}$ $Y = \begin{pmatrix} \dfrac{5}{2} & \dfrac{1}{2} \\ \dfrac{5}{7} & 5 \end{pmatrix}$ $Z = \begin{pmatrix} \dfrac{43}{18} & \dfrac{31}{18} \\ \dfrac{19}{18} & -\dfrac{11}{18} \end{pmatrix}$ **3 a** -30 **b** 44 **c** 78 **d** 190

4 a $\cos^2\theta + \sin^2\theta = 1$ **b** $-\sin 2\theta$ **c** -2 **d** $2abc$ **e** $\cos^3\theta + \sin^3\theta - 3\sin^2\theta + \tan^3\theta$ **f** $-y^2 + y - 2$ **g** $a^3 - a^2b - ab^2 + b^3$

5 a 4 units^2 **5 b** 4 units^2 **c** 20 units^2 **6 a** Collinear **b** Not collinear **c** Collinear **8** $k = -1.22, 1.08$ **10** $c = -5, 4.5$

11 $y = -3.36, 3.86$ **12** $x = y = 1$ or $x = y = -1$ **13 a** $\dfrac{1}{3}B^{-1}(2A-B)$ **b** $X = I$ **14 b** $\dfrac{1}{8}\begin{pmatrix} -x-3 & x-1 \\ x+1 & 3-x \end{pmatrix}$

Chapter 11 · Exercise 3

1 a $x = \dfrac{15}{7}, y = -\dfrac{17}{7}$ **b** $p = -21, q = -14$ **c** $x = \dfrac{5}{6}, y = \dfrac{2}{9}$ **d** $x = 7, y = 17$ **2 a** $x = \dfrac{17}{4}, y = -\dfrac{11}{20}$ **b** $a = -\dfrac{61}{11}, b = -\dfrac{9}{11}$

2 c $x = -\dfrac{5}{7}, y = \dfrac{22}{7}$ **d** $x = -\dfrac{16}{5}, y = -\dfrac{69}{5}$ **3 a** $x = -\dfrac{1}{3k+1}, y = -\dfrac{(5k+2)}{3k+1}, k \neq -\dfrac{1}{3}$ **b** $x = \dfrac{3(3-k)}{3-k^2}, y = \dfrac{3(k-1)}{3-k^2}, k \neq \pm\sqrt{3}$

3 c $x = \dfrac{12}{11k-5}, y = \dfrac{-13k+7}{11k-5}, k \neq \dfrac{5}{11}$ **d** $x = \dfrac{4k+14}{k^2+k+3}, y = \dfrac{10k-12}{k^2+k+3}, k \in \mathbb{R}$ **4 a** Unique solution **b** Unique solution

4 c No unique solution **d** $k = -\dfrac{5}{3}, 1$ **5 a** $c = 1, 0$. Lines are parallel. **b** $c = -2$. Lines are parallel.

6 a Consistent. Lines intersect giving unique solution.　**b** Consistent. Same line giving infinite solutions.　**c** Consistent. Same line giving infinite solutions.

7 $p = 3$　**8** $\lambda = -6, x = 1, y = -1$

Chapter 11　　Exercise 4

1 a $x = -1, y = 9, z = -13$　**b** $x = 10, y = 10, z = -36$　**c** $x = 2, y = -4, z = -3$　**d** $x = 2, y = -1, z = 2$　**2 a** $x = -3, y = 2, z = 4$

2 b $x = -2, y = 1, z = 4$　**c** $x = -1, y = 1, z = 2$　**d** $x = 4, y = -5, z = 2$　**3 a** $x = 4, y = \frac{1}{3}, z = -\frac{2}{3}$　**b** $x = \frac{55}{47}, y = \frac{40}{47}, z = -\frac{64}{47}$

3 c $x = \frac{1}{2}, y = 1, z = -2$　**d** $x = 2, y = 3, z = -2$　**4 a** Determinant = 6. Unique solution.　**b** Determinant = 0. No unique solution.

4 c Determinant = 0. No unique solution.　**d** Determinant = 0. No unique solution.　**5 a** $x = -1, y = -1, z = 1$　**b** $x = -\frac{3}{19}, y = -\frac{59}{19}, z = \frac{28}{19}$

5 c $x = -1, y = -2, z = 4$　**d** $x = \frac{1}{4}, y = \frac{1}{2}, z = 2$　**6 a** $x = \frac{74}{19}, y = -\frac{3}{19}, z = -\frac{9}{19}$　**b** $x = 2, y = -1, z = -1$

6 c $x = \frac{35}{66}, y = -\frac{43}{66}, z = -\frac{13}{33}$　**d** $x = 0, y = 2, z = 3$　**7 a** $x = \frac{5 - 4\lambda}{2}, y = \lambda, z = \frac{-2\lambda - 3}{4}$　**b** $x = -\frac{2}{5} + \lambda, y = \lambda, z = \frac{8}{5} - \lambda$

7 c $x = \frac{19 - 5\lambda}{13}, y = \lambda, z = \frac{7\lambda - 11}{13}$　**d** $x = \lambda, y = \mu, z = \frac{4 - 2\lambda - \mu}{3}$　**e** $x = \frac{7}{5}, y = 0, z = \frac{2}{5}$　**f** No solution.　**g** No solution.

7 h $x = 4, y = 4, z = 6$　**8　a** $x = 2\lambda - 3, y = \lambda, z = \frac{5\lambda - 10}{2}$　**b** $x = -\frac{5}{44}, y = \frac{15}{44}, z = -\frac{7}{22}$　**c** No solution.

8 d No solution.　**e** $x = \frac{44 - 4\lambda}{11}, y = \lambda, z = \frac{3\lambda}{11}$　**f** $x = 1, y = 2, z = -3$　**9 a** $\begin{pmatrix} \frac{1}{5} & \frac{3}{5} & -\frac{2}{15} \\ -\frac{1}{10} & \frac{1}{5} & -\frac{1}{10} \\ -\frac{1}{6} & 0 & \frac{1}{18} \end{pmatrix}$　**b** $x = \frac{7}{15}, y = -\frac{2}{5}, z = -\frac{1}{9}$

10 a 0　**b** $c = 3$　**c** $x = \lambda, y = \lambda, z = -2\lambda$

12 $a = \frac{4b - 38}{b - 20}$　**13 a** $-2k^2 - 11k + 37$　**b** $k = -7.86, 2.36$　**14** $a = 36$

Chapter 11　　Review Exercise

1 R is an $n \times p$ matrix　S is an $m \times p$ matrix　**2** $\lambda = 1$ or 6　**3 a** $\frac{1}{k^2 + 1}\begin{pmatrix} k & 1 \\ -1 & k \end{pmatrix}$　**b** $x = 1, y = -k$　**4** $x = 1, y = 8$

5 a $\begin{pmatrix} 1 & k + 3 & 5 \\ 1 & 3 & k + 1 \\ 1 & 1 & k \end{pmatrix}\begin{pmatrix} x \\ y \\ z \end{pmatrix} = \begin{pmatrix} 0 \\ k + 2 \\ 2k - 1 \end{pmatrix}$　**b** $k = \frac{8}{3}$　**c** $z = \frac{k^2 - 5k - 4}{8 - 3k}$　**d** If $k = \frac{8}{3}$ there is no solution.

5 d Otherwise $x = \frac{-k^3 - 2k^2 + 29k - 22}{8 - 3k}, y = \frac{k^2 - 6k + 14}{8 - 3k}, z = \frac{k^2 - 5k - 4}{8 - 3k}$　**6** $a = -1, b = 3$　**7 a** $3p - 3q - r = 0$

7 b Solution is not unique. $x = \frac{17 + 3\lambda}{51}, y = \lambda, z = -\frac{5\lambda + 11}{17}$　**8 b** $k = -3$　**9** $p = 3, q = -5$　**10 a** $c = -2.5, 0.5, 2$

10 b $\begin{pmatrix} 36 & 39 & 50 \\ 14 & 16 & 20 \\ 15 & 18 & 22 \end{pmatrix}$　**c** Since M is singular, A must be singular.　**11** $k = 5$　**12 b** $c = -3$　**c** $x = -\frac{1 + 7\lambda}{2}, y = \lambda, z = \frac{11\lambda + 7}{2}$

13 a $a = 7, b = 2$　**b** $x = -1, y = 2, z = -1$　**15** $a = 1$　**16** $a = 4, b = -1$　**17** $y_1 = 16z_1 + 36z_2 - 58z_3$

17 $y_2 = -18z_1 + 11z_2 - 3z_3$　$y_3 = 13z_1 + 12z_2 - 17z_3$

Chapter 12　　Exercise 1

1 a $a = 1$　$b = \frac{1}{2}$　$c = 4$　**b** $a = \frac{1}{3}$　$b = 0, \frac{1}{2}$　$c = -8$　**c** $a = \frac{8}{7}$　$b = -3$　**2 a** $PQ = i - 4j$　**b** $|PQ| = \sqrt{17}$

3 a $AB = \begin{pmatrix} -3 \\ 2 \end{pmatrix}$　**b** $|AB| = \sqrt{13}$　**4** $\begin{pmatrix} 1 \\ 3 \end{pmatrix}$　**5 a** $\sqrt{34}$　**b** $\sqrt{53}$　**c** $\sqrt{90}$　**d** $\sqrt{29}$　**e** $\sqrt{21}$　**f** $\sqrt{57}$　**6 a** Parallel　**b** Parallel

6 c Not parallel　**d** Not parallel　**7 a** $c = 6$　**b** $c = -7$　**c** $c = 6$　**8** $\begin{pmatrix} \frac{13}{2} \\ \frac{13\sqrt{3}}{2} \end{pmatrix}$　**9** $\begin{pmatrix} \frac{-5}{\sqrt{62}} \\ \frac{-6}{\sqrt{62}} \\ \frac{1}{\sqrt{62}} \end{pmatrix}$　**10 a** Not parallel　**b** Not parallel　**c** Parallel

11 $PQ = \begin{pmatrix} 2 \\ 3 \end{pmatrix}$ $QR = \begin{pmatrix} -4 \\ -6 \end{pmatrix}$ $PR = \begin{pmatrix} -2 \\ -3 \end{pmatrix}$ $|PQ| = \sqrt{13}$ $|QR| = \sqrt{52}$ $|PR| = \sqrt{13}$ **12 a** $R = \begin{pmatrix} 3 \\ 3 \\ 5 \end{pmatrix}$

12 b $PQ = \begin{pmatrix} 4 \\ -2 \\ -1 \end{pmatrix}$ $QR = \begin{pmatrix} -1 \\ 4 \\ 2 \end{pmatrix}$ $SR = \begin{pmatrix} 4 \\ -2 \\ -1 \end{pmatrix}$ $RP = \begin{pmatrix} -3 \\ -2 \\ -1 \end{pmatrix}$ **c** $|PQ| = \sqrt{21}$ $|QR| = \sqrt{21}$ $|SR| = \sqrt{21}$ $|RP| = \sqrt{14}$

12 d $PQ = \begin{pmatrix} \frac{4}{\sqrt{21}} \\ -\frac{2}{\sqrt{21}} \\ \frac{-1}{\sqrt{21}} \end{pmatrix}$ $QR = \begin{pmatrix} \frac{-1}{\sqrt{21}} \\ \frac{4}{\sqrt{21}} \\ \frac{2}{\sqrt{2}} \end{pmatrix}$ $SR = \begin{pmatrix} \frac{4}{\sqrt{21}} \\ -\frac{2}{\sqrt{21}} \\ \frac{-1}{\sqrt{21}} \end{pmatrix}$ $RP = \begin{pmatrix} \frac{-3}{\sqrt{14}} \\ \frac{-2}{\sqrt{14}} \\ \frac{-1}{\sqrt{14}} \end{pmatrix}$

14 $SU = -a - b$ **15** $a = -1, b = 1$ **16** $p = -\frac{4}{7}, q = \frac{4}{7}, r = -\frac{6}{7}$ **17** $x = -\frac{69}{53}, y = \frac{75}{53}$

Chapter 12 — Exercise 2

1 a $3i + 4j - k$ **b** $6i - 2j - 9k$ **c** $-2i + 11j + 2k$ **d** $17i - 21j - 28k$ **e** $-3i - 6j + 49k$ **f** $20i - 25j - 84k$

1 g $13mi + 117mj - 91mk$ **2 a** $17i + 16j$ **b** $\sqrt{421}$ **c** $74.7°$ **3** $q = 1$ Ratio is 1:2 **4 a** $AB = b - a$ **b** $AC = \frac{1}{2}(b - a)$

4 c $CB = \frac{1}{2}(b - a)$ **d** $OC = \frac{1}{2}(a + b)$ **5 a** $\begin{pmatrix} -5 \\ 20 \end{pmatrix}$ **b** $\sqrt{425}$ **c** $\begin{pmatrix} 3 \\ 2 \\ -6 \end{pmatrix}$ **d** $\begin{pmatrix} 2 \\ -8 \end{pmatrix}$ **6 a** $\begin{pmatrix} 6 \\ 10 \end{pmatrix}$ **b** $\sqrt{136}$ **c** $\begin{pmatrix} -4 \\ -8 \end{pmatrix}$ **d** $\begin{pmatrix} \frac{-25}{4} \\ \frac{-47}{4} \end{pmatrix}$

7 a $\frac{1}{1 + k}(b - a)$ **b** $\frac{k}{1 + k}(a - b)$ **c** $(a - b)$ **d** $\frac{ka + b}{1 + k}$

8 a i $CD = -a$ **ii** $CA = -b - a$ **iii** $BD = b - a$ **iv** $AX = \frac{1}{3}b$ **v** $XD = \frac{2}{3}b$ **b** $AC = \begin{pmatrix} 6k \\ 4c \end{pmatrix}$

9 a $BE = \frac{mb}{m + n + 3}$ **7 b** $EF = \frac{nb}{m + n + 3}$ **c** $CF = \frac{-3b}{m + n + 3}$ **d** $AF = a + \frac{(m + n)b}{m + n + 3}$ **e** $ED = -a + \frac{(2m + 3n + 9)b}{m + n + 3}$

10 a $BC = b$ **b** $FH = b - a$ **c** $AH = b + c$ **d** $AG = a + b + c$ **e** $BH = b + c - a$ **12 a** $DG = d - a - b$ **b** $AH = d$

12 c $FA = -b - c - d$ **14 a** $AB = c, BC = -a, AC = c - a, OB = a + c$ **b** They are perpendicular.

Chapter 12 — Exercise 3

1 a 11 **b** -1 **c** 29 **d** 2 **e** 7 **f** 54 **g** 40 **h** 25 **2 a** -5 **b** 6 **c** 30 **d** -9 **e** 3 **f** 55 **g** -26 **h** 1 **3 a** $58.7°$ **b** $86.6°$

3 c $24.8°$ **d** $129°$ **e** $54.0°$ **f** $50.0°$ **4** $p \cdot q = 14$, $\cos \theta = \sqrt{\frac{7}{19}}$ **5** a and d, a and f, b and c, b and e. **6 a** $-\frac{3}{2}$ **b** -11 **c** $\frac{15}{2}$ **d** $-3, 2$

8 $-\frac{5}{\sqrt{35}}i + \frac{3}{\sqrt{35}}j + \frac{1}{\sqrt{35}}k$ **9** $x = 17.9$ or $x = 6.5$ **10** $70.5°$ **12** It is not a rhombus

16 It is a rectangle since $A\hat{B}C$ is 90° but we do not know if $AB = BC$

Chapter 12 — Exercise 4

1 a $-14i - 5j - 8k$ **b** $-14i - 5j - 8k$ **c** $14i + 5j + 8k$ **d** $28i + 10j + 16k$ **e** $14i + 5j + 8k$ **f** $42i + 15j + 24k$ **g** 0

2 a $\frac{21\sqrt{3}}{2}$ **b** $\frac{9\sqrt{39}}{2}$ **c** 6 **7 a** $-\frac{18}{\sqrt{817}}i + \frac{3}{\sqrt{817}}j + \frac{22}{\sqrt{817}}k$ **b** $\sqrt{\frac{817}{986}}$ **8 a** $-\frac{2}{\sqrt{6}}i - \frac{1}{\sqrt{6}}j + \frac{1}{\sqrt{6}}k$ **b** $\sqrt{\frac{54}{55}}$

9 $\frac{-3}{\sqrt{19}}i + \frac{3}{\sqrt{19}}j - \frac{1}{\sqrt{19}}k$ **10** $\frac{\sqrt{74}}{2}$ units² **11** $\frac{\sqrt{341}}{2}$ units² **12** $\sqrt{6}$ units² **13** $4\sqrt{189}$ or $\sqrt{3024}$ units² **14** $\sqrt{850}$ units²

15 $\sqrt{234}$ units² **16** $PQ = 25i - 5j + 10k$ and $PS = -6i - 14j + 2k$ Area $= 10\sqrt{1734}$ units²

Chapter 12 — Review Exercise

1 b $\frac{3}{2}\sqrt{146}$ or $\frac{1}{2}\sqrt{1314}$ units² **c** $-10.8i + 9.6j - 1.2k$ **d** $\frac{137}{\sqrt{24817}}$ **2 a** $6i - 12j + (2p + 1)k$ **b** $p = 4$ **4** $\sqrt{2 - 2\cos\theta}$ **5** $\begin{pmatrix} 5\cos\theta + 3 \\ 5\sin\theta + 2 \end{pmatrix}$

6 0 **8** 0.7021 **10** $\alpha = \frac{\pi}{2} - 2\theta$ **11 a** A has coordinates $(2, 4, 6)$ B has coordinates $(6, -3, 0)$ C has coordinates $(4, -7, -6)$

11 b 50.2 units² **11 c** $\begin{pmatrix} 3 \\ -\frac{3}{2} \\ 0 \end{pmatrix}$ **d** $96.3°$ **e** 22.6 units² **12** $m = \frac{10}{3}$

Chapter 13 — Exercise 1

1 a $r = \begin{pmatrix} 0 \\ 2 \\ -3 \end{pmatrix} + \lambda\begin{pmatrix} 1 \\ -2 \\ -1 \end{pmatrix}$ **b** $r = i - 2j + \lambda(i - 4j - 2k)$ **c** $r = \begin{pmatrix} 4 \\ 4 \\ 3 \end{pmatrix} + \lambda\begin{pmatrix} 0 \\ -5 \\ 12 \end{pmatrix}$ **d** $r = 5i + 2j + k + \lambda(3i + 6j - k)$ **e** $r = -3i - j + \lambda(2i - j)$

1 f $r = \begin{pmatrix} -5 \\ 1 \end{pmatrix} + \lambda\begin{pmatrix} 4 \\ -7 \end{pmatrix}$ **2 a** $r = \begin{pmatrix} 2 \\ 1 \\ 2 \end{pmatrix} + \lambda\begin{pmatrix} 4 \\ -3 \\ -1 \end{pmatrix}$ **b** $r = \begin{pmatrix} -3 \\ 1 \\ 0 \end{pmatrix} + \lambda\begin{pmatrix} 7 \\ -2 \\ 2 \end{pmatrix}$ **c** $r = \begin{pmatrix} 2 \\ -2 \\ 3 \end{pmatrix} + \lambda\begin{pmatrix} -2 \\ 9 \\ -6 \end{pmatrix}$ **d** $r = \begin{pmatrix} 3 \\ 4 \\ -2 \end{pmatrix} + \lambda\begin{pmatrix} -1 \\ -9 \\ 1 \end{pmatrix}$

2 e $r = \begin{pmatrix} 4 \\ -3 \end{pmatrix} + \lambda\begin{pmatrix} -3 \\ 0 \end{pmatrix}$ **3 a** $r = i - 2j - 4k + \lambda(3i + j - 5k)$ $x = 1 + 3\lambda, y = -2 + \lambda, z = -4 - 5\lambda$ $\dfrac{x - 1}{3} = y + 2 = \dfrac{z + 4}{-5}$

3 b $r = \begin{pmatrix} -3 \\ -2 \\ 3 \end{pmatrix} + \lambda\begin{pmatrix} 4 \\ -7 \\ 3 \end{pmatrix}$ $x = -3 + 4\lambda, y = -2 - 7\lambda, z = 3 + 3\lambda$ $\dfrac{x + 3}{4} = \dfrac{y + 2}{-7} = \dfrac{z - 3}{3}$

3 c $r = j + k + \lambda(i - 3k)$ $x = \lambda, y = 1, z = 1 - 3\lambda$ $y = 1, x = \dfrac{1 - z}{3}$

3 d $r = \begin{pmatrix} 4 \\ 1 \\ 0 \end{pmatrix} + \lambda\begin{pmatrix} -1 \\ 2 \\ 2 \end{pmatrix}$ $x = 4 - \lambda, y = 1 + 2\lambda, z = 2\lambda$ $4 - x = \dfrac{y - 1}{2} = \dfrac{z}{2}$

4 a $x = 1 - 2\lambda, y = -1 + 3\lambda, z = 2 - 2\lambda$ $\dfrac{1 - x}{2} = \dfrac{y + 1}{3} = \dfrac{2 - z}{2}$ **b** $x = 2 + 3\mu, y = -5 - \mu, z = -1 + 4\mu$ $\dfrac{x - 2}{3} = \dfrac{y + 5}{-1} = \dfrac{z + 1}{4}$

4 c $x = 2 + 4m, y = 8 - 7m, z = -1 + 6m$ $\dfrac{x - 2}{4} = \dfrac{8 - y}{7} = \dfrac{z + 1}{6}$ **d** $x = 1 + 2n, y = -1 - 3n, z = 7 - n$ $\dfrac{x - 1}{2} = \dfrac{y + 1}{-3} = 7 - z$

4 e $x = 4 + 3s, y = 6 - 5s$ $\dfrac{x - 4}{3} = \dfrac{6 - y}{5}$ **f** $x = 1 + 2t, y = -6 - 5t$ $\dfrac{x - 1}{2} = \dfrac{y + 6}{-5}$

5 a $r = \begin{pmatrix} -7 \\ 6 \\ 4 \end{pmatrix} + \lambda\begin{pmatrix} 3 \\ 1 \\ 2 \end{pmatrix}$ **b** $r = \begin{pmatrix} -4 \\ -5 \\ 1 \end{pmatrix} + \mu\begin{pmatrix} -1 \\ 3 \\ 5 \end{pmatrix}$ **c** $r = \begin{pmatrix} 4 \\ 0 \\ -3 \end{pmatrix} + m\begin{pmatrix} 5 \\ 3 \\ 4 \end{pmatrix}$ **d** $r = \begin{pmatrix} -4 \\ -1 \\ 5 \end{pmatrix} + n\begin{pmatrix} 0 \\ 1 \\ -1 \end{pmatrix}$ **6 a** $r = \begin{pmatrix} 3 \\ -5 \\ -1 \end{pmatrix} + \lambda\begin{pmatrix} 4 \\ 3 \\ -3 \end{pmatrix}$

6 b $r = \begin{pmatrix} \frac{5}{2} \\ -1 \\ -1 \end{pmatrix} + \lambda\begin{pmatrix} 2 \\ -\frac{4}{3} \\ 2 \end{pmatrix}$ **c** $r = \begin{pmatrix} \frac{2}{5} \\ -\frac{5}{3} \\ -\frac{7}{2} \end{pmatrix} + \lambda\begin{pmatrix} \frac{4}{5} \\ -2 \\ \frac{3}{2} \end{pmatrix}$ **d** $r = \begin{pmatrix} -\frac{1}{6} \\ \frac{4}{3} \\ 4 \end{pmatrix} + \lambda\begin{pmatrix} 1 \\ 1 \\ -9 \end{pmatrix}$ **e** $r = \begin{pmatrix} \frac{5}{3} \\ -\frac{2}{3} \\ \frac{1}{3} \end{pmatrix} + \lambda\begin{pmatrix} 1 \\ 2 \\ 3 \end{pmatrix}$ **f** $r = \begin{pmatrix} \frac{5}{3} \\ \frac{3}{7} \\ 2 \end{pmatrix} + \lambda\begin{pmatrix} 49 \\ -4 \\ 0 \end{pmatrix}$

7 a No **b** Yes **c** No **d** Yes **e** No **f** Yes **8** $r = \begin{pmatrix} 3 \\ 7 \\ -1 \end{pmatrix} + \lambda\begin{pmatrix} 6 \\ -9 \\ 2 \end{pmatrix}$ Position vector is $\begin{pmatrix} 6 \\ 5 \\ 2 \\ 0 \end{pmatrix}$

9 Crosses the xy plane at $(-9, -7, 0)$ Crosses the yz plane at $\left(0, 11, -\dfrac{9}{2}\right)$ Crosses the xz plane at $\left(-\dfrac{11}{2}, 0, -\dfrac{7}{4}\right)$

10 a $r = 2i - j + 5k + \lambda(3i - j + k)$ Crosses the xy plane at $(-13, 4, 0)$ Crosses the yz plane at $\left(0, -\dfrac{1}{3}, \dfrac{13}{3}\right)$ Crosses the xz plane at $(-1, 0, 4)$

10 b $r = \begin{pmatrix} 2 \\ 6 \\ 7 \end{pmatrix} + \lambda\begin{pmatrix} 1 \\ -1 \\ -1 \end{pmatrix}$ Crosses the xy plane at $(9, -1, 0)$ Crosses the yz plane at $(0, 8, 9)$ Crosses the xz plane at $(8, 0, 1)$

11 Crosses the xy plane at $\left(-\dfrac{29}{15}, \dfrac{14}{5}, 0\right)$ Crosses the yz plane at $\left(0, \dfrac{17}{4}, -\dfrac{29}{12}\right)$ Crosses the xz plane at $\left(-\dfrac{17}{3}, 0, \dfrac{14}{3}\right)$

Chapter 13 — Exercise 2

1 a Skew **b** Intersect at the point $(2, 1, 6)$ **c** Parallel **d** Skew **e** Skew **f** Parallel **g** Skew

2 a $\dfrac{\sqrt{210}}{6}$ **b** $\sqrt{17}$ **c** 1 **d** $\dfrac{5}{\sqrt{2}}$ **e** Lies on line. **f** $\dfrac{2\sqrt{3}}{\sqrt{7}}$

3 $AB: r = \begin{pmatrix} 0 \\ 1 \\ -2 \end{pmatrix} + \lambda\begin{pmatrix} 3 \\ 4 \\ 7 \end{pmatrix}$ $\dfrac{x}{3} = \dfrac{y - 1}{4} = \dfrac{z + 2}{7}$ $AD: r = \begin{pmatrix} 0 \\ 1 \\ -2 \end{pmatrix} + \lambda\begin{pmatrix} 1 \\ 2 \\ -1 \end{pmatrix}$ $\dfrac{y - 1}{2} = -z - 2$ Coordinates of C are $(4, 7, 4)$

4 a 60.5° **b** 36.3° **c** 71.2° **d** 88.4° **e** 62.8° **5** $a = -2$ Point of intersection is $(1, 5, -3)$ **6** $r = \begin{pmatrix} -3 \\ 8 \end{pmatrix} + t\begin{pmatrix} -3 \\ 2 \end{pmatrix}$

7 a $r = 2i - 3j + k + \lambda(i + 8j - 3k)$ **b** $p = 34$

Chapter 13 — Exercise 3

1 a $r.\begin{pmatrix} \frac{1}{\sqrt{18}} \\ \frac{4}{\sqrt{18}} \\ \frac{-1}{\sqrt{18}} \end{pmatrix} = 3$ **b** $r.\left(\frac{1}{\sqrt{35}}i - \frac{3}{\sqrt{35}}j + \frac{5}{\sqrt{35}}k\right) = \frac{-8}{\sqrt{35}}$ **c** $r.\begin{pmatrix} \frac{2}{\sqrt{5}} \\ 0 \\ \frac{1}{\sqrt{5}} \end{pmatrix} = 6$ **d** $r.\left(\frac{-1}{\sqrt{18}}i - \frac{4}{\sqrt{18}}j + \frac{1}{\sqrt{18}}k\right) = \frac{-15}{\sqrt{18}}$

1 e $r.\left(\frac{1}{\sqrt{74}}i - \frac{8}{\sqrt{74}}j + \frac{3}{\sqrt{74}}k\right) = \frac{-19}{\sqrt{74}}$ **f** $r.\left(\frac{-3}{\sqrt{19}}i + \frac{3}{\sqrt{19}}j + \frac{1}{\sqrt{19}}k\right) = \frac{3}{\sqrt{19}}$ **g** $r.\left(\frac{10}{\sqrt{1133}}i - \frac{3}{\sqrt{1133}}j + \frac{32}{\sqrt{1133}}k\right) = \frac{-63}{\sqrt{1133}}$

1 h $r.\begin{pmatrix} \frac{7}{\sqrt{354}} \\ \frac{17}{\sqrt{354}} \\ \frac{4}{\sqrt{354}} \end{pmatrix} = \frac{32}{\sqrt{354}}$ **i** $r.\begin{pmatrix} \frac{20}{\sqrt{557}} \\ \frac{6}{\sqrt{557}} \\ \frac{11}{\sqrt{557}} \end{pmatrix} = \frac{86}{\sqrt{557}}$ **j** $r.\left(\frac{-3}{\sqrt{26}}i - \frac{4}{\sqrt{26}}j + \frac{1}{\sqrt{26}}k\right) = 0$ **k** $r.\left(\frac{3}{\sqrt{13}}i - \frac{2}{\sqrt{13}}j\right) = \frac{15}{\sqrt{13}}$

1 l $r.\left(\frac{-7}{\sqrt{65}}j + \frac{4}{\sqrt{65}}k\right) = 0$ **2 a** $r.(5j + 2k) = -2$ **b** $r.(5i - 4j + 15k) = 42$ **c** $r.\begin{pmatrix} -15 \\ 3 \\ 11 \end{pmatrix} = -3$ **d** $r.(i + 11j - 9k) = 16$

3 a $x - 2y + 7z = 9$ **b** $4x - y = -6$ **c** $15x + 13y - 8z = -38$ **d** $-4x + 3y + 8z = 21$ **4** $r.\begin{pmatrix} 1 \\ -3 \\ 7 \end{pmatrix} = 4$

5 a $\frac{1}{\sqrt{30}}i + \frac{2}{\sqrt{30}}j - \frac{5}{\sqrt{30}}k$ **b** $\frac{4}{\sqrt{42}}i - \frac{1}{\sqrt{42}}j + \frac{5}{\sqrt{42}}k$ **c** $\frac{-3}{7}i - \frac{6}{7}j + \frac{2}{7}k$ **d** $\frac{5}{\sqrt{45}}i - \frac{2}{\sqrt{45}}j + \frac{4}{\sqrt{45}}k$ **7** $r.(13i - 4j - 11k) = 31$

8 The direction normals are equal. Distance $\frac{8}{\sqrt{19}}$ units.

9 b $r.(i + 5j + k) = 1$ **c** Distance of π_1 to origin is $\frac{8}{\sqrt{27}}$ units Distance of π_2 to origin is $\frac{1}{\sqrt{27}}$ units Distance between π_1 and π_2 is $\frac{7}{\sqrt{27}}$ units

10 $P_1 = r.(2i + 5j - 6k) = -41$. Distance of P_1 to origin is $\frac{41}{\sqrt{65}}$ units Distance of P_2 to origin is $\frac{14}{\sqrt{65}}$ units Distance between P_1 and P_2 is $\frac{55}{\sqrt{65}}$ units

11 $r.(i - k) = 1$ r_1 is not contained in the plane. r_2 is contained in the plane.

12 $r.\left(\frac{8}{\sqrt{89}}i - \frac{5}{\sqrt{89}}k\right) = \frac{-17}{\sqrt{89}}$ Distance of plane from the origin is $\frac{17}{\sqrt{89}}$ units.

13 $r.\begin{pmatrix} 2 \\ -3 \\ -2 \end{pmatrix} = -13$ Distance of plane from the origin is $\frac{13}{\sqrt{17}}$ units. **14** $r = (4i + 3j + 7k) + \lambda(2i + 2j - 5k)$

15 b $r.(i + j - k) = -9$ **c** Distance of P_1 to origin is $\frac{8}{\sqrt{3}}$ units Distance of P_2 to origin is $\frac{9}{\sqrt{3}}$ units Distance between P_1 and P_2 is $\frac{17}{\sqrt{3}}$ units

16 a The line and the plane intersect. **b** The line and the plane are parallel. **c** The line and the plane intersect. **d** The line and the plane intersect.

Chapter 13 — Exercise 4

1 a $\left(9, \frac{1}{2}, -\frac{2}{3}\right)$ **b** $\left(4, \frac{1}{2}, -2\right)$ **c** $(3, -2, -9)$ **d** $\left(\frac{51}{7}, \frac{1}{7}, 1\right)$ **e** $\left(\frac{20}{3}, \frac{-13}{9}, \frac{4}{3}\right)$ **f** $(16, -10, -7)$

2 a $45.0°$ **b** $23.2°$ **c** $11.0°$ **d** $64.5°$ **e** $90°$ **f** $55.9°$ **3 a** $11.7°$ **b** $57.8°$ **c** $17.6°$ **d** $32.5°$ **e** $34.1°$ **f** $5.51°$

4 a $r = (4i + j) + \lambda(14i + 17j + 13k)$ **b** $r = \begin{pmatrix} 2 \\ -2 \\ 0 \end{pmatrix} + \lambda\begin{pmatrix} 7 \\ -1 \\ 4 \end{pmatrix}$ **c** $r = (7i + 3j) + \lambda(2i - k)$ **d** $r = (i - 6j) + \lambda(-3i - 2j + k)$

4 e $r = (-4i - 22j) + \lambda(7i + 31j + 2k)$ **f** $r = \left(\frac{56}{25}i + \frac{29}{25}j\right) + \lambda(36i - 26j - 25k)$ **6** $48.5°$ **7 a** $r = \begin{pmatrix} 1 \\ 2 \\ -3 \end{pmatrix} + \lambda\begin{pmatrix} 3 \\ 1 \\ -3 \end{pmatrix}$ **b** $(-2, 1, 0)$

8 a $a = -3$ **b** $74.2°$ **c** 13.9 units.

Chapter 13 — Review Exercise

1 i $\frac{3}{\sqrt{178}}i - \frac{12}{\sqrt{178}}j + \frac{5}{\sqrt{178}}k$ **ii** $r.\left(\frac{3}{\sqrt{178}}i - \frac{12}{\sqrt{178}}j + \frac{5}{\sqrt{178}}k\right) = \frac{7}{\sqrt{178}}$ **iii** $73.5°$ **2 b** $\frac{\sqrt{30}}{2}$ units² **c** $\frac{1}{\sqrt{30}}i - \frac{2}{\sqrt{30}}j - \frac{5}{\sqrt{30}}k$

2 d AD has equation $x - 1 = \dfrac{1 - y}{3} = \dfrac{1 - z}{4}$ BD has equation $2x - 4 = \dfrac{4 + 2y}{-9} = \dfrac{6 + 2z}{-7}$ **e** 16.4° **3 a ii** $r = \begin{pmatrix} 2 \\ 1 \\ 1 \end{pmatrix} - \lambda \begin{pmatrix} 1 \\ 2 \\ 1 \end{pmatrix}$ **b** $3x - 2y + z = 5$

3 c $r = \begin{pmatrix} 2 \\ 1 \\ 1 \end{pmatrix} + \lambda \begin{pmatrix} -1 \\ -2 \\ -1 \end{pmatrix}$ **4 b** $\left(\dfrac{18}{5}, \dfrac{-36}{5}, 4 \right)$ **c** $r . \begin{pmatrix} -2 \\ 4 \\ 9 \end{pmatrix} = 0$ **d** $\left(-33, \dfrac{-51}{2}, 4 \right)$ **5 a** $\left(1, -\dfrac{13}{5}, \dfrac{12}{5} \right)$ **b** $(6, -1, 4)$ **c** $r = \begin{pmatrix} 6 \\ -1 \\ 4 \end{pmatrix} + \lambda \begin{pmatrix} 25 \\ 8 \\ 8 \end{pmatrix}$

5 d 62.7° **6 a** $\dfrac{x - 2}{1} = \dfrac{y - 5}{1} = \dfrac{z + 1}{1}$ **b** $\left(\dfrac{1}{3}, \dfrac{10}{3}, \dfrac{-8}{3} \right)$ **c** $\left(\dfrac{-4}{3}, \dfrac{5}{3}, \dfrac{-13}{3} \right)$ **d** 8.52 **7** $r = \begin{pmatrix} -2 \\ 3 \\ 0 \end{pmatrix} + \lambda \begin{pmatrix} -1 \\ 1 \\ 1 \end{pmatrix}$ **8 a** $AB = \begin{pmatrix} -1 \\ -3 \\ 1 \end{pmatrix}$ $BC = \begin{pmatrix} 1 \\ 1 \\ 0 \end{pmatrix}$

8 b $-i + j + 2k$ **c** $\dfrac{\sqrt{6}}{2}$ **d** $r . \begin{pmatrix} -1 \\ 1 \\ 2 \end{pmatrix} = 3$ **e** $x = 2 - \lambda$ $y = -1 + \lambda$ $z = -6 + 2\lambda$ **f** $3\sqrt{6}$ **g** $\dfrac{-i}{\sqrt{6}} + \dfrac{j}{\sqrt{6}} + \dfrac{2k}{\sqrt{6}}$ **h** $(-4, 5, 6)$

9 a A lies in the plane. B does not lie in the plane **b** $\dfrac{2 - x}{2} = \dfrac{y + 3}{4} = z - 8$ **c** 43.6° **d** 3.16 units. **e** $r . (i + 3j - 10k) = -87$

10 a $-2i + 2j - k$ **b i** $n_1 = 6i + 3j - 2k$ and $n_2 = -2i + 2j - k$ **ii** 79.0° **d i** $(8, -20, -12)$

11 a $r = (2i + j + 4k) + \lambda(3i + j)$ **b** $(-4, -1, 4)$ **c** 0.716 **d** $(0, -13, 18)$ **e** $\left(-\dfrac{13}{3}, 0, \dfrac{28}{3} \right)$ **f** $r = \begin{pmatrix} 0 \\ -13 \\ 18 \end{pmatrix} + \lambda \begin{pmatrix} -\dfrac{13}{3} \\ 13 \\ -\dfrac{26}{3} \end{pmatrix}$

12 a ii $r = (-j - k) + \lambda(3i + 11j + k)$ **b** $c = -2$. Line of intersection. **c ii** $\dfrac{7\sqrt{2}}{6}$ **13** $r = \begin{pmatrix} 7 \\ 0 \\ -4 \end{pmatrix} + \lambda \begin{pmatrix} -7 \\ 1 \\ 5 \end{pmatrix}$

14 a $r = (2i + 3j + 7k) + t(3i + j + 3k)$ **b** $(8, 5, 13)$ **c** $2x + 3y - 4z + 4 = 0$ **e i** $i + j - k$ **ii** $PO = i - 2j - 4k$ **iii** $\dfrac{12}{\sqrt{14}}$

15 b $\dfrac{4}{\sqrt{21}}$ **c** $r . (i + 4j + 2k) = -9$ **d** $\dfrac{13}{\sqrt{21}}$ **16 a** $\left(4, \dfrac{5}{2}, \dfrac{13}{2} \right)$ **b** $x + 2y = 9$ **c** $\left(\dfrac{37}{5}, \dfrac{4}{5}, \dfrac{9}{5} \right)$ **d** $(6.63, 3.50, 4.65)$

16 $(6.87, 3.99, 4.15)$ or $(4.53, -0.694, 4.15)$

Chapter 14   Exercise 1

1 $5x$ **2** $10x$ **3** $-2x$ **4** $2x^2$ **5** $6x^2$ **6** x^3 **7** x^4 **8** x^5 **9** $3x^3$ **10** $4x^{-1}$

Chapter 14   Exercise 2

1 $x^2 - x + c$ **2** $\dfrac{1}{3}x^3 + c$ **3** $\dfrac{1}{4}x^4 + c$ **4** $\dfrac{1}{5}x^5 + c$ **5** $2x^3 - 5x + c$ **6** $2x^4 + 2x^2 - 3x + c$ **7** $\dfrac{5}{3}x^3 - 4x + c$ **8** $-x^{-1} + c$

9 $\dfrac{2}{3}x^{\frac{3}{2}} + c$ **10** $3x^{\frac{1}{3}} + c$ **11** $7x + 2x^{-2} + c$ **12** $2x^{\frac{1}{2}} + c$ **13** $\dfrac{2}{7}x^7 - x^5 + c$ **14** $2x^{\frac{5}{2}} + 2x^{-2} + c$ **15** $x^4 + 2x^2 - 9x + c$

16 $x - x^2 + 2x^3 - \dfrac{1}{4}x^4 + c$ **17** $-3x^{-2} + c$ **18** $y = -\dfrac{1}{2}x^{-4} + c$ **19** $y = \dfrac{2}{3}x^{\frac{3}{2}} - 2x^{\frac{1}{2}} + c$ **20** $y = 8x - \dfrac{9}{2}x^{\frac{2}{3}} + c$ **21** $y = 4x^4 - 12x^2 + c$

22 $y = \dfrac{2}{3}x^3 - \dfrac{21}{2}x^2 + 27x + c$ **23** $y = 3x^3 - 12x^2 + 16x + c$ **24** $y = -\dfrac{1}{2}x^{-2} + \dfrac{5}{4}x^{-4} + c$ **25** $y = \dfrac{8}{7}x^{\frac{7}{2}} - \dfrac{14}{3}x^{\frac{3}{2}} + c$ **26** $y = \dfrac{14}{19}x^{\frac{19}{2}} - 2x^{\frac{3}{2}} + c$

27 $y = -6p^{-2} + c$ **28** $y = \dfrac{32}{9}k^{\frac{9}{4}} + c$ **29** $y = \dfrac{1}{6}z^6 - \dfrac{1}{2}z^2 + c$ **30** $y = -t^{-1} - 3t^{-2} - 3t^{-3} + c$ **31** $y = \dfrac{2}{3}t^{\frac{3}{2}} - \dfrac{4}{9}t^3 + c$

Chapter 14   Exercise 3

1 $y = 6x - 4$ **2** $y = 2x^2 + 3$ **3** $y = 4x^2 - 3x - 18$ **4** $y = -x^2 + 5x$ **5** $y = x^4 - 2x^3 + 7x + 3$ **6** $y = \dfrac{4}{3}x^3 - 6x^{-1} - \dfrac{509}{6}$

7 $y = 16x^{\frac{1}{2}} - 46$ **8** $y = \dfrac{1}{7}t^7 - \dfrac{3}{5}t^5 - \dfrac{4}{3}t^3 + \dfrac{818}{105}$ **9** $Q = \dfrac{2}{21}p^{\frac{7}{2}} - \dfrac{8}{33}p^{\frac{11}{2}} + 2$

Chapter 14   Exercise 4

1 $\dfrac{1}{4}x^4 - 2\ln|x| + c$ **2** $4e^x - \cos x + c$ **3** $5\ln|x| - \sin x + c$ **4** $-6\cos x - \dfrac{6}{5}x^5 + c$ **5** $8\cos x + 7e^x + c$

6 $5e^x + 2\cos x + 3\ln|x| + c$ **7** $\dfrac{1}{3}e^x - \dfrac{5}{2}\ln|x| - 7\cos x + c$ **8** $\dfrac{e^x}{15} - 10x^{\frac{3}{2}} + \sin x + c$

Chapter 14 — Exercise 5

1 $-\dfrac{1}{5}\cos 5x + c$ 2 $\dfrac{1}{6}\sin 6x + c$ 3 $-\dfrac{1}{2}\cos 2x + c$ 4 $-2\cos\dfrac{1}{2}x + c$ 5 $2\sin 4x + c$ 6 $2\cos 3x + c$ 7 $-\dfrac{5}{2}\sin 2x + c$ 8 $\dfrac{1}{6}e^{6x} + c$

9 $\dfrac{1}{5}e^{5x} + c$ 10 $e^{4x} + c$ 11 $\dfrac{4}{3}e^{6t} + c$ 12 $-\dfrac{5}{6}e^{6p} + c$ 13 $4x^2 - \dfrac{1}{2}e^{2x} + c$ 14 $-2e^{-2x} + c$ 15 $y = \dfrac{1}{2}\ln|2x - 3| + c$

16 $y = \dfrac{1}{8}\ln|8x + 7| + c$ 17 $y = 2\ln|2x - 5| + c$ 18 $y = \dfrac{1}{18}(3x - 1)^6 + c$ 19 $y = \dfrac{1}{28}(4x - 7)^7 + c$ 20 $y = -\dfrac{1}{8}(4x + 3)^{-2} + c$

21 $y = -\dfrac{1}{10}(3 - 2x)^5 + c$ 22 $y = -\dfrac{2}{3}(3x - 2)^{-2} + c$ 23 $y = -\dfrac{3}{2}(2t - 1)^{-1} + c$ 24 $y = \dfrac{4}{3}\ln|3x - 1| + c$ 25 $y = 2\ln|3x - 5| + c$

26 $y = -8\ln|4 - p| + c$ 27 $y = -3\ln|6 - t| + c$ 28 $\dfrac{3}{2}e^{4x} + c$ 29 $-\dfrac{1}{3}\cos 3x - 2x^2 + c$ 30 $-\dfrac{1}{2}e^{-8x} - 2\sin 2x + c$

31 $\dfrac{1}{2}\ln|2x - 1| + \dfrac{1}{18}(3x + 4)^6 + c$ 32 $2x^3 - \dfrac{2}{3}\ln|3x + 2| + c$

Chapter 14 — Exercise 6

1 3 2 38 3 20 4 0 5 $\dfrac{2}{3}$ 6 $-\dfrac{8}{3}$ 7 201 8 216468 9 1490 10 $\dfrac{2}{3}$ 11 0 12 $\dfrac{1 - 3\pi}{3}$ 13 21 14 -200 15 312.6

16 0.490 17 0.0429 18 0.549 19 1.85 20 -3.47 21 8.56 22 0, because both functions are odd 23 $2\ln|2k + 1|$

Chapter 14 — Exercise 7

1 32 2 $\dfrac{343}{6}$ 3 50 4 70.4 5 2 6 6.39 7 5.55 8 0.619 9 22 10 1.69 11 0.825 12 2.19 13 27 14 10.4 15 2 16 2.27

17 $\dfrac{4}{3}$ 18 9 19 $\dfrac{4}{3}$ 20 1.48 21 $\dfrac{2}{3}\ln\left|\dfrac{3p + 5}{8}\right|$ 22 $\dfrac{1}{2}(1 - e^{-2p}) + \dfrac{1}{3}p^3$ 23 $k = 4$ 24 $a = 1$

Chapter 14 — Exercise 8

1 $\dfrac{64}{3}$ 2 61 3 16 4 $\dfrac{85}{4}$ 5 $\dfrac{253}{12}$ 6 $\dfrac{2401}{16}$ 7 $\dfrac{863}{6}$ 8 22.1 9 23 10 408 11 $\dfrac{21}{4}$ 12 2 13 6 14 15.3

Chapter 14 — Exercise 9

1 $\dfrac{1}{6}$ 2 $\dfrac{1}{8}$ 3 $\dfrac{1}{2}$ 4 $\dfrac{1}{3}$ 5 $\dfrac{125}{6}$ 6 $\dfrac{407}{4}$ 7 36 8 6.43 9 $\dfrac{160}{3}$

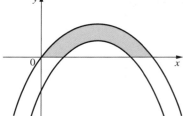

10 1.60 11 1.85 12 3.08 13 $\dfrac{5}{2}$ 14 3.92 15 $\dfrac{8}{3}$ 16 $\dfrac{32}{3}$

17 $\dfrac{256}{3}$ 18 18 19 $\dfrac{10\sqrt{5}}{3}$ 20 3.62 21 4.53 22 3.21

Chapter 14 — Review Exercise

1 a $\dfrac{4}{3}x^3 - 7x + c$ b $3x^3 + 2x^2 - 5x + c$ c $-4x^{-2} + c$ d $-\dfrac{1}{6}(3 - 2x)^3 + c$ 2 a $y = -\dfrac{1}{x} + \dfrac{3}{2}x^{-4} + c$ b $y = p^3 - \dfrac{1}{8}p^8 + c$

2 c $y = \dfrac{3}{8}t^2 - \dfrac{1}{2}t^{\frac{1}{2}} + c$ 3 $y = -\dfrac{3}{2}x^2 + 8x - 2$ 4 a $4e^x + \cos x + c$ b $7\sin x - 4\ln|x| + c$ c $\dfrac{1}{3}e^{6x} - 5\ln|x| - 4\cos x + c$

5 a $3\sin 2x + c$ b $2e^{2x} + c$ c $\dfrac{1}{2}\ln|4x - 3| + c$ d $\dfrac{1}{21}(3x - 2)^7 + c$ e $\dfrac{7}{3}e^{3x} + \dfrac{1}{3}(3x - 4)^{-4} + c$ 6 $\dfrac{4x^{\frac{5}{2}}}{5} + \dfrac{3}{x} + c$

7 a $\dfrac{382}{25}$ b $\dfrac{1}{2} + \dfrac{1}{4}\pi$ c $3\sin 2k$ 8 a $e^5 - e^2$ b 36 c $\dfrac{2}{3}$ 9 $\dfrac{3}{2}\ln(2p - 5)$

10 $\dfrac{407}{4}$ 11 a 0.753 b 2.45 c 1.78 12 a $\dfrac{9}{2}$ b 24.3 13 a 13.3 b 1.93 14 30.2

Chapter 15 — Exercise 1

1 $-\cos\left(\theta - \dfrac{3\pi}{4}\right) + c$ 2 $\dfrac{1}{3}\sin\left(3x + \dfrac{\pi}{4}\right) + c$ 3 $\dfrac{1}{32}e^{32x - 7} + c$ 4 $2e^{\sin 2} - 2$ 5 $\dfrac{1}{4}\ln|8x - 9| + c$ 6 $\dfrac{1}{54}(x^6 - 9)^9 + c$

7 $\dfrac{1}{16}[(2p^4 + 1)^2 - 1]$ 8. $\dfrac{1}{3}(1 + x^2)^{\frac{3}{2}} + c$ 9 $\dfrac{1}{12}(4x^2 - 3)^{\frac{3}{2}} + c$ 10 $\dfrac{1}{12}(3\tan x + 4)^4 + c$ 11 $(x^2 - 1)^{\frac{1}{2}} + c$ 12 $\dfrac{1}{10}[1 - (2\cos 0.5 - 1)^5]$

13 $\dfrac{1}{24}[(6e^{2a} - 7)^2 + 1]$ 14 $-\dfrac{1}{3}(\cos 2x - 1)^{\frac{3}{2}} + c$ 15 $\dfrac{1}{12}(x^2 + 2x - 4)^6 + c$ 16 $\ln|x^2 - 3x + 5| + c$ 17 $\dfrac{3}{2(\cos x + 8)^2} + c$

18 $\frac{1}{2}\ln|2e^x - 4| + c$ **19** $\frac{1}{3}\ln|3\sin x - 12| + c$ **20** $\frac{2}{15}(3x^3 + 6x - 19)^{\frac{5}{2}} + c$ **21** $\frac{1}{21}(1 - 3\cos 2x)^{\frac{7}{2}} + c$ **22** $\frac{1}{6}\ln|3\tan 2x - 7| + c$

23 $\frac{15}{128}$ **24** $\frac{1}{3}\ln|3x^2 - 3x + 4| + c$ **25** $-\frac{1}{9(3x^2 - 3x + 4)^3} + c$ **26** $(\ln|p|)^2$ **27** $\ln|e^{2x} + 1| + c$ **28** $\frac{1}{6}\ln|2p^2 + 2p - 5|$

Chapter 15   Exercise 2

1 $\frac{1}{3}\tan^{-1}\frac{x}{3} + c$ **2** $\sin^{-1}\frac{x}{5} + c$ **3** $\cos^{-1}\frac{x}{6} + c$ **4** $3\tan^{-1}\frac{x}{3} + c$ **5** $2\sin^{-1}\frac{x}{2\sqrt{2}} + c$ **6** $\frac{\sqrt{3}}{3}\tan^{-1}\frac{p\sqrt{3}}{3} - \frac{\pi\sqrt{3}}{18}$ **7** 0.0203

8 $\frac{\sqrt{3}}{3}\sin^{-1}x\sqrt{3} + c$ **9** $\sin^{-1}(x - 1) + c$ **10** $\frac{1}{8}\tan^{-1}\left(\frac{x + 1}{2}\right) + c$ **11** $\frac{1}{3}\tan^{-1}\left(\frac{x + 3}{3}\right) + c$ **12** $5\cos^{-1}\left(\frac{x + 2}{3}\right) + c$ **13** $\frac{1}{6}\tan^{-1}\left(\frac{3x + 1}{2}\right) + c$

14 $\frac{2}{\sqrt{7}}\tan^{-1}\left(\frac{2x + 3}{\sqrt{7}}\right) + c$ **15** $4\sin^{-1}\left(\frac{2x - 3}{7}\right) + c$ **16** $\frac{1}{3}\sin^{-1}\left(\frac{x - 1}{2\sqrt{3}}\right) + c$ **17** 0.0623 **18** $\sin^{-1}\left(\frac{p + 3}{\sqrt{3}}\right) - \sin^{-1}\left(\frac{\sqrt{3}}{3}\right)$

Chapter 15   Exercise 3

1 $\sin x - \frac{1}{3}\sin^3 x + c$ **2** $-\frac{1}{2}\cos 2x + \frac{1}{6}\cos^3 2x + c$ **3** $-\cos x + \frac{2}{3}\cos^3 x - \frac{1}{5}\cos^5 x + c$ **4** $\frac{1}{2}\tan 2p - p$ **5** $\frac{x}{2} + \frac{\sin 4x}{8} + c$

6 $\frac{x}{2} - \frac{\sin 4x}{8} + c$ **7** $\frac{1}{32}(12x - 8\sin 2x + \sin 4x) + c$ **8** $-\cos x + \frac{4}{3}\cos^3 x - \frac{6}{5}\cos^5 x + \frac{4}{7}\cos^7 x - \frac{1}{9}\cos^9 x + c$ **9** $\frac{1}{6}\tan^2 3x + \frac{1}{3}\ln|\cos 3x| + c$

10 $\frac{1}{32}(4x - \sin 4x) + c$ **11** $\frac{1}{4}\tan^4 x + \frac{1}{6}\tan^6 x + c$ **12** $-\frac{1}{6}\cos^3 2x + \frac{1}{10}\cos^5 2x + c$ **13** $\frac{1}{3}\sin^3 p - \frac{1}{5}\sin^5 p$

Chapter 15   Exercise 4

1 $\frac{(x + 2)^5}{5} + c$ **2** $\frac{(2 + 7x)^4}{28} + c$ **3** $-\sqrt{1 - 2x} + c$ **4** $\frac{-3}{4(2x + 1)^2} + \frac{(1 + 2x)^{\frac{3}{2}}}{3} + c$ **5** $\frac{(3 + 5x)^{\frac{3}{2}}}{30} + c$

6 $\frac{-2}{3}(1 - x)^{\frac{3}{2}} - 2(1 - x)^{\frac{1}{2}} - \frac{1}{1 - x} + c$ **7** $\frac{3}{4}\sin\left(4x - \frac{\pi}{2}\right) + c$ **8** $\tan^{-1} 2x + c$ **9** $\frac{1}{4}\ln|3 - 4\cos x| + c$ **10** $-\frac{1}{2}\tan\left(\frac{\pi}{3} - 2x\right) + c$

11 $-\frac{2}{3}\cos(3x + \alpha) + c$ **12** $\frac{1}{4}e^{4x + 1} + c$ **13** $\frac{2^x}{\ln 2} + c$ **14** $\frac{1}{3}\ln|3x + 1| + c$ **15** $\ln|x^2 + 4| + c$ **16** $\frac{1}{2}\ln|x^2 + 2x + 3| + c$

17 $\frac{1}{4}\ln|x^4 + 3| + c$ **18** $\sin^{-1} 3x + c$ **19** $\frac{3}{4}\tan^{-1} 2x + c$ **20** $\frac{1}{14}(x^2 + 6x - 8)^7 + c$ **21** $\frac{1}{10}(\sin 2x + 3)^5 + c$ **22** $2e^{1 - \cot\frac{x}{2}} + c$

23 $\frac{1}{12}(1 + x^{\frac{2}{3}})^8 + c$ **24** $-\frac{1}{4}\cos^4 x + c$ **25** $\frac{1}{2(\cot x - 3)^2} + c$ **26** $\ln|e^x + 2| + c$ **27** $2(e^x + 2)^{\frac{1}{2}} + c$ **28** $\frac{1}{64}(24x + 8\sin 4x + \sin 8x) + c$

29 $\frac{1}{2}\ln|x^2 + 2x + 3| + c$ **30** $\sqrt{2}\tan^{-1}\left(\frac{x + 1}{\sqrt{2}}\right) + c$ **31** $2\sin^{-1}\left(\frac{x - 2}{3}\right) + c$ **32** $2(-x^2 + 4x + 5)^{\frac{1}{2}} + c$ **33** $\frac{1}{5}(\sin^2 x + 3)^5 + c$

Chapter 15   Exercise 5

1 $\frac{(x^2 + 3)^6}{12} + c$ **2** $\frac{1}{4}\ln|6x^2 + 4x - 13| + c$ **3** $-\sqrt{1 - \sin 2x} + c$ **4** $\frac{2}{15}(x - 2)^{\frac{3}{2}}(3x + 4) + c$ **5** $\frac{(2p - 1)^{\frac{3}{2}}(p + 1) - 2}{3}$

6 $\frac{(2x + 1)^{\frac{3}{2}}(x + 8)}{3} + c$ **7** $\tan^{-1} x^2 + c$ **8** $\frac{2(3x - 4)^{\frac{3}{2}}(9x + 38)}{135} + c$ **9** $\frac{(p - 2)^4(8p - 1) - 7}{20}$ **10** $\frac{1 - 6x}{24(2x - 1)^3} + c$ **11** $\frac{1 - 5x}{10(x - 3)^5} + c$

12 $\frac{2(p - 2)^{\frac{3}{2}}}{3} + 4(p - 2)^{\frac{1}{2}} - \frac{26\sqrt{7}}{3}$ **13** $\frac{(x + 5)^2}{2} - 15(x + 5) + 75\ln|x + 5| + \frac{125}{x + 5} + c$ **14** $\frac{2(5p + 2)^{\frac{3}{2}}(15p - 4)}{375} - \frac{416\sqrt{3}}{125}$

15 $\frac{x^2 - 4x + 8}{x - 2} + c$ **16** $\frac{1}{2}\tan^{-1}(2\tan x) + c$ **17** $\frac{3}{2}(\sin^{-1} x + 4x\sqrt{1 - x^2}) + c$ **18** $\frac{1}{2}\tan^{-1}\left(2\tan\frac{x}{2}\right) + c$

19 $2\sin^{-1}\frac{p}{2} + p\cos\left(\sin^{-1}\frac{p}{2}\right) - 2\sin^{-1}\frac{1}{4} + \frac{1}{2}\cos\left(\sin^{-1}\frac{1}{4}\right)$ **20** $\frac{1}{\sqrt{5}}\tan^{-1}(\sqrt{5}\tan x) + c$ **21** $3\tan^{-1}\left(\frac{5\tan\frac{1}{2}p + 3}{4}\right) - 3\tan^{-1}\frac{3}{4}$

22 $\frac{1}{32}\tan 2x + c$ **23** $\frac{8}{3n}\tan^{-1}\sqrt{x^n - 1} + c$

Chapter 15   Exercise 6

1 $x\sin x + \cos x + c$ **2** $\frac{e^{2x}}{4}(2x - 1) + c$ **3** $\frac{x^5}{25}(5\ln x - 1) + c$ **4** $\frac{-x\cos 2x}{2} + \frac{\sin 2x}{4} + c$ **5** $\frac{(p + 1)^{10}}{110}(10p - 1) - \frac{4608}{55}$

6 $-x^2\cos x + 2x\sin x + 2\cos x + c$ **7** $\frac{e^{2x}}{4}(2x^2 - 2x + 1) + c$ **8** $\frac{x^3}{9}(3\ln 3x - 1) + c$ **9** $x^3\ln 8x - \frac{x^3}{3} + c$ **10** $\frac{-e^{-3x}}{27}(9x^2 + 6x + 2) + c$

11 $\dfrac{e^x}{2}(\cos x + \sin x) + c$ **12** $x \sin^{-1} x + (1 - x^2)^{\frac{1}{2}} + c$ **13** $x \tan^{-1} x - \dfrac{1}{2}\ln|1 + x^2| + c$ **14** $e^{2x}(x - 1) + c$ **15** $\dfrac{e^{3x}}{10}(\sin x + 3 \cos x) + c$

16 $\dfrac{e^{2x}}{13}(2 \sin 3x - 3 \cos 3x) + c$ **17** $\dfrac{e^x}{5}(\sin 2x - 2 \cos 2x) + c$ **18** $\dfrac{x^{n+1}}{(n+1)^2}[(n+1)\ln x - 1] + c$ **19** $\dfrac{e^{ax}}{a^2 + b^2}(a \sin bx - b \cos bx) + c$

20 $\dfrac{(2p+1)^{n+1}(np - 1) + 1}{2(n+1)(n+2)}$

Chapter 15   Exercise 7

1 $-(1 - x^2)^{\frac{1}{2}} + \sin^{-1} x + c$ **2** $\dfrac{3}{2}\ln|x^2 + 4| + 2 \tan^{-1}\dfrac{x}{2} + c$ **3** $\dfrac{1}{2}\ln|x^2 + 3| + \dfrac{5\sqrt{3}}{3}\tan^{-1}\dfrac{x\sqrt{3}}{3} + c$ **4** $2 \ln|x^2 + 4x + 8| - \dfrac{1}{2}\tan^{-1}\left(\dfrac{x + 2}{2}\right) + c$

5 $\ln|x^2 + 4x + 6| - \dfrac{\sqrt{2}}{2}\tan^{-1}\left(\dfrac{x + 2}{\sqrt{2}}\right) + c$ **6** $2(-x^2 - 6x - 4)^{\frac{1}{2}} + \sin^{-1}\left(\dfrac{x + 3}{\sqrt{5}}\right) + c$ **7** $x - 7 \ln|x + 4| + c$ **8** $\dfrac{x^2}{2} - 3x + 10 \ln|x + 3| + c$

Chapter 15   Exercise 8

1 $\dfrac{\sin 3x}{3} - \dfrac{\cos 2x}{2} + c$ **2** $\dfrac{1}{3}(8x^{\frac{3}{2}} + 8(x + 1)^{\frac{3}{2}} + (1 - 3x)^4) + c$ **3** $\dfrac{1}{3}\ln|3x^2 + 1| + c$ **4** $2 \tan^{-1} 2x + c$ **5** $\dfrac{1}{2}\tan\left(2x - \dfrac{\pi}{3}\right) + c$

6 $2\sqrt{1 + \sin x} + c$ **7** $2e^{x^2 + x - 5} + c$ **8** $\dfrac{1}{(n - 1)\cos^{n-1} x} + c \; n \neq 1,\; \ln|\cos x| + c$ for $n = 1$ **9** $\tan^{-1} 2x(+ 1) + c$ **10** $2 \sin^{-1}\left(\dfrac{x + 4}{5}\right) + c$

11 $\dfrac{(1 - x)^8}{36}(7 - 8x) + c$ **12** $x + 5 - 10 \ln|x + 5| - \dfrac{25}{x + 5} + c$ **13** $\dfrac{4(3x - 4)^{\frac{3}{2}}}{135}(9x + 8) + c$ **14** $\dfrac{1 - 5p}{10(p - 3)^5} - \dfrac{1}{80}$ **15** $\tan^{-1} e^x + c$

16 $x(1 - x^2)^{\frac{1}{2}} + \sin^{-1}x + c$ **17** $\dfrac{2\sqrt{3}}{3}\tan^{-1}\left(\dfrac{\tan\dfrac{x}{2}}{\sqrt{3}}\right) + c$ **18** $-\dfrac{1}{4}\cot\left(\sin^{-1}\dfrac{x}{2}\right) + c$ **19** $\dfrac{3}{2}\tan^{-1} x^2 + c$

20 $\sqrt{x - 2} + \dfrac{\sqrt{2}}{2}\tan^{-1}\sqrt{\dfrac{x - 2}{2}} + c$ **21** $\dfrac{-1}{x}(\ln|x| + 1) + c$ **22** $\dfrac{e^{3x}}{9}(3x - 1) + c$ **23** $x \sin\left(x + \dfrac{\pi}{6}\right) + \cos\left(x + \dfrac{\pi}{6}\right) + c$

24 $\dfrac{e^{-2x}}{4}(\sin 2x - \cos 2x) + c$ **25** $-2x^2 \cos\dfrac{x}{2} + 8x \sin\dfrac{x}{2} + 16 \cos\dfrac{x}{2} + c$ **26** $x \ln|2x + 1| - x + \dfrac{1}{2}\ln|2x + 1| + c$ **27** $x \tan^{-1}\dfrac{1}{x} + \dfrac{1}{2}\ln|x^2 + 1| + c$

28 $\dfrac{e^{ax}}{a^2 + 4}(a \sin 2x - 2 \cos 2x) + c$ **29** $\dfrac{1}{3}\sin^3 x + c$ **30** $-\dfrac{1}{3}\cos^3 x + c$ **31** $\dfrac{1}{4}\tan^4 x + c$ **32** $\dfrac{\theta}{2} + \dfrac{\sin 2\theta}{4} + c$ **33** $\dfrac{x}{2} - \dfrac{\sin 6x}{12} + c$

34 $\dfrac{\sin 2x}{2} - \dfrac{1}{6}\sin^3 2x + c$ **35** $-4 \cos\dfrac{x}{4} + \dfrac{8}{3}\cos^3\dfrac{x}{4} - \dfrac{4}{5}\cos^5\dfrac{x}{4} + c$ **36** $\dfrac{2}{3}\tan^3\dfrac{x}{2} - 2 \tan\dfrac{x}{2} + x + c$ **37** $\dfrac{1}{4}\left(x - \dfrac{\sin 4ax}{4a}\right) + c$

38 $\dfrac{1}{3}\tan^3 x + \dfrac{1}{5}\tan^5 x + c$ **39** $\dfrac{1}{4}\left(\dfrac{3}{2}x + 6 \sin\dfrac{x}{3} + \dfrac{3}{4}\sin\dfrac{2x}{3}\right) + c$ **40** $-2(1 - x^2)^{\frac{1}{2}} - \sin^{-1} x + c$

41 $\ln\left|\dfrac{3x^2}{2} - 2x + 3\right| - \dfrac{2\sqrt{14}}{7}\tan^{-1}\left(\dfrac{\sqrt{14}}{14}(3x - 2)\right) + c$ **42** $\dfrac{x^3}{3} + \dfrac{x^2}{2} + x + 2 \ln|x - 1| + c$ **43** $\dfrac{x^2}{4} + \dfrac{x}{4} + \dfrac{13 \ln|2x - 1|}{8} + c$

Chapter 15   Exercise 9

1 $\dfrac{(x - 3)^4}{4} + c$ **2** $\dfrac{2}{9}(3x - 5)^{\frac{3}{2}} + c$ **3** $\dfrac{1}{4}e^{4x - 5} + c$ **4** $\dfrac{12x^{\frac{7}{3}}}{7} - \dfrac{192x^{\frac{23}{12}}}{23} + \dfrac{32x^{\frac{3}{2}}}{3} + c$ **5** $-\dfrac{1}{4}\operatorname{cosec} 4x + c$ **6** 0.0791

7 $\dfrac{-e^{-2x}}{4}(2x^2 + 2x + 1) + c$ **8** 0.169 **9** $\ln|x^2 + 1| - 3 \tan^{-1} x + c$ **10** $\dfrac{1}{2}\sin^{-1}\dfrac{2x}{5} + c$ **11** $\dfrac{-1}{4}(2e^x + 1)^{-2} + c$ **12** $-\dfrac{1}{b}\ln\left|\dfrac{a + b \cos p}{a + b}\right|$

13 $\dfrac{-(1 + 5x)}{25(2 + 5x)^2} + c$ **14** $\dfrac{3}{4}\tan^{-1}\left(\dfrac{x - 3}{4}\right) + c$ **15** $\dfrac{1}{32}(12 + 8 \sin 2x + \sin 4x) + c$ **16** -0.142 **17** 16.5

18 $\dfrac{-3x^2}{4}\cos 2x + \dfrac{3x}{4}\sin 2x + \dfrac{3}{8}\cos 2x + c$ **19** $\sin^{-1}\left(\dfrac{x + 2}{\sqrt{33}}\right) + c$ **20** $\dfrac{x}{\ln 4}(\ln x - 1) + c$ **21** $\sin^{-1}(x - 1) + c$ **22** $\dfrac{x^5}{25}(5 \ln 2x - x) + c$

23 2.44 **24** 3.16 **25** $\dfrac{e^{-3x}}{10}(\sin x - 3 \cos x) + c$ **26** $\dfrac{\sqrt{x}}{25}(2x^2 + 70) + c$ **27** $\dfrac{1}{5}\tan^5 x + \dfrac{1}{7}\tan^7 x + c$ **28** $\dfrac{1}{2}\sin^{-1} 2x + 2\sqrt{1 - 4x^2} + c$

29 $\dfrac{2}{\sqrt{5}}\tan^{-1}\left(\sqrt{5}\tan\dfrac{x}{2}\right) + c$ **30** 1.44 **31** $\dfrac{-(1 + 3a)}{3(a + 1)^3} + \dfrac{1}{3}$ **32** $x \cos^{-1} 2x - \dfrac{\sqrt{1 - 4x^2}}{2} + c$ **33** 1 **34** $\dfrac{\pi a^4}{16}$ **35** -0.0280 **36** 1.50

37 $-\dfrac{1}{4}\ln|\cos 2x| + c$ **38** 0

Chapter 15   Exercise 10

1 0.215 **2** 4.48 **3** 0.148 **5** 9.42 **6** $2(4 - x^2)^{\frac{1}{2}}$ **7** $0.227a^2$ **10** $\dfrac{4 - e^3}{2}$ **11** $\dfrac{2(2a - 1)^{\frac{3}{2}}(3a + 1)}{15}$ **12** 5.66 **13** 11.7 **14** 3.03

Chapter 15 ⬤ Review Exercise

1 $e - e^k + 1$　**2** $\dfrac{4}{15}\left(\dfrac{x+2}{2}\right)^{\frac{3}{2}}(3x-4)+c$　**3** $\dfrac{(k^2-4)^{\frac{3}{2}}}{3\sqrt{5}}$　**4** $x\arctan x - \dfrac{1}{2}\ln|1+x^2|+k$　**5** 0.307　**6** $a=25,\ b=2,\ \sin^{-1}\left(\dfrac{x-2}{5}\right)+c$

7 $a=1.07$　**8** $-\dfrac{e^{-2x}}{4}(2x^2+2x+1)+k$　**9 a** $y=\dfrac{ex}{2}$　**b** $\dfrac{1}{10}$　**10** $\dfrac{1}{2}\arctan\left(\dfrac{x+3}{2}\right)+k$　**11** $-\dfrac{a}{b}\ln(3-b\sin x)+k$　**12** 0.690

13 a $(0,1)$ is a maximum　**b** $y=0$　**d** $\dfrac{\pi}{2}-1$　**14** 1　**15 b i** $\dfrac{2\pi}{9}$　**ii** $\dfrac{4\pi}{9}$　**iii** $\dfrac{6\pi}{9}$　**c** $\dfrac{n\pi}{9}(n+1)$

Chapter 16 ⬤ Exercise 1

1 $y=\dfrac{x^3}{3}-\cos x+k$　**2** $y=\dfrac{(3x-7)^5}{15}+k$　**3** $y=-\dfrac{2}{3}(1-x^2)^{\frac{3}{2}}+k$　**4** $y=-x\cos x+\sin x+k$

5 $y=-\ln|1-\sin x|+k$　**6** $y=\dfrac{3}{2k}e^{\frac{2kx}{3}}\left[1-\dfrac{3}{2k}\right]+c$　**7** $y=-\dfrac{1}{3}(1-15x^2)^{\frac{1}{2}}+k$　**8** $y=\dfrac{1}{8}(4x-\sin 4x)+k$　**9** $y=\dfrac{4}{135}(3x+2)^{\frac{5}{2}}$

10 $y=-\ln|\cos x|+kx+c$　**11** $y=\dfrac{x^4}{24}\ln x-\dfrac{13x^4}{288}+\dfrac{kx^2}{2}+cx+d$

12 $y=x\cos x-4\sin x+\dfrac{kx^3}{6}+\dfrac{cx^2}{2}+dx+e$　**13** $y=\dfrac{1}{2}\ln|4x^2+3|-\dfrac{1}{2}\ln 19$　**14** $y=\dfrac{1}{2}\ln\left(\dfrac{4x^2+3}{19}\right)$

15 $y=\dfrac{(2x-1)^6}{120}+2x+\dfrac{239}{240}$　**16** $y=2x\tan^{-1}x-\ln|1+x^2|-2x\tan^{-1}\left(\dfrac{\pi}{4}\right)+3x+5$

Chapter 16 ⬤ Exercise 2

1 $\dfrac{y^2}{2}=-\ln|\cos x|+k$　**2** $\tan^{-1}y=\dfrac{1}{2}\ln|x|+k$　**3** $\dfrac{1}{2}\ln|3+2y|=-\dfrac{1}{3}\ln|4-3x|+k$　**4** $y=\ln|\cos x+\sin x|+k$

5 $y^4=2x^2\ln x-x^2+k$　**6** $y=\ln\left(\dfrac{5}{c-6\ln|x|}\right)$　**7** $\ln y=4\sin^{-1}\dfrac{x}{2}+k$　**8** $s^3=3(t\sin^{-1}t+(1-t^2)^{\frac{1}{2}}+k)$

9 $y=\dfrac{(3x-1)^{10}(30x+1)}{110}+c$　**10** $v^2=\dfrac{\sin 2at}{2a}+t+k$　**11** $y=\ln\left(\dfrac{-e^{-2x}}{2}+k\right)$　**12** $\ln|y|=\dfrac{3}{2}\ln|x^2+2x|+k$

13 $\ln y=\dfrac{-(3-x)^5}{5}+\ln 4+\dfrac{1}{5}$　**14** $\dfrac{3}{2}y^{\frac{2}{3}}=\dfrac{-e^{-2x}}{2}+\dfrac{3}{2}+\dfrac{1}{2e^2}$　**15** $-\cot y=\ln x-1-\ln 4$

16 $\dfrac{1}{2}\ln|2y^2+3|=\dfrac{3}{8}\ln|4x^2-1|+\dfrac{1}{2}\ln 131-\dfrac{3}{8}\ln 3$　**17** $\dfrac{\theta^3}{3}=\dfrac{e^{2t}}{5}(2\sin t-\cos t)+\dfrac{\pi^3}{24}+\dfrac{1}{5}$　**18** $2\sin^{-1}3s=3(t^2-1)+\pi$

Chapter 16 ⬤ Exercise 3

1 $y=-\ln|\cos x|+k$　**2** $x=\dfrac{\omega^2 t^3}{6}+kt+c$　**3** $x=\dfrac{-\omega\sin nt}{n^2}+kt+c$　**4** $\dfrac{y^3}{3}=\ln|1+x^3|+9$　**5** $\tan y=\tan x+k$

6 a $\dfrac{dz}{dx}=1+\dfrac{dy}{dx}$　$\tan^{-1}z=x+k$　**c** $y=\tan(x+k)-x$　**7** $y=\dfrac{Bx^4}{24A}-\dfrac{Bx^3}{12A}+\dfrac{Bx}{24A}$　**8** $V=-4\cos\left(t+\dfrac{\pi}{4}\right)+6$

9 $r=4t\ln t-4t+24-20\ln 5$

Chapter 16 ⬤ Exercise 5

1 $v=\dfrac{4t^3}{3}+t$　$s=\dfrac{t^4}{3}+\dfrac{t^2}{2}$　**2 a** $v=\dfrac{-3t^4}{2}+10$　**2 b** $-6134\ \text{ms}^{-1}$　**2 c** $s=\dfrac{-3t^5}{10}+10t$　**2 d** $-29900\ \text{m}$　**3** 26.6m

4 a $v=\dfrac{\sin 3t}{3},\ s=-\dfrac{\cos 3t}{9}+\dfrac{11}{18}$　**4 b** $\dfrac{\pi}{3}$ seconds　**5** $s=-\dfrac{t^2}{2}-t-2\ln|t-1|+14$　**6** $v=\sqrt{4-2\cos\left(s+\dfrac{\pi}{4}\right)}$

7 $v=\pm\sqrt{\dfrac{2se^{2s}-e^{2s}+9}{2}}$　**8** $v=\pm\sqrt{\dfrac{(2s-1)^5+223}{5}}$　**9 a** $7\dfrac{3}{4}\text{ms}^{-1}$　**b** $\dfrac{1}{t}+7\dfrac{1}{2}\to 7\dfrac{1}{2}\ \text{ms}^{-1}$ as $t\to\infty$　**10 b** $\dfrac{e^8-137}{4}$

11 a 2.60　**11 b** $v=220e^{-2.60t}$　**c** 108m　**12 a** $v=\dfrac{2}{\omega}$　**12 b** $\dfrac{2\pi}{\omega}$　**13** $v=\dfrac{g}{k}-\dfrac{g}{ke^{kt}},$ Yes $v\to\dfrac{g}{k}$

Answers

1 a 176 **1 b** 107 **1 c** 57.4 **1 d** 8.90 **1 e** 104 **1 f** 2060 **1 g** 327 **1 h** 1.23 **1 i** 1.08 **1 j** 274 **2 a** 8π **2 b** 5.98 **2 c** 0.592

2 d 0.622 **2 e** 113 **2 f** 0.965 **2 g** 12.8 **3** $\dfrac{\pi a^5}{5}$ **4** 145 **5** 1940 **6** 84.2 **7** 5cm, 196 cm² **8**

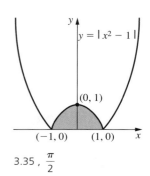

9 $\dfrac{49\pi}{16}$ **10** $\pi^2 b^2 a$ **11** $\pi^3(n^2 + (n-1)^2)$

12 a $\dfrac{1}{4}(2\sin^{-1}x + 2x\sqrt{1-x^2}) + k$ **b.** $\dfrac{\pi}{4}(\pi - 2\sin^{-1}a + 2a\sqrt{1-a^2})$

3.35 , $\dfrac{\pi}{2}$

1 $\dfrac{\pi}{12}$ **2** $y = \sqrt{\dfrac{x}{2} - 1}$ **3** $4m$ **4** $\dfrac{\pi(e^{8a} - 1)}{4e^{4a}}$ **5** $y = Ce^{\frac{5x^2}{2}}$

6 $v = \sqrt{3\ln\left(\dfrac{s^2 + 1}{2}\right) + 4}$. At $t = 5$, $v = \sqrt{3\ln 13 + 4}$

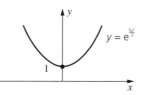

7 a Correct **b i** $A = 78$, $k = \dfrac{1}{15}\ln\left(\dfrac{8}{13}\right)$ **ii** 45.3 mins

8 $\cos^{-1}\left(\dfrac{1}{\sqrt{T}}\right) + \dfrac{\sqrt{T-1}}{T} - \dfrac{\pi}{4} - \dfrac{1}{2}$ **9** Correct **10** 169.8 years **11a** $v = \dfrac{3}{2\pi}\left(1 - \cos\dfrac{2\pi t}{3}\right)$, $s = \dfrac{3}{2\pi}\left(t - \dfrac{3}{2\pi}\sin\dfrac{2\pi t}{3}\right)$

b $t = 3$ seconds **12 a** Correct **b** 0, 3, 6 seconds **c i** distance $= \displaystyle\int_0^3 t\sin\dfrac{\pi t}{3}\,dt - \int_3^6 t\sin\dfrac{\pi t}{3}\,dt$ **ii** $\dfrac{36}{\pi}$ metres **13 a** $v = v_0 e^{-kt}$ **b** $t = \dfrac{1}{k}\ln 2$

14 $\dfrac{\pi}{4}(\pi + 2)$ **15 a**

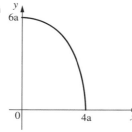

b $V_x = 72\pi a^3$ **c** $\dfrac{V_y}{V_x} = \dfrac{32}{45}$

1 a 18i **b** 38i **c** 112i **d** 60i **2 a** 12i **b** 15i **c** $-24i$ **d** $-52i$ **3 a** 240i **b** -32 **c** 45i **d** 96i **e** $-72i$ **f** 35 **g** $-90i$
4 a $-\dfrac{15}{2}$ **b** 2 **c** $\dfrac{5}{2}i$ **d** $-16i$ **5 a** 9 **b** -1 **6 a** 9i **b** $-\dfrac{5}{3} + \dfrac{7i}{3}$ **c** $-\dfrac{1}{2}i$ **d** 1 **e** $1 - \dfrac{3}{4}i$

1 a $8 + 16i$ **b** $18 + 28i$ **c** $7 + i$ **d** $14 - 21i$ **e** $-7 + 11i$ **2 a** $3 - 5i$ **b** $13 + 22i$ **c** $-20 + 10i$ **d** $7 + 13i$

3 a $-4 + 19i$ **b** $47 + 35i$ **c** $7 + 2i$ **d** $115 - 111i$ **e** 306 **f** $a^2 + b^2$ **g** $1 + 70i$ **h** $117 + 44i$ **i** $x^2 - y^2 + 2ixy$

3 j $m(m^2 - 3n^2) + in(3m^2 - n^2)$ **k** $(m^3 - 18m^2 - 12m + 24) + i(3m^3 + 6m^2 - 36m - 8)$ **4 a** $\dfrac{3 + i}{5}$ **b** $\dfrac{-15 + 10i}{13}$ **c** $\dfrac{7 + 26i}{29}$

4 d $\dfrac{-10 - 33i}{29}$ **e** $6 - 2i$ **f** $\dfrac{2x^2 - y^2 + 3ixy}{4x^2 + y^2}$ **g** $\dfrac{3 + 4i}{5}$ **h** $-\dfrac{3}{2} + \dfrac{i}{2}$ **i** $\dfrac{(2x - 5y) + i(5x + 2y)}{x^2 + y^2}$ **j** $\dfrac{x(y^2 + 3x^2 + 2ixy)}{x^2 + y^2}$

5 a -4 **b** $-11 - 2i$ **c** $-237 - 3116i$ **d** $-8432 - 5376i$ **6 a** $x = 15, y = -7$ **b** $x = 8, y = 0$ **c** $x = 0, y = -3$

6 d $x = 5, y = 12$ **e** $x = -6, y = -3$ **f** $x = -\dfrac{5}{17}, y = -\dfrac{14}{17}$ **g** $x = -21, y = 20$ **h** $x = y = \pm\sqrt{\dfrac{15}{2}}$ **i** $x = -\dfrac{72}{25}, y = \dfrac{29}{25}$

6 j $x = -3, y = 1$ **k** $x = -\dfrac{33}{169}, y = \dfrac{2591}{169}$ **7 a** $\text{Re}(z) = 27, \text{Im}(z) = -8$ **b** $\text{Re}(z) = -\dfrac{53}{185}, \text{Im}(z) = -\dfrac{89}{185}$ **c** $\text{Re}(z) = \dfrac{117}{145}, \text{Im}(z) = \dfrac{41}{145}$

7 d $\text{Re}(z) = \dfrac{72}{65}, \text{Im}(z) = -\dfrac{61}{65}$ **e** $\text{Re}(z) = \dfrac{2a}{4 + b^2} - \dfrac{12}{16 + a^2}, \text{Im}(z) = -\dfrac{ab}{4 + b^2} - \dfrac{3a}{16 + a^2}$ **f** $\text{Re}(z) = 0, \text{Im}(z) = -\dfrac{2xy}{x^2 + y^2}$

7 g $\text{Re}(z) = -597, \text{Im}(z) = 122$ **h** $\text{Re}(z) = \cos\dfrac{2\pi}{3}, \text{Im}(z) = \sin\dfrac{2\pi}{3}$ **i** $\text{Re}(z) = -128, \text{Im}(z) = -128\sqrt{3}$

7 j $\text{Re}(z) = \dfrac{x}{1 + y^2} + \dfrac{12}{16 + 9x^2}, \text{Im}(z) = -\dfrac{xy}{1 + y^2} + \dfrac{9x}{16 + 9x^2}$ **8 a** $1 + 4i, -1 - 4i$ **b** $1.10 + 0.455i, -1.10 - 0.455i$

c $2.12 - 0.707i, -2.12 + 0.707i$ **8 d** $3.85 + 1.69i, -3.85 - 1.69i$ **e** $1.92 + 1.30i, -1.92 - 1.30i$ **f** $0.734 - 0.454i, -0.734 + 0.454i$

8 g $0.704 - 0.369i, -0.704 + 0.369i$ **h** $1.59 + 1.42i, -1.59 - 1.42i$ **i** $0.541 + 0.0416i, -0.541 - 0.0416i$ **9 a** $x = -3 \pm i$

9 b $x = \dfrac{-1 \pm i\sqrt{3}}{2}$ **c** $x = \dfrac{-3 \pm i\sqrt{51}}{2}$ **d** $x = \dfrac{-3 \pm i\sqrt{6}}{3}$ **e** $x = \dfrac{-3 \pm i\sqrt{95}}{4}$ **10 a** $x^2 - 4x + 13 = 0$ **b** $x^2 - 6x + 10 = 0$

10 c $x^2 - 8x + 25 = 0$ **d** $x^3 - 3x^2 + 7x - 5 = 0$ **e** $x^3 - 8x^2 + 25x - 26 = 0$ **f** $x^4 - 10x^3 + 20x^2 + 90x - 261 = 0$

11 a $x^2 - 4x + 53 = 0$ **b** $x^2 - 8x + 25 = 0$ **c** $x^2 - 14x + 85 = 0$ **d** $x^2 - 2ax + a^2 + b^2 = 0$ **12** $x^4 - 10x^3 + 42x^2 - 82x + 65 = 0$

13 $3, 1 - i$ **14 a** $3, -1 + 2i, -1 - 2i$ **b** $2, -1 + 3i, -1 - 3i$ **15** $\dfrac{-214 + 735i}{53}$ **16** $-8 + 2i, -8 - i$ **17** $\dfrac{348 - 115i}{13}$

18 $z_1 = \dfrac{26 - 2i}{17}, z_2 = \dfrac{21 + i}{17}$ **19** $\dfrac{3 - i\sqrt{3}}{2}$ **20** $p = -\dfrac{3}{5}, q = \dfrac{9}{5}$ **21** $\dfrac{88 - 966i}{25}$

Chapter 17   Exercise 3

1 a

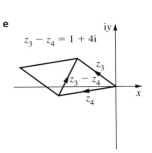

$z_1 + z_3 = -3 + i$

b $z_2 + z_3 = -2 + 7i$

c $z_1 - z_4 = 6 - i$

d $z_1 + z_4 = -4 - 3i$

e $z_3 - z_4 = 1 + 4i$

2 a $2\left(\cos\dfrac{\pi}{3} + i\sin\dfrac{\pi}{3}\right)$ **b** $\sqrt{8}\left[\cos\left(-\dfrac{3\pi}{4}\right) + i\sin\left(-\dfrac{3\pi}{4}\right)\right]$ **c** $\sqrt{26}[\cos(2.94) + i\sin(2.94)]$

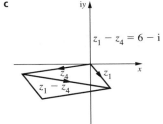 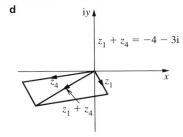

2 d $\sqrt{41}[\cos(-0.896) + i\sin(-0.896)]$ **e** $10(\cos 0 + i\sin 0)$ **f** $6\left(\cos\dfrac{\pi}{2} + i\sin\dfrac{\pi}{2}\right)$

3 a $4\sqrt{2}e^{i\frac{\pi}{4}}$ **b** $5e^{-2.21i}$ **c** $\sqrt{53}e^{1.85i}$ **d** $\sqrt{82}e^{-1.46i}$ **e** $8e^{0i}$ **f** $2e^{i\frac{\pi}{3}}$

4 a $1 + i\sqrt{3}$ **b** $-\dfrac{\sqrt{15}}{2} + i\dfrac{\sqrt{5}}{2}$ **c** $5\sqrt{2} - 5i\sqrt{2}$ **d** $3.74 - 1.00i$ **e** $-\dfrac{3\sqrt{2}}{2} + \dfrac{3\sqrt{2}}{2}i$ **f** $-\dfrac{\sqrt{5}}{2} + i\dfrac{\sqrt{15}}{2}$ **g** $\dfrac{15\sqrt{3}}{2} - \dfrac{15}{2}i$

4 h $1.67 - 4.03i$ **5 a** $r = \sqrt{313}, \theta = 0.825$ **b** $r = \sqrt{701}, \theta = 1.38$ **c** $r = 2\sqrt{370}, \theta = 0.487$ **d** $r = 4\dfrac{\sqrt{370}}{5}, \theta = 1.41$

6 a Root 1 $r = 2.65, \theta = 2.17$ Root 2 $r = 2.65, \theta = -2.17$ **b** Root 1 $r = \sqrt{5}, \theta = 1.11$ Root 2 $r = \sqrt{5}, \theta = -1.11$

6 c Root 1 $r = \sqrt{7}, \theta = 0.714$ Root 2 $r = \sqrt{7}, \theta = -0.714$ **7 a i** $z_1 = 13e^{-1.18i}$ **ii** $z_2 = 5e^{2.21i}$ **iii** $z_3 = 25e^{-0.284i}$ **iv** $z_4 = 2e^{i\frac{\pi}{3}}$

7 b i $r = 65, \theta = 1.04$ **ii** $r = 325, \theta = -1.46$ **iii** $r = 26, \theta = -0.129$ **iv** $r = 2.5, \theta = 1.17$ **v** $r = 5, \theta = -2.50$ **vi** $r = 6.5, \theta = -2.22$

7 vii $r = 25, \theta = -1.33$ **viii** $r = 150, \theta = 0.763$

8 a

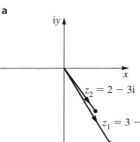

and **b**

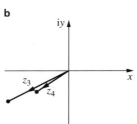

c Rotation of $\dfrac{\pi}{2}$ radians clockwise.

9 a $z_1 = \sqrt{7}[\cos(-0.714) + i\sin(-0.714)]$ $z_2 = \sqrt{2}\left[\cos\left(-\dfrac{3\pi}{4}\right) + i\sin\left(-\dfrac{3\pi}{4}\right)\right]$

9 b $|z_1z_2| = \sqrt{14}$, $\arg(z_1z_2) = -3.07$, $\left|\dfrac{z_1}{z_2}\right| = \sqrt{\dfrac{7}{2}}$, $\arg\left(\dfrac{z_1}{z_2}\right) = 1.64$

10 $z_1 = -\dfrac{1}{53} + \dfrac{23}{53}i$, $z_2 = \dfrac{4}{13} + \dfrac{7}{13}i$ **11 b** $z = 1 + 0i, \dfrac{-1 + i\sqrt{3}}{2}, \dfrac{-1 - i\sqrt{3}}{2}$ **12 b i** 16 **ii** 10

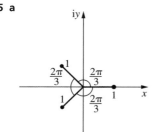

14 a $1[\cos(0.841) + i\sin(0.841)]$, $1[\cos(-0.841) + i\sin(-0.841)]$ **b** 1.68 radians

15 a $r = \dfrac{3}{2}$, $\theta = 1.23$ **15 b** 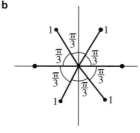 **c** No.

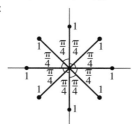

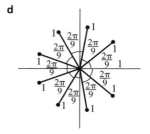

Chapter 17   Exercise 4

1 a $1024(\cos 10\theta + i\sin 10\theta)$ **b** $\cos 25\theta + i\sin 25\theta$ **c** $\dfrac{1}{243}[\cos(-5\theta) + i\sin(-5\theta)]$ **d** $\cos(-9\theta) + i\sin(-9\theta)$ **e** $\cos\dfrac{1}{2}\theta + i\sin\dfrac{1}{2}\theta$

1 f $\dfrac{1}{\sqrt[3]{4}}\left[\cos\left(-\dfrac{1}{3}\theta\right) + i\sin\left(-\dfrac{1}{3}\theta\right)\right]$ **g** $\cos 0 + i\sin 0$ **h** $\cos\left(-\dfrac{\pi}{2}\right) + i\sin\left(-\dfrac{\pi}{2}\right)$ **i** $\cos\left(\dfrac{3\pi}{4}\right) + i\sin\left(\dfrac{3\pi}{4}\right)$ **j** $\cos\left(\dfrac{\pi}{10}\right) + i\sin\left(\dfrac{\pi}{10}\right)$

2 a $(\cos\theta + i\sin\theta)^7$ **b** $4(\cos\theta + i\sin\theta)^{\frac{1}{2}}$ **c** $6(\cos\theta + i\sin\theta)^{-3}$ **d** $(\cos\theta + i\sin\theta)^{-\frac{1}{4}}$ **e** $(\cos\theta + i\sin\theta)^{-2}$ **f** $(\cos\theta + i\sin\theta)^{-\frac{1}{8}}$

3 a $\cos 8\theta + i\sin 8\theta$ **b** $\cos\dfrac{5}{2}\theta + i\sin\dfrac{5}{2}\theta$ **c** $\cos 3\theta + i\sin 3\theta$ **d** $\cos\theta - i\sin\theta$ **e** $\cos 11\theta + i\sin 11\theta$ **f** $\cos 3\theta - i\sin 3\theta$

3 g $\cos\dfrac{1}{6}\theta - i\sin\dfrac{1}{6}\theta$ **h** $\cos\dfrac{3}{4}\pi - i\sin\dfrac{3}{4}\pi$ **i** $\cos\dfrac{3}{4}\pi + i\sin\dfrac{3}{4}\pi$ **j** $\cos\dfrac{\pi}{12} + i\sin\dfrac{\pi}{12}$ **4 a** $2 + 3i, -2 - 3i$ **b** $0.644 - 1.55i, -0.644 + 1.55i$

4 c $2^{\frac{1}{6}}\left[\cos\left(-\dfrac{3\pi}{4}\right) + i\sin\left(-\dfrac{3\pi}{4}\right)\right], 2^{\frac{1}{6}}\left[\cos\left(-\dfrac{\pi}{12}\right) + i\sin\left(-\dfrac{\pi}{12}\right)\right], 2^{\frac{1}{6}}\left[\cos\left(\dfrac{7\pi}{12}\right) + i\sin\left(\dfrac{7\pi}{12}\right)\right]$ **d** $1.69 - 0.606i, -0.322 + 1.77i, -1.37 - 1.16i$

4 e $1.456 + 0.347i, -0.347 + 1.456i, 0.347 - 1.456i, -1.456 - 0.347i$ **f** $1.54 - 0.640i, 1.09 + 1.27i, -0.872 + 1.42i,$

4 f $-0.132 - 1.67i, -1.62 - 0.389i$ **g** $2^{\frac{1}{8}}\left[\cos\left(-\dfrac{35\pi}{36}\right) + i\sin\left(-\dfrac{35\pi}{36}\right)\right], 2^{\frac{1}{8}}\left[\cos\left(-\dfrac{23\pi}{36}\right) + i\sin\left(-\dfrac{23\pi}{36}\right)\right], 2^{\frac{1}{8}}\left[\cos\left(-\dfrac{11\pi}{36}\right) + i\sin\left(-\dfrac{11\pi}{36}\right)\right],$

4 g $2^{\frac{1}{8}}\left[\cos\left(\dfrac{\pi}{36}\right) + i\sin\left(\dfrac{\pi}{36}\right)\right], 2^{\frac{1}{8}}\left[\cos\left(\dfrac{13\pi}{36}\right) + i\sin\left(\dfrac{13\pi}{36}\right)\right], 2^{\frac{1}{8}}\left[\cos\left(\dfrac{25\pi}{36}\right) + i\sin\left(\dfrac{25\pi}{36}\right)\right]$

5 a 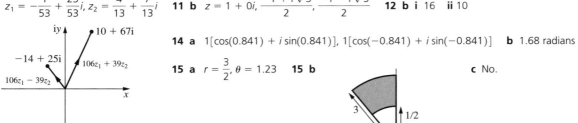 **b** **c** **d**

6 a $16\left[\cos\left(\dfrac{\pi}{2}\right) + i\sin\left(\dfrac{\pi}{2}\right)\right]$ **b** $1.85 + 0.765i,\ -0.765 + 1.85i,\ 0.765 - 1.85i,\ -1.85 - 0.765i$ **7 a** $2\left[\cos\left(\dfrac{\pi}{3}\right) + i\sin\left(\dfrac{\pi}{3}\right)\right]$

7 b Real part is -2^{15} Imaginary part is $-2^{15}\sqrt{3}$ **9 b** $\tan\left(-\dfrac{15\pi}{16}\right), \tan\left(-\dfrac{7\pi}{16}\right), \tan\left(\dfrac{\pi}{16}\right), \tan\left(\dfrac{9\pi}{16}\right)$ **10** $-\dfrac{1}{2} - i\dfrac{\sqrt{3}}{2}$

11 a $z_2 = 5[\cos(-0.927) + i\sin(-0.927)]$ **b** $\dfrac{4}{25}$ **12** Product $= 4$, Sum $= -2$

15 a $z_1 = 2\left(\cos\dfrac{2\pi}{7} + i\sin\dfrac{2\pi}{7}\right)$ **b** $z_1^2 = 4\left(\cos\dfrac{4\pi}{7} + i\sin\dfrac{4\pi}{7}\right),\ z_1^3 = 8\left(\cos\dfrac{6\pi}{7} + i\sin\dfrac{6\pi}{7}\right)$

$z_1^4 = 16\left[\cos\left(-\dfrac{6\pi}{7}\right) + i\sin\left(-\dfrac{6\pi}{7}\right)\right], z_1^5 = 32\left[\cos\left(-\dfrac{4\pi}{7}\right) + i\sin\left(-\dfrac{4\pi}{7}\right)\right]$ $z_1^6 = 64\left[\cos\left(-\dfrac{2\pi}{7}\right) + i\sin\left(-\dfrac{2\pi}{7}\right)\right],\ z_1^7 = 128(\cos 0 + i\sin 0)$

15 c

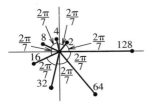

15 d Rotate $\dfrac{2\pi}{7}$ anticlockwise. Enlargement scale factor 2.

16 b $\cos\left(\dfrac{\pi}{3}\right) + i\sin\left(\dfrac{\pi}{3}\right), \cos\left(-\dfrac{\pi}{3}\right) + i\sin\left(-\dfrac{\pi}{3}\right)$ **c** $\cos\left(-\dfrac{2\pi}{3}\right) + i\sin\left(-\dfrac{2\pi}{3}\right)$

17 c $z^6 + 6z^4 + 15z^2 + 20 + \dfrac{15}{z^2} + \dfrac{6}{z^4} + \dfrac{1}{z^6}$ **d** $a = \dfrac{1}{32}, b = \dfrac{3}{16}, c = \dfrac{15}{32}, d = \dfrac{5}{16}$

Chapter 17   Review Exercise

1 $r = \sqrt{14.8}, \theta = -2.06$ **2** $k = -2$ **3** $a = -2, b = 5$ **4** 4 **5 b** 4^{16} **6** $x = -\dfrac{47}{65}, y = -\dfrac{1}{65}$ **7** $x^3 - 5x^2 + 10x - 12 = 0$

8 Real part $= \dfrac{x^3 + xy^2 + x}{x^2 + y^2}$, Imaginary part $= \dfrac{x^2y + y^3 - y}{x^2 + y^2}$ **9** $\sqrt{3}$ **10 a** $|z| = 2, \arg(z) = \dfrac{\pi}{3}$ **b** $|z^2| = 4, \arg(z^2) = \dfrac{2\pi}{3}$

10 c i

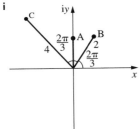

ii $2\sqrt{3}$ **iii** $2\sqrt{3} - \dfrac{3}{2}$ **11 b** $1, \dfrac{-1 + i\sqrt{3}}{2}, \dfrac{-1 - i\sqrt{3}}{2}$

11 c $\cos 0 + i\sin 0, \cos\dfrac{2\pi}{3} + i\sin\dfrac{2\pi}{3}, \cos\left(-\dfrac{2\pi}{3}\right) + i\sin\left(-\dfrac{2\pi}{3}\right)$

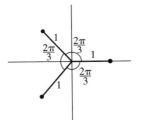

11 d Each side has length $\sqrt{3}$. Area of triangle $= \dfrac{3\sqrt{3}}{4}$ **12 a** $a = -\dfrac{5}{3}, b = \dfrac{16}{9}$ **b** $a = -\dfrac{17}{4} - \dfrac{31i}{4}, b = -\dfrac{17}{4} + \dfrac{31i}{4}$

13 d i $a = 32, b = -32, c = 6$ **ii** 6 **14** $z = 5 + i, \omega = 6 - i$ **15 a i** $\cos^3\theta - 3\cos\theta\sin^2\theta + i(3\cos^2\theta\sin\theta - \sin^3\theta)$ **c** $\dfrac{23\sqrt{2}}{20}$

16 a i 1 **ii** $\dfrac{2\pi}{3}$ **c** $\dfrac{3}{2} + \dfrac{3\sqrt{3}}{2}i$ **17** $|z| \le 5, \dfrac{\pi}{3} \le \arg(z) \le \pi, -2 \le \text{Re}(z) \le \dfrac{5}{2}$ **18 a** $\dfrac{10k + i(k^2 - 21)}{k^2 + 49}$ **b** $k = \pm\sqrt{21}$

19 a $1, \dfrac{-1 + i\sqrt{3}}{2}, \dfrac{-1 - i\sqrt{3}}{2}$ **c** $\begin{pmatrix} 3 & 0 & 0 \\ 0 & 3 & 0 \\ 0 & 0 & 3 \end{pmatrix}$ **d** $x = -1, y = z = 2$ **21 a** $2 - i, -2 + i$ **b**

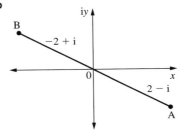

21 c $(\sqrt{3}, 2\sqrt{3}), (-\sqrt{3}, -2\sqrt{3})$

Chapter 18 — Exercise 3

Answers are given when asked to form a conjecture

1 $D^n = \begin{pmatrix} 1 & 2^n - 1 \\ 0 & 2^n \end{pmatrix}$ **2** $\sum_{r=1}^{n} 3r + 2 = \frac{1}{2}n(3n + 7)$ **3** $\sum_{r=1}^{n} 4r - 7 = n(2n - 5)$ **4** Any value $\geq 18p$

Chapter 18 — Review Exercise

5 sum of the first n odd numbers $= n^2$ **6** $n^2 + 2n + 2$ **7** $n^2 + 4$

Chapter 19 — Exercise 1

8 $(n + 1)! - 1$ **16** $M^n = \begin{pmatrix} n + 1 & -n \\ n & 1 - n \end{pmatrix}$ **1 a** Continuous   **b** Discrete   **c** Continuous   **d** Continuous   **2**
mode = 4., median = 4, mean = 4.15   **3** mode = Blue

4 b 81-90   **c** 78.2   **5** 16.7   **6 a** 1.58m.   **b** There is no information about the ages or gender of the students.   **7** 0.927

8 a People below this height are not allowed on the ride.   **b** Mean = 1.79m

Chapter 19 — Exercise 2

1
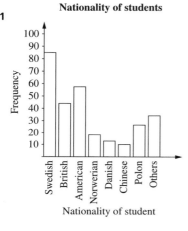
Nationality of students

2
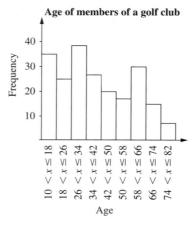
Age of members of a golf club

3
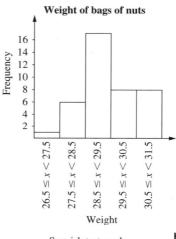
Weight of bags of nuts

4
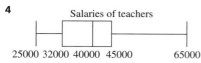
Salaries of teachers

5
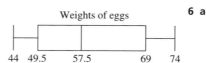
Weights of eggs

6 a
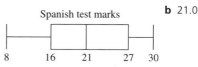
Spanish test marks   **b** 21.0

7
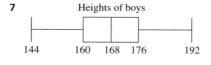
Heights of boys

8
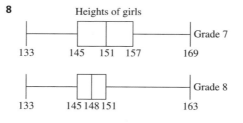
Heights of girls

9 median = 13,

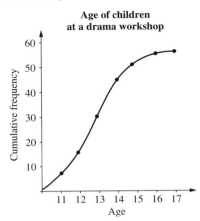

Age of children at a drama workshop

10 estimate median = 25

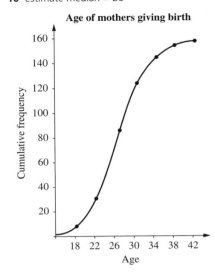

Age of mothers giving birth

11 estimate $Q_1 = 12$, $Q_2 = 21$, $Q_3 = 24$

12 estimate $Q_1 = 20$, $Q_2 = 20.5$, $Q_3 = 21.25$, 10th percentile = 19

13 estimate $Q_1 = 35m$, $Q_2 = 42m$, $Q_3 = 48m$, 35th percentile = 38m, 95th percentile = 54m

Chapter 19   Exercise 3

1 a $Q_2 = 10$, $IQ = 8$ **b** $Q_2 = 59.5$, $IQ = 6.5$ **c** $Q_2 = 67$, $IQ = 39$ **d** $Q_2 = 176$, $IQ = 89$ **e** $Q_2 = 40$, $IQ = 3$

2 The two sets have a similar spread as $IQ = 3$ for both sets. The average age for set B is less as the median is 18 and the median for set A is 19.

3 Box and whisker plots. Medicine $IQ = 1$, median = 2. Law $IQ = 2$, median = 3. Law students rated the lecturing higher but there was a greater spread of opinion among this group.

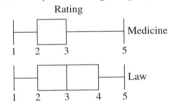

4 a 2.28   **b** 3.74   **c** 33.4   **d** 5.50   **e** 9.17

5 Mean = 19.01, standard deviation = 6.28, estimated standard deviation = 6.31

6 $\bar{x} = 501$, variance = 2.61

7 a $Q_2 = 105$, $IQ = 16$ **b** $\bar{x} = 110$, $s = 16.2$

Chapter 19   Exercise 4

1 a i $Q_1 = 8.3$, $Q_2 = 9.9$, $Q_3 = 11.9$ **ii** $\bar{x} = 9.8$, $s = 1.79$ **b i** $Q_1 = 183$, $Q_2 = 263$, $Q_3 = 298$ **ii** $\bar{x} = 238$, $s = 60.2$

1 c i $Q_1 = 34000$, $Q_2 = 45500$, $Q_3 = 57250$ **ii** $\bar{x} = 44500$, $s = 14300$ **d i** $Q_1 = 0.62$, $Q_2 = 0.755$, $Q_3 = 0.845$ **ii** $\bar{x} = 0.724$, $s = 0.189$

2 Graph $IQ = 3$ **3** Daniel Graph, median = 185.5, range = 167 Paul Graph, median = 198.5, range = 71

4 $\bar{x} = 40.2$, $s = 29.1$, New mean = 46.2, $s = 33.5$ **5** $\bar{x} = 16.6$, $s = 1.44$

Chapter 19 — Review Exercise

1 a Continuous **b** Discrete **c** Discrete **d** Continuous **2** Red **3** $\bar{x} = 76.0, s = 16.4$ **4** 1.49

5

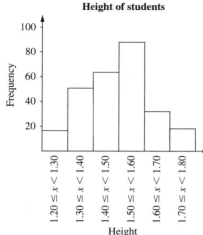

Height of students

6 a

| | | | | | | |
12   33.5 39.5   51   60

b 17.5 **c** 40.9

7 estimate $Q_1 = 18$, $Q_2 = 24$, $Q_3 = 28$ **8 a** $Q_2 = 1, IQ = 1$ **b** $\bar{x} = 1.63, s = 1.51$

9 $\bar{x} = £264, s = £41.70$. After bonus, $\bar{x} = £296, s = £46.70$

10 a 29.9 **b** 0.183 **11 a** median = 135 **b** $IQ = 11$ **12 a** 31.3 **b** 9.84

13 a 156 **b** $x = 44$ minutes

$\bar{x} = £264, s = £41.70$. After bonus, $\bar{x} = £296, s = £46.70$

Chapter 20 — Exercise 1

1 a $\frac{1}{4}$ **b** $\frac{1}{2}$ **c** $\frac{1}{2}$ **2 a** $\frac{1}{10}$ **b** $\frac{1}{2}$ **c** $\frac{3}{10}$ **d** $\frac{2}{5}$ **3 a** $\frac{1}{4}$ **b** $\frac{13}{20}$ **c** $\frac{9}{10}$ **d** $\frac{13}{20}$ **e** 0 **4 a** 0.48

4 b Because the probability of either a novel or a mathematics book is 1. **5** No. $P(A \cup B) \neq 1$ **6 a** $\frac{7}{11}$ **b** $\frac{922}{1155}$ **7** $\frac{1}{2}$ **8 a** $\frac{1}{5}$ **b** $\frac{9}{15}$

8 c $\frac{13}{15}$ **d** $\frac{13}{15}$ **e** $\frac{2}{3}$ **9 a** $\frac{1}{4}$ **b** $\frac{5}{8}$ **c** $\frac{1}{8}$ **d** 1 **e** $\frac{3}{4}$ **f** $\frac{1}{2}$ **g** 0 **10 a** $\frac{1}{9}$ **b** $\frac{5}{6}$ **c** $\frac{1}{6}$ **d** $\frac{1}{6}$ **e** $\frac{1}{9}$ **f** $\frac{1}{4}$ **g** $\frac{2}{3}$

11 a 0.1 **b** 1 **c** 0.5 **12** $\frac{1}{10}$ **13** $\frac{1}{4}$ **14** 0.72 **15 a** $\frac{1}{3}$ **b** $\frac{13}{18}$ **c** $\frac{5}{18}$ **d** $\frac{7}{9}$ **e** $\frac{7}{18}$

15 f Because it is not possible to have one die showing a 5 and for the sum to be less than 4. **16** $\frac{23}{32}$

17 a $\frac{1}{2}$ **b** $\frac{9}{16}$ **c** 1

18 a Events X and Y are not mutually exclusive because 2 fish of type A and 2 fish of type B fit both.

18 b Events X and Z are not mutually exclusive because 2 fish of type A, 1 fish of type B and 1 fish of type C fit both. **c** Events Y and Z are mutually exclusive because the event Y does not allow a fish of type C and event Z does.

19 0.15 **20 a** $\frac{83}{500}$ **b** $\frac{1}{25}$

Chapter 20 — Exercise 2

1 a $\frac{5}{42}$ **b** $\frac{3}{7}$ **2 a** $\frac{9}{15}$ **b** $\frac{1}{6}$ **c** $\frac{1}{2}$ **3 a** $\frac{7}{16}$ **b** $\frac{1}{4}$ **c** $\frac{2}{7}$ **d** $\frac{1}{2}$

4 a $\frac{1}{14}$ **b** $\frac{1}{6}$ **c** $\frac{2}{7}$ **5 a** $\frac{2}{5}$ **b** $\frac{1}{2}$ **c** $\frac{1}{2}$ **6** 0.0768 **7 a** $\frac{181}{208}$

7 b No, because $P(A \cup B) \neq 1$ **8** $\frac{14}{17}$ **9 a** **b** $\frac{2}{3}$ **c** $\frac{1}{2}$

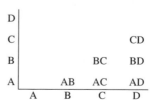

Chapter 20 — Exercise 3

1 a 0.3 **b** 0.5 **2 a**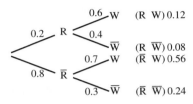

b $\frac{1}{36}$ **c** $\frac{25}{36}$ **d**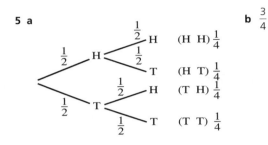

e $\frac{1}{2}$ **f** $\frac{3}{4}$

3 a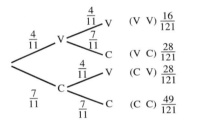

b 0.12 **c** 0.32 **d** 0.3 **e** 0.176 **4 a** $\frac{1}{8}$ **b** $\frac{3}{8}$ **c** $\frac{1}{4}$ **d** $\frac{7}{8}$

5 a

b $\frac{3}{4}$

5 c

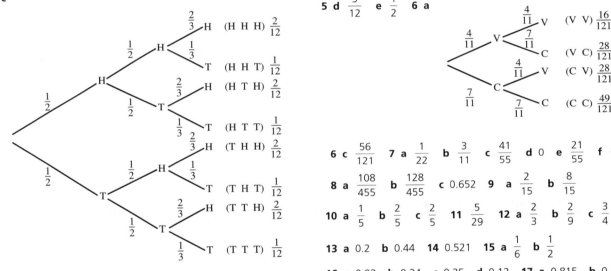

5 d $\frac{5}{12}$ **e** $\frac{1}{2}$ **6 a**

b $\frac{49}{121}$

6 c $\frac{56}{121}$ **7 a** $\frac{1}{22}$ **b** $\frac{3}{11}$ **c** $\frac{41}{55}$ **d** 0 **e** $\frac{21}{55}$ **f** $\frac{7}{15}$

8 a $\frac{108}{455}$ **b** $\frac{128}{455}$ **c** 0.652 **9 a** $\frac{2}{15}$ **b** $\frac{8}{15}$

10 a $\frac{1}{5}$ **b** $\frac{2}{5}$ **c** $\frac{2}{5}$ **11** $\frac{5}{29}$ **12 a** $\frac{2}{3}$ **b** $\frac{2}{9}$ **c** $\frac{3}{4}$

13 a 0.2 **b** 0.44 **14** 0.521 **15 a** $\frac{1}{6}$ **b** $\frac{1}{2}$

16 a 0.03 **b** 0.34 **c** 0.25 **d** 0.12 **17 a** 0.815 **b** 0.149

Chapter 20 — Exercise 4

1 a 0.974 **b** 6 **c** 6 **2 a** 0.832 **b** 24 **3** 0.985 **4** $\frac{3}{4}$ **5** $\frac{2}{3}$ **6** $\frac{45}{53}$ **7** $\frac{32}{45}$ **8 a i** $\frac{11}{609}$ **ii** 0.0212 **iii** 0.0153 **b** 0.0839

9 0.84 **10 a** 0.222 **b** 0.074 **c** 0.144 **d** 6 **11 a** 0.00877 **b** 0.322 **c** 0.632 **d** 0.561 **12 a** 0.3 **b** 0.35 **c** 0.075 **d** 0.3

13 a $\frac{1}{16}$ **b** $\frac{1}{64}$ **c** 19 **14 a** 0.0191 **b i** 0.00459 **ii** 0.0255 **15 a** $\frac{1}{27000}$ **b** 90 **c** $\frac{1}{900}$

Answers

Chapter 20 — Exercise 5

1 720 **2** 16065 **3** 24 **4** 60 **5** 4838400 **6** 604800 **7** 241920 **8** 2177280 **9 a** 36 **b** 6 **c** 12 **d** 24 **10** 70 **11 a** 90720

b 5040 **11 c** 2520 **12 a** 40320 **b** 5040 **c** 20160, 2520 **13 a** 2494800 **b** 635040 **14 a** 5040 **b** 4320 **c** 720 **15 a** 36 **b** 10

16 a 9000000 **16 b** Increases by 16000000 **17 a** 831600 **b** 176400 **c** 151200 **18** 119 **19** 38760 **20 a** 4 **b** 48 **c** 24 **d** 256

21 a 60 **b** 30 **22 a** 28 **b** 28 **c** 35 **23 a** 120 **b** 90 **24** 756756 **25** 70 **26** 210 **27** 3185325 **28** 27

Chapter 20 — Exercise 6

1 a 100 **b** $\frac{1}{2}$ **c** $\frac{1}{5}$ **d** $\frac{1}{5}$ **2 a** 362880 **b i** $\frac{2}{9}$ **ii** $\frac{7}{9}$ **c** $\frac{1}{189}$ **3 a** 15 **b** $\frac{2}{3}$ **4 a** 360 **b** $\frac{2}{3}$ **c** $\frac{8}{15}$ **5 a** 3838380

5 b $\frac{3}{20}$ **c** $\frac{5}{39}$ **6 a** 10440 **b** $\frac{12}{29}$ **c** $\frac{1}{15}$ **7 a** 16 **b i** $\frac{1}{2}$ **ii** $\frac{3}{8}$ **iii** $\frac{3}{16}$ **8 a** 3150 **b** $\frac{1}{30}$ **c** $\frac{1}{45}$ **d** $\frac{1}{6}$ **9 a** 720 **b** $\frac{1}{15}$

9 c $\frac{1}{30}$ **10 a** 3628800 **b** $\frac{1}{30}$ **c** $\frac{7}{15}$ **d** $\frac{3}{28}$

Chapter 20 — Review Exercise

1 a $\frac{1}{6}$ **b** Events are not independent since $P(Black) \times P(Brown) = \frac{1}{3}$ and $P(Black \cap Brown) = \frac{1}{6}$

Events are not mutually exclusive because $P(Black \cap Brown) \neq 0$ Events are exhaustive since $P(Black \cup Brown) = 1$

2 a 0.984 **b** 7 **3 a** 151200 **b** 10080 **4 a** $\frac{1}{6}$ **b** 0.0670 **c** $\frac{1}{6} \times \left(\frac{5}{6}\right)^{2n-2}$ **5** $\frac{10}{13}$ **6** 62 **7 a** 453600

7 b 90720 **c** 362880 **8** $\frac{19}{30}$ **9 a** 15 **b** $\frac{8}{15}$ **c** $\frac{4}{15}$ **10** 0.888 **11 a** 252 **b** 250 **c** 126 **d** 32 **12** $\frac{4}{5}$ **13 a** 360

13 b 216 **14** 0.048 **15 a** 20160 **b** $\frac{1}{4}$ **c** $\frac{1}{56}$ **16 a** $\frac{17}{42}, \frac{9}{17}$

16 b ii $\frac{5}{6}$ **iii** $\frac{7}{25}$ **iv** $\frac{28}{51}$ **17 a** 252 **b** 196 **c** 186

18 a $\frac{3n+1}{2}$ **b** $\frac{n+1}{3n+1}$ **20 a** 0.549 **b** 0.368 **c** 0.439

21 a 0.581 **b** 0.0918 **c** 0.0663

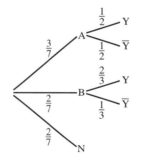

b i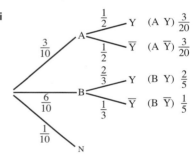

Chapter 21 — Exercise 1

1 a $b = 0.2$ **b** 0.6 **c** 0.65 **d** 0.8 **e** 2 **2 a** $a = 0.33$ **b** 0.87 **c** 0.75 **d** 0.73 **e** 0.25 **f** 7

3 a

X	0	1	2	3
P(X = x)	$\frac{1}{8}$	$\frac{3}{8}$	$\frac{3}{8}$	$\frac{1}{8}$

b

X	0	1	2	3
P(X = x)	$\frac{27}{343}$	$\frac{108}{343}$	$\frac{144}{343}$	$\frac{64}{343}$

c

X	0	1	2	3
P(X = x)	$\frac{125}{216}$	$\frac{75}{216}$	$\frac{15}{216}$	$\frac{1}{216}$

3 d

X	2	3	4	5	6	7	8	9	10	11	12
P(X = x)	$\frac{1}{36}$	$\frac{2}{36}$	$\frac{3}{36}$	$\frac{4}{36}$	$\frac{5}{36}$	$\frac{6}{36}$	$\frac{5}{36}$	$\frac{4}{36}$	$\frac{3}{36}$	$\frac{2}{36}$	$\frac{1}{36}$

e

X	0	1	2	3
P(X = x)	0.512	0.384	0.096	0.008

4 a

X	0	1	2	3	4	5
P(X = x)	0	$\frac{1}{55}$	$\frac{4}{55}$	$\frac{9}{55}$	$\frac{16}{55}$	$\frac{25}{55}$

b

X	1	2	3	4	5	6
P(X = x)	$\frac{1}{21}$	$\frac{2}{21}$	$\frac{3}{21}$	$\frac{4}{21}$	$\frac{5}{21}$	$\frac{6}{21}$

c

X	7	8	9	10
P(X = x)	$\frac{6}{30}$	$\frac{7}{30}$	$\frac{8}{30}$	$\frac{9}{30}$

4 d

X	12	13	14	15
P(X = x)	$\frac{9}{42}$	$\frac{10}{42}$	$\frac{11}{42}$	$\frac{12}{42}$

5 a $k = \frac{1}{9}$

X	3	4	5
P(X = x)	$\frac{2}{9}$	$\frac{3}{9}$	$\frac{4}{9}$

b $k = \frac{1}{74}$

X	4	5	6
P(X = x)	$\frac{15}{74}$	$\frac{24}{74}$	$\frac{35}{74}$

5 c $k = \frac{1}{225}$

Y	1	2	3	4	5
P(Y = y)	$\frac{1}{225}$	$\frac{8}{225}$	$\frac{3}{225}$	$\frac{64}{225}$	$\frac{5}{9}$

d $k = 35$

B	1	2	3	4	5
P(B = b)	$\frac{5}{35}$	$\frac{6}{35}$	$\frac{7}{35}$	$\frac{8}{35}$	$\frac{9}{35}$

6 a

X	0	1	2	3
$P(X = x)$	$\dfrac{6}{504}$	$\dfrac{108}{504}$	$\dfrac{270}{504}$	$\dfrac{120}{504}$

b $\dfrac{249}{252}$ **7 a**

X	0	1	2	3
$P(X = x)$	$\dfrac{27}{64}$	$\dfrac{27}{64}$	$\dfrac{9}{64}$	$\dfrac{1}{64}$

b $\dfrac{5}{32}$

8 a

Y	0	1	2	3	4
$P(Y = y)$	$\dfrac{1}{126}$	$\dfrac{20}{126}$	$\dfrac{60}{126}$	$\dfrac{40}{126}$	$\dfrac{5}{126}$

b $\dfrac{125}{126}$

Chapter 21   Exercise 2

1 a $b = 0.3$ $E(X) = 4.7$ **b** $b = 0.15$ $E(X) = 3.35$ **c** $b = 0.28$ $E(X) = 3.06$ **d** $b = 0.1$ $E(X) = 2.6$

2 $E(X) = \dfrac{1}{2}$ **3 a** $c = \dfrac{12}{91}$ **b** $E(X) = \dfrac{441}{91}$ **4 a** $E(X) = \dfrac{x - 3}{4}$ **b** $x = 3$

5

X	−1	1
$P(X = x)$	0.3	0.7

6

X	−1	1	3
$P(X = x)$	0.15	0.25	0.6

7

Y	0	2	4	6
$P(Y = y)$	$\dfrac{4}{10}$	$\dfrac{1}{10}$	$\dfrac{1}{10}$	$\dfrac{4}{10}$

8 $E(X) = \dfrac{25}{9}$ **9** $E(X) = 1$ **10 a** 2.05 **b** 5.45 **c** 3.10 **d** 8.15 **11 a** 3 **b** 14 **c** 7 **d** 8

12 a $\dfrac{691}{400}$ **b** 5.8 **c** 6 **d** 0.986 **13 a** 7 **b** $\dfrac{329}{6}$ **c** $\dfrac{35}{6}$ **14 a** $\dfrac{9}{7}$ **b** $\dfrac{15}{7}$ **c** $\dfrac{24}{49}$ **15 a** $\dfrac{1}{14}$ **b** 2.53 **c** 1.65

16 $E(Y) = \dfrac{28}{15}$, $Var(Y) = 0.782$

Y	0	1	2	3	4
$P(Y = y)$	$\dfrac{2}{39}$	$\dfrac{56}{195}$	$\dfrac{84}{195}$	$\dfrac{8}{39}$	$\dfrac{1}{39}$

17 a $\dfrac{1}{2}$ **b** $\dfrac{26}{7}$ **c** 1.44

18 a can be a probability density function   Mean $= \dfrac{110}{35}$ Standard deviation $= 1.05$

19 a 0, 1, 2, 4 **b**

X	0	1	2	3	4
$P(X = x)$	$\dfrac{15}{24}$	$\dfrac{4}{24}$	$\dfrac{4}{24}$	0	$\dfrac{1}{24}$

c Mean $= \dfrac{2}{3}$ Variance $= 1.06$

20 a 10, 11, 12, 13, 14 **b**

X	10	11	12	13	14
$P(X = x)$	$\dfrac{6}{90}$	$\dfrac{24}{90}$	$\dfrac{30}{90}$	$\dfrac{24}{90}$	$\dfrac{6}{90}$

c $E(X) = 12$, $Var(X) = \dfrac{16}{15}$ **d** $\dfrac{83}{225}$

21 a $\dfrac{1}{14}$ **b** $E(X) = \dfrac{4}{7}$, $Var(X) = 0.816$ **c** 0, 1, 2, 3, 4, 5, 6 **d**

X	0	1	2	3	4	5	6
$P(X = x)$	$\dfrac{81}{196}$	$\dfrac{54}{196}$	$\dfrac{27}{196}$	$\dfrac{24}{196}$	$\dfrac{7}{196}$	$\dfrac{2}{196}$	$\dfrac{1}{196}$

e $E(Y) = \dfrac{8}{7}$, $Var(Y) = 1.63$

Chapter 21   Exercise 3

1 a 0.27 **b** 0.532 **c** 0.0556 **2 a** 0.201 **b** 0.833 **c** 0.834 **3 a** 0.208 **b** 0.0273 **c** 0.973 **d** 0.367 **4** 0.751 **5 a** 2.4 **b** 1.44 **c** 2

6 a 2.4 **b** 1.68 **c** 2 **7 a** $n = 7$, $p = \dfrac{1}{4}$, $q = \dfrac{3}{4}$ **b** 0.445 **c** 1 or 2 **8 a** 0.00345 **b** 0.982 **c** 0.939 **9 a** 0.0872 **b** 0.684

9 c 0.684 **d** 0.847 **10 a** 3.52×10^{-5} **b** 0.0284 **c** 0.683 **d** 0.163 **11 a** 0.238 **b** 0.0158 **12 a** 0.245 **b** 0.861 **c** 0.997

13 a 0.060 **b** 0.00257 **c** 0.00191 **d** 0.230 **e** 0.129 **14 a** 10 **b** 2.74 **c** 0.416 **15 a** $\dfrac{4}{7}$ **b** 17 **16 a** 0.0258 **b** 0 **c** $\dfrac{1}{3}$ **d** 4

16 e 0.00258 **f** 0.00858 **17 a** Mean $= 2.1$ Variance $= 1.81$ **b** 2 **c** 0.204 **d** 0.148 **e** 0.473 **18 a** $X \sim Bin\left(8, \dfrac{1}{3}\right)$ **b** 0.156

18 c $\dfrac{8}{3}$ **d** 0.961 **19 a** 8 **b** 0.822 **c** 8 **d** 0 0.0108

Chapter 21   Exercise 4

1 a 0.224 **b** 0.423 **c** 0.353 **d** 3 **2 a** 0.134 **b** 0.151 **c** 0.554 **d** 6 **3 a** 0.125 **b** 0.332 **c** 0.933 **d** 10

4 a 1.68 **b** 0.0618 **c** 0.910 **5 a** 2.48 **b** 0.213 **c** 0.763 **d** 0.0404 **6 a** 2.69 **b** 0.0799 **c** 0.136 **d** 0.505 **e** 0.944

7 a 2.10 **b** 0.0991 **c** 0.0204 **d** 0.350 **8 a** 7.62 **b** 0.996 **9 a** 1 **b** 0.981 **10 a** 0.905 **b** 0.00468 **c** 0.000151

11 a 0.874 **b** 0.191 **c** 0.223 **d** 0.0426 **12 a** 0.0804 **b** 0.751 **c** 2 **d** 0.182 **e** 0.454 **13 a** 0.0149 **b** 0.223

14 a 0.0183 **b** 0.215 **c** 0.927 **d** Mean $= 80$ Variance $= 80$ **e** 0.849 **f** 0.0262 **15 a** 0.0324 **b** 0.992 **c** 0.112 **d** 0.868 **e** 0.654

16 a i 0.839 **ii** 21.4, 21.4 **iii** 21 **b** 0.000241 **17 a** 0.195 **b** 0.785 **c** 0.152 **d** 0.166 **e** 8

18 a 0.0823 **b** 0.0499 **c** 0.00363 **d** 0.0000314

Chapter 21 — Review Exercise

1 a $\dfrac{1}{10}$ **b** 3 **c** 1 **2 a** $\lambda = 2.99$ **b** 0.424 **3 a** 0.225 **b** 3 or 4 **c** 17.5 **4 a** 20 **b** 12.4 **c** 42.2 **5 a** 0.191 **b** 0.246

6 a 0.176 **b** 0.905 **c** 7 **d** 6.5 **e** 0.0158 **7 a** $\dfrac{12}{25}$ **b** $\dfrac{48}{25}$ **8 a** 0.175 **b** 0.141 **9 a** 0.160 **b** 4 **c** 0.271 **d** 0.0808

10 a 0.0729 **b** 22 **11 a i** $\dfrac{1}{9}$ **ii** $\dfrac{1}{81}$ **b i** $\dfrac{73}{648}$ **ii** $\dfrac{575}{1296}$ **c ii**

X	1	2	3	4	5	6
$P(X = x)$	$\dfrac{1}{1296}$	$\dfrac{15}{1296}$	$\dfrac{65}{1296}$	$\dfrac{175}{1296}$	$\dfrac{369}{1296}$	$\dfrac{671}{1296}$

iii $\dfrac{6797}{1296}$

12 a 0.0105 **b** 0.0226 **c** 1 **d** 10 **e** 0.0232 **13** 30 **14 a** 0.222 **b** 0.939 **c** 0.104 **d** 0.00370 **e** 0.332 **f** 0.0145 **g** 0.995

15 a $P(A|B) = \dfrac{P(A \cap B)}{P(B)}$ **b** $P(A_1 \cup A_2) = P(A_1) + P(A_2)$ **c** $P(E_1 \cap E_2) = P(E_1) \times P(E_2)$

15 d This is a distribution that deals with events that either occur or do not occur, i.e. there are two complementary outcomes. We are usually told the number of times an event occurs and we are given the probability of the event happening or not happening.

15 e i $^4C_k \theta^k (1 - \theta)^{4-k}$ **ii** $E(X) = 4\theta$, $Var(X) = 4\theta(1 - \theta)$ **iii** $6\theta^2(1 - \theta)^2 + 4\theta^4(1 - \theta) + \theta^4$ **iv** 0.688, 0.996 **v** $\dfrac{^4C_j \theta^j (1 - \theta)^{4-j}}{6\theta^2(1 - \theta)^2 + 4\theta^4(1 - \theta) + \theta^4}$

15 e vi 0.545, 0.0488 **vii** 0.969 **16 a**

No heads	0	1	2	3	4	5	6
Amount received in Euros X	30	25	15	12	18	25	40
$P(X = x)$	0.00137	0.0165	0.0823	0.219	0.329	0.263	0.0878

16 b Gain of 20.3 Euros **c** 59.5 **d** 305 **17** $\alpha = \dfrac{1}{7}$, $E(X) = \dfrac{123}{49}$ **18 b i** 0.0977 **ii** 0.885

Chapter 22 — Exercise 1

1 a $k = 1$ **b**
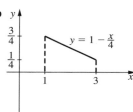
c $\dfrac{1}{2}$ **d** $\dfrac{27}{32}$ **2 a** $c = 4$ **b**
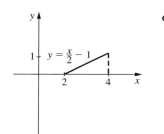
c $\dfrac{3}{16}$ **d** 0

3 a $k = \sqrt{2}$ **b**
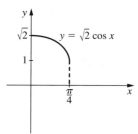
c 0.707 **d** 0.634 **4 a** $k = 1.08$ **b**
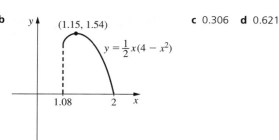
c 0.306 **d** 0.621

5 a $\dfrac{3}{269}$ **b**
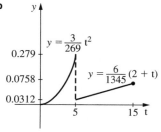
c 0.494 **d** 0.375

Chapter 22 — Exercise 2

1 a $k = \dfrac{1}{2}$ **b** $\dfrac{4}{3}$ **c** $\dfrac{2}{9}$ **2 a** $\dfrac{1}{\ln 3}$ **b** $\dfrac{2}{\ln 3}$ **c** $\dfrac{4}{\ln 3} - \dfrac{4}{(\ln 3)^2}$ **3 a** $c = 0.755$ **b** 0.530 **4 a** $k = \dfrac{1}{16}$ **b** $\dfrac{23}{6}$ **c** 1.97 **d** 4 **e** $\dfrac{15}{32}$

5 a $k = \dfrac{1}{9}$ **b** 2.25 **c** $\dfrac{27}{80}$ **d** 2.38 **6 a** $c = \dfrac{1}{(e-1)}$ **b** $\dfrac{1}{(e-1)}$ **7 a** $k = 1$ **b** 0.571 **c** 0.141 **d** $\dfrac{\pi}{6}$ **e** 0.169

8 a $k = 1.56$ **b** 0.616 **c** 0.422 **d** 0.546 **e** 0.616 **9 a** $0 \le x \le \dfrac{1}{2}$ **b** $0 < x < 0.421$ **c** 0.115 **d** 0.0209 **10 a** 1 **b** 0

10 c 0.273 **12 a i** 7.71 **ii** 0.947 **b** 0.318 **13 a** $k = \dfrac{e^2}{3(e^2-1)}$ **b** 0.148 **c** 0.00560 **14 a**

14 b $\sqrt{135}$ **c** 12.4 **15 a** $c = 0$, $k = \dfrac{3}{8}$ **b** 0.15 **c** 1.587 **d** 0.305

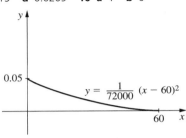

Chapter 22 — Exercise 3

1 a 0.775 **b** 0.589 **c** 0.633 **d** 0.0392 **e** 0.9234 **f** 0.0973 **g** 0.203 **h** 0.562 **i** 0.841 **j** 0.5392 **2 a** 0.121 **b** 1.53
2 c -0.396 **d** -0.678 **e** 1.69 **f** -0.485 **g** 0.999 **h** 1.56 **i** 0.509 **j** 0.813 **3 a** 0.00332 **b** 0.901 **c** 0.00332 **d** 0.968
4 a 0.0912 **b** 0.997 **c** 0.952 **d** 0.122 **e** 0 **5 a** 0.106 **b** 0.809 **c** 0.998 **d** 0.101 **e** 0 **6 a** 0.275 **b** 0.00139
6 c 0.683 **d** 0.0279 **7 a** 0.840 **b** 0.0678 **c** 0.683 **d** 0.997 **8 a** 40.6 **b** 38.9 **c** 41.9 **d** 39.2 **9 a** 93.9 **b** 84.6 **c** 86.6
9 d 82.2 **10 a** 5.89 **b** 13.7 **c** 18.1 **d** 4.18 **11** Upper quartile $Z \ge 0.674$ Lower quartile $Z \le -0.674$ **12** 0.935 **13** 0.912 **14** 0.999
15 0.939 **16** 0.134

Chapter 22 — Exercise 4

1 15.1 **2** 75.6 **3** 30.5 **4** 39.0 **5** 6.81 **6** 7.81 **7** 14.1 **8** 11.2 **9** $\mu = 11.6, \sigma = 4.53$ **10** $\mu = 46.3, \sigma = 4.26$

11 $\mu = 290, \sigma = 11.1$ **12 a** $\mu = 23.6, \sigma = 6.13$ **b** 0.432 **13** 11.7

Chapter 22 — Exercise 5

1 a 0.453 **b** 2.29 kg **2 a** 10.6% **b** $m = 589g, n = 600g$ **3 a** 0.309 **b** 0.227 **c** 0.440 **4 a** 0.106 **b** 0.734 **c** 0.599 **d** 0.159

4 e 0.606 **f** 0.292 **5 a** 440 **b** 82.3 kg **6** $\left(\mu - \sigma, \dfrac{1}{\sigma\sqrt{2e\pi}}\right), \left(\mu + \sigma, \dfrac{1}{\sigma\sqrt{2e\pi}}\right)$ **7** 0.886 **8 a** 5 **b** 57.4 **9 a** 7.93 **b** 48.9

9 c 7 **10** $\mu = 64.56, \sigma = 11.36$ **11 a** 0.235 **b** $564g$ **c** 114 **12 a** 19.0 **b** 117 **c** 98.2 **13 a** 90.9% **b** 94.7 **14** 0.338

15 a $194 \le X \le 303$ **b** 0.423 **16 a** $\mu = 28.5, \sigma = \sqrt{1.49}$ **b** 0.587 **17 a** 126 **b** $280g$ **18 a** 1.43 **b** 0.0146 **19** 4.14

20 a 32.8 **b** 0.161 **21** 0.00123 **22** $\mu = 59.3, \sigma = 18.6$ **23 a** 4.82 **b** 0.0173

Chapter 22 — Review Exercise

1 a $\mu = 34.5, \sigma = 3.93$ **b** 0.996 **2 a** 0.0668 **b** 142 cm **c** $q = 140$ cm, $r = 180$ cm **d** 0.121 **e** 0.332 **3 a** 0.946 **b** 0.798

3 c 0.000109 **d** 0.999851 **4** $2\dfrac{1}{2}$ **5 a** 99.9% **b** 11.4 **c** 3.96% **d** 0.00110 **6 a** 0.0327 **b** 8.00 **c** Day 1: 2620. Day 2: 2610. **7 b** 0

7 c 0.268 **d** 0.350 **e** 0.348 **8 a** $\dfrac{1}{4}$ **b** $E(X) = \pi$, $Var(X) = 2.93$ **c** 0.323

9 a 1.63 **c** 0.434 **d** $\$6610$ **10 a** $\mu = 28.6, \sigma = 14.3$ **b** 12.6% **c** $x = 0$ Model is not perfect. **d i** $\dfrac{8}{125}$

10 d ii $\dfrac{36}{125}$ **iii** $\dfrac{98}{125}$ **10 e** $< 0.4^3$ **f** Either the events are not independent or the distribution is not continuous

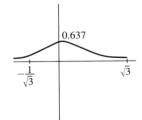

11 a i $E(X) = \dfrac{1}{12}\displaystyle\int_0^2 x(8x - x^3)\, dx$ **ii** 1.24 **b ii** 1.29 **c** 1.63 **12 a i** 1.355 **ii** 110.37 **b** $A = 108.63, B = 112.11$ **13 a** $\dfrac{4}{81}$ **b** 0.6

13 c 0.24 **d** 9000 cents. **14** 0.783 **15 b** $e^{\frac{1}{4}} - e^{\frac{1}{2}} + \dfrac{1}{4}e$ **c** $E(X) = \dfrac{e}{2} - 1$, $Var(X) = 1 + \dfrac{e}{3} - \dfrac{e^2}{4}$ **d** 0.290 **e** 0.0243 **f** 0.179

Index

Index